Lecture Notes in Nanoscale Science and Technology

Volume 42

Lecture Notes in Nanoscale Science and Technology (LNNST) aims to report latest developments in nanoscale science and technology research and teaching – quickly, informally and at a high level. Through publication, LNNST commits to serve the open communication of scientific and technological advances in the creation and use of objects at the nanometer scale, crossing the boundaries of physics, materials science, biology, chemistry, and engineering. Certainly, while historically the mysteries in each of the sciences have been very different, they have all required a relentless step-by-step pursuit to uncover the answer to a challenging scientific question, but recently many of the answers have brought questions that lie at the boundaries between the life sciences and the physical sciences and between what is fundamental and what is application. This is no accident since recent research in the physical and life sciences have each independently cut a path to the edge of their disciplines. As both paths intersect one may ask if transport of material in a cell is biology or is it physics? This intersection of curiosity makes us realize that nanoscience and technology crosses many if not all disciplines. It is this market that the proposed series of lecture notes targets.

Shrikaant Kulkarni • Susanta Das
Editors

Ionic Liquids, Nanofluids and Nanotechnology

Innovations, Applications and Sustainability

Springer

Editors

Shrikaant Kulkarni
School of Engineering
Sanjivani University
Kopargaon, Maharashtra, India

Susanta Das
School of Engineering
Ajeenkya D Y Patil University
Pune, Maharashtra, India

ISSN 2195-2159 ISSN 2195-2167 (electronic)
Lecture Notes in Nanoscale Science and Technology
ISBN 978-3-032-04007-7 ISBN 978-3-032-04008-4 (eBook)
https://doi.org/10.1007/978-3-032-04008-4

This Springer imprint is published by the registered company Springer Nature Switzerland AG
The registered company address is: Gewerbestrasse 11, 6330 Cham, Switzerland

Preface

In the relentless quest to address global challenges through science and innovation, the intersection of ionic liquids, nanofluids, and nanotechnology has emerged as a vibrant and transformative domain. These materials—individually powerful and collectively synergistic—are shaping the future of energy, medicine, electronics, and environmental sustainability. What once seemed the subject of theoretical fascination is now taking form in real-world applications, from intelligent cooling systems and targeted drug delivery to carbon capture and adaptive smart materials.

Ionic Liquids, Nanofluids, and Nanotechnology: Innovations, Applications, and Sustainability offers a deep, multidisciplinary examination of this exciting frontier. The book explores the properties, design strategies, functional integration, and real-life applications of these advanced materials. It moves fluidly across the foundational science and the forward-leaning technological landscapes that are redefining materials engineering, biotechnology, and chemical sciences.

A defining characteristic of the twenty-first-century scientific revolution is the emergence of ML and AI as key enablers of innovation. These computational tools are accelerating the pace of discovery, allowing researchers to predict material behavior, optimize synthesis protocols, and extract insights from complex, high-dimensional datasets. When paired with the adaptive nature of nanomaterials and ionic systems, AI and ML open the door to novel capabilities—such as real-time performance modeling, autonomous material design, and precision tuning of physical and chemical properties.

Moreover, this volume goes beyond theoretical discourse by addressing real-world challenges and ethical considerations—from nanoparticle toxicity and long-term environmental impact to regulatory gaps and equitable technology access. Case studies and educational perspectives included in this book provide a pragmatic view of how these innovations are being integrated into curriculum design, research collaborations, and industrial protocols.

The journey begins with Chap. 1, which lays the groundwork by offering an integrative introduction to ionic liquids, nanofluids, nanomaterials, and nanotechnology. It emphasizes the interconnectivity and synergistic potential of these materials, revealing their centrality to modern scientific inquiry and engineering solutions.

Chapter 2 delves deeper into the properties and characterization techniques that define the behavior and performance of these materials. By detailing experimental methodologies such as spectroscopy, electron microscopy, and calorimetry, it offers insights into the structure–property relationships that underpin functionality and application.

Chapter 3 provides a focused exposition on nanofluids, discussing their synthesis, thermophysical behavior, and theoretical underpinnings. The chapter identifies key challenges, such as stability and environmental safety, while also highlighting the vast application potential in energy and process industries.

The foundational understanding is reinforced in Chap. 4, which revisits the core concepts but with a wider lens on multidisciplinary integration. It presents the materials and technologies not as isolated innovations, but as interconnected platforms critical for future-oriented scientific and industrial breakthroughs.

Chapter 5 explores ionic liquids as sustainable alternatives to traditional solvents in organic reactions. Their negligible volatility, tunable reactivity, and alignment with green chemistry principles position them as transformative agents in eco-friendly chemical synthesis and the circular economy.

The emergence of bionanofluids—discussed in Chap. 6—represents an exciting development where biotechnology and nanoscience intersect. By using bio-derived nanoparticles, researchers are opening new pathways in sustainable energy systems, biomedical engineering, and environmentally friendly heat transfer technologies.

Chapter 7 addresses the hybridization of ionic liquids and nanofluids, examining how IL-based nanofluids bring together the best of both systems. Their enhanced thermal and electrochemical properties unlock powerful capabilities in energy storage, solar thermal systems, and high-performance electrochemical devices.

The biomedical horizon is explored in Chap. 8, which investigates novel clinical applications of ionic liquids (ILs), nanofluids (NFs), and nanoparticles (NPs). Their biocompatibility and functional tunability enable advances in targeted drug delivery, cancer therapy, diagnostic imaging, and tissue engineering.

A broader perspective on biomedical, biotechnology, and nanobiotechnology applications is presented in Chap. 9. Here, the material highlights how ILs) and nanofluids intersect with modern medicine, from enzyme catalysis and biosensing to regenerative medicine and precision drug delivery, while also considering essential safety and regulatory challenges.

Chapter 10 delves into the historical context and future trajectory of nanotechnology and its biomedical relevance. It ties together the advancements and challenges associated with ILs, NFs, and NPs, with a call for sustainable, biocompatible, and clinically translatable materials that bridge the gap between research laboratories and real-world applications.

In Chap. 11, the interdisciplinary utility of ionic liquids, nanofluids, and nanoparticles in biomedical, biotechnology, and clinical settings is explored. It emphasizes their emerging roles in enzyme stabilization, drug delivery, and therapeutic strategies—offering new frontiers in disease treatment and diagnostics.

Chapter 12 delves into the catalytic potential of ionic liquid nanoparticles, highlighting their effectiveness in both homogeneous and heterogeneous systems.

The chapter outlines their roles in organic synthesis, biomass conversion, and environmental remediation, all while advocating sustainable and green chemical practices.

In Chap. 13, nanomaterials take center stage in the development of advanced biosensors. With a focus on refractive index-based detection inspired by dual self-referenced sensing methods, this chapter showcases the promise of quasi-crystalline photonic structures in real-time malaria diagnostics and optical sensing innovations.

Chapter 14 extends the discussion to energy applications, surveying how nanomaterials transform devices like batteries, supercapacitors, and fuel cells. The chapter underscores their potential in smart materials and the ethical and practical considerations in advancing energy sustainability.

Chapter 15 offers a detailed analysis of nanofluids in energy storage, especially in solar thermal systems. It emphasizes the thermophysical benefits of nanofluids in improving heat transfer, energy density, and equipment miniaturization—key components of next-generation storage solutions.

In Chap. 16, the nuanced role of magnetic control in double-diffusive convection within magnetic nanofluids is studied. Through a sophisticated mathematical and numerical framework, this chapter provides a deeper understanding of transport phenomena relevant to thermal management systems.

Chapter 17 showcases the synthesis and antimicrobial potential of silver nanoparticle-based nanofluids. These multifunctional materials are discussed in the context of biomedical, environmental, and consumer product applications, emphasizing their scalability and practical integration.

Chapter 18 investigates the role of ionic liquids in the sustainable pretreatment of lignocellulosic biomass for biorefineries. It presents ionic liquids as powerful tools for biomass dissolution, analyzing their molecular interactions and sustainability from both ecological and economic lenses.

Chapter 19 confronts the growing concerns surrounding the toxicity of ionic liquids, nanofluids, and nanomaterials. As these materials proliferate across applications, understanding their environmental fate and biological interactions is crucial for developing safer alternatives and regulatory frameworks.

In Chap. 20, readers will encounter a forward-looking analysis of how ML and AI enhance the precision and efficiency of ILs and nanofluids design. These advanced systems now support the prediction of thermophysical properties, optimize synthesis pathways, and promote sustainability across industries such as green chemistry and biomedicine.

Chapter 21 delves deeper into the modeling techniques and predictive systems such as neural networks and ensemble learning. It introduces tools like the Committee Machine Intelligent System (CMIS) and explains their ability to optimize nanofluid formulations and performance in complex applications like renewable energy systems and biomedical devices.

In Chap. 22, the fusion of experimental data with AI models is explored, illustrating how the convergence of these domains enables cost-effective and scalable solutions. The chapter emphasizes AI's role in the rapid design and customization of nanomaterials for use in areas like smart lubrication and nanomedicine delivery.

Chapter 23 continues this narrative by showcasing the predictive modeling capabilities of ANNs and SVR, establishing how these approaches outperform traditional methods in understanding nonlinear behaviors in nanofluids. The chapter underscores AI's transformative role in accelerating the discovery of next-generation nanomaterials.

In Chap. 24, readers are introduced to the real-time prediction potential of ML tools such as MLP and RBF-ANN, and their applications in optimizing heat exchangers and improving energy efficiency. This chapter also highlights the ability of AI to assess toxicity, improve nanoparticle synthesis, and personalize treatments in clinical settings, while also recognizing current limitations in data quality and model interpretability.

Finally, Chap. 25 reflects on the broader landscape—presenting both the challenges and opportunities of this evolving field. It discusses the critical need for standardized protocols, explores ethical concerns, and showcases successful case studies ranging from CO_2 capture to biomedical imaging. The chapter also outlines how educational reform and interdisciplinary research are essential to cultivating the next generation of researchers and practitioners in the field.

This book is crafted to function both as a comprehensive academic resource and a practical guide for researchers, graduate students, and professionals engaged in chemical engineering, materials science, nanotechnology, and computational modeling. Whether addressing challenges in sustainability, biomedical innovation, or energy systems, the chapters within offer a forward-looking perspective on how data-driven, intelligent approaches are shaping the future of ionic liquids, nanofluids, and advanced materials.

More than a repository of current knowledge, this volume is a call to action—a reflection of the urgent need to balance scientific progress with environmental stewardship, ethical responsibility, and societal impact. It represents a collaborative effort to bridge foundational science with real-world applications, encouraging innovation that is not only cutting-edge but also responsible and sustainable.

We hope this work will serve as both a catalyst and a compass: sparking new ideas, guiding research endeavors, and fostering interdisciplinary collaboration across academia and industry. As we stand at the convergence of materials science and intelligent technologies, may this book inspire the next wave of discoveries and solutions in the dynamic field of nanotechnology.

Our sincere thanks go to all the contributors whose dedication, expertise, and vision have shaped this work. May it be a valuable stepping stone toward new breakthroughs and a more sustainable, intelligent, and equitable future.

Kopargaon, Maharashtra, India Shrikaant Kulkarni
Pune, Maharashtra, India Susanta Das

Acknowledgments We extend our heartfelt gratitude to all the contributors whose dedication, expertise, and vision have been instrumental in shaping this volume. Their scholarly efforts and commitment to advancing the fields of ionic liquids, nanofluids, and nanotechnology have made this work possible. We hope that this book serves as a meaningful stepping stone toward future breakthroughs that are sustainable, intelligent, and equitable.

We are also deeply thankful to our families, friends, and colleagues for their unwavering support and encouragement throughout this journey—from the initial stages of drafting the proposal for the edited volume *Ionic Liquids, Nanofluids, and Nanotechnology: Innovations, Applications, and Sustainability* to the completion of this ambitious project. Without their patience, understanding, and belief in our vision, this endeavor would not have come to fruition.

Finally, we sincerely thank Springer Nature for providing us with the platform to bring this work to a global audience. Their support and collaboration have been invaluable in transforming this vision into reality.

Introduction

In the face of escalating global challenges—from climate change and energy insecurity to healthcare demands and sustainable development—the scientific community has increasingly turned to novel materials and interdisciplinary innovation. At the forefront of this transformative journey lie ionic liquids, nanofluids, and nanotechnology. Individually, each represents a powerful leap in material science. Collectively, they form a synergistic triad capable of revolutionizing entire industries and reshaping our approach to energy, medicine, electronics, and environmental sustainability.

Ionic Liquids, Nanofluids, and Nanotechnology: Innovations, Applications, and Sustainability offers a comprehensive and forward-looking exploration of this rapidly evolving domain. By integrating foundational science with real-world applications, this volume traverses diverse disciplines—from chemical engineering and biotechnology to materials science and computational modeling. The book serves as both an academic reference and a practical guide, shedding light on the properties, synthesis strategies, functional integration, and emerging uses of these materials across conventional and cutting-edge contexts.

A defining hallmark of twenty-first-century scientific progress is the rise of AI and ML as accelerators of innovation. These technologies are now instrumental in predicting material behavior, optimizing synthesis pathways, and deciphering complex, high-dimensional datasets. When coupled with the adaptive potential of ILs and nanomaterials, AI and ML are enabling unprecedented capabilities: from real-time performance modeling to autonomous design of sustainable systems and precision biomedical tools.

This book does not stop at technical breakthroughs—it also confronts the ethical, environmental, and societal implications of advancing nanotechnology. Issues such as nanoparticle toxicity, regulatory oversight, equitable access, and long-term sustainability are critically examined. Case studies, educational insights, and real-world implementations help bridge theory with practice, providing readers with both inspiration and a roadmap for responsible innovation.

Structured into 25 chapters, this volume begins with foundational insights into the interconnected roles of ILs, NFs, and NT (Chaps. 1–4), followed by deep dives into specific material properties, synthesis techniques, and characterizations. The book then transitions to application-driven chapters addressing energy systems, biomedical technologies, catalysis, sensors, and environmental solutions (Chaps. 5–19). The final section (Chaps. 20–25) highlights the convergence of AI and materials science, showcasing how intelligent algorithms are transforming the design, optimization, and deployment of ILs and nanofluids across diverse sectors.

Each chapter contributes a unique perspective while reinforcing the unifying themes of innovation, application, and sustainability. Whether discussing silver nanoparticle-based antimicrobial fluids, AI-driven toxicity modeling, or the use of ionic liquids in biomass conversion, the narrative remains focused on the shared mission: to engineer smarter, safer, and more sustainable materials for the future.

Ultimately, this book represents a collaborative vision—uniting researchers, engineers, educators, and thought leaders who are actively shaping the next generation of advanced materials. As we move further into an era defined by digital tools and material convergence, we hope this work serves as both a catalyst for discovery and a compass for responsible progress.

We express our sincere gratitude to the contributors whose expertise and commitment have made this endeavor possible. It is our hope that this volume inspires interdisciplinary dialogue, sparks innovative research, and supports the development of technologies that not only push scientific boundaries but also uphold our collective responsibility to people and the planet.

Contents

Abbreviations

AAMC	Association of American Medical Colleges
ACBV	AC breakdown voltage
AFM	Atomic force microscopy
Ag	Silver
AgNPs	Silver nanoparticles
AI	Artificial intelligence
Al_2O_3	Aluminum oxide
$AlCl_3$	Aluminum chloride
ANN	Artificial Neural Network
APIL	Aprotic ionic liquid
Au	Gold
AuNPs	Gold nanoparticles
AuNRs	Gold nanorods
Au-SiO_2	Gold-silica
BMIMCl	1-butyl-3-methylimidazolium chloride
BTMS	Battery thermal management systems
CAMD	Computer-aided molecular design
CDs	Carbon dots
CdSe/ZnS	Cadmium selenide/zinc sulfide
CNN	Convolutional Neural Networks
CNT	Carbon nanotube
$CoFe_2O_4$	Cobalt ferrite
CRISPR	Clustered regularly interspaced short palindromic repeats
CSP	Concentrating solar power
CT	Computed tomography
Cu	Copper
CuO	Copper oxide
CVD	Chemical vapor deposition
DA	Detection accuracy
DCM	Dichloromethane
DDAB	Didodecyldimethylammonium bromide

DESs	Deep eutectic solvents
DFT	Density functional theory
DILs	Dicationic ionic liquids
DL	Deep learning
DLS	Dynamic light scattering
DMF	Dimethyl formamide
DMSO	Dimethyl sulfoxide
DNA	Deoxyribonucleic acid
DNN	Deep Neural Networks
EAN	Ethylammonium nitrate
EDLRF	Electrical double layer repulsive force
EDX	Energy dispersive X-ray spectroscopy
EG	Ethylene glycol
EMA	European Medicines Agency
EPA	Environmental Protection Agency
eV	Electron volts
F/g	Farads per gram
FDA	Food and Drug Administration
Fe_3O_4	Iron (II, III) oxide
FR	Fenton reaction
FT-IR	Fourier transform infrared spectroscopy
GMOs	Genetically modified organisms
GPR	Gaussian process regression
HAADF-STEM	High-angle annular dark-field scanning transmission electron microscopy
HEBM	High-energy ball milling
HER	Hydrogen evolution reaction
HF	Hydrofluoric acid
HHMI	Howard Hughes Medical Institute
HTFs	Heat transfer fluids
IL	Ionic liquids
IL-6	Interleukin-6
ILNPs	Ionic liquid nanoparticles
INFs	Ionanofluids
IoNT	Internet of Nano-things
IoT	Internet of Things
LCB	Lignocellulosic biomass
LCCs	Lignin-carbohydrate complexes
LCD	Liquid-crystal display
LFCS	Lipid Formulation Classification System
LHW	Liquid hot water treatment
LiBs	Lithium-ion batteries
LiF	Lithium fluoride
LNPs	Liposomal NPs
LOC	Lab-on-a-chip

LTO	Lithium titanate oxide
MAE	Mean absolute error
mAh/g	Milliampere-hours per gram
MEMS	Micro electro mechanical systems
MFS	Mononuclear phagocyte system
MGS	Multilayered graphite sheets
MH	Magnetic hyperthermia
ML	Machine learning
MLP	Multilayer perceptron
MoS_2	Molybdenum disulfide
MRI	Magnetic resonance imaging
mRNA	Messenger ribonucleic acid
MWCNT	Multiwalled carbon nanotube
NASEM	National Academies of Sciences, Engineering, and Medicine
NB	Naïve Bayes
NDMA	N-Nitrosodimethylamine
NFs	Nanofluids
nm	Nanometers
NMR	Nuclear magnetic resonance
NPs	Nanoparticle
NT	Nanotechnology
N-UNCD	Nitrogen-doped ultrananocrystalline diamond
PCM	Phase change materials
PEC	Photoelectrochemical
PEG	Polyethylene glycol
PIL	Protic ionic liquid
PILs	Polymeric ionic liquids
PLGA	Poly(lactic-co-glycolic acid)
PQCs	Photonic quasicrystals
PTT	Photothermal therapy
PVT	Photovoltaic thermal systems
PZT	Lead zirconate titanate
QDs	Quantum dots
QF	Quality factor QF
RBC	Red blood cells
RBF-ANN	Radial Basis Function Neural Networks
REACH	Registration, Evaluation, Authorisation and Restriction of Chemicals
RF	Random forest
RFA	Radio frequency ablation
RI	Refractive index
RL	Reinforcement learning
RNN	Recurrent Neural Networks
ROS	Reactive oxygen species
RPC	Ring parasite cells

RT	Room temperature
SAL	Surface active liquids
$ScCO_2$	Supercritical carbon dioxide
SCs	Supercapacitors
SEM	Scanning electron microscopy
SERS	Surface-enhanced Raman scattering
SLNs	Solid lipid nanoparticles
SML	Supervised machine learning
SPC	Schizont parasite cells
SPIONs	Superparamagnetic iron oxide nanoparticles
SPR	Surface plasmon resonance
SVM	Support vector machine
SVR	Support vector regression
SWCNT	Single-walled carbon nanotube
TCE	Thermal conductivity enhancement
TEM	Transmission electron microscopy
TES	Thermal energy system
TGA	Thermogravimetric analysis
THF	Tetrahydrofuran
TiO_2	Titanium dioxide
TLS	Thermal lens spectrometry
TMM	Transfer matrix method
TN	Twisted nematic
TPC	Trophozoite parasite cells
TPPs	Tamm plasmon polaritons
TSIL	Task-specific ionic liquid
UHV	Ultrahigh vacuum
US	Ultrasound-assisted pretreatment
UV-Vis	UV-visible spectroscopy
VOCs	Volatile organic compounds
VRFBs	Vanadium redox flow batteries
WEKA	Waikato Environment for Knowledge Analysis
ZIL	Zwitterionic ionic liquid

An Introduction to Ionic Liquids, Nanofluids, Nanomaterials, and Nanotechnology

B. Namratha, Sushma, and Santosh L. Gaonkar

1 Introduction

The discipline of materials science has experienced significant progress in recent years. A multitude of advancements is currently being realized in the domain of material synthesis. The materials can be categorized into three broad classifications: structural, functional, and smart materials, based on their inherent properties. Structural materials are characterized by their stiffness, toughness, and ability to bear loads. Examples of such materials encompass ceramics, polymers, metals, and composites. These materials are utilized in various sectors, including engineering, automotive, and construction. Functional materials are engineered to exhibit specific electrical, chemical, or physical characteristics. They are meticulously designed to execute particular tasks. Silicon, ionic liquids, nanofluids, nanomaterials, and nitinol represent a selection of these materials. Smart materials possess the capability to respond to a range of external triggers. In response to alterations in their environment, they can modify their size, shape, properties, and functionalities. Examples of smart materials include electrochromic materials and self-healing materials. While all smart materials are classified as functional materials, not all functional materials exhibit the characteristics of smart materials [1–3].

B. Namratha
Department of Chemistry, The Yenepoya Institute of Arts, Science, Commerce and Management, Yenepoya (Deemed to be University), Mangaluru, Karnataka, India

Sushma
Department of Chemistry, Cambridge Institute of Technology, Bangalore, Karnataka, India

S. L. Gaonkar (✉)
Department of Chemistry, Manipal Institute of Technology, Manipal Academy of Higher Education, Manipal, Karnataka, India
e-mail: sl.gaonkar@manipal.edu

S. Kulkarni, S. Das (eds.), *Ionic Liquids, Nanofluids and Nanotechnology*,
Lecture Notes in Nanoscale Science and Technology 42,
https://doi.org/10.1007/978-3-032-04008-4_1

This chapter primarily emphasizes the properties and applications of functional materials, including ionic liquids, nanofluids, and nanomaterials, as well as their interrelation with nanotechnology.

1.1 Definition and Importance of Functional Materials

Materials that are specifically engineered to fulfill designated functions, as opposed to serving structural purposes, are designated as functional materials [4]. The significance of functional materials is increasingly recognized due to their applicability across a broad spectrum of scientific fields, facilitating the production of innovative and advanced materials; these materials exhibit exceptional optical, magnetic, thermal, and electrical properties that can be tailored to meet specific applications. The properties of these materials can be influenced by external stimuli, as well as by their composition and structural attributes; consequently, they find applications in multiple domains, including electronics, water purification, medicine, energy storage, heat transfer, and computation. The various categories of functional materials are enumerated in Table 1.

1.2 Evolution and Growth of Advanced Materials

The development and progression of materials have profoundly impacted humanity [22, 23]. A plethora of materials has been developed in response to human needs, spanning from the prehistoric era to the contemporary age. Following the Stone Age, Bronze Age, Iron Age, and the era of polymers, we now enter the age of functional materials. This era commenced with the discovery of superconductivity in mercury within U-shaped capillaries in the year 1911 by Heike Kamerlingh Onnes [24]. In 1912, Erwin Schrödinger posited that certain liquids could exhibit self-polarization during the solidification process, a phenomenon that was later designated as ferroelectric materials; however, this concept was initially dismissed. By the 1950s, researchers had identified numerous oxide-based ferroelectric materials belonging to 30 distinct families [25].

In subsequent research, Paul Walden elucidated in 1914 that certain ionic substances, designated as ionic liquids, exhibit melting points that fall below 100 °C [26]. In 1954, pioneering work at Bell Laboratories culminated in the development of the inaugural silicon-based solar cells, which were deemed practical for utilization [27]. The phenomenon known as the shape memory effect was investigated in nitinol in 1962 [28]. The liquid-crystal display (LCD) was innovatively developed in 1964 at RCA Laboratories in Princeton, New Jersey. The commercial viability of LCD technology was first realized in 1970 with the advent of the twisted-nematic (TN) operational mode [29]. The serendipitous discovery of fullerene transpired in 1985, thereby laying the groundwork for the exploration of carbon nanomaterials

Table 1 List of functional materials

Category of functional material	Functions	Examples	Applications	References
Superconductors	These materials at low temperatures show no electrical resistance, making magnetic levitation and effective power transfer possible	Hg, PB, alloys like Nb-Ge, MgB_2 (intermetallic compound), insulators like graphite	MRI, NMR, induction heaters, magnetic separation, cables, electric aircraft	[5]
Piezoelectric materials	They generate electric charges under mechanical stress	Barium titanate, lead zirconate titanate (PZT)	Sensors, ultrasonic cleaning, ultrasound imaging, actuators	[6]
Ferroelectrics	They possess spontaneous electric polarization	$BiFeO_3$, $BaTiO_3$, MXenes	Gas adsorption, photocatalysis, energy storage, water purification	[7]
Shape memory alloys	These substances can revert to their initial form after undergoing deformation	Nitinol	Medical applications like implants	[8]
Semiconductors	Materials whose conductivity lies between those of conductors and insulators	Silicon, germanium	Solar cells, electronic devices like CD players, cell phones, and telecommunication systems like receivers and transistors	[9]
Dielectric	It is an insulator with poor conductivity of electricity	Ceramics, glass	PCBs, dielectric antenna, transformers, sensors	[10]
Ferromagnetic	This material, which is attracted by the magnetic field and retains magnetism even after the magnetic field is removed	Fe, Co, and Ni and their alloys	Spintronic devices, biomedical devices, artificial neural networking	[11]
Ferrites	Ceramic-like materials which has oxides of iron with one or more metals that are ferrimagnetic in nature	Manganese zinc ferrites	Inductors, telecommunications	[12]
Magneto resistive	Tendency of the material that changes the electrical resistance under the influence of an external magnetic field	Non-magnetic materials, semiconductors, and magnetic metals	Hard disk drives, biosensors, magnetic field sensors	[13]

(continued)

Table 1 (continued)

Category of functional material	Functions	Examples	Applications	References
Luminescent material	Materials that emit light when exposed to different forms of energy	Lanthanide doped nanophosphor materials, quantum dots	Photovoltaics, light-emitting diodes	[14]
Thermoelectric	They transform thermal energy into electrical energy or the other way around	Bismuth chalcogenides, germanium chalcogenides,	Portable power generation, consumer electronics	[15]
Hydrogels	3D network capable of absorbing and retaining water	Starch-based hydrogels	Food packaging, cosmetic industry, and drug delivery	[16]
Electrochromic materials	Materials that alter their color when subjected to an electric field	Tungsten oxide, Prussian blue	Display systems	[17]
Nanomaterials	Materials whose size is in 1–100 nm	Activated carbon, clay	Water purification	[18]
Metamaterials	An engineered material with unusual electromagnetic properties, like a negative refractive index	Terahertz metamaterials, gold metamaterials	Aerospace and near space	[19]
Pyroelectrics	Generate a charge when the temperature changes	Gallium nitride, tourmaline	Sensors, dye degradation	[20]
Photovoltaic materials	They convert light energy into electricity	Copper indium gallium selenide (CIGS), silicon	Solar cells, BIPV	[21]

[30]. In 1991, Iijima and Ichihashi were credited with the initial identification of the single-walled, carbon nanotube [31]. In 1995, endeavors were initiated to distribute nanoparticles within a base fluid, subsequently referred to as nanofluids. These innovative fluids are posited as the future of heat transfer mediums [32]. In 2004, a research team from Manchester, led by Andre Geim, identified single-atom-thick crystallites within graphite. Ultimately, this finding substantiated the existence of genuine two-dimensional crystals, such as graphene, which had been previously assumed to be incapable of sustaining a planar configuration [33].

The concept of metamaterials, characterized by their advanced properties, was initially proposed by the Russian theorist Veselago in 1968, who articulated the theoretical possibility of fabricating materials possessing both negative permeability and permittivity. Nevertheless, it was Lucent Technology that first integrated innovative metamaterial enhancements for the production of resonant antennas in 2002. These materials were subsequently employed to achieve unanticipated electromagnetic characteristics in 2005, for the development of artificial media structures in 2010, and for the recent commercial availability of metamaterial-based satellite antennas in 2017 [34].

The identification of quasicrystals, wherein atomic arrangements span extensive distances but lack the periodic repetition characteristic of conventional crystals, was recognized in 2011. Traditionally, researchers upheld the belief that any crystalline atomic arrangement must exhibit a flawless repeating pattern in all spatial directions [35]. The emergence of carbon dots (CDs) in 2004 garnered significant interest, as these carbon-based nanomaterials, with dimensions under 10 nm and a generally spherical morphology, have captivated the scientific community since their initial discovery [36].

Recently, perovskite materials have attracted considerable scholarly attention, particularly concerning perovskite solar cells. Researchers are presently evaluating tandem solar cells, which integrate silicon and perovskite within their architecture. During this period, materials encompassing semiconductors, ferroelectric substances, and nanomaterials with distinctive electrical, optical, and magnetic characteristics were developed. Collectively, these materials have played a pivotal role in advancing the technologies prevalent in contemporary society, whether within the telecommunications sector, the healthcare industry, consumer electronics such as computers, smartphones, and televisions, or the transportation domain encompassing vehicles, aircraft, and maritime vessels. Through the innovation of technologies that incorporate these materials, there has been a substantial enhancement in healthcare, a reduction in mortality rates, and an elevation in living standards [37–39].

2 Ionic Liquids

2.1 Introduction of Ionic Liquids

Green chemistry represents a progressive domain within the field of chemistry that emphasizes the development of products that are environmentally benign and devoid of hazardous substances. Every chemical reaction necessitates the use of solvents, which are often detrimental and can lead to significant ecological challenges. Numerous strategies have been implemented to mitigate environmental hazards and reduce toxicity. The adoption of environmentally sustainable solvents (water, supercritical fluids, and ionic liquids) constitutes one approach to diminishing toxicity. Among these sustainable solvents, ionic liquids are particularly noteworthy due to their distinctive characteristics [40].

An ionic liquid (IL) is defined as a salt that exists in a liquid state at ambient temperature. Ionic liquids (ILs) are typically defined as substances composed entirely of ions and having a melting temperature below 100 °C. They include organic anions (such as triflate, benzoate, alkyl sulfates, alkyl carbonates, and various carboxylates) and inorganic anions (such as halides, nitrate, perchlorate, sulfate, nitrite, hexafluorophosphate, tetrafluoroborate, and azide), along with organic cations (including phosphonium, pyridinium, pyrrolidinium, and alkylammonium). In general, these ions are bonded together through different types of interactions. For a cation to remain liquid at room temperature, it should be asymmetrical, meaning the alkyl groups should be distinct from one another [41].

By carefully choosing specific pairs of cations and anions, it is possible to adjust both the polarity and the hydrophilic or hydrophobic characteristics of ionic liquids. Paul Walden first recognized an ionic liquid, namely ethyl ammonium nitrate, in 1914; however, he could not predict that ionic liquids would become a significant research focus nearly a century later. In 1940, while working at the Rice Institute in Texas, Tom Weir and Frank Hurley performed experiments that combined aluminum chloride with heated organic salt powder (alkyl pyridinium chloride), leading to the discovery of a clear, colorless liquid now known as ionic liquids [26].

Due to their unique attributes, ILs serve a broad spectrum of industries, including battery manufacturing, nanoparticle production, organic chemical processes, pharmaceutical delivery systems, the pharmaceutical industry, coatings, cleaning agents, and lubricants [42].

2.2 Classification

There are multiple classification methods for ionic liquids, which are discussed in the following sections [43–53]. Figure 1 displays a flowchart outlining the various categories of ionic liquids.

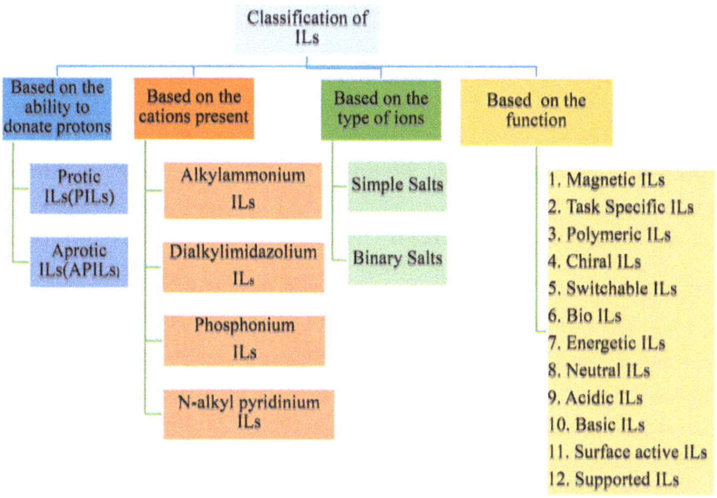

Fig. 1 Classification of ionic liquids

Based on their ability to donate protons, ionic liquids are usually divided into two main types.

Protic ionic liquids (PILs): Protons are transferred from a Brønsted acid to a Brønsted base to form these substances. Usually produced through a neutralization reaction, PILs typically exhibit improved conductivity, reduced melting points, and increased fluidity in comparison to aprotic ionic liquids. They are inexpensive, easy to produce, and do not generate byproducts. For example, ethyl ammonium nitrate serves as an illustration.

Aprotic ionic liquids (APILs): In these ionic liquids, proton transfer is absent due to the inherent characteristics of the ions making up the salt. They are produced through a multi-step process that results in higher melting points, increased costs, and lower fluidity and conductivity. An example would be 1-ethyl-3-methylimidazolium tetrafluoroborate.

Based on their cation segment: Ionic liquids can be categorized into four types—alkylammonium-ILs, dialkylimidazolium-ILs, phosphonium-based ILs, and N-alkyl pyridinium-based ILs.

Based on the Type or Number of Ions

Simple Salts: This category includes ionic liquids that consist of a single cation and an anion, conforming to the formula $[A^+]$ $[X^-]$, where A symbolizes the cation and X stands for the anion. An example is $[EtNH_3]$ $[NO_3]$.

Binary ionic liquids: These are created through the combination of two ionic liquid salts, involving an equilibrium process. For example, the mole fractions of aluminum (III) chloride and 1,3-dialkylimidazolium chloride.

Based on their Functionality

Magnetic ionic liquids (MILs): These ionic liquids are developed by infusing a paramagnetic atom, generally a d- or f-transition metal, into the anion and/or cation. With the distinct advantage of inherent magnetic mobility, MILs maintain all the unique properties of ionic liquids, such as 1-butyl-3-methylimidazolium tetrachloroferrate (III).

Task-specific ionic liquids (TSILs): These salts exhibit the typical characteristics of ionic liquids, where a functional group with a particular application is covalently bonded to either the cation or anion. The range of functional groups employed includes urea, thiourea, thioether, phosphoryl, ethylene glycol, maltolate, or thioglycolate. As a result, the type and placement of the introduced functional group within the structure can affect the properties of TSILs, in addition to the combination of cation and anion; for instance, a TSIL containing thiourea when paired with MWCNTs.

Polymeric ionic liquids (PILs): This new category of materials merges the beneficial characteristics of polymers and ionic liquids and is created by polymerizing ionic liquid monomers. An example is poly(1-vinyl-3-butylimidazolium bromide).

Chiral ILs: This type of ionic liquid includes a chiral group that is enantioselective, either in the cationic part, the anionic part, or both. For example, 1-butyl-3-methylimidazolium lactate.

Switchable ILs: Under mild conditions, these ionic liquids can modify their physical properties by introducing or removing chemical triggers such as CO_2, CS_2, or COS. These molecular triggers facilitate solute separation and help address the recycling challenges associated with conventional ionic liquids. One example is the equimolar combination of 1,8-diazabicyclo-[5.4.0]-undec-7-ene and 1-hexanol.

Biocompatible ionic liquids (Bio-ILs): The cations and anions found in these compounds come from biocompatible sources, including carboxylic acids, amino acids, glucose, and various others, which are chosen for the synthesis of these materials. Due to the fact that the cations and/or anions sourced from proteins are nontoxic, readily biodegradable, and biocompatible, most studies on bio-ILs have concentrated on these elements. Furthermore, these substances are economically viable and plentiful in nature, which signifies an increasing interest in their possible uses.

Energetic ionic liquids (EILs): These ionic compounds are characterized by a melting point below 100 °C and are known to release significant energy, with their ions potentially utilized in propellants, pyrotechnics, and explosives. For example, 1-butyl-3-methylimidazolium dicyanoborohydride.

Neutral ionic liquids (NILs): These are organic salts that typically provide cations along with stable inorganic anions (such as BF_4^- or PF_6^-) in the composition of neutral ionic liquids.

Acidic ionic liquids: These liquids serve as catalysts due to their high acidity, which is indicated by their relatively high Guttmann acceptor numbers. They can be classified into Brønsted acidic ILs and Lewis acidic ILs.

Basic ionic liquids: Typically, these consist of an abundance of nucleophilic anions (like Cl^-, OH^-, etc.) paired with organic salts (including tertiary ammonium cations, imidazolium, or other heterocyclic cations). Their primary use is in base-catalyzed organic reactions to create ionic liquid crystals.

Surface active liquids (SAL): These ionic liquids are characterized by having a long-chain structure attached to either a cation, an anion, or both. Their amphiphilic nature allows them to effectively lower the interfacial tension between water and oil.

Supported ILs (SILs): This category of ionic liquids is fixed onto a solid support. This approach reduces the amount of ILs needed while enhancing their effectiveness and selectivity in chemical reactions. Grafting agents are used to facilitate the attachment of supported ILs to the solid substrate. Consequently, these ILs exhibit greater stability and can be recycled multiple times, such as 1-butyl-3-methylimidazolium hexafluorophosphate supported on silica.

2.3 *Physicochemical Properties*

The various elements that influence a specific ionic liquid encompass the structural characteristics of the cation and anion, temperature, water content, and chain length. In comparison to organic solvents, ionic liquids exhibit a higher density. Relative to water, ionic liquids are regarded as exhibiting exceptionally high viscosity. The ions present in ionic liquids possess mobility, thereby facilitating their capability to conduct electricity. Ionic liquids are characterized by an exceedingly low vapor pressure. Consequently, these liquids remain in a liquid state at room temperature and resist evaporation even under ultrahigh vacuum (UHV) conditions. The majority of ionic liquids are devoid of color; however, any contamination can result in a distinct orange or yellow coloration. The predominant factors contributing to the coloration of ionic liquids are excessive heating and the utilization of colored raw materials. The enhanced hydrophilicity of anionic components leads to improved water solubility of ionic liquids. Ionic liquids consist of highly stable ions that exhibit no reactivity at temperatures reaching as high as 400 °C. The excellent solvent properties of ionic liquids arise from their weak coordinating ability, which minimizes interference with chemical reactions. Furthermore, they are recyclable and biodegradable and may even function as catalysts [54–56].

2.4 *Synthesis of ILS*

In most cases, the synthesis of ionic liquids requires just a single step. There are various synthesis methods for ionic liquids, including both traditional and environmentally friendly techniques.

$$NR_3 + R'X \longrightarrow [R_3R'N]^+X^-$$

Where, NR_3 is organic base/tertiary amine [R is alkyl/benzyl group]

R'X is alkyl halide/haloalkane [R' is alkyl group; X is halide like Cl^-, Br^-, I^-]

Fig. 2 Synthetic route of direct quaternization

Traditional methods/chemical methods: This category encompasses either single-step or two-step reactions. Direct quaternization yields a quaternary halide salt, which can subsequently be treated with silver and ammonium salts or Lewis acids to generate ionic liquids. Certain halide salts are available commercially and can be employed for metathesis or Lewis acid-mediated reactions. In instances where halide salts are not accessible, the direct quaternization approach should be implemented [57–59].

Direct quaternization: This reaction constitutes a one-step process. In this methodology, an organic base is combined with a haloalkane to synthesize quaternary ammonium salts. The synthetic pathway for this method is illustrated in Fig. 2.

Metathesis reaction: This reaction can take place when the salt generated from direct quaternization is engaged with silver or the ammonium salt of the target anion. The addition of silver improves both the quantity and quality of the resulting ionic liquids. This method is especially beneficial for creating ionic liquids that show solubility in water.

Lewis-acid-based reactions: This reaction occurs when the salt generated via direct quaternization is reacted with a Lewis acid, such as $AlCl_3$, among others.

Direct alkylation: This method entails the interaction between a nitrogen-containing base and an alkyl halide, resulting in the formation of ionic liquids. Ionic liquids based on imidazole are predominantly synthesized utilizing this technique.

Green methods of synthesis: The principal drawback of traditional methodologies lies in extended reaction times, suboptimal yields, and challenges associated with the handling of chemical substances. Green methodologies have been introduced primarily to mitigate the shortcomings inherent in conventional synthesis techniques. Solvent-free synthesis, sonochemical reactions, microwave-mediated reactions, and microfluidic reactions exemplify classic instances of this approach [60–63].

2.5 Applications of ILs

As electrolytes in energy storage: The deployment of ionic liquids (ILs) in energy applications represents a paramount area of research concerning their utilization. ILs can serve as solvents in the fabrication of perovskite solar cells, thereby enhancing the kinetics associated with perovskite crystal formation, which results in larger grain sizes, reduced defect densities, and superior film quality. Additionally, ILs can function as dopants to modify the properties of hole or electron transport layers. They can additionally act as carbon sources for catalysts employed in fuel cell electrodes while also being used as electrode materials for batteries and supercapacitors, as well as electrolyte solutions for different battery types. Furthermore, ILs exhibit a propensity to self-organize on the porous electrodes of supercapacitors. [EMIM] [BF_4] serves as an example of this phenomenon [64–66].

In separation technologies: The extraction of metals constitutes a principal application within the domain of separation technology. The incorporation of ILs in extraction methodologies and environmentally sustainable technologies is substantiated by their low volatility and elevated selectivity ratios. A quintessential illustration of this application is the extraction of metals from electronic waste. ILs have been investigated as supported liquid membranes, stationary phases in chromatography, and additives in various processes [67].

In biomass processing: IL-based biomass processing encompasses the pretreatment of lignocellulosic biomass, the dissolution of whole biomass, and subsequent deconstruction or fractionation processes. When lignocellulosic biomass is subjected to pretreatment with ILs, the plant cell wall typically experiences a series of physical and chemical modifications. Numerous ILs have been employed for biomass pretreatment, including methylimidazolium, pyrrolidinium, morpholinium, and choline cations in conjunction with carboxylate, triflate, methane sulfonate, amino acid, and chloride anions. Dissolution represents another prevalent technique utilized in the processing of lignocellulosic biomass. Cations like choline, quaternary ammonium, and methylimidazolium, when combined with carboxylate, chloride, amino acid, and phosphonium anions, have been found to fully dissolve a variety of herbaceous and woody feedstocks. In the regeneration stage, ILs can also be utilized to form thermo-reversible cross-links between the lignocellulosic components, providing distinct opportunities for modifying the structural and chemical characteristics of the resulting matrix [68].

Catalysis and organic synthesis: This domain continues to be a pivotal area of inquiry within the discipline of chemistry due to its relevance across numerous scientific fields. A pertinent illustration of Friedel-Craft acylation, which necessitates the presence of any Lewis acid to facilitate the acylation of aromatic rings, exemplifies the utilization of ILs. It is well-established, however, that for complete conversion to occur, a substantial excess of $AlCl_3$ is generally required. Moreover, post-reaction, $AlCl_3$ cannot be recovered or repurposed for subsequent reactions. To mitigate this challenge, various Lewis acids—including $ZnCl_2$, $FeCl_3$, and $AlCl_3$—have been investigated for their capacity to enhance the reusability of imidazolium salts. These salts are posited to offer greater reusability and environmental safety while maintaining comparable catalytic efficacy. Ionic liquids are utilized in various other organic reactions as well [69–73].

3 Nanofluids

3.1 Definition and Characteristics

A nanofluid is defined as a fluid that contains a notable concentration of nanoparticles suspended within a base fluid, which can include substances like water, oil, or ethylene glycol. The creation of a nanofluid involves the addition of metal or non-metal nanoparticles (ranging from 1 to 100 nm) and their oxides into a base fluid. The inclusion of these nanoparticles improves the thermal conductivity and heat transfer performance of the base fluid. Choosing the right base fluid is essential when formulating nanofluids. The characteristics of a nanofluid, such as its viscosity and thermal conductivity, are affected by the properties of the base fluid [74–76].

3.2 Categories of Nanofluids

Single or mono nanofluids: They result from blending a specific kind of nanoparticle with a base fluid. These can be classified into various categories: Pure metal-based nanofluids include Cu, Ag, and Au; metal oxide-based nanofluids consist of CuO, MgO, and ZnO; carbon-based nanofluids encompass SWCNT and MWCNT; nitride-based nanofluids include BN, SiN, and AlN; while carbide nanofluids comprise TiC and SiC.

Hybrid nanofluids: These are formed by combining multiple types of nanoparticles with a base fluid, such as CuO combined with MWCNTs, Al with Cu, or Ag with MgO.

3.3 Formulation of Nanofluids

The creation of nanofluids involves the dispersion of nanoparticles into a base liquid. This procedure necessitates the attainment of a consistent, resilient, and persistent suspension, ensuring that there are no chemical alterations to the base liquid and reducing the nanoparticles' coalescence. The microstructure of a nanofluid can be modified by varying synthesis parameters, such as temperature, acidity, concentrations of reactants and additives, as well as the order in which additives are introduced to the solution [77–79].

Nanofluids are generally synthesized using two primary procedural methods:

One-step method: In this method, the production of nanoparticles and their mixing in the base fluid take place at the same time in a single step. Methods like direct evaporation, physical vapor deposition, and liquid chemical direct evaporation are utilized to generate the nanofluid in this one-step process. This involves the thermal breakdown of nanoparticles under reduced pressure, simplifying the transformation of nanoparticles into a liquid state without multiple preparatory steps. There is no requirement for drying, storage, transportation, dispersion, or oxidation, which enhances stability and reduces particle agglomeration.

Two-step method: Nanoparticle synthesis initially takes place as a dry powder, and in the subsequent step, the nanoparticles are suspended in the base fluid. Continuous-flow microfluidic reactions, phase transfer techniques, submerged arc nanoparticle synthesis systems (SANSS), and direct evaporation methods are some of the innovative techniques reported for nanofluid preparation.

Furthermore, synthetic techniques can also be classified into chemical, physical, and hybrid methods. The chemical method encompasses the in situ production of nanoparticles, which are then dispersed in a base fluid to prevent agglomeration, or it may involve the individual synthesis of nanoparticles that are subsequently dispersed in base fluids. Physical methods utilize techniques such as grinding or laser ablation to distribute nanoparticles. These physical synthesis techniques provide benefits like ease of execution, scalability, and lower contamination risks when compared to certain chemical methods. The hybrid method is a dual process that merges both chemical and physical synthetic approaches to achieve a customized synthesis of nanofluids with particular attributes. This approach aims to harness the advantages of different synthesis types, tackling issues related to scalability, consistency, and control over the characteristics of nanoparticles. In hybrid synthesis, the process typically starts with a chemical phase, where nanoparticles are created through chemical reactions, sol-gel processes, or precipitation methods. This is followed by the use of physical techniques, such as sonication or high-pressure homogenization, to adjust, refine, or disperse these nanoparticles within the base fluid [80–82].

3.4 Applications of Nanofluids

Thermal Management and Cooling Systems [83, 84]

Electronics cooling: Utilized for regulating heat in microelectronics and microprocessors in compact high-power devices. Enhances cooling in liquid-cooled heat sinks for CPUs and GPUs.

Automotive sector: Nanofluids, such as Al_2O_3/ethylene glycol in radiator coolants, enhance the engine's cooling efficiency. There are potential advantages in mitigating overheating risks, which can lead to better fuel efficiency.

Heating, ventilating, and air conditioning (HVAC): Lowers energy consumption in air conditioners, chillers, and refrigeration systems. Employed in solar collectors to boost thermal absorption and heat exchange.

Renewable energy systems: PVT solar panels and other solar thermal systems, along with geothermal systems, have shown improved heat transfer. They enhance the efficiency of thermal energy storage units that utilize phase change materials (PCMs).

Nuclear reactors and power plants: Increases the heat removal rate and safety in the coolant loops of nuclear reactors. Compact heat exchangers in fossil fuel or solar thermal plants can benefit from this.

Industrial heat exchangers: Achieved in process plants, power generation, and chemical processes in oil refining. They allow for compact and high-efficiency designs due to increased heat transfer rates.

Solar energy applications: Nanofluids—suspensions of nanoparticles in base fluids like water or oil—are being more thoroughly examined to enhance the operation of solar energy systems. Their excellent thermal properties make them highly effective for solar thermal applications where efficient heat capture, transfer, and storage are crucial [76].

Role in the medical field: Nanoparticles can be employed across a diverse spectrum of medical applications and are currently under investigation as agents for drug delivery and diagnostics. One of the most promising applications of nanoparticle drug delivery systems lies in their ability to enhance the bioavailability, biocompatibility, and biodistribution profiles of numerous pharmaceuticals when integrated into a nanocarrier, particularly in instances where drug solubility represents a critical limitation to bioavailability. Lipid nanoparticles have been extensively utilized to augment the delivery of drugs that exhibit poor water solubility, facilitating the dissolution of the drug within the nanoparticle matrix to form a colloidal suspension in aqueous environments. Following administration, lipid nanoparticles are capable of fusing with and traversing cellular membranes or, alternatively entering cells via mechanisms such as endocytosis, thereby enabling the direct delivery of therapeutic agents into the cellular interior [84, 85].

4 Nanomaterials

Nanomaterials refer to substances that have dimensions in the range of 10^{-9} nm to 1–100, or can be described as particles that measure between 1 nm and 100 nm. These materials can be naturally occurring, synthetically produced, or formed by accident [86].

4.1 Classification of Nanomaterials

Nanomaterials can be classified in various ways [86–89].

Classification by Geometry

Nanospheres: Nanospheres are a type of nanoparticle characterized by their spherical shape, with sizes ranging from 10 to 200 nm. They can be either biodegradable or non-biodegradable. Biodegradable nanospheres naturally break down and are made from materials like starch or albumin, making them environmentally friendly. Examples include modified starch nanospheres, albumin nanospheres, gelatin nanospheres, polypropylene dextran nanospheres, and polylactic acid nanospheres. In contrast, non-biodegradable nanospheres do not decompose, as they consist of materials such as polymers and metals, and may negatively impact the environment. Examples in this category include polymeric nanospheres and metal nanospheres.

Nanorods: Nanorods are nanoscale materials characterized by a rod-like shape, with dimensions varying from 1 to 100 nm. They are comparable to nanotubes and are derived from semiconducting materials or metals. Gold nanorods, zinc oxide nanorods, and carbon nanorods are some examples. These materials are utilized in drug delivery, electronics, and optoelectronic devices.

Nanocapsule: Nanocapsules are tiny capsules that encapsulate another substance within. They are used in drug delivery systems where the capsules are filled with medication and coated with antibodies that attach to cancer cells to release the drugs specifically at those targeted locations. Examples include polymeric capsules, lipid capsules, and protein capsules.

Nanopatches: Nanopatches represent a needle-free, painless vaccination method designed to target various types of cells.

Nanomagnets: Nanomagnets are employed in the extraction of heavy metals and excess steroid drugs from the bloodstream.

Nanocrystalline silver: Nanocrystalline silver acts as an antimicrobial substance in the treatment of wounds. It is highly attracted to negatively charged groups found in microbial cells. Silver assaults several areas within the cell, disrupting essential physiological processes such as the formation of cell walls, membrane transport, nucleic acid synthesis, translation, protein folding and functioning, as well as electron transport.

Nanoprism: Nanoprism structures are flat and triangular, possessing anisotropic properties. They display distinct structural and crystallographic characteristics that significantly affect their surface plasmon resonance (SPR) properties. These characteristics include sharp edges, corners, and specific facet orientations, which determine their optical behavior. Their applications include chemotherapy, radiotherapy, neurotherapeutics, and antiviral and antibacterial treatments. Examples are gold nanoprisms and silver nanoprisms.

Nanowires: Nanowires are slender, thread-like structures resembling ultrafine electrical wires. Most research focuses on elements from groups III to V for nanowire systems. Their dimensions range from 10 nm to 100 nm, and they can be either conductive or non-conductive. Conductive nanowires are made of metals such as copper, gold, and silver, while non-conductive nanowires consist of SiO_2 and polymers.

Nanoplates: Flat, thin, plate-like nanoparticles.

Nanodots: Small spherical particles.

Nanocages: Nanocages are miniature structures that resemble boxes with a cage-like design. They are hollow and porous in nature.

Based on Behavior

Nanomaterials such as nanoparticles, nanoenzymes, micelles, and xenocells fall into this category.

Classification by Structural Configuration

This category includes nanobots, bimetallic structures, core-shell systems, metal oxide frameworks, biological materials, Janus particles, aerogels/hydrogels, and functionalized nanoparticles.

Based on Structural Configuration

Carbon-based: These substances are made entirely of carbon atoms, including fullerenes, multiwalled carbon nanotubes, and graphene. They demonstrate excellent thermal conductivity and corrosion resistance, making them ideal for use in nanofluids for heat and mass transfer, along with gas absorption. For example, quantum dots.

Organic: These compounds are formed from organic materials. Examples include ferritin, micelles, liposomes, and dendrimers. Micelles and liposomes are non-toxic, biodegradable, thermo- and photo-responsive nanomaterials. The surfaces of dendrimers have chain ends capable of participating in multiple chemical transformations. They are utilized in molecular detection, nano-enabled sensing, and light harvesting. Dendrimers with three dimensions possess empty interiors capable of hosting additional molecules, making them useful in drug delivery applications.

Metallic/inorganic: These do not possess carbon atoms in their backbone structure. They are metal-based/metal oxide-based nanoparticles. Metal-based nanoparticles consist solely of pure metals, offering excellent optical, electrical, thermal, antibacterial, and UV-visible properties, along with tunable catalytic characteristics.

Metal oxide nanoparticles: These comprise positive metal ions paired with negative oxygen ions, such as in zinc oxide and silicon dioxide. They are generally more stable than metallic nanoparticles and are employed in solar cells, biofuels, LEDs, optoelectronic packaging, and transistors.

Metal sulphide nanoparticles: These contain metal ions combined with sulfur. They serve applications in optics and imaging, as well as sensors, for example, ZnS and CdS.

Ceramic-based: Ceramic nanoparticles are inorganic solids composed of carbonates, carbides, phosphates, and oxides of metals and metalloids such as titanium and calcium. They are usually formed by heating and then slowly cooling and can be found in amorphous, polycrystalline, dense, porous, or hollow forms. Their primary use is in biomedical fields because of their extraordinary steadiness and high load-bearing capacity; nonetheless, they are also employed in catalysis, dye degradation, photonics, and optoelectronics.

Lipid based: These incorporate lipid components and are effectively used in numerous biomedical applications.

Polymeric: These generally appear as nanospheres or nano-capsular shapes. Nanospheres are matrix particles with an overall solid mass, while nanocapsules completely encapsulate the solid mass within the particle.

Semiconductor nanoparticles: Composed of semiconductor materials, these have properties that lie between those of metals and non-metals. They possess distinct wide band gaps and demonstrate considerable alterations in their properties when bandgap tuning is utilized, in contrast to bulk semiconductor materials, resulting in uses in photocatalysis, optics, and digital devices.

Classification by Dimensions

0 dimension: These have all three dimensions within the nanoscale range, for instance, quantum dots and fullerenes.

1 dimension: These possess one dimension that exceeds the nanoscale range, such as nanorods and nanotubes.

2 dimensions: These materials have two dimensions beyond the nanoscale range, such as nanosheets.

3 dimensions: They are not restricted to the nanoscale in any dimension, like arrays of nanotubes and nanowires.

Categorized by Pore Dimensions

Micro: These materials contain extremely small pores. They are capable of holding only minute molecules, such as gases or linear structures, and generally demonstrate slow diffusion rates along with strong interaction properties. An example of this includes Na-Y and naturally occurring clay minerals.

Meso: These materials have pores that are sufficiently large to accommodate larger molecules, such as aromatic compounds or significant polymeric monomers. The rates at which adsorbed molecules diffuse are typically influenced by capillary action, beginning with an initial interaction with the pore surfaces and then progressing to the filling of the pores. Examples include MCM-41, MCM-48, SBA-15, and carbon mesoporous materials, among others.

Macro: These materials are mainly employed as matrices for storing functional substances, as scaffolds for attaching functional groups, such as catalytic sites, and as sensing materials, owing to the quick diffusion of chemical species throughout the pore structure. Examples include carbon micro-tubes, porous gels, and porous glasses. The pores found in these substances can accommodate very large molecules, and the interactions with the pore walls are typically less significant.

Categorized by Origin

Natural: Nanomaterials found in nature are referred to as natural nanomaterials. Examples include viruses, protein molecules, minerals like clay, natural colloids such as milk and blood, gelatin, fog, insect wings, volcanic ash, ocean spray, gecko feet, spider silk, lotus leaves, etc.

Artificial: Materials that are synthesized in laboratories or are man-made are termed artificial nanomaterials. Examples: Quantum dots, fullerenes, CNTs, etc.

Based on Shape

Nanomaterials can take on various forms, including spherical, cylindrical, conical, tubular, hollow, core, and spiral shapes.

4.2 Properties of Nanomaterials

Electronic properties: In macroscopic materials, the conduction of electrons is characterized by delocalization. However, when the dimensions are reduced to the nanoscale, quantum mechanical effects become predominant. In the case of zero-dimensional nanomaterials, all spatial dimensions are effectively constrained to the nanoscale, resulting in the confinement of electrons within a three-dimensional framework, thereby eliminating any possibility of the freedom of movement of electrons. Conversely, in one-dimensional nanomaterials, electrons are restricted to a two-dimensional plane, which permits electron delocalization along the longitudinal axes of structures such as nanotubes, nanorods, or nanowires. This confinement of electrons induces a transformation from continuous energy bands to discrete energy states, thereby causing conducting materials to exhibit characteristics akin to either semiconductors or insulators.

Optical properties: The phenomenon of quantum confinement in nanomaterials allows for the modulation of visible light emission through the alteration of nanoscale parameters. Empirical observations have demonstrated that a reduction in the size of nanomaterials results in a shift of the emission peak toward shorter wavelengths, commonly referred to as a blue shift.

Magnetic properties: The dimensions of magnetic nanoparticles exert a profound influence on their magnetization characteristics. Notably, magnetization experiences a significant enhancement when the particle size is diminished to below 20 nm.

Consequently, by decreasing the particle size of granular magnetic materials, the performance of magnets fabricated from such materials can be substantially improved.

Catalytic activity: Nanomaterials exhibit a propensity for rapid degradation or oxidation upon environmental exposure. Their electronic architecture and the quantity of surface atoms exhibit considerable divergence from those of bulk materials, which serves as a critical determinant of their elevated reactivity. Nanomaterials have long been acknowledged as highly promising candidates for heterogeneous catalysis, owing to their substantial surface-to-volume ratio and the capability to optimize active facets through morphological alterations of the nanocrystals.

4.3 Nanomaterial Synthesis

Nanomaterials can primarily be synthesized through two main approaches: the top-down approach and the bottom-up approach (Fig. 3).

Top-down approach: The methodology employed for the synthesis of nanoparticles intended for functional materials involves a systematic diminution in dimensions from bulk substances, utilizing a variety of analytical techniques, which may include lithography, milling, electrospinning, sputtering, laser ablation, and chemical vapor deposition. These methodologies exhibit a significant dependency on energy consumption, and the resultant product typically demonstrates a less homogeneous size distribution in contrast to that generated by the bottom-up approach. The ultimate dimensions of the synthesized product are influenced by numerous variables, including rotational speed, thermal regulation, ambient noise levels, and the potential for contamination [90].

Bottom-up approach: The strategy for nanoparticle synthesis is centered on the direct assembly of particles from atomic and molecular constituents in either gaseous or liquid phases. The processing methodologies encompass a combination of both chemical and physical techniques. Initial reactants undergo dissolution in a liquid medium or vaporization, followed by subsequent condensation to yield particles. The synthesis methodologies comprise supercritical synthesis, co-precipitation, sol-gel techniques, laser pyrolysis, chemical reduction, solvothermal and hydrothermal methods, chemical vapor deposition, molecular condensation, self-assembly, and environmentally friendly synthesis. The dimensions of the final product are contingent upon a variety of operational parameters during the synthesis process, such as the spatial separation between the target and substrate in chemical vapor deposition. Influential factors such as temperature, the solvent employed, and pressure play a pivotal role in hydrothermal and solvothermal approaches, while the co-precipitation technique is influenced by pH, reaction duration, precursor concentration, stirring speed, and temperature. This methodological approach offers significant advantages, including the capacity to generate fine nanoparticles with a uniform size distribution and the flexibility to modify deposition parameters to achieve diverse morphologies and dimensions. There exists substantial potential for

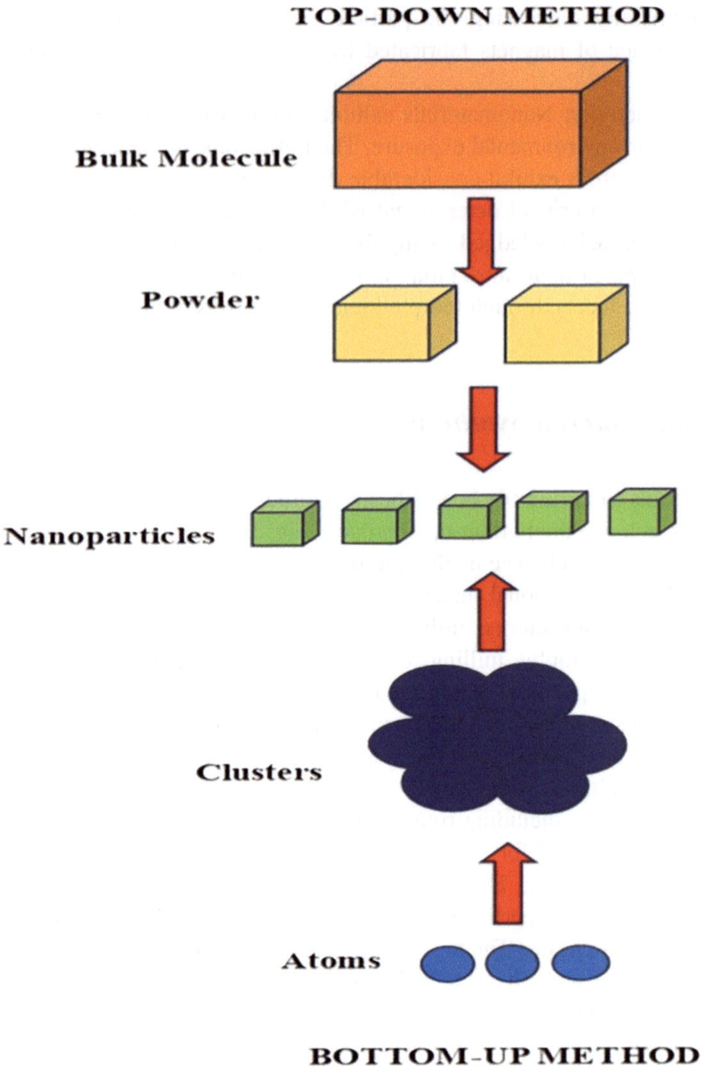

Fig. 3 Illustration of nanoparticle synthesis using top-down and bottom-up methods

parameter manipulation, given that synthesis can be executed in either the gas or liquid phase. Nonetheless, notable limitations include the requirement for chemical purification and the constraints associated with large-area deposition [91, 92].

Nanoparticle synthesis from waste: The economic feasibility of processes involving nanoparticles poses a considerable obstacle to the pervasive industrial application of nanomaterials. The application of various top-down and bottom-up methodologies for the synthesis of nanoparticles derived from polymeric materials, rubber, metallic scrap, and natural waste can effectively mitigate this issue.

Furthermore, the extraction of noble metals from electronic waste and solar panel components through pyro-/hydrometallurgical processes can contribute to waste minimization and enhance processing efficiency. Nanoparticles synthesized from coconut shells via mechanical milling can also function as particle reinforcements and fillers in a range of aesthetic uses [93].

4.4 Applications of Nanomaterials

Medical field: Nanoparticles (NPs) have increasingly attracted scholarly attention across the spectrum of medical disciplines due to their ability to facilitate drug delivery within an optimal dosage range, which frequently results in enhanced therapeutic efficacy, diminished adverse effects, and improved patient compliance. Iron oxide nanoparticles, such as magnetite (Fe_3O_4) and its oxidized variant maghemite (Fe_2O_3), rank among the most commonly employed materials for biomedical applications. Superparamagnetic iron oxide nanoparticles, characterized by appropriate surface chemistry, can be utilized in various in vivo applications, including the enhancement of MRI contrast, the promotion of tissue repair, the execution of immunoassays, the detoxification of biological fluids, the induction of hyperthermia, the targeted delivery of pharmaceuticals, and the separation of cells [94, 95].

Electronics: There has been an escalating scholarly interest in the development of printed electronics in contemporary years, attributable to their advantageous characteristics in comparison to conventional silicon methodologies, in addition to the prospective capabilities for economically viable, large-area electronic systems that are amenable to flexible displays and sensors. The advancement of printed electronics that integrate a variety of functional inks comprising nanoparticles such as metallic nanoparticles, organic electronic compounds, carbon nanotubes (CNTs), and ceramic nanoparticles is projected to progress expeditiously as a feasible large-scale production technique for pioneering electronic appliances [96].

Energy harvesting: Recent investigations have underscored the inherent shortcomings and imminent exhaustion of fossil fuel reserves, owing to their non-renewable nature. In light of this, researchers are recalibrating their methodologies to cultivate renewable energy sources derived from abundantly available materials at minimal costs. They have discerned that nanoparticles present themselves as optimal candidates for such endeavors, attributed to their substantial surface area, distinctive optical properties, and remarkable catalytic functionalities. NPs are particularly prevalent in photocatalytic applications for producing energy through photoelectrochemical (PEC) processes and electrochemical water splitting [97].

Water purification: Various methods, such as adsorption and photocatalysis, can be employed for water purification using nanoparticles like ZnO, CuO, NiO, and Ag. Nanotechnology represents a superior option for eliminating contaminants, such as heavy metals, from water [98–100].

5 Nanotechnology: An Integrative Perspective

5.1 Fundamentals of Nanotechnology

The discipline of nanotechnology encompasses the manipulation of substances at the atomic and molecular dimensions, typically ranging from 1 to 100 nm. While the designation "nanotechnology" was first coined by Norio Taniguchi in 1974, it was the physicist Richard Feynman who laid the foundational theoretical framework for this domain. By its very nature, nanotechnology encompasses various disciplines, merging materials science, engineering, biology, chemistry, and physics. This leads to the creation of new materials and devices with distinctive properties, such as enhanced strength, lighter weight, or improved chemical reactivity. The ability to manipulate matter at the nanoscale paves the way for innovative developments across numerous fields, including environmental science, electronics, energy, and healthcare [101, 102].

5.2 The Impact of Nanotechnology on Functional Materials

Improving Material Attributes

Nanotechnology plays a crucial role in improving the properties of materials, resulting in the emergence of functional materials utilized in diverse industries [103, 104]:

Electronics: Nanoscale transistors and components allow for miniaturization, resulting in faster and more efficient electronics.

Energy: Nanomaterials enhance the performance of fuel cells, batteries, and solar panels by optimizing surface area and charge transport.

Healthcare: Nanoparticles in drug delivery systems facilitate targeted delivery to specific cells, enhancing effectiveness and minimizing side effects.

Environment: Due to their enhanced surface interface and superior reactive properties, nanomaterials are employed for environmental monitoring, reducing pollution, and purifying water.

Current Developments and Future Opportunities

Recent advancements in nanotechnology include the creation of nanomachines capable of being programmed for specific functions, such as environmental cleanup and targeted medication delivery. For example, researchers are exploring the use of light-activated nanomachines to treat specific cancers and combat drug-resistant diseases. Future advancements in nanotechnology could result in breakthroughs in areas like targeted therapy, quantum computing, and sustainable energy. Nonetheless, these opportunities will need to overcome challenges related to cost, scalability, and integration with existing technologies [105].

5.3 *Challenges and Ethical Implications*

Concerns About Environmental Impact and Toxicity

Due to their unique characteristics, nanomaterials raise concerns regarding potential toxicity and environmental effects. The production, use, or breakdown of nanoparticles can lead to their release into the environment, impacting ecosystems and possibly human health. For instance, the antibacterial properties of nanosilver used in textiles may leach into water systems, affecting aquatic organisms. Moreover, the diminutive dimensions and significant reactivity of nanoparticles may engender unforeseen interactions with biological systems, thereby necessitating comprehensive hazard analysis and safety review [106].

Regulatory Challenges and Safety Issues

The regulations governing nanotechnology are complex and differ by country and region. It is possible that the specific characteristics and risks associated with nanomaterials are not sufficiently addressed by existing laws concerning chemical and product safety. For example, current thresholds and exemptions may not apply to nanomaterials, which could fall under multiple regulatory categories (chemicals, cosmetics, and food). Regulatory agencies are working on establishing comprehensive guidelines and risk assessment protocols to ensure the safe development and utilization of nanotechnology. Global harmonization initiatives are aimed at enhancing consistency and collaboration in nanotechnology regulation, encouraging innovation while protecting the environment and public health [107].

6 Synergies Between Ionic Liquids, Nanofluids, and Nanomaterials

Ionic liquids (ILs) are characterized by their remarkable thermal stability, low volatility, and customizable solvation properties. When combined with nanomaterials, ILs serve various roles as functionalizing agents, reaction media, stabilizers, and solvents. They provide a unique reaction environment conducive to the controlled synthesis of nanoparticles, particularly those composed of metals and metal oxides. Furthermore, ILs enhance the conductivity and ionic mobility of nanostructured electrodes, thereby rendering them suitable for applications in fuel cells, batteries, and supercapacitors.

Nanofluids are engineered suspensions that incorporate nanoparticles, typically with dimensions less than 100 nm, dispersed within base fluids such as water, ethylene glycol, and oils. Ionic liquids can be utilized as either base fluids or additives within nanofluids to develop hybrid nanofluid systems that exhibit enhanced electrochemical, rheological, and thermal properties. Hybrid nanofluids demonstrate superior heat transfer capabilities, which are critical for high-performance cooling systems in the automotive, aerospace, and electronics industries. Under the influence of external magnetic or electric fields, nanofluids containing ionic liquids and functional nanoparticles exhibit tunable conductivity and viscosity.

The amalgamation of ionic liquids, nanomaterials, and nanofluids is facilitating the advancement of more efficient and green energy solutions. By augmenting thermal storage and heat transfer, IL-based nanofluids enhance the efficiency of solar collectors. Likewise, lithium-ion, sodium-ion, and redox-flow batteries utilize IL-nanoparticle electrolytes, which improve cycle life, safety, and electrochemical stability. In the realm of hydrogen storage for fuel cell applications, nanomaterials modified with ionic liquids have shown considerable promise in the efficient adsorption and release of hydrogen.

Research is ongoing into perovskites and IL-functionalized quantum dots to enhance charge transport and light absorption. Hybrid materials that combine nanostructured semiconductors with ionic liquids exhibit promising thermoelectric conversion efficiency. The synergistic application of ionic liquids and nanomaterials is facilitating advancements in drug delivery, biosensing, and diagnostics.

The intersection of ionic liquids, nanofluids, and nanomaterials signifies a transformative frontier in material science and technology. Their hybridization leverages the distinct advantages of each component, yielding synergistic benefits for applications ranging from smart manufacturing and healthcare to renewable energy and environmental remediation. These hybrid systems possess the potential to redefine performance benchmarks in biomedical and industrial sectors, provided that challenges related to toxicity, cost, and scalability of production are adequately addressed [108].

7 Conclusion and Future Directions

This chapter has examined the application of ionic liquids, nanofluids, nanomaterials, and their interrelations with nanotechnology. Each section presents a foundational overview and delineates their applications across diverse domains such as electronics, energy, and material synthesis.

These materials are instrumental in the development of smart materials and devices applicable in drug delivery systems, electrochromic materials, and self-healing substances. By employing the principles of green chemistry, it is feasible to engineer products that are environmentally sustainable. The synthesis and application of two-dimensional materials in electronic contexts are also highlighted.

These materials have the potential to be integrated with artificial intelligence to facilitate the discovery of novel processes and products. Moreover, there exists an opportunity for the synthesis of environmentally friendly biodegradable materials that serve multifunctional purposes [108, 109].

References

1. A. Nande, N.T. Kalyani, A. Tiwari, S.J. Dhoble, Exploring the world of functional materials, in *Functional Materials from Carbon, Inorganic, and Organic Sources*, (Elsevier, Amsterdam, 2023), pp. 1–19. https://doi.org/10.1016/B978-0-323-85788-8.00014-8
2. F. Arab Hassani, Q. Shi, F. Wen, T. He, A. Haroun, Y. Yang, Y. Feng, C. Lee, Smart materials for smart healthcare—Moving from sensors and actuators to self-sustained nanoenergy nanosystems. Smart Mater. Med. **1**, 92 (2020). https://doi.org/10.1016/j.smaim.2020.07.005
3. A.H. Wilson, Functional materials, in *Pergamon Materials Series*, vol. 5, (Elsevier, Amsterdam, 2001), pp. 253–304. https://doi.org/10.1016/S1470-1804(01)80010-9
4. F. Tiwari, Exploring the world of functional materials revolutionizing technology and beyond. Adv. Mater. Sci. Res. **6**, 58 (2023)
5. C. Yao, Y. Ma, Superconducting materials: Challenges and opportunities for large-scale applications. iScience **24**, 102541 (2021). https://doi.org/10.1016/j.isci.2021.102541
6. B. Chandra Sekhar, B. Dhanalakshmi, B. Srinivasa Rao, S. Ramesh, K. Venkata Prasad, P.S.V. Subba Rao, B. Parvatheeswara Rao, Piezoelectricity and its applications, in *Multifunctional Ferroelectric Materials*, ed. by D. Ranjan Sahu, (IntechOpen, London, 2021). https://doi.org/10.5772/intechopen.96154
7. W. Ding, J. Lu, X. Tang, L. Kou, L. Liu, Ferroelectric materials and their applications in activation of small molecules. ACS Omega **8**, 6164 (2023). https://doi.org/10.1021/acsomega.2c06828
8. S.C. Bhatia, Energy resources and their utilisation, in *Advanced Renewable Energy Systems*, (Elsevier, Amsterdam, 2014), pp. 1–31. https://doi.org/10.1016/B978-1-78242-269-3.50001-2
9. T. Duerig, A. Pelton, D. Stöckel, An overview of nitinol medical applications. Mater. Sci. Eng. A **273**, 149 (1999). https://doi.org/10.1016/S0921-5093(99)00294-4
10. M.A. Rahman, A review on semiconductors including applications and temperature effects in semiconductors. Am. Sci. Res. J. Eng. Technol. Sci. **7**, 50 (2014)
11. M. Hussain, H. Zahra, S.M. Abbas, Y. Zhu, Flexible dielectric materials: Potential and applications in antennas and RF sensors. Adv. Electron. Mater. **10**, 2400240 (2024). https://doi.org/10.1002/aelm.202400240
12. R. Valenzuela, Novel applications of ferrites. Phys. Res. Int. **2012**, 1 (2012). https://doi.org/10.1155/2012/591839
13. O. Vitayaya, P.Z.Z. Nehan, D.R. Munazat, M.T.E. Manawan, B. Kurniawan, Magnetoresistance (MR) properties of magnetic materials. RSC Adv. **14**, 18617 (2024). https://doi.org/10.1039/D4RA01989J
14. A. Kumar Soni, B. Pratap Singh, Luminescent materials in lighting, display, solar cell, sensing, and biomedical applications, in *Luminescence - OLED Technology and Applications*, ed. by S. Pyshkin, (IntechOpen, London, 2020). https://doi.org/10.5772/intechopen.82123
15. M.R. Shankar, A.N. Prabhu, A review on structural characteristics and thermoelectric properties of mid-temperature range chalcogenide-based thermoelectric materials. J. Mater. Sci. **58**, 16591 (2023). https://doi.org/10.1007/s10853-023-09028-8
16. M. Qamruzzaman, F. Ahmed, M.I.H. Mondal, An overview on starch-based sustainable hydrogels: Potential applications and aspects. J. Polym. Environ. **30**, 19 (2022). https://doi.org/10.1007/s10924-021-02180-9

17. C. Gu, A.-B. Jia, Y.-M. Zhang, S.X.-A. Zhang, Emerging electrochromic materials and devices for future displays. Chem. Rev. **122**, 14679 (2022). https://doi.org/10.1021/acs.chemrev.1c01055
18. P. Jangid, M. Prabhu Inbaraj, Applications of nanomaterials in wastewater treatment. Mater Today Proc **43**, 2877 (2021). https://doi.org/10.1016/j.matpr.2021.01.126
19. R. Liu, C. Ji, Z. Zhao, T. Zhou, Metamaterials: Reshape and rethink. Engineering **1**, 179 (2015). https://doi.org/10.15302/J-ENG-2015036
20. Y. Liang, X. Tang, Q. Zhu, J. Han, C. Wang, A review: Application of tourmaline in environmental fields. Chemosphere **281**, 130780 (2021). https://doi.org/10.1016/j.chemosphere.2021.130780
21. M. Siva Ramkumar, R. Felshiya Rajakumari, N. Kannan, R. Premkumar, S. Mohanasundaram, S. Purushotham, D. Ramya, K. Rajan, Semiconductor materials for solar PV technology and challenges towards electrical engineering. Adv. Mater. Sci. Eng. **2022**, 1 (2022). https://doi.org/10.1155/2022/7272489
22. L. Kennedy, The prehistoric ages: How humans lived before written records, https://www.history.com/articles/prehistoric-ages-timeline
23. D. Feldman, Polymer history. Des. Monomers Polym. **11**, 1 (2008). https://doi.org/10.1163/156855508X292383
24. D. Van Delft, P. Kes, The discovery of superconductivity. Europhys. News **42**, 21 (2011). https://doi.org/10.1063/1.3490499
25. N. Author, Ferroelectricity: 100 years on, https://physicsworld.com/a/ferroelectricity-100-years-on/
26. Z. Lei, B. Chen, Y.-M. Koo, D.R. MacFarlane, Introduction: Ionic liquids. Chem. Rev. **117**, 6633 (2017). https://doi.org/10.1021/acs.chemrev.7b00246
27. L.L. Kazmerski, The 70th anniversary of the bell telephone laboratories "solar battery" discovery. Solar Compass **10**, 100073 (2024). https://doi.org/10.1016/j.solcom.2024.100073
28. A. Cladera, B. Weber, C. Leinenbach, C. Czaderski, M. Shahverdi, M. Motavalli, Iron-based shape memory alloys for civil engineering structures: An overview. Constr. Build. Mater. **63**, 281 (2014). https://doi.org/10.1016/j.conbuildmat.2014.04.032
29. H. Kawamoto, The history of liquid-crystal display and its industry, in *2012 Third IEEE History of Electro-Technology Conference (HISTELCON)*, (IEEE, Pavia, 2012), pp. 1–6
30. S. Goodarzi, T. Da Ros, J. Conde, F. Sefat, M. Mozafari, Fullerene: Biomedical engineers get to revisit an old friend. Mater. Today **20**, 460 (2017). https://doi.org/10.1016/j.mattod.2017.03.017
31. J. Chen, S. Wei, H. Xie, A brief introduction of carbon nanotubes: History, synthesis, and properties. J. Phys.: Conf. Ser. **1948**, 012184 (2021). https://doi.org/10.1088/1742-6596/1948/1/012184
32. L. Zheng, X. Zhang, Differential transform method, in *Modelling and Analysis of Modern Fluid Problems*, (Elsevier, Amsterdam, 2017), pp. 179–251
33. Discovery of Graphene - Graphene - The University of Manchester, https://www.graphene.manchester.ac.uk/learn/discovery-of-graphene/
34. R. Kumar, M. Kumar, J.S. Chohan, S. Kumar, Overview on metamaterial: History, types and applications. Mater. Today Proc. **56**, 3016 (2022). https://doi.org/10.1016/j.matpr.2021.11.423
35. D. Monroe, Nobel prize–Discovery of quasicrystals. Physics **28**, 14 (2011)
36. J.D.G. Da Rocha, M.A.P. Cechinel, L.F. Rocha, H.G. Riella, N. Padoin, C. Soares, Exploring the potential of rare earth doped carbon dots: Concepts and applications. Chem. Eng. J. Adv. **17**, 100583 (2024). https://doi.org/10.1016/j.ceja.2024.100583
37. J.-A. Alberola-Borràs, R. Vidal, I. Mora-Seró, Evaluation of multiple cation/anion perovskite solar cells through life cycle assessment. Sustain. Energy Fuels **2**, 1600 (2018). https://doi.org/10.1039/C8SE00053K
38. E.O. Pyzer-Knapp, J.W. Pitera, P.W.J. Staar, S. Takeda, T. Laino, D.P. Sanders, J. Sexton, J.R. Smith, A. Curioni, Accelerating materials discovery using artificial intelligence, high performance computing and robotics. Npj Comput. Mater. **8**, 84 (2022). https://doi.org/10.1038/s41524-022-00765-z

39. J. Joy, E. George, S. Anas, S. Thomas, Applications of self-healing polymeric systems, in *Self-Healing Polymer-Based Systems*, (Elsevier, Amsterdam, 2020), pp. 495–513. https://doi.org/10.1016/B978-0-12-818450-9.00016-7
40. P. Anastas, N. Eghbali, Green chemistry: Principles and practice. Chem. Soc. Rev. **39**, 301 (2010). https://doi.org/10.1039/B918763B
41. J.L. Shamshina, O. Zavgorodnya, R.D. Rogers, Ionic liquids, in *Reference Module in Chemistry, Molecular Sciences and Chemical Engineering*, (Elsevier, Amsterdam, 2018), pp. 218–225. https://doi.org/10.1016/B978-0-12-409547-2.13931-9
42. G. Kaur, H. Kumar, M. Singla, Diverse applications of ionic liquids: A comprehensive review. J. Mol. Liq. **351**, 118556 (2022). https://doi.org/10.1016/j.molliq.2022.118556
43. W. Silva, M. Zanatta, A.S. Ferreira, M.C. Corvo, E.J. Cabrita, Revisiting ionic liquid structure-property relationship: A critical analysis. IJMS **21**, 7745 (2020). https://doi.org/10.3390/ijms21207745
44. K. Ghandi, A review of ionic liquids, their limits and applications. GSC **4**, 44 (2014). https://doi.org/10.4236/gsc.2014.41008
45. H. Niedermeyer, J.P. Hallett, I.J. Villar-Garcia, P.A. Hunt, T. Welton, Mixtures of ionic liquids. Chem. Soc. Rev. **41**, 7780 (2012). https://doi.org/10.1039/C2CS35177C
46. M.H. Zainal-Abidin, M. Hayyan, J. Matmin, A.M. Al-Fakih, N. Jamaluddin, W.M.A. Wan Mahmood, R. Abdul Wahab, F. Abdullah, Greening industrial applications with magnetic-based deep eutectic solvents: A promising future. J. Ind. Eng. Chem. **124**, 1 (2023). https://doi.org/10.1016/j.jiec.2023.04.011
47. B.A.D. Neto, A.A.M. Lapis, R.Y. Souza, Task-specific ionic liquids: Design, properties and applications, in *Encyclopedia of Ionic Liquids*, ed. by S. Zhang, (Springer, Singapore, 2019), pp. 1–11. https://doi.org/10.1007/978-981-10-6739-6_33-1
48. L. Sola, Column technology for capillary electromigration methods, in *Capillary Electromigration Separation Methods*, (Elsevier, Amsterdam, 2018), pp. 45–68. https://doi.org/10.1016/B978-0-12-809375-7.00003-4
49. R. Ratti, Ionic liquids: Synthesis and applications in catalysis. Adv. Chem. **2014**, 1 (2014). https://doi.org/10.1155/2014/729842
50. Z. Liu, P. Hu, X. Meng, R. Zhang, H. Yue, C. Xu, Y. Hu, Synthesis and properties of switchable polarity ionic liquids based on organic superbases and fluoroalcohols. Chem. Eng. Sci. **108**, 176 (2014). https://doi.org/10.1016/j.ces.2013.12.040
51. R. Santiago, I. Díaz, M. González-Miquel, P. Navarro, J. Palomar, Assessment of bio-ionic liquids as promising solvents in industrial separation processes: Computational screening using COSMO-RS method. Fluid Phase Equilib. **560**, 113495 (2022). https://doi.org/10.1016/j.fluid.2022.113495
52. A. Singh, N. Kaur, A. Parmar, H.K. Chopra, Structure and properties of ionic liquids: Green aspects, in *Ionic Liquids in Analytical Chemistry*, (Elsevier, Amsterdam, 2022), pp. 1–32. https://doi.org/10.1016/B978-0-12-823334-4.00004-7
53. I.A. Berezianko, S.V. Kostjuk, Ionic liquids in cationic polymerization: A review. J. Mol. Liq. **397**, 124037 (2024). https://doi.org/10.1016/j.molliq.2024.124037
54. P. Sharma, S. Sharma, H. Kumar, Introduction to ionic liquids, applications and micellization behaviour in presence of different additives. J. Mol. Liq. **393**, 123447 (2024). https://doi.org/10.1016/j.molliq.2023.123447
55. Y. Cao, S. Yao, X. Wang, Q. Peng, H. Song, *The Physical and Chemical Properties of Ionic Liquids and its Application in Extraction* (Nova Science Publisher, Hauppauge, 2012)
56. O. Soleimani, Properties and applications of ionic liquids. J. Chem. Rev. **2**, 169 (2020). https://doi.org/10.33945/SAMI/JCR.2020.3.4
57. R. Ferraz, C. Prudencio, M. Vieira, R. Fernandes, J.P. Noronha, Z. Petrovski, ChemInform abstract: Ionic liquids synthesis - methodologies. ChemInform **46** (2015). https://doi.org/10.1002/chin.201538264
58. G. Zbancioc, I.I. Mangalagiu, C. Moldoveanu, Ultrasound assisted synthesis of imidazolium salts: An efficient way to ionic liquids. Ultrason. Sonochem. **23**, 376 (2015). https://doi.org/10.1016/j.ultsonch.2014.10.028

96. A. Sharma, D. Manocha, A. Arora, A. Kumar, C.M. Srivastava, V. Rawat, H. Kim, M. Verma, G.K. Rao, Applications of green nanomaterials in electronic and electrical industries, in *Green Nanomaterials for Industrial Applications*, (Elsevier, Amsterdam, 2022), pp. 397–421. https://doi.org/10.1016/B978-0-12-823296-5.00016-2
97. M.Z.A. Mahmud, A concise review of nanoparticles utilized energy storage and conservation. J. Nanomater. **2023**, 1 (2023). https://doi.org/10.1155/2023/5432099
98. J. Joy, J. Mathew, S.C. George, Nanomaterials for photoelectrochemical water splitting—Review. Int. J. Hydrog. Energy **43**, 4804 (2018). https://doi.org/10.1016/j.ijhydene.2018.01.099
99. S. Kamath, V.G.V. Ramanjaneyalu, S. Kamila, Application of ZnO Nano rods for the batch adsorption of Cr(VI): A study of kinetics and isotherms. Am. J. Appl. Sci. **16**, 101 (2019). https://doi.org/10.3844/ajassp.2019.101.112
100. M. Pinto, P.S.F. Ramalho, N.F.F. Moreira, A.G. Gonçalves, O.C. Nunes, M.F.R. Pereira, O.S.G.P. Soares, Application of magnetic nanoparticles for water purification. Environ. Adv. **2**, 100010 (2020). https://doi.org/10.1016/j.envadv.2020.100010
101. B.A. Omran, Fundamentals of nanotechnology and nanobiotechnology, in *Nanobiotechnology: A Multidisciplinary Field of Science*, (Springer International Publishing, Cham, 2020), pp. 1–36. https://doi.org/10.1007/978-3-030-46071-6_1
102. N. Battard, Convergence and multidisciplinarity in nanotechnology: Laboratories as technological hubs. Technovation **32**, 234 (2012). https://doi.org/10.1016/j.technovation.2011.09.001
103. S. Malik, K. Muhammad, Y. Waheed, Nanotechnology: A revolution in modern industry. Molecules **28**, 661 (2023). https://doi.org/10.3390/molecules28020661
104. N. Chausali, J. Saxena, R. Prasad, Recent trends in nanotechnology applications of bio-based packaging. J. Agric. Food Res. **7**, 100257 (2022). https://doi.org/10.1016/j.jafr.2021.100257
105. S. Mahendra Kumar, Y. Rajni, T. Sandeep Prasad, Recent advances in nanotechnology. Int. J. Nanomater. Nanotechnol. Nanomed. **9**, 15 (2023). https://doi.org/10.17352/2455-3492.000053
106. P.A. Schulte, F. Salamanca-Buentello, Ethical and scientific issues of nanotechnology in the workplace. Environ. Health Perspect. **115**, 5 (2007). https://doi.org/10.1289/ehp.9456
107. S. Wasti, I.H. Lee, S. Kim, J.-H. Lee, H. Kim, Ethical and legal challenges in nanomedical innovations: A scoping review. Front. Genet. **14**, 1163392 (2023). https://doi.org/10.3389/fgene.2023.1163392
108. Y. Wen, A review of applications and future prospects of nanotechnology. ACE **56**, 135 (2024). https://doi.org/10.54254/2755-2721/56/20240640
109. M. Sahoo, S. Vishwakarma, C. Panigrahi, J. Kumar, Nanotechnology: Current applications and future scope in food. Food Front. **2**, 3 (2021). https://doi.org/10.1002/fft2.58

Properties and Characterization of Ionic Liquid, Nanofluid, and Nanomaterials

Ashwini Patil and Lovleen Marwaha

1 Structural and Morphological Characterization

Knowledge of the structural and morphological features of ionic liquids, nanofluids, and nanomaterials is the cornerstone for the design of material properties for a diversity of applications, including catalysis, energy storage, and biomedical devices. Specific characterization of this area requires a thorough examination of crystallinity, particle size and shape, and external and internal morphology of material particles [1]. All these factors are very important in determining overall physical, chemical, and mechanical behaviors of these advanced materials (Fig. 1).

Electron microscopy type approaches, in particular scanning electron microscopy (SEM) and transmission electron microscopy (TEM), qualify as the primary approaches to morphological analysis. SEM makes use of the concentrated electron beam that sweeps across the sample surface, producing a high-resolution image that shows the topographical details and surface texture [1]. This technique is especially beneficial in the detection of the external characteristics of nanomaterials and nanofluids, where the phenomena of surface roughness and particle aggregation can quite substantially affect the stability and activity of the fluids [2]. Contrary to this, with TEM, there is the capability of probing the internal setup of materials; this involves the passing of electrons through ultra-thin sections. TEM not only gives an understanding of the shape and size of individual nanoparticles but also enables one to see crystallographic defects, lattice fringes, and grain boundaries at near-atomic levels [3]. Combined use of SEM and TEM therefore gives complementary information on the morphology of the surface from SEM and the internal structures from TEM, which is very essential even for a complete understanding of nanomaterials.

A. Patil · L. Marwaha (✉)
Department of Zoology, School of Bioengineering and Biosciences, Lovely Professional University, Phagwara, Punjab, India

S. Kulkarni, S. Das (eds.), *Ionic Liquids, Nanofluids and Nanotechnology*, Lecture Notes in Nanoscale Science and Technology 42,
https://doi.org/10.1007/978-3-032-04008-4_2

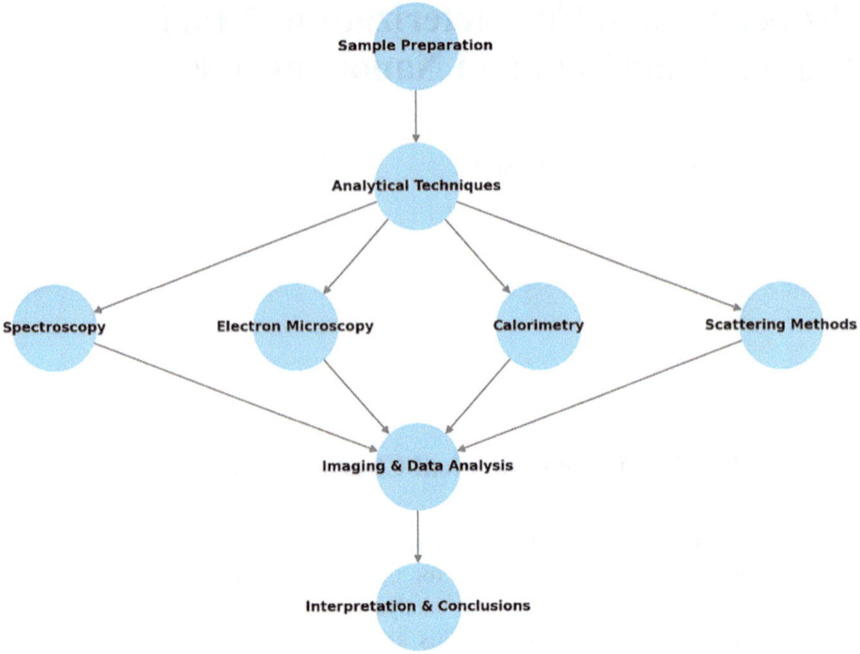

Fig. 1 Schematic diagram of characterization workflow. A flowchart outlining the steps from sample preparation to imaging and data interpretation

Another key method used to determine crystalline nature of materials is the X-ray diffraction (XRD). Using the analysis of diffraction patterns obtained for X-rays on interaction with the atomic planes with a crystalline pattern, the XRDs can be used for identifying the crystalline phases and determination of the crystal sizes. In the case of nanomaterials, where the effects of the quantum confinement are usually involved, accurate determination of lattice parameters and crystallinity is important [4]. For instance, a high degree of order in crystalline nanoparticles may be reflected through well-defined diffraction peaks, whereas a reduction in the size of particles or increase in the strain in the lattice may be indicated through broadening of peaks [5]. Notably, the ionic liquids may pose problems for XRD analysis since most of them do not exhibit long-range order found in crystalline solids. Instead, they demonstrate diffuse scattering patterns that signify the disorderness of their ionic networks [6].

The atomic force microscopy (AFM) adds more value to the morphological characterization toolkit by providing three-dimensional surface profiling with sub-nanometer resolution. In contrast to the conventional electron microscopy techniques, AFM is conducted at ambient temperatures without necessitating the use of conductive coatings, which makes it an ideal technique to be used in the analysis of soft or delicate materials such as ionic liquids [7]. AFM is capable of generating fine topographical maps that reveal the differences in the roughness of the surface and nanostructured components of surfaces after scanning the sharp tip over the surface. This is an extremely useful ability when dealing with nanofluids, where interactions between the dispersed nanoparticles and the host fluid determine variables such as viscosity and thermal conductivity [8].

Dynamic light scattering (DLS) is widely applied for determining the size distribution of nanoparticles being suspended in a solution [9]. This approach is based on the detection of fluctuations of the light intensity scattered by Brownian particles, which is correlated to the hydrodynamic radius of the Brownian particles. DLS is especially beneficial for the study of dispersion stability of the nanofluids since it provides rapid in situ readings on particle size, aggregation, and its behavior. Time-dependent monitoring of size distribution is crucial for the optimization of synthesis protocols and to keep nanofluid in a stable and uniformly dispersed state, which has direct consequence for its thermal and rheological properties.

The integration of those advanced methods allows understanding the structural and morphological characteristics of materials in a multidimensional context. For example, even though SEM and TEM provide critical information on surface and internal structures, respectively, XRD and AFM are able to confirm by providing quantification of crystallinity and surface structure. DLS, however, provides quantitative data on particle size distribution, which is one of the significant determinants of the functional performance of nanofluids and nanomaterials [10]. This in-depth description is not only necessary for the basic scientific knowledge, but also for the applied engineering of materials with specific properties.

Recent studies emphasize the importance of multiprocedural approach to address the complexity of materials on nanoscale. Lee et al. proved that the catalytic efficiency of the metallic nanoparticles is strongly associated with their exact size and surface morphology, as revealed by a combined TEM, AFM, and DLS analysis. In a similar line, investigations on ionic liquids have shown that very small differences in molecular arrangement, which can be seen by XRD and AFM, can profoundly affect the electrochemical behavior. The above-described results emphasize that the comprehensive structural and morphological descriptions are imperative for rational design and synthesis of next-generation materials.

Characterization of ionic liquids, nanofluids, and nanomaterials with respect to their structure and morphology is a multidisciplinary process that benefits from the synergy among SEM, TEM, XRD, AFM, and DLS (Table 1). Individually, each of the techniques brings in unique insights from the range of surface topology, internal structure to crystallinity, and particle dispersion that cumulatively give an insight into the behavior of material at nanoscale. With the research continuing to progress, the use of these methodologies will still be an important part of designing and optimizing materials that perform exceptionally and have superior function.

Table 1 Comparison of structural characterization techniques [11]

Technique	Principle	Resolution	Applications	Limitations
Scanning electron microscopy (SEM)	Uses electron beams to scan the surface morphology of a sample	High (10 nm to 1 μm)	Surface morphology, particle size analysis	Limited to surface imaging, requires vacuum conditions
Transmission electron microscopy (TEM)	Uses electron transmission to reveal internal nanostructures	Very high (0.1 nm to 1 nm)	Nanostructure imaging, material defects analysis	Sample preparation is complex, potential radiation damage
X-ray diffraction (XRD)	Uses X-ray diffraction patterns to determine crystallographic structure	Moderate (100 nm to 1 μm)	Crystal structure determination, phase identification	Limited to crystalline materials, requires pure samples
Atomic force microscopy (AFM)	Measures surface topography using a mechanical probe	High (sub-nanometer)	Surface roughness, mechanical property evaluation	Slow scanning, limited area of analysis
Fourier transform infrared spectroscopy (FTIR)	Measures molecular vibrations to identify chemical bonds	Moderate (depends on sample preparation)	Identification of organic and polymeric compounds	Requires extensive sample preparation
Raman spectroscopy	Uses inelastic scattering of light to analyze molecular structure	Moderate (depends on material type)	Material composition, chemical bonding analysis	Interference from fluorescence and other background signals

2 Thermal Properties and Behavior

The ability of a material to be viable in many different cutting-edge technologies depends upon the thermal performance of the material. Particularly, thermal conductivity, heat capacity, phase transition behavior, and thermal stability are key properties that determine the way in which materials handle and distribute heat [12]. These thermal properties in nanostructured materials and ionic liquids are strongly affected by size-induced effects, interfacial interactions, and liquids nature, which is ionic in nature and results in heat transfer processes that are different from the bulk materials. Similar thermophysical behavior has also been observed for graphene-based ionanofluids [13]. Furthermore, heat-capacity enhancement in ionic-liquid nanofluids has been reported [14].

Nanostructuring may result in a significant change in thermal conductivity because of the increased contribution of phonon scattering at the interfaces and the grain boundaries. With the proximity of the dimensions of a material toward the nanoscale, the phonon mean free path is seen to be similar to the size of the material, which leads to an increase in boundary scattering events [15]. This can result in thermal conductivity of a bulk material being decreased. Nevertheless, in certain cases, nanostructuring itself can also create pathways for greater heat transfer efficiency if the interfaces are designed to accommodate phonon coupling over grain boundaries [16]. In the same way, an ionic aspect of some liquids brings in complicated intermolecular interactions that determine the energy storage and transport mechanisms. The fact that ionic liquids also contain ionic aggregates and long-range electrostatic interactions may change vibrational modes in the fluid, leading to changes in the heat capacity and thermal conductivity (Fig. 2).

In order to evaluate these thermal properties, systematically different advanced characterization techniques are used for that purpose. Differential scanning calorimetry (DSC) is one of the extensively applied techniques for exploring phase transitions as well as heat capacity. By keeping the heat flow into or out of a sample under scanning as a function of temperature, DSC can determine endothermic and exothermic changes like melting, crystallization, and glass transition. These phase transitions in nanomaterials may take place under temperatures that are greatly different from the bulk form as a result of the fact that the size is confined and there are increased surface energy effects. DSC not only quantifies such transitions but also measures specific heat capacity, an important parameter that is needed in thermal management applications [17].

Monitoring the mass change of a material as it is being heated under controlled conditions provides complementary insights offered by thermogravimetric analysis (TGA). This method is very useful in analyzing thermal stability and decomposition pattern. In nanostructured systems, such high surface area to volume ratio can cause enhanced oxidation or thermal degradation processes, which TGA may note as a function of polymer material temperature. Such analyses are very important for defining the highest working limits of materials in a high-temperature environment and finding the temperatures at which volatile elements or structural changes take place [7].

Table 2 Detailed thermal properties comparison of nanofluid, ionic liquid, and nanomaterial [21]

Material/system	Thermal conductivity (W/m·K)	Heat capacity (J/g·K)	Decomposition temperature (°C)	Measurement conditions	Remarks
Nanofluid A	0.65 ± 0.03	4.2 ± 0.1	200 ± 5	Measurements conducted over 25–100 °C at ambient pressure using laser flash analysis (LFA) for conductivity and differential scanning calorimetry (DSC) for heat capacity	Exhibits moderate thermal conductivity and high heat capacity, making it well-suited for applications requiring efficient heat storage and moderate thermal management
Ionic liquid B	0.15 ± 0.01	1.8 ± 0.05	350 ± 10	Measurements performed between 20 and 120 °C at ambient pressure; DSC was used for assessing phase transitions and heat capacity, while thermogravimetric analysis (TGA) was used to determine thermal stability	Displays lower thermal conductivity but excellent thermal stability, ideal for high-temperature processes where stability is paramount
Nanomaterial C	3.20 ± 0.15	0.9 ± 0.03	600 ± 15	Measurements taken over 25–300 °C under an inert atmosphere to prevent oxidation; LFA and DSC were employed to assess thermal conductivity and heat capacity, respectively, and TGA confirmed decomposition behavior	Offers significantly higher thermal conductivity with lower heat capacity, making it optimal for applications that demand rapid heat dissipation and stability at elevated temperatures

3.2 *Types of Flow Behavior*

1. Newtonian flow: A fluid where viscosity remains constant irrespective of shear rate (e.g., water, simple oils). Certain CNT-based nanofluids maintain Newtonian flow at moderate operating temperatures [26].
2. Shear-thinning (pseudoplastic) behavior: Viscosity decreases with increasing shear rate, common in polymer solutions and some nanofluids.
3. Shear-thickening (dilatant) behavior: Viscosity increases with shear rate, observed in dense suspensions.
4. Bingham plastic behavior: The fluid behaves as a solid until a critical shear stress is applied, after which it flows with a nearly constant viscosity.
5. Thixotropic and rheopectic fluids: Time-dependent viscosity changes, where thixotropic fluids become less viscous over time under shear, while rheopectic fluids become more viscous.

3.3 *Shear Stress Response*

The shear stress (τ) in a fluid is given by:

$$\tau = \eta \cdot \dot{\gamma} \quad \tau = \eta \cdot \dot{\gamma}$$

where η is the viscosity, $\gamma^{\cdot}$ is the shear rate.

For non-Newtonian fluids, this relationship is not linear. Depending on the nature of the nanofluid, the addition of nanoparticles or colloidal structures alters this relationship by affecting interparticle interactions, electrostatic forces, and hydrodynamic resistance.

Non-Newtonian Behavior of Nanofluids

Nanofluids often exhibit complex viscoelastic behavior, where elasticity and viscosity are both significant [27]. This is particularly influenced by:

- Particle size and distribution: Smaller particles enhance Brownian motion, influencing viscosity.
- Particle concentration: Higher concentrations lead to increased interactions, sometimes resulting in gel-like structures.
- Surface chemistry and coating: Modifications to nanoparticle surfaces influence fluid structure and viscosity.

Impact of Ionic Liquids on Rheology

Ionic liquids (ILs) are emerging as effective rheological modifiers in nanofluids and colloidal suspensions. Their addition can:

- Reduce viscosity by enhancing lubrication and disrupting particle aggregation.
- Increase thermal conductivity without significantly increasing flow resistance.
- Enhance shear stability, making nanofluids more suitable for industrial applications.

The presence of ILs influences electrostatic and steric interactions, leading to better dispersion of nanoparticles and controlling flow properties under different conditions.

3.4 *Key Techniques for Rheological Characterization*

To accurately assess the rheological behavior of nanofluids and colloidal suspensions, the following techniques are commonly employed:

Rheometry

Rheometers measure viscosity, shear stress, and elasticity under different shear rates (Fig. 3). Rotational and oscillatory rheometers are commonly used to determine:

- Flow curves (shear stress vs. shear rate).
- Viscoelastic properties (storage and loss modulus).
- Yield stress and thixotropy.

Viscometry

Viscometers provide viscosity measurements under controlled conditions. They can be classified into:

- Capillary viscometers (measure flow time through a narrow tube).
- Rotational viscometers (measure torque required for fluid movement).
- Falling ball viscometers (measure the time taken for a sphere to descend through the fluid).

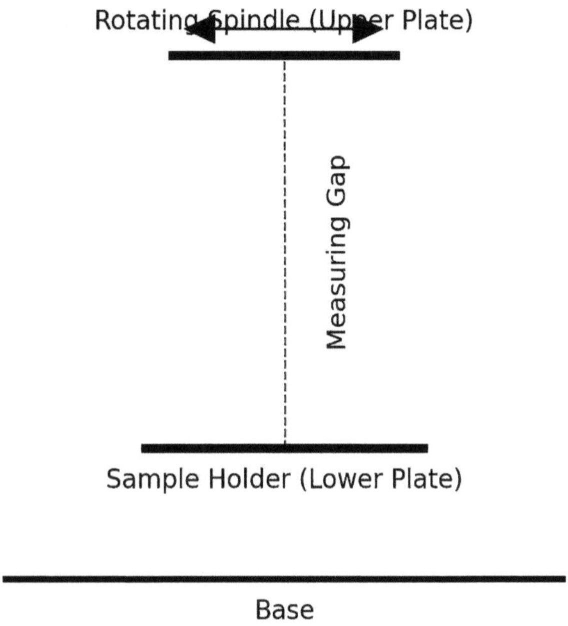

Fig. 3 Schematic of a typical rheometer setup used to measure the rheological properties of fluids, such as viscosity, elasticity, and flow behavior

Shear Rate Vs. Viscosity Measurements

Plotting shear rate vs. viscosity allows for the classification of shear-thinning, shear-thickening, or Newtonian behavior. These measurements help in:

- Understanding fluid stability and dispersion quality.
- Optimizing formulations for industrial applications.
- Predicting performance in flow systems such as heat exchangers and lubrication systems.

4 Electrical and Optical Characterization

Characterization of electronics and optics is an important step in realizing the intrinsic properties of materials, especially when applied in electronics, optoelectronics, and photonics. These descriptions are essential to determining how materials react to electric field and electromagnetic radiation which directly influences their response in the devices such as sensors, transistors, capacitor, optical filter, solar cells, and displays [28]. Electrical characterization entails the characterization of a material in terms of conductivity, resistivity, dielectric property, and impedance behavior which would make it possible to determine how charge is transported and how polarization effects and the phenomena of energy dissipation exist in the material. Conversely, the optical characterization gives indications about light absorption, transmission, scattering, and refractive index [29]. These optical properties are of great interest in applications such as solar energy harvesting, thin-film coatings, optical waveguides, and biomedical imaging technologies.

Different electrical and optical actions are displayed by materials based on their atomic structure and composition, and moreover, the concentration of free charge carriers. For example, metals possess high electrical conductivity on account of their free electrons, while semiconductors have bandgap-dependent conductivity and are, therefore, suitable for controlled electronic and photonic applications [30]. Dielectric materials, on the other hand, have a low electric conductivity and yet are very important in energy storage devices because they are able to polarize under an external field. Similarly, the optical properties like absorption and reflection govern interactions of material with the light and influence their usage in mirrors, anti-reflective coatings, and photodetectors. In order to understand more about these properties, so many advanced methods of characterization such as dielectric and impedance spectroscopy, UV-visible and IR spectroscopy, and ellipsometry are used.

4.1 Electrical Characterization

Electrical Conductivity and Dielectric Properties

Electric conductivity is one of the most elementary characteristics of a material since it defines the performance of this material in terms of the flux of electric charge within this material if exposed to an external electric field. It is quantitatively described as the reciprocal of the electrical resistivity (ρ) as:

$$\sigma = 1\,\rho \quad \sigma = \rho 1$$

where σ refers to the electrical conductivity. Electrical behavior also differs significantly with different materials in terms of structuring, nature of atomic binding, and carrier charge levels. Conductors like metals have high conductivity of electricity because of their delocalized electron clouds; this allows electrons to conduce freely under electric field. Semiconductors like silicon and gallium arsenide have semiconductivity that can be adjusted by doping or other external stimulation such as temperature and light. Insulators, which include ceramics and polymers, have an extremely low conductivity as there are no free electrons; however, they can be useful where construction of the electrical insulation and dielectric properties are required [31].

Dielectric materials are of special interest in electrical characterization as they govern the charge storage, energy dissipation, and polarization effects. When we place a dielectric material in an electric field, it is polarized, i.e., the positions of its positive and negative charge centers change slightly causing the development of an internal electric field that acts against the external one. It is the dielectric loss that establishes the energy dissipation properties of the material, while for its ability to store the electrical energy, the dielectric constant (εr) is responsible. The knowledge of these parameters is important for designing capacitors, insulating materials, and high-frequency electronic devices [32]. Furthermore, dielectric behavior is a very strong function of frequency; therefore, materials must be analyzed at a wide range of frequencies to determine how suitable they are to various applications.

Dielectric Spectroscopy

Dielectric spectroscopy is one of the important techniques for the determination of the dielectric properties of material since it involves the determination of how the material reacts to stimuli of an alternating electric field with varying frequencies. This technique is a necessity in studying the mechanisms of polarization, charge storage, and loss behavior of materials. Dielectric spectroscopy is used to classify different types of polarization mechanisms including electronic polarization, which is present at high frequencies; ionic polarization, which involves the movement of ions; and dipolar polarization, which is observed in materials that possess permanent dipoles. Furthermore, interfacial polarization that occurs at interfaces of the heterogeneous materials is very crucial in composite materials, nanostructures, and polymer-based dielectrics.

One of the main benefits of the dielectric spectroscopy is its capacity to deliver frequency-dependent information on how a certain material reacts to the applied electric field to enable researchers know if the material is suitable for high-speed electronic application, energy storing device, or microwave communication system. Materials having high dielectric constant and low dielectric loss are suitable for capacitors, while those with a high loss are used in the energy dissipation technology like the electromagnetic interference (EMI) shielding.

Impedance Spectroscopy

Impedance spectroscopy is another strong technique for electrical characterization for studying mechanism of charge transport, interface polarization, and resistive and capacitive behavior of materials. Impedance (ZZ) is a complex quantity that has a real part (resistive behavior) and imaginary part (capacitive or inductive behavior) represented mathematically as follows:

$$Z = Z' + jZ'' \quad Z = Z' + jZ''$$

where Z' stands for material's resistance and Z'' stands for reactance. This technique is of great importance in the understanding of the conductivity of nanomaterials, electrolytes, and electrochemical interfaces that are of crucial importance in applications such as supercapacitors, batteries, and fuel cells. From analyzing impedance spectra, useful parameters like charge transfer resistance, ion diffusion rates, and dielectric relaxation times can be retrieved and as such, the behavior of material in electronic and energy storage applications can be better understood.

4.2 Optical Characterization

UV-Visible and Infrared (IR) Spectroscopy

Techniques of optical spectroscopy like UV-visible and infrared spectroscopy are valuable in revealing the electronic transitions, vibrations of molecules, and composition structure of materials. It is a common approach that is used to investigate optical absorption and transmission properties of materials that are important in the development of the photonic devices, sensors, and solar energy harvesting system. Particle morphology has a strong influence on optical absorption and photothermal conversion efficiency [33]. The absorption spectrum reveals about the bandgap energy, which is an important parameter of semiconductor materials that find applications in the photovoltaic and optoelectronics. The materials with direct bandgap have good optical absorption and hence are applicable to light emitting diodes (LEDs) and photodetectors, while indirect bandgap materials are used in application where slower optical transition is accepted.

On the other hand, infrared spectroscopy is largely applied to the study of the vibrations of the molecules and determination of functional group. This approach gives information about the chemical bonding, the composition of structure, and interaction of molecules in a material. It is very useful for the characterization of polymers, biomaterials, and nanocomposites, since it enables researchers to detect alterations on bonding and molecular structure when the subject is subjected to modifications or external factors like temperature and pressure.

Ellipsometry

Ellipsometry is a very precise optical method applied to determining the refractive index, thickness, and the optical constants for thin films and surfaces. This technique involves an analysis of the change in the state of polarization of light on reflection from a surface of some material. The refractive index (n) and extinction coefficient (k) of a material are calculated through ellipsometry to determine the manner in which light interacts with the material at the various wavelengths. This method is of particular significance in the manufacturing of semiconductor devices, the creation of optical coatings, and the biosensing operations that require the accurate control of the thickness of the film and the optical properties. Ellipsometry is nondestructive and able to deliver high-resolution data, thus being a technique of choice for nanoscale films and multilayer structures study.

5 Stability, Dispersion, and Aggregation Behavior

The stability, dispersion, and aggregation behavior of ionic liquids, nanofluids, and nanomaterials represent the most important reasons why they are useful and have a good performance in different applications. These materials are currently being used more and more in areas such as catalysis, energy storage, biomedical engineering, and environmental science since they possess unique physicochemical properties. But their activity depends largely on their ability to maintain stability and good dispersion in various media in time. Stability means the ability of such materials to remain intact, while possessing useful characteristics even at the varied conditions in the surrounding environment, such as change of temperature, pH, or solvent composition. Dispersion, on the other hand, is the even distribution of the particles or molecules in a medium, while aggregation simply implies that particles tend to clump together, hence leading to sedimentation or phase separation. Knowledge and mastery of such behaviors are critical to the improvement in providing ionic liquids, nanofluids, and nanomaterials on a real-world application.

The stability of these materials depends on a variety of factors such as the surface charge, particle size, and nature of medium surrounding them. The zeta potentiality is one of the most important parameters in the evaluation of the stability and it denotes the effective surface charge of particles in the colloidal system. Zeta potential is a cornerstone of indicating electrostatic contact between the particles because it demonstrates the degree of repulsive power that prevents an aggregation. High zeta potential value (usually above ± 30 mV) represents good electrostatic repulsion and improves stability of dispersion. On the other hand, poor zeta potential values mean poor repulsion force and the system is easily subjected to aggregation and sedimentation. Apart from the electrostatic stabilization, the steric stabilization also plays an important role in stability maintenance. This means, use of polymers, surfactants, or other stabilizing agents, which adsorb onto the surface area of particles and act as a physical barrier, thereby preventing them from being in close contact. Steric stabilization is especially applicable in systems where electrostatic repulsion is not strong enough to bring about stability, like in high-ionic-strength media where electrostatic forces are screened.

Aggregation behavior of the second group is the other important factor that should be regulated in a very attentive manner in order to achieve the efficient use of ionic liquids, nanofluids, and nanomaterials. Aggregation is brought about when the particles come close because of VdW forces or other attractive interactions and hence form larger clusters [34]. This may lead to sedimentation, decrease of surface area, and loss of functionality. Particle concentration, pH, ionic strength, and temperature may have a major effect on the aggregation behavior. For instance, adding ionic strength to a medium can squeeze the electric double layer surrounding particles and decrease zeta potential as well as lead to the aggregation of particles. Similarly, modified surface charge for particles can be brought about by the alteration of the pH which would influence interaction and stability in particles. The knowledge of these factors is very essential in designing systems in which aggregation is reduced and the long-term stability is ensured.

5.1 Key Techniques

"Fundamental characterization of the stability, dispersion, aggregation tendencies of Ionic Liquids, Nanofluids and Nanomaterials": According to the title, several sophisticated analytical techniques are used to study the stability, dispersion, aggregation behavior of ionic liquids, nanofluids, and nanomaterials. These techniques yield useful information about the physicochemical properties of these materials and direct researchers to devise ways in which they can improve their performance.

5.2 Zeta Potential Measurements

Zeta potential is a very common stability-evaluating parameter of colloidal systems. It is conducted by applying an electric field to dispersion and determining electrophoretic mobility of particles. The zeta potential value represents the amount of surface charge, as well as the force with which the repulsive nature of particles acts. High values of zeta potential (commonly > ±30 mV) suggest good electrostatic repulsion, and, consequently, inhibited aggregation and improved dispersant stability. On the other hand, small values of zeta potential imply week forces of repulsion that makes the system more vulnerable to aggregation and sedimentation [35]. Zeta potential measurements are quite effective in the optimization of the formulations of nanofluids and nanomaterials as they give a quantitative readout of the stability in different states. For instance, the researchers can use the zeta potential measurements in order to find out the effect of pH or ionic strength on the stability of a given dispersion or stabilizing agents. Such information may then be used for designing systems that are more stable and perform better.

Dynamic Light Scattering (DLS) for Aggregation Analysis

Dynamic light scattering (DLS) is a useful tool in the determination of size distribution and aggregation properties of particles in a dispersion. DLS does its measurements based on the fluctuations in scattered light brought about by a phenomenon known as Brownian motion of particles. These changes are applied when determining the hydrodynamic diameter of particles to obtain the information about particle size distribution and the existence of aggregates. DLS is very helpful in determining how the particle size may change with time, an inference to the onset of aggregation [36]. For instance, an observed increase of the average particle size detected by DLS could imply the formation of aggregates. DLS can also be used to determine how different factors (e.g., pH, temperature, or the presence of stabilizing agents) influence their aggregation. By supplying real-time information on the sizes and aggregations of particles, DLS allows researchers to optimize on the formulations of ionic liquids, nanofluids, and nanomaterials in order to reduce the aggregations and increase the stability.

Sedimentation Tests

Sedimentation tests are quite a simple but reliable way to investigate the long-term stability of dispersions. These tests occur through tracking the settling of particles as long as gravity or centrifugation. The rate of sedimentation gives important information regarding stability of the system because stable dispersions show little or no sedimentation, whereas unstable systems show rapid settling of particles. Sedimentation tests might be performed under some different conditions, including different pH, ionic strength, or temperature, to check for their influence on stability. For instance, the sedimentation tests can be used by the researchers for comparison of the stability of different formulation of nanofluids or to assess the efficiency of stabilizing agents. Other sophisticated methods that can be used to quantify sedimentation and to reveal more details on the stability include turbidity measurements or laser diffraction. The sedimentation tests are especially helpful in cases in which long-term stability is of utmost importance, for instance, in the creation of nanofluids to use for heat transfer or nanomaterials to be used in drug delivery.

Chemical Composition and Surface Analysis

Chemical composition and surface studies are the basis to the study and optimization of the performance of advanced materials such as nanomaterials, ionic liquids, and nanofluids. These techniques offer pivotal understanding of the chemical structure, elemental context, and surface alterations of materials; this determines their reactivities, the stabilities, and functions in such a vast range of applications: catalysis, energy storage, biomedical engineering, and environmental purification [37]. In-depth knowledge of such properties enables researchers to customize materials at the molecular level so that they can suit specific needs of the intended applications.

Fourier transform infrared spectroscopy (FTIR) is one of the major methods employed in order to reveal chemical bonds in a material. Exposure of a sample to infrared radiation causes absorption of certain wavelengths on the basis of the vibrational modes of the chemicals in the sample. As a result of this absorption, a spectrum that is used as the molecular fingerprint to identify the functional groups present in the material is produced. For example, the presence of the hydroxyl (–OH), carbonyl (C=O), or amine ($-NH_2$) groups can be identified by the characteristic absorption bands. There is FTIR application which is specifically useful for monitoring processes associated with surface functionalization; the spectrum of a material either before or after chemical modification can indicate successful addition of new functional groups [38]. For instance, a change in the absorption peak at around 3400 cm^{-1} can suggest the presence of hydrophilic groups, which are able to increase dispersibility of the material in aqueous media or contribute to its catalytic activity. This capability to be sensitive to movement of molecules has placed FTIR as an indispensable tool for analysis of materials, both of qualitative nature and semiquantitative nature.

As a complement to FTIR, Raman spectroscopy gives additional information about the molecular structure through the analysis of vibrational modes with the help of inelastic scattering of light. In contrast to FTIR that uses absorption, Raman spectroscopy uses energy variations that occur when a laser light exchanges interactions with vibrations on a molecule. This method is very useful for characterizing materials that show very strong Raman-active modes like the carbon-based nanomaterials. In these materials, Raman spectra usually have characteristic features including D and G bands—indicators of structural disorder and graphitic content, respectively. The ratio of these bands can provide for a quantitative measure of the defects density as well as the overall crystallinity of the sample (*I_D/I_G*). Vibrational modes, which may be either weak or simply missing from FTIR spectra, get often detected by the Raman spectroscopy, thus completing the picture of molecular structure and corroborating subtle changes caused by surface modifications [39]. This synergetic application of FTIR and Raman guarantees investigators with a strong tool for describing the nature and environments of the functional groups present in a material.

X-ray photoelectron spectroscopy (XPS) provides a surface-sensitive technique for studying the elemental composition and chemical state of atoms on the top several nanometer layer of a material. By exposing a substance to X-rays, XPS induces the emission of core electrons, which are studied depending on their kinetic energy. This kind of analysis produces highly specific binding energy values to various elements and oxidation states. The high-level sensitivity of the XPS makes it an invaluable tool for the confirmation of the surface functionalization. For instance, the appearance of a new nitrogen peak when an XPS spectrum was recorded for the functionalized material after its coverage with amine groups testifies to a high degree of the success of the chemical modification. Also, variations in the binding energies of carbon, oxygen, or any other element may reflect changes in the chemical environment, and may reveal about the bonding arrangements and the amount of oxidation on the surface. As a quantitative technique, XPS not only confirms the presence of certain elements in the system but also allows for estimation of their relative concentrations, and that is absolutely necessary for the complete comprehension of the surface reactions and optimization of the same.

Energy dispersive X-ray spectroscopy (EDX) is usually combined with the techniques of electron microscopy, such as scanning electron microscopy (SEM), to provide the analysis of elements with spatial resolution. When electrons are bombarded on a material, the typical X-rays are emitted from the constituent elements. From the analysis of these X-rays, EDX can obtain elemental composition of the material with high spatial resolutions; as a result, one could map out distribution of elements on the surface. This is especially essential in heterogeneous materials for which the uniform character of surface functionalization may make a difference in the properties of materials of up to 50% [40]. For example, in a composite material, whereby a catalytic metal is present in the matrix, EDX mapping can determine whether the metal is homogeneously distributed or there is agglomeration existing of the metal, which may reduce catalytic efficacy. The correlation between these chemical revelations and physical structure represented with SEM images is a complete way of knowing the chemistry of the surface and the architecture of the material.

6 Stability, Dispersion, and Aggregation Behavior

Stability, dispersion, and aggregation behavior are important parameters that determine their capability and reliability of the ionic liquids, nanofluids, and nanomaterials in practical applications. These materials are frequently used in areas where constant thermal conductivity, viscosity, and optical property are necessary. In most cases, the success of the system relies on the system's efficiency in remaining well dispersed over time without aggregations or settling down that could also lead to a compromise of the material's functioning. To overcome these challenges, researchers apply a set of analytical techniques that help in obtaining the microscopic and macroscopic understanding of the behavior of the colloidal dispersions.

Among the major techniques used can be identified zeta potential measurement, which delivers quantitative evaluation of surface charge of particles in a dispersion. Zeta potential is an expression of the relationship between attractive forces (e.g., van der Waals interactions) and repulsive forces (normally caused by electrostatic charges). A high absolute value of zeta potential—positive or negative—indicates strong electrostatic repulsion, thus promoting particle separation and preventing particle from aggregating, amplifying its translation from paste to its implantation in the cornea. This measurement is normally done using electrophoretic light scattering, where mobility of the charged particles under an applied electric field is determined. By varying other quantities like the pH of the solution or the concentration of soap (surfactants) etc., a researcher can be able to tune zeta potential for a more stable dispersion. For instance, when the magnitude of the zeta potential is raised from 20 mV to 40 mV, it can drastically minimize the possibility of agglomeration of particles, hence improving long-term stability of a nanofluid [41].

Enhancing zeta potential measurements, dynamic light scattering (DLS) continues to be a very common technique to study aggregation behavior of nanoparticles from suspension. DLS measures the variations in the intensity of the scattered light due to Brownian motion of particles. Such fluctuations correlate directly with the hydrodynamic diameter of particles, hence permitting the definition of particle size distribution. Possibly even slight variations in the mean particle size can be picked up early, which may indicate that the aggregation has started. Since DLS could be repeated repeatedly in time, it is a useful monitoring system for the dispersion's stability in real time. This is quite important in applications such as drug delivery or electronic cooling systems, where the small changes in the size of particles may affect significantly the performance. Besides, through correlation of DLS data with results showing zeta potential, researchers can conclude both as to the mechanisms underpinning the maintenance of dispersion, either steric hindrance or electrostatic repulsions.

Sedimentation tests provide the more macroscopic determinant of stability in that it bases the settling behavior of particles upon the effect of gravity. Normal sedimentation test involves the setting of the dispersion aside undisturbed, whereby the changes in the concentration or turbidity of the supernatant are calculated with time. When a layer of sediments is formed at the bottom of the container, it means that the forces of repulsion are unable to prevent the gravity settling. This type of testing is essential for imitating real life, where the stability of materials is needed for long periods of time. A dispersion in which it has settled little after several days is robust but if it has settled quickly, this would mean the formulation needs more stabilization. Sedimentation tests can be improved by a visual or instrumental analysis (e.g., changes of optical density) to quantify settling rate and extent.

Integrating zeta potential measurements, DLS, and sedimentation tests, researchers form a multidimensional perception on the stability, dispersion/aggregation behavior of advanced materials. At nanometer scale, zeta potential and DLS observations offer detailed insight on forces at play as well as size distribution of the particles in suspension, while sedimentation tests take a practical approach to forming a conclusion on long-term stability. This all-rounded approach allows the optimization of material formulations by altering the ionic strength, pH, and by adding the appropriate stabilizer or surfactant. Finally, such a thorough investigation involves a guarantee that materials are able to be designed to contain the optimal combination of steric and electrostatic stabilization, which will help further enable their use for high performance in the fields of energy systems and biomedical devices, among others.

7 Conclusion

Finally, properties and characterization of ionic liquids, nanofluids, and nanomaterials display a complicated interplay of structural, thermal, rheological, electrical, chemical, and stability parameters that control the entire performance of these systems in advanced uses. Comprehensive structural and morphological studies, with the help of the SEM, TEM, XRD, AFM, and DLS methods, give important insights about the particle size, shape, crystallinity, and surface features, which are determinants in the study of the behavior of the materials on the nanoscale. Thermal characterization by DSC, TGA, and LFA emphasizes the huge effect of nanostructuring and the peculiar nature of ions on mechanisms of heat transfers, phase transitions, and thermal behaviors. These thermal properties are very important for engineering materials for energy storage, cooling systems, and other heat-sensitive technologies. Also, the rheological and flow properties resulting from viscometry and shear testing elucidate how diluents and dispersion of nanoparticles affect the dynamics of movement of fluids, which is important for heat transfer applications and stability of fluids.

The chemical composition and surface analysis build on the understanding further by exposing molecular bonds, functional groups, and distribution of elements using FTIR, Raman spectroscopy, XPS, and EDX. Such techniques not only demonstrate the success of surface modifications but also demonstrate the highly complex interplay between the chemistry of the surface and the functional performance of the material. Lastly, stability, dispersion, and aggregation behavior evaluation by methods of zeta potential measurements, DLS, and sedimentation tests guarantees that these materials possess predetermined properties over time, which is a decisive factor for the long-term credibility of the materials in real-world scenarios.

These integrated methods of comprehensive characterization provide a multidimensional system that aids in the fine-tuning of ionic liquids, nanofluids, and nanomaterials. This holistic approach allows the researchers not only to gain an understanding of intrinsic properties of these advanced materials but also to engineer them to perform high-tech functions especially in other technological disciplines. Further studies of this foundation will keep the bar of material functionality high, improving these systems' utility in catalysis, energy conversion, and biomedicine among others.

References

1. A.G.M. Ferreira, P.N. Simões, A.F. Ferreira, M.A. Fonseca, M.S.A. Oliveira, A.S.M. Trino, Transport and thermal properties of quaternary phosphonium ionic liquids and IoNanofluids. J. Chem. Thermodyn. **64**, 80–92 (2013)
2. Y. Javed, K. Ali, K. Akhtar, Jawaria, M.I. Hussain, G. Ahmad, T. Arif, TEM for atomic-scale study: Fundamental, instrumentation, and applications in nanotechnology, in *Handbook of Materials Characterization*, (Springer, Cham, 2018), pp. 147–216
3. Y.U. Paulechka, A.G. Kabo, A.V. Blokhin, G.J. Kabo, M.P. Shevelyova, Heat capacity of ionic liquids: Experimental determination and correlations with molar volume. J. Chem. Eng. Data **55**(8), 2719–2724 (2010)
4. A. Rahimi, M. Rahjoo, S.S. Hashemi, M.R. Sarlak, M.H. Malekshah, E.H. Malekshah, Combination of dual-MRT lattice Boltzmann method with experimental observations during free convection in enclosure filled with MWCNT-MgO/water hybrid nanofluid. Thermal Sci. Eng. Progr. **5**, 422–436 (2018)
5. M.P. Shevelyova, Y.U. Paulechka, G.J. Kabo, A.V. Blokhin, A.G. Kabo, T.M. Gubarevich, Physicochemical properties of imidazolium-based ionic nanofluids: Density, heat capacity, and enthalpy of formation. J. Phys. Chem. C **117**(9), 4782–4790 (2013)
6. O. Lebedeva, D. Kultin, L. Kustov, Electrochemical synthesis of unique nanomaterials in ionic liquids. Nano **11**(12), 3270 (2021)
7. V.M.B. Nunes, M.J.V. Lourenço, F.J.V. Santos, M.L.S.M. Lopes, C.A. Nieto de Castro, Accurate measurement of physicochemical properties on ionic liquids and molten salts, in *Molten Salts and Ionic Liquids: Never the Twain?* (Wiley, Hoboken, 2010), pp. 229–263
8. E.A. Franceschini, *Nanostructured Multifunctional Materials: Synthesis, Characterization, Applications and Computational Simulation* (Routledge, London, 2021)

9. K.N. Marsh, J.A. Boxall, R. Lichtenthaler, Room temperature ionic liquids and their mixtures—A review. Fluid Phase Equilib. **219**(1), 93–98 (2004)
10. M.A. Marcos, N.E. Podolsky, D. Cabaleiro, L. Lugo, A.O. Zakharov, V.N. Postnov, K.N. Semenov, MWCNT in PEG-400 nanofluids for thermal applications: A chemical, physical and thermal approach. J. Mol. Liq. **294**, 111616 (2019)
11. P. Navarro, M. Larriba, E. Rojo, J. Garcia, F. Rodríguez, Thermal properties of cyano-based ionic liquids. J. Chem. Eng. Data **58**(8), 2187–2193 (2013)
12. J.M. França, C.A. Nieto de Castro, M.M. Lopes, V.M. Nunes, Influence of thermophysical properties of ionic liquids in chemical process design. J. Chem. Eng. Data **54**(9), 2569–2575 (2009)
13. J. Alizadeh, M.K. Moraveji, An experimental evaluation on thermophysical properties of functionalized graphene nanoplatelets ionanofluids. Int. Commun. Heat Mass Transf. **98**, 31–40 (2018)
14. K. Oster, C. Hardacre, J. Jacquemin, A.P.C. Ribeiro, A. Elsinawi, Understanding the heat capacity enhancement in ionic liquid-based nanofluids (ionanofluids). J. Mol. Liq. **253**, 326–339 (2018)
15. G. Huminic, A. Huminic, Heat transfer capability of ionanofluids for heat transfer applications. Int. J. Thermophys. **42**, 1–17 (2021)
16. R.L. Gardas, J.A. Coutinho, A group contribution method for heat capacity estimation of ionic liquids. Ind. Eng. Chem. Res. **47**(15), 5751–5757 (2008)
17. T.C. Paul, A.K.M.M. Morshed, E.B. Fox, A.E. Visser, N.J. Bridges, J.A. Khan, Thermal performance of ionic liquids for solar thermal applications. Exp. Thermal Fluid Sci. **59**, 88–95 (2014)
18. A.A. Minea, S.S. Murshed, A review on development of ionic liquid based nanofluids and their heat transfer behavior. Renew. Sust. Energ. Rev. **91**, 584–599 (2018)
19. K. Oster, C. Hardacre, J. Jacquemin, A.P. Ribeiro, A. Elsinawi, Thermal conductivity enhancement phenomena in ionic liquid-based nanofluids (ionanofluids). Aust. J. Chem. **72**(2), 21–33 (2018)
20. A. Singh, V.A. Chavan, P. Ariya, B.V.S.S. Udaynadh, E. Adithi, Nanomaterial characterization techniques. Fut. Trends Chem. Mater. Sci. Nano Technol. **3**(1), 1–19 (2024)
21. S. Aman, I. Khan, Z. Ismail, M.Z. Salleh, Q.M. Al-Mdallal, Heat transfer enhancement in free convection flow of CNTs Maxwell nanofluids with four different types of molecular liquids. Sci. Rep. **7**(1), 2445 (2017)
22. S.K. Das, N. Putra, W. Roetzel, Pool boiling characteristics of nano-fluids. Int. J. Heat Mass Transf. **46**(5), 851–862 (2003)
23. E.B. Fox, A.E. Visser, N.J. Bridges, J.W. Amoroso, Thermophysical properties of nanoparticle-enhanced ionic liquids (NEILs) heat-transfer fluids. Energy Fuel **27**(6), 3385–3393 (2013)
24. D.P. Soman, S. Karthika, P. Kalaichelvi, T.K. Radhakrishnan, Impact of viscosity of nanofluid and ionic liquid on heat transfer. J. Mol. Liq. **291**, 111349 (2019)
25. D. Tomida, S. Kenmochi, T. Tsukada, K. Qiao, Q. Bao, C. Yokoyama, Viscosity and thermal conductivity of 1-hexyl-3-methylimidazolium tetrafluoroborate and 1-octyl-3-methylimidazolium tetrafluoroborate at pressures up to 20 MPa. Int. J. Thermophys. **33**, 959–969 (2012)
26. G. Yellapu, C.C. Vishal, M.P. Kandoth, P. Saha, R.R. Bojja, S. Gandham, R. Kanaparthi, Functionalized multi-walled carbon nanotubes based Newtonian Nano fluids for medium temperature heat transfer applications. Thermal Sci. Eng. Progr. **12**, 13–23 (2019)
27. C.A. Nieto de Castro, S.I. Vieira, M.J. Lourenco, S.M. Murshed, Understanding stability, measurements, and mechanisms of thermal conductivity of nanofluids. J. Nanofluids **6**(5), 804–811 (2017)
28. C. Hermida-Merino, A.B. Pereiro, J.M. Araújo, C. Gracia-Fernández, J.P. Vallejo, L. Lugo, M.M. Piñeiro, Graphene IoNanofluids, thermal and structural characterization. Nano **9**(11), 1549 (2019)

29. S.U. Ilyas, R. Pendyala, M. Narahari, Experimental investigation of natural convection heat transfer characteristics in MWCNT-thermal oil nanofluid. J. Therm. Anal. Calorim. **135**, 1197–1209 (2019)
30. V.S. Patil, A. Cera-Manjarres, D. Salavera, C.V. Rode, K.R. Patil, C.A. Nieto de Castro, A. Coronas, Ru-imidazolium halide ionanofluids: Synthesis, structural, morphological and thermophysical properties. J. Nanofluids **5**(2), 191–208 (2016)
31. T.J. Choi, S.P. Jang, M.A. Kedzierski, Effect of surfactants on the stability and solar thermal absorption characteristics of water-based nanofluids with multi-walled carbon nanotubes. Int. J. Heat Mass Transf. **122**, 483–490 (2018)
32. N. Sezer, M. Koç, Stabilization of the aqueous dispersion of carbon nanotubes using different approaches. Thermal Sci. Eng. Progr. **8**, 411–417 (2018)
33. L. Zhang, L. Chen, J. Liu, X. Fang, Z. Zhang, Effect of morphology of carbon nanomaterials on thermo-physical characteristics, optical properties and photo-thermal conversion performance of nanofluids. Renew. Energy **99**, 888–897 (2016)
34. S.H. Lee, B.H. Jun, Silver nanoparticles: Synthesis and application for nanomedicine. Int. J. Mol. Sci. **20**(4), 865 (2019)
35. B. Bakthavatchalam, K. Habib, R. Saidur, B.B. Saha, K. Irshad, Comprehensive study on nanofluid and ionanofluid for heat transfer enhancement: A review on current and future perspective. J. Mol. Liq. **305**, 112787 (2020)
36. S. Azarfar, S. Movahedirad, A.A. Sarbanha, R. Norouzbeigi, B. Beigzadeh, Low cost and new design of transient hot-wire technique for the thermal conductivity measurement of fluids. Appl. Therm. Eng. **105**, 142–150 (2016)
37. A.P.C. Ribeiro, S.I.C. Vieira, J.M. França, C.S. Queirós, E. Langa, M.J.V. Lourenço, C.N. de Castro, *Thermal Properties of Ionic Liquids and Ionanofluids* (IntechOpen, London, 2011)
38. A. Kakavandi, M. Akbari, Experimental investigation of thermal conductivity of nanofluids containing of hybrid nanoparticles suspended in binary base fluids and propose a new correlation. Int. J. Heat Mass Transf. **124**, 742–751 (2018)
39. H.W. Xian, N.A.C. Sidik, R. Saidur, Impact of different surfactants and ultrasonication time on the stability and thermophysical properties of hybrid nanofluids. Int. Commun. Heat Mass Transf. **110**, 104389 (2020)
40. S.K. Pramanik, S. Sreedharan, R. Tiwari, S. Dutta, N. Kandoth, S. Barman, J.A. Thomas, Nanoparticles for super-resolution microscopy: Intracellular delivery and molecular targeting. Chem. Soc. Rev. **51**(24), 9882–9916 (2022)
41. X. Li, G. Zeng, X. Lei, The stability, optical properties and solar-thermal conversion performance of SiC-MWCNTs hybrid nanofluids for the direct absorption solar collector (DASC) application. Sol. Energy Mater. Sol. Cells **206**, 110323 (2020)

Harnessing Nanofluids: A Comprehensive Guide to Theory, Synthesis, Properties, and Utilization

Swagata Chetia, Sneha Sonari, Hirendra Das, and Pranjal Saikia

1 Introduction to Nanofluids

Nanofluids represent specially designed colloidal suspensions in a biphasic system, where nanoparticles exist in the solid phase and the base fluid (BF) in the liquid phase. The nanoparticles utilized in nanofluids include metals (e.g., copper, silver, aluminum), metal oxides (e.g., alumina, titania, copper oxide, iron oxide, silica), and carbon-based materials (e.g., carbon nanotubes, graphene, silicon carbide, calcium carbonate, titanium nanotubes). These nanoparticles are incorporated into the base fluids to enhance specific characteristics like heat conduction capability, viscosity, rheological and tribological characteristics, and general thermal transfer efficiency performance in comparison to conventional technological fluids. Even at modest nanoparticle concentrations, significant improvements could be attained in both thermal and transfer of heat coefficients, which will lead to the energy efficiency enhancement in numerous applications. In the beginning of 1990s, "nanofluids" was first termed in at Argonne National Laboratory, U.S.A. by Choi and Eastman [1]. Initially, efforts were focused on creating suspensions by incorporating solid particles of micrometer or even millimeter sizes into base fluids. However, these early approaches faced considerable challenges, such as surface wear, obstruction of microchannels, pipeline degradation, and increased pressure drop, which severely limited their practical applications. These drawbacks were overcome when Choi and Eastman introduced nanoscale metallic particle and carbon nanotube suspensions in

S. Chetia · S. Sonari · P. Saikia
Department of Applied Sciences (Chemical Science Division), Gauhati University, Guwahati, Assam, India

H. Das (✉)
Department of Electronics and Communication Technology, Gauhati University, Guwahati, Assam, India

S. Kulkarni, S. Das (eds.), *Ionic Liquids, Nanofluids and Nanotechnology*, Lecture Notes in Nanoscale Science and Technology 42,
https://doi.org/10.1007/978-3-032-04008-4_3

various base fluids, yielding promising results. Since then, extensive research has been conducted on different aspects of nanofluids, such as heat transfer efficiency, flow resistance, mass per unit volume, specific heat, various heat transfer modes, pressure drop, pumping power, and other fundamental properties [2].

Traditional thermal management liquids like water, oil, kerosene, glycols, and refrigerants typically exhibit low thermal conductivity. However, their widespread use in applications such as engine cooling, heating and cooling systems, fuel cells, automotive components, electronics, and industrial processes necessitates improvements in their thermal performance to enhance efficiency [3]. Relative to traditional cooling agents, nanofluids possess significantly higher thermal conductivities. Additionally, they offer the advantage of minimal obstruction of flow channels and negligible pressure drop, making them highly beneficial for heat transfer applications.

Apart from these advantages, nanofluids also possess the following key benefits [4]:

- Enhanced specific surface area, facilitating greater heat transfer between particles and fluids.
- Excellent dispersion stability, driven by dominant Brownian motion of nanoparticles.
- Lower pumping power requirements compared to conventional liquids for equivalent heat transfer enhancement.
- Reduced particle blockage, minimizing system clogging and supporting miniaturization.
- Customizable properties, such as surface wettability and thermal conductivity, by adjusting nanoparticle concentration for different applications.

Research on nanofluids has focused on their implementation across numerous applications, particularly for cleaner and more efficient energy generation and utilization. Although not all applications directly influence energy transmission, they hold potential to reduce dependence on nonrenewable energy sources. Examples of nanofluid applications include engine cooling, electronics cooling, welding cooling, transformer cooling oil, solar water heating, diesel-electric generator jacket water coolant, and recovery of waste gases from boiler exhaust flues, among others [5, 6]. Moreover, nanofluids demonstrated considerable potential in biomedical applications such as cell labeling, cell separation, cellular regeneration, drug transport, magnetic resonance imaging (MRI), cancer treatment, and sensing-imaging technologies [7]. Consequently, the large-scale, cost-effective production of well-dispersed industrial nanofluids has become a major focus of research and development. As the field of nanofluids continues to evolve, it is essential to understand its foundational principles, which are deeply rooted in various scientific and engineering disciplines. Figure 1 presents a conceptual depiction of traditional and modern domains relevant to nanofluid research. These fields encompass a broad spectrum of studies, reflecting the interdisciplinary nature of nanofluid research. Furthermore, advancements in analytical techniques have significantly enhanced nanofluid research, aided by sophisticated tools such as scanning electron

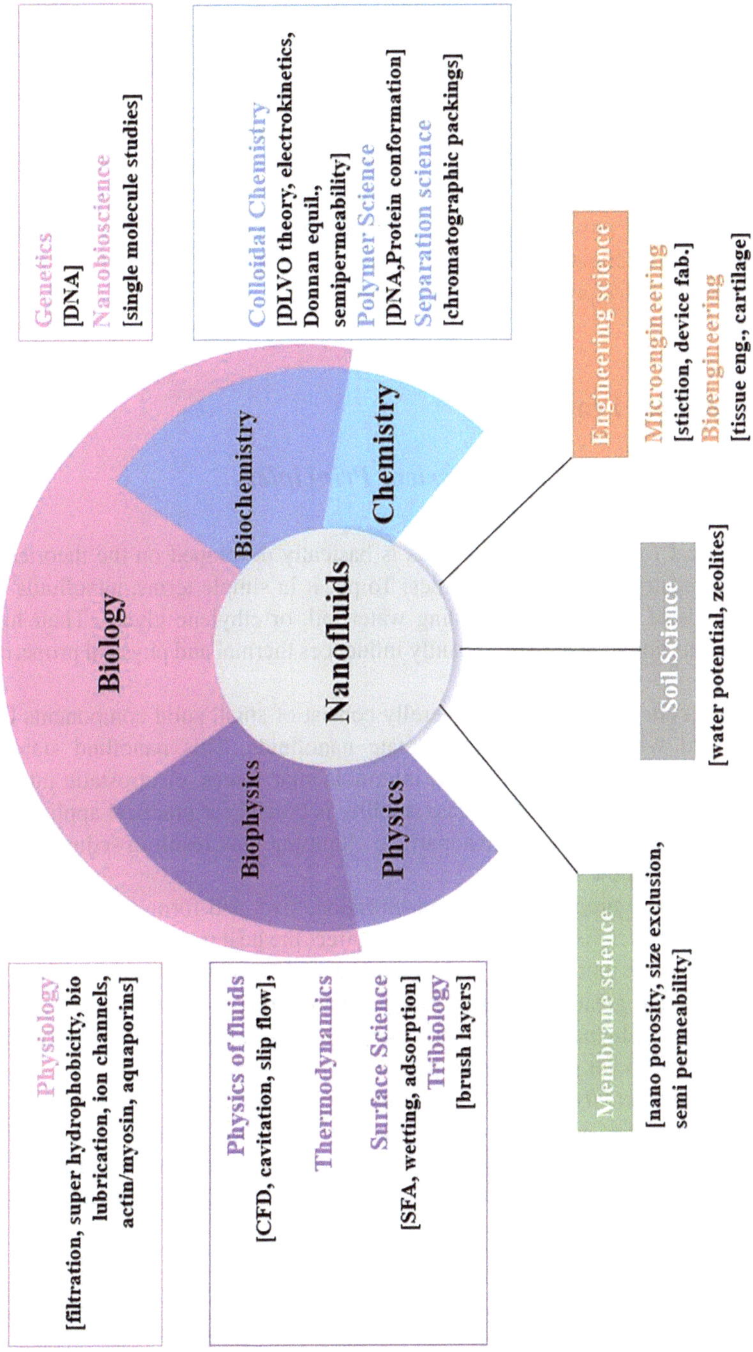

Fig. 1 Disciplines related to nanofluids

microscopy (SEM) and atomic force microscopy (AFM), which enable precise characterization of nanostructures.

By modification of microscale physics, nanofluid research and engineering aims to improve the characteristics at the megascale and macroscale levels of the fluid. Research should be concentrated in the area that deals with issues like: (1) effective manipulation and control over nanoparticle dispersion and stability, (2) comprehending the interaction of physical phenomena at various scales, and (3) optimizing the microscale physics to maximize the best macroscale performance. In this chapter, we cover these basic fundamental aspects, along with synthesis procedures, characterization techniques, applications, and the prospective research pathways required for progress of nanofluid technology.

2 Theoretical Framework

2.1 Thermodynamic and Physical Principles

The theoretical foundation of nanofluids is basically developed on the theories of physical chemistry and thermodynamics. To put it in simple terms, nanofluids are special blends of base fluids, including water, oil, or ethylene glycol. Their high surface area-to-volume ratio significantly influences thermal and physical properties [8].

Colloidal systems: Nanofluids generally consist of small solid components that are combined with a base fluid to create nanofluids. This nanofluid stability completely depends on forces such as intermolecular forces, electrostatic interactions, and Brownian motion. Adequate stability is crucial for practical applications because the presence of excess nanoparticle clumping may result in sedimentation and obstruction in flow systems.

Thermodynamic properties: The concentration, size, and form of nanoparticles as well as the characteristics of the base fluid affect the heat transfer capability, flow resistance, mass per unit volume, and specific heat of nanofluids. Nanoparticles typically demonstrate greater heat efficiency compared to base fluids, enhancing heat transfer. However, higher nanoparticle concentrations increase viscosity, affecting power requirements for pumping and pressure drop in applications [9].

Thermal conductivity enhancement: Enhanced nanofluid thermal conductivity arises from Brownian motion, liquid layering effects near nanoparticles, ballistic phonon transport, and nanoparticle clustering. Brownian motion facilitates heat transfer within the base fluid, significantly improving thermal conductivity [10, 11].

2.2 Particle Suspension Dynamics

The behavior of nanoparticles in a fluid is governed by classical fluid dynamics and nanoscale phenomena, influenced by particle size, shape, concentration, and base fluid properties.

Brownian motion: The random movement of nanoparticles due to collisions with fluid molecules contributes to thermal conductivity enhancement. The Brownian diffusion coefficient is given by the Stokes-Einstein equation:

$$D_B = k_B T 6\pi \mu \mathrm{r}$$

where k_B is the Boltzmann constant, T is absolute temperature, μ is the dynamic viscosity of the base fluid, and r is the nanoparticle radius.

Aggregation and sedimentation: Nanoparticles tend to form clusters due to interparticle forces, leading to sedimentation, which affects viscosity and thermal conductivity. Density differences between nanoscale particles and the foundational liquid can cause gradual settling, reducing stability and performance.

Zeta potential: Zeta potential determines the electrostatic charge at the nanoparticle surface. A zeta potential exceeding ± 30 mV indicates excellent stability, as electrostatic repulsion prevents aggregation [12].

2.3 Heat Transfer Mechanisms

Heat transfer in nanofluids involves a combination of conduction and convection mechanisms, with nanoparticles enhancing the heat efficiency of the base fluid.

Enhanced thermal conductivity: The Maxwell–Garnett model describes effective thermal conductivity:

$$k_{eff} = \frac{2k_{bf} + k_p - 2\varnothing\left(k_{bf} - k_p\right)}{2k_{bf} + k_p + \varnothing\left(k_{bf} - k_p\right)} k_{bf}.$$

where k_{eff} is the effective thermal conductivity, k_{bf} is the base fluid conductivity, k_p is the nanoparticle conductivity, and ϕ is the proportion of the volume occupied by nanoparticles.

Brownian motion and microconvection: Brownian motion induces micro convection, enhancing heat transfer. This effect strengthens at higher temperatures due to increased nanoparticle activity.

Liquid layering: A nanolayer of liquid forms around nanoparticles, exhibiting different thermal properties than the bulk fluid. This enhances heat transfer by improving the thermal interface between nanoparticles and the surrounding fluid medium.

Thermophoresis: Nanoparticles move in response to temperature gradients, altering their spatial distribution within the fluid. The thermophoretic velocity is given by:

$$V_{th} = -0.26\left(\frac{k_{bf}}{2k_{bf} + k_p}\right)\frac{\mu}{\rho}\frac{\Delta T}{T_k}$$

where μ is the dynamic viscosity, ρ is the density, ΔT is the temperature gradient, and T_k is the absolute temperature [13–15].

3 Properties of Nanofluids

Nanofluids are specially designed colloidal dispersions to reduce ecological effects while improving the thermal and flow characteristics of operational fluids. They exhibit improved thermophysical attributes like heat conduction ability, thermal diffusivity, viscosity, and convective thermal transfer coefficients in comparison to base fluids such as oil or water [16]. Conventional liquids have limitations in thermal conductivity, environmental impact, and worker safety, whereas nanofluids serve as smart fluids. By incorporating nanoparticles into the base fluid (BF), its thermal properties can be significantly enhanced. This section provides a comprehensive and up-to-date review of the thermophysical characteristics of nanofluids.

3.1 Thermal Conductivity

Generally, the most important characteristic of nanofluids lies in the superior thermal conductivity. As the concentration of the nanoparticle increases, the enhancement of the thermal capability also increases to a certain extent, after which the rise shows degradation, or in other words, a reduction can be observed. This reduction is dependent on the type of nanoparticles. The measurement of the thermal conductivity can be done by using the following techniques:

- Transient hot-wire method
- Temperature oscillation method
- 3-ω method
- Steady-state parallel plate method

The primary factor affecting the heat transfer ability of nanofluids is volume concentration [17]. Thermal conductivity enhancements of 30.6%, 64%, and 64% for MXene, graphene, and MWCNT, respectively, were observed when these nanoparticles were dispersed in water, indicating superior heat transfer capabilities of CNT and graphene nanoparticles [18]. A systematic investigation into the effective thermal conductivity (λ_{ef}) of nanofluids was conducted by varying base fluids and particle types, including silicon dioxide (SiO_2), titanium dioxide (TiO_2), and copper oxide (CuO), along with polymers such as polystyrene (PS) and polymethylmethacrylate (PMMA) [19]. Studies by Zhang et al. and Shahsavar et al. showed that as the thermal conductivity of nanofluids improves with rising temperature and nanoparticle concentration, indicating that temperature plays a dominant role as the heat transport efficiency increases [20, 21].

3.2 Viscosity

Viscosity determines the pumping power, pressure drop in laminar flow, and convective heat transfer. There are discrepancies in reported viscosity values for nanofluids due to variations in experimental conditions and measurement techniques [22]. Nanofluids containing TiO_2, Al_2O_3, and Al_2O_3 nanoparticles are unsuitable as enhanced oil recovery (EOR) additives due to their low dispersibility in biopolymer solutions, attributed to the low density of silanol groups on their surfaces [23]. Nguyen et al. [24] studied the dynamic viscosity of nanofluids at different temperatures and particle sizes, where they reported that the nanoparticles bearing a large size have the ability to exhibit high viscosity in comparison to nanoparticles having a small size. Much before this finding, researchers discovered that the increase in concentration of the nanoparticles gradually leads to higher viscosity without any change in it [25, 26].

3.3 Density

As proved by scientific theories, there is a linear change of density of nanofluids along with temperature and concentrations [27]. Nanoparticles typically have much higher densities than base fluids, significantly increasing the overall nanofluid density. Sharifpur et al. examined the density of SiOx-ethylene glycol/water, SiO_2-water, CuO-glycerol, and MgO-glycerol nanofluids experimentally within the temperature spectrum of 10–40 °C, with volume fractions between 1% and 6% [28]. Selvakumar et al. proposed an effective density prediction model that accounted for clusters surrounded by interfacial layers and primary nanoparticles [29].

3.4 Specific Heat

The thermal performance of nanofluids is influenced by their specific heat capacity (C_p). The specific heat of nanoparticles depends on their type, concentration, size, shape, and the properties of base fluids. A nanofluid composed of molten solar salt (MSS) with 1.0% SiO_2 nanoparticles exhibited a 15.65% increase in specific heat compared to pure MSS. An enhancement of 25.03% was observed in mixtures containing $NaNO_3$ nanostructures bonded with SiO_2 nanoparticles [30].

3.5 Electrical Conductivity

Nanofluids containing conductive nanoparticles, such as carbon nanotubes or metals, can enhance electrical conductivity. The conductivity rises with the concentration of conductive nanoparticles (e.g., copper, silver, or carbon-based materials), though excessive conductivity may cause short-circuiting or undesired electrical behavior. Research has demonstrated that increasing the proportion of volume fraction and temperature generally leads to higher electrical conductivity in nanofluids [31]. Electrical conductivity can also indicate nanofluid stability. Akilu et al. reported data on the electrical conductivity of several ethylene glycol- and propylene glycol-based β-SiC nanofluids [32]. Zyla et al. prepared ethylene glycol-based nanodiamond nanofluids, finding a nonlinear progression in electrical conductivity with increasing volume fraction [33].

3.6 Optical Properties

The optical properties of nanofluids, such as light absorption and scattering, vary depending on nanoparticle concentration and type. These properties significantly influence heat absorption in nanofluids. Karami et al. investigated the optical properties of CuO nanofluids and found an increase in absorption capacity at very low volume concentrations [34]. Ahmad et al. reviewed the optical properties, including absorption, transmittance, scattering, and extinction coefficient of nanofluids containing metal, metal oxide, carbon nanotubes, graphite, and graphene [35].

3.7 Rheological Properties

Rheological properties describe the deformation and flow behavior of materials. The rheology of nanofluids determines their flow characteristics and pressure drop in flow systems. Typically, fluid flow behavior is characterized by the relationship between shear stress (τ) and shear rate (γ). Nanofluid rheology directly impacts pressure drop. Richmond et al. found that adding TiO_2 to SiO_2/water nanofluids altered the flow behavior from Newtonian to non-Newtonian [36]. This shear-thinning effect is characteristic of non-Newtonian fluids, where the viscosity decreases with increasing shear rate. Such behavior is particularly important for applications in heat exchangers, where variable flow rates and shear conditions are encountered. The ability to control the rheological properties enables the fine-tuning of flow behavior, which in turn affects the pressure drop, heat transfer efficiency, and overall system performance. This understanding is a key factor in advancing the use of nanofluids in various industrial and technological applications, from cooling systems to biomedical devices. Table 1 summarizes thermophysical properties of various nanofluids.

4 Synthesis Processes of Nanofluids

Proper synthesis of nanofluids is crucial, as inadequate preparation can lead to reduced thermal efficiency, biphasic heat transfer behavior, and equipment issues like clogging and damage [16]. Nanofluids are generally synthesized using two main approaches: the two-step method and the one-step method. Figure 2 illustrates different approaches to nanoparticle synthesis.

4.1 Two-Step Method

For the production of nanofluids on large scale, a two-step technique is widely used. Usually, the technique proceeds by first synthesizing nanoparticles in the form of dry powders through physical or chemical procedures and then gradually dispersion of it in a base fluid with the help of either physical or ultrasonic agitation, high-shear mixing, homogenization, or ball milling [43]. Common nanoparticle synthesis techniques include transition metal salt reduction, ligand reduction, organometallic displacement, microwave synthesis, and co-precipitation [44]. A major challenge in nanofluid production is nanoparticle agglomeration due to van der Waals forces and weakened Brownian motion, leading to sedimentation and reduced homogeneity [45]. This can be mitigated using surfactants, dispersants at

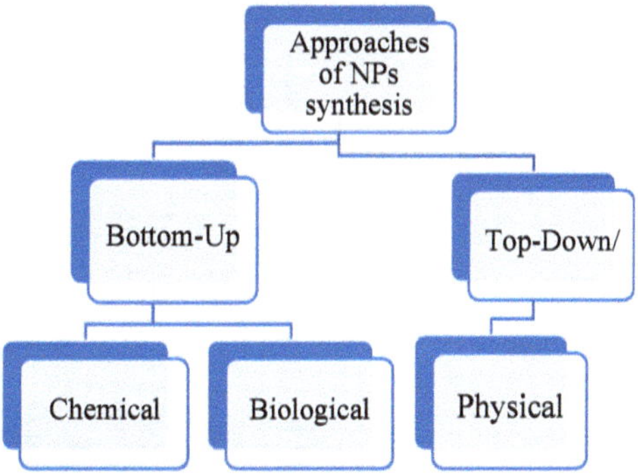

Fig. 2 Approaches of nanoparticle synthesis

Table 1 Thermophysical properties of various nanofluids

Nanofluid	Thermal conductivity ($Wm\ K^{-1}$)	Electrical conductivity ($\mu S\ cm^{-1}$)	pH	Particle size	Volume fraction (%)	Temperature (°C)	References
Al_2O_3-water	0.661	–	8.0	15–50	0.2	25	[37]
CuO-water	0.679	–	9.5	25–60	0.2	25	[37]
Al_2O_3-EG	–	10.26	–	80	1	30	[38]
CuO-EG	–	18.21	–	80	1	30	[38]
MgO-EG	0.354	–	–	20	5	30	[39]
Al_2O_3-EG	0.328	–	–	20	5	30	[39]
MgO-EG	0.328	–	–	60	5	25	[40]
Al_2O_3-glycerol	–	0.15	4.21	20–30	2	30	[41]
Al_2O_3-water	–	2.9	5.5	20	0.01	25	[42]

critical micelle concentrations, and dispersion techniques such as ultrasonic baths, magnetic stirrers, high-pressure homogenizers, and ultrasonic jammers [46].

Various studies have explored synthesis techniques and their impact on nanofluid properties. Suresh et al. used chemical precipitation to synthesize Al_2O_3 nanoparticles, studying dispersion and thermal behavior [47]. Hashemi and Behabadi investigated heat transfer and pressure variations in nanofluids via direct synthesis [48]. Kumar et al. enhanced MgO nanoparticle heat transfer capability for improved nanofluid performance [49]. Lee et al. applied the two-step method to alumina-based nanofluids, analyzing thermal conductivity and viscosity [50]. Murshed et al. with the help of this technique prepared TiO_2-water nanosuspensions, to study the effect that particle size and concentration has on thermal energy movement [51].

4.2 One-Step Method

The one-step method simultaneously synthesizes and disperses nanometer-sized particles dispersed in a carrier fluid, avoiding drying, storage, and transportation, which minimizes agglomeration and enhances stability [52]. With the help of this technique, a highly stable nanolubricants can be synthesized. However, there are two main setbacks: first is that it is of higher cost in comparison to the two-step method, and second is the difficulty in controlling nanoparticle size and morphology. Consequently, to achieve the desired fluid properties, residual reactants must be removed. Out of many, the top most one-step synthesis techniques include thermal decomposition, physical vapor condensation, chemical reduction, and polyol synthesis [44]. Lo et al. synthesized CuO/water nanofluids using Cu wire as an electrode, achieving a stable, agglomeration-free suspension [53]. Zhu et al. employed microwave-assisted reduction of $CuSO_4{\cdot}5H_2O$ with sodium hypophosphite in ethylene glycol, demonstrating enhanced dispersion and faster reaction rates. Mohammadpoor et al. compared single-step and two-step Cu-based nanofluids, concluding that single-step-prepared nanofluids exhibited superior stability [54].

Despite cost and particle control challenges, the one-step method remains promising for producing uniformly dispersed nanofluids. Research efforts are needed to improve its scalability and cost-effectiveness. While both one-step and two-step methods have advantages and limitations (Fig. 3), they enhance nanoparticle dispersion and fluid stability in suitable base fluids, as summarized in Table 2.

4.3 Chemical Methods

Chemical vapor deposition (CVD) is a sophisticated technique where gaseous precursors react on a substrate to form deposits. Zhang et al. used a two-step rotary

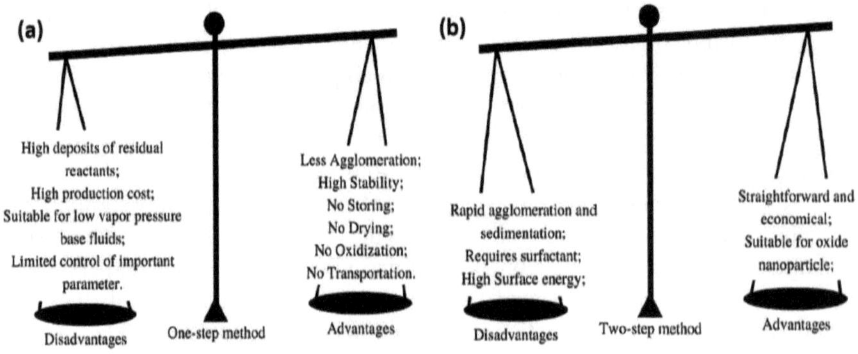

Fig. 3 Advantages and disadvantages of (**a**) one-step method and (**b**) two-step method. Reproduced with permission (CC-BY 4.0) from [55]

Table 2 Summary of nanofluid synthesis studies using one-step and two-step methods

NPs	Base	Method	Stability	References
CuO	Water	Two-step	Optimal stability with 95:5 CTAB–SDS	[56]
SiO_2	Water	Two-step	0–8 days	[57]
ZnO	DIW	Two-step	1–60 days	[58]
MWCNT	Water	Two-step	1 h	[59]
SiO_2/TiO_2	Tetraethyl orthosilicate (TEOS)titanium butoxide (TNBT), ammonia ($NH_3 \cdot H_2O$) and alcohol	One-step	20 min	[60]
WO_3–silver	TO	One-step	N/A	[61]

CVD to mix Cu nanoparticles with Al_2O_3, forming Al_2O_3-Cu nanocomposites [62]. Similarly, rf-CVD was employed to synthesize biocompatible graphene/HA/Au nanocomposites using acetylene and methane as carbon sources.

The sol-gel method involves forming a sol, transitioning into a gel, and calcining it to obtain crystalline powders. Behnajady et al. outlined the process in four stages [63], while Thiagarajan and Ovejero synthesized metal oxide nanostructures and Au-Fe_3O_4 hybrid nanoparticles, demonstrating improved stability and biocompatibility [64, 65]. This method has also been adapted for SiO_2 and CuO nanoparticles. An elevated temperature and pressure are used for hydrothermal and solvothermal synthesis to control particle size and morphology, successfully producing ethylenediamine- and L-phenylalanine-doped carbon dots [66]. Precipitation and co-precipitation rely on supersaturation-induced nucleation for nanoparticle formation. This approach has been applied to synthesize CeO_2 from cerium nitrate and NaOH. Magnetic nanofluids (ferrofluids) have gained attention due to their tunable properties. Su et al. synthesized Fe_3O_4-based ferrofluids via co-precipitation to study their thermal properties, as illustrated in Fig. 4 [67].

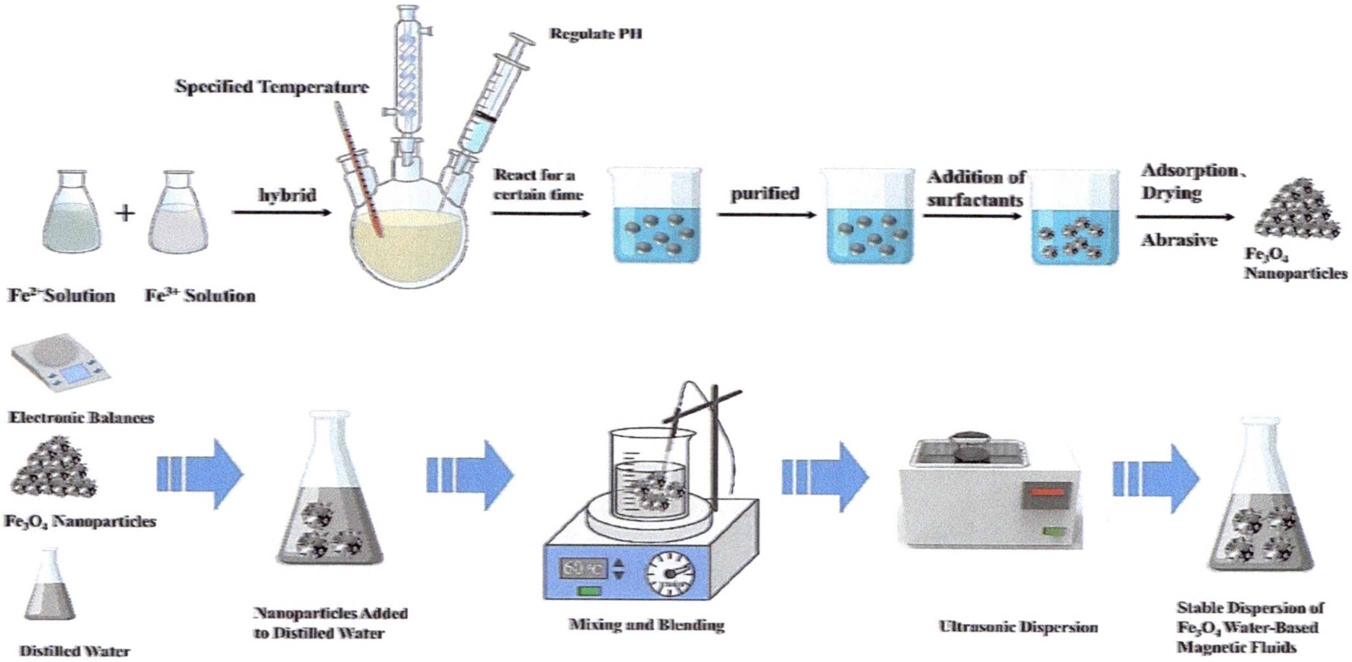

Fig. 4 Schematic representation of water-Fe_3O_4 magnetic nanofluid synthesis process. Reproduced with permission (CC-BY 4.0) from [67]

4.4 Physical Methods

Physical methods for nanofluid synthesis encompass advanced techniques to produce nanoparticles with controlled properties.

- Laser ablation in liquid (LAL) is a cutting-edge technique that uses an Nd:YAG laser to ablate a target material in a liquid medium, forming nanoparticles. For instance, CuNPs-EG nanofluid was synthesized using this method under ambient conditions [68]. Similarly, Riahi et al. demonstrated the versatility of pulsed LAL by preparing Al_2O_3-water nanofluids, highlighting its broad applicability [69].
- Ball milling and high-energy ball milling (HEBM) are widely used mechanical techniques for reducing particle size and synthesizing nanocomposites. HEBM has been employed to refine ZnO microcrystalline powder into nanoparticles [70] and is also utilized in oxide- and carbide-strengthened aluminum alloys and wear-resistant coatings [69].
- Electrospinning is an advanced method primarily used for fabricating polymer-based nanofibers. It has been employed to synthesize one-dimensional $Na_{0.5}Bi_{0.5}TiO_3$ (NBT) nanofibers for energy storage applications [71]. Figure 5 illustrates the electrospinning-based synthesis of palladium-decorated polyethyleneimine/polycaprolactone composites reported by Curiu et al. [72].
- Magnetron sputtering is another key technique where high-energy particles bombard a solid surface, leading to thin film deposition. This approach has been successfully implemented to fabricate WSe_2 layered nanofilms on SiO_2 and carbon paper substrates [73].

Although these synthesis techniques are efficient for stabilizing and dispersion nanoparticles, a gap remains due to which agglomeration occurs, affecting fluid flow behavior [74]. Stability evaluation methods include:

- Sedimentation (qualitative): Simple but time-consuming.
- Centrifugation (accelerated sedimentation): Singh et al. used this method to assess Ag nanofluid stability in ethanol synthesized via microwave reduction of $AgNO_3$ with PVP [75].
- UV Light Absorbance Rate and Zeta Potential (quantitative): Zeta potential offers outstanding insights into colloidal stability, with high absolute values indicating better dispersion and lower values leading to coagulation or flocculation [76].

Thus, while traditional sedimentation and centrifugation provide qualitative stability assessments, zeta potential offers a more precise evaluation of dispersion stability in nanofluids.

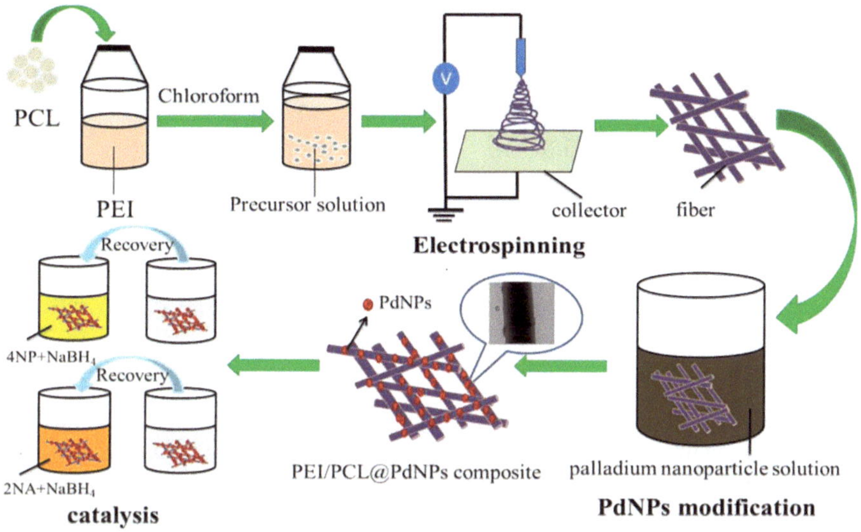

Fig. 5 Schematic representation of water-Fe_3O_4 magnetic nanofluid synthesis process using electrospinning process. Reproduced with permission (CC-BY 4.0) from [72]

5 Characterization of Nanofluids

Nanofluids are characterized by their physical and chemical properties, as well as their structure, shape, size, and composition. The most used characterization techniques include zeta potential analysis, sedimentation method, and electron microscopy. The analysis methods are frequently employed by researchers to characterize nanofluids. The nanofluids are characterized by the following techniques: SEM, TEM, XRD, FT-IR, DLS, TGA, and zeta potential analysis. Microscopic techniques can be used to characterize nanomaterials and nanofluids.

5.1 Microscopic Techniques

Electron microscopy is a powerful technique that utilizes an electron beam to generate high-resolution images, offering details of the morphology, stress, crystallinity, and magnetic characteristics of nanomaterials in depth. This technique enables imaging at high resolution and magnification, making it a vital tool for nanomaterial characterization. For example, silver nanoparticles synthesized through electrical explosion of wires in liquids without any additives were examined using transmission electron microscopy (TEM) and field-emission scanning electron microscopy (FESEM) [77]. In another experiment, Mohamed Hosny et al. synthesized platinum nanoparticles using *P. salicifolium* extract and revealed that these nanoparticles

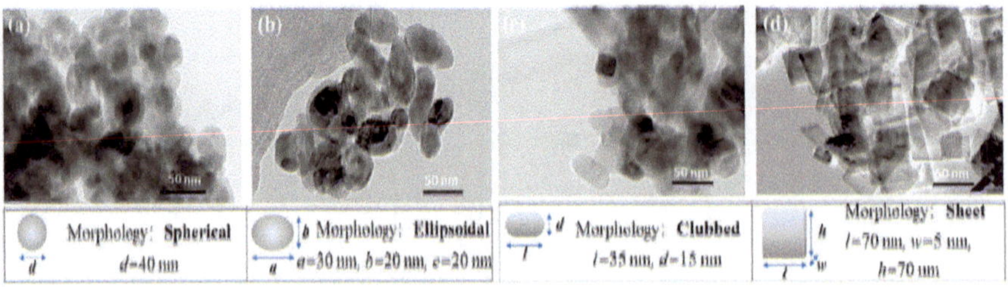

Fig. 6 TEM image showing the morphology of TiO_2 nanoparticles of different shapes: (**a**) spherical, (**b**) elliptical, (**c**) cubed, (**d**) sheet. Reproduced with permission (CC-BY 4.0) from reference [79]

are spherical, with measurements ranging from 1 to 3 nm, through TEM analysis [78]. Additionally, Cui et al. explored the shapes of different TiO_2 nanoparticles with various shapes and demonstrated that TiO_2-water nanofluids with clubbed and sheet-shaped nanoparticles had a high relative thermal conductivity (RTC) in comparison to other shapes. Remarkably, the sheet-shaped TiO_2 nanofluids achieved the highest RTC at 600 K with a 4% concentration of nanoparticles [79]. Figure 6 illustrates the morphology of different TiO_2 nanoparticles as designed by Cui et al., highlighting their impact on thermal conductivity.

SEM is a widely used technique for observing surface phenomena of materials even at the nanoscale; however, it has certain limitations. Ijaz et al. reported the fabrication of silver nanoparticles through a green extract and characterized them microscopically and spectroscopically to evaluate their antibacterial activity [80]. SEM offers high-resolution imaging capabilities, achieving a resolution of up to 1 nm, making it an essential tool for detailed surface analysis. Additionally, Dahy et al. developed an intelligent optimization model to classify nanoparticles with an accuracy ranging from 97% to 98% [81]. In another study, helium ion microscopy was employed to investigate the effects of TiO_2 nanoparticles on *E. coli* [82]. Scanning tunneling microscopy, which generates three-dimensional images by scanning a sample with a stylus, is another powerful tool for surface imaging at the atomic level. Furthermore, the microstructure and morphology of α-SiC and β-SiC particles were evaluated using SEM and TEM techniques, providing valuable insights into their structural characteristics [83].

5.2 Spectroscopic Techniques

Spectroscopic techniques share the fundamental principle of obtaining a characteristic spectrum of a sample based on its interaction with an energy source such as laser, infrared light, X-rays, or UV-Vis light. These techniques provide both qualitative and, in certain instances, numerical data regarding the chemical makeup of a sample. Infrared (IR) and Raman spectroscopy, both nondestructive and efficient, utilize the inelastic scattering of light to characterize materials. A UV-Vis spectrophotometer was used to study the degree of γ-Al_2O_3 nanofluid dispersion in nanofluids [84–86]. Additionally, photoluminescence (PL) spectroscopy examines the electronic excitation of a sample followed by a relaxation process. FTIR spectroscopy is an additional extensively employed method that determines a sample's molecular composition and structure by analyzing its absorbance of infrared light energy at various wavelengths. In the green synthesis of silver nanoparticles, FTIR analysis revealed that polyphenols and flavanols, extracted from wine production residues, played a key role in reducing $AgNO_3$ to form silver nanostructures.

5.3 *Thermal Technique*

Thermal measurements provide insights into both physical and chemical adsorption, including chemisorption, thermal potting, and solid–gas reactions. TGA is widely used to assess the thermal durability of nanofluids. The TGA results of MWCNT/ethylene glycol nanofluid demonstrated maximum thermal stability at 140 °C, along with a highest zeta potential value of 61.8 mV, indicating ethylene glycol as the optimal base liquid for achieving enhanced thermophysical properties and long-term stability [87]. Thermal lens spectrometry (TLS) is a technique that measures the amount of light absorbed by a specimen, offering understanding into its thermal properties. Dual-beam thermal lens spectrometry has been used to characterize the thermal diffusivity of aqueous dispersions of silicon oxide nanofluids. From this study it was proven that with an increase in thermal diffusivity, the concentration of SiO_2 nanoparticles decorated with Au in the nanofluid increases. Similarly, the dual-beam thermal lens method was utilized to investigate the thermal diffusivity of ZnO/rGO nanofluid, revealing an enhancement in thermal diffusion relative to water [88]. Additionally, ion nanofluids, formulated by incorporating nanocarbons such as diesel soot (DS), camphor soot (CS), CNTs, and graphene into an ionic liquid base (1-butyl-3-methylimidazolium methyl sulfate, BMMS), were analyzed using nondestructive photopyroelectric calorimetry (PPE). Furthermore, the thermal effectiveness of silver nanofluids was investigated using a front-photopyroelectric technique that employed a metalized polyvinylidene difluoride (PVDF) pyroelectric sensor.

6 Applications of Nanofluids

6.1 *Biomedical*

Nanotechnology has a significant impact on the advancement of targeted drug delivery, enabling nanoparticles to transport drugs directly to a specific cell.

Since the 1990s, research on biomicro-electro-mechanical systems (MEMS) has driven commercialization and technological advancements in drug delivery systems. Biological nanoforms such as albumin, gelatin, and phospholipid-based vesicles, along with the chemical agents like polymers and nanoparticles in nanofluids, have been extensively studied. Nanofluids hold an upper hand on biomedical applications, including targeted drug delivery and biological process regulation, for the comparable size of nanoparticles to biomolecules. Additionally, nanofluids have the ability to enhance surgical safety by providing effective cooling at the surgical site, reducing the risk of organ damage and improving patient survival rates. Magnetic nanoparticles in biological fluids play the role of transport agents for drugs or radiation, offering innovative cancer therapy approaches. These nanoparticles absorb more energy than microparticles when exposed to alternating current (AC) magnetic fields, which are harmless to the human body. Moreover, they showcase a higher affinity towards the tumor cells than normal cells, making AC-stimulated magnetic nanoparticles a promising agent for cancer treatment [89–91].

6.2 Renewable Energy

Solar energy, a sustainable and environmentally friendly power source, has a significant impact on electricity generation, thermal heating, and chemical processing. However, conventional flat-plate solar collectors, though cost-effective, face challenges related to low efficiency and limited output temperatures, restricting their broader applicability. To address these constraints, scientists have explored alternative designs and advanced heat transfer mechanisms. In a significant study, Tyagi et al. conducted a theoretical analysis comparing a non-concentrating direct absorption solar collector (DAC) with a conventional flat-plate collector [92]. In this system, solar radiation interacted with a thin film of flowing nanofluid, leading to enhanced absorption characteristics.

Building upon this research, scientists have further investigated innovative techniques to improve energy efficiency in solar devices. The integration of nanoparticles into conventional heat transfer fluids has been identified as a promising strategy for enhancing thermal properties. Studies suggest that by dispersing nanoparticles within base fluids, heat transfer rates can be improved by up to 100 times compared to conventional fluids [93]. This substantial enhancement in thermal conductivity enables more conversion of thermal energy from solar sources, which builds the path for a high-performance solar collector.

6.3 Application as a Coolant

Conventional cutting fluids have been used to mitigate the heat issue, but environmental concerns have necessitated the adoption of minimal lubrication techniques.

While increased surface area improves heat dissipation, it also raises energy consumption. Nanofluids have emerged as a promising alternative to traditional cooling and lubricating fluids, offering energy savings and reduced emissions. Routbort et al. demonstrated that replacing conventional cooling liquids with nanofluids could save up to one billion BTUs of energy [94]. In the rubber and tire industry, nanofluids effectively enhance cooling during manufacturing.

The development of energy-efficient central heating systems has also benefited from nanotechnology. CNT dispersions in water form nanofluids with significantly improved convective heat transfer, up to four times higher than conventional fluids. When incorporated into commercial water boilers, nanofluids enhance efficiency by 10% and modify flow characteristics (laminar or turbulent). However, their effectiveness depends on a marginal increase in viscosity alongside enhanced thermal conductivity. While laminar flow applications benefit from nanofluids, large-scale turbulent flow applications may not be as advantageous. Nonetheless, in small-scale systems, nanofluids provide superior cooling speeds and improve heat exchange performance at ambient temperatures.

6.4 Nanofluids in Sensors for Exploration

Beyond these applications, nanofluids continue to play a crucial role in various emerging technologies, as illustrated in Fig. 7.

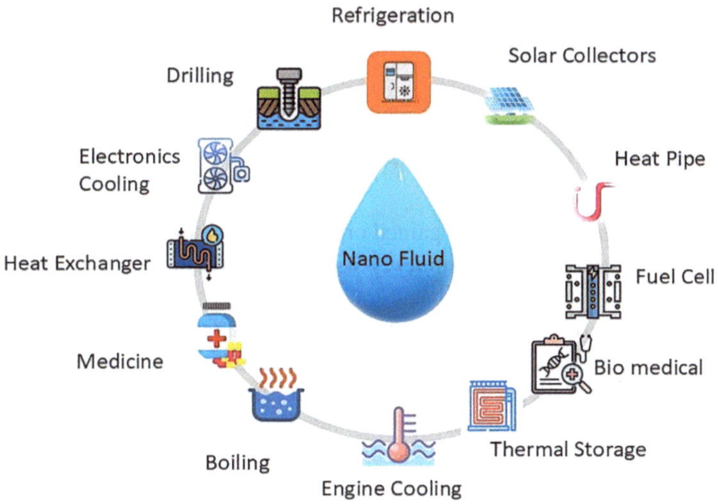

Fig. 7 Various applications of nanofluids

7 Challenges and Future Prospects

Despite the remarkable properties of nanofluids, several factors continue to hinder their widespread adoption. Key challenges include:

- Inconsistent experimental outcomes across different studies.
- Unsatisfactory characterization techniques, leading to discrepancies in reported suspension properties.
- Insufficient theoretical models to fully explain the mechanisms governing property modifications.

To advance the field, addressing the following critical issues is essential:

7.1 *Stability Issues*

Nanoparticles tend to form aggregates due to strong van der Waals forces, which reduces their effectiveness over time. Several methods, likewise surfactant incorporation, surface modification of nanoparticles, and high-pressure treatment, have been explored to enhance stability. Eastman et al. demonstrated that thermal conductivity in ethylene glycol-based nanofluids containing 0.3% copper nanoparticles diminishes over time [95, 96]. Their research proved that nanoparticles which are developed freshly exhibit higher thermal conductance power in comparison to those stored for several months. Similarly, Lee and Mudawar observed a decline in stability over a 30-day storage period in Al_2O_3 nanofluids, further reinforcing the relationship between stability and thermal performance.

7.2 *Environmental and Health Impacts*

A prior knowledge has to be also gained on the potential toxicity and environmental impact of nanofluids. Many commonly used nanoparticles such as Al_2O_3, TiO_2, CuO, and Fe_3O_4 are classified as hazardous. Safety data records reveal that alumina, titania, and cupric oxide pose acute and chronic aquatic toxicity, with long-lasting environmental effects. Thus, future research should focus on eco-friendly nanofluid development using biodegradable or bio-inspired nanomaterials, as well as rigorous toxicity assessments to ensure safety in biomedical and industrial applications.

8 Future Directions in Nanofluid Research

Current development in research on nanofluids primarily focuses on their characterization and convection properties, which includes stability, temperature flow, and heat transfer performance. However, deep investigations are required to develop a comprehensive understanding of nanofluid behavior. Studies indicate that oxide nanofluids are generally inefficient for heat transfer applications, whereas certain CNT-based nanofluids demonstrate enhanced thermal performance. However, existing theoretical frameworks struggle to fully explain these observations, necessitating new models and methodologies. Researchers have justified various emerging aspects of nanofluid behavior under different conditions. One such significant area of study involves the electromagnetized entropy generation in nanofluid peristaltic flow, particularly under Reynolds number constraints and long-wavelength approximations.

Notably, CNT nanofluids showcase significant variation in thermal conductivity on the basis of tube morphology. Long, monodisperse CNTs show a high thermal conductivity, whereas shorter and irregularly shaped CNTs exhibit little to no improvement. Similarly, the heat transfer capabilities are dependent on the particle shape of oxide nanofluids. However, a major drawback in this field is the lack of precisely controlled nanoparticle specifications, which holds back the reproducibility and optimization. By achieving greater control over nanoparticle, researchers can put an effort to develop next-generation nanofluids with superior thermal properties for industrial and technological applications.

9 Conclusion

Remarkable advancements in nanofluid research in recent years have revealed their unique properties. This chapter emphasizes the historical evolution, synthesis methods, applications, and future prospects of nanofluids. Importance is also driven on the properties of nanofluids, like thermal conductivity, viscosity, and specific heat have been analyzed, along with mathematical models for electrical conductivity and thermophoretic viscosity. Influences of nanofluids on heat transfer, cooling systems, biomedical applications, and various industries, including fossil fuel boilers, cancer therapeutics, electronics, mechanical systems, and nuclear reactors, are also stated. However, challenges such as particle agglomeration and stability remain as a gap that continues to hinder their practical implementation.

Hence, future research must draw attention to improve techniques for synthesis techniques, stability enhancement, and refining theoretical models to unlock their full potential. Despite the existing challenges, advancements in research for nanofluid mark a significant step toward innovative and efficient technological solutions.

References

1. S.U. Choi, J.A. Eastman, *Enhancing Thermal Conductivity of Fluids with Nanoparticles (No. ANL/MSD/CP-84938; CONF-951135-29)* (Argonne National Lab. (ANL), Argonne, 1995)
2. I.M. Mahbubul, R. Saidur, M.A. Amalina, Latest developments on the viscosity of nanofluids. Int. J. Heat Mass Transf. **55**(4), 874–885 (2012)
3. R. Saidur, K.Y. Leong, H.A. Mohammed, A review on applications and challenges of nanofluids. Renew. Sust. Energ. Rev. **15**(3), 1646–1668 (2011)
4. S.U. Choi, Nanofluids: A new field of scientific research and innovative applications. Heat Transfer Eng. **29**(5), 429–431 (2008)
5. D. Elcock, *Potential Impacts of Nanotechnology on Energy Transmission Applications and Needs (no. ANL/EVS/TM/08-3)* (Argonne National Lab. (ANL), Argonne, 2007)
6. R. Ravisankar, V.S.K. Venkatachalapathy, N. Alagumurthy, K. Thamizhmaran, A review on oxide and metallic form of nanoparticle in heat transfer. Int. J. Eng. Sci. Technol. **6**(3), 63 (2014)
7. A.K. Gupta, M. Gupta, Synthesis and surface engineering of iron oxide nanoparticles for biomedical applications. Biomaterials **26**(18), 3995–4021 (2005)
8. M.J. Uddin, K.S. Al Kalbani, M.M. Rahman, M.S. Alam, N. Al-Salti, I. Eltayeb, Fundamentals of nanofluids: Evolution, applications and new theory. Int. J. Biomath. Syst. Biol. **2**(1), 1–32 (2016)
9. A.T. Baheta, A.D. Woldeyohannes, Effect of particle size on effective thermal conductivity of nanofluids. Asian J. Sci. Res. **6**(2), 339–345 (2013)
10. P. Keblinski, S.R. Phillpot, S.U.S. Choi, J.A. Eastman, Mechanisms of heat flow in suspensions of nano-sized particles (nanofluids). Int. J. Heat Mass Transf. **45**(4), 855–863 (2002)
11. J.A. Eastman, S.R. Phillpot, S.U.S. Choi, P. Keblinski, Thermal transport in nanofluids. Annu. Rev. Mater. Res. **34**(1), 219–246 (2004)
12. S.A. Angayarkanni, J. Philip, Review on thermal properties of nanofluids: Recent developments. Adv. Colloid Interf. Sci. **225**, 146–176 (2015)
13. L. Waldmann, Uber die Kraft eines inhomogenen Gases auf kleine suspendierte Kugeln. Z. Naturforsch. **14**, 589–599 (1959)
14. B.V. Derjaguin, S.P. Bakanov, Dokl. Akad. Nuuk SSSR **117**, 959 (1957)
15. E. Mason, J. Chapman, J. Chem. Phys. **36**, 627 (1962)
16. H.B. Bacha, N. Ullah, A. Hamid, N.A. Shah, A comprehensive review on nanofluids: Synthesis, cutting-edge applications, and future prospects. Int. J. Thermofluids **22**, 100595 (2024)
17. I. Gonçalves, R. Souza, G. Coutinho, J. Miranda, A. Moita, J.E. Pereira, A. Moreira, R. Lima, Thermal conductivity of nanofluids: A review on prediction models, controversies and challenges. Appl. Sci. **11**(6), 2525 (2021)
18. K.V. Wong, O. De Leon, Applications of nanofluids: Current and future. Adv. Mech. Eng. **2**, 519659 (2010)
19. F.E. Berger Bioucas, T.M. Koller, A.P. Fröba, Effective thermal conductivity of nanofluids containing silicon dioxide, titanium dioxide, copper oxide, polystyrene, or polymethylmethacrylate nanoparticles dispersed in water, ethylene glycol, or glycerol. Int. J. Thermophys. **46**(2), 27 (2025)
20. H. Zhang, S. Yan, H. Gao, X. Yuan, T. Ming, M.H. Ahmadi, X. Zhao, Experimental investigation and prediction of changes in thermal conductivity of carbon nanotube nanofluid. Heat Mass Transf. **127**, 105526 (2021)
21. A. Shahsavar, A. Shaham, M.A. Mirzaei, M. Jamei, F. Seifikar, S. Azizian, Assessment of thermal conductivity of polyethylene glycol-carbon dot nanofluid through a combined experimental-data mining investigation. Mater. Res. Technol. **19**, 2695–2704 (2022)
22. P.C. Mishra, S. Mukherjee, S.K. Nayak, A. Panda, A brief review on viscosity of nanofluids. Nano Lett. **4**(4), 109–120 (2014)

23. R.H. Castro, L.M. Corredor, S. Llanos, M.A. Causil, A. Arias, E. Pérez, F.B. Cortés, Experimental investigation of the viscosity and stability of scleroglucan-based nanofluids for enhanced oil recovery. Nano **14**(2), 156 (2024)
24. C.T. Nguyen, F. Desgranges, G. Roy, N. Galanis, T. Maré, S. Boucher, H. Angue Mintsa, Temperature and particle-size dependent viscosity data for water-based nanofluids—Hysteresis phenomenon. Heat Fluid Flow **28**(6), 1492–1506 (2007)
25. G. Yalçın, S. Öztuna, A.S. Dalkılıç, S. Wongwises, Effect of particle size on SiO_2 nanofluid viscosity determined by a two-step method. Thermal Anal. Calorimetry **149**, 13681–13696 (2024)
26. Z. Said, L.S. Sundar, A.K. Tiwari, H.M. Ali, M. Sheikholeslami, E. Bellos, H. Babar, Recent advances on the fundamental physical phenomena behind stability, dynamic motion, thermophysical properties, heat transport, applications, and challenges of nanofluids. Phys. Rep. **946**, 1–94 (2022)
27. D. Yadav, A. Nirala, R. Kumar, P. Kumar Singh, Density variation in nanofluids as a function of concentration and temperature. Mater. Today Proc. **46**, 6576–6580 (2021)
28. M. Sharifpur, S. Yousefi, J.P. Meyer, A new model for density of nanofluids including nanolayer. Int. Commun. Heat Mass Transfer **78**, 168–174 (2016)
29. R.D. Selvakumar, J. Wu, A comprehensive model for effective density of nanofluids based on particle clustering and interfacial layer formation. J. Mol. Liq. **292**, 111415 (2019)
30. F.M. Abir, D. Shin, Specific heat capacity of solar salt-based nanofluids: Molecular dynamics simulation and experiment. Materials **17**(2), 506 (2024)
31. G. Kumar Poongavanam, S. Duraisamy, V.S. Vigneswaran, V. Ramalingam, Review on the electrical conductivity of nanofluids: Recent developments. Mater. Today Proc. **39**, 1532–1537 (2021)
32. S. Akilu, A.T. Baheta, K. Kadirgama, E. Padmanabhan, K.V. Sharma, Viscosity, electrical and thermal conductivities of ethylene and propylene glycol-based β-SiC nanofluids. Mol. Liq. **284**, 780–792 (2019)
33. G. Żyła, J.P. Vallejo, J. Fal, L. Lugo, Nanodiamonds—Ethylene glycol nanofluids: Experimental investigation of fundamental physical properties. Heat Mass Transf. **121**, 1201–1213 (2018)
34. M. Karami, M.A. Akhavan-Behabadi, M. Raisee Dehkordi, S. Delfani, Thermo-optical properties of copper oxide nanofluids for direct absorption of solar radiation. Sol. Energy Mater. Sol. Cells **144**, 136–142 (2016)
35. S.H.A. Ahmad, R. Saidur, I.M. Mahbubul, F.A. Al-Sulaiman, Optical properties of various nanofluids used in solar collector: A review. Renew. Sust. Energ. Rev. **73**, 1014–1030 (2017)
36. W.R. Richmond, R.L. Jones, P.D. Fawell, The relationship between particle aggregation and rheology in mixed silica–titania suspensions. Chem. Eng. J. **71**(1), 67–75 (1998)
37. W. Xian-Ju, L. Xin-Fang, Influence of pH on nanofluids' viscosity and thermal conductivity. Chin. Phys. Lett. **26**(5), 056601 (2009)
38. K.K. Sarojini, S.V. Manoj, P.K. Singh, T. Pradeep, S.K. Das, Electrical conductivity of ceramic and metallic nanofluids. Colloids Surf. A Physicochem. Eng. Asp. **417**, 39–46 (2013)
39. H. Xie, W. Yu, W. Chen, MgO nanofluids: Higher thermal conductivity and lower viscosity among ethylene glycol-based nanofluids containing oxide nanoparticles. J. Exp. Nanosci. **5**(5), 463–472 (2010)
40. M. Hemmat Esfe, P.M. Behbahani, A.A.A. Arani, M.R. Sarlak, Thermal conductivity enhancement of SiO_2–MWCNT (85: 15%)–EG hybrid nanofluids: ANN designing, experimental investigation, cost performance and sensitivity analysis. J. Therm. Anal. Calorim. **128**, 249–258 (2017)
41. S.A. Adio, M. Sharifpur, J.P. Meyer, Investigation into effective viscosity, electrical conductivity, and pH of γ-Al2O3-glycerol nanofluids in Einstein concentration regime. Heat Transfer Eng. **36**(14-15), 1241–1251 (2015)
42. E.V. Timofeeva, A.N. Gavrilov, J.M. McCloskey, Y.V. Tolmachev, S. Sprunt, L.M. Lopatina, J.V. Selinger, Thermal conductivity and particle agglomeration in alumina nanofluids: Exper-

iment and theory. Physical review E—Statistical, nonlinear, and soft matter. Physics **76**(6), 061203 (2007)
43. W. Yu, H. Xie, A review on nanofluids: Preparation, stability mechanisms, and applications. J. Nanomater. **2012**(1), 435873 (2012)
44. A. Angayarkanni, J. Philip, *Synthesis of Nanoparticles and Nanofluids* (Royal Society of Chemistry, London, 2022)
45. W.N. Septiadi, I.A.N.T. Trisnadewi, N. Putra, I. Setyawan, Synthesis of hybrid nanofluid with two-step method. E3S Web Conf. **67**, 03057 (2018)
46. J.R. Babu, K.K. Kumar, S.S. Rao, State-of-art review on hybrid nanofluids. Renew. Sust. Energ. Rev. **77**, 551–565 (2017)
47. M. Chandrasekar, S. Suresh, A.C. Bose, Experimental studies on heat transfer and friction factor characteristics of Al2O3/water nanofluid in a circular pipe under laminar flow with wire coil inserts. Exp. Thermal Fluid Sci. **34**(2), 122–130 (2010)
48. S.M. Hashemi, M.A. Akhavan-Behabadi, An empirical study on heat transfer and pressure drop characteristics of CuO–base oil nanofluid flow in a horizontal helically coiled tube under constant heat flux. Int. Commun. Heat Mass Transfer **39**(1), 144–151 (2012)
49. J. Sathish Kumar, G. Senthilkumar, S. Ramachandran, Novel strategy of mixing MgO in CuO/water nanofluid for thermal conductivity improvement: Experimental study. Case Stud. Thermal Eng. **52**, 103723 (2023)
50. S. Lee, S.U.-S. Choi, S. Li, J.A. Eastman, Measuring thermal conductivity of fluids containing oxide nanoparticles. J. Heat Transf. **121**(2), 280–289 (1999)
51. S.M.S. Murshed, K.C. Leong, C. Yang, Enhanced thermal conductivity of TiO_2—Water based nanofluids. Int J Therm Sci **44**(4), 367–373 (2005)
52. Y. Li, J. Zhou, S. Tung, E. Schneider, S. Xi, A review on development of nanofluid preparation and characterization. Powder Technol. **196**(2), 89–101 (2009)
53. C.-H. Lo, T.-T. Tsung, L.-C. Chen, C.-H. Su, H.-M. Lin, Fabrication of copper oxide nanofluid using submerged arc nanoparticle synthesis system (SANSS). J. Nanopart. Res. **7**(2-3), 313–320 (2005)
54. M. Mohammadpoor, S. Sabbaghi, M.M. Zerafat, Z. Manafi, Investigating heat transfer properties of copper nanofluid in ethylene glycol synthesized through single and two-step routes. Int. J. Refrig. **99**, 243–250 (2019)
55. M. Ho, C. Oon, L. Tan, Y. Wang, Y. Hung, A review on nanofluids coupled with extended surfaces for heat transfer enhancement. Results Eng. **17**, 100957 (2023)
56. I. Fazeli, M.R. Sarmasti Emami, M.H. Nazeri, Comparing the effect of single and mixture surfactants on the improvement in stability and thermal conductivity of CuO nanofluid: An experimental study. J. Therm. Anal. Calorim. **146**(2), 967–980 (2021)
57. T. Zhang, Q. Zou, Z. Cheng, Z. Chen, Y. Liu, Z. Jiang, Effect of particle concentration on the stability of water-based SiO2 nanofluid. Powder Technol. **379**, 457–465 (2021)
58. A. Qamar, Z. Anwar, H. Ali, S. Imran, R. Shaukat, M.M. Abbas, Experimental investigation of dispersion stability and thermophysical properties of ZnO/DIW nanofluids for heat transfer applications. Alex. Eng. J. **61**(5), 4011–4026 (2022)
59. A. Asadi, I.M. Alarifi, Effects of ultrasonication time on stability, dynamic viscosity, and pumping power management of MWCNT-water nanofluid: An experimental study. Sci. Rep. **10**(1), 15182 (2020)
60. X. Zhao, L. Huang, J. Hou, Z. Ding, L. Ma, J. Wu, N. Hao, Microfluidic one-step and large-scale production of silica and titania nanofluids toward phase-change heat transfer intensification of power electronic devices. Chem. Eng. J. **503**, 158227 (2025)
61. S. Aberoumand, A. Jafarimoghaddam, Tungsten (III) oxide (WO_3)–silver/transformer oil hybrid nanofluid: Preparation, stability, thermal conductivity and dielectric strength. Alex. Eng. J. **57**(1), 169–174 (2018)
62. K.A. Altammar, A review on nanoparticles: Characteristics, synthesis, applications, and challenges. Front. Microbiol. **14**, 1155622 (2023)

63. M.A. Behnajady, H. Eskandarloo, N. Modirshahla, M. Shokri, Investigation of the effect of sol–gel synthesis variables on structural and photocatalytic properties of TiO2 nanoparticles. Desalination **278**(1–3), 10–17 (2011)
64. S. Thiagarajan, A. Sanmugam, D. Vikraman, Facile methodology of sol-gel synthesis for metal oxide nanostructures, in *Recent Applications in Sol-Gel Synthesis*, (IntechOpen, London, 2017), pp. 1–17
65. J.G. Ovejero, M.A. Garcia, P. Herrasti, Self-assembly of au-Fe3O4 hybrid nanoparticles using a sol–gel pechini method. Molecules **26**(22), 6943 (2021)
66. Y. Newman Monday, J. Abdullah, N.A. Yusof, S. Abdul Rashid, R.H. Shueb, Facile hydrothermal and solvothermal synthesis and characterization of nitrogen-doped carbon dots from palm kernel shell precursor. Appl. Sci. **11**(4), 1630 (2021)
67. Z. Su, Y. Cheng, Z. Liu, J. Zhou, D. Li, Y. Li, Experimental study on thermal conductivity of water-based magnetic fluid loaded with different nanoparticles. Nano **13**(22), 2952 (2023)
68. V.E. Manqueros-Avilés, H. Coto-Fuentes, K.V. Guevara-Amatón, F. Valdés-Perezgasga, J. Alonso-Chamarro, Multilayered manufacturing method for microfluidic systems using low-cost, resin-based three-dimensional printing. Sensors **25**(3), 694 (2025)
69. A. Riahi, S. Khamlich, M. Balghouthi, T. Khamliche, T.B. Doyle, W. Dimassi, M. Maaza, Study of thermal conductivity of synthesized Al2O3-water nanofluid by pulsed laser ablation in liquid. J. Mol. Liq. **304**, 112694 (2020)
70. T.P. Yadav, R.M. Yadav, D.P. Singh, Mechanical milling: A top down approach for the synthesis of nanomaterials and nanocomposites. Nanosci. Nanotechnol. **2**(3), 22–48 (2012)
71. N. Salah, S.S. Habib, Z.H. Khan, A. Memic, A. Azam, E. Alarfaj, S. Al-Hamedi, High-energy ball milling technique for ZnO nanoparticles as antibacterial material. Int. J. Nanomedicine **6**, 863–869 (2011)
72. Y. Liu, H. Luo, Z. Gao, H. Xie, R. Guo, F. Wang, D. Zhang, Electrospinning synthesis of NaO 5BiO 5TiO3 nanofibers for dielectric capacitors in energy storage application. Nano **12**(6), 906 (2022)
73. C. Wang, J. Yin, S. Han, T. Jiao, Z. Bai, J. Zhou, Q. Peng, Preparation of palladium nanoparticles decorated polyethyleneimine/polycaprolactone composite fibers constructed by electrospinning with highly efficient and recyclable catalytic performances. Catalysts **9**(6), 559 (2019)
74. P. Ayyub, R. Chandra, P. Taneja, A.K. Sharma, R. Pinto, Synthesis of nanocrystalline material by sputtering and laser ablation at low temperatures. Appl. Phys. A **73**, 67–73 (2001)
75. M.A. Rahman, S.M. Hasnain, S. Pandey, A. Tapalova, N. Akylbekov, R. Zairov, Review on nanofluids: Preparation, properties, stability, and thermal performance augmentation in heat transfer applications. ACS Omega **9**(30), 32328–32349 (2024)
76. A.K. Singh, V.S. Raykar, Microwave synthesis of silver nanofluids with polyvinylpyrrolidone (PVP) and their transport properties. Colloid Polym. Sci. **286**, 1667–1673 (2008)
77. E. Ponticorvo, M. Iuliano, C. Cirillo, A. Maiorino, C. Aprea, M. Sarno, Fouling behavior and dispersion stability of nanoparticle-based refrigeration fluid. Energies **15**(9), 3059 (2022)
78. E. Ju Park, S. Won Lee, I.C. Bang, H.W. Park, Optimal synthesis and characterization of Ag nanofluids by electrical explosion of wires in liquids. Nanoscale Res. Lett. **6**, 1–10 (2011)
79. M. Hosny, M. Fawzy, E.M. El-Fakharany, A.M. Omer, E.M. Abd El-Monaem, R.E. Khalifa, A.S. Eltaweil, Biogenic synthesis, characterization, antimicrobial, antioxidant, antidiabetic, and catalytic applications of platinum nanoparticles synthesized from Polygonum salicifolium leaves. J. Environ. Chem. Eng. **10**(1), 106806 (2022)
80. W. Cui, Z. Cao, X. Li, L. Lu, T. Ma, Q. Wang, Experimental investigation and artificial intelligent estimation of thermal conductivity of nanofluids with different nanoparticles shapes. Powder Technol. **398**, 117078 (2022)
81. I. Ijaz, A. Bukhari, E. Gilani, A. Nazir, H. Zain, R. Saeed, Green synthesis of silver nanoparticles using different plants parts and biological organisms, characterization and antibacterial activity. Environ. Nanotechnol. Monit. Manage. **18**, 100704 (2022)

82. R.E. Abouzeid, M.E. Owda, S. Dacrory, Effective adsorption of cationic methylene blue dye on cellulose nanofiber/graphene oxide/silica nanocomposite: Kinetics and equilibrium. J. Appl. Polym. Sci. **139**(25), e52377 (2022)
83. G. Dahy, M.M. Soliman, H. Alshater, A. Slowik, A.E. Hassanien, Optimized deep networks for the classification of nanoparticles in scanning electron microscopy imaging. Comput. Mater. Sci. **223**, 112135 (2023)
84. B.W. Arey, V. Shutthanandan, Y. Xie, A. Tolic, N. Williams, G. Orr, Investigation of cellular interactions of nanoparticles by helium ion microscopy. Microsc. Res. Tech. **88**(5), 1599 (2011)
85. N. Nikkam, M. Saleemi, E.B. Haghighi, M. Ghanbarpour, R. Khodabandeh, M. Muhammed, M.S. Toprak, Fabrication, characterization and thermophysical property evaluation of SiC nanofluids for heat transfer applications. Nano-Micro Lett. **6**, 178–189 (2014)
86. M.S. Liu, M.C.C. Lin, I.T. Huang, C.C. Wang, Enhancement of thermal conductivity with carbon nanotube for nanofluids. Int. Commun. Heat Mass Transfer **32**(9), 1202–1210 (2005)
87. R. Sadeghi, S.G. Etemad, E. Keshavarzi, M. Haghshenasfard, Investigation of alumina nanofluid stability by UV–vis spectrum. Microfluid. Nanofluid. **18**, 1023–1030 (2015)
88. B. Bakthavatchalam, K. Habib, R. Saidur, S. Shahabuddin, B.B. Saha, Influence of solvents on the enhancement of thermophysical properties and stability of multi-walled carbon nanotubes nanofluid. Nanotechnology **31**(23), 235402 (2020)
89. S. Mathew, F. Francis, S.A. Joseph, M.S., K., Enhanced thermal diffusivity of water based ZnO nanoflower/rGO nanofluid using the dual-beam thermal lens technique. Nanostruct. Antimicrob. Peptides **28**, 100784 (2021)
90. V. Sridhara, B.S. Gowrishankar, Snehalatha, L.N. Satapathy, Nanofluids—A new promising fluid for cooling. Trans. Indian Ceramic Soc. **68**(1), 1–17 (2009)
91. W. Yu, D.M. France, S.U. Choi, J.L. Routbort, *Review and Assessment of Nanofluid Technology for Transportation and Other Applications (no. ANL/ESD/07-9)* (Argonne National lab. (ANL), Argonne, 2007)
92. S.M. Moghimi, A.C. Hunter, J.C. Murray, Long-circulating and target-specific nanoparticles: Theory to practice. Pharmacol. Rev. **53**(2), 283–318 (2001)
93. H. Tyagi, P. Phelan, R. Prasher, Predicted efficiency of a low-temperature nanofluid-based direct absorption solar collector. J. Sol. Energy Eng. **131**(4), 041004 (2009)
94. R.R. Srikant, D.N. Rao, M.S. Subrahmanyam, V.P. Krishna, Applicability of cutting fluids with nanoparticle inclusion as coolants in machining. Proc. Inst. Mech. Eng. **223**(2), 221–225 (2009)
95. J.A. Eastman, S.U.S. Choi, S. Li, W. Yu, L.J. Thompson, Anomalously increased effective thermal conductivities of ethylene glycol-based nanofluids containing copper nanoparticles. Appl. Phys. Lett. **78**(6), 718–720 (2001)
96. M.N. Pantzali, A.A. Mouza, S.V. Paras, Investigating the efficacy of nanofluids as coolants in plate heat exchangers (PHE). Chem. Eng. Sci. **64**(14), 3290–3300 (2009)

Foundations and Applications of Ionic Liquids, Nanofluids, Nanomaterials, and Nanotechnology: A Comprehensive Introduction

Kirti Deshpande

1 Introduction: Overview of Ionic Liquids, Nanofluids, Nanomaterials, and Nanotechnology

In recent years, the scientific domain has faced a rapid evolution in ionic liquids, nanofluids, nanomaterials, and nanotechnologies, each having set a stamp of transformation in the various industrial and research domains. Ionic liquids (ILs) are organic salts with melting points below a certain threshold and possessing exotic properties such as near-zero vapor pressure, high thermal stability, and tunable solvation ability; their applicability runs through green chemical technologies and electrochemical applications [1, 2]. Nanofluids, created by dispersing nanoparticles in base fluids, promise to offer solutions to cooling and biomedical applications, having higher thermal conductivity and convective heat transfer performance [3]. Nanomaterials are said to have their novel mechanical, electrical, and optical properties endowed because of their nanometer-scale structures, through which they revolutionize sensor design, drug delivery, and structural materials [4]. The remainder of the domain is connected under nanotechnology, which deals with the manipulation and engineering of matter at the atomic and molecular level with the utmost precision [5]. This interdisciplinary field draws upon physics, chemistry, biology, and materials science to come up with novel products and systems featuring enhanced functionalities.

K. Deshpande (✉)
School of Pharmacy, Vishwakarma University, Pune, India

S. Kulkarni, S. Das (eds.), *Ionic Liquids, Nanofluids and Nanotechnology*,
Lecture Notes in Nanoscale Science and Technology 42,
https://doi.org/10.1007/978-3-032-04008-4_4

1.1 Importance and Scope of the Chapter

Ionic liquids occur in an interesting interdisciplinary context, coupling with the knowledge domains of nanofluids, nanomaterials, and nanotechnologies: a new conglomerate of areas of study that are very much relevant in today's scientific and industrial scenario. The different physicochemical nature of ionic liquids permits greener processing alternatives and better electrochemical systems [2]. Nanofluids, having high thermal capacity, are being investigated for cooling in advanced systems and biomedical systems [3]. Similar to this, nanomaterials provide nanometer-size-based properties of sensors, structural materials, and biomedical delivery systems, thus offering materials with a series of functionalities never achieved before [4]. Nanotechnology is an umbrella that enables atomic manipulation and brings together different scientific disciplines [5]. This chapter tries to bring together the basic principles, the characterization methods, and the applications that will serve as a banner for students, researchers, and professionals to get an understanding and contribute to these disruptive technologies.

2 Ionic Liquids: Fundamentals and Applications

Ionic liquids (ILs) are specially crafted solvents that boast a wide range of applications thanks to their remarkable tunability and stability. They work as effective media in catalytic processes and green technologies. It is helping to reduce waste and boost reaction efficiency [6]. Further, ILs serve as cutting-edge electrolytes in energy storage systems and provide enhanced safety, electrochemical performance, and conductivity [7].

2.1 Physicochemical Properties of Ionic Liquids

Generally, any salt that is liquid below 100 °C is classified as an IL. Their unique physicochemical features have made ILs the primary focus of research for diverse scientific and industrial applications [1, 2]. Their defining characteristic is perhaps their almost negligible vapor pressure, which means they tend to decrease volatile organic compound (VOC) emissions, being thus safer alternatives in solvent-based processes. They further have very strong thermal and chemical stability, allowing them to be used in extreme conditions, such as catalysis at high temperatures or electrochemical systems [6]. Another key aspect that can be mentioned is tenability—the physicochemical characteristics of the ILs, such as polarity, viscosity, and hydrophobicity, can be very finely tuned by altering the exact cation-anion combinations involved. Ionic liquids also exhibit very good ionic conductivity, which makes them available for use in batteries and in fuel cells (e.g., electrolyte for lithium-ion batteries) [7].

2.2 Applications in Catalysis and Green Technologies

Ionic liquids (ILs) have become standout players as effective solvents and catalysts in a variety of chemical processes, especially in the realm of green chemistry, thanks to their non-volatile and recyclable characteristics [1, 8]. Their knack for dissolving a broad spectrum of both organic and inorganic compounds makes them incredibly useful in catalytic systems, such as acid/base reactions, organometallic chemistry, and biocatalysis.

BASF, a global chemical company, developed an ionic-liquid-based solvent system as part of its OASE® blue technology. This system is designed for CO_2 capture from industrial flue gases, such as those produced by power plants and refineries. The captured CO_2 is then utilized in the production of urea, methanol, or cyclic carbonates, which are value-added products [9].

2.3 Energy Storage and Electrochemical Applications

Due to their gigantic electrochemical stability, broad electrochemical windows, and high ionic conductivity, ionic liquids represent ideal electrolytes for energy storage systems [7]. Their usage has been successfully accepted in lithium-ion batteries, supercapacitors, and dye-sensitized solar cells. On the other hand, it is generally accepted that the low volatility of IL electrolytes avoids the risk of leakage and degradation of electrochemical devices. Their flexibility and ionic nature allow tuning of the electrochemical properties to meet the required specifications of future energy systems.

For example, ILs such as N-methyl-N-propylpyrrolidinium bis(fluorosulfonyl)imide ([Pyr_{13}][FSI]) enhance battery safety due to their non-flammability and thermal stability [10].

2.4 Ionic Liquids in Biomedical and Pharmaceutical Applications

The unique combination of enhanced thermal performance and biocompatibility positions nanofluids at the forefront of innovation in both industrial and medical fields. While challenges such as stability, biotoxicity, and scalability persist, ongoing research continues to push boundaries, especially with the advent of smart nanofluids that respond to external stimuli such as temperature, pH, or magnetic fields [11, 12].

ILs as Solubilizing Agents for Poorly Water-Soluble Drugs

A major hurdle in drug development is hardly ever the aqueous solubility of drugs. ILs can greatly enhance the solubility and bioavailability of hydrophobic drugs by offering them another medium of solvation [13].

For example:

- Lidocaine-based ILs could be used to increase the solubility of poorly soluble anti-inflammatory drugs such as ibuprofen, thus offering an improved transdermal delivery system [14].
- The choline-based ILs have improved solubilization of curcumin, a natural compound with anticancer properties requiring a good degree of solubility in water [15].

ILs in Drug Delivery Systems

ILs can act as solvents or co-solvents, or they can even form active pharmaceutical ingredients (API-ILs) when the drug itself is part of the ionic liquid structure [13]. This opens pathways for enhanced delivery across biological membranes.

API-ILs: Some ILs are made upon pairing pharmaceutically active cations and anions. For example, the IL consisting of the salicylic acid anion paired with the lidocaine cation possesses the dual commercial properties of analgesic and anti-inflammatory effects with enhanced skin permeation characteristics [14].

Lipid and polymeric carriers based on ILs have been studied for the targeted delivery of anticancer drugs; basically, liposomal formulations using ILs have shown improvements in encapsulation efficiency as well as stability [16].

Antimicrobial and Antiviral Properties of ILs

Some ionic liquids may encompass broad-spectrum antimicrobial activity, as they may be used as antiseptic formulations or potentially incorporated as additives into medical devices. Based on studies involving imidazolium-based ILs, this type of IL exhibited substantial bactericidal activity against diverse Gram-positive as well as Gram-negative microorganisms (e.g., *Staphylococcus aureus*, *Escherichia coli*) [17].

ILs derived from long alkyl chain cations can damage microbial cell membranes and could theoretically serve as surface disinfectants as well as in wound dressings [18].

ILs in Enzyme Stabilization and Biocatalysis

Moreover, ILs can also serve as characteristic microenvironments in which to preserve the structure and activity of enzymes in a process, which could prove beneficial, particularly in drug synthesis and diagnostic applications [19]. For example, lipases showed higher catalytic activity when in the presence of IL(s) such as 1-Butyl-3-methylimidazolium hexafluorophosphate [BMIM][PF6] instead of conventional organic solvents that are typically used in drug synthesis, which ultimately allowed for the enantioselective synthesis of intermediates used in pharmaceuticals [20]. ILs are also being used as part of bioanalytical devices, or biosensors, to modify the conductivity and stability of biomolecules, which also promotes use in biomolecular devices in a biosensor [21].

ILs for Nucleic Acid and Gene Delivery

Additionally, some ILs and their charged constituents and amphiphilicity can also have potential utility to modulate interactions with nucleic acids (e.g., DNA or RNA), as this is a growing area of investigation for genetic therapies [22].

The cholinium-based ILs show promise slightly in demarcating DNA compaction as well as protecting from enzyme degradation, which subsequently could amplify gene delivery [23].

3 Nanofluids: Enhanced Thermal Properties and Applications

Nanofluids are specially created mixtures dissolving nanoparticles in common conventional base fluids to improve thermal characteristics for advanced applications. Their performance also depends on their stability and composition. These fluids enhance heat transfer and thermal conductivity; owing to such features, they find applications in advanced biomedical technologies and high-performance cooling systems [24].

3.1 Composition and Characteristics of Nanofluids

The nanofluid is usually considered as a base fluid in which fine nanoparticles are dispersed. The nanoparticles of metals such as copper or aluminum, metal oxides such as aluminum oxide (AlO_3) and titanium oxide (TiO_2), or carbon-based materials including graphene and carbon nanotubes [25]. The nanoparticle suspensions are generally referred to as nanofluids. When surface treatment or surfactants are used, nanoparticles remain stable if their diameter is less than 100 nm, thus preventing any form of settling or coalescing [26].

The concentration of nanoparticles plays an important role in determining the physical and chemical properties of nanofluids, along with their size, shape, composition, and type of base fluid. Enhanced stability is one of their striking properties; for example, TiO_2 nanofluids have been reported to remain stable in water agitated ultrasonically for several weeks while using surfactants to prevent sedimentation [27]. They can be tailored for viscosity, surface tension, and optical applications. These properties are realized at the application level; for instance, cuprous oxide (CuO) nanoparticles suspended in transformer oil increase the dielectric strength and heat transfer of the oil, thereby improving the reliability of electrical equipment [28]. Meanwhile, hybrid nanofluids of metal oxides and graphene are used in solar collectors due to their optical absorption and thermal stability [29]. This, together with their intriguing properties, makes nanofluids a promising new contender as materials for electronics and energy applications.

3.2 Thermal Conductivity and Heat Transfer Enhancement

One of the main advantages of nanofluids over conventional base fluids is their increased heat conductivity. The enhanced surface area of nanoparticles, their Brownian motion, and the formation of nanolayers at the solid–liquid interface are credited with improving the energy transfer mechanisms within the fluid [30]. A nanofluid containing 0.3% volume fraction of CuO nanoparticles in ethylene glycol demonstrated a more than 40% improvement in thermal conductivity when compared to the base fluid [31]. Nanofluids enhance both static thermal conductivity and convective heat transfer. For dynamic systems like heat exchangers, radiators, and microchannel cooling devices, they are therefore perfect [32]. A practical example is the use of AlO_3–water nanofluids in automobile radiators, which resulted in reduced fluid pumping power and increased engine cooling efficiency [33].

3.3 Applications in Thermal Management and HVAC Systems

Nanofluids have been gaining considerable attention in the field of thermal management systems (e.g., Heating, Ventilation, and Air Conditioning (HVAC)) due to their better heat transfer characteristics when compared to traditional heat transfer fluids [34]. Nanofluids are made by suspending nanoparticles (typically metals, oxides, carbides, and carbon-based materials) in a base fluid (e.g., water, ethylene glycol, and oils). Their thermal conductivity, convection heat transfer, and specific heat capacity all make them potentially useful products to improve the energy efficiency in HVAC systems.

Enhanced Heat Transfer in HVAC Systems

Conventional HVAC systems essentially have limited thermal and energy efficiency because of the thermal properties of the working fluids. However, nanofluids can overcome many of these thermal limitations by increasing the thermal conductivity. Enhancing the thermal conductivity of the base fluid was observed in nanofluids with nanoparticle volume fractions as small as <1% volume fraction [35]. Alumina-water nanofluids showed enhancement of thermal conductivity by up to 30% [36]. For convective heat transfer enhancement, nanoparticles disturb the boundary layers, thereby producing micro-convection. This, in a way, will help in improving heat transfer rates in heat exchangers and coils [37]. Such enhancements will lessen the temperature drop across an HVAC coil and thereby shorten the cooling and/or heating time, lessen compressor workload, and reduce energy consumption [38].

Applications in Specific HVAC Components

Nanofluids can be used in an air-conditioner evaporator and condenser to enhance heat absorption and transfer. TiO_2 and CuO nanofluids have been used within refrigerant circuits and found to reduce compressor work and increase cooling capacity [39]. Nanofluids improve heat exchange performance for fan coil units in more compact designs with lower material costs. Experiments have shown that SiO_2-ethylene glycol nanofluids increase the outlet air temperature of heating coils quickly, thus improving indoor thermal comfort without saving energy [40].

Chilled Water and Cooling Tower Systems.

In cold water loops, nanofluids cool more rapidly. This decreases the chiller unit load and cycle time. Nanofluids of ZnO- and Fe_3O_4-based composition have been investigated to be applied in cooling towers to improve heat rejection rates and reduce make-up water demand through lower operating temperatures [41].

Energy Efficiency and Environmental Benefits.

Green building technologies and green energy practices are supportive of the application of nanofluids in HVAC systems. Based on the type of nanoparticles and the system design, HVAC energy consumption can be lowered by 10–20% [42]. Efficient heat transfer helps in reducing carbon footprint [38].

3.4 *Applications in Biomedical Fields*

Nanofluids provide many benefits for diagnosis, treatment, and drug delivery because of their enhanced thermal conductivity, ability to topically target functionality, and minimally invasive properties [43]. The range of nanofluids is created from tailoring the chemical, size, and shape of nanomaterials to allow custom-tailored nanofluids for many great medical applications.

Cancer Hyperthermia Therapy

One of the most promising biomedical applications of nanofluids is in magnetic fluid hyperthermia (MFH). Magnetic nanofluids, typically superparamagnetic iron oxide nanoparticle (SPION)-containing nanofluids, when used in MFH, locally heat tumors [44]. Magnetic hydrothermotherapy (MHHT) is a new form of nanofluid hyperthermia whereby magnetic field resonance allows for the elimination of tumors.

Alteration of the magnetic field resonates with the SPIONs and produces heat in very localized areas to cause fatal apoptosis or necrosis to tumor cells without harming neighboring cells. SPION-based nanofluids have been effective in treating glioblastomas and prostate cancers, where exerting local, precise control of the temperature is critical.

Drug Delivery Systems

The second use of nanofluids is as carriers for controllably and intentionally targeted drug delivery that naturally leads to localized therapy and decreased systemic side effects.

For example, thermosensitive nanofluids can be designed to find any biological site, where, upon reaching a certain temperature, they can release medications and other compounds such as metabolic enhancers, molecular graphics, and organ drive [45]. Lipid-based nanofluids consisting of gold nanoparticles (AuNPs) use LTIR-visible light-generated heating to cause local tumor heating followed by drug release [46]. Magnetically sensitive analyzed nanofluids can be spatially located to specific biological tissues using an external or internal magnet that keeps drugs attached to the NP/biological solid carrier [4, 44].

Biomedical Imaging and Contrast Enhancement

Nanofluids are used in imaging tools such as magnetic resonance imaging (MRI), computed tomography (CT), and fluorescence imaging due to their optical and magnetic properties. SPION-based nanofluids act as MRI contrast agents, so images are sharper [44].

Quantum dot nanofluids enhance tissue imaging for fluorescence-guided surgery because of their photostability and high fluorescence [4]. Gold nanofluids have increased x-ray absorption in CT imaging and can replace iodine-based contrast agents with a lesser toxic level [46].

Tissue Engineering and Regenerative Medicine

Nanofluids can enhance cell growth and differentiation by simulating the natural extracellular environment. Hydrogel systems with nanofluid suspension of bioactive nanoparticles (e.g., hydroxyapatite or silica) aid bone and cartilage regeneration [47]. In wound healing, nanofluids with zinc oxide nanoparticles or silver nanoparticles can provide antibacterial protection and promote regeneration [17, 18]. Their manipulated rheological and temperature properties make them advantageous as injectable scaffolds or for 3D bioprinting [48].

Cryopreservation and Organ Transport

The importance of nanofluids in cryopreservation is promising for cells, tissues, and organs to be stored below freezing [49]. Nanofluids improve thermal homogeneity in cryoprotective solutions that reduce ice crystal formation, which destroys biological structure. Nanofluids with graphene oxide or TiO_2 have the potential to maintain structural stability during the freeze–thaw process, thus improving organ transport or transplantation [24].

Antimicrobial and Antiviral Applications

Some nanofluid formulations (particularly with silver, copper, or zinc oxide nanoparticles) possess broad-spectrum antibacterial and antiviral properties [17, 18]. Nanofluid applications are being considered in topical formulations, catheter coatings, and surgical dressings to reduce hospital-acquired infections (HAIs). Nanofluids have multiple mechanisms of action. They can disrupt microbial cell membranes, act on processes to affect replication, and prevent biofilm development [50].

Biosensing and Diagnostics

Nanofluids are a featured component of biosensor platforms for the detection of biomolecules, such as glucose, cholesterol, or a disease marker, in biological samples [21]. Metallic nanoparticles (such as gold and silver) added to sensing fluids improve signal sensitivity with surface plasmon resonance (SPR) and electrochemical sensors. Nanofluids allow for rapid response to biomolecular changes and provide diagnostic feedback on a real-time basis [51].

4 Nanomaterials: Engineering at the Nanoscale

Nanomaterials are specially designed structures that usually measure less than 100 nm. They stand out because of their unique physical, chemical, and mechanical properties, which arise from their large surface area and quantum effects [52]. These materials can be prepared using the top-down method and the bottom-up method. They lead to a variety of forms like nanoparticles, nanotubes, and nanowires. Thanks to their improved structural strength, nanomaterials are perfect for lightweight composites, cutting-edge sensors, and targeted drug delivery systems in both medical and industrial fields [4, 52].

4.1 Types and Synthesis Methods of Nanomaterials

Nanomaterials can be grouped into four main categories: zero-dimensional (like nanoparticles), one-dimensional (such as nanowires and nanotubes), two-dimensional (including graphene and nanosheets), and three-dimensional (which covers nanocomposites and nanostructured bulk materials) [53]. The way we create these materials typically falls into two approaches: top-down or bottom-up. The top-down method breaks down larger materials into nanoscale structures using techniques like ball milling, lithography, or laser ablation [54]. On the flip side, the bottom-up approach constructs nanostructures molecule by molecule or atom by atom through methods like the sol-gel method, co-precipitation, and chemical vapor deposition [55].

For example, carbon nanotubes can be made using CVD by breaking down hydrocarbons over metal catalysts at high temperatures [4]. Likewise, metal oxide nanoparticles like ZnO and TiO_2 are frequently created through sol-gel and hydrothermal methods because these techniques allow for precise control over size and shape [24, 30]. The choice of synthesis method plays a crucial role in determining the particle size, surface characteristics, and crystalline structure, all of which significantly affect how well the nanomaterials perform in various applications [53].

4.2 Structural Properties and Engineering Applications

Nanomaterials have structural characteristics that make them most valuable in many engineering sectors [5]. At the nanoscale, materials usually gain superior mechanical, electrical, or optical properties due to their surface area-to-volume ratio or quantum effects. Nanoparticles, for example, would show different optical characteristics from bulk composition; thus, they are of use in sophisticated sensors and optics [56]. Besides, with their high surface area, they promote catalytic activity, further aiding environmental remediation and energy conversion technologies [57]. Nanomaterials, such as carbon nanotubes and graphene, in structural engineering promise to change lightweight composites and high-performance materials for aerospace and automotive industries due to their superior strength and conductivity [4, 52]. One trick of synthesizing nanomaterial calls for precise control to keep the consistent properties, and safe handling becomes quite important, owing to their possible health or environmental considerations [58]. These issues keep perpetually sustaining new research so that nanomaterials can really be utilized in engineering [59].

4.3 Lightweight Materials, Sensors, and Drug Delivery Systems

Nanomaterials are a key component in engineering lightweight material systems, along with the development of advanced sensors and efficacious drug delivery systems [4]. Lightweight composites are increasingly employed in aerospace, automotive, and sports applications. These materials provide high mechanical strength but do so with a great reduction in weight, which is very much desirable for fuel economy and performance enhancement [52].

In sensor technologies, ZnO nanowires and gold nanoparticles are just some of those nanomaterials used because of their favorable properties of surface area and sensitivity [60]. Carbon nanomaterials, for instance, are used in biosensors to quickly and accurately detect the levels of glucose, toxins, or pathogens [21, 51]. Because of their rapid electron conduction and distinctive sensing properties, they can monitor in real time, whether in environmental, biomedical, or industrial settings [21].

Nanomaterials also change drug delivery systems. Lipid nanoparticles and polymeric nano-carriers such as dendrimers and micelles can be used to encapsulate drugs and deliver them to specific targeted cells, thereby reducing side effects [4]. An ideal example is that of lipid nanoparticles used for mRNA-based COVID-19 vaccines that protect the mRNA and ensure their delivery into viable cells [61]. In general, coinciding with these properties, nanomaterials, because of size, surface properties, and tunable function, have found the eldorado of opportunities in lightweight engines [59].

4.4 *Environmental Applications of Nanomaterials*

Nanomaterials, due to their distinct surface properties, reactivity, and dimensions at the nanoscale, are being increasingly applied to a wide variety of environmental applications [57]. The high surface area-to-volume ratio, adjustable porosity, and potential for adsorbing, degrading, or transforming contaminants provide a significant advantage in achieving pollution control and water and air purification, as well as sustainable energy technologies [62]. These applications contribute to a global effort to address environmental issues stemming from industrialization, urbanization, and climate change.

Water Purification and Wastewater Treatment

Nanomaterials present novel solutions to the problems surrounding the treatment of polluted water sources, including the removal of heavy metals, organic matter, pathogens, and colorants.

Adsorption of heavy metals: Carbon nanotubes (CNTs), graphene oxide (GO), and iron oxide nanoparticles adsorb toxic metals such as lead (Pb^{2+}), arsenic (As^{5+}), and mercury (Hg^{2+}) with excellent efficiency [57]. Functionalized GO sheets can adsorb metal ions using surface interactions, with high efficiency at low concentrations [63].

Degradation of organic pollutants: Titanium dioxide (TiO_2) nanoparticles are photodegradative of organic pollutants, including pesticides, dyes, and pharmaceutical residues, when activated with UV or visible light to photodegrade. ZnO nanostructures demonstrate similar photocatalytic activity and are also incorporated into advanced oxidation processes (AOPs) for wastewater treatment [24].

Antimicrobial activity: Silver (Ag) and copper (Cu) nanoparticles are effective antimicrobial agents for water disinfection and reducing the microbial load of wastewater. They primarily inactivate pathogens by disrupting cell membranes and interfering with metabolic transitions [17, 50].

Air Pollution Control

Nanomaterials are integrated using air cleaning systems, catalysts, and sensor devices to combat airborne pollutants, primarily particulate matter, volatile organic compounds, and greenhouse gases [64].

Gas adsorption and capture: Activated carbon nanomaterials and metal–organic frameworks (MOFs) are useful to adsorb CO_2, SO_2, and NO_2 to minimize industrial emissions. MOFs have high selectivity and tunability to make them ideal for carbon capture and storage (CCS) technology [65].

Photocatalytic air purification: TiO_2-based nanocoatings on buildings and filters can degrade VOCs like formaldehyde and benzene using solar or artificial light

[62]. Air cleaning devices with embedded nanocatalysts are being adopted in urban building environments to improve indoor and outdoor air quality [64].

Soil remediation: Nanomaterials are able to immobilize, degrade, or transform the contaminants stagnant in soils, including hydrocarbons, heavy metals, and persistent organic pollutants (POPs). Zero-valent iron nanoparticles (nZVI) have been extensively studied for their ability to reduce chlorinated hydrocarbons (e.g., trichloroethylene) to harmless compounds. Clay-based nanocomposite materials with metal oxides and nanocomposites increase sorption and stabilization of metals in contaminated soils [66].

Environmental sensing and monitoring: Nanomaterials are built into nanosensors that detect pollutants with unprecedented sensitivity and specificity. Gold and silver nanoparticles and quantum dots are used in optical and electrochemical sensors to detect pesticides, pathogens, and toxic gases [4, 51]. These sensors are utilized for real-time environmental monitoring for the rapid, on-site detection of contamination events [64].

5 Nanotechnology: Principles and Applications

Nanotechnology, a science of designing and engineering functional systems at the nanometer scale, mainly focuses on controlled manipulation of atoms and molecules to develop structures with enhanced or entirely new properties. Based upon the basic principles, it enables precise fabrication and integration of nanostructures, supporting breakthroughs in fields like diagnostics, energy systems, and smart materials [5].

5.1 Manipulation at Atomic and Molecular Scales

Manipulating materials at atomic and molecular scales is essentially at the heart of nanotechnology to create materials and devices with a desired level of precision and properties. At this scale, matter behaves differently from its bulk counterpart, factoring in quantum effects and an increased surface-to-volume ratio [53]. Scientists modify and design the secondary physical, chemical, and biological characteristics of materials by bringing them to life through the manipulation of atoms and molecules for specific applications.

Having the ability to view and manipulate atoms on surfaces using scanning-tunneling microscopy (STM) and atomic-force microscopies (AFM) has been considered a breakthrough, allowing scientists to position atoms in desired ways; this is the preliminary step in forming molecular machines or nanoscale circuits [67]. On the other hand, molecules may undergo self-assembly into predetermined structures without the need for outside intervention, which is the basis for constructing nanostructured materials [68].

Electronic manipulation of atoms has brought about nanometer-scale transistors with greater processing power and less size and energy consumption. In material science, carbon nanotubes and graphene are designed atomically in order to make lightweight structures that are ultra-strong and highly conductive [69].

Engineering matter at atomic precision not only aids in the progress of technology but also opens pathways for cures in medicine, like targeted drug delivery and nanoscale diagnostics, revolutionizing conventional approaches to healthcare and material design [70].

5.2 Interdisciplinary Foundations: Physics, Chemistry, Biology, and Engineering

Nanotechnology is inherently interdisciplinary, drawing principles of physics, chemistry, biology, and engineering toward the design and manipulation of matter at the nanoscale level. From physics, one learns of quantum mechanics and surface phenomena that dominate the behavior of materials at the nanometer scale. Chemistry contributes methods for molecular synthesis, self-assembly, and surface functionalization, which are necessary for preparing nanostructures that are stable and reactive in any given environment [71]. Biology serves as a guide to nanoscale organization, such as DNA nanostructures and protein folding, inspiring biomimetic designs for applications in drug delivery and biosensing [72]. Engineering combines these to realize production methods, devices, and systems suitable for mass production in the actual environment. These include photolithographic and nanoimprint nanofabrication approaches that link physics and engineering to realize nanoscale electronics and sensors [73]. Therefore, crossing traditional boundaries paves the road to innovation from materials science to medical diagnostics, and therefore, cross-disciplinarity is essential in the advancement of nanotechnology research and applications.

5.3 Design, Fabrication, and Control of Nanostructures

This topic is concerned with nanodesign, fabrication, and control with extremely refined techniques to obtain materials and devices with nanoscale properties. Design begins from materials and structural configurations that match the intended function—directly into such properties as conductivity, strength, or biocompatibility. Fabrication techniques include top-down processing, in which patterns are etched onto substrates using electron-beam lithography, and bottom-up methods such as CVD and molecular self-assembly, whereby the construction occurs atom by atom or molecule by molecule [74].

One must exercise extreme control over size, shape, and surface properties, as these parameters directly influence the behavior of the nanostructures; e.g., gold nanoparticles present size-dependent optical behaviors useful in imaging and diagnostics [75]. Meanwhile, surface-functionalized nanoparticles may be created for targeted drug delivery or environmental sensing. Examples of visualization and manipulation methods at the atomic level would include atomic-force microscopy (AFM) and scanning-tunneling microscopy (STM), wherein structural fidelity is greatest. This, then, is the set of tools with which one goes about designing sophisticated nanosystems for electronics, medicine, energy, and environmental technologies [76].

5.4 *Nanotechnology in Renewable Energy and Sustainability*

Nanotechnology is truly revolutionizing the way we tackle global energy issues by boosting the efficiency, affordability, and eco-friendliness of renewable energy systems. At the heart of this change is the ability to manipulate materials on a nanoscale, enhancing their physical, chemical, optical, and electrical properties—key factors for energy conversion, storage, and conservation [77]. By integrating nanotechnology into renewable energy, we're not just promoting sustainability; we're also reducing our reliance on fossil fuels, cutting down greenhouse gas emissions, and paving the way for a low-carbon future. Table 1 further provides the details of various sectors where nanotechnology is applied.

Table 1 Details of nanotechnology in renewable energy

Sector	Applications	Key technologies
Solar energy	Nanostructured solar cells, anti-reflective coatings, and self-cleaning coatings	Quantum dots, perovskite nanoparticles, and carbon nanotubes (CNTs)
Wind and hydro energy systems	Advanced coatings, lubricants, and bearings	Nanocoatings, nanocomposites, and nanoparticle-infused lubricants
Energy storage	Lithium-ion batteries, solid-state batteries, supercapacitors, and hydrogen storage	Nano-sized electrode materials and solid-state electrolytes
Bioenergy and fuel cells	Enzyme immobilization, catalysis, and PEM fuel cells	Nanostructured catalysts and graphene/CNT-supported nanocatalysts
Energy efficiency and smart grids	Thermoelectric materials, smart coatings, building materials, and sensors	Nanostructured thermoelectrics, smart coatings, and nanosensors
Environmental and economic sustainability	Resource conservation, lifecycle benefits, and carbon footprint reduction	Nano-enhanced materials and extended operational life

6 Characterization Methods of Ionic Liquids, Nanofluids, Nanomaterials, and Nanotechnology

Characterization of ionic liquids, nanofluids, nanomaterials, and nanotechnology depends on advanced analytical techniques applicable in assessing physical, chemical, and structural attributes. Such modern equipment sets the nanoscale environment precisely, allowing for advances in material sciences and engineering. The more it develops, the more minute it becomes in manifestation, enabling even deeper insight into nanoscale performance [78].

6.1 Analytical Techniques and Instrumentation

Analytical techniques are essential for assessing the structural, chemical, thermal, and mechanical properties of ionic liquids, nanofluids, and nanomaterials. The techniques must ensure that one locates their behavior accurately so that essential conditions are placed for optimization of their application in electronics, energy storage, and biomedical systems [79]. Table 2 gives collective information on various analytical techniques.

6.2 Advances in Characterization and Measurement

Recent advances in characterization techniques have helped improve the analysis of nanomaterials, nanofluids, and ionic liquids in terms of precision, resolution, and speed. Instrumentation advances now allow real-time analysis in situ, leading to an enhanced understanding of nanoscale behavior during physical, chemical, or thermal processes [80].

Key Advances and Applications:

1. In situ electron microscopy: Current generation TEMs and SEMs permit one to observe a dynamic real-time process, e.g., the growth, aggregation, and phase transformation of nanoparticles [81].
2. High-resolution spectroscopy: Surface-Enhanced Raman Spectroscopy (SERS) is used for ultrasensitive detection of molecules, particularly of surface-functionalized nanoparticles and ionic liquids [82].
3. X-ray techniques: Synchrotron X-ray diffraction and spectroscopy work at atomic-level resolution to expose crystallographic changes and electronic structure [83].
4. Lab-on-a-chip systems: Microfluidic-integrated sensors are now able to rapidly analyze nanofluids and ionic liquids with a very small amount of sample required [84].

Table 2 Details of analytical techniques

Technique	Property measured	Application example	Material/system studied
Fourier transform infrared spectroscopy (FTIR)	Functional groups	Identification of bonding in ionic liquids	Ionic liquids, surface-modified nanoparticles
UV-visible spectroscopy	Optical absorbance, plasmon resonance	Monitoring particle size and aggregation of gold and silver nanoparticles	Metal nanoparticles (e.g., Au, Ag)
Nuclear magnetic resonance (NMR)	Chemical environment, molecular structure	Analysis of ionic liquid structure and dynamics	Ionic liquids
Scanning electron microscopy (SEM)	Surface morphology	Imaging surface features of nanoparticles or coatings	Nanomaterials, nanocoatings
Transmission electron microscopy (TEM)	Internal structure, morphology, crystallinity	Determining the graphene layer number or nanoparticle dispersion in nanofluids	Graphene, nanofluids
Atomic-force microscopy (AFM)	Surface topology, roughness	Mapping nanostructure surfaces at atomic resolution	Thin films, nanocoatings
Thermo gravimetric analysis (TGA)	Thermal stability, weight loss	Evaluating degradation behavior of nanofluids at elevated temperatures	Nanofluids, polymer nanocomposites
Differential scanning calorimetry (DSC)	Phase transitions, thermal behavior	Studying the melting/crystallization behavior of ionic liquids	Ionic liquids, polymer blends

6.3 Computational Methods in Nanotechnology Characterization

Computational approaches have gained relevance in nanotechnology, mainly to characterize, model, and predict the behavior of nanomaterials and nanostructures. Unlike the conventional experimental approaches, these computational methods give information at atomic and molecular levels, thus comprehensively analyzing structure-property relationships observed experimentally, predicting behaviors of materials under different conditions, and creating virtual prototypes of nanosystems [85].

7.1 Innovations in Ionic Liquid Applications

Innovative applications of ionic liquids in various domains are listed below in Table 3 with relative examples.

7.2 Emerging Nanofluid Technologies

Nanofluid technologies are undergoing an evident metamorphosis from traditional thermal application systems toward multifunctional systems oriented for advanced industries. Currently, nanofluids are engineered with specialized optical, electronic, mechanical, and biocompatible properties so that their synthesis would be considered in energy harvesting, smart materials, additive manufacturing, biomedicine, and sustainable technologies. These efforts go beyond single-function fluids and, therefore, push nanofluids into complex aerospace, electronics, environmental, and medical arenas for a platform-based approach [27]. Table 4 includes a descriptive overview of various nanofluid technologies.

7.3 Nanomaterials in Next-Generation Technologies

Nanomaterials are speeding through development, essentially destroying the well-trodden development path to a surefire stone-lined, sustainable solution that promises performance in a single package. At such a scale, materials can be endowed with fantastic mechanical strength, thermal and electrical conductivity, light responsiveness, and chemical reactivity applications that grassroots cannot even dream of using conventional materials. Nanomaterials signify scientific advances stretching from high-speed electronics to life-saving biomedical devices to green environmental systems [97]. As of modern times, they forever redefine the philosophies of miniaturization, efficiency, responsiveness, and adaptability that one puts into modern technology. Table 5 provides details of various application domains with their respective uses.

7.4 Future Directions in Nanotechnology Research

The future of nanotechnology research holds many opportunities across research fields with the continuous developments that are anticipated. As nanomaterials undergo evolution, sustainability, scalability, and multifunctionality become a focus. Future research might aim at producing greener nanomaterials, with the goal that such materials would not pollute the environment as much and would be highly

Table 3 Details of innovations in ionic liquid applications

Domain	Innovative applications	Key functional attributes of ILs	Example/impact
Advanced separation technologies	– Selective gas separations (e.g., CO_2/CH_4, CO_2/N_2) – Liquid–liquid extraction of rare earth elements and toxic metals – Chiral separations in pharmaceutical processes	– High selectivity – Tunable solvation properties – Low volatility – Functionalization with selective moieties	Enables cleaner and more efficient recovery of critical resources; replaces volatile organic solvents [94]
Photoresponsive and stimuli-responsive ILs	– UV/visible light-triggered changes in polarity/viscosity – Magnetically responsive rheological control – pH-sensitive drug delivery systems	– External field responsiveness – Dynamic reversibility – Environmental sensitivity (e.g., pH, light, and magnetism)	Facilitates smart materials for diagnostics, controlled release, and adaptive devices
Sensor and analytical platforms	Electrochemical biosensors (glucose, urea, heavy metals) selective gas sensors (e.g., NH_3, NOx) wearable electrolyte/pH sensors	Broad electrochemical windows – High ionic conductivity – Compatibility with biomolecules and analytes	Enables real-time monitoring in healthcare and environmental analysis with high sensitivity and stability [95]
Additive manufacturing and material design	– Conductive inks for printed electronics – Rheology modifiers in 3D printing – Templates for MOFs and porous nanomaterials	– Thermal stability – Tunable viscosity – Compatibility with polymers and nanoparticles	Enhances material precision and functionality in next-gen manufacturing and energy systems
Biotechnological interfaces and biomass valorization	– Dissolution and conversion of biopolymers (cellulose, chitin) – Enzyme immobilization in IL-based matrices – Algae lipid extraction	– Biocompatibility – Chemical inertness – Enhanced enzyme performance – Preservation of biological structures	Supports eco-friendly biofuel production, biodegradable plastics, and reusable enzyme-based processes [96]

Table 4 Descriptive tabular overview of emerging nanofluid technologies

Innovation area	Advanced functionalities	Key materials and features	Application examples
Photothermal and optical-grade nanofluids	Solar-to-thermal energy conversion, laser energy absorption, spectral control	Plasmonic nanoparticles (e.g., Au, Ag), wavelength-tuned absorption, thermal stability	Photothermal energy harvesting, microfabrication via laser processing, and cooling in photovoltaic systems
Electrically tunable nanofluids	Field-responsive viscosity and conductivity, adaptive mechanical damping	Dielectric or conductive nanoparticles in insulating base fluids, electrorheological behavior	Lab-on-chip microfluidics, aerospace vibration dampers, adaptive biomedical actuators
Additive manufacturing applications	Printable conductive inks, integrated thermal management, flexible electronics	Silver/graphene-based nanofluids, thermally conductive fillers, rheology control	3D-printed wearable circuits, printed heat spreaders for electronics
Eco-responsive and biocompatible nanofluids	Environmentally safe heat transfer, biocompatibility for medical use	Plant-based oils, cellulose nanocrystals, and biodegradable surfactants	Organ cooling in transplantation, green cooling fluids for HVAC systems
Energy storage and conversion nanofluids	Charge storage, enhanced heat capacity, multifunctional energy carriers	Redox-active nanoparticles, phase-change nanomaterials, and high surface area carriers	Flow battery electrolytes, nanofluid-based thermal energy storage in solar thermal plants

efficient in their purpose. These green nanomaterials, depending on sectors like energy, medicine, and environmental remediation, would be fabricated either from renewable resources or via biological processes.

In medicine, nanoparticles will continue providing ways for personalized therapeutics wherein drugs are delivered to specific cells or tissues at a molecular scale, thus giving treatments that have fewer side effects. Integration of nanotechnology with artificial intelligence will create systems for enhanced diagnostics and therapeutics, allowing the use of predictive medicine and real-time monitoring of health conditions.

8 Conclusion

Nanotechnology implies a convergence of several scientific disciplines to manipulate matter at the atomic and molecular levels. This would include the fabrication

Table 5 Details of nanomaterials in technologies

Application domain	Functional contributions of nanomaterials	Representative nanomaterials	Key use cases and innovations
Advanced electronics and photonics	Miniaturization, high-speed data transfer, tunable optical response	Graphene, TMDs (e.g., MoS_2), quantum dots, CNTs	High-frequency transistors, photodetectors, quantum communication devices, wearable and flexible electronics
Energy storage and conversion	Enhanced energy density, faster charge-discharge rates, improved conversion efficiency	Graphene oxide, silicon nanowires, MOFs, nanocatalysts	Lithium-ion batteries, supercapacitors, PEM fuel cells, redox flow batteries
Healthcare and biomedicine	Targeted drug delivery, molecular diagnostics, image enhancement	Gold nanoparticles, polymeric micelles, magnetic nanoparticles, liposomes	Cancer therapeutics, MRI contrast agents, smart biosensors, non-invasive disease screening
Environmental sustainability and remediation	Pollutant degradation, oil absorption, CO_2 capture	Carbon nanotube composites, graphene oxide, MOFs, nanocatalysts	Wastewater treatment, oil spill clean-up, air purification, carbon sequestration
Smart and autonomous systems	Responsiveness to stimuli, self-repair, and sensory integration	CNT-based sensors, nanofibers, nanocomposites with microcapsules	Self-healing coatings, adaptive robotics, smart garments with biosignal monitoring, tactile feedback interfaces

of materials whose properties are enhanced mechanically, electrically, thermally, or optically. The main principles are nanoscale manipulation using sophisticated fabrication techniques, ranging from lithography, chemical vapor deposition, and molecular self-assembly, to control over the structure and function of something.

Characterization tools such as TEM, AFM, FTIR, and SERS give knowledge about nanoscale properties, with recent advances, like in situ electron microscopy and microfluidic sensors, adding to improvements in accuracy and real-time analysis. These tools need to be harnessed to characterize nanomaterials, nanofluids, and ionic liquids enlisted for myriad applications. For example, gold nanoparticles find their use in cancer diagnostics, graphene is employed in flexible electronics, and nanofluids are used in thermal management systems.

In the future, trends forecast smart nanomaterials built through AI-assisted nanodesign with eco-friendly approaches to synthesis. Self-healing coatings, sensors embedded in garments, and target-specific nanomedicine are just other examples of this vast domain of nanotechnology. It is truly an intensely interdisciplinary field with enormous transformative potential. The interplay and fusion of scientists and engineers allow the birth and development of energy-, electronics-, environment-, and health-care-related technologies, with nanotechnology being one of the platforms for the construction of next-generation technology.

References

1. T. Welton, Room-temperature ionic liquids: Solvents for synthesis and catalysis. Chem. Rev. **99**, 2071–2083 (1999)
2. N.V. Plechkova, K.R. Seddon, Applications of ionic liquids in the chemical industry. Chem. Soc. Rev. **37**, 123–150 (2008)
3. S.U.S. Choi, J.A. Eastman, Enhancing thermal conductivity of fluids with nanoparticles. ASME Int. Mech. Eng. Congr. Expo. **66**, 99–105 (1995)
4. Z. Liu, S.M. Tabakman, Z. Chen, H. Dai, Preparation of carbon nanotube bioconjugates for biomedical applications. Nat. Protoc. **4**, 1372–1381 (2009)
5. K.E. Drexler, *Nanosystems: Molecular Machinery, Manufacturing, and Computation* (John Wiley & Sons, New York, 1992)
6. P. Wasserscheid, W. Keim, Ionic liquids—New "solutions" for transition metal catalysis. Angew. Chem. Int. Ed. **39**, 3772–3789 (2000)
7. J. Wang, Y. Yamada, K. Sodeyama, C.H. Chiang, Y. Tateyama, A. Yamada, Superconcentrated electrolytes for a high-voltage lithium-ion battery. Nat. Energy **1**, 16020 (2016)
8. D. Zhao, M. Wu, Y. Kou, E. Min, Ionic liquids: Applications in catalysis. Catal. Today **74**, 157–189 (2002)
9. C. Kohlpaintner, M. Schulte, J. Falbe, Carbon monoxide, in *Ullmann's Encyclopedia of Industrial Chemistry*, (Wiley, Weinheim, 2000)
10. G.B. Appetecchi, M. Montanino, M. Carewska, S. Passerini, Ionic liquids and polymer ionic liquids for advanced batteries. J. Power Sources **269**, 845–851 (2014)
11. J. Sarkar, P. Ghosh, A. Adil, A review on smart nanofluids for heat transfer applications. J. Mol. Liq. **195**, 135–143 (2015)
12. O. Mahian, A. Kianifar, S.A. Kalogirou, I. Pop, S. Wongwises, A review of the applications of nanofluids in solar energy. Int. J. Heat Mass Transf. **57**, 582–594 (2013)
13. M. Moniruzzaman, M. Goto, Ionic liquids for drug delivery and pharmaceutical applications. Int. J. Pharm. **400**, 219–232 (2011)
14. J. Stoimenovski, D.R. MacFarlane, K. Bica, R.D. Rogers, Crystalline vs. ionic liquid salt forms of active pharmaceutical ingredients: A position paper. Pharm. Res. **27**, 521–526 (2010)
15. J.L. Shamshina, P.S. Barber, R.D. Rogers, Ionic liquids in drug delivery. Expert Opin. Drug Deliv. **10**, 1367–1381 (2013)
16. M.C. Gutiérrez, M.L. Ferrer, F.D. Monte, Lipid and polymeric ionic liquid carriers for enhanced anticancer drug delivery. Angew. Chem. Int. Ed. **49**, 2158–2162 (2010)
17. J. Pernak, K. Sobaszkiewicz, I. Mirska, Antimicrobial activities of ionic liquids. Green Chem. **5**, 52–56 (2003)
18. J. Ranke, A. Othman, P. Fan, A. Müller, K. Kümmerer, Influence of ionic liquid structure on biodegradability and antimicrobial activity. Chemosphere **76**, 1002–1007 (2009)
19. R.A. Sheldon, Biocatalysis in ionic liquids. Green Chem. **7**, 267–278 (2005)
20. F. van Rantwijk, R.M. Lau, R.A. Sheldon, Biocatalytic transformations in ionic liquids. Trends Biotechnol. **21**, 131–138 (2003)
21. Q. Liu, F. Wu, Y. Lu, Application of ionic liquids in electrochemical biosensors. Anal. Chim. Acta **746**, 1–10 (2012)
22. W.L. Hough, M. Smiglak, H. Rodríguez, R.P. Swatloski, S.K. Spear, D.T. Daly, J. Pernak, J.E. Grisel, R.D. Carliss, M.D. Soutullo, J.H. Davis, R.D. Rogers, The third evolution of ionic liquids: Active pharmaceutical ingredients. New J. Chem. **31**, 1429–1436 (2007)

23. S. Saini, P. Bansal, V. Rana, Ionic liquid-mediated gene delivery: Current insights and future perspectives. Drug Discov. Today **25**, 1275–1287 (2020)
24. S.K. Das, S.U.S. Choi, W. Yu, T. Pradeep, *Nanofluids: Science and Technology* (Wiley, Hoboken, 2007)
25. J. Buongiorno et al., A benchmark study on the thermal conductivity of Nanofluids. J. Appl. Phys. **106**, 094312 (2009)
26. S.M.S. Murshed, K.C. Leong, C. Yang, Stability of nanofluids. Int J Therm Sci **47**, 569–578 (2008)
27. R. Saidur, K.Y. Leong, H.A. Mohammad, A review on applications and challenges of nanofluids. Renew. Sust. Energ. Rev. **15**, 1646–1668 (2011)
28. A.G.A. Nnanna, Experimental model of nanofluid thermal conductivity for energy applications. J. Energy Resour. Technol. **129**, 84–89 (2007)
29. M. Afrand, M. Akbari, N. Sina, Effect of hybrid nanoparticles on thermal conductivity of nanofluids. J. Therm. Anal. Calorim. **123**, 1399–1406 (2016)
30. J.A. Eastman, S.U.S. Choi, S. Li, W. Yu, L.J. Thompson, Enhanced thermal conductivity through the development of nanofluids. Appl. Phys. Lett. **78**, 718–720 (2001)
31. S. Lee, S.U.S. Choi, S. Li, J.A. Eastman, Measuring thermal conductivity of fluids containing oxide nanoparticles. J. Heat Transf. **121**, 280–289 (1999)
32. M.P. Beck, Y. Yuan, P. Warrier, A.S. Teja, The thermal conductivity of alumina nanoparticles in ethylene glycol-based fluids. J. Appl. Phys. **103**, 1–5 (2009)
33. W. Duangthongsuk, S. Wongwises, Heat transfer and pressure drop characteristics of TiO_2–water nanofluid in a double-tube counter flow heat exchanger. Int. J. Heat Mass Transf. **52**, 2059–2067 (2009)
34. V. Bianco, O. Manca, S. Nardini, Nanofluids applications in HVAC systems—A review. Appl. Therm. Eng. **30**, 1448–1461 (2011)
35. Y. Xuan, Q. Li, Investigation on convective heat transfer and flow features of nanofluids. Int. J. Heat Mass Transf. **46**, 4319–4327 (2003)
36. B.X. Wang, L.P. Zhou, X.F. Peng, A fractal model for predicting the effective thermal conductivity of liquid with suspension of nanoparticles. Int. J. Heat Mass Transf. **46**, 2665–2672 (2003)
37. H.A. Mintsa, G. Roy, C.T. Nguyen, D. Doucet, New temperature dependent thermal conductivity data for water-based nanofluids. Int J Therm Sci **48**, 363–371 (2009)
38. A. Turgut, I.H. Tavman, M. Chirtoc, H.P. Schuchmann, C. Sauter, S. Tavman, Investigation of viscosity and thermal conductivity of suspensions containing nanoparticles. Int. J. Thermophys. **30**, 1213–1226 (2009)
39. D.P. Kulkarni, D.K. Das, G.A. Chukwu, Heat transfer and pressure drop in nanofluid flow through a parallel plate channel. J. Heat Transf. **130**, 052401 (2008)
40. W.M. El-Maghlany, M.A. Habib, Numerical simulation of forced convection heat transfer in nanofluid flow over a heated pipe. Int. J. Heat Fluid Flow **41**, 74–85 (2013)
41. L. Huang, Y. Zhang, Z. Wang, Experimental study on heat rejection performance of a hybrid nanofluid used in cooling towers. Energ. Buildings **76**, 202–209 (2014)
42. S.K. Verma, A.K. Tiwari, Progress of nanofluid application in solar collectors: A review. Energy Convers. Manag. **100**, 324–346 (2015)
43. J.M. Khodadadi, S. Hosseinizadeh, Nanoparticle-enhanced phase change materials (NEPCM) with great potential for improved thermal energy storage. Int. Commun. Heat Mass Transf. **34**, 534–543 (2007)
44. R. Gupta, D. Sharma, Magnetic nanoparticles for hyperthermia applications. J. Magn. Magn. Mater. **416**, 252–259 (2016)
45. M.V. Yigit, A. Moore, Z. Medarova, Magnetic nanoparticles for cancer diagnosis and therapy. Pharm. Res. **29**, 1180–1188 (2012)
46. X. Huang, I.H. El-Sayed, W. Qian, M.A. El-Sayed, Cancer cell imaging and photothermal therapy in the near-infrared region by using gold nanorods. J. Am. Chem. Soc. **128**, 2115–2120 (2006)

47. E.S. Place, N.D. Evans, M.M. Stevens, Complexity in biomaterials for tissue engineering. Nat. Mater. **8**, 457–470 (2009)
48. D. Chimene, K.K. Lennox, R.R. Kaunas, A.K. Gaharwar, Advanced bioinks for 3D printing: A materials science perspective. Ann. Biomed. Eng. **44**, 2090–2102 (2016)
49. B. Wowk, Thermodynamic aspects of vitrification. Cryobiology **60**, 11–22 (2010)
50. M. Rai, S.D. Deshmukh, A.P. Ingle, A.K. Gade, Silver nanoparticles: The powerful nanoweapon against multidrug-resistant bacteria. J. Appl. Microbiol. **112**, 841–852 (2012)
51. P. Baptista, E. Pereira, P. Eaton, G. Doria, A. Miranda, I. Gomes, P. Quaresma, R. Franco, Gold nanoparticles for the development of clinical diagnosis methods. Anal. Bioanal. Chem. **391**, 943–950 (2008)
52. M.C. Roco, The long view of nanotechnology development: The national nanotechnology initiative at 10 years. J. Nanopart. Res. **13**, 427–445 (2011)
53. B. Bhushan, *Springer Handbook of Nanotechnology*, 3rd edn. (Springer, Berlin, 2010)
54. H. Gleiter, Nanostructured materials: Basic concepts and microstructure. Acta Mater. **48**, 1–29 (2000)
55. S. Yang, P.J. Rivero, J. Goicoechea, F.J. Arregui, Layer-by-layer fabrication of nanostructured materials for chemical and biological sensing. Sensors **19**, 975 (2019)
56. S.A. Maier, *Plasmonics: Fundamentals and Applications* (Springer, New York, 2007)
57. N. Savage, M.S. Diallo, Nanomaterials and water purification: Opportunities and challenges. J. Nanopart. Res. **7**, 331–342 (2005)
58. A. Nel, T. Xia, L. Mädler, N. Li, Toxic potential of materials at the nanolevel. Science **311**, 622–627 (2006)
59. R.P. Feynman, There's plenty of room at the bottom. Eng. Sci. **23**, 22–36 (1960)
60. J. Wang, Electrochemical glucose biosensors. Chem. Rev. **108**, 814–825 (2008)
61. N. Pardi, M.J. Hogan, F.W. Porter, D. Weissman, mRNA vaccines – A new era in vaccinology. Nat. Rev. Drug Discov. **17**, 261–279 (2018)
62. M.N. Chong, B. Jin, C.W. Chow, C. Saint, Recent developments in photocatalytic water treatment technology: A review. Water Res. **44**, 2997–3027 (2010)
63. F. Perreault, A.F. De Faria, M. Elimelech, Environmental applications of graphene-based nanomaterials. Chem. Soc. Rev. **44**, 5861–5896 (2015)
64. L. Zhang, P. Pornwongthong, Y. Zhang, T. Emrick, T.P. Russell, Nanoengineered air filters with enhanced particle capture performance using nanofiber coating. Nano Lett. **16**, 3642–3649 (2016)
65. J.R. Li, R.J. Kuppler, H.C. Zhou, Selective gas adsorption and separation in metal–organic frameworks. Chem. Soc. Rev. **38**, 1477–1504 (2009)
66. B. Karn, T. Kuiken, M. Otto, Nanotechnology and in situ remediation: A review of the benefits and potential risks. Environ. Health Perspect. **117**, 1823–1831 (2009)
67. G. Binnig, H. Rohrer, Scanning tunneling microscopy. Helv. Phys. Acta **55**, 726–735 (1982)
68. G.M. Whitesides, B. Grzybowski, Self-assembly at all scales. Science **295**, 2418–2421 (2002)
69. S. Iijima, Helical microtubules of graphitic carbon. Nature **354**, 56–58 (1991)
70. M. Ferrari, Cancer nanotechnology: Opportunities and challenges. Nat. Rev. Cancer **5**, 161–171 (2005)
71. G. Schmid, *Nanoparticles: From Theory to Application* (Wiley-VCH, Weinheim, 2004)
72. N.C. Seeman, DNA in a material world. Nature **421**, 427–431 (2003)
73. S.Y. Chou, P.R. Krauss, P.J. Renstrom, Imprint lithography with 25-nanometer resolution. Science **272**, 85–87 (1996)
74. Y. Cui, C.M. Lieber, Functional nanoscale electronic devices assembled using silicon nanowire building blocks. Science **291**, 851–853 (2001)
75. M.A. El-Sayed, Some interesting properties of metals confined in time and Nanometer space of different shapes. Acc. Chem. Res. **34**, 257–264 (2001)
76. R. Garcia, R. Perez, Dynamic atomic force microscopy methods. Surf. Sci. Rep. **47**, 197–301 (2002)
77. Q. Zhang, E. Uchaker, S.L. Candelaria, G. Cao, Nanomaterials for energy conversion and storage. Chem. Soc. Rev. **42**, 3127–3171 (2013)

78. D.A. Skoog, F.J. Holler, S.R. Crouch, *Principles of Instrumental Analysis* (Cengage Learning, Boston, 2018)
79. J.C. Vickerman, I.S. Gilmore, *Surface Analysis: The Principal Techniques* (Wiley, Hoboken, 2009)
80. L. Zhang, Y. Zhang, D. Yang, X. Jiang, Recent advances in characterization techniques for nanomaterials. Anal. Bioanal. Chem. **411**, 6205–6224 (2019)
81. F.M. Ross, Opportunities and challenges in liquid cell electron microscopy. Science **350**, aaa9886 (2015)
82. R. Le, P.G. Etchegoin, *Principles of Surface-Enhanced Raman Spectroscopy* (Elsevier, Amsterdam, 2009)
83. P.A. Lee, L. Yang, X-ray absorption and synchrotron radiation in nanomaterials. Chem. Soc. Rev. **41**, 2310–2330 (2012)
84. G.M. Whitesides, The origins and the future of microfluidics. Nature **442**, 368–373 (2006)
85. D. Srivastava, Multiscale modeling of nanosystems. Science **303**, 168–169 (2004)
86. M.P. Allen, D.J. Tildesley, *Computer Simulation of Liquids* (Oxford University Press, Oxford, 2017)
87. R.O. Jones, O. Gunnarsson, The density functional formalism, its applications and prospects. Rev. Mod. Phys. **61**, 689–746 (1989)
88. O.C. Zienkiewicz, R.L. Taylor, *The Finite Element Method* (McGraw-Hill, London, 2000)
89. D.P. Landau, K. Binder, *A Guide to Monte Carlo Simulations in Statistical Physics* (Cambridge University Press, Cambridge, 2014)
90. P. Keblinski, J.A. Eastman, D.G. Cahill, Nanofluids for thermal transport. Mater. Today **8**, 36–44 (2005)
91. K.T. Butler, D.W. Davies, H. Cartwright, O. Isayev, A. Walsh, Machine learning for molecular and materials science. Nature **559**, 547–555 (2018)
92. J. Fish, *Multiscale Methods: Bridging the Scales in Science and Engineering* (Oxford University Press, Oxford, 2009)
93. L. Zhang, J. Liu, C. Huang, C. Wang, Emerging trends in nanotechnology and its impact on society. Nano Today **30**, 100837 (2020)
94. D. Mecerreyes, Polymeric ionic liquids: Broadening the properties and applications of polyelectrolytes. Prog. Polym. Sci. **36**, 1629–1648 (2011)
95. F. Arduini, A. Amine, D. Moscone, G. Palleschi, Ionic liquids in biosensing and electrochemical biosensors. TrAC Trends Anal. Chem. **79**, 114–126 (2016)
96. C. Wu, J. Zhang, H. Wang, Ionic liquids for biomass conversion: An overview. ACS Sustain. Chem. Eng. **6**, 5911–5927 (2018)
97. M.C. Roco, C.A. Mirkin, M.C. Hersam, Nanotechnology research directions for societal needs in 2020. J. Nanopart. Res. **13**, 897–919 (2011)

Ionic Liquids: Alternative to Conventional Molecular Solvents for Organic Reactions

Amandeep Singh, Shikha Baskar, and Chinnappan Baskar

1 Introduction

Concerns have emerged within the field related to chemicals and its processing because of the substantial amounts of flammable as well as hazardous chemicals or solvents used daily. The persistent application of these solvents adversely affects all forms of life, leading to the deterioration of vital organs [1]. Despite the contribution of organic solvents, for instance, dimethyl sulfoxide (DMSO), acetone, dimethylformamide (DMF), and other aromatic compounds like toluene and benzene, to environmental degradation, their usage remains prevalent. The main objective of the green chemistry is to minimize solvent usage or have another lesser harmful chemicals/solvent option to substitute them [2]. Green solvents include supercritical fluids, water, ionic liquids (ILs), and other nontoxic macromolecules/polymers, for instance, polyethylene glycol (PEG) considered for the low toxicity, availability, and their recyclability. Among these, ILs are particularly favored by researchers due to their remarkable properties, including excellent biodegradability, good stability, lesser toxicity, and low vapor pressure. These attributes position ILs as superior substitutes for conventional chemicals, thereby serving as effective greener chemicals or solvents.

A. Singh (✉)
Indian Institute of Technology (IIT), Ropar, Rupnagar, Punjab, India

S. Baskar
Faculty of Science and Technology, State Christian Institute (IAKN), Ambon, Maluku, Indonesia

C. Baskar
Department of Chemistry, Central University of Kashmir, Ganderbal, Jammu and Kashmir, India

Department of Chemistry, Pattimura University, Ambon, Maluku, Indonesia

S. Kulkarni, S. Das (eds.), *Ionic Liquids, Nanofluids and Nanotechnology*,
Lecture Notes in Nanoscale Science and Technology 42,
https://doi.org/10.1007/978-3-032-04008-4_5

Solvents have an important character in various household and industrial processes. The consequences of the loss of solvents and their emissions have prompted initiatives aimed at reducing or completely eliminating them. Since the 1990s, this issue has gained significant attention within the realm of the green chemistry, leading to the emergence of the concept of "green" solvents. This notion resulted in a considerable body of chemical literature and the innovation of what are termed neoteric solvents [3]. A thorough review of existing publications reveals that only a limited number of novel materials have achieved extensive application as solvents or chemicals. The pursuit of solvents with reduced environmental impact is rendered ineffective if conducted without careful consideration of the specific conditions under which these solvents will be applied on an industrial level. Broader sustainability issues, predominantly the utilization of sources to be considered as nonfossil in the production of solvents, take precedence over the inherent "greenness" of the solvents themselves. While the concept of solvency is universally applicable, the idea of a universal solvent, or alkahest, remains an elusive ideal [4].

One of the foundational understandings of the green chemistry emphasizes on the necessity to minimize the overall utilization of assisting constituents, for instance, separation agents and solvents, and ensures that any substances or chemicals used are harmless [5]. The hazardous and toxic characteristics of numerous solvents, specifically chlorinated hydrocarbons, raise significant ecological issues including harmful releases and water pollution. It is acknowledged that adoption of nontraditional chemicals as substitutes for ecologically detrimental conventional chemicals can significantly decrease the generation of waste solvents, thereby mitigating their adverse environmental effects [6]. Among these innovative solvent systems are water, ILs, supercritical carbon dioxide, solvent-free operations, and fluorinated processes [7]. Notably, ILs have appeared as a particularly capable substitute among these nontraditional solvents. The ILs are considered as salts whose m. p. remains lower than b.p. of water [8]. The ILs are referred to various terms including ionic fluids, neoteric solvents, designer solvents, molten salts. Typically, ILs contain a segment of organic positive ion along with another segment of inorganic negative ion [9]. For an IL to remain in a liquid state at RT, it is preferable for the cation to be asymmetrical, meaning that attached -R groups should differ. The hydrophilicity/hydrophobicity, as well as polarity of ILs, can be adjusted through the suitable selection of positive ion and negative ion, a feature that has led to their designation as designer solvents. Contemporary society relies significantly on progress in polymer science; however, growing environmental issues related to plastic waste necessitate the pursuit of sustainable alternatives. ILs have surfaced as noninnocent solvents, showcasing considerable promise for polymer chemistry in synthesis, as well as depolymerization processes [10, 11]. Also, the relationship between ILs and circular economy is formed through multiple pathways that promote sustainability, improve resource efficiency, and reduce waste [12].

2 Green Chemistry

The green chemistry, like circular or sustainable chemistry, represents a branch of chemical engineering and chemistry devoted to develop processes and products to reduce and eradicate the overall applications and production of destructive materials. In contrast to environmental chemistry, which examines the repercussions of pollutants on the environment, green chemistry emphasizes the ecological implications of chemical practices, including the reduction of nonrenewable resource consumption and the implementation of technologies aimed at pollution prevention [13]. The primary objectives of green chemistry, specifically, the creation of added resource-efficient and fundamentally harmless designs for materials, molecules, processes, and products, should be applied across the diverse variety of circumstances. In 1998, formulated by Paul A. and John W., the subsequent list delineates a framework aimed at creating more environmentally friendly chemicals, processes, or products [14]. Green chemistry principles are described in Fig. 1.

Efforts are underway not only to assess the environmental sustainability of a chemical process but also to incorporate additional factors, i.e., yield of chemicals, cost of reactants, safety considerations in chemical management, equipment requirements, energy consumption, and the simplicity of product recovery and purification. A quantitative analysis [15] revealed that the synthesis of aniline from nitrobenzene scored 64 points of 100, categorizing this as a satisfactory synthesis, while the amide synthesis using sodium bis(trimethylsilyl)amide was rated merely adequate with a total of 32 points.

Converting waste into valuable products represents a significant advancement in promoting environmental sustainability. The ultimate objective in the realm of

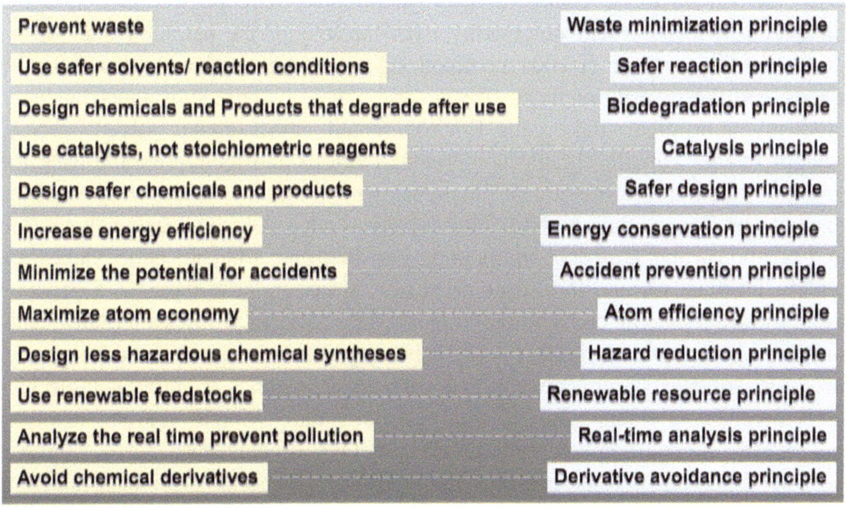

Fig. 1 Green chemistry principles

green chemistry is to prioritize source prevention. To mitigate the toxic effects of a substance, it is imperative to avoid its generation from the outset. Solvents are extensively used across various chemical and related industries, becoming integral to modern life in the twenty-first century. In 1991, various solvents including toluene, methanol, methyl ethyl ketone, xylene, and dichloromethane (DCM) accounted for five of the top ten chemicals disposed of. Given the substantial volumes of solvent usage, the associated hazards and safety concerns have consistently influenced the development and the choice of the solvents. Due to the increasing attention for the environment and health linked to hazardous properties of solvents, alternative solvents are meticulously evaluated for their potential risks, leading to the creation of new solvent systems designed to be more environmentally friendly. Numerous solvents, particularly halogenated ones like carbon tetrachloride and chloroform, have been identified as toxic and potential carcinogens, while others exhibit neurotoxic effects. To enhance the greenness and environmental compatibility of separation and extraction processes, alternative solvents such as ILs and supercritical fluids offer unique advantages, particularly their tunability. ILs are considered innovative chemicals and are viewed as green alternatives for several conventional chemicals and solvents due to their recyclability and nonvolatility.

3 Conventional Molecular Solvents

Traditional organic solvents, including various halogenated hydrocarbons, hydrocarbons, and carbon-based chemicals, are characterized by their high vapor pressure and immiscibility with water. These volatile organic compounds (VOCs), primarily sourced from petrochemicals, exhibit an extensive variety of chemical features. They are known to be explosive, toxic, flammable, narcotic, carcinogenic, and mutagenic, which present noteworthy risks throughout the handling and storing processes. Contact with mucous membranes, skin, or eyes can lead to serious health issues, and the severity of these effects can escalate if these substances enter the bloodstream or affect vital organs. Environmentally, most organic solvents used to categorize into volatile organic compounds (VOCs) and their presence back the pollution, including formation of photochemical smog and the degradation of the ozone layer. Despite their high toxicity, solvents such as DMSO, DMF, halogenated hydrocarbons, and aromatic compounds like benzene, toluene, and xylenes are widely used in various industrial processes. Consequently, there is a growing trend toward replacing conventional organic solvents with greener alternatives that minimize environmental impact while achieving similar results. The development of green solvents or nonconventional solvents to synthesize various molecules has emerged as a viable solution, with environmentally friendly options including ILs, water, supercritical carbon dioxide ($scCO_2$), organic carbonates, nontoxic polymer fluids, and deep eutectic solvents (DESs), among others. All these materials are reusable, nonvolatile, environmentally friendly, biocompatible, and represent comparatively novel resources that function effectively as reaction media, solvents, and catalysts for various organic reactions as shown in Table 1.

4 Green Solvents

Eco-friendly solvents, known as green solvents, are chemical agents that align to green chemistry principles. Their significance surged in 2015 when the United Nations established a sustainability-oriented development agenda comprising 17 goals, highlighting the importance of solvents and green chemistry in fostering the sustainability in real world [16]. These solvents are formulated to be more environmentally benign, often sourced from agricultural products or other sustainable processes, serving as alternatives to traditional petrochemical solvents. Anticipated attributes of green solvents encompass recyclability, biodegradability, and minimal toxicity [17].

The predominant use of various solvents in different processes like paints and coatings accounts for about 46% of their overall application. Other, less significant uses comprise degreasing, cleaning, chemical synthesis, and adhesives [18]. Conventional solvents/chemicals are frequently halogenated or toxic, whereas green solvents tend to be less detrimental to the environment and to health, and are sustainable by nature. Preferably, these chemicals originate from the renewable materials and get biodegraded into harmless substances [19, 20]. However, the production of various solvents from different biomass can sometimes pose greater environmental risks than producing the same chemicals from the fossils [21]. Therefore, an environmental implication of chemical/solvent production should be taken care while selecting a chemical/solvent [22]. Additionally, the post-use fate of the solvent is crucial. In scenarios where chemical/solvent processing and recycling are practical, the overall energy costs and the environmental impacts related to their recycling must be evaluated. In such cases, water which requires significant energy for purification would not be the most environmentally friendly option. Conversely, solvents used in consumer products are discharged in the atmosphere during application, making the atmospheric influence of the chemical/solvent critical as compared to the cost of energy and recycling effects; in these instances, water is supposed to be more sustainable selection. In summary, it is essential to assess ecological effects of the solvent during its whole lifespan, from production to disposal (or from production to reuse if it is recycled). Therefore, the most thorough definition of green solvents is articulated as follows: "a green solvent is one who minimizes ecological influence of a product or process across its complete life cycle" [23].

Consequently, a solvent may be considered environmentally friendly for one specific use, due to its lower ecological impact compared to alternative solvents, while simultaneously being deemed unsuitable for another application. A prime illustration of this is water, which serves as an eco-friendly solvent for household products like toilet bowl cleaners, yet fails to qualify as a green solvent in the

Table 1 Conventional molecular solvents, their uses in organic reactions, and associated harmful effects

Solvent	Chemical formula	Common uses in organic reactions	Harmful effects
DCM	CH_2Cl_2	Extraction, halogenation, Friedel-crafts reactions	Suspected carcinogen, CNS depression, liver toxicity
Chloroform	$CHCl_3$	Solvent for spectroscopy, organic synthesis	Carcinogenic, hepatotoxic, nephrotoxic, CNS effects
Toluene	$C_6H_5CH_3$	Nitration, sulfonation, Grignard reactions	Neurotoxic, respiratory irritation, possible reproductive toxicity
Benzene	C_6H_6	Friedel-crafts alkylation/acylation, solvent	Known human carcinogen (leukemia), bone marrow suppression
Carbon tetrachloride	CCl_4	Solvent, chlorination reactions	Hepatotoxic, nephrotoxic, possible human carcinogen
Diethyl ether	$(C_2H_5)_2O$	Grignard reactions, extractions	Extremely flammable, CNS depression, forms explosive peroxides
Tetrahydrofuran (THF)	C_4H_8O	Polymerization, lithium-based reactions	Peroxide former, CNS effects, respiratory irritation
Acetone	$(CH_3)_2CO$	Cleaning, nucleophilic substitution	Eye and respiratory irritation, CNS effects in high doses
DMF	$HCON(CH_3)_2$	Nucleophilic substitution, peptide coupling	Reproductive toxicity, liver damage, skin absorption hazard
DMSO	$(CH_3)_2SO$	Polar aprotic solvent, organometallics	Skin penetrant (carries toxins), mild hepatotoxicity in high doses
Acetonitrile	CH_3CN	SN_2 reactions, electrochemistry	Toxic by inhalation and ingestion, eye and skin irritant
Hexane	C_6H_{14}	Nonpolar solvent, extraction, chromatography	Neurotoxic, possible reproductive toxicity, flammable
1,4-Dioxane	$C_4H_8O_2$	Solvent in oxidation reactions, polymerization	Probable human carcinogen, forms explosive peroxides

production of polytetrafluoroethylene. In this case, using water necessitates the incorporation of perfluorinated surfactants known for their environmental persistence. Conversely, supercritical carbon dioxide emerges as the most environmentally benign solvent for this process as it operates effectively without the need for surfactants [23]. Therefore, it is essential to recognize that a solvent cannot be universally classified as a "green solvent" without specifying the context of its application. Notable green solvents include water, ethanol, glycerol (glycerin), $scCO_2$, ILs, dimethyl carbonate (DMC), limonene, ethyl lactate as given in Table 2.

5 Ionic Liquids

ILs represent a category of environmentally friendly solvents extensively used in chemical processing, attributed to their distinctive characteristics, i.e., nonflammability, nonvolatility, and remarkable thermal/chemical stability. These substances are salts, typically composed of an organic positive ion paired with an inorganic negative ion, remaining in a fluid state at temperatures below 100 °C. Chemical and physical attributes of ionic liquids are readily altered through modifying combinations of cations and anions in their molecular structures. They are frequently employed in the pre-concentration of pesticides found in food and environmental samples, and ongoing research is exploring their potential applications in innovative sensing technologies.

Ionic solvents are applicable in various industrial applications such as cleaning, extraction, and synthesis. The substitution of traditional solvents with eco-friendly alternatives allows industries to minimize their ecological footprint and enhance the sustainability of their operations.

Frequently used cations include imidazolium, ammonium, phosphonium, and pyridinium, while anion counterparts consist of halides, hexafluorophosphate, nitrate, and tetrafluoroborate. According to a research, these ILs exhibit nonflammability and demonstrate stability in chemical, electrochemical, and thermal contexts [24]. Such characteristics render ILs suitable as environmentally friendly solvents, as their low volatility significantly reduces VOC emissions in comparison with traditional solvents. However, concerns regarding ecotoxicity and limited biodegradability of ILs are acknowledged, primarily due to the reliance on nonrenewable resources for their synthesis such as halogenated alkanes and imidazole sourced from the petroleum. Recently, ILs derived from biodegradable and renewable resource are developed, yet the production remains constrained by elevated costs [4].

Among the diverse categories of ILs, the ILs having N-heterocyclic positive ions are the utmost prevalent. Imidazolium salts signify the furthermost significant within the domain with several commercially available options. The solvent characteristics including m.p., viscosity, and solubility are finely adjustable across a broad spectrum by modifying the moiety over N atoms and altering opposite ions. The versatility positions ILs as true tailor-made chemicals/solvents. Despite

Table 2 Green solvents along with their uses in organic reactions

Green solvent	Type	Common uses in organic reactions	Green advantages
Water	Polar protic solvent	Hydrolysis, condensation, enzymatic and aqueous-phase organic reactions	Nontoxic, nonflammable, abundant, environmentally benign
Ethanol (EtOH)	Polar protic alcohol	Esterification, fermentation, extractions, phase-transfer reactions	Biodegradable, renewable (bioethanol), low toxicity
Methanol (MeOH)	Polar protic alcohol	Transesterification, methylation, reductions	Low toxicity (relatively), biodegradable
Acetone	Polar aprotic ketone	Aldol condensations, SN_2 reactions, solvent for cleaning and crystallization	Readily biodegradable, relatively low toxicity
Ethyl acetate	Polar aprotic ester	Extractions, chromatography, esterification	Biodegradable, low toxicity, pleasant odor
$scCO_2$	Nonpolar (supercritical fluid)	Hydrogenations, extractions, polymerizations	Inert, recyclable, nontoxic, replaces halogenated solvents
2-Methyltetrahydrofuran (2-MeTHF)	Cyclic ether (bio-based)	Grignard reactions, Diels-Alder, Suzuki couplings	Derived from biomass, less toxic than THF, recyclable
Glycerol	Polyol	Biocatalysis, esterification, phase-transfer reactions	By-product of biodiesel, biodegradable, low vapor pressure
DMC	Carbonate ester	Methylation, transesterification, as replacement for phosgene or methyl halides	Nontoxic, biodegradable, produced from CO_2
Cyrene™	Dihydrolevoglucosenone	Diels-Alder, SNAr, amidation, as replacement for NMP and DMF	Bio-derived, low toxicity, good solvating power
Propylene carbonate (PC)	Cyclic carbonate	Nucleophilic substitution, electrochemical reactions	Biodegradable, low vapor pressure, nontoxic
ILs (e.g., [EMIM][OAc])	Ionic compound	Solvents and catalysts for organic synthesis and biomass conversion	Nonvolatile, recyclable, tunable properties
DES (e.g., choline chloride: urea)	Hydrogen-bonded mixture	Biocatalysis, esterification, oxidation, biomass processing	Easy to prepare, biodegradable, nontoxic

the widespread applications of imidazolium salts in the organic synthesis and catalysis, ILs do present certain drawbacks. A notable limitation is their failure to act as inert solvents/chemicals under the influence of a strong base by undergoing deprotonation resulting in the formation of N-heterocyclic carbenes. In contrast, 1,2,3-triazolium does not have any acidic ring-C adjacent to the N atoms which may confer advantages in this regard. Interestingly, salts of 1,2,3-triazolium are largely overlooked for an excellent ionic liquid until now even though the first instances were identified as early as 1887, yet their IL characteristics remained unexplored. Before the initial study regarding the stability of 1,3,4-trialkyl-1,2,3-triazolium-based ILs, these were primarily of interest when these compounds contained 1-amino functional groups which were associated with O-rich negative ions, thus being applicable as highly energetic fuels. Some common ILs are given in Table 3.

The imidazolium-based ionic liquids have a large contribution due to their thermal stability and tunability. Anion choice (e.g., BF_4^-, PF_6^-, NTf_2^-, Ac^-) significantly affects hydrophobicity, viscosity, and catalytic activity. Deep eutectic solvents (DESs), though not traditional ILs, are considered green IL analogs with similar properties, some of them are shown in Fig. 2. Many ILs are recyclable and nonvolatile and can enhance selectivity and reaction rates.

6 Uses of Ionic Liquids in Organic Reactions

Cost-effective IL N-methylimidazolium p-toluenesulfonate facilitates an electrochemical denitrative cyanation of nitroarene feedstocks under mild circumstances without transition metals. This efficient method demonstrates an extensive acceptance for different functional groups and is applicable for the modification of bioactive compounds [25] as shown in Eq. (1).

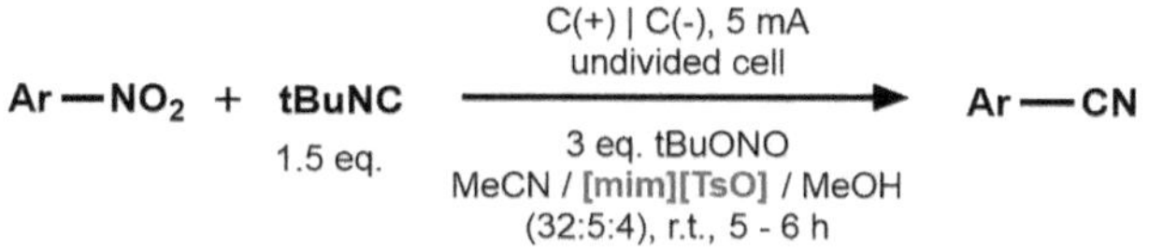

(1)

The Brønsted acidic IL, [dsim][Cl], facilitates sulfonation reaction of any aromatic moiety through the in situ production of H_2SO_4 at a temperature of 50 °C in a slight aqueous environment [26] as shown in Eq. (2).

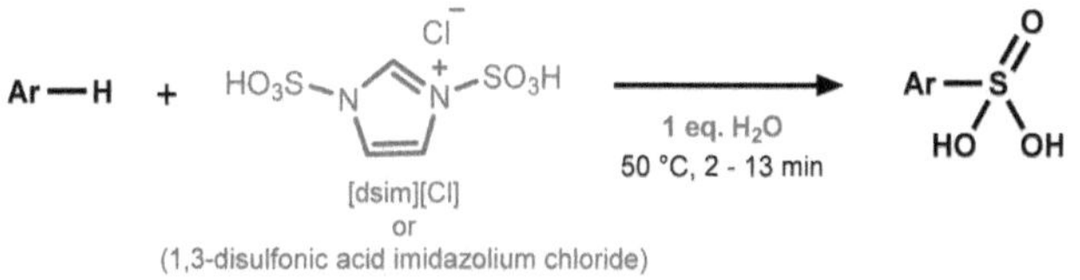

(2)

Table 3 ILs and their common uses

IL	Cation-anion	Type	Common uses in organic reactions
$[BMIM^+][BF_4^-]$	1-butyl-3-methylimidazolium tetrafluoroborate	Imidazolium-based IL	Solvent for Diels-Alder, alkylation, and transition-metal catalyzed reactions
$[BMIM^+][PF_6^-]$	1-butyl-3-methylimidazolium hexafluorophosphate	Imidazolium-based IL	Biphasic catalysis, hydrogenation, nucleophilic substitution
$[EMIM^+][BF_4^-]$	1-ethyl-3-methylimidazolium tetrafluoroborate	Imidazolium-based IL	Electrochemical applications, solvent in cycloaddition and oxidation reactions
$[EMIM^+][OAc^-]$	1-ethyl-3-methylimidazolium acetate	Imidazolium-based IL	Biomass dissolution, catalysis of esterification and transesterification
$[HMIM^+][cl^-]$	1-hexyl-3-methylimidazolium chloride	Imidazolium-based IL	Solvent for biocatalysis, polymerization
$[BMIM][NTf_2]$	1-butyl-3-methylimidazolium bis(trifluoromethylsulfonyl)imide	Hydrophobic IL	Acid/base-catalyzed reactions, fluorination, IL-based biphasic systems
[DBU][ac]	1,8-diazabicyclo[5.4.0]undec-7-ene acetate	Protic/basic IL	Knoevenagel condensation, Michael addition, green base catalyst
[choline][cl] + urea (deep eutectic solvent)	Choline chloride + urea	DES	Solvent for aldol condensation, esterification, biomass processing
$[P_{14,6,6,6}{}^+][cl^-]$	Trihexyl(tetradecyl)phosphonium chloride	Phosphonium-based IL	Phase-transfer catalyst, IL-based extraction systems
$[N_{4,4,4,4}{}^+][OH^-]$	Tetrabutylammonium hydroxide	Ammonium-based IL	Base-catalyzed reactions (e.g., aldol, Claisen-Schmidt), CO_2 capture

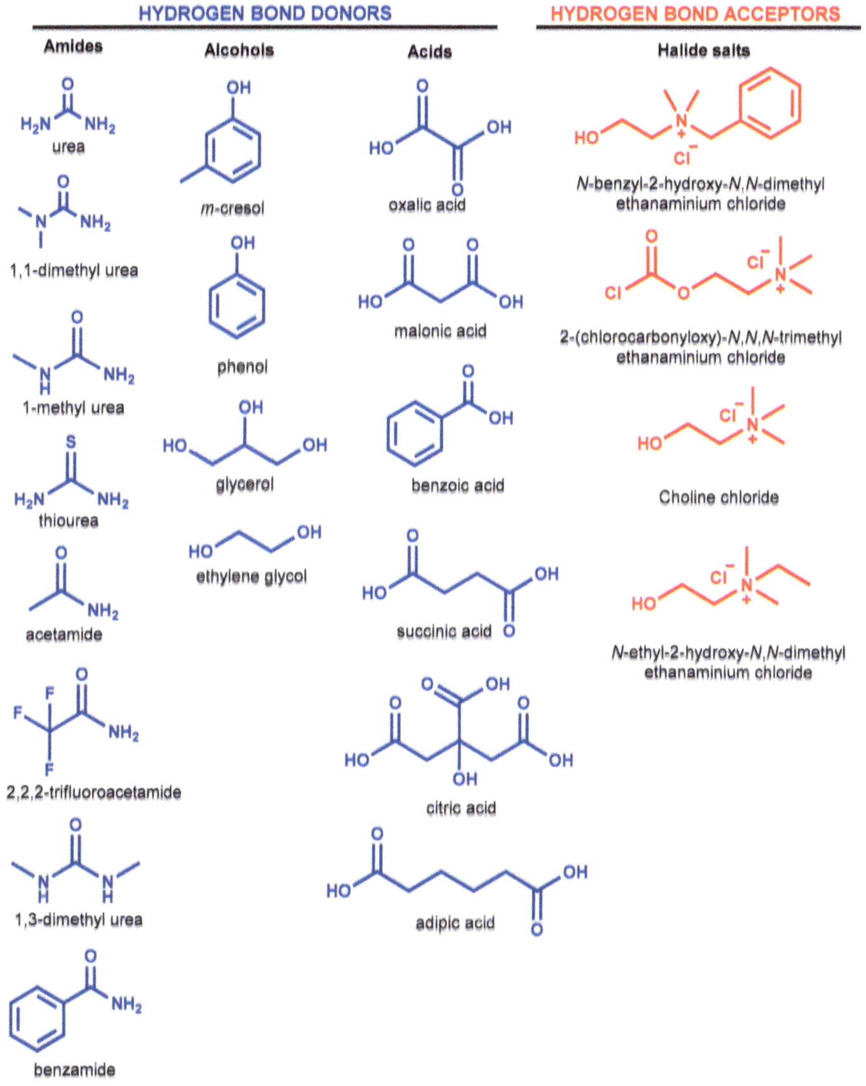

Fig. 2 Hydrogen bond donors and acceptors for ILs

Trifluoroacetyl nitrate ($CF_3CO_2NO_2$) and triflyl nitrate ($TfONO_2$) produced through metathesis in easily accessible ethylammonium nitrate (EAN) IL serve as highly effective electrophilic nitrating agents for various heteroaromatic and aromatic compounds. The relative studies on nitration reveal that EAN/Tf_2O outperforms EAN/TFAA in the nitration of significantly deactivated molecules [27]. For this, refer to Eq. (3).

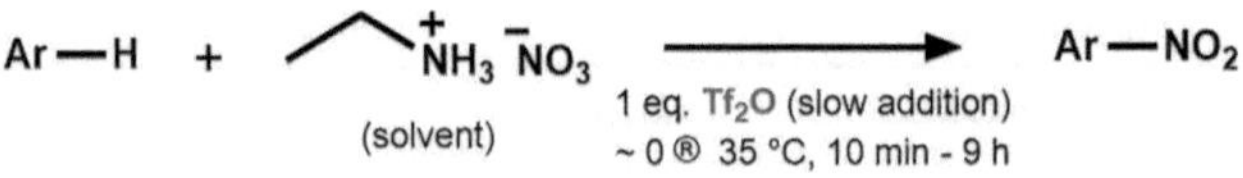

(3)

A collection of 3,4-dimethyl-3-phospholene 1-oxides, 1-alkoxy-3-methyl-3-phospholene 1-oxides, and 1-alkoxy-3-methylphospholane 1-oxides was synthesized with high production through a microwave-equipped transesterification process using IL from respective methyl or ethyl esters [28] as shown in Eq. (4).

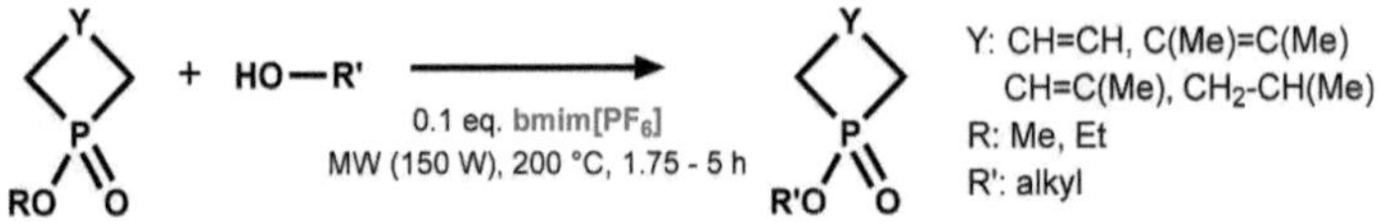

(4)

An organotelluride catalyst, supported by a recyclable IL, facilitates phosphite esters' aerobic oxidation into phosphate esters. The approach yields high conversion and enables the straightforward separation of resultant compound with added benefit of reusability [29] as shown in Eq. (5).

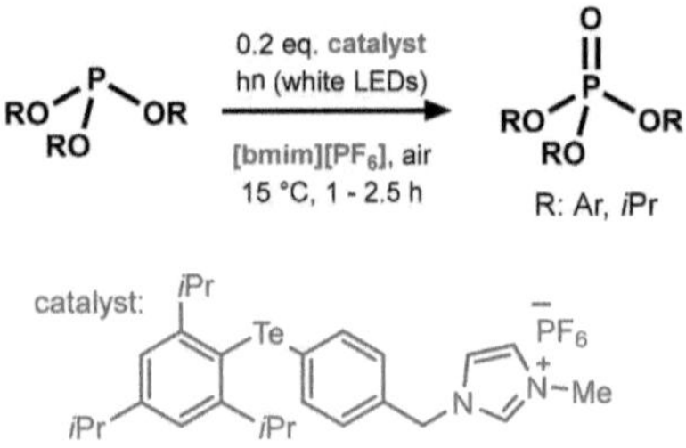

(5)

Alcohols were effectively transformed into alkyl halides through the utilization of 1-n-butyl-3-methylimidazolium halide ILs, in conjunction with Brønsted acids at normal conditions. Resulting alkyl halide compounds were readily separated from reaction mixture by means of straightforward decantation or extraction [30] as shown in Eq. (6).

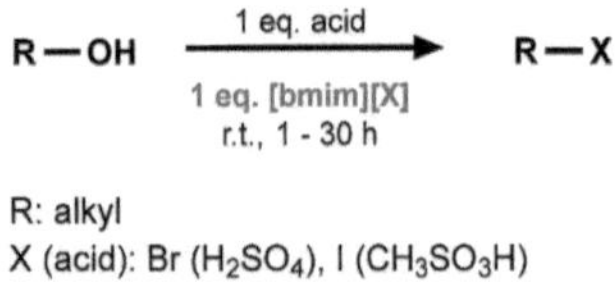

(6)

A straightforward microwave-aided condensation reaction of 2-aminothiophenol with aromatic aldehydes using cost-effective IL yields 2-arylbenzothiazoles without the need for solvents or catalysts. Furthermore, this IL is amenable to recycling for the future use [31] as shown in Eq. (7).

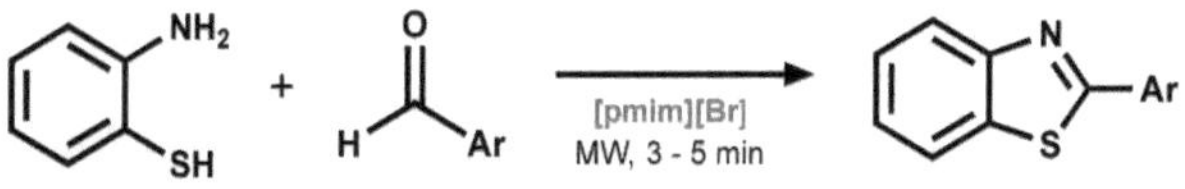

(7)

The H_2O_2 serves as the sustainable oxidizing reagent by facilitating hydroxylation of arylboronic acids effectively in an aerobic environment devoid of metals and bases using a room-temperature IL [32] as shown in Eq. (8).

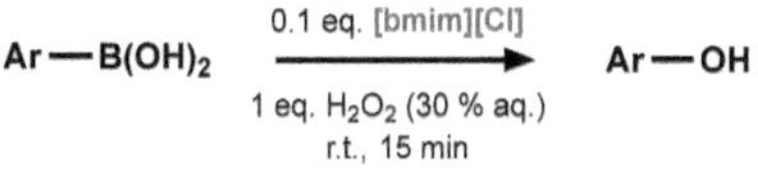

(8)

Another IL serves as a nitrating reagent. In this reaction, different arylboronic acids were subjected for the nitration in the presence of IL. It was found that IL was facilitating the nitro-Hunsdiecker process involving different α,β-unsaturated acid derivatives, resulting in the formation of diverse nitroarenes and nitroolefins, all achieved without the need for cocatalysts or solvents under mild reaction conditions [33] as shown in Eq. (9).

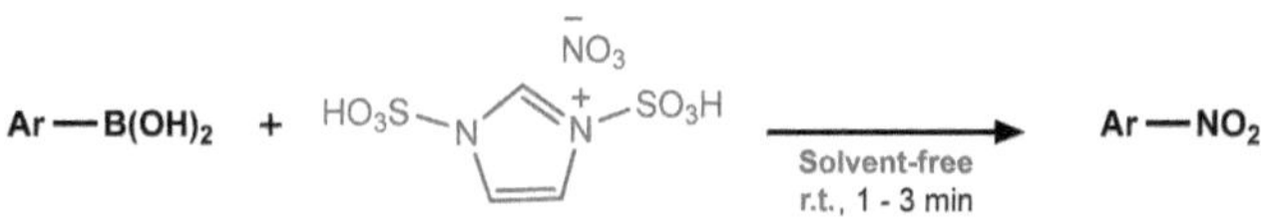

(9)

IL of the form [bmim][X] (where X represents I, Br, Cl, SCN, OAc) serves as an effective substance for nucleophilic substitution reactions involving sulfonate esters obtained using 1° and 2° alcohols (refer to Eq. (10)). This recently established methodology is notably ecological as it employs stoichiometric quantities of ILs as the exclusive reagents, eliminating the need for supplementary solvents and activating agents. Furthermore, these ILs are easily recyclable [34].

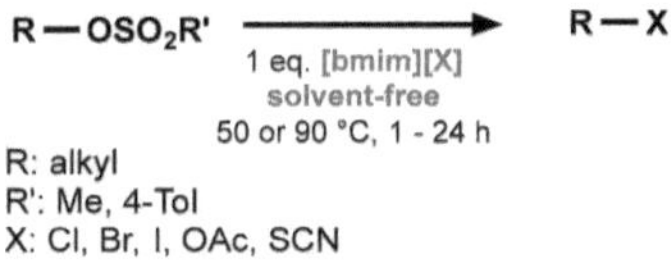

(10)

A one-pot practical and an efficient approach to synthesize the terminal vinylphosphonates through nitromethane, trialkyl phosphites, and aldehydes is achieved using the Henry-Michael reaction. This was then followed by the nitro elimination using 5-hydroxypentylammonium acetate as a task-specific IL resulting in favorable yields [35] as shown in Eq. (11).

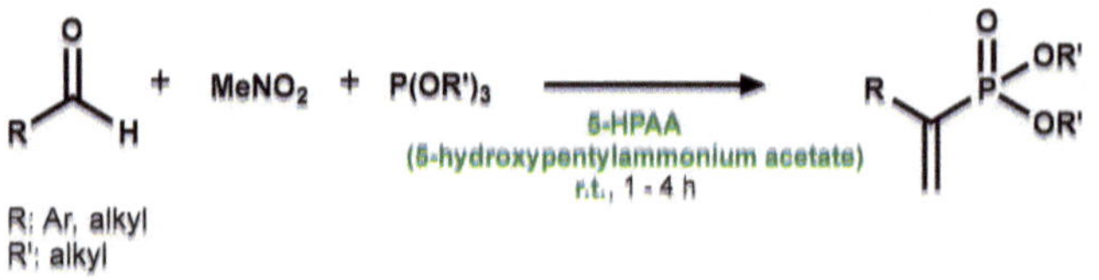

(11)

The research conducted by Jessop's team demonstrated the effective asymmetric hydrogenation reaction of tiglic acid (refer to Eq. (12)) and its predecessor molecule for anti-inflammatory medication of an ibuprofen (refer to Eq. (13)) catalyst [36]. The resultant compounds were isolated through $scCO_2$ extraction.

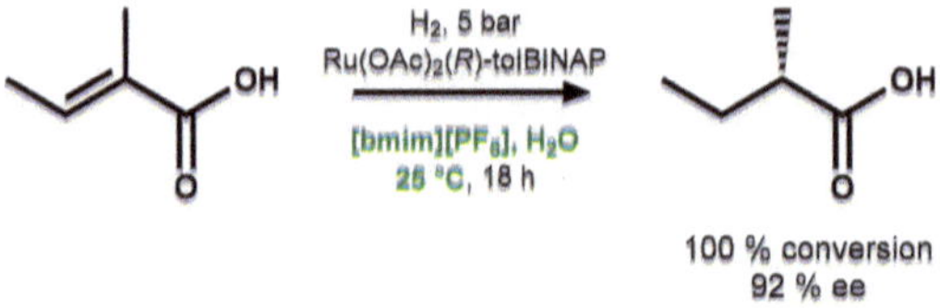

(12)

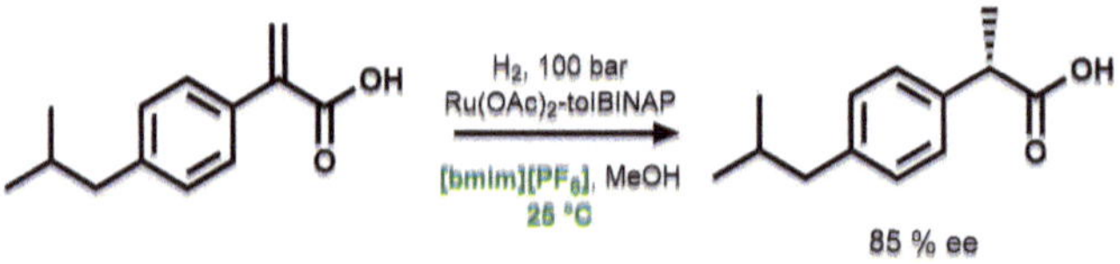

(13)

Despite the high stability of ILs and their assessment as solvents for the oxidation reactions, there is a startling lack of focus on conducting catalytic oxidation reactions. In that direction, another research involved the uses of asymmetric Jacobsen-Katsuki epoxidations using NaOCl present in [bmim][PF_6], facilitated by a chiral manganese complex known as Jacobsen's catalyst [37] as shown in Eq. (14).

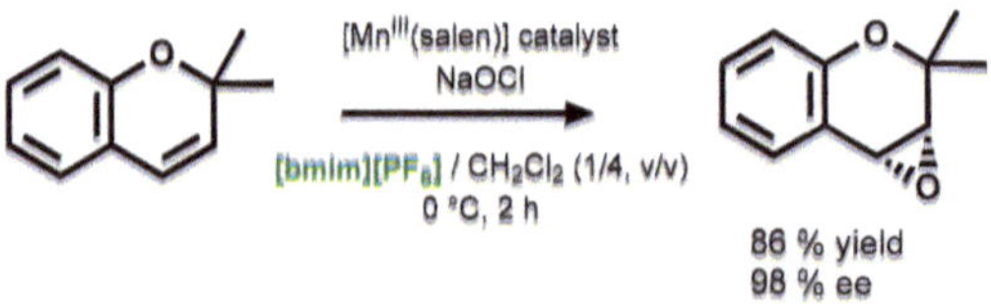

(14)

Another study showed the hydroformylation reaction of 1-octane catalyzed by platinum in RT IL [bmim][$SnCl_3$] [38] as shown in Eq. (15). The biphasic structure provided the benefit of easy product separation and facilitated the recovery of Pt catalyst.

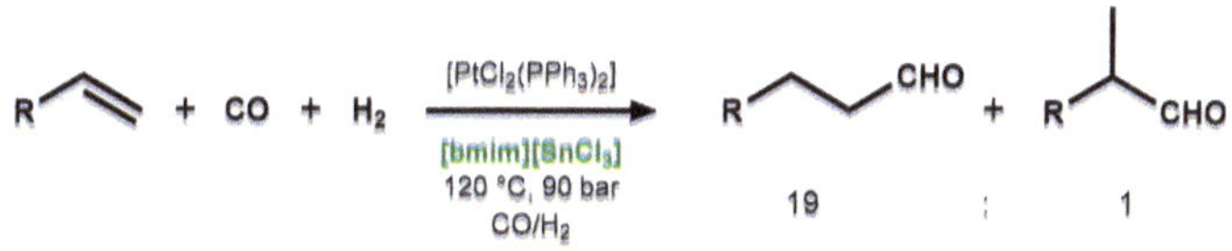

(15)

Also, the hydroformylation reaction of 1-pentene catalyzed by rhodium catalyst in RT IL [bmim][PF_6] (refer to Eq. (16)) can be accomplished in a greener way [39].

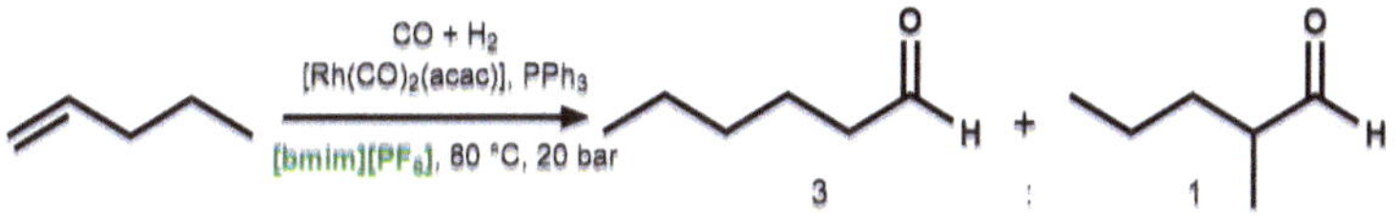

(16)

In another research, Heck couplings were accomplished using [bmim][PF_6] or n-hexylpyridinium hexafluorophosphate using $PdCl_2$ (refer to Eq. (17)) or $Pd(OAc)_2/Ar_3P$ (refer to Eq. (18)) as catalytic agents [40]. This process contains a workup procedure using the three-phase system comprising [bmim][PF_6], H_2O, and hexane. The obtained products exhibited solubility in the organic phase, whereas catalyst remained within the ionic layer. The by-product salt, [Hbase]X, was found to be soluble in water layer.

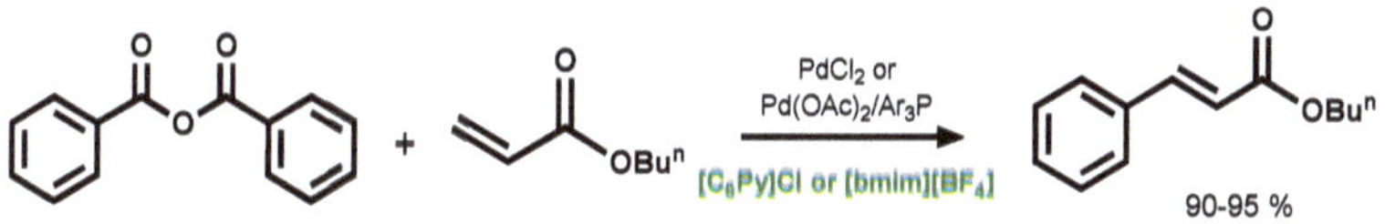

(17)

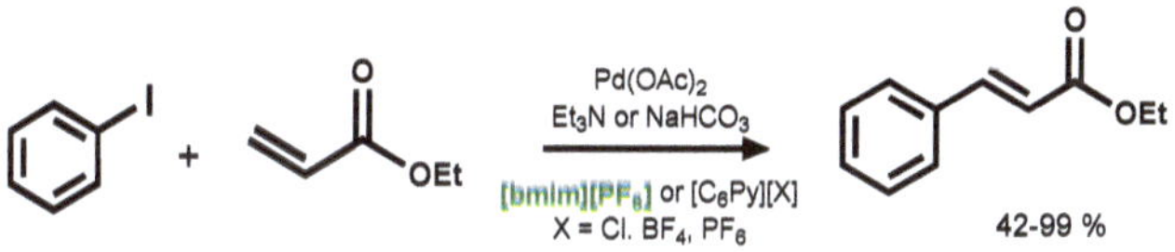

(18)

A cascade reaction, involving the reduction of sulfoxides and C–S bond metathesis with alkyl bromides in IL, yields valuable sulfides without any base as shown in Eq. (19). This method uses the traditional Kornblum oxidation in order to facilitate the sulfoxide reduction using RBr. The method is characterized by its exceptional acceptance for different groups, minor conditions for the reaction, potential for scalability, and the use of an environmentally friendly solvent [41].

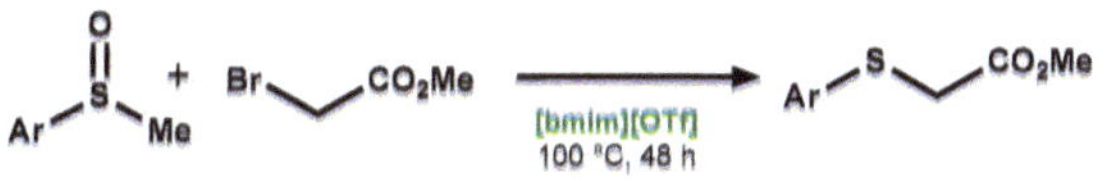

(19)

The protection of amine and carbonyl groups is crucial in the realm of organic synthesis. In an investigation, researchers protect the carbonyls through the formation of ketals or acetals using IL [42] as shown in Eqs. (20) and (21). The resultant product was efficiently isolated from the reaction mixture allowing for IL to be reused after the removal of moisture.

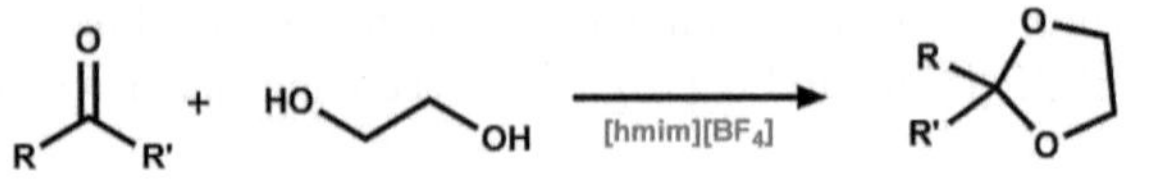

(20)

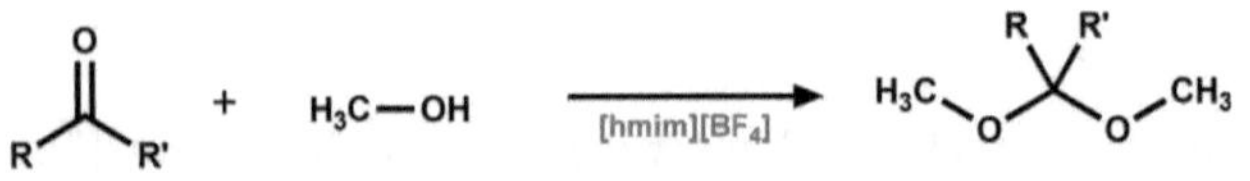

(21)

The Henry reaction is a highly valuable method for forming C–C bonds widely used in organic synthesis to produce 2-nitroalcohols through the interaction of carbonyl compounds with nitroalkanes using an alkaline catalyst as shown in Eq. (22). In this direction, researchers have established an effective method for the Henry reaction employing the IL [43].

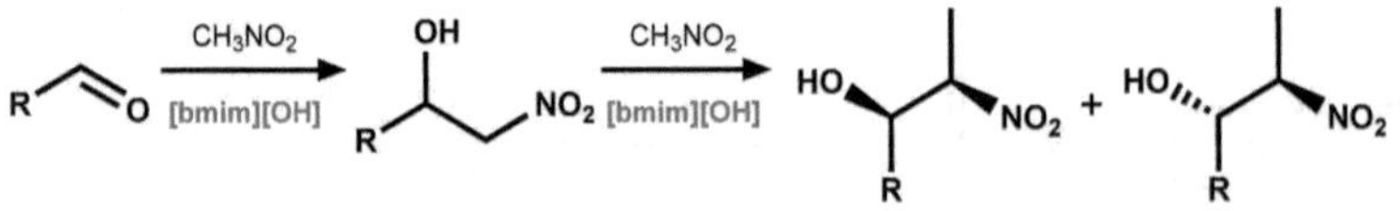

(22)

Furthermore, researchers have created an efficient and convenient method for the Markovnikov addition reaction to synthesize vinyl esters using N-heterocycle precursor in bmim[OH] IL eliminating the need for additional catalysts or organic solvents [44] (refer to Eq. (23)).

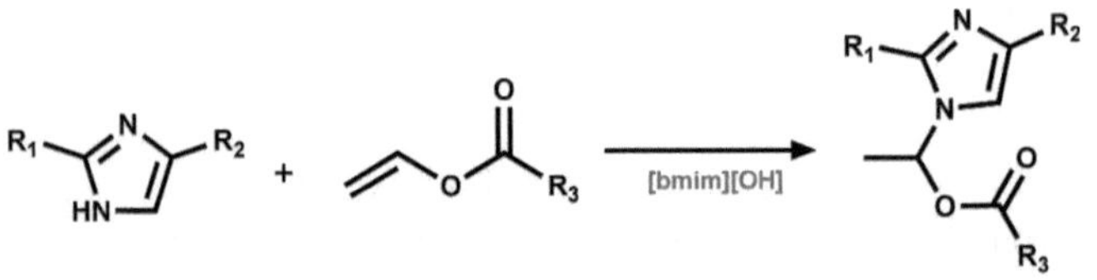

(23)

Furthermore, researchers have established an effective protocol for synthesizing acetamides by the use of CaL B in an IL which serves as an environmentally friendly medium [45]. This method is suitable for a diverse range of acetic acids, aliphatic esters, and amines, yielding outstanding results in the production of the corresponding acetamides. The catalyst demonstrates exceptional activity and can be reused for as many as four successive cycles as shown in Eq. (24).

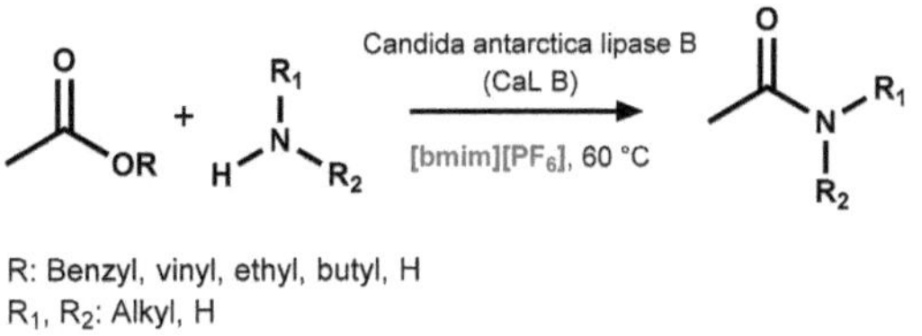

(24)

The IL served both as a solvent and as a catalyst facilitating the effective transformation of aldehydes and ketones into amines with high yields and minimal side product formation. The described method was characterized by its mild conditions, efficiency, and its user- and environmentally friendly nature (refer to Eq. (25)).

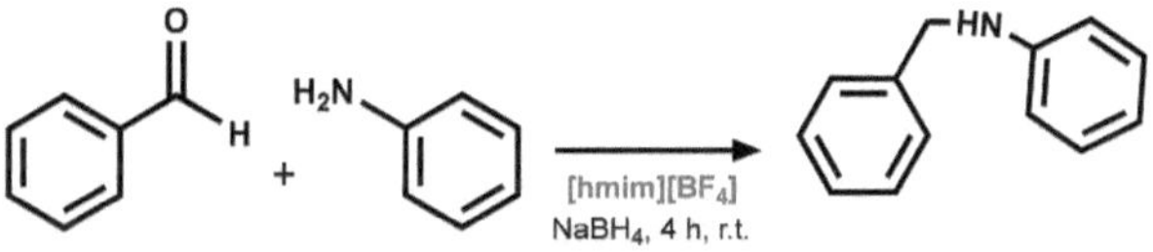

(25)

The Friedlander reaction or annulation process includes the interaction between o-aminoarylaldehyde and various methylene compounds to synthesize the quinoline. In that direction, researchers have investigated the application of BAILs as environmentally friendly media and catalysts to develop active quinolines and other associated polyheterocycles employing the Friedlander heteroannulation method without the inclusion of any additional catalyst [46] as shown in Eq. (26).

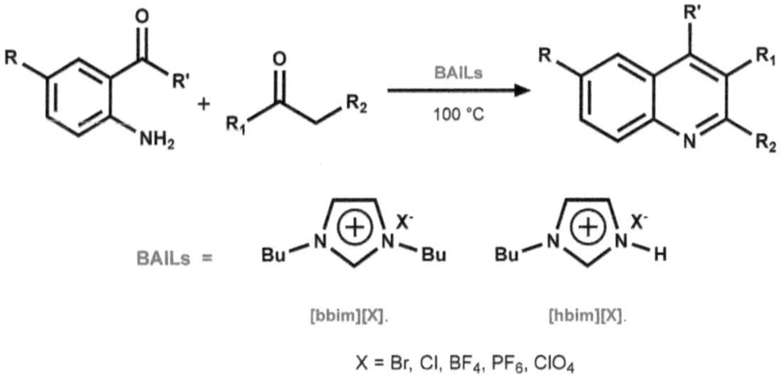

(26)

In addition, researchers examined the effectiveness of other supported IL phase catalysts. This catalyst was used to carry out hydroformylation in a regioselective manner using H_2O as medium [47]. Following this, the hydroformylation products underwent hydrogenation with a Ru/PPh_3/SILP catalyst resulting in the formation of 1,4-butanediol. This heterogenized homogeneous catalyst system demonstrated exceptional conversion rates and selectivity for n-aldehyde (refer to Eq. (27)).

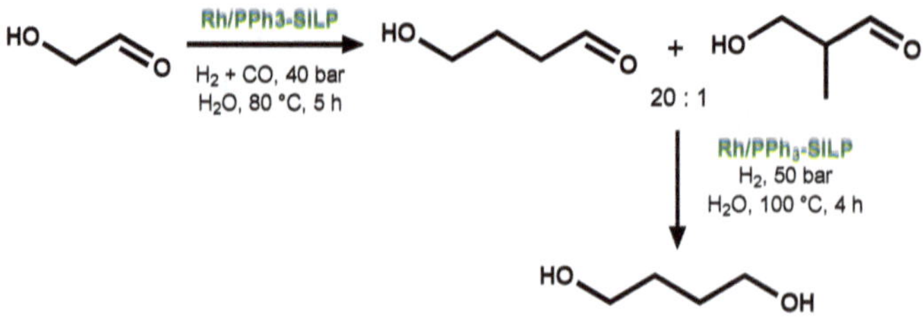

(27)

The process of Pd-catalyzed allylic substitution using carbon nucleophiles represents a valuable synthetic approach to form carbon-carbon bonds. Such reactions were conducted in the ILs employing monophasic and biphasic systems using [bmim][BF_4] IL [48] as shown in Eq. (28).

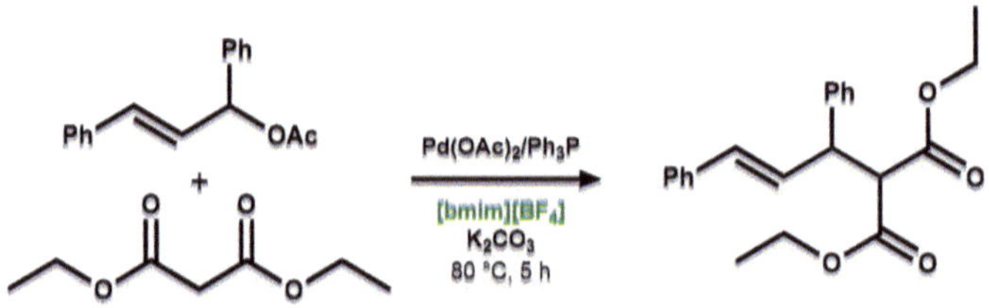

(28)

The same reaction was also achieved in ILs employing monophasic and biphasic systems using another IL [bmim][Cl]-methylcyclohexane [49], as shown in Eq. (29).

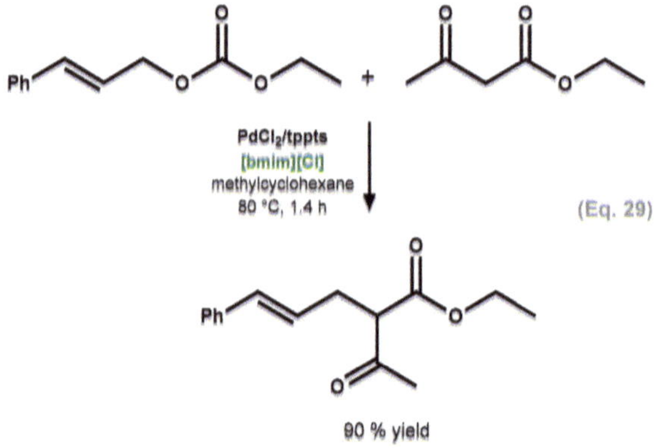

(29)

Recent research uses Pickering emulsion droplets for a continuous cyanosilylation reaction of ketones. The catalysts were immobilized through emulsification which exhibited remarkable stability throughout the continuous process. This methodology employed two ILs; [C_4mim][Cl] served as a catalyst encapsulated within the [C_4mim][BF_4] droplets which acted as the medium [50] as shown in Eq. (30). Following the optimization of continuous-flow parameters, various ketones were applied as precursors to yield corresponding products with outstanding efficiency.

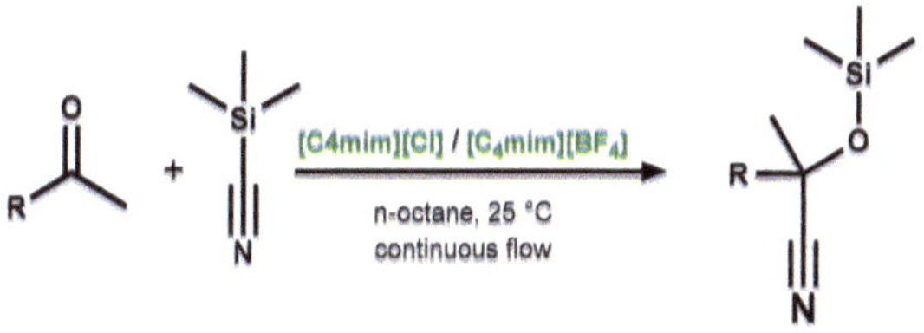

(30)

In another research, a continuous process using a water/IL biphasic system was developed [51] (refer to Eq. (31)). The study explored various hydrophobic ILs based on [C4mim] which are known to create biphasic systems with water. This configuration facilitated the effective parting and the catalyst recycling. Furthermore, desired compounds were achieved with an impressive yield of 90% at a temperature of 60 °C within half an hour.

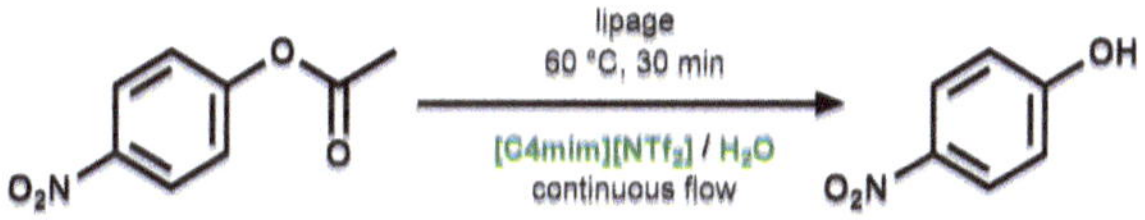

(31)

7 Conclusion

ILs constitute a highly promising category of solvents, exhibiting an extensive array of uses in the organic synthesis. Their distinctive characteristics present opportunities to tackle the limitations of conventional solvents and advance the establishment of sustainable synthetic methodologies. Ongoing research in the domain is critical to realizing complete capabilities of ILs and enabling their incorporation into standard practices of organic synthesis. Due to their capacity to liquefy various metal complexes, these liquids are increasingly being used in current years for numerous reactions, thereby improving both reaction rates and the selectivity. While they were originally employed mainly as solvents, their applications are now expanding into diverse areas such as catalysis, electrochemistry, spectroscopy, and materials science. ILs meet the criteria of green chemistry. The continuous investigation into environmentally friendly, biodegradable ILs is decisive for their complete integration with the principles of a circular economy. Their connection to the circular economy is established through various avenues that enhance sustainability, optimize resource use, and minimize waste.

8 Challenges and Future Scopes

Despite their considerable promise, the broad implementation of ILs in organic synthesis encounters numerous obstacles, including issues related to cost, toxicity, and scalability in the synthesis and recovery processes. Additional research is essential to overcome these challenges and refine the properties of ILs for targeted applications. Prospective advancements in the domain of ILs for organic reactions may encompass: (1) The creation of innovative ILs with improved characteristics, such as reduced viscosity, enhanced recyclability, and minimized toxicity. (2) The incorporation of ILs into continuous-flow methodologies to boost efficiency, scalability, and process management. (3) The investigation of synergistic interactions between ILs and alternative solvent systems to improve reaction results and selectivity. (4) The assessment of the environmental implications and life cycle analysis of IL-based methodologies to guarantee their sustainability.

Most research efforts have focused on N-alkylimidazolium, N-alkylpyridinium, N-alkyl thiazolium, tetraalkylammonium, and N,N-dialkyl pyrazolium cations paired with poorly coordinating anions, i.e., PF_6^-, BF_4^-, $(F_3CSO_2)_2N^-$, $F_3CSO_3^-$, $CF_3SO_3^-$, and $ZnCl_3^-$. Numerous simulations and experimental investigations primarily address the structure of imidazolium-based ILs revealing that these function as micro-biphasic consisting of both polar and nonpolar areas [52]. Nonetheless, the application of these ILs presents several drawbacks including toxicity due to potential leaching into soil or water bodies which could lead to persistent pollution and environmental hazards, as well as cost issues that render them less feasible for large-scale industrial uses such as metal electroplating, electrodeposition, and biocatalysis. Consequently, the primary challenges in the research application of conventional ILs revolve around their availability and associated costs. Various concerns including toxicity and accessibility will restrict their practical implementation in larger-scale applications involving other metals and biomaterials. Although numerous protocols for synthesizing traditional ILs have been documented, not all laboratories engaged in applied research possess the requisite expertise, operational practices, and equipment to perform synthesis. Furthermore, achieving pure, dried traditional ILs or conducting post-synthesis purification can often be problematic. Additionally, the commercial availability of certain traditional ILs is limited to small quantities and the pricing of many of these liquids remains prohibitive for applied engineering research [53–55].

References

1. Y. Xiong, Y. Huang, K. Du, Health risk-oriented source apportionment of hazardous volatile organic compounds in eight Canadian cities and implications for prioritizing mitigation strategies. Environ. Sci. Technol. **56**(17), 12077–12085 (2022). https://doi.org/10.1021/acs.est.2c02558
2. S. Panda, S. Gorantla, Green analytical approaches and eco-friendly solvents: Advancing industrial applications and environmental sustainability: A comprehensive review. Orient. J. Chem. **41**(2), 616–627 (2025). https://doi.org/10.13005/ojc/410231
3. M. Shaibuna, L.V. Theresa, K. Sreekumar, Neoteric deep eutectic solvents: History, recent developments, and catalytic applications. Soft Matter **18**(14), 2695–2721 (2022). https://doi.org/10.1039/D1SM01797G

4. N. Winterton, The green solvent: A critical perspective. Clean Techn. Environ. Policy **23**(9), 2499–2522 (2021). https://doi.org/10.1007/s10098-021-02188-8
5. P.T. Anastas, T.C. Williamson, *Green Chemistry: Frontiers in Benign Chemical Syntheses and Processes* (Oxford University Press, Oxford, 1998). https://books.google.co.in/books/about/Green_Chemistry.html?id=OrNTAAAAMAAJ&redir_esc=y#:~:text=This%20topical%20volume%20provides%20an%20in-depth%20profile%20of,likely%20future%20developments%20in%20environmentally%20benign%20chemical%20synthesis. Accessed 15 May 2025
6. D.J. Adams, P.J. Dyson, S.J. Tavener, *Chemistry in Alternative Reaction Media* (Wiley, Hoboken, 2003). https://doi.org/10.1002/0470869666
7. P. Pollet, E.A. Davey, E.E. Ureña-Benavides, C.A. Eckert, C.L. Liotta, Solvents for sustainable chemical processes. Green Chem. **16**(3), 1034–1055 (2014). https://doi.org/10.1039/C3GC42302F
8. W. Peter, W. Thomas, Front matter and index, in *Ionic Liquids in Synthesis*, ed. by W. Peter, W. Thomas, 2nd edn., (Wiley, Hoboken, 2008), pp. 1–776. https://doi.org/10.1002/3527600701.fmatter_indsub
9. T. Welton, Room-temperature ionic liquids. Solvents for synthesis and catalysis. Chem. Rev. **99**(8), 2071–2084 (1999). https://doi.org/10.1021/cr980032t
10. R. Salas et al., Ionic liquids in polymer technology. Green Chem. **27**(6), 1620–1651 (2025). https://doi.org/10.1039/D4GC05445H
11. V. Kumar, A. Singh, Polypropylene clay nanocomposites. Rev. Chem. Eng. **29**(6), 14 (2013). https://doi.org/10.1515/revce-2013-0014
12. A. Singh, S.L. Banerjee, K. Kumari, P.P. Kundu, Recent innovations in chemical recycling of polyethylene terephthalate waste: A circular economy approach toward sustainability, in *Handbook of Solid Waste Management*, (Springer, Singapore, 2022), pp. 1149–1176. https://doi.org/10.1007/978-981-16-4230-2_53
13. R. Cernansky, Chemistry: Green refill. Nature **519**(7543), 379–380 (2015). https://doi.org/10.1038/nj7543-379a
14. T.A. Paul, C.W. John, *Green Chemistry: Theory and Practice* (Oxford University Press, New York, 2000). https://global.oup.com/academic/product/green-chemistry-9780198506980?lang=en&cc=us#. Accessed 14 May 2025
15. K. Van Aken, L. Strekowski, L. Patiny, EcoScale, a semi-quantitative tool to select an organic preparation based on economical and ecological parameters. Beilstein J. Org. Chem. **2**, 3 (2006). https://doi.org/10.1186/1860-5397-2-3
16. D. Bégin, C. Couture, M. Gérin, M. Debia, *Solvants verts: fondements, santé, sécurité, environnement et substitution* (Institut de Recherche Robert-Sauvé en Santé et en Sécurité du Travail, Montréal, 2020). https://search.worldcat.org/en/title/1162187204. Accessed 14 May 2025
17. Anonymous, 3.2 tools of green chemistry. Bharathidasan University, Tiruchirappalli. https://www.bdu.ac.in/cde/SLM/M.Sc.%20Chemistry/Chemistry%20II%20Year/Gree%20Chemistry%20&%20Nono%20Science/Unit%203.2%20Green%20Chemistry.pdf. Accessed 14 May 2025
18. B. Torok, T. Dransfield, *Green Chemistry: An Inclusive Approach* (Elsevier, Boston, 2017). https://books.google.co.in/books/about/Green_Chemistry.html?id=i00ADQAAQBAJ&redir_esc=y. Accessed 14 May 2025
19. J. Sherman, B. Chin, P.D.T. Huibers, R. Garcia-Valls, T.A. Hatton, Solvent replacement for green processing. Environ. Health Perspect. **106**, 253 (1998). https://doi.org/10.2307/3433925
20. D. Prat et al., Sanofi's solvent selection guide: A step toward more sustainable processes. Org. Process. Res. Dev. **17**(12), 1517–1525 (2013). https://doi.org/10.1021/op4002565
21. V. Isoni, L.L. Wong, H.H. Khoo, I. Halim, P. Sharratt, Q-SA$\surd$ESS: A methodology to help solvent selection for pharmaceutical manufacture at the early process development stage. Green Chem. **18**(24), 6564–6572 (2016). https://doi.org/10.1039/C6GC02440H
22. C.J. Clarke, W.-C. Tu, O. Levers, A. Bröhl, J.P. Hallett, Green and sustainable solvents in chemical processes. Chem. Rev. **118**(2), 747–800 (2018). https://doi.org/10.1021/acs.chemrev.7b00571

23. P.G. Jessop, Green/alternative solvents, in *Encyclopedia of Sustainable Technologies*, (Elsevier, Amsterdam, 2017), pp. 611–619. https://doi.org/10.1016/B978-0-12-409548-9.10253-2
24. M. Cvjetko Bubalo, S. Vidović, I. Radojčić Redovniković, S. Jokić, Green solvents for green technologies. J. Chem. Technol. Biotechnol. **90**(9), 1631–1639 (2015). https://doi.org/10.1002/jctb.4668
25. Y. Chen et al., Electrochemically driven denitrative cyanation of nitroarenes. Org. Lett. **26**(36), 7555–7559 (2024). https://doi.org/10.1021/acs.orglett.4c02552
26. A. Moosavi-Zare, M. Zolfigol, E. Noroozizadeh, Regioselective sulfonation of aromatic compounds over 1,3-disulfonic acid imidazolium chloride under aqueous media. Synlett **27**(11), 1682–1684 (2016). https://doi.org/10.1055/s-0035-1561938
27. G. Aridoss, K.K. Laali, Ethylammonium nitrate (EAN)/Tf_2O and EAN/TFAA: Ionic liquid based systems for aromatic nitration. J. Org. Chem. **76**(19), 8088–8094 (2011). https://doi.org/10.1021/jo201374a
28. N. Harsági, N.Z. Kiss, G. Keglevich, P-chloride-free synthesis of phosphoric esters: Microwave-assisted esterification of alkyl- and dialkyl phosphoric ester-acids obtained from phosphorus pentoxide. Synthesis **54**(13), 3047–3054 (2022). https://doi.org/10.1055/a-1811-8586
29. A. Mihoya, Y. Shibuya, A. Ito, A. Toyoda, M. Oba, S. Koguchi, Aerobic oxidation of phosphite esters to phosphate esters by using an ionic-liquid-supported organotelluride reusable catalyst. Synlett **31**(20), 2043–2045 (2020). https://doi.org/10.1055/s-0040-1706068
30. R.X. Ren, J.X. Wu, Mild conversion of alcohols to alkyl halides using halide-based ionic liquids at room temperature. Org. Lett. **3**(23), 3727–3728 (2001). https://doi.org/10.1021/ol016672r
31. B.C. Ranu, R. Jana, S.S. Dey, An efficient and green synthesis of 2-arylbenzothiazoles in an ionic liquid, [pmIm]Br under microwave irradiation. Chem. Lett. **33**(3), 274–275 (2004). https://doi.org/10.1246/cl.2004.274
32. E.-J. Shin, G.-T. Kwon, S.-H. Kim, Room-temperature ionic liquids (RTILs) as green media for metal- and base-free ipso-hydroxylation of arylboronic acids. Synlett **30**(15), 1815–1819 (2019). https://doi.org/10.1055/s-0037-1611894
33. M. Zarei, E. Noroozizadeh, A.R. Moosavi-Zare, M.A. Zolfigol, Synthesis of nitroolefins and nitroarenes under mild conditions. J. Org. Chem. **83**(7), 3645–3650 (2018). https://doi.org/10.1021/acs.joc.7b03289
34. Y. Liu, Y. Xu, S. Jung, J. Chae, A facile and green protocol for nucleophilic substitution reactions of sulfonate esters by recyclable ionic liquids. Synlett **23**(18), 2692–2698 (2012). https://doi.org/10.1055/s-0032-1317473
35. S. Sobhani, M. Honarmand, A simple and efficient method for one-pot, three-component synthesis of terminal vinylphosphonates using a task-specific ionic liquid. Synlett **24**(2), 236–240 (2013). https://doi.org/10.1055/s-0032-1317965
36. R.A. Brown, P. Pollet, E. McKoon, C.A. Eckert, C.L. Liotta, P.G. Jessop, Asymmetric hydrogenation and catalyst recycling using ionic liquid and supercritical carbon dioxide. J. Am. Chem. Soc. **123**(6), 1254–1255 (2001). https://doi.org/10.1021/ja005718t
37. C.E. Song, E.J. Roh, Practical method to recycle a chiral (salen)Mn epoxidation catalyst by using an ionic liquid. Chem. Commun. **10**, 837–838 (2000). https://doi.org/10.1039/b001403f
38. P. Wasserscheid, H. Waffenschmidt, Ionic liquids in regioselective platinum-catalysed hydroformylation. J. Mol. Catal. A Chem. **164**(1-2), 61–67 (2000). https://doi.org/10.1016/S1381-1169(00)00259-4
39. Y. Chauvin, L. Mussmann, H. Olivier, A novel class of versatile solvents for two-phase catalysis: Hydrogenation, isomerization, and hydroformylation of alkenes catalyzed by rhodium complexes in liquid 1,3-dialkylimidazolium salts. Angew. Chem. Int. Ed. Engl. **34**(23-24), 2698–2700 (1996). https://doi.org/10.1002/anie.199526981
40. A.J. Carmichael, M.J. Earle, J.D. Holbrey, P.B. McCormac, K.R. Seddon, The heck reaction in ionic liquids: A multiphasic catalyst system. Org. Lett. **1**(7), 997–1000 (1999). https://doi.org/10.1021/ol9907771

41. C. Liu, D. Chen, Y. Fu, F. Wang, J. Luo, S. Huang, Sulfoxide reduction/C(sp^3)–S metathesis cascade in ionic liquid. Org. Lett. **22**(14), 5701–5705 (2020). https://doi.org/10.1021/acs.orglett.0c02089
42. H.-H. Wu, F. Yang, P. Cui, J. Tang, M.-Y. He, An efficient procedure for protection of carbonyls in Brønsted acidic ionic liquid [Hmim]BF4. Tetrahedron Lett. **45**(25), 4963–4965 (2004). https://doi.org/10.1016/j.tetlet.2004.04.138
43. H. Wu, F. Zhang, Y. Wan, L. Ye, An efficient protocol for henry reaction using basic ionic liquid [bmIm]OH as catalyst and reaction medium. Lett. Org. Chem. **5**(3), 209–211 (2008). https://doi.org/10.2174/157017808783955817
44. J.-M. Xu, B.-K. Liu, W.-B. Wu, C. Qian, Q. Wu, X.-F. Lin, Basic ionic liquid as catalysis and reaction medium: A novel and green protocol for the markovnikov addition of N-heterocycles to vinyl esters, using a task-specific ionic liquid, [bmIm]OH. J. Org. Chem. **71**(10), 3991–3993 (2006). https://doi.org/10.1021/jo0600914
45. K.P. Dhake, Z.S. Qureshi, R.S. Singhal, B.M. Bhanage, Candida Antarctica lipase B-catalyzed synthesis of acetamides using [BMIm(PF6)] as a reaction medium. Tetrahedron Lett. **50**(23), 2811–2814 (2009). https://doi.org/10.1016/j.tetlet.2009.03.170
46. S.S. Palimkar, S.A. Siddiqui, T. Daniel, Rajgopal, J. Lahoti, K.V. Srinivasan, Ionic liquid-promoted regiospecific friedlander annulation: Novel synthesis of quinolines and fused polycyclic quinolines. J. Org. Chem. **68**(24), 9371–9378 (2003). https://doi.org/10.1021/jo035153u
47. A.G. Panda, M.D. Bhor, S.R. Jagtap, B.M. Bhanage, Selective hydroformylation of unsaturated esters using a Rh/PPh3-supported ionic liquid-phase catalyst, followed by a novel route to pyrazolin-5-ones. Appl. Catal. A Gen. **347**(2), 142–147 (2008). https://doi.org/10.1016/j.apcata.2008.06.008
48. J. Ross, W. Chen, L. Xu, J. Xiao, Ligand effects in palladium-catalyzed allylic alkylation in ionic liquids. Organometallics **20**(1), 138–142 (2001). https://doi.org/10.1021/om000712y
49. C. de Bellefon, E. Pollet, P. Grenouillet, Molten salts (ionic liquids) to improve the activity, selectivity and stability of the palladium catalysed Trost–Tsuji C–C coupling in biphasic media. J. Mol. Catal. A Chem. **145**(1-2), 121–126 (1999). https://doi.org/10.1016/S1381-1169(99)00032-1
50. Z. Meng, M. Zhang, H. Yang, Pickering emulsion droplets hosting ionic liquid catalysts for continuous-flow cyanosilylation reaction. Green Chem. **21**(3), 627–633 (2019). https://doi.org/10.1039/C8GC03585G
51. Q. Deng, N.N. Tran, M. Razi Asrami, L. Schober, H. Gröger, V. Hessel, Ionic liquid/water continuous-flow system with compartmentalized spaces for automatic product purification of biotransformation with enzyme recycling. Ind. Eng. Chem. Res. **59**(48), 21001–21011 (2020). https://doi.org/10.1021/acs.iecr.0c02785
52. L.M.N.B.F. Santos et al., Ionic liquids: First direct determination of their cohesive energy. J. Am. Chem. Soc. **129**(2), 284–285 (2007). https://doi.org/10.1021/ja067427b
53. J.-P.T. Mikkola, P.P. Virtanen, K. Kordás, H. Karhu, T.O. Salmi, SILCA—Supported ionic liquid catalysts for fine chemicals. Appl. Catal. A Gen. **328**(1), 68–76 (2007). https://doi.org/10.1016/j.apcata.2007.05.030
54. K. Kant, P. Naik, N.A. Jyoti, C.C. Malakar, Ionic liquids in C–H activation: synthesis and functionalization of heterocycles and carbocycles. Org. Biomol. Chem. **23**(18), 4260–4305 (2025). https://doi.org/10.1039/D4OB02109F
55. X. Xue et al., Synthesis of poly(ionic liquid)s-functionalized carbon dots with superior tribological stability and potential-controlled friction effect under oil lubrication. Langmuir (2025). https://doi.org/10.1021/acs.langmuir.5c01309

Bionanofluids: Introduction, Properties, Characterization, and Applications

Sheetal S. Samant, Shruti V. Kolambkar, Faizan A. Beerwala, Pramod J. Hurkadale, and Vijay A. Jagtap

1 Introduction

1.1 Definition and Concept of Bionanofluids

Bionanofluids are multifaceted systems that include carbon-based nanomaterials with well-tuned surface characteristics, which allow excellent dispersion in different fluids. These nanofluids may also contain hybrid nanoparticles (NPs) or polymeric nanostructures to enhance the targeting specificity of the nanofluid to particular biological entities. Bionanofluids, containing suspended nanoparticles smaller than 100 nm, have applications in various domains. From the medical point of view, these fluids are promising in medical imaging, therapeutic applications, and mass transfer processes. The diversity of bionanofluids can be realized in the applications including medical imaging, disinfection, and therapeutic procedure having implications in both healthcare and industrial sectors [1]. The thermophysical properties of bionanofluids are markedly different when in contact with biological fluids, such as blood, and can contribute to an increased heat transfer and fluid dynamic in the cardiovascular system. The pioneering work of nanofluids by Choi in 1995 has opened the doors for exploring their applications in advanced heat transfer operations. Nanofluids enable localized delivery of therapeutic agents and also assist in producing implantable medical devices. This again highlights the multidimensional characteristics of nanofluids beyond typical engineering areas. In addition, nanofluids have a great potential in mass transfer processes enhancement

S. S. Samant (✉) · V. A. Jagtap
Yashwantrao Bhonsale College of Pharmacy, Sawantwadi, Maharashtra, India

S. V. Kolambkar · F. A. Beerwala · P. J. Hurkadale
KLE College of Pharmacy, JNMC Campus, Belagavi, Karnataka, India

S. Kulkarni, S. Das (eds.), *Ionic Liquids, Nanofluids and Nanotechnology*,
Lecture Notes in Nanoscale Science and Technology 42,
https://doi.org/10.1007/978-3-032-04008-4_6

and they are suitable to be applied in heat exchangers and energy systems [2]. The carbon nanomaterials dispersed in a fluid can be denoted as a bionanofluid. For the compatibility of such particles, their surfaces are modified by polar groups for polar fluids and non-polar groups for non-polar fluids. They are also functionalized with targeting ligands for specific conjugation to selected biological entities. Hybrid bionanofluids go one step ahead by including hybrid nanoparticles including alloys, transition metals, semiconductors, semi-metals, and polymers, also functionalized for selective ligand-receptor binding. Such advanced fluids can be incorporated into a variety of formulations, such as hydrogels, foams, creams, sprays, and dried materials, to provide new applications. Unique features of bionanofluid and hybrid bionanofluids used in multimodal imaging, such as photoluminescence, luminescence, photoacoustic imaging, MRI, and ultrasound, are also beneficial compared to traditional contrast agents [1]. Bionanofluids promise a remarkable potential in enhancing mass transfer operations. The significant improvement in transport efficiency may be due to complex particles and fluid interaction, which may be facilitated by dynamic motion [3].

1.2 Historical Background and Evolution of Bionanofluids

Bionanofluids, being an exclusive category of nanofluids, utilize a biological ingredient for enhancing their properties that impart exclusive advantages over traditional nanofluids. Despite the efforts of both types of fluids to improve heat transfer characteristics, bionanofluids experience unique phenomena, such as bioconvection, that can greatly enhance mass and heat transfer. It is this property that makes them very attractive for high surface area applications that require optimal heat and mass transfer. The most interesting of all is that the viscosity of bionanofluids is generally lower than that of conventional nanofluids, improving the rheological property and heat transfer performance. Conversely, traditional nanofluids are often found to have limitations in their thermophysical properties, leading to limited heat transfer rates. A thorough screening indicates that bionanofluids due to their innovative composition may potentially contribute higher thermal conductivities and stabilized applications, and hence, they can be regarded as an effective alternative to the nanofluids in various thermal engineering applications [4]. Bionanofluids can be considered as a significant hybrid of nanotechnology and medicine. The idea of nanofluid was introduced in the late twentieth century to enhance thermal properties with nanoparticles dispersing into the base fluid. With further scientific development, a variety of nanoparticles such as metallic, oxide, and carbon-based nanoparticles have been proposed to enhance fluid properties [5]. Subsequent studies resulted in the development of bionanofluids, in which the nanoparticles were integrated into a biological environment via complex interactions (e.g., dynamic activity of microorganisms) to enhance the efficiency of mass transfer [3].

2 Synthesis and Stabilization of Bionanofluids

2.1 Steps 1: Nanoparticle Synthesis

Green Synthesis (Biogenic Methods)

It uses biological systems (plants, microorganisms, and fungi) for synthesizing nanoparticles. This technique can thus decrease the use of dangerous chemicals and the associated environmental damage. Different biological agents are used in green synthesis, such as plant extracts and microorganisms. Phytochemicals present in plants aid in the reduction of metal ions to nanoparticles. It is a relatively inexpensive, rapid, and easy approach to be used in the synthesis of nanoparticles, resulting in stability of the particles. Microorganisms such as bacteria, fungi, and algae may serve as biological factories where they use enzymes to convert metal ions to nanoparticles via intracellular or extracellular mechanisms. The green synthesis is compatible with the issues of environmental safety and human health and avoids poisons. The utilization of renewable resources in these processes is contributing to the sustainable development [6].

Chemical Methods

Bionanofluids can be prepared by different chemical approaches.

1. Co-precipitation: The co-precipitation method is used to synthesize nanoparticles from a single solution. This method is simple and cost-effective and used for the tailored synthesis of particular properties of nanoparticles.
2. Chemical vapor deposition (CVD): CVD leads to transformation of gaseous precursors into solid materials on a substrate surface. This approach is very convenient and efficient to synthesize high-quality nanoparticles with superior purity and can be adjusted to various compositions, making it a good candidate in bionanofluid preparation.
3. Chemical reduction: Metal ions present in the solution are reduced to form nanoparticles. The size and shape of nanoparticles can be tuned by varying reaction conditions.
4. Microemulsion: It involves the preparation of a stable mixture of oil, water, and surfactants which can encapsulate nanoparticles. Through microemulsions, nanoparticles can be synthesized where we have fine control over the dimensions, as well as the dispersion.
5. Laser pyrolysis: This is the technique of the pulsed-laser ablation, where laser is used to break down the precursors for nanoparticle formation. This method provides for nanoparticles of very high purity and uniform size, leading to superior efficacy of bionanofluids.

Surface Oxidation

Surface oxidation is a technique to introduce oxygen-containing groups to the surfaces of materials such as carbon nanofibers to make them more hydrophilic and enhance their interaction with liquids. This change is very important to develop stable nanofluids with good thermal properties. Nanoparticle agglomeration is reduced by surface oxidation, thereby enhancing suspension and optimum performance [10].

Surfactant Addition

Surfactants such as polyvinylpyrrolidone (PVP) may be employed as an additive in nanofluids to enhance its stability through reducing interparticle forces by hindering the aggregation [10].

pH Control

Best stabilization of a nanofluid is obtained when the pH is kept within a certain range (usually 4–9), since it modifies the ionic environment and interparticle repulsive forces [11].

2.3 Step 3: Nanoparticle Dispersion into a Base Fluid

Bionanofluids are prepared by suspending the nanoparticles in base fluid. Bionanofluids have unique advantages and are used above all in mass transfer operations, taking advantage of the bioconvection driven by the microorganisms. The performance of the base fluid can be remarkably enhanced by adding nanoparticles that have different shapes or compositions, such as nanotube, nanosheet. There are generally two methods, i.e., one-step and two-step methods. The one-step process consists in preparing nanoparticles directly into the base fluid, while in the two-step method, nanoparticles are synthesized and then dispersed in the base fluid. Long-term stability of the dispersion is a challenge as agglomeration of nanoparticles can induce sedimentation and lose of efficacy. Stability is also influenced by other factors like NP size, shape, base fluid properties, and presence of surfactant. Nanofluids have higher thermal conductivities, and heat transfer coefficients are suitable for cooling applications [12].

3 Thermophysical Properties of Bionanofluids

Thermophysical properties of bionanofluids are especially important in addressing their potential for heat transfer systems. These are dependent on the material, temperature, and concentration.

3.1 Thermal Conductivity of Bionanofluids

The heat transfer capability of a body is described by its thermal conductivity. For bionanofluids, the thermal conductivity of the bionanofluid is an important factor for effective heat transfer. The inclusion of nanoparticles like multiwalled carbon nanotubes (MWCNTs) can enhance the heat transfer for these fluids substantially [13].

3.2 Viscosity of Bionanofluids

Most of the bionanofluids tend to have lower viscosity as compared to conventional fluids, which leads to better flow and heat transfer behavior. Viscosity is a vital factor in the enhancement of the bionanofluid as it regulates the flow of the fluid. Viscosity is closely related to nanoparticle concentration and generally increases with concentration because of the stronger interaction of particles. Viscosity is also affected by temperature, and higher temperatures in general cause particles to have greater kinetic energy and lead to lower viscosity due to an increased ability to flow [14].

3.3 Density and Specific Heat Capacity of Bionanofluids

The density and the specific heat of bionanofluids depend strongly on the biomaterial employed. The density (the mass per unit volume) is an important property since it influences the consequent behavior and performance of the fluid in thermal applications. It is obvious that the density depends on the biomaterial of which bionanofluid is comprised. A substance's specific heat capacity determines the amount of thermal energy needed to increase its temperature by one degree relative to water. This property significantly impacts the fluid's ability to transfer heat, making it a crucial factor in heat transfer applications. In the case of bionanofluids, specific heat is not constant, but temperature-dependent primarily, and finally, it becomes a function of temperature at a higher range. This simply means that the hotter the fluid, the more thermal energy it can hold [14].

4 Rheological Behavior of Bionanofluids

The flow of bionanofluids, a unique class of nanofluids comprising nanoparticles and bioingredients, depends on a number of parameters, such as nanoparticle characteristics, base fluid characteristics, and environmental factors. It is therefore of great interest to study these effects to optimize bionanofluids for various applications ranging from heat transfer to other industrials. To enhance the performance of the bionanofluid, the in-depth information related to its rheology is critical.

4.1 Effects of Particle Size and Concentration on Rheology

Bionanofluids are colloidal suspensions of nanoparticles dispersed in a base fluid. Nanoparticles have sizes that afford new and desirable surface properties improving base fluid performance. The concentration of nanoparticles in bionanofluid determines the properties and effectiveness of bionanofluids. In hybrid nanofluids, the rheology is dominated to a larger extent by the nanoparticle concentration. With the increase in the concentration of nanoparticles, interparticle interactions would become more significant that might affect viscosity and flow behavior [15]. The nanoparticle size (generally lower than 100 nm) and its shape largely affect the viscosity and the flow characteristics of the nanofluid. Spherical particles typically exhibit Newtonian fluid behavior, while nonspherical particles (such as cylindrical nanotubes) more commonly possess non-Newtonian behaviors [16]. When nanoparticle content in bionanofluid is increased, the viscosity increases which in turn changes the behavior of the flow of fluid [15].

4.2 Rheological Models for Bionanofluids

The study of flow property is necessary for the optimization of bionanofluids for a particular application. Taking into consideration of temperature, NP concentration, and microorganism presence, these models allow us to have an understanding of interactions between NPs and base fluid.

Herschel-Bulkley Models

The power-law and Herschel-Bulkley models among others have been popular models to describe the non-Newtonian features of nanofluids particularly when the nanomaterial concentration is not so high. The Herschel-Bulkley model is a mathematical model that describes the flow of materials that do not follow Newton's law of viscosity and that have a yield stress. With the help of this model, the behaviors of different fluids (e.g., nanofluids) under different shear rates can be understood. In the context of nanofluids, the Herschel-Bulkley model is one of the most important models to describe the rheological behavior of fluids that include nanoscale particles in it. The Herschel-Bulkley model is characterized by three parameters, namely, the present yield stress, the consistency index, and the flow behavior index. Yield stress is the lowest stress required for a fluid to flow; when force is applied below that point, the fluid behaves more like a solid. The flow index, on the other hand, is indicative of the viscosity of the fluid at a given shear rate. At the same time, the flow behavior index describes the viscosity-shear relationship, whether the fluid is shear-thinning or shear-thickening. It should be highlighted that the rheology of the nanofluid experiences a marked variation at some shear rate threshold, modifying its flow pattern. This behavior is determined by different parameters such as the concentration, the size, and the properties of the nanoparticles. The performance of the nanofluid in various applications such as cooling systems and lubrication is quite important, and having an insight on flow behavior under different conditions is quite essential [17].

Ostwald-de Waele Model

The Ostwald-de Waele model is used for bioconvection flows to describe how gyrotactic microorganisms change the flow. The Ostwald model and the de Waele model are both power-law models and are used to explain the flow behavior of non-Newtonian fluids, which exhibit viscosity that changes in accordance with a shear rate. In the context of this study, we apply these models to analyze bioconvective nanofluid flow with gyrotactic microorganisms and nanoparticles. In particular, the Ostwald model is suitable for pseudoplastic fluids that have a lower viscosity with the increase in the shear rate. The fluid's viscosity decreases as it flows more rapidly, making way for easier movement. It is this property that is particularly advantageous when an application will benefit from an improved flow of fluid, such as bioengineering nanofluid fuel cells. On the other hand, the de Waele model is also used for dilatant fluids, i.e., for those which solidify as the shear rate is high. The Ostwald and de Waele models provide a foundation for evaluating complex bioconvection nanofluid flow structures, which are important in developing bioengineering and nanofluid fuel cell devices [18].

4.3 Non-Newtonian Behavior of Bionanofluids

Bionanofluids also present unique non-Newtonian behavior, where the rheological characteristics deviate from the Newtonian viscosity law. This property is highly desirable for these biomedical engineering applications or thermal-control systems, where precise manipulation of biomicro/nanoparticles is required. In the vast majority of such systems, the bionanofluid behaves as shear-thinning or viscoplastic and could be analyzed by the power-law model or the Bingham plastic model [19, 20]. These fluids have enhanced heat transfer properties, thereby making them suitable for use in thermal management, e.g., as heat exchangers and coolants [21].

5 Characterization Techniques for Bionanofluids

Characterization techniques are of great significance for understanding the properties of bionanofluids and their further application in specific cases. Characterization of such nanoparticle suspensions enables to study structural, thermal, and mechanical properties.

5.1 Microscopic Techniques (TEM, SEM, AFM)

Scanning Electron Microscopy (SEM)

It permits the visualization of surface morphology and structure of materials at microscopic scales. SEM is an imaging technique in which a focused electron beam is scanned over the surface of a sample to form an image that is enlarged and displays surface topology and composition. The interaction between electron beam and the sample generates signals that contain information about the sample structure; these signals are converted into an image. This enables the investigation of surface morphology, texture, and composition and is indicative of the size and shape of the nanoparticles [22].

Transmission Electron Microscopy (TEM)

TEM is a sophisticated imaging technique that enables researchers to examine the internal structure of materials at the nanoscale. By providing detailed images of the internal arrangement and morphology of nanoparticles, TEM facilitates understanding of the properties and potential applications of these materials. TEM allows us to study the internal structure of materials at a nanometer scale. This process enables a clear view of individual nanoparticles internal structure, as well as nanoparticle morphology, which can be useful in establishing the properties and the applications of the materials. Such visualization of the nanoscale distribution and interaction of constituents provides strong indication to understand the mechanical and thermal properties of the material. TEM, as an atomic-scale probe, is capable of viewing high-resolution microstructure features. When used with EDX, TEM can yield data about the elemental composition of the materials of interest.

Atomic Force Microscopy (AFM)

AFM is an advanced characterization method that provides high-resolution topographical information and has been applied in order to study particle interactions and surface properties of materials. AFM maps topography at the nanoscale through scanning a sharp tip across the sample surface; therefore, it unveils surface roughness and morphology [23].

5.2 *Thermal and Mechanical Characterization*

Thermogravimetric Analysis (TGA)

It provides information on material composition and thermal stability by measuring the weight change as a function of temperature. This method consists of heating a sample at a regulated constant rate and measuring the mass of the sample during the process, to provide data of the material properties at different temperature levels [4, 22].

Dynamic Light Scattering (DLS)

It is a powerful tool to assess particle size distribution and stability for colloidal suspensions, e.g., nanofluids. The fluctuation of scattering light intensity due to the particle Brownian motion is measured, from which the information on particle size and distribution can be accessed. Upon laser beam interaction with particles, the intensity of the scattered light from the particles fluctuates through time and those fluctuations are related to particle size. The smaller particles cause rapid fluctuations, whereas the larger particles produce slower changes. For this, the analysis of these fluctuations allows determination of the size distribution of suspended particles in DLS. The small particles lead to fast fluctuations, while the large one drives slower variation. These fluctuations in intensity can be analyzed by DLS to establish the size distribution of the particles in suspension. This approach is necessary as it forms the basis for describing nanoparticle dispersion, as well as stability in the base fluid, which is responsible for achieving the desired improvement in thermal behavior [24].

5.3 *Spectroscopic Techniques (FT-IR, UV-Vis, NMR)*

Fourier Transform Infrared (FT-IR)

FT-IR spectroscopy is a useful technique to study the surface chemistry of the particles in different phases: solid and liquid. Information may be given on the

ment, and bioimaging. Bionanofluids also offer prospects for tissue engineering, antimicrobial surfaces, and thermal therapy within medical applications. The coalescence of computational modeling and green synthesis approaches allows the design of greener and safer bionanofluids. Bionanofluid research is an exciting new area of study; its interdisciplinary nature integrates the advantages of nanotechnology, biology, and engineering in the development of innovation [3].

8 Conclusion

Bionanofluids have enormous potential in different fields. They have attractive thermophysical, optical, and rheological properties. They have created opportunities for applications in biomedical, energy, and environmental areas. The distinct features of bionanofluids such as enhanced heat transfer, selective delivery, and biocompatible material led to their application in numerous fields. The medical applications of bionanofluids include drug delivery, cancer therapy, and imaging. Their superior thermal conductivity also qualifies them for heat transfer applications in electronics and renewable energy devices. The issues of stability, toxicity, and upscaling must be solved to take full advantage of bionanofluids. Altogether, bionanofluids have great potential to revolutionize a broad feature of industries ranging from medical to energy. Because of their unique characteristics and a range of potential applications, bionanofluids are expected to play a significant role in the future of technology and medicine.

References

1. M. Trifiro, M. Paliouras, P.J.R. Roche, I. Dotan, L.K. Beitel, C. Alvarado, et al., *Bionanofluid for Use as a Contrast, Imaging, Disinfecting and/or Therapeutic Agent* (Google, Mountain View, 2014). https://patents.google.com/patent/EP3068442A1/en
2. M. Ghassemi, A. Shahidian, *Bio-Nanofluid Simulation* (Elsevier, Amsterdam, 2017), pp. 89–112. https://doi.org/10.1016/B978-0-12-803779-9.00005-4
3. B. Kanjilal, N. Sharifi, A. Masoumi, I. Noshadi, I. Noshadi, *Bionanofluids and Mass Transfer Characteristics* (Elsevier, Amsterdam, 2022), pp. 217–244. https://doi.org/10.1016/B978-0-12-823996-4.00008-2
4. A.A. Okafor, C.O. Mgbemena, B.U. Oreko, Preparation, characterization, stability and thermophysical properties of bio, non-bio (metallic and non-metallic) and hybrids nanofluids: A review. J. Nanofluids **11**(6), 803–818 (2022). https://doi.org/10.1166/jon.2022.1896
5. S.Z. Heris, Y. Mohammadfam, R. Javadpour, H. Pourpasha, *Nanofluids* (Elsevier, Amsterdam, 2024), pp. 27–56. https://doi.org/10.1016/b978-0-323-95511-9.00019-6
6. N.S. Alsaiari, F.M. Alzahrani, A. Amari, H. Osman, H.N. Harharah, N. Elboughdiri, M.A. Tahoon, Plant and microbial approaches as green methods for the synthesis of nanomaterials: Synthesis, applications, and future perspectives. Molecules **28**(1), 463 (2023). https://doi.org/10.3390/molecules28010463
7. R.P. Chauhan, Processes for nanoparticle synthesis, in *Advances in Chemical and Materials Engineering*, (IGI Global, London, 2024), pp. 69–100. https://doi.org/10.4018/979-8-3693-4397-5.ch003

8. R. Zanella, Metodologías para la síntesis de nanopartículas: controlando forma y tamaño. Mundo Nano **5**(1), 69–81 (2014). https://doi.org/10.22201/CEIICH.24485691E.2012.1.45167
9. K. Diao, Q. Jiang, M. Yang, J.-L. Lu, Y. Zhu, Enhancing the dynamic thermal stability of metal oxide oil-based nanofluids through synchronous chemical modification and physical coating. Surf. Interfaces (2024). https://doi.org/10.1016/j.surfin.2024.104119
10. N. Saidi, N. Abdullah, M.N. Norizan, N. Janudin, N.A. Kasim, M.J. Osman, I.S. Mohamad, M.A. Rosli, Surface-oxidised carbon nanofibre-based nanofluids: Structural, morphological, stability and thermal properties. Nano **12**(21), 3922 (2022). https://doi.org/10.3390/nano12213922
11. J. Wang, X. Yang, J.J. Klemeš, K. Tian, T. Ma, B. Sunden, A review on nanofluid stability: Preparation and application. Renew. Sustain. Energy Rev. (2023). https://doi.org/10.1016/j.rser.2023.113854
12. A.S.N. Husainy, P.M. Chougule, K.R. Badade, S.A. Hasure, A.S. Patil, S.A. Tukshetti, A glance on preparation, stability, properties and applications of nanofluids. Asian Rev. Mech. Eng. **7**(2), 30–41 (2018). https://doi.org/10.51983/arme-2018.7.2.2456
13. N.K. Das, R. Hussain, A.K. Vatti, T. Banerjee, Thermal performance of bio-nanofluid dihydrolevoglucosenone: Experimental and atomistic simulation investigations. AICHE J. (2024). https://doi.org/10.1002/aic.18718
14. J.L.T. Chen, A.N. Oumer, A.A. Azizuddin, A review on thermo-physical properties of bio, non-bio and hybrid nanofluids. J. Mech. Eng. Sci. **13**(4), 5875–5904 (2019). https://doi.org/10.15282/JMES.13.4.2019.12.0468
15. Z. Said, *Rheological Behavior of Hybrid Nanofluids* (Elsevier, Amsterdam, 2022), pp. 111–129. https://doi.org/10.1016/b978-0-323-85836-6.00005-3
16. A.K. Sharma, A.K. Tiwari, A.R. Dixit, Rheological behaviour of nanofluids: A review. Renew. Sustain. Energy Rev. **53**(53), 779–791 (2016). https://doi.org/10.1016/J.RSER.2015.09.033
17. Y. Rudyak, Rheology of nanofluids: Experiments and molecular dynamics modeling. J. Phys. Conf. Ser. **2585**, 012009 (2023). https://doi.org/10.1088/1742-6596/2585/1/012009
18. A.M. Obalalu, S.O. Salawu, O.A. Olayemi, C. Odetunde, A.O. Akindele, Computational study of bioconvection rheological nanofluid flow containing gyrotactic microorganisms: A model for bioengineering nanofluid fuel cells. Int. J. Model. Simul. (2023). https://doi.org/10.1080/02286203.2023.2204209
19. L. Yang, L. Yang, K. Du, K. Du, A comprehensive review on the natural, forced, and mixed convection of non-Newtonian fluids (nanofluids) inside different cavities. J. Therm. Anal. Calorim. **140**(5), 2033–2054 (2020). https://doi.org/10.1007/S10973-019-08987-Y
20. G. Kefayati, A.P. Bassom, A lattice Boltzmann method for single- and two-phase models of nanofluids: Newtonian and non-Newtonian nanofluids. Phys. Fluids **33**(10), 102008 (2021). https://doi.org/10.1063/5.0067744
21. M. Tawalbeh, I. Shomope, A. Al-Othman, Comprehensive review on non-Newtonian nanofluids, preparation, characterization, and applications. Int. J. Thermofluids (2024). https://doi.org/10.1016/j.ijft.2024.100705
22. S.M. Alay-e-Abbas, K. Mahmood, A. Ali, M.I. Arshad, N. Amin, M.S. Hasan, *Characterization Techniques for Bionanocomposites* (Elsevier, Amsterdam, 2020), pp. 105–144. https://doi.org/10.1016/B978-0-12-816751-9.00005-2
23. Ç.K. Kurukavak, Surface and mechanical characterization of bionanocomposites for advanced applications: A review. Proc. Inst. of Mech. Eng. Part L: J. Mater. Des. Appl. **236**(10), 1949–1966 (2022). https://doi.org/10.1177/14644207221096937
24. G.S. Reddy, V. Sumalatha, Nanofluids: Bridging nanotechnology and fluid dynamics for enhanced thermal performance. J. Phys. **2837**(1), 012053 (2024). https://doi.org/10.1088/1742-6596/2837/1/012053
25. H. Rabbi, A.Z. Sahin, B.S. Yilbas, A. Al-Sharafi, Methods for the determination of nanofluid optical properties: A review. Int. J. Thermophys. **42**(1), 1–42 (2021). https://doi.org/10.1007/S10765-020-02762-0
26. Z. Ji, R. Dai, Z. Zhang, Characterization of fine particulate matter in ambient air by combining TEM and multiple spectroscopic techniques–NMR, FTIR and Raman spectroscopy. Environ. Sci.: Processes Impacts **17**(3), 552–560 (2015). https://doi.org/10.1039/C4EM00678J

27. M. Sheikhpour, M. Arabi, A. Kasaeian, A. Rabei, Z. Taherian, Role of nanofluids in drug delivery and biomedical technology: Methods and applications. Nanotechnol. Sci. Appl. **13**, 47–59 (2020). https://doi.org/10.2147/NSA.S260374
28. D. Mandal, K. Anand, P. Prabhakar, *Biomedical Applications of Nanofluids in Drug Delivery* (Springer, Cham, 2023), pp. 105–124. https://doi.org/10.1007/978-981-99-5281-6_5
29. R. Asif, R. Yasmin, M. Mustafa, J. Nisar, A. Ambreen, A. Rehman, S. Unmbreen, *Antimicrobial Potential, Drug Delivery and Therapeutic Applications of Bio-Nanoparticles in Medicine* (Springer, Cham, 2023), pp. 115–130. https://doi.org/10.1007/978-981-99-4149-0_6
30. Z. Wang, Z. Wang, M. Zhang, J. Cui, M. Xie, Y. Jiang, *Biological Applications of Nanofluids: Antimicrobial Activity and Drug Delivery* (Springer, Cham, 2022), pp. 19–45. https://doi.org/10.1007/978-981-19-8050-3_2
31. S. Esfandeh, *Application of Nanofluids and Future Directions* (Elsevier, Amsterdam, 2024), pp. 289–309. https://doi.org/10.1016/b978-0-443-21576-6.00006-6

Ionic Liquids in Nanofluids: Design and Applications

Farhat Aisha Ansari and Amita Somya

Abbreviations

[C2mim][Ac]	1-Ethyl-3-methylimidazolium acetate
[C2mim][$C(CN)_3$]	1-Ethyl-3-methylimidazolium tricyanomethanide
[C2mim][C_2SO_4]	1-Ethyl-3-methylimidazolium ethylsulfate
[C2mim][CH_3SO_3]	1-Ethyl-3-methylimidazolium methanesulfonate
[C2mim][Dca]	1-Ethyl-3-methylimidazolium dicyanamide
[C2mim][SCN]	1-Ethyl-3-methylimidazolium thiocyanate
[C4mim][BF_4]	1-Butyl-3-methylimidazolium tetrafluoroborate
[C4mim][$C(CN)_3$]	1-Butyl-3-methylimidazolium
[C4mim][CF_3SO_3]	1-Butyl-3-methylimidazolium Trifluoromethanesulfonate
[C4mim][Dca]	1-Butyl-3-methylimidazolium dicyanamide
[C4mim][NTf2]	1-Butyl-3-methylimidazolium bis (trifluoromethylsulfonyl)imide
[C4mim][PF_6]	1-Butyl-3-methylimidazolium hexafluorophosphate
[C4mim][SCN]	1-Butyl-3-methylimidazolium thiocyanate
[C4mmim][NTf2]	1-Butyl-2,3-dimethylimidazolium bis (trifluoromethylsulfonyl)imide
[C6mim][BF4]	1-Hexyl-3-methyl imidazolium tetrafluoroborate
[C6mim][NTf2]	1-Hexyl-3-methylimidazolium bis (trifluoromethylsulfonyl)imide

F. A. Ansari (✉)
Department of Chemistry, Integral University, Lucknow, Uttar Pradesh, India

A. Somya
Department of Chemistry, Amity University, Bengaluru, Karnataka, India

S. Kulkarni, S. Das (eds.), *Ionic Liquids, Nanofluids and Nanotechnology*,
Lecture Notes in Nanoscale Science and Technology 42,
https://doi.org/10.1007/978-3-032-04008-4_7

[C6mim][PF6]	1-Hexyl-3-methylimidazolium hexafluorophosphate
[C8mim][NTf2]	1-Methyl-3-octylimidazolium bis (trifluoromethylsulfonyl)imide

1 Introduction: Ionic Liquid-Based Nanofluids

Ionic liquids (ILs) constitute a distinctive and adaptable category of chemicals formed exclusively from ions. Ionic liquids (ILs) have garnered substantial scientific and industrial attention related to their exceptional physicochemical characteristics, including low vapour pressure, enhanced heat and chemical resistivity, and the ability to respond to certain conditions through the choice of diverse positive and negative ion combinations. Ionic fluids are a novel category of fluids consisting only of ions, exhibiting melting temperature less than 100 °C [1]. In recent years, the formation of normal temperature ionanofluids has garnered significance from scientists, industry professionals owing to comparatively lesser melting points (<30 °C). Compared to typical molecular liquids, the density of ionic liquids exhibits a modest drop, predominantly in a linear fashion, with rising temperature. Scientists studied how an elevation in temperature from 15 °C to 37 °C results in a reduction of the molecular density of [BMIM][NTf2] from 1446 to 1375.7 kg/m^3 (4.86%) [1, 2]. The thickness of ionic liquids (ILs) generally exceeds that of traditional thermal transfer fluids, for example, ethylene glycol.

Nevertheless, similar to other fluids, the thickness of ILs decreases significantly (non-linearly) with rising temperature as reported by research [3]. The decrease in fluidity of ionised fluids is especially essential for thermal applications at higher temperatures, since it can substantially decrease the required pumping power. Ionic liquids (ILs) are fluids at normal temperature, consisting of ionic mixtures. It has attracted attention as heat exchange liquids because of their superior thermophysical characteristics [4]. These features encompass a low pressure gradient and fluctuation, significant heat capacity, a comparatively lesser melting point, considerable ionic conductivity, and strong solution capacity. In addition to their utilisation as heat transfer fluids, ionic liquids have numerous industrial uses. It has several uses, particularly as solutions in chemical, oil-gas industries, processing of coal, and pharmaceutical preparations [5–7].

The integration of nanomaterials into various ILs (ionic liquids) creates a new domain, referred to as 'ionanofluids'. The idea that the use of nanofluids enhances the thermodynamic characteristics is due to the dispersal of a small amount or mass fraction of particles in the ionised liquids (ILs). Multiple groups of researchers [8–11] are investigating ILs-based nanofluids, and various efforts have been undertaken to evaluate their thermophysical characteristics, including temperature stability, rheological actions, electrical conductivity, heat retention, and thermal stability under both ambient and induced convection circumstances.

The integration of ionic liquids with nanoparticles creates an effective platform for the preparation and stability of metal oxides and other nanomaterials (Fig. 1). Ionic liquids serve as solvents, stabilisers, and surface modifications,

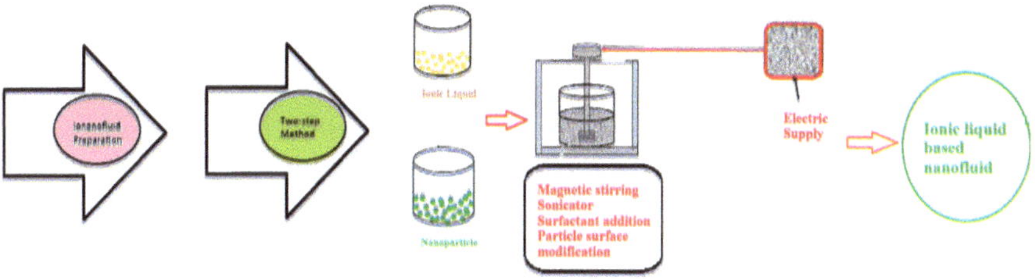

Fig. 1 Rationale for integrating ionic liquids with nanoparticles

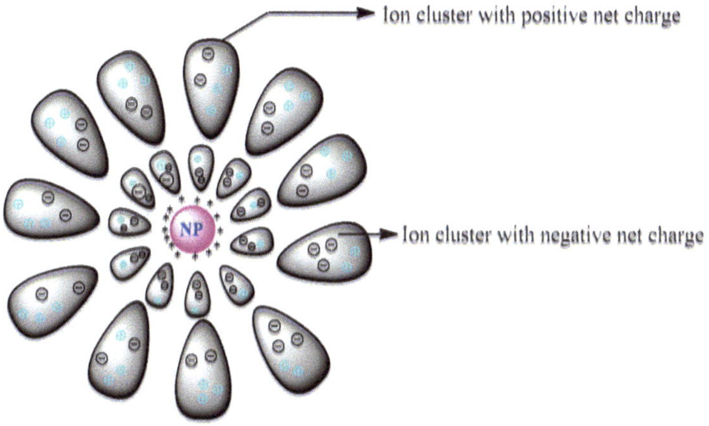

Fig. 2 Schematic representation of clusters of ions encircling nanoparticles. The ion clusters create a safeguarding electric dual layer

utilising electrostatics, steric, and hydrogen-bonding effects to regulate nanoparticle dimensions, morphology, and dispersion stability. The adaptability of ionic liquid features, particularly alkyl chain size, enables customised nanoparticle-ionic liquid combined systems with potential uses in catalysis, electrochemical equipment, and materials research. The titania nanotube scattering an ionic liquid-infused polymeric framework demonstrated substantial electrical conductance of 10^{-3} S cm^{-1} at 120 °C, enhanced thermal endurance at temperatures exceeding 250 °C, and enhanced shape, characterised by a decrease in crystallised area volume and a rise in the non-crystal phase, indicating significant prospects for several uses [12]. This review examines the prospective applications of new hybrids made of nanomaterial in ionic fluids, while considering the possible connections between the form and function of the nanomaterials. This will encompass diverse ionic fluid and nanoparticle mixed substances, such as gels, colloid glasses, suspensions, and ionic fluid-impregnated nanomaterials.

Fukushima et al. [13] pioneered the incorporation of high concentrations of composite nanotubes in ionic fluids at the ambient temperature, resulting in the formation of liquids referred to as 'bucky gels', which may have potential uses in various technical and chemical methods (Fig. 2). These 'bucky gels' are fluid, representing a high accumulation of CNT ionanofluids. Sheveylyova et al. [14] investigated [C4mim][BF_4] and [C4mim][PF_6] ionanofluids containing MWCNT (multi-walled carbon nanotubes) at approximately 12 wt% amount, revealing an optimal heat transfer improvement of 8%. Moganty et al. [15] examined a new type of production of silica ionogels created by dispersion of nanoparticles of silica in SpmImTSFI, which were subsequently disseminated in the underlying ionic liquid BmpyrTFSI. The study concentrated on an extensive thermal span, and the incorporation of SiO_2-SpmImTSFI retarded the crystallisation of the host ionic liquid. They proposed a novel method involving blended ionic liquids with enhanced thermo-mechanical features.

Researchers [16] investigated that [C4mim][$N(SO_2C_2F_5)_2$] combined with nanomaterial showed the most substantial rise in its viscosity attributing this phenomenon to the relationship between those nanoparticles and the charged ions of the ionic liquid. Other research appeared to be inconsistent regarding the rheometrical behaviour of ionanofluids. In contrast to the work on its viscosity, there has been a greater focus on the measurement of heat conductance, with results showing promising improvements in heat conductance due to incorporating nanoparticles into the fundamental ionic liquid [11, 17]. In another investigation, Deb et al. [18] examined the effect of ionic-liquid-linked aluminium oxide nanoparticles on the non-isothermal low temperature crystallisation in ionic-fluid-based ionanofluids, concluding that the attached nanoparticles function as diverse nucleation agents, thereby reducing the inherent energy barrier to nucleation process.

Nanofluids, which consist of fluids with distributed nanoparticles, can be synthesised mainly using two approaches: the single-step procedure and the double-step technique. Every approach possesses unique methods, benefits, and limitations.

2 Double-Step Process

In a dual-step process, nanoparticles are first produced in powdered form, utilising either a physical or chemical process. The synthesised nanomaterials are subsequently distributed in a liquid by mechanical or ultrasound agitation, high-shear blending, homogenisation, or through ball milling techniques. Nanoparticles are synthesised through the decrease of transition-metal salts, ligand diminution, expulsion from organometallic compounds, microwave synthesis [19] (Fig. 3), and the co-precipitation [20].

Nanoparticles exhibit a propensity to agglomerate owing to their elevated area of coverage and interface reactivity. To combat this, nanoparticles are frequently stabilised with surfactants. Bonnemann et al. designed an approach to prepare palladium (Pd) particles measuring 2.2 nm in diameter. Copper nanofluids can be synthesised by diffusing Cu nanoparticles in water and oil from transformers using sonication with stabilisers.

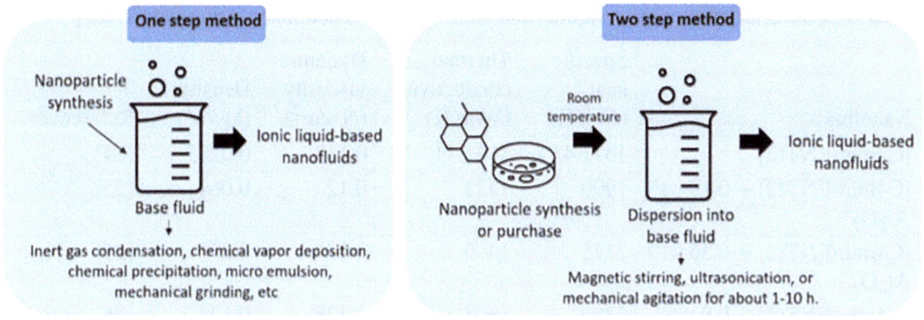

Fig. 3 Two-step synthesis procedure for nanofluids, nanoparticles synthesised via a microwave-based synthesiser are disseminated in a base medium via a mechanical agitator

3 Single-Step Method

Nanoparticle synthesis and dissemination in the base medium are executed concurrently in a single-step approach [21]. A single-step method can therefore be employed to make evenly distributed nanoparticles firmly in a base medium [22]. The techniques used to synthesise nanofluids through a single-step method encompass thermal degradation, physical vapour condensate, chemical reduction, and polyol formation.

4 Physical Properties of Ionic Liquid Nanofluids

Ionic liquid-based nanofluids (INFs) integrate the distinctive characteristics of ionic liquids (ILs) with nanoparticles (NPs) to produce enhanced heat transfer fluids (HTFs) exhibiting exceptional thermophysical capabilities. The following is an examination of their principal characteristics and uses.

To obtain valid computational results, it is essential to depend on existing data from experiments about heat conduction, viscosity, density, and specific heat. Table 1 highlights the accessible parameters acquired from the research activity.

5 Applications of Ionanoliquids

Ionanocolloids have distinctive physiochemical properties characteristic of ionic liquids, notably their minimal pressure of vapour, elevated thermal resistance, and significant conductive nature [29]. Moreover, they are environmentally friendly materials and ensure a more secure use by decreasing the hazards of inadvertent leakage [30]. The exceptional characteristics of ionanocolloids render them viable fluids for utilisation in several energy -associated domains (Fig. 4).

Table 1 Characteristics of ionic liquids and their nanofluids at ambient temperature [23]

Nanofluids	Specific heat (J/kg K)	Thermal conductivity (W/m K)	Dynamic viscosity (N s/m^2)	Density (kg/m^3)	References
[C4mim][NTf2]	1372.44	1441.34	0.117	0.0285	[24]
[C4mim][NTf2] + 0.18 wt% Al_2O_3	1900	1423	0.12	0.064	[25]
[C4mim][NTf2] + 0.36 wt% Al_2O_3	2175	1460	0.124	0.073	[25]
[C4mim][NTf2] + 0.9 wt% Al_2O_3	2380	1508	0.128	0.124	[25]
[C4mim][NTf2] + 1 wt% MWCNT	1396.03	1423	0.129	0.03158	[26]
[C2mim][$EtSO_4$]	1614.96	1241.37	0.175	0.05001	[24, 25]
[C2mim][$EtSO_4$] + 1 wt% MWCNT	1642.72	1226.06	0.189	0.05398	[26]
[HMIM][BF_4]	2266	1145.4	0.166	0.25	[27]
[HMIM][BF_4] + 0.03 wt% MWCNT	2258	1146	0.173	0.18	[27]
[HMIM][BF_4] + 0.06 wt% MWCNT	2227	1147	0.187	0.19	[27]
[HMIM][BF_4] + 0.03 wt% graphene	2258	1097	220.181	0.1804	[28]
[HMIM][BF_4] + 0.06 wt% graphene	2236	1092	0.192	0.185	[28]

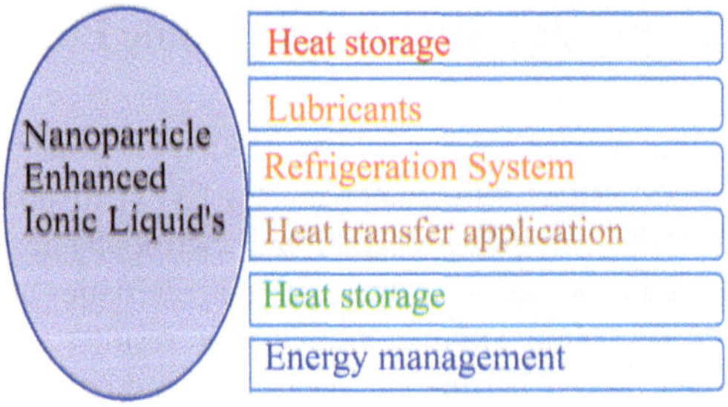

Fig. 4 Applications of ionanocolloids in thermal energy

5.1 *Thermal Conductivity Enhancements*

Numerous studies were conducted to assess the ionanofluids' thermally conductive performance.

In general, the base fluid's thermal conduction can be raised by up to 68% by adding small amounts of nanoparticles. Jozwiak et al. [31] assessed the impact of adding various nanoparticles of carbon at varying concentrations and observed that the nanoparticle dispersion had a greater impact on the conductivity of heat than the quantity of the nanoparticles. The [C2mim][SCN] thermal efficiency was most enhanced at a concentration of 1 wt% of commercial SWCNT, but the synthesised

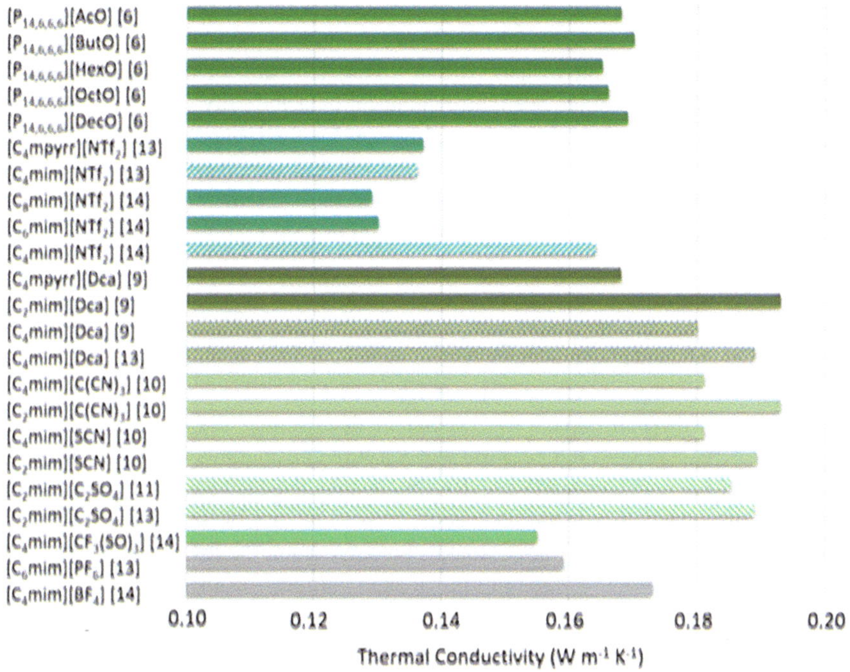

Fig. 5 Representation of the thermal conductance of ionanoliquids 1 wt% of MWCNT

MWCNT also demonstrated remarkable performance, raising the degree of thermal conductivity by 44% [31].

Conversely, in the study by Fox et al. [32], the distribution of 0.5 wt% of SWCNT marginally reduced the heat conductance of [C4mmim][NTf2], whereas the incorporation of an equivalent weight of MWCNT enhanced thc thermal conductivity by approximately 4% [32]. The researchers assessed the impact of nanomaterial morphology utilising 0.5 wt% of alumina spheres and whiskers, noting significant disparities; round shapes enhanced thermal conductivity by 3%, while hair-like strands increased it by 7%. Furthermore, the influence of temperature is more significant in hair-like than in spheres, with elevated temperatures resulting in increased conductivity of heat [32]. Although a definitive link between cation or anionic type and the heat conductivity of ionanocolloids was not established, França et al. observed that the heat conductivity of [C2mim] + ionanofluids typically exceeds that of [C4mim] + based nanofluids [33]. Figure 5 illustrates the ability to conduct heat of various ionanocolloids, comparing different positive and negative ions under same circumstances (1 wt% of MWCNT at the ambient temperature).

The comprehension of the primary mechanisms underlying the improvement in thermal conductivity in base fluids with nanoparticle inclusion remains ambiguous because of insufficient and commonly accepted empirical evidence to corroborate the proposed theories. Researchers also regard several mechanisms described and examined for nanofluids as accountable for the aforementioned improvement of heat conductance in ionanocolloids. The clear summary of these processes includes the nature of heat transfer in nanoparticles, fragmentation of the liquid -particle interaction at the molecular level, Brownian particle mobility, and the effects of nanoparticle accumulation. The principal reason for the ionanocolloids' enhanced thermal efficiency is the layering of the fluid at the nanoparticle interface. The particles in this particular layer exhibit significantly greater order, resulting in superior thermal conductivity relative to most fluids [34].

5.2 *Heat Transfer Performance*

The particular thermal capacity is crucial in determining the heat transfer rate, typically increasing with temperature and varying considerably based on the kind of ionic liquids and nanoparticles. Recent results indicate that the most recent factor is particularly influential on specific heat capacity [35], and the relationship with nanoparticle quantity varies among authors. Some scientists noted that the introduction of minimal concentrations of nanoparticles resulted in a decrease in the particular heating ability of the ionic liquid [28, 36].

This rise may be related to the development of interface nanolayers on the outer layer of the nanoparticles by ionic fluid molecules [37]. Figure 6 illustrates the particular thermal properties of ionanocolloids (INC) in respect to their base

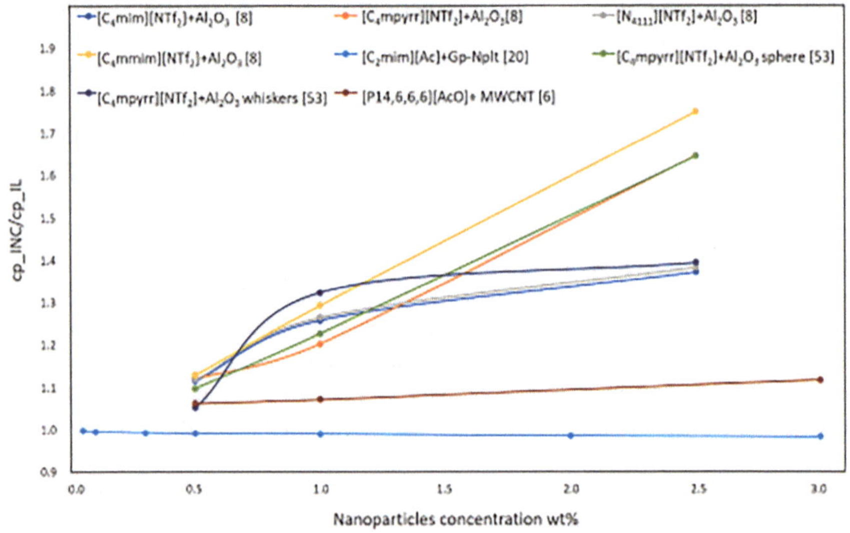

Fig. 6 Comparative thermal conductivity of ionanoliquids and ionic liquids (ILs) with varying concentrations of nanoparticles

fluid (IL) at varying concentrations of nanoparticles. A number of systems demonstrated a nearly straight-line rise in specific heat capacity increasing nanoparticle concentration, whereas others displayed a parabolic pattern, likely attributable to the clustering of nanoparticles.

5.3 *Viscosity*

A lot of things affect the viscosity of ionanocolloids, which is essential to their real-world uses for power generation and thermal control. The parameters include the composition of the base fluid, its properties, and the quantity of the nanoparticles, temperature, dispersion duration, shear velocity, and vibration frequency. These factors can have such complicated and important effects on the way a fluid behaves that, for example, a rise in the quantity of nanoparticles could make ionanocolloids a liquid that is not Newtonian [37, 38]. This happens because the nanoparticles are interacting with each other more, which causes them to change between an ordered and an unorganised state depending on the amount of pressure [37, 38].

This process is elucidated by the enhanced interactions among the nanoparticles, resulting in a change between ordered and disordered states in response to the applied stress. In such situations, the attractive forces prevail over the repulsive Brownian forces, clusters form, impeding the movement of the nanoparticles and requiring more power for movement, hence causing a considerable increase in viscosity [37, 38]. Nanoparticles make the viscosity higher, and there are just a few cases where the thickness went down. These used [P6,6,6,14][Phosph], [P6,6,6,14][NTf2], and [C6mim][BF_4] as base fluids. It was found that adding MWCNT to phosphonium-based ionic fluids makes it less thick, by as much as 75.4% [3]. The researchers said that such tendency could be due to an unconventional shear dispersion (or pseudo-plastic) flow, and that could be linked to the MWCNT becoming more useful as a result of the relationship among the nanoparticles and the charged particles in the ionic fluid [3].

5.4 *Density*

In thermal transfer methods, the density of ionanoliquids is an additional feature to speculate about, mostly for technical reasons, because it affects frictional forces, factor, pressure fall, and transporting capacity of the equipment [39]. It depends on the kind of ionic fluid, its composition and amount of nanoparticles, and the ambient heat. The framework of the ionic liquid tells you a lot regarding its density, with the composition of the anion playing a big role in determining this attribute [40]. The density of pure ionic fluids typically ranges from 1 to 2.3 g cm^{-3} for fluorinated ionic liquids; however, phosphonium-based ionic liquids frequently have densities below 1 g cm^{-3}. This study assessed literature indicating that the densities of the pure ionic liquids employed as base fluids ranged from 0.881 g cm^{-3} for [P14,6,6,6][DecO] to 1.437 g cm^{-3} for [C4mim][NTf2] at ambient temperature [41].

6 Challenges and Future Perspectives

Although ionanoliquids have advantageous characteristics, their potential for use in industry as substitutes for traditional liquids in thermal energy contexts remains constrained by several aspects that shall be examined in this section. These constraints present numerous potential for enhancement and advancement in this domain and provide a review of the primary factors that should be examined for the potential use of this novel category of heat-transferring liquids.

6.1 Stability, Scalability, and Cost Issues

The primary problem associated with ionanocolloids is its acquired price, with purest ionic liquids priced up to five times costlier than Therminol VLT, Dowtherm MX, or Dowtherm A [42]. Due to ionic liquids being a relatively novel category of fluids, their massive production remains underdeveloped; however, advancements are being made to facilitate their availability in substantial numbers in the coming years. The procedure's growth is effectively established, the procurement prices will significantly decrease and become equivalent to those of conventional thermal transfer liquids.

A further constraint is the expenditures associated with ILs, which exhibit significant viscosity and need substantial mechanical energy, hence increasing the overall expense of the entire procedure. However, as the ambient temperature increases, viscosity significantly decreases, hence reducing total operating costs. While the viscosity is elevated at ambient temperature, an increase in temperature significantly reduces the consistency of the ionic liquids, rendering it comparable to that of commercial fluids. An elevation in temperature of 100 K may decrease the operational cost of processes utilising ionic liquids by 50%, a phenomenon not observed with market fluids, which maintain comparable fluidity as well as cost at elevated temperatures [42].

6.2 Emerging Trends and Research Directions for Ionanofluids

The significant advancements in technical sectors stimulate a growing need for more durable, effective, and adaptable heat transfer fluids. In this context, ionanocolloids present themselves as highly potential options; therefore, a thorough investigation of their primary thermophysical attributes, together with those of corresponding pure base liquids (ILs), is required for practical uses.

The properties of the ions of the ionic fluid significantly impact their thermophysical properties, with the anion having the biggest impact. Imidazolium is the predominant ionic liquid and is extensively utilised for ionanocolloids, whereas the NTf2 anion is the most frequently referenced in scientific research. In general, the incorporation of nanoparticles promotes heat conductivity, exhibiting a practically linear improvement with greater nanoparticle content. The sort of nanoparticle has a more substantial influence than concentration, with SWCNT and MWCNT exhibiting the highest efficacy. Despite the scarcity of research conducted, the morphology and dimensions of the nanoparticles have been observed to affect thermal conductivity outcomes.

The temperature significantly influences the fluidity of ionanocolloids, with a rise in temperature markedly decreasing viscosity, which is advantageous for their heat and power applications at high temperatures. The amount of density diminishes with the elongation of the cation and escalates with the mass in molecules of the anion associated with the ionised base liquid. Nanoparticles facilitate an increase in the volume of the base fluid, with this expansion being directly related to their concentration. Heat only marginally reduces density, owing to volumetric growth.

The durability of ionanoliquids surpasses that of ordinary nanofluids; yet, it remains an issue that significantly affects the efficiency of heating systems. Functionalisation of nanoparticles or alterations in pH may represent the most likely approaches to enhance the endurance of ionanocolloids while minimally affecting their thermophysical properties.

Significantly more meticulous and organised computational and experimental investigations should be undertaken to expand the types of ionic liquids and nanoparticles utilised, and more attention should be dedicated to elucidating the key processes involved and developing coherent theories regarding the thermophysical properties of ionanocolloids.

References

1. M. Kanakubo, K.R. Harris, Density of 1-Butyl-3-methylimidazolium Bis (trifluoromethanesulfonyl) amide and 1-Hexyl-3-methylimidazolium Bis (trifluoromethanesulfonyl) amide over an extended pressure range up to 250 MPa. J. Chem. Eng. Data **60**, 1408–1418 (2015)
2. A.P. Fröba, H. Kremer, A. Leipertz, Density refractive index interfacial tension and viscosity of ionic liquids [EMIM][$EtSO_4$], [EMIM][NTf_2], [EMIM][$N(CN)_2$], and [OMA][NTf_2] in dependence on temperature at atmospheric pressure. J. Phys. Chem. B **112**, 12420–12430 (2008)
3. A.G.M. Ferreira, P.N. Simoes, A.F. Ferreira, M.A. Fonseca, M.S.A. Oliveira, A.S.M. Trino, Transport and thermal properties of quaternary phosphonium ionic liquids and IoNanofluids. J. Chem. Thermodyn. **64**, 80–92 (2013)
4. R.D. Rogers, K.R. Seddon, Ionic liquids—Solvents of the future? Science **302**, 792–793 (2003)

5. P. Kubisa, Application of ionic liquids as solvents for polymerization processes. Prog. Polym. Sci. **29**, 3–12 (2004)
6. J.F. Wishart, Energy applications of ionic liquids. Energy Environ. Sci. **2**, 956–961 (2009)
7. P. Singh, K. Kumari, A. Katyal, R. Kalra, R. Chandra, Copper nanoparticles in ionic liquid: An easy and efficient catalyst for selective carba-Michael addition reaction. Catal. Lett. **127**, 119–125 (2009)
8. N.J. Bridges, A.E. Visser, E.B. Fox, Potential of nanoparticle-enhanced ionic liquids (NEILs) as advanced heat-transfer fluids. Energy Fuel **25**, 4862–4864 (2011)
9. A. Wittmar, D. Ruiz-Abad, M. Ulbricht, Dispersions of silica nanoparticles in ionic liquids investigated with advanced rheology. J. Nanopart. Res. **14**, 1–10 (2012)
10. J.M.P. Francìa, S.I.C. Vieira, M.J.V. Lourencìo, S.M.S. Murshed, C.A.N. de Castro, Thermal conductivity of [C4mim][(CF3SO 2)2N] and [C2mim][EtSO4] and their ionanofluids with carbon nanotubes: Experiment and theory. J. Chem. Eng. Data **58**, 467–476 (2013)
11. S. Atashrouz, M. Mozaffarian, G. Pazuki, Modeling the thermal conductivity of ionic liquids and ionanofluids based on a group method of data handling and modified Maxwell model. Ind. Eng. Chem. Res. **54**, 8600–8610 (2015)
12. M. Ishikawa, T. Sugimoto, M. Kikuta, E. Ishiko, M. Kono, Pure ionic liquid electrolytes compatible with a graphitized carbon negative electrode in rechargeable lithium-ion batteries. J. Power Sources **162**, 658–662 (2006)
13. T. Fukushima, T. Aida, Ionic liquids for soft functional materials with carbon nanotubes. Chem. Eur. J. **13**, 5048–5058 (2007)
14. M.P. Shevelyova, Y.U. Paulechka, G.J. Kabo, A.V. Blokhin, A.G. Kabo, T.M. Gubarevich, Physicochemical properties of imidazolium-based ionic nanofluids: Density, heat capacity, and enthalpy of formation. J. Phys. Chem. C **117**, 4782–4790 (2013)
15. S.S. Moganty, S. Srivastava, Y. Lu, J.L. Schaefer, S.A. Rizvi, L.A. Archer, Ionic liquid tethered nanoparticle suspensions: A novel class of ionogels. Chem. Mater. **24**, 1386–1392 (2012)
16. A.E. Visser, N.J. Bridges, B.L. Garcia-Diaz, J.R. Gray, E.B. Fox, Al_2O_3-based nanoparticle enhanced ionic liquids (NEILs) for advanced heat transfer fluids, in *Ionic Liquids: Science and Applications, ACS Symposium Series*, vol. 1117, (ACS, New York, 2012), pp. 259–270
17. Z. He, P. Alexandridis, Nanoparticles in ionic liquids: Interactions and organization. Phys. Chem. Chem. Phys. **17**, 18238–18261 (2015)
18. D. Deb, S. Bhattacharya, Influence of ionic-liquid-tethered Al_2O_3 nanoparticle on the nonisothermal cold crystallization in ionic-liquid-based nanofluids. J. Phys. Chem. C **121**, 6962–6976 (2017)
19. I. Fernández-Barahona, M. Muñoz-Hernando, F. Herranz, Microwave-driven synthesis of iron-oxide nanoparticles for molecular imaging. Molecules **24**(7), 1224 (2019). https://doi.org/10.3390/molecules24071224
20. T. Ahn, J.H. Kim, H.-M. Yang, J.W. Lee, J.-D. Kim, Formation pathways of magnetite nanoparticles by coprecipitation method. J. Phys. Chem. C **116**(10), 6069–6076 (2012). https://doi.org/10.1021/jp211843g
21. B. Pant, M. Park, S.-J. Park, One-step synthesis of silver nanoparticles embedded polyurethane nano-fiber/net structured membrane as an effective antibacterial medium. Polymers **11**, 1185 (2019). https://doi.org/10.3390/polym11071185
22. S. Borsley, W. Edwards, I.K. Mati, G. Poss, M. Diez-Castellnou, N. Marro, E.R. Kay, A general one-step synthesis of alkanethiyl-stabilized gold nanoparticles with control over core size and monolayer functionality. Chem. Mater. **35**(15), 6168–6177 (2023). https://doi.org/10.1021/acs.chemmater.3c01506
23. A.A. Minea, S.M. Sohel Murshed, A review on development of ionic liquid based nanofluids and their heat transfer behavior. Renew. Sust. Energ. Rev. **91**, 584–599 (2018). https://doi.org/10.1016/j.rser.2018.04.021
24. C.A. de Castro, S.M.S. Murshed, M.J.V. Lourenço, F.J.V. Santos, M.L. Matos Lopes, J.M.P. França, Ionanofluids: new heat transfer fluids for green processes development, in *Green Solvents I: Properties and Applications in Chemistry*, ed. by M. Ali, Inamuddin, (Springer, Cham, 2012)

25. T.C. Paul, A.K.M. Morshed, E.B. Fox, J.A. Khan, Thermal performance of Al_2O_3 nanoparticle enhanced ionic liquids (NEILs) for concentrated solar power (CSP) applications. Int. J. Heat Mass Transf. **85**, 585–594 (2015)
26. S.M.S. Murshed, C.A. Nieto de Castro, M.J.V. Lourenço, J. França, A.P.C. Ribeiro, S.I.C. Vieira, C.S. Queirós, Ionanofluids as novel fluids for advanced heat transfer applications. World Acad. Sci. Eng. Technol. **52**, 795–798 (2011)
27. F. Wang, L. Han, Z. Zhang, X. Fang, J. Shi, W. Ma, Surfactant-free ionic liquid-based nanofluids with remarkable thermal conductivity enhancement at very low loading of graphene. Nanoscale Res. Lett. **7**, 314–319 (2012)
28. J. Liu, F. Wang, L. Zhang, X. Fang, Z. Zhang, Thermodynamic properties and thermal stability of ionic liquid-based nanofluids containing graphene as advanced heat transfer fluids for medium-to-high-temperature applications. Renew. Energy **63**, 519–523 (2014)
29. T.C. Paul, A.K.M.M. Morshed, E.B. Fox, J.A. Khan, Enhanced thermophysical properties of NEILs as heat transfer fluids for solar thermal applications. Appl. Therm. Eng. **110**, 1–9 (2017). https://doi.org/10.1016/j.applthermaleng.2016.08.004
30. A.A. Minea, Overview of ionic liquids as candidates for new heat transfer fluids. Int. J. Thermophys. **41**, 151 (2020). https://doi.org/10.1007/s10765-020-02727-3
31. B. Jozwiak, G. Dzido, E. Zorebski, A. Kolanowska, R. Jedrysiak, J. Dziadosz, et al., Remarkable thermal conductivity enhancement in carbon-based ionanofluids: Effect of nanoparticle morphology. ACS Appl. Mater. Interfaces **12**, 38113–38123 (2020). https://doi.org/10.1021/acsami.0c09752
32. E.B. Fox, A.E. Visser, N.J. Bridges, J.W. Amoroso, Thermophysical properties of nanoparticle-enhanced ionic liquids (NEILs) heat-transfer fluids. Energy Fuel **27**, 3385–3393 (2013). https://doi.org/10.1021/ef4002617
33. J.M.P. Franca, M.J.V. Lourenco, M. Sohel Murshed, A.A.H. Padua, C.A. Nieto de Castro, Thermal conductivity of ionic liquids and ionanofluids and their feasibility as heat transfer fluids. Ind. Eng. Chem. Res. **57**, 6516–6529 (2018). https://doi.org/10.1021/acs.iecr.7b04770
34. P. Keblinski, S.R. Phillpot, S.U.S. Choi, J.A. Eastman, Mechanisms of heat flow in suspensions of nano-sized particles (nanofluids). Int. J. Heat Mass Tran. **45**, 855–863 (2002). https://doi.org/10.1109/icosp.2004.1441590
35. L. Sang, T. Liu, The enhanced specific heat capacity of ternary carbonates nanofluids with different nanoparticles. Sol. Energy Mater. Sol. Cells **169**, 297–303 (2017). https://doi.org/10.1016/j.solmat.2017.05.032
36. F.F. Zhang, F.F. Zheng, X.H. Wu, Y.L. Yin, G. Chen, Variations of thermophysical properties and heat transfer performance of nanoparticle enhanced ionic liquids. R. Soc. Open Sci. **6**, 182040 (2019). https://doi.org/10.1098/rsos.182040
37. B. Jozwiak, S. Boncel, Rheology of ionanofluids—A review. J. Mol. Liq. **302**, 112568 (2020). https://doi.org/10.1016/j.molliq.2020.112568
38. F. Pinto, M. Meo, Design and manufacturing of a novel shear thickening fluid composite (STFC) with enhanced out-of-plane properties and damage suppression. Appl. Compos. Mater. **24**, 643–660 (2017). https://doi.org/10.1007/s10443-016-9532-1
39. B. Bakthavatchalam, K. Habib, R. Saidur, B.B. Saha, K. Irshad, Comprehensive study on nanofluid and ionanofluid for heat transfer enhancement: A review on current and future perspective. J. Mol. Liq. **305**, 112787 (2020). https://doi.org/10.1016/j.molliq.2020.112787
40. R. Alcalde, M. Atilhan, S. Aparicio, Intermolecular forces in 1-butyl-3-methylimidazolium bis(trifluoromethylsulfonyl)imide + ethanol mixtures. J. Mol. Liq. **258**, 1–9 (2018). https://doi.org/10.1016/j.molliq.2018.02.064
41. C.P. Fredlake, J.M. Crosthwaite, D.G. Hert, S.N.V.K. Aki, J.F. Brennecke, Thermophysical properties of imidazolium-based ionic liquids. J. Chem. Eng. Data **49**, 954–964 (2004). https://doi.org/10.1021/je034261a
42. K. Oster, C. Hardacre, J. Jacquemin, A.P.C. Ribeiro, A. Elsinawi, Ionic liquid-based nanofluids (ionanofluids) for thermal applications: An experimental thermophysical characterization. Pure Appl. Chem. **91**, 1309–1340 (2019). https://doi.org/10.1515/pac-2018-1114

Innovative Biomedical and Clinical Approaches in Ionic Liquids, Nanofluids, and Nanoparticles

A. Ashwini

1 Introduction on Convergence of Ionic Liquids, Nanofluids, and Nanoparticles in Biomedicine

Nanotechnology integration along with innovative chemical systems has established new possibilities for diagnostic technologies and drug delivery systems and therapeutic treatments in biomedicine. Three main likely materials in biomedical investigation include ionic fluids, nanofluids, and nanoparticles, which present different physicochemical properties adjustable for progressive biomedical uses [1]. This interdisciplinary grouping of material science, nanotechnology, and chemistry with medication creates an actual approach to resolving long-standing healthcare glitches. These liquid combinations demonstrate medical value because they soften different biomolecules while increasing enzyme roles and work as both solvent movers and antimicrobial representatives.

Engineered colloidal interruptions named nanofluids contain base fluid precipitations, which contain nanoparticles. Such fluids validate superior abilities in thermal conductivity as well as chemical stability landscapes, together with attractive or optical behaviors that surpass conventional fluid possessions. Bioengineers practice nanofluids to develop groundbreaking biomedical devices that change drugs to specific positions as well as progress MRI imaging and use heat to fight tumor cells. Nanoparticles—particles with dimensions on the nanometer scale—are already well established in biomedical research due to their size-dependent properties, large surface area-to-volume ratios, and ease of surface functionalization. Scientists utilize them for three types of biomedical applications: delivering drugs to specific

A. Ashwini (✉)
Department of Electronics and Communication Engineering, Vel Tech Rangarajan Dr. Sagunthala R&D Institute of Science and Technology, Chennai, Tamil Nadu, India

S. Kulkarni, S. Das (eds.), *Ionic Liquids, Nanofluids and Nanotechnology*, Lecture Notes in Nanoscale Science and Technology 42,
https://doi.org/10.1007/978-3-032-04008-4_8

areas, creating imaging devices, performing genetic treatments, and building tissue engineering platforms.

Multiple advancements in technology emerge when ionic liquids (ILs) merge with nanofluids while integrating nanoparticles. Organic medicinal scientists use ILs to sustain nanoparticles and nanofluids, which create optimized systems for managing therapeutic agent distribution. Lab-diagnostic equipment based on nanoparticles within ionic liquid nanofluids demonstrates highly refined diagnostic capabilities [2]. Medically safe IL-magnetic nanoparticle hybrid nanofluid systems enable the creation of new minimally invasive treatment methods through magnetic hyperthermia. Scientists use biocompatible ionic liquids with nanoparticles' structural properties to create new scaffolds, which serve as platforms in regenerative medicine.

The recent research puts emphasis on sustainable green synthesis systems developed with biodegradable ionic liquids and bionanoparticles to reduce toxic effects while decreasing environmental harm, which represents a critical issue within nanomaterial clinical adoption [3]. A revolutionary new phase emerges in biomedicine because of the integration between ionic liquids and nanofluids and nanoparticles. The combined effects of these elements will create advanced biomedical tools, which lead to personalized medicine based on precise approaches.

2 Ionic Liquids in Drug Delivery and Therapeutics

The liquid state salts known as "designer solvents" refer to ionic liquids (ILs) which remain liquid at temperatures under 100 °C. These solutions only consist of ions that integrate organic or inorganic anions in conjunction with bulky organic cations. ILs show considerable potential in biomedical solutions because their unique characteristics include base vapor pressure values, chemical and thermal resistance levels, variable polarities, and superior solvent abilities. The chemical modification of IL components through cations or anions permits the customizing of their biological behavior and functionality, which enables developers to elaborate sophisticated delivery solutions.

The application of ionic liquids in drug delivery methods enhances the solubility as well as stability and improves bioavailability for medicines that have limited water solubility. The therapeutic effectiveness of numerous pharmaceutical compounds becomes constrained by their insufficient solubility rates [4]. Pure and stable drug compounds become accessible through ILs because they dissolve hydrophilic as well as hydrophobic pharmaceutical ingredients. ILs function as protective agents to preserve unstable drugs because they create an environment that prevents hydrolytic and enzymatic transformations. Some drug molecules are transformed into active pharmaceutical ingredient-ionic liquids (API-ILs) through pairing them with suitable counterions. Boiling drug molecules into ILs improves both their physical attributes while simultaneously enabling controlled release methods and delivery system targeting ability [5].

The study of ionic liquids continues for therapeutic uses while these compounds help pharmaceutical drugs become more soluble and stable. Ionic liquids demonstrate bioactivity themselves because they conduct antimicrobial, anticancer, and anti-inflammatory properties, which enables them to serve both as carriers together with therapeutic agents. Developing ionic liquids allows them to detect specific external stimuli, which enables targeted drug release while maintaining full control over the delivery process—this essential skill is particularly useful for cancer treatment when exact drug targeting strategies are required [6]. Cell membranes can interact with ILs at adjustable strength levels to enhance drug delivery through membranes and decrease adverse side effects in the system. Research continues to determine biodegradable and non-toxic ionic liquids for clinical applications because proper design and toxicity assessment are essential due to the fact that not all ionic liquids are biocompatible.

3 Nanofluids for Enhanced Medical Diagnostics and Imaging

A nanofluid describes engineered suspensions composed of nanoparticles dispersed inside base fluids, which can include water besides oils and biological solutions. The unique combination of properties emerges from nanoparticles such as metals, including gold or silver, together with metal oxides, including iron oxide, then carbon-based materials, including graphene, in nanofluids. Medical diagnostics and imaging benefit from nanofluids because these fluids improve signal sensitivity and targeted imaging while providing enhanced contrast features [7]. Smaller particles in these agents allow proper dispersion and stability while providing effortless tissue penetration for successful non-invasive medical examination procedures.

Nanofluids function as vital contrast agents throughout medical imaging procedures, especially magnetic resonance imaging, computed tomography, and ultrasound imaging. Superparamagnetic agents that comprise iron oxide nanofluids serve as enhanced MRI contrast agents by manipulating tissue-level magnetic fields. CT scanning contrast detection improves by using nanofluids containing gold nanoparticles because the high atomic number of gold increases X-ray absorption [8]. Through ligand-functionalization of nanoparticles, healthcare professionals can detect biomarkers and cancer cells specifically, which provides precise diagnosis in the early stages.

Experts in the biomedical sector apply increasing attention to nanoparticle structures used with metal elements and ceramics or polymers, since they demonstrate strong potential for medical treatments through gene therapy, drug delivery, and regeneration procedures. Nanoparticle interaction methods inside biological organisms enable the delivery of treatments through different biological membrane barriers, including the blood–brain barrier [9]. The chapter describes personalized medicine through genetic profiles as an upcoming medical development, which relies on functionalized nanoparticles. Advanced treatment development faces fundamental roadblocks due to toxicity-related safety concerns, biocompatibility issues, and regulatory hurdles, despite its exceptional potential for development. This study links research in material science and biotechnology alongside medical findings to create future health innovations out of these novel materials.

Nanofluids are subject to current research investigations regarding multi-modal imaging capabilities that leverage singular nanoparticle systems to enhance dual image detection simultaneously. Special nanofluid design would merge MRI with fluorescence imaging to create a system that shows disease information at anatomical and molecular levels. Photoacoustic imaging techniques use specific nanofluids that convert absorbed light into ultrasound waves, which provide high-resolution imaging of deep tissues. Nanofluids will become the key to diagnostic medical advancements through their ability to deliver functions that include rapid detection with safety plus precise and individualized imaging abilities.

4 Nanoparticles in Precision Medicine and Targeted Therapy

Medical advances in precision medicine and targeted therapy primarily result from nanoparticle characteristics, including their small dimensions and surface capabilities and functional tailoring capabilities [10]. Through precision medicine, patients receive personalized medical treatments which utilize nanoparticles as ideal platforms for achieving customized care. The nanometer-scale measurements (1–100 nm) of nanoparticles let them easily pass through tissues and biological barriers along with circulating through the body. Research has exposed that nanoparticles can be fortified with ligands or antibodies or aptamers that pursue distinct cellular indicators to reach exact unhealthy cells, thereby dipping injury to strong tissues.

The distribution of drugs and genes or other therapeutic features occurs through nanoparticles during directed therapy practices [11]. Nanoparticles originate equipped with landscapes that make them issue their substance gratified only after particular activates such as pH differences and enzyme responses and various forms of external stimulus from heat or light bases. Tumor microenvironments uphold lower pH levels than usual tissues which nanoparticles use through spatial drug release tools at their target site. Targeted drug distribution schemes improve both therapeutic consequences and minimize standard chemotherapy and radioactivity side effects as of their directed approach. Liposomes, collected with dendrimers, gold nanoparticles and polymeric nanoparticles, characterize the advanced nanoparticle types which have established clinical support to treat tumor along with cardiac and infectious diseases.

Precision diagnostics uses nanoparticles as imaging contrast mediators and biosensors to transport highly beneficial analytical capabilities. A diagnostic method enabled through nanoparticles combines a therapeutic role with diagnostic landscapes, thus letting physicians to track drugs inside patients, measure treatment accomplishment, and spot disease declines in real time. Nanoparticles signify recent technical advancements because they function as distribution vehicles to provide medications while detecting cellular environments [12]. Treatment methods become imaginable to adapt rapidly through biological feedback checking, which constitutes an important component of accurate medical systems. The healthcare background of these days progresses toward more customized, robust actions because of nanoparticles' wide-ranging effect.

4.1 Liposomes

Natural cell membranes counterpart the construction of spherical vesicles recognized as liposomes, which consist of lipid bilayers. These nanoparticles keep a high level of compatibility with biological surroundings while being able to switch both kinds of drugs that soften in water and in fat. Liposomes turn into precision medication by shielding drugs from bloodstream ruin prevention and directing delivery to precise areas, which results in reduced injury to surrounding healthy materials. Liposomes become better at perceiving precise cell types after surface alterations consist of ligand or antibody accumulation. Research scientists established Doxil as a liposomal preparation that covers the chemotherapy drug DOXOrubicin which delights both ovarian growth and Kaposi's tumor.

4.2 Polymeric Nanoparticles

Polymeric nanoparticles originate their arrangement from ecological synthetic polymers, with polylactic acid (PLA). These nanoparticles deliver researchers with durable, versatile capabilities and allow the release of medicine at measured rates over prolonged extents. Drug targeting routine and stability inside the body become improved after surface alterations to these drug distribution schemes. Polymeric nanoparticles of targeted therapy receive engineering to permit them to collect primarily in unhealthy areas like tumors over passive directing features from enhanced permeability and maintenance or active levelling from surface ligand procedure. Scientists widely utilize these materials for developing drug delivery platforms of anticancer drugs as well as genetic sequences and vaccines.

4.3 Gold Nanoparticles

The popularity of AuNPs as precision medicine research tools stems from their key benefit set, including outstanding bioavailability, simple modification capabilities, and distinct optical signal characteristics. The solid nanoparticles demonstrate exceptional applications for imaging and therapy because they effectively absorb and scatter visible light. The combination of targeted therapy uses gold nanoparticles that guide through the bloodstream towards tumors, where photothermal therapy produces heat to precisely destroy cancer cells. Medical researchers use these nanoparticles to transfer drugs together with genetic material to distinct bodily tissues. Their minimal degradation, along with their detectability, makes them outstanding candidates for diagnostic testing and treatment in a single approach known as theranostics.

4.4 Magnetic Nanoparticles

Iron oxide nanoparticles such as Fe_3O_4 positioned as magnetic nanoparticles possess magnetic features that enable their dual application in imaging and therapeutic modalities. The external magnetic field serves as a guidance system to lead magnetic nanoparticles precisely to desired locations inside the body, thus enhancing treatment delivery to target sites. The diagnostic accuracy in magnetic resonance imaging is enhanced through the application of magnetic nanoparticles as contrast agents. Agents composed of magnetic nanoparticles produce heating effects for cancer cell elimination during magnetic hyperthermia therapy by subjecting them to alternating magnetic fields.

4.5 Dendrimers

The synthetic macromolecule dendrimers use branched tree architecture combined with a defined structure. Multicomponent drug delivery is enabled by their distinct structural arrangement, which provides accessible binding locations for drugs and targeting ligands as well as imaging agents. Dendrimers serve as carriers of multiple therapeutic compounds while enabling controlled drug delivery and diminished harmful side effects. The surface of dendrimers allows scientists to attach targeting molecules such as antibodies, which can find and connect with particular disease markers. The use of dendrimers signifies a promising beneficial method for cancer conducts alongside genetic factor therapy as well as dealing with infectious diseases outstanding to their detailed and skillful nature.

4.6 Silica Nanoparticles

The pore arrangement of mesoporous silica nanoparticles (MSNs) permits considerable drug integration because of their high order preparation. Experts adjust these nanoparticles with easiness to offer directed delivery abilities in addition to using ecological triggers to switch the drug release proportion. Silica nanoparticles find applications in precision medicine because they serve to transport chemotherapeutic agents and imaging agents and genetic material. The combination of stability and biocompatibility with tunable size makes silica nanoparticles useful agents for diagnostic and therapeutic purposes.

4.7 Carbon-Based Nanoparticles

The fantastic mechanical and electrical properties combined with surface modification ability make carbon nanotubes, graphene oxide, and fullerenes attractive for various applications. The field of precision medicine employs these nanoparticles to supply medications while also featuring them for biosensing operations and photothermal therapeutic applications. Through functionalization, carbon nanotubes possess excellent cellular penetration abilities to deliver medical substances directly into cellular cytoplasm [13]. The major clinical barriers to using these materials lie in their biocompatibility evaluation alongside toxicity reduction.

4.8 Exosomes

Exosomes represent naturally occurring cell-derived vesicles that function as biological communication messengers of the human body. The vesicles transport both proteins and lipids and RNA as well as DNA materials between different cells. The exploration of exosomes grows as medical investigators study them as biological delivery agents of drugs and genes in precision medicine. Cells serve as their origin, making exosomes less likely to start immune reactions while also allowing engineers to load drug agents directly for target tissue delivery. The ability to create personalized treatments for real-time monitoring has made exosomes become one of the most promising nanoparticle types in medical applications.

5 Ionic Liquid-Based Biosensors for Clinical Diagnostics

Ionic liquids demonstrate special salt characteristics through their capability to maintain liquid form at or near room temperature while offering high ionic conductivity alongside wide electrochemical windows and chemical and thermal stability as well as adjustable solubility. The properties of ionic liquids enable their development as exceptional materials for building next-generation biosensors used in clinical diagnosis tests. Traditional solvents dissipate, but ionic liquids protect biological substances such as enzymes, DNA, and antibodies alongside maintaining their functional integrity [14]. The stabilization property serves as a decisive factor in clinical biosensors because it allows preservation of biological recognition element performance for accurate diagnostic outcomes. Figure 1 shows the types of ionic liquids.

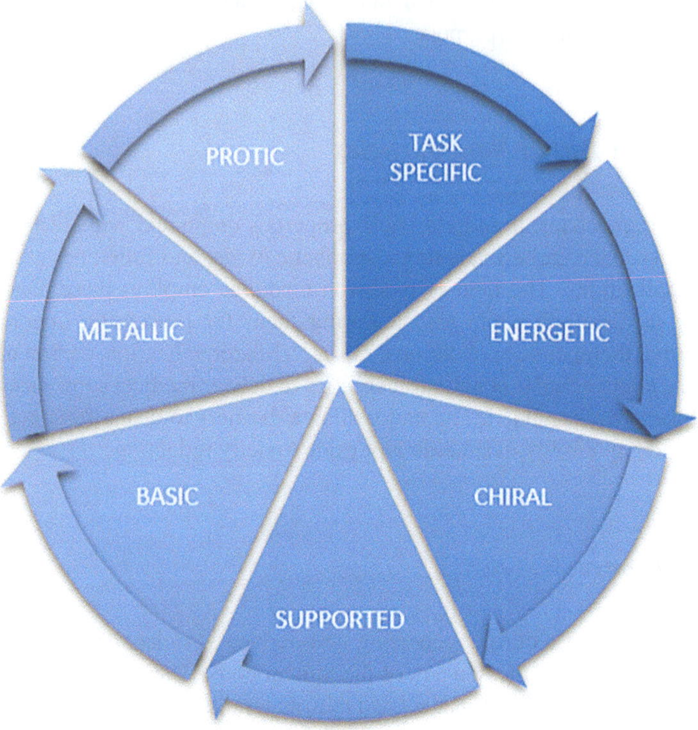

Fig. 1 Types of ionic liquids

The detection of diverse biomarkers such as glucose, cholesterol, cancer biomarkers, infectious agents, and nucleic acids exists through IL-based biosensors in clinical diagnostic settings. The biological element gets immobilized through the medium or matrix, which is composed of ionic liquids. A highly biocompatible microenvironment combination offered by ionic liquids enhances the sensor device operations by providing faster responses while increasing sensitivity levels and maintaining stability. The ion-conducting properties of ILs promote rapid and dependable electrochemical signals among biological components and electrodes, which drives the development of point-of-care diagnostic devices.

The main benefit of ionic liquid-based biosensors arises from their ability to enable miniature diagnostic devices that integrate flexibly and can be worn. Scientists are now developing highly sensitive portable biosensors through combining ionic liquids with nanomaterials, including carbon nanotubes, graphene, and metal nanoparticles. Biomarker detection through these sensors reaches a level that enables early disease assessment capabilities. The adjustable properties of ionic liquids enable precise customization for medical applications through pH-sensitive substances for wound monitoring as well as temperature-sensitive materials for fever detection [15]. Current advances in biosensor research indicate they will become essential for both personal medicine approaches and live patient healthcare tracking in the upcoming future.

6 Nanofluids in Regenrative Machine and Tissue Engineering

Various medical applications employ nanofluids either in scaffolding designs to reconstruct tissues or use them as delivery solutions for tissue restoration. The application of nanofluids that contain hydroxyapatite nanoparticles serves bone regeneration purposes because these particles share a similar chemical composition with bone mineral. The nanosized fluid possesses increased mechanical properties and encourages bone tissue growth while accelerating tissue-to-surrounding integration. Nanofluids containing magnetic nanoparticles benefit from external magnetic field applications, which let users direct and regulate tissue developmental patterns while achieving spatial precision in the healing process [16]. The ability proves most beneficial for creating neural networks and blood vessels as complex structures, as shown in Fig. 2.

Biomedical investigators utilize nanofluids within regenerative medicine to develop controlled release systems that transport growth factors and genes along

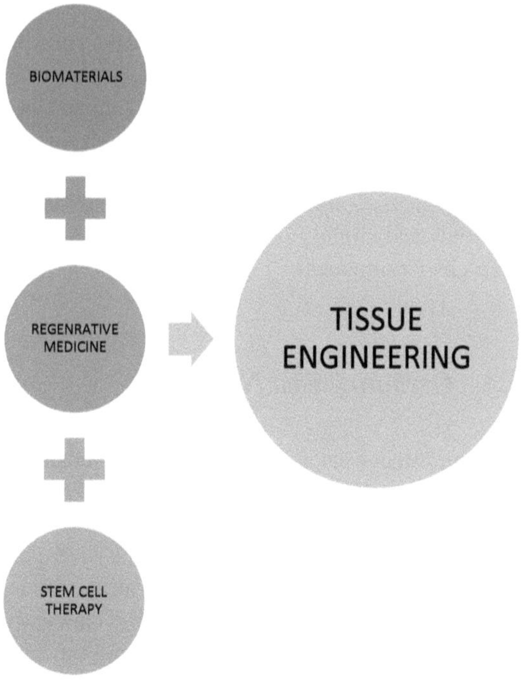

Fig. 2 Nano in regenerative machine and tissue engineering

with therapeutic agents directly to damaged tissue locations. Nanofluid systems with smart functionality respond to stimuli, including temperature, pH, and enzymatic activity, to release therapeutic agents at precise locations and particular times. Nanofluids improve stem cell actions by making stem cells live longer and decision-making stem cell alteration pathways while also providing systems to display transplant cell spots noninvasively.

6.1 Metal and Metal Oxide Nanofluids

Investigation on tissue engineering often contains metal-based nanofluids with gold (Au) or silver (Ag) or titanium dioxide (TiO_2) or zinc oxide (ZnO) nanoparticles. During tissue regeneration bacterial defense becomes imaginable through the usage of these nanoparticles. The grouping of silver nanoparticles in nanofluids used for wound remedial scaffold jobs to combat pollution together with refining tissue recovery time [17]. Titanium dioxide nanofluids display both excellent compatibility with biological schemes and better capabilities for pleasant bone cells recognized as osteoblasts, resulting in their standing for bone muscle growth.

6.2 Hydroxyapatite Nanofluids

Nanofluids made with hydroxyapatite (HA) grasp a special rank for bone and dental tissue manufacturing due to their usual presence and their inert characteristics as calcium apatite. HA nanofluids spare human bone and tooth mineral arrangement, thus manufacturing robust bone scaffolds for incorporation into normal tissues. Osteoconductivity progresses from these nanofluids because they permit cells to join and increase while creating new bone tissue more successfully. 3D-printed bone scaffolds and slightly offensive injectable bone fillers comprise these materials as their chief component.

6.3 Carbon-Based Nanofluids

The third major category of nanofluids includes carbon-based nanoparticles which include graphene oxide, carbon nanotubes, and fullerenes. These materials present high levels of mechanical durability combined with electronic capabilities and pliability suitable for developing tissues that need electrical and mechanical stimulation such as muscle as well as nerve and cardiac tissues. Research has demonstrated the use of graphene nanofluids for building conductive scaffolds that guide the proper organization of nerves cells for promoting neural tissue growth. Carbon-based nanofluids help guide stem cells toward selecting specific cells by both transmitting biological indicators and physical stimuli.

6.4 *Magnetic Nanofluids*

Researchers use dispersed iron oxide (Fe_3O_4)-based magnetic nanoparticles for producing magnetic nanofluids commonly referred to as ferrofluids within fluids. The usage of such materials shows great potential in the field of magnetically guided tissue engineering. Cell organization and scaffold migration occur better thanks to magnetic nanofluids during external magnetic field use, which results in enhanced structure formation of new tissues [18]. Magnetic nanofluids provide a useful tool for hyperthermia therapy because they generate localized heating, which enables tissue repair alongside healthy tissue protection.

6.5 *Polymer-Based Nanofluids*

The regenerative site requires a targeted and extended release of growth factors and drugs and genetic material through PLGA-poly(lactic-co-glycolic acid) polymeric nanoparticles, which constitute essential components of nanofluids. The polymer-based nanofluids maintain biocompatible and biodegradable properties for ensuring they do not generate adverse responses. To support proper tissue maturation in soft tissue engineering applications, these nanofluids enable controlled bioactive molecule release in cartilage repair besides playing a role in vascular tissue engineering and skin regeneration.

7 Ionic Liquids and Nanotechnology in Antimicrobial and Antiviral Treatments

The antimicrobial and antiviral development relies heavily on ionic liquids (ILs) because they possess distinct physical and chemical attributes. The liquid state property of IL salts at room or near-room temperatures allows them to break microbial membranes and prevent essential protein structures and enzyme processes. Antimicrobial ionic liquids show strong microbial activity against bacteria (including Gram-positive and Gram-negative strains), fungi, and viral pathogens because they are derived from imidazolium, pyridinium, and phosphonium cation structures. Research scientists achieve better ionic liquid performance through structural modification techniques which comprise alkyl chain length variations and functional group integration [19]. The structural modifications applied to ILs enable their use as biocompatible agents in medical wound dressings alongside their application in disinfectants and bactericide drug delivery carriers as well as medical equipment coatings.

Antimicrobial and antiviral treatment methods have experienced a breakthrough with nanotechnology thanks to the development of nanoparticles (NPs) that demonstrate intrinsic biological properties or function as carriers for therapeutic compounds. Nanoparticles made from silver, gold, copper oxide, and zinc oxide demonstrate superb antimicrobial activity through the process of reactive oxygen species production, membrane breakdown, and DNA copying interference. Nanoparticles used in antiviral applications function by stopping viruses from entering host cells and by reinforcing antiviral drug delivery while blocking the viral replication sequence. The deactivation of HIV along with influenza and multiple enveloped viruses occurs through silver nanoparticles, which adhere to viral proteins to stop host cell attachment. The nanoparticle scale enables deep tissue penetration compared to standard antibiotic drugs, thus presenting a major advantage over other antibiotics as well as antivirals. Researchers have developed IL-functionalized nanoparticles through combining nanotechnology with ionic liquids to achieve better synergy between antimicrobial and antiviral effectiveness [20]. The incorporation of ionic liquids as surface coatings enhances the dispersion stability as well as improves biocompatibility and specific biological interaction capabilities of nanoparticles in these systems.

In tool growth for clinicians, inventers focus on creating boundaries that offer flat integration within their workspace and uphold simplicity. The treatment involves of electronic health record arrangements as well as analytical tools and platforms powered by AI expertise. The tools used by clinicians for difficult quick choices need to be simple to use and require basic training because difficult interfaces diminish work efficiency. The design of user-friendly interfaces with low navigation complexity represents usability in interface creation. An effective AI diagnostic tool design must present vital information using unassuming displays that label essential data points in a straightforward and readable manner [21]. The accessibility of tools refers to making these resources easily available through different clinical platforms to offer ready access for healthcare professionals. The integration with mobile devices and computers ensures doctors can obtain patient data and generate choices both in stressful situations and when employed with portable plans.

8 Applications of Ionic Liquid-Based Drug Screening

Research in drug detection now trusts seriously in ionic liquids as gifted tools as of their unique properties since these salts exist in a molten state at or near room disease. Ionic materials offer high flexibility because their physicochemical possessions can be tuned together with ionic conductivity and divergence and solubility features. IL characteristics allow their claim as drug cleaners in adding to biomolecule keys and high-throughput broadcast platforms.

8.1 Applications in Drug Screening

Solvent and Medium for Drug Screening

Ionic liquids donate to outstanding drug screening as they effectively run watery and oily compounds concurrently, which allows evaluation of varied pharmaceutical applicants that other dissolving agents cannot handle [22]. The transmission of composites in liquid media develops possible with ILs as these solvents combine their high ionic conductivity with low instability properties to provide evaporation-free situations that overpower limitations created with traditional solvents. They successfully dissolve intricate biomolecules comprising proteins along with nucleic acids and lipids, which advances their welfare for bioassays.

Stability and Reusability

The durability of ILs starts them as appropriate agents for drug transmission stages because drug screening procedures rest on constant solution steadiness throughout the assay series. Monitoring drug detections becomes more effective after ILs validate the ability to operate in showing assays for numerous cycles deprived of performance degradation [23]. They also display high thermal stability, which is beneficial when showing drugs under circumstances that may need elevated temperatures (for instance, enzyme inhibition studies or thermal stability assays).

High-Throughput Screening (HTS) in Pharmaceutical Research

High-throughput transmission systems progress more efficiently by using Ionic liquids as they permit rapid analysis of several drug candidates for biotic activity. Investigators can complete thousands of bioassays through automatic screening stages by utilizing ILs as an intermediate solution. The ionic liquid succeeds in interacting with biological molecules while their structure remains intact, which provides a better drug screening process with reduced adverse reactions compared to traditional organic solvents.

Ionic Liquids in Enzyme Inhibition Assays

The enzyme inhibition studies require effective solvents, and ILs demonstrate their suitability to perform tests for drug candidate assessment against specific enzymes. Derivatives with this structure provide conditions that stabilize enzymes together with testing elements, which leads to precise outcome analysis. Such features prove essential when screening drug candidates that act on metabolic disorder enzymes as well as cancer and infectious disease enzymes.

10 Sustainability and Green Chemistry in Biomedical Applications

Biomedical applications hold sustainability along with green chemistry plans due to their dual purposes of sinking environmental effects and increasing medical treatment safety and efficiency. Green chemistry signifies an approach to progress chemical products and industrial processes which diminish hazardous substance practice and waste generation in order to guard human health as well as ecological welfare [28]. The biomedical field chains safety and efficacy of materials utilization with renewable resources management over energy-efficient procedures that reduce final waste.

10.1 Key Principles of Green Chemistry in Biomedical Applications

Use of Renewable Resources

Medical applications of bio-based polymers include their use in medical devices together with drug delivery systems and tissue scaffolds through the utilization of polylactic acid (PLA) and polyhydroxyalkanoates (PHA). The production of these materials happens from sustainable resources, including corn and sugarcane, and they decompose naturally, which improves their environmental impact over conventional petroleum-based plastics. Medical devices have decreased dependency on fossil fuels because of their adoption, which results in reduced carbon emissions [29].

Energy Efficiency and Green Manufacturing Processes

Green chemistry supports operational methods in manufacturing that both decrease resource consumption and eliminate dangerous substances from production requirements. The biomedical device production process currently uses additive manufacturing and 3D printing to build patient-tailored prosthetic implants. The production method results in significantly less waste than conventional machining does and enables complete material precision. Solvent-free synthesis and microfluidic-based synthesis allow manufacturers to operate without organic solvents that commonly prove dangerous to the environment.

Nontoxic and Biodegradable Materials

Nontoxic materials that possess biocompatibility and biodegradability levels are vital needs when implementing biomedical products. Green chemistry promotes the development of substances that protect living beings while dissolving into body tissues and decomposing into harmless byproducts. Nanotechnology firms use biodegradable nanoparticles and environmentally friendly manufacturing methods to produce metal nanoparticles, which serve as platforms for pharmaceutical delivery systems and medical imaging mechanisms and tissue regeneration methods. Controlled drug delivery systems use nanoparticles made of natural biopolymers including, chitosan and albumin, because these do not produce long-term toxicity problems.

Waste Minimization and Closed-Loop Systems

The principles of green chemistry work to minimize waste outcomes that occur from production through the complete product lifespan, starting with manufacturing and ending at disposal. Most elements of biomedical production can embrace waste reduction techniques for developing medical products, including surgical instruments and diagnostic equipment. Healthcare institutions can establish sustainable practices through closed-loop systems, which either recycle waste materials or put them back into service, such as reusing bioplastics and reconditioning medical devices. The implementation of atom-efficient chemical reactions together with catalysts in drug synthesis produces efficient therapeutics by reducing by-product generation.

10.2 Examples of Green Chemistry in Biomedical Applications

Green Synthesis of Nanoparticles

Nanoparticle production through conventional methods requires toxic chemicals in addition to dangerous substances that use high amounts of energy during synthesis processes. Green synthesis of nanoparticles occurs through environmentally friendly techniques that employ plant extracts and microbial agents and bio-waste materials. Green tea extract combined with aloe vera extract functions as both a stabilizing and reducing agent to produce silver nanoparticles and gold nanoparticles through green synthesis. These ecofriendly nanoparticles provide sustainable drug delivery possibilities and antibacterial treatments and diagnostic solutions instead of conventional options.

Sustainable Drug Delivery Systems

Drug delivery systems experience revolutionary changes through green chemistry because researchers develop controlled-release systems using degradable biopolymers. Green chemistry principles enable the synthesis of three drug delivery systems including polymeric micelles as well as liposomes and hydrogels, which provide both safety and effectiveness to the environment [30]. Drug delivery systems equipped with these polymers enable targeted medication delivery, which decreases side effects until they degrade completely to nonharmful products.

Biodegradable Implants and Scaffolds

Tissue engineering and regenerative medicine produce the most important advantage of green chemistry within biomedicine. Healthcare practitioners utilize biodegradable scaffolds alongside implants as components for bone as well as cartilage and skin regeneration. Scientists fabricate new PLGA, PCL, and collagen-based scaffolds through environment-friendly manufacturing methods. The implants break down naturally through time, so patients do not need additional surgical procedures, and the environment does not face long-term disposal issues.

Green Chemistry in Diagnostic Devices

Diagnostics research for environmentally friendly chemistry has produced sustainable diagnostic tools through the development of biosensors together with microfluidic devices. The diagnostic devices make use of green solvents simultaneously with biodegradable materials to attain lower environmental impact for disposable tests. Point-of-care diagnostic tools that utilize paper-based biosensors provide healthcare access to all population segments because these sensors are both affordable and easy to make with environmentally friendly materials.

10.3 Challenges and Future Outlook

The implementation of sustainable and green chemistry in biomedical applications shows great promise, yet challenges remain unsolved. The production scale-up of bio-materials requires cost reduction, while safety standards must be strengthened to test new green materials and sustainable technologies need regulatory authorization. Patient outcomes together with environmental protection remain achievable in the biomedical industry when resources are allocated to developing sustainable materials while building efficient power facilities and waste management systems.

11 Conclusion

Patient-focused accountable medicine becomes achievable through the coordination of sustainable technologies and green chemistry approaches in clinical healthcare practice. The biomedical industry reaches environmental sustainability through its adoption of renewable resource systems using nontoxic materials while executing energy-saving methods for waste reduction purposes to improve device safety and drug effects and diagnostic performance. The vital role of nanotechnology in biomedical product development produces innovations that establish superior performance through its artificial intelligence components in implants as well as delivery systems and nano-forward diagnostic instruments together with biodegradable frameworks. Other than supplying innovative health solutions, these technologies adopt green chemistry principles, which ensure their products become eco-friendly while sustaining over time. Complete achievement of green chemistry's biomedicine capabilities depends on ongoing science-based research together with developmental work and united efforts between material scientists and biotechnology specialists and sustainable production experts. The fields of sustainability together with green chemistry in biomedicine no longer represent future trends because they have evolved into foundational cornerstones for contemporary healthcare development.

References

1. R.A. Khan, H.A. Mohammed, G.M. Sulaiman, A.A. Subaiyel, A. Karuppaiah, H. Rahman, Y.E. Choonara, Molecule(s) of interest: I. Ionic liquids–gateway to newer nanotechnology applications: Advanced nanobiotechnical uses', current status, emerging trends, challenges, and prospects. Int. J. Mol. Sci. **23**(22), 14346 (2022)
2. A.M. Curreri, S. Mitragotri, E.E. Tanner, Recent advances in ionic liquids in biomedicine. Advanced Science **8**(17), 2004819 (2021)
3. R. Ravichandran, Nanoparticles in drug delivery: Potential green nanobiomedicine applications. Int. J. Green Nanotechnol. Biomed. **1**(2), B108–B130 (2009)
4. S.K. Zandu, H. Chopra, I. Singh, Ionic liquids for therapeutic and drug delivery applications. Curr. Drug Abuse Rev. **12**(1), 26–41 (2020)
5. S.N. Pedro, C.S. Freire, A.J. Silvestre, M.G. Freire, Ionic liquids in drug delivery. Encyclopedia **1**(2), 324–339 (2021)
6. L. Jiang, Y. Sun, A. Lu, X. Wang, Y. Shi, Ionic liquids: Promising approach for oral drug delivery. Pharm. Res. **39**(10), 2353–2365 (2022)
7. M. Sheikhpour, M. Arabi, A. Kasaeian, A. Rokn Rabei, Z. Taherian, Role of nanofluids in drug delivery and biomedical technology: Methods and applications. Nanotechnol. Sci. Appl. **13**, 47–59 (2020)
8. G. Nobrega, R.R. de Souza, I.M. Gonçalves, A.S. Moita, J.E. Ribeiro, R.A. Lima, Recent developments on the thermal properties, stability and applications of nanofluids in machining, solar energy and biomedicine. Appl. Sci. **12**(3), 1115 (2022)
9. M. Malika, S.S. Sonawane, Few more significant applications of nanofluids, in *Nanofluid Applications for Advanced Thermal Solutions*, (Elsevier, Amsterdam, 2023), pp. 267–286

10. M.A. Alghamdi, A.N. Fallica, N. Virzì, P. Kesharwani, V. Pittalà, K. Greish, The promise of nanotechnology in personalized medicine. J. Personal. Med. **12**(5), 673 (2022)
11. A. Sengar, Precision in practice: Nanotechnology and targeted therapies for personalized care. Int. J. Adv. Nano Comput. Anal. **3**(2), 56–67 (2024)
12. M.T. Manzari, Y. Shamay, H. Kiguchi, N. Rosen, M. Scaltriti, D.A. Heller, Targeted drug delivery strategies for precision medicines. Nat. Rev. Mater. **6**(4), 351–370 (2021)
13. R.D. Munje, S. Muthukumar, B. Jagannath, S. Prasad, A new paradigm in sweat based wearable diagnostics biosensors using room temperature ionic liquids (RTILs). Sci. Rep. **7**(1), 1950 (2017)
14. R. Ghanbari, D. Khorsandi, A. Zarepour, M. Ghomi, Z. Tavakkoliamol, A. Fahimipour, A. Zarrabi, Ionic liquid-based sensors. Mater. Chem. Horizons **1**(2), 123–135 (2022)
15. D.N. Unal, S. Sadak, B. Uslu, A review on electrochemical and optical sensing platform based on ionic liquids for different molecules determination. Crit. Rev. Anal. Chem. **53**(4), 798–824 (2023)
16. S. Yuanlei, D. Andriukaitis, V. Pham, A. Karimipour, Z. Li, Study of the bio-nanofluid heat transfer rate and thermo-physical properties of fused deposition modeling of bone scaffold dip-coated by polyvinyl alcohol/nano-diopside. J. Therm. Anal. Calorim. **149**(17), 9893–9901 (2024)
17. P.K. Gupta, Tissue engineering and regenerative medicine, in *Nanotoxicology in Nanobiomedicine*, (Springer International Publishing, Cham, 2023), pp. 125–141
18. A. Shafiq, A.B. Çolak, T.N. Sindhu, Development of an intelligent computing system using neural networks for modeling bioconvection flow of second-grade nanofluid with gyrotactic microorganisms. Numer. Heat Transf. B Fundam. **85**, 1749–1766 (2023)
19. M. Amaral, A.B. Pereiro, M.M. Gaspar, C.P. Reis, Recent advances in ionic liquids and nanotechnology for drug delivery. Nanomedicine **16**(1), 63–80 (2021)
20. M. Abdollahi, A. Baharian, M. Mohamadhoseini, M. Hassanpour, P. Makvandi, M. Habibizadeh, N. Nikfarjam, Advances in ionic liquid-based antimicrobial wound healing platforms. J. Mater. Chem. B **38**, 9478–9507 (2024)
21. Z. Fallah, E.N. Zare, M.A. Khan, S. Iftekhar, M. Ghomi, E. Sharifi, R.S. Varma, Ionic liquid-based antimicrobial materials for water treatment, air filtration, food packaging and anticorrosion coatings. Adv. Colloid Interf. Sci. **294**, 102454 (2021)
22. H.W. Khan, A.A. Elgharbawy, M.A. Bustam, M. Goto, M. Moniruzzaman, Ionic liquid-based green emulsion liquid membrane for the extraction of the poorly soluble drug ibuprofen. Molecules **28**(5), 2345 (2023)
23. D.M. Correia, L.C. Fernandes, M.M. Fernandes, B. Hermenegildo, R.M. Meira, C. Ribeiro, S. Lanceros-Méndez, Ionic liquid-based materials for biomedical applications. Nano **11**(9), 2401 (2021)
24. M. De Boeck, S. Missotten, W. Dehaen, J. Tytgat, E. Cuypers, Development and validation of a fast ionic liquid-based dispersive liquid–liquid microextraction procedure combined with LC–MS/MS analysis for the quantification of benzodiazepines and benzodiazepine-like hypnotics in whole blood. Forensic Sci. Int. **274**, 44–54 (2017)
25. K. Gulati, D. Chopra, N.A. Kocak-Oztug, E. Verron, Fit and forget: The future of dental implant therapy via nanotechnology. Adv. Drug Deliv. Rev. **199**, 114900 (2023)
26. A. Abouchenari, N. Tajbakhsh, A. Shahbaz, G. Alamdari-Mahd, Advancements in dental implant technology: The impact of smart polymers utilized through 3D printing. J. Synth. Sinter. **4**(2), 4 (2024)
27. M.R. Karim, M.I.H. Siddiqui, A.K. Assaifan, M.O. Aijaz, I.A. Alnaser, Nanotechnology and prosthetic devices: Integrating biomedicine and materials science for enhanced performance and adaptability. J. Disab. Res. **3**(3), 20240019 (2024)
28. Y. Zhang, K. Poon, G.S.P. Masonsong, Y. Ramaswamy, G. Singh, Sustainable nanomaterials for biomedical applications. Pharmaceutics **15**(3), 922 (2023)
29. E.S. Beach, Z. Cui, P.T. Anastas, Green chemistry: A design framework for sustainability. Energy Environ. Sci. **2**(10), 1038–1049 (2009)
30. F. Kurul, B. Doruk, S.N. Topkaya, Principles of green chemistry: Building a sustainable future. Discover Chem. **2**(1), 68 (2025)

Bridging Materials Science and Biomedicine: Clinical Innovations Using Ionic Liquids, Nanofluids, and Nanoparticles

Jamal Ahmad, Insha Khan, and Mohd Kamran Khan

1 Introduction

Advances in materials science have opened new frontiers in biomedicine by leveraging ionic liquids (ILs), nanofluids (NFs), and nanoparticles (NPs) [1–3]. Ionic liquids (ILs) are salts made entirely of cations and anions, which remain liquid at temperatures near or below 100 °C. Their adjustable structures (Fig. 1) result in unique physical and chemical properties, including a very low vapor pressure, a broad range of liquid temperatures, high stability upon heating, and the ability to dissolve various substances. Initially explored as "green" solvents, ILs are now being tailored as bioactive materials and pharmaceutical excipients owing to their ability to dissolve biomolecules and modulate biological activity [1, 5]. Nanofluids are defined as colloidal suspensions of nanometer-sized particles in a base fluid, which gives the fluid improved thermal, magnetic, or optical properties compared to traditionally used fluids [2, 6]. In the fields of medicine and biomedical engineering, where there is a need, such as in injectable therapeutics, contrast agents, or heat mediators, nanofluids can be employed by dispersing functional nanoparticles, such as metallic or magnetic particles (biocompatible fluids). Particles within the range of 1–100 nanoparticles (NPs) are considered

J. Ahmad (✉)
Department of Bioengineering, King Fahd University of Petroleum and Minerals, Dhahran, Saudi Arabia
e-mail: g202417540@kfupm.edu.sa

I. Khan
Department of Biotechnology, Jamia Hamdard University, New Delhi, India

M. K. Khan
Department of Chemistry, King Fahd University of Petroleum and Minerals, Dhahran, Saudi Arabia

S. Kulkarni, S. Das (eds.), *Ionic Liquids, Nanofluids and Nanotechnology*,
Lecture Notes in Nanoscale Science and Technology 42,
https://doi.org/10.1007/978-3-032-04008-4_9

nanoparticles. At these dimensions, advancements in bulk materials can be made for drug delivery and diagnostics, posing a significant leap forward with materials behaving on a nanoscale. These novel materials with their unique optical, magnetic, and chemical properties can surpass traditional methods in targeted drug delivery, imaging contrast, and biosensing. Many frameworks of nanomedicine, liposomes, polymeric NPs, or even iron oxide nanocrystals have advanced into clinical phases, which showcase the feasibility of nanoscale techniques [3, 7].

Although they may seem ILs, nanofluids and nanoparticles differ from each other, they all enable novel biomedical and clinical functionalities through engineered small-scale physicochemical properties. For instance, ILs can be tailored to biocompatible solvents for drug formulation, or even as active pharmaceutical ingredients (API-ILs) [8], biomolecules. Nanoparticles can be integrated into fluids (yielding nanofluids) to assist magnetic hyperthermia treatment or drug delivery via injection [9]. Even more, ILs are used for the synthesis and stabilization of nanoparticles [1], making use of the ionic liquid's tunability coupled with the nanomaterial's functionality. Given these complementary strengths, advanced materials are more intertwined in biomedical research. In this chapter, we outline all their physicochemical properties as well as current and novel biotechnological and biomedical applications, clinical prospects, safety considerations, and the associated risks. Some foundational elements are depicted in Fig. 1. A technological survey highlights the innovations shaping the future of diagnostics, therapeutics, and regenerative medicine with ILs, nanofluids, and nanoparticles.

2 Chemical and Physical Properties of Nanoparticles, Ionic Liquids, and Nanofluids

2.1 Ionic Liquids

The unique characteristics of ionic liquids (ILs) arise from their low-temperature physical states. Structurally, ILs consist of bulky organic cations, which are "1-alkyl-3-methylimidazolium" and "tetraalkylammonium," and paired with an anion that is either inorganic or organic (Fig. 1). The reduction of the melting point and the formation of fluid salts are caused by charge delocalization and asymmetry. Other critical characteristics of ILs are negligible vapor pressure, meaning ILs do not evaporate even at elevated temperature, which makes them nonvolatile and thermally stable. The stability up to 300 °C makes ionic liquids thermally stable and to remain in the liquid state over a wide range of temperatures. This allows ILs to be used in processes where traditional organic solvents would boil or decompose. This characteristic, combined with being nonvolatile, makes ILs useful in high-temperature processes and "green" chemistry, focusing on the minimization of air emissions [1]. Their polarity and solvation properties are highly tunable by ion selection; ILs can dissolve a remarkable range of substances, from inorganic salts to

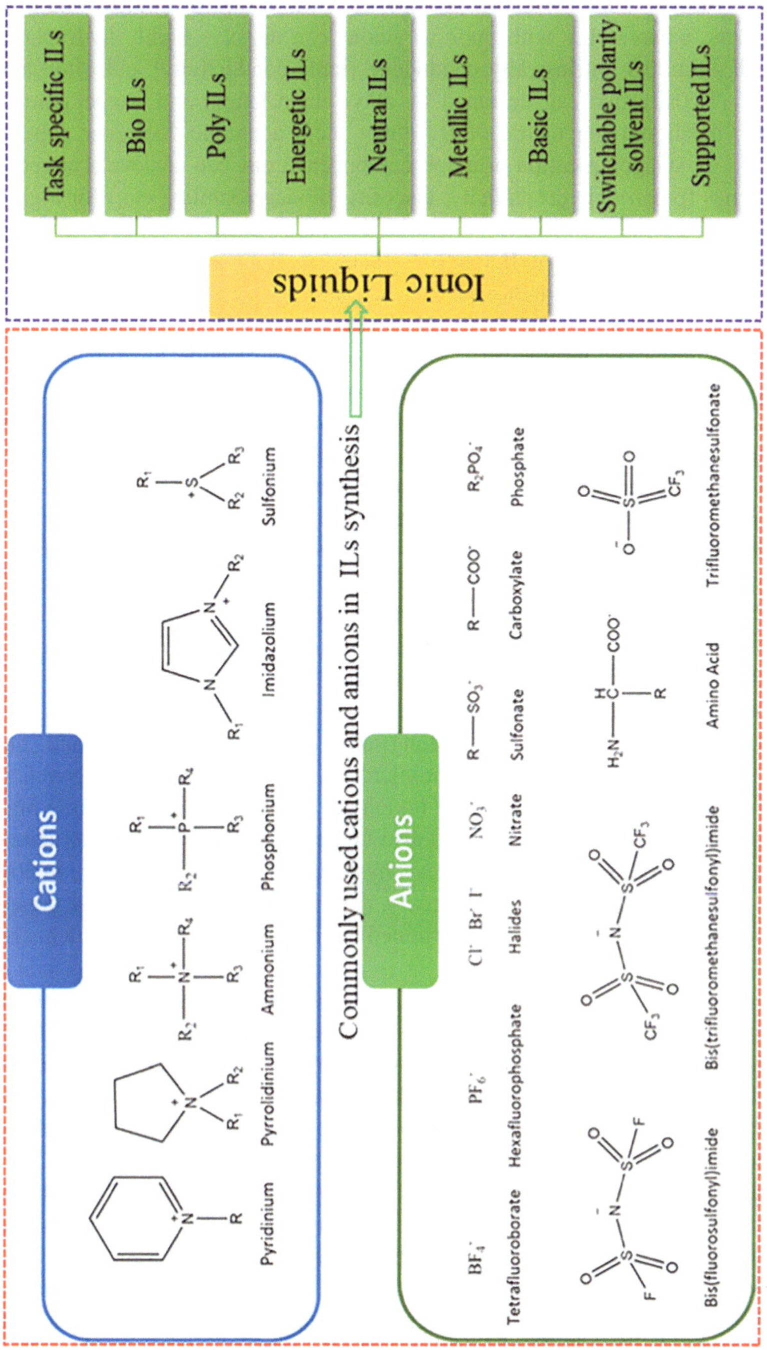

Fig. 1 Grouping ILs by the numerous ways ions can be combined [4]

biomolecules, by appropriate choice of cation/anion combinations [1]. For example, ILs with hydrogen-bonding anions (acetate, etc.) can dissolve cellulose and other biopolymers, whereas ILs with fluorous anions can dissolve highly hydrophobic solutes. ILs are also intrinsically conductive (ionic conductivity) and often have moderate viscosity that can be tuned by alkyl chain length and temperature. In summary, the liquid–ion nature of ILs gives them a unique "designer solvent" capability: by swapping cation or anion components, one can achieve a target set of properties (polarity, hydrophilicity, viscosity, biodegradability, etc.) for a given application [1].

It is useful to categorize ILs into "generations." First-generation ILs were developed for inert, stable industrial solvents and tend to have inert anions like BF_4^- or PF_6^-, which impart stability but can be poorly biodegradable [10]. Second-generation ILs introduced functional ions (e.g., having functional groups for task-specific behaviors), and third-generation ILs refer to those that incorporate biologically active ions, effectively merging the IL with a drug or biomolecule [8]. This latter category includes the concept of API-ILs, where either the cation or anion is itself a therapeutic agent (for instance, an IL formed from an active drug anion paired with an organic cation). In such cases, the IL not only serves as a solvent or carrier but also contributes directly to pharmacological activity. The immense combinatorial space of possible cation–anion pairs (often cited as "tunability") means ILs can be tailored to possess a desired hydrophilicity, miscibility with water or lipids, and biocompatibility [5]. As an illustration, cholinium derivatives ILs, which are based on choline, a less toxic nutrient, have better biodegradability, and this makes them viable for incorporation into pharmaceutical preparations [11]. Imidazolium or pyridinium ILs with long alkyl chains, on the other hand, may possess considerable cytotoxicity or antimicrobial activity, which can be beneficial (as antiseptic agents) or undesirable (if toxicity to human cells is excessively high) [11]. Therefore, it is very important to also understand and control the structure–property relationships of ILs. Some key physicochemical parameters are melting point, viscosity, density, ionic conductivity, cop miscibility with water or organic solvents, and thermal stability, all determined by the specific ionic constituents and their interactions.

In biomedical contexts, two additional considerations are important: biocompatibility and stability in biological environments. Many early ILs (e.g., 1-butyl-3-methylimidazolium tetrafluoroborate) were found to be cytotoxic or ecotoxic at significant levels [11, 12]. Research has shown that toxicity tends to correlate with certain structural features—for instance, long alkyl chains increase hydrophobicity and membrane disruption, leading to higher toxicity [11]. As a result, the focus on developing biocompatible or "task-specific" ionic liquids has emerged, which centers on their therapeutic application by minimizing toxicity [1]. Examples include ILs derived from amino acids, choline, or bioactive anions such as salicylate, which serve as both solvents and potential therapeutics [11, 13]. It is also possible that these ILs are bio-ILs designed to biodegrade into nontoxic metabolites. Generally, ionic liquids present a highly customizable platform that can be precisely tailored to bolster their application in drug delivery, biomaterials, and medical interfacing devices.

2.2 *Nanofluids*

Nanofluids are a new type of engineered colloids that consist of solid particles of nanometer size (1–100 nm) dispersed homogeneously in a carrier liquid, water, oils, or glycols [2, 6]. The physical properties of a fluid change drastically with the addition of nanoparticles, even at small volume fractions, typically less than 5%. Thermal conductivity is one of the most studied attributes, and suspensions of different nanoparticles are known to greatly exceed base fluid values. Classic work has shown that incorporating metallic or ceramic nanoparticles into liquids enhances their thermal conductivity significantly greater than the base vector fluid alone [14]. This property makes nanofluids attractive for cooling and hyperthermia applications (as depicted in Fig. 2). In biomedical settings, magnetic nanofluids (also known as ferrofluids, containing iron oxide or other magnetic nanoparticles) are of special interest because an external magnetic field can manipulate and heat these fluids. Nanofluids containing superparamagnetic iron oxide nanoparticles (SPIONs) can be remotely heated by alternating magnetic fields, enabling controlled thermal therapies (hyperthermia) for cancer when delivered to tumor sites [9]. The fluid nature allows injection or perfusion into target tissue, and the nanoparticles then act as localized heat sources under the field, as shown in Fig. 2.

Apart from thermal properties, nanofluids can exhibit altered rheology (viscosity behavior) and optical properties. Some nanofluids demonstrate plasmonic behavior if the nanoparticles support surface plasmons (e.g., gold nanoshells in a fluid); such fluids can be used as photoresponsive agents, absorbing light and converting it to heat for photothermal therapy (PTT) [16]. About rheology, the incorporation of nanoparticles usually increases viscosity modestly while also introducing non-Newtonian behavior, particularly if the particles are covalently bonded to polymers or if they undergo self-assembly to form structured networks. However, several biomedical nanofluids are engineered to maintain relatively low viscosity to facilitate easy injection. Stability is a critical issue: nanoparticles tend to aggregate with other nanoparticles due to van der Waals forces. A stable nanofluid requires preventing aggregation over its intended shelf-life and during use. This is achieved by using surfactants or surface coatings on the nanoparticles to provide electrostatic or steric repulsion. For instance, biocompatible polymers like polyethylene glycol (PEG) are often grafted to nanoparticle surfaces in nanofluids to ensure stability and "stealth" properties (reducing opsonization by the immune system) [6, 17]. The size and surface chemistry of the nanoparticles strongly influence nanofluid behavior: smaller particles (e.g., 10–20 nm) remain suspended more easily (Brownian motion counteracting sedimentation) and can even behave like macromolecules in solution, whereas larger or heavier particles may settle over time under gravity. To mitigate this, magnetic nanofluids can be formulated as gels or with density-matched media

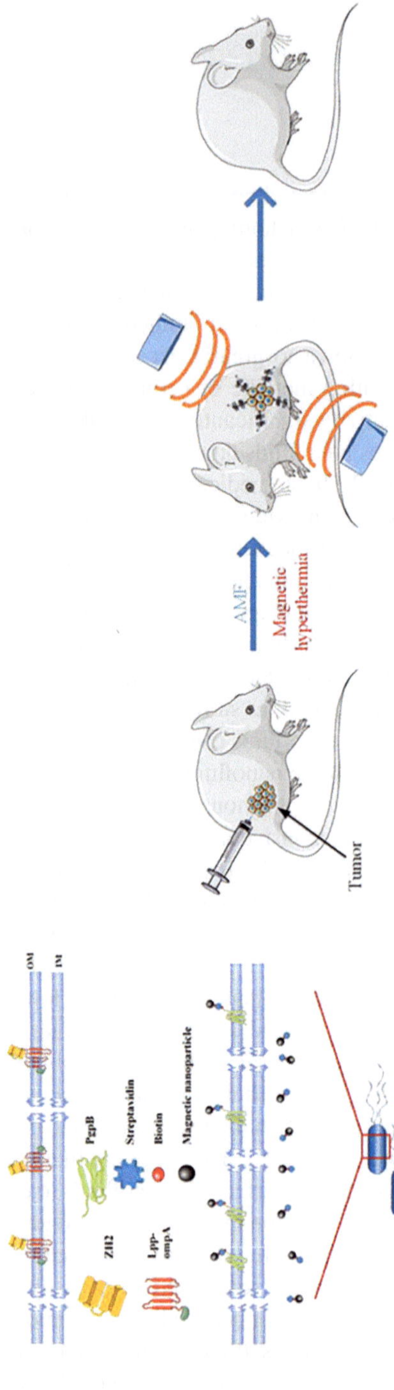

Fig. 2 Schematic illustration of the strategy applied in this study. Refer to the text for details [15]

if needed, but typically sub-100 nm SPIONs in water with proper coatings are stable for extended periods [5].

From a materials perspective, the nanoparticles in nanofluids can be metals (gold, iron, etc.), metal oxides (Fe_3O_4 and ZnO), or even polymeric or carbon-based nanoparticles. Each type confers different properties: metallic particles (e.g., Au) give plasmonic optical features and high X-ray absorption, metal oxides can be magnetic (Fe_3O_4) or photocatalytic (TiO_2), and quantum dots (QDs) provide fluorescence. Magnetic nanofluids are noteworthy for biomedical use because they combine magnetic manipulability with fluid injectability. These fluids typically have magnetite or maghemite nanoparticles ~10 nm in diameter, often coated with dextran or citrate for biocompatibility, suspended in water or saline. They respond to magnetic fields, allowing targeted movement (magnetic targeting) and induction heating (magnetic hyperthermia). Another emerging class is nanofluids for imaging, such as iron oxide nanofluids used as magnetic resonance imaging (MRI) contrast agents (where the fluid can be injected and the particles provide T2 contrast). Indeed, several iron oxide nanoparticle fluids were clinically approved as MRI agents (e.g., Feridex and Resovist), although some were later withdrawn; their formulation can be considered a nanofluid [3]. Overall, nanofluids inherit a combination of fluid-like transport properties and nanoparticle-induced functional properties, making them appealing for biomedical applications that require a therapeutic or imaging function distributed in a liquid form (e.g., injectable drug nanosuspensions, coolants for medical devices, and contrast agents) [6].

Stability must also be examined colloidally in vivo: when injected into tissues or blood, nanofluid particles will encounter proteins (opsonins) which could bind to them and either aggregate or be cleared by the mononuclear phagocyte system [17]. As noted, surface functionalization such as PEGylation is pivotal to retaining colloidal stability in the dynamic and complex multisystem of living organisms. Another property, osmolarity, should also be considered for the polymerized nanofluids meant for injection—the addition of nanoparticles may increase effective solute osmolarity; tissue and fluid balance must be maintained in a borderline state of isotonicity to avert sustaining cellular injury. To conclude, nanofluids are capable of exhibiting the flow properties of liquids and the functional characteristics of nanomaterials. Careful consideration of the size, composition, and surface chemistry of the nanoparticles enables the engineering instability of nanofluids whilst bestowing altered thermal, magnetic, or optical functionalities within a biomedical framework [2].

2.3 *Nanoparticles*

Nanoparticles serve as the basic building blocks for numerous nanofluids and nanomedicines. More distinct and identifying features of a nanoparticle are its high surface area to volume ratio and the quantum effects. When the size of a particle approaches a certain scale, such as the de Broglie wavelength or phonon mean free path, classical behavior is dominated by quantum effects. With metals such as gold, nanoparticles with a size smaller than 100 nm exhibit localized surface plasmon resonance (LSPR). LSPR is the oscillation of conduction electrons in response to certain light frequencies. This gives rise to intense optical absorbance bands (e.g., ~520 nm for 20–30 nm AuNPs), which can be captured for imaging and photothermal therapy. Also, semiconductor nanoparticles, or quantum dots, have fluorescence that is dependent on size, emitting bright light of color far from the particle size because of quantum confinement [3]. Magnetic nanoparticles, for example, ferrites, transition to a single-domain superparamagnetic state below a certain size (approximately 20 nm for magnetite). Here, they have strong magnetization but no permanent magnetism unless a field is applied. This is advantageous for biomedical applications because it means that magnetic NP dispersions do not agglomerate when the outside magnetic field is off, but can be freely altered, agitated, or heated under alternating field [9].

Dimension aside, nanoparticles derive many properties from material composition. Metallic nanoparticles, like gold and silver, are photonic active and are used as radiopaque agents. Silica, alumina, and calcium phosphates are ceramic nanoparticles that can have chemical stability and reinforce composites or act as delivery matrices. Polymers, including Polylactic acid (PLA) and poly lactic-co-glycolic acid (PLGA), are typically designed to be biodegradable carriers that release drugs over time. An important subset are lipid-based nanoparticles (like solid lipid nanoparticles [LNPs] and liposomes), which, while not solid inorganic particles, are often ~50–100 nm vesicles used in drug delivery—notably, the mRNA vaccines for COVID-19 use lipid nanoparticles (LNPs) to encapsulate and protect the mRNA [18, 19]. These LNPs are essentially nanoscopic droplets of ionizable lipids and have proven remarkably effective in human vaccination, underscoring how nanoparticle technology has become clinically indispensable. Indeed, as of 2021, dozens of nanomedicine products (liposomal drugs, polymer–drug conjugates, and NP suspensions) have been approved globally [3, 7].

Physiochemically, nanoparticles can be considered as an intermediate state of matter, bridging single molecules and bulk materials. They often possess high surface energy, which can drive phenomena like protein adsorption in biological fluids (the formation of a "protein corona" around the NP) [10, 12]. The functional features, charge, and hydrophobic properties of a given colloidal NP determine its stability as well as its interaction with cells [10]. A charged NP, for instance, is likely to stick to the cell membranes, which contain negative charges, facilitating its internalization, yet may also result in toxicity [10, 20], for example, with PEGylated neutral hydrophilic NPs. The size of nanoparticles affects their biodistribution: typically, particles under ~5 nm can be rapidly cleared renally (like small molecule drugs), 10–100 nm particles often accumulate in liver and spleen if not actively targeted (via the enhanced permeability and retention [EPR] effect in tumors or via phagocytes), and larger ones (>200 nm) might be trapped in the vasculature or filtered by the spleen [2, 3, 6]. Hence, controlling the NP size during fabrication is essential in achieving the biological outcome of interest [6].

One more essential characteristic is surface modification. Nanoparticles are often coated or functionalized to provide the desired activity or biocompatibility [14]. For example, receptors for specific cells can be targeted with antibodies or ligands covalently bonded to the surface of nanoparticles (this is a part of a drug delivery or imaging system) [11, 13]. Nanoparticles may have an enzyme or aptamer covalently bonded to them, thus producing sensors that detect certain compounds with a signal output [21]. The broad range of surface chemistry allows engineering nanoparticles as multifunctional "theranostic" platforms that integrate both imaging and therapeutic agents [7, 13]. For instance, a magnetic core-gold shell nanoparticle. The iron oxide core permits MRI or magnetic guidance, the gold shell enables optical imaging and photothermal treatment, and on its surface, a chemotherapeutic drug and a targeting peptide can be covalently attached—all in one construct. Such theranostic nanoparticles illustrate how the nanoscale features integrate diagnosis and therapy [13, 22].

To recap, nanoparticles have optical, electrical, and magnetic properties that are vastly size dependent [7, 12, 13]. They also have a highly tunable surface chemistry and an exceptionally high surface area, which makes them ideal candidates for biomedical applications [10, 12]. Their use in biomedicine is one of their most prominent and rapidly growing applications [11]. One major difference between a molecular drug and a nanoparticle is that the latter does not have a single, precisely defined molecular formula. Instead, it is described by several physicochemical distributions, such as size distribution, surface charge, shape, and hydrophobicity [7, 12].

The ability to fabricate reproducible and uniform nanoparticles remains one of the more complex challenges in nanomaterials science [12, 23]. Despite these challenges, modern synthesis techniques such as chemical reduction, self-assembly, and hydrothermal synthesis have enabled the production of monodisperse nanoparticles at scale [12, 13, 23]. For successful clinical translation, it is essential to design nanoparticles with reproducibility, stability, and biocompatibility, which in turn requires a detailed understanding of nanoscale behavior, particularly from a physicochemical and nanotechnological engineering standpoint [14, 22].

A detailed table (e.g., Table 1 in this chapter) contains examples of medical products incorporating nanoparticles, illustrating how these engineered materials are already improving human health [16, 17].

3 Biomedical and Biotechnology Applications

3.1 *Drug Delivery and Pharmaceutical Formulations*

The use of nanoparticles and ionic liquids to aid drug delivery systems in medicine is one of the earliest examples of their application in biomedicine [1, 3, 12]. The applications of ionic liquids (ILs) in pharmaceutics highlight their potential as novel

Table 1 Selected nanoparticle-based therapeutics and imaging agents in clinical use

Product (Company)	Nanoparticle type	Indication/use	Status
1. Doxil (Johnson & Johnson)	PEGylated liposomal doxorubicin (≈100 nm liposomes)	Metastatic cancers (ovarian, Kaposi's sarcoma, etc.)—improved drug delivery and reduced cardiotoxicity	FDA-approved (1995)
2. Abraxane (Celgene)	Albumin-bound paclitaxel nanoparticles (~130 nm)	Breast, lung, and pancreatic cancers—solvent-free chemo delivery with tumor targeting via albumin pathway	FDA-approved (2005)
3. Ferumoxytol (Feraheme) (AMAG)	Superparamagnetic iron oxide (SPION) ~30 nm, carboxymethyl dextran-coated (intravenous nanofluid)	Iron replacement therapy for anemia; also used off-label as an MRI contrast agent for vascular imaging (T1/T2 contrast)	FDA-approved (2009)
4. Visudyne (Bausch and Lomb)	Liposomal verteporfin (≈100 nm liposomes)	Photodynamic therapy for age-related macular degeneration—targeted light-activated therapy in the eye	FDA-approved (2000)
5. Onpattro (Alnylam)	Lipid nanoparticle encapsulating siRNA (≈80 nm LNP)	Treatment of hereditary transthyretin amyloidosis—delivers siRNA to the liver to silence the disease gene	FDA-approved (2018)
6. Comirnaty/Spikevax (Pfizer/BioNTech and Moderna)	Lipid nanoparticle mRNA vaccine (≈100 nm ionizable LNPs carrying mRNA)	mRNA vaccine for COVID-19—LNP protects mRNA and facilitates uptake into cells for antigen production	FDA-approved (2020)
7. NanoTherm (MagForce) (EU)	Magnetic iron oxide nanoparticle fluid (nanofluid injected into the tumor)	Focal thermal ablation therapy for glioblastoma (used with external AMF to heat the tumor)—minimal invasive thermotherapy	CE-marked in the EU (2010); in clinical use in Europe
8. Resovist® (Bayer) (EU/Japan)	Superparamagnetic iron oxide (SPION) ~60 nm, carboxydextran coated (intravenous)	MRI contrast agent for liver imaging (T2*)—enhances detection of liver lesions by accumulating in Kupffer cells	Approved in EU/Japan (2001)—later withdrawn in US (development halted)

solvents and carriers for difficult-to-formulate drugs. The limited water solubility of many active pharmaceutical ingredients (APIs) often hampers bioavailability. However, for ionizable APIs, forming an IL with a suitable counterion can transform the drug into a more soluble or stable ionic liquid form [12]. For example, lidocaine and prilocaine have been converted into an IL that is liquid at room temperature and offers improved transdermal delivery [24]. In other examples, choline-based or ammonium-based ILs have shown a remarkable ability to dissolve hydrophobic APIs and can be incorporated into polymeric systems to yield nanoparticles or hydrogels [24]. Because IL polarity and hydrophobicity are tunable via ion selection, the solvent environment can be "matched" to a drug's chemical nature. ILs also reduce drug crystallization, maintaining APIs in an amorphous and bioavailable state [25]. Additionally, ILs have been used as transdermal permeation enhancers, where specific ILs transiently disrupt skin lipids to facilitate drug absorption [26].

Some ILs offer additional therapeutic properties. For instance, ILs containing salicylic acid exhibit anti-inflammatory and antimicrobial effects [19]. Egorova and colleagues demonstrated that pairing therapeutic anions (e.g., salicylate and penicillin) with biocompatible cations (e.g., choline) results in bifunctional IL drugs that act as both solvent and active pharmaceutical agent [27]. These API-ILs can prevent crystallization and enhance tissue permeability. One such example is choline fenofibrate, which improves water solubility and oral absorption of fenofibrate [5]. ILs are also useful in biomacromolecule stabilization. For example, choline-based deep eutectic solvents (DESs) help preserve enzyme activity and are being explored in enzyme therapy and in situ biocatalysis [10].

In contrast, nanoparticles have a long-established role in drug delivery. Nanocarriers can protect payloads from degradation, prolong circulation, and release drugs in a controlled fashion [28]. Multiple NP-based therapies are already U.S. Food and Drug Administration (FDA)-approved or in clinical development: e.g., Doxil (liposomal doxorubicin), Abraxane (albumin-bound paclitaxel), and NanoTherm (iron oxide NPs for thermotherapy) [21, 29, 30]. Nanoparticles benefit from their small size, which facilitates tissue penetration. In oncology, the enhanced permeability and retention (EPR) effect enables passive accumulation of NPs in tumors due to leaky vasculature [16, 20]. PEGylated liposomes, like those in Doxil, enhance circulation time and targeting, while reducing side effects such as cardiotoxicity [31]. Similarly, Abraxane's 130 nm albumin–paclitaxel NPs bypass solubility issues and improve tumor targeting without toxic solubilizers [27].

In addition, polymeric NPs (e.g., PLGA) are being used for encapsulation of small molecules, proteins, and nucleic acids, enabling sustained release over days to weeks [23, 24]. These systems are currently being tested in vaccines, cancer therapy, and insulin delivery [25].

Nanofluids, in this context, refer to suspensions of drug-loaded nanoparticles. Many "nanosuspension" drugs (e.g., sirolimus and fenofibrate) are delivered intravenously or orally as nanocrystals, improving dissolution and bioavailability [27]. In targeted delivery, magnetic nanofluids have been explored, where drug-functionalized iron oxide nanoparticles are directed to target sites (e.g., tumors) by external magnets, achieving site-specific drug accumulation and release [18, 30]. In preclinical rabbit tumor models, this technique—called magnetic drug targeting—resulted in significant tumor remission [29]. Although clinical translation is still limited, ongoing trials using magnetic thermotherapy combined with radiotherapy (e.g., for glioblastoma) highlight its potential for localized, low-toxicity cancer treatment [28].

ILs also play a role in biopharmaceutical formulation and purification. They help stabilize hydrolysis-prone APIs by replacing water with biocompatible IL media [1]. For example, certain vaccine components and enzymes may benefit from IL-based storage to prevent denaturation, although this remains under investigation [32]. Similarly, deep eutectic ILs have been used to extract and formulate plant-derived APIs and natural products [33].

Together, ILs, nanofluids, and NPs address the challenges in modern drug delivery: solubilization, protection, and targeting. ILs enhance solubility and formulation versatility; NPs enable precise targeting and controlled release; and nanofluids allow injectable or perfusable forms of these nanosystems. All three technologies together are pushing the boundaries of precision medicine [9, 12, 14]. Table 1 in this chapter summarizes the current clinical applications of such systems.

3.2 Antimicrobial and Biosensing Applications

Other than maintaining the drug delivery methodologies, ionic liquids and nanoparticles have been extensively used in biosensors, diagnostic assays, and antimicrobial platforms within biomedical applications [9, 14, 16]. Various ionic liquids, particularly those containing long alkyl chains, demonstrate antimicrobial properties similar to cationic surfactants such as benzalkonium chloride [12, 13]. Imidazolium-based and quaternary-ammonium-based ILs are capable of disrupting microbial membranes, enabling effective antibacterial and antifungal activity when dosed appropriately [17, 28]. These ILs have been studied as disinfectants and preservatives in pharmaceutical and medical contexts [31, 34].

In vitro studies show ILs can inactivate pathogens while exhibiting lower toxicity toward mammalian cells, especially when optimized for cation/anion structure [23, 34]. A notable example includes choline-based ILs synthesized using geranic acid, yielding Choline-geranate (CAGE) ILs with potent topical antiseptic activity [21]. The benefit of ILs in this regard lies in their fluidic nature, which permits deeper penetration into biofilms and crevices, and their intrinsic solvent action, enabling dissolution of the biofilm matrix [18, 30].

Furthermore, ILs can be incorporated into polymers, including wound dressings or electroconductive biopolymers, to offer sustained antimicrobial action. For instance, a chitosan hydrogel doped with ILs has demonstrated antibacterial wound healing capability while preserving biocompatibility [28, 32]. The bioactivity and specificity of ILs can also be modulated by selecting natural amino acid-based or organic acid-based anions, targeting specific microbial species [28, 32]. That said, the toxicity of ILs to human cells must be carefully managed, hence the need to favor biocompatible cations, such as choline or amino acid-based systems, to ensure effective antimicrobial action without systemic side effects [28, 32].

Nanoparticles, in parallel, show excellent utility in biosensing and diagnostics, driven by their optical, magnetic, and electrochemical properties [28, 32]. Gold nanoparticles (AuNPs) are particularly well known for colorimetric biosensing due to their localized surface plasmon resonance, which gives them a red color in solution (~520 nm) that shifts to purple/blue upon aggregation [28, 32]. This behavior underpins lateral flow assays (e.g., home pregnancy tests, COVID-19 antigen kits) that use AuNP-conjugated antibodies to generate visual results [28, 32].

In more advanced biosensors, AuNPs functionalized with DNA or antibodies aggregate upon target recognition, causing a measurable color shift, forming the basis of signal amplification assays [28, 32]. AuNPs are also extensively used as surface-enhanced Raman scattering (SERS) tags. When conjugated with Raman reporter molecules, they allow the detection of trace-level analytes, including tumor markers, with enhancements of 10^5–10^6-fold [32, 33]. Qian et al. notably demonstrated in vivo cancer detection using SERS-labeled AuNPs functionalized with antibodies, allowing highly specific imaging of tumor-associated biomarkers [20].

Beyond gold, quantum dots (QDs), such as CdSe/ZnS core-shell particles, are also utilized for biosensing and bioimaging. QDs emit bright, size-tunable fluorescence and are more photostable than traditional fluorophores, enabling cell labeling, long-term imaging, and molecular diagnostics [22, 35]. They are frequently incorporated into fluorescence resonance energy transfer (FRET) biosensors, which change fluorescence upon target binding (e.g., DNA hybridization, antigen–antibody recognition) [17].

Diagnostics are also supported by magnetic nanoparticles (typically iron oxide NPs), especially for magnetic separation and magnetic transduction in biosensing [7, 10]. These SPIONs can be coated with antibodies or aptamers to capture specific targets from complex biological samples (e.g., blood), allowing enrichment of bacterial pathogens or nucleic acids [7, 22].

Ionic Liquids in Biosensing

ILs have been of particular use in the area of electrochemical sensors. Numerous ionic liquids are electrically conductive and nonvolatile, which makes them highly suitable as electrolytes in biosensor devices [1, 13, 20]. In particular, electrochemical biosensors (e.g., for glucose or cholesterol monitoring) have incorporated ILs into electrode matrices to both preserve enzyme stability and enhance electron transfer [16, 32].

For instance, incorporating an ionic liquid into a carbon paste electrode creates a conductive microenvironment that supports oxidase enzymes like glucose oxidase in facilitating direct electron transfer from the redox-active center of the enzyme to the electrode surface [17, 21]. This enables stable and sensitive amperometric glucose detection.

ILs can also immobilize enzymes without denaturing them, particularly when used to form ionogels (ionic gels), which trap the enzymes in a quasi-liquid matrix, preserving their activity for biosensing applications [22] As ILs are "designable," they can be functionalized with specific chemical groups (e.g., nitrile and hydroxyl), which improve selectivity and signal quality by tailoring interactions with the target analyte [33, 35].

Moreover, ILs are being used in wearable and flexible biosensors, particularly as electrolytes in nonvolatile electrochemical cells that can continuously monitor sweat electrolytes or lactate [12, 16]. ILs' low vapor pressure prevents evaporation, and their ionic conductivity ensures stable signal output over time, both critical for continuous and long-term monitoring in wearable devices [12].

In antimicrobial applications, silver nanoparticles (AgNPs) are among the most well-documented nanomaterials [10, 29]. AgNPs exhibit broad-spectrum antibacterial activity primarily due to the release of Ag^+ ions, which disrupt microbial membranes, proteins, and DNA via oxidative stress and ion toxicity [29]. They are used in commercial wound dressings and as coatings for medical devices to reduce the risk of infection [24]].

Because of their nanoscale size, AgNPs offer a high surface area for sustained Ag^+ release, which improves antimicrobial efficacy even at lower concentrations compared to bulk silver [25, 26]. Emerging work has explored IL–nanoparticle hybrid systems (e.g., IL-capped AgNPs) that synergize the membrane-disrupting action of ILs with the ion-release properties of silver, resulting in enhanced antimicrobial activity [18, 36].

Overall, ionic liquids (ILs) and nanoparticles (NPs) form an indispensable foundation for modern biosensing and antimicrobial platforms. ILs allow the development of robust and sensitive electrochemical biosensors, as well as liquid-phase antimicrobial formulations or coatings [18, 22]. Nanoparticles amplify optical, electrochemical, and magnetic biosensors, support point-of-care diagnostics (e.g., colloidal gold-based lateral flow devices), and are deployed as nanoscale antimicrobials [10, 23, 24].

Surface functionalization of NPs enables highly specific targeting of diverse biomarkers, from glucose and lactate to cancer markers and infectious disease antigens, often achieving femtomolar detection sensitivity in advanced nano-enabled assays [10, 23].

Looking ahead, smart theranostic systems could emerge that integrate ionic liquid sensors with nanoparticle carriers, triggering on-demand drug delivery upon detecting pathological changes, thereby enabling closed-loop diagnostic–therapeutic responses [10, 22, 34].

3.3 Bioimaging

Bioimaging

Ionic liquid (IL), nanofluid, and nanoparticle (NP) innovations have significantly contributed to bioimaging, the visualization of biological structures and processes, by enhancing contrast, specificity, and multimodal functionality [10, 11]. Conventional imaging techniques (e.g., MRI, computed tomography [CT], PET, fluorescence imaging, and ultrasound) often require contrast agents to improve image quality. Among modern contrast agents, nanoparticles have emerged as robust enhancers across multiple imaging modalities [7, 10].

For instance, iron oxide nanoparticles, particularly superparamagnetic iron oxide nanoparticles (SPIONs), serve as effective T_2 MRI contrast agents, causing signal loss (darkening) in tissues such as the liver, spleen, and tumors [4, 7, 10]. Several SPION-based formulations have reached clinical stages or are currently under clinical trials for liver tumor detection and lymph node imaging [13, 17]. Recent advances in iron oxide nanoclusters with targeted polymer coatings have led to the development of next-generation MRI contrast agents with increased relaxivity and tumor/inflammation specificity [10, 17].

Likewise, gold nanoparticles (AuNPs) ranging from 10 to 30 nm are under investigation as computed tomography (CT) contrast agents due to their high X-ray attenuation, far surpassing that of traditional iodinated contrast media [10]. Preclinical animal studies have demonstrated that AuNPs (nanospheres and nanorods) can act as long-circulating blood pool contrast agents, enabling prolonged imaging windows and passive tumor targeting through the EPR effect, thereby enhancing cancer visualization in CT scans [10].

In the realm of fluorescence imaging, quantum dots (QDs) enable multicolor detection, thanks to their size-dependent, narrow-band emission spectra [10, 23, 24]. QDs are uniquely suited for simultaneous labeling of multiple biomolecular targets using distinct emission wavelengths. This capability is widely applied in cell imaging, tumor marker tracking, and increasingly explored for intraoperative margin detection in oncology [10, 23]. Their photostability, brightness, and bioconjugation compatibility make them valuable tools in both basic research and translational bioimaging applications.

It is a niche role that ionic liquids (ILs) are playing in imaging, particularly as enabling components in formulation and signal modulation [2, 22]. Certain ILs are being investigated as solvent systems for poorly soluble imaging agents, such as gadolinium chelates or hydrophobic fluorophores, which require high-concentration, injectable solutions for MRI or optical imaging [1, 2, 18]. Some ILs not only dissolve imaging agents but also stabilize them, improving shelf life and safety profiles [2, 21].

In ultrasound imaging, nanofluids composed of perfluorocarbon nanodroplets act as acoustic contrast agents, improving echogenicity. While not ionic liquids themselves, these droplets are low-volatility liquids with similar physicochemical properties [7, 10]. In this context, IL-formulated emulsions or ionic surfactants may help stabilize such nanodroplets for injectable use.

Additionally, ionic liquid crystals are being explored for their anisotropic optical properties, such as field-responsive birefringence, making them candidates for switchable optical contrast agents [22]. Although these systems are in early conceptual development, their dynamic light modulation potential makes them attractive for future adaptive imaging applications.

Photothermal and Photodynamic Theranostics

Nanoparticles have facilitated a lot of advancements in the field of photothermal and photodynamic theranostics [10, 20]. Gold nanoshells and nanorods are highly effective at converting light into heat through plasmonic resonance, making them suitable for localized thermal ablation of tumors [10, 33]. In clinical trials for prostate tumors, silica–gold nanoshells (Auroshell) have been used for MRI-guided photothermal ablation, where gold nanoparticles serve both as thermal mediators (therapy) and imaging agents via CT and photoacoustic imaging (diagnosis) [34]. Similarly, photoacoustic imaging, which detects ultrasound waves generated by pulsed laser-induced heating of nanoparticles, benefits significantly from gold nanorods, nanoshells, and carbon nanotubes, which absorb near-infrared (NIR) light efficiently [10, 23]. These nanomaterials can be functionalized with targeting ligands (e.g., antibodies or peptides) to direct them to tumor sites, allowing real-time imaging and simultaneous photothermal therapy (PTT) for killing cancer cells with minimal off-target damage [10, 23].

Theranostic hyperthermia is also achieved using magnetic nanoparticles, specifically superparamagnetic iron oxide nanoparticles (SPIONs). Upon delivery into a tumor, these particles can be imaged by MRI and then heated by an alternating magnetic field, resulting in localized hyperthermia and tumor cell death [7, 13]. This modality is being used clinically in Germany for recurrent glioblastoma under the NanoTherm® therapy developed by MagForce, where a SPION-based nanofluid is injected directly into the tumor and later activated by alternating magnetic fields, with the particles monitored using MRI for precise localization and coverage [13, 17]. The ability to both visualize and thermally ablate the tumor through the same nanomaterial is a model example of a theranostic approach.

Another promising theranostic strategy involves radiotherapy enhancement using high atomic number (Z) nanoparticles. For example, hafnium oxide nanoparticles (NBTXR3) have been developed as radiosensitizers and also function as radiopaque materials for CT imaging [10, 12]. Upon intratumoral injection, these nanoparticles amplify the local radiation dose during X-ray therapy, while simultaneously enabling CT-based visualization to confirm their distribution and ensure accurate treatment planning [22]. NBTXR3 has reached clinical trials in patients with locally advanced soft tissue sarcomas, showing promise in improving radiation efficacy and achieving localized imaging-guided therapeutic outcomes [32].

Biosafety and Toxicity

This section will address it, but in the context of applications, any theranostic or imaging agent, when it comes to the use of nanoparticles or IL components, needs to be thoroughly screened. For example, the quantum dots are useful but tend to contain cadmium, that concerns about toxicity issues. Cadmium-free QDs (such as indium phosphide) are being worked on and developed. On the same note, ionic liquids, if used in vivo, have to be demonstrated to be nontoxic at the necessary doses. Potentially efficacious ILs under some particular lab tests may be too hemolytic or inflammatory for systemic administration and will only be utilized in topical applications.

Finally, it should be stressed that convergence of diagnostics and therapy at the nanoscale level of theranostics is significantly powered by ionic liquids, nanofluids, and nanoparticles. ILs are flexible media and stabilizers for contrast agents, and on some occasions, can directly participate in imaging or therapy chemistry. Nanoparticles are multifunctional particles that are imaged by way of different modalities (optical, magnetic, acoustic, or ionizing) while simultaneously performing a therapy role (drug delivery, heat generation, etc.). Such synergy holds more personalized and real-time monitored therapies. For instance, one could imagine an IL nano-particle drip to the tumor which will report when a picture of the size of the tumor exists the nanoparticle, will deploy a drug in response to some stimulus, and will allow the physician to watch the reaction of the tumor. Such capabilities are an example of why the biomedical society is eager to see the unification of these technologies to build their smart theranostic systems that lead to better results and image-guided interventions.

4 Regenerative Medicine and Tissue Engineering

Regenerative medicine strives to repair or replace damaged tissues or organs, where, in many cases, engineered scaffolds, cells, and bioactive factors are used. In this case, ILs, nanofluids, and nanoparticles have emerging roles in the development of a better biomaterial and promoting the growth and integration of tissue [23–25].

4.1 Ionic Liquids in Tissue Engineering

Remarkably, ionic liquids (ILs) find application in tissue engineering, particularly for polymer processing and the fabrication of biocompatible scaffolds [1, 12, 18]. Many conventional solvents are inadequate at dissolving structural biopolymers like cellulose, chitin, or silk, which limits their utility in scaffold fabrication. However, certain ILs, such as 1-ethyl-3-methylimidazolium acetate ([EMIM][Ac]), can directly dissolve cellulose or chitin, enabling the formation of porous biopolymer scaffolds via IL-assisted dissolution and regeneration using antisolvent techniques [12, 22].

The IL here acts as a removable processing medium, which can be completely washed out postfabrication, resulting in biocompatible, structured scaffolds suitable for cell adhesion and tissue growth [12]. Furthermore, polymer networks incorporating ILs have been used to create ionogels, soft, flexible, and ionically conductive gels with embedded IL [18, 24]. These materials are particularly relevant in neural and cardiac tissue engineering, where ionic conductivity enhances electrophysiological compatibility [24, 27].

For example, choline-based ILs have been incorporated into hydrogel matrices to produce conducting scaffolds, which stimulate electrically active cells in neural tissues [27]. IL–polymer composite scaffolds not only support cell proliferation and adhesion, but also enable drug release, electrical cue delivery, or antimicrobial protection, depending on the ions used [18, 24].

The biological effects of ILs within scaffolds vary depending on their chemical structure. Some ILs contain functional groups (e.g., carboxylates and amino acids) as (in Fig. 3), that interact with cell membranes or receptors, potentially enhancing cell adhesion or modulating cell signaling pathways [18, 22]. Others may be metabolized or biodegraded into nontoxic byproducts, particularly when using natural ions such as choline, amino acids, or organic acids [21, 24]. That said, IL toxicity must be carefully considered. Many early generation ILs (especially those with long alkyl chains or fluorinated anions) show cytotoxic or membrane-disrupting activity [5, 14]. Therefore, ILs intended for biomedical use, particularly those embedded in scaffolds, are typically selected for low toxicity, and are often used at subtoxic concentrations or based on choline-derived systems as shown in Fig. 3 [1, 21].

4.2 Nanoparticles in Tissue Scaffolds

Nanoparticles are widely combined with the methods of tissue engineering to either reinforce scaffolds or bind biomolecules, offering enhanced mechanical, biological, and signaling properties (as depicted in Fig. 4) [6, 10, 25]. Bioceramic nanoparticles, especially hydroxyapatite (HAp), are frequently embedded into bone scaffolds to mimic the mineral phase of bone and stimulate osteogenesis [21, 29].

Due to their high surface area, nanoscale HAp particles act as nucleation sites for calcium phosphate deposition, vital for new bone formation [21]. Several studies show that polymer-based scaffolds (e.g., collagen, PLGA, and PCL) reinforced with nano-HAp improve osteoblast attachment, enhance biocompatibility, and stimulate stem cell differentiation into bone-forming cells (osteoblasts) as shown in Fig. 4 [21, 29].

Silver nanoparticles (AgNPs) are widely used for skin and soft tissue regeneration due to their broad-spectrum antimicrobial activity. When incorporated into scaffolds (e.g., chitosan matrices), AgNPs release Ag^+ ions, which disrupt bacterial

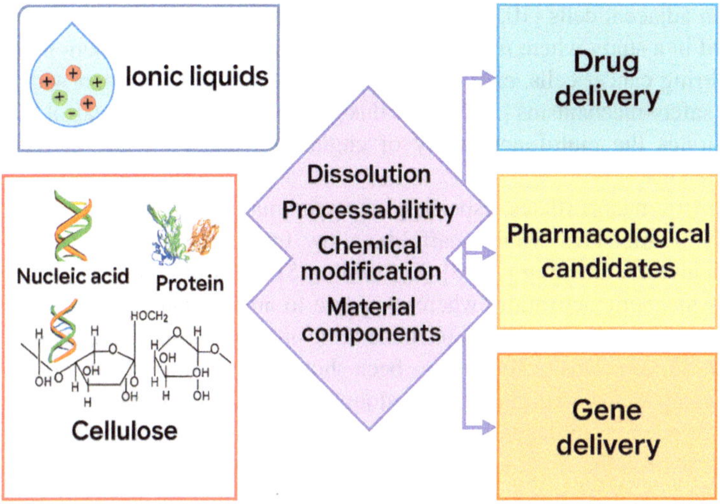

Fig. 3 Strategies and application of IL/biopolymer materials

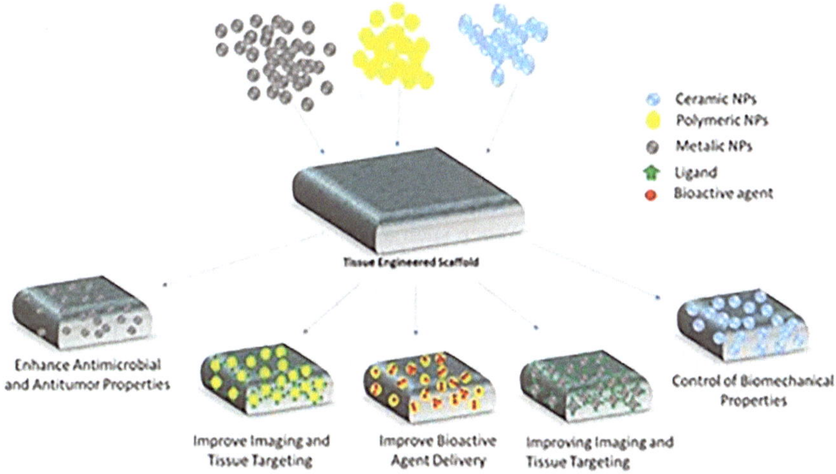

Fig. 4 Different kinds of nanoparticles—ceramic, polymeric, metallic—that can be used for several uses in Tissue engineering and regenerative medicine (TERM), including tissue targeting and imaging, bioactive chemical administration, altering mechanical properties of scaffolds, and offering antibacterial and anticancer qualities [37]

membranes and suppress infection during healing [29, 30]. Such antimicrobial scaffolds not only reduce bioburden but also accelerate wound closure in burns and ulcers [29].

Gold nanoparticles (AuNPs) have also been integrated into tissue engineering applications, particularly in cardiac and neural scaffolds, where their electroconductive properties help enhance cell signaling and promote electrical communication between adjacent cells [10, 24]. A fascinating use of nuclear-targeted AuNPs was reported in a study where implanted gold NPs enabled thermal ablation of residual or recurring cancer cells, effectively combining scaffold integration with potential cancer safety mechanisms [28]. While this goes beyond classical scaffold functions, it illustrates the multifunctionality of engineered nanomaterials in regenerative environments.

Magnetic nanoparticles, especially superparamagnetic iron oxide nanoparticles (SPIONs), embedded within scaffolds, allow for remote mechanical stimulation and noninvasive tracking [7, 13]. Embedding SPIONs in hydrogels or elastomers enables magnetic activation, where exposure to an external magnetic field causes the particles to vibrate or rotate, generating mechanostimulation at the cellular level [7, 13, 25]. This technique has been shown to promote mechanotransduction, encouraging stem cell differentiation along muscle or bone lineages [25].

Moreover, SPION-laden scaffolds provide MRI contrast, allowing clinicians to monitor scaffold position, integration, and degradation in vivo [7, 17]. The SPIONs act as T_2-weighted negative contrast agents, producing a localized signal drop where scaffolds reside, helping visualize cellularized implants without invasive procedures.

4.3 *Nanofluids in Regenerative Medicine*

They can be found in injectable tissue engineering. There are thermoresponsive nanofluids that change to gel at body temperature, used in an injectable form, which then solidifies in situ [19, 25]. When such a nanofluid has nanoparticles that either carry growth factors or provide structural cues, it can serve as a minimally invasive tissue engineering strategy [25]. For example, a suspension of peptide nanofibers (which can be regarded as a nanofluid until gelation) has been used for forming a scaffold inside the body that promotes cell infiltration and cartilage regeneration [19, 27]. Some researchers have developed "magneto-responsive" hydrogels where a nanofluid of magnetic particles and polymer is injected, and then a magnetic field is applied in the shaping or strengthening of the scaffold within the defect site [7, 25].

Again, nanoparticles, in the case of stem cell therapy, act as tracking agents and as carriers of therapeutic cargo. Stem cells can be tagged with superparamagnetic nanoparticles, allowing MRI tracking after transplantation (the clinician can observe cell migration and localization) [7, 13, 17]. Such labeling must be done carefully so as not to interfere with stem cell function; however, studies using iron oxide-labeled mesenchymal stem cells confirm feasibility and effectiveness in monitoring distribution and therapeutic engagement [13, 25]. In addition, nanoparticles can be used to deliver genes or gene-editing tools into stem cells before implantation or in vivo. For instance, PLGA nanoparticles carrying plasmid DNA encoding growth factors have been incorporated into scaffolds to achieve prolonged gene release and in situ transfection of host cells, effectively programming cells at the injury site to express required regenerative signals (i.e., gene-activated matrices) [10, 11, 19].

Ionic liquids (ILs) are less directly applied to cell therapy, but one fascinating application is their use in cryopreservation of cells or tissues. Certain ILs lower freezing points and form eutectic mixtures, making them potential cryoprotectants [23]. Protic ILs have been studied as alternatives to DMSO/glycerol in freezing solutions due to their ability to form hydrogen-bond networks, stabilizing cell membranes during freezing [23]. Initial results show that some ILs preserve cells with lower toxicity compared to DMSO, though this remains an emerging field [23].

To view the convergence of these technologies, consider an example in bone regeneration: a collagen scaffold is reinforced with hydroxyapatite nanoparticles to mimic bone mineral. A cholinium-based IL with an osteogenic drug is introduced into the scaffold matrix. The scaffold is seeded with stem cells. Once implanted, the HAp provides mechanical stimulation and nucleation for mineralization, the IL releases the drug, inducing stem cells to differentiate into osteoblasts, and the scaffold degrades over time as new bone forms. Concurrently, gold nanoparticles added to the scaffold could facilitate CT-based tracking, while silver nanoparticles could provide antimicrobial protection [21, 29]. Although theoretical, this is a plausible hybrid strategy grounded in current experimental trends.

The intersection of ionic liquids and nanoparticles is particularly interesting in regenerative medicine. For example, IL-functionalized nanoparticles, where a nanoparticle is coated with an IL, can be dispersed more uniformly in a polymer scaffold and may present cell-adhesive functional groups [18, 24]. A novel concept is bio-ionic liquids (Bio-ILs), which are not only biocompatible but also act as bio-signaling agents, mimicking natural osmolytes or ECM components. These bio-ILs could generate scaffold environments with ionic compositions that simulate native tissue and promote cell functionality [24].

A study reported a GelMA (gelatin methacrylate) hydrogel containing a choline-based bio-IL, which enhanced cardiac cell encapsulation and functionality in 3D, illustrating how ILs can be designed to tune scaffold conductivity, mechanical compliance, and cell signaling properties [27].

Tissue engineering bioreactors may also be configured to exploit nanofluids. For instance, perfusing a developing construct with a nanofluid containing magnetic nanoparticles, followed by magnetic field application, can be used to "exercise" muscle fibers or to align collagen networks, showcasing a biophysical stimulation method based on these material interfaces [7, 25].

In short, ionic liquids and nanomaterials advance regenerative medicine by enabling advanced material processing, enhanced scaffold functionality, and responsive cell environments. ILs improve biomaterial solubility, conductivity, and biocompatibility, while nanoparticles serve as mechanical reinforcers, gene carriers, and imaging agents [10, 18, 24]. Ensuring long-term biocompatibility and regulatory compliance remains essential, especially for implantable platforms. Nonetheless, research is encouraging, most IL-based scaffolds show low cytotoxicity, and several nanoparticle-infused scaffolds (e.g., bone fillers with nano-HAp) are already used clinically [4, 12, 29]. In the future, scaffolds may be smart, integrated systems that monitor force, deliver drugs on-demand, and communicate wirelessly via embedded nano sensors, in line with the precision goals of next-generation regenerative therapies [10, 24].

5 Clinical Applications and Toxicity/Biocompatibility Considerations

Translation of ionic liquids, nanofluids, and nanoparticles from bench to clinic involves rigorous assessments of biocompatibility, safety, and regulatory acceptability [1, 5, 7]. This section evaluates the current clinical status and reviews what is known about their interactions with biological systems.

5.1 *Clinical Applications and Approvals*

Nanoparticle formulations have made the most clinical headway. As outlined in Table 1, numerous FDA-approved nanomedicines and imaging agents currently exist [4, 13]. Liposomal drugs (e.g., Doxil, Visudyne) and polymeric nanoparticle drugs (e.g., Abraxane) have been commercially available for years and are proven to improve therapeutic outcomes in oncology and ocular diseases [4].

Iron oxide nanoparticle suspensions were approved for clinical use as MRI contrast agents (e.g., Ferumoxides and Ferumoxytol) and for iron supplementation, although some (e.g., Feridex) were withdrawn due to market-related or safety concerns such as anaphylaxis [13, 17]. Nevertheless, Ferumoxytol remains widely used off-label as a vascular MRI contrast agent, given its long circulation time and T1/T2-weighted imaging capability [13].

Although gold nanoparticles (AuNPs) have not yet been incorporated into routine clinical use, a gold nanoshell-based photothermal therapy (AuNS) for tumors has completed clinical trials, demonstrating both safety and therapeutic potential through localized ablation of cancer tissues [10, 31].

In 2020–21, the COVID-19 mRNA vaccines brought lipid nanoparticles (LNPs) to the forefront, marking the first large-scale use of LNPs (~100 nm) in healthy human populations [25, 27]. The vaccines developed by Pfizer/BioNTech and Moderna demonstrated high safety and efficacy, reinforcing the credibility of nanocarrier-based platforms [25, 33]. This breakthrough is expected to accelerate clinical translation of other nanoparticle-based therapeutics, such as siRNA-loaded LNPs (e.g., Onpattro), which was approved in 2018 for hereditary transthyretin-mediated amyloidosis, and many more that are in the clinical pipeline [4, 25, 27].

On the other hand, ionic liquids (ILs) have not yet received regulatory approval as therapeutic agents but are gradually progressing through preclinical and early phase clinical studies [1, 5, 18]. Some IL formulations have been tested in compounding pharmacies or early trials. For instance, topical IL-based antimicrobials are being investigated for skin infections, and transdermal IL systems for pain management have also been evaluated [13, 21].

A notable candidate, cholinc geranate (CAGE), a cholinium-based IL, was reported in 2017 to enhance transdermal delivery of antibiotics and showed favorable outcomes in MRSA-infected wound models during preclinical evaluation [21].

Additionally, deep eutectic solvents (DESs), often made from choline chloride and natural components like glycerol or amino acids, are structurally similar to ILs and are seen as more favorable for clinical use due to their GRAS (generally recognized as safe) status [15, 22]. DESs are being studied particularly for topical drug delivery and biomolecular stabilization.

Regulatory bodies such as the FDA and EMA will likely treat ILs as new excipients or active pharmaceutical entities, depending on their use case, which means they require extensive toxicology profiling and ADME studies [5, 7]. One challenge is that ILs do not neatly fall into existing drug classifications (e.g., not purely small molecules or polymers, but rather ionic assemblies) [1, 7].

Despite that, the field is advancing by developing biocompatible ILs, which are designed to biodegrade into known safe metabolites [5]. For instance, an IL formulated from choline and glycine would break down into choline (a nutrient) and glycine (a nontoxic amino acid), both familiar and well-tolerated molecules, which make such systems attractive from a safety standpoint [1, 21].

5.2 *Toxicity and Biocompatibility*

The same properties that grant these materials their new properties (small size for NPs, charge for ILs) can also generate different toxicological profiles [5, 7, 9]. Nanoparticles tend to interact with proteins and cells in unanticipated ways: they could trigger immune responses, cause oxidative stress, or, in the worst cases, accumulate and cause organ damage [9, 17, 29]. It's well known that some nanoparticles induce oxidative stress either by the formation of reactive oxygen species (ROS) or by releasing toxic ions (e.g., silver NPs release Ag^{+}) [9, 29, 30].

Rodent research has shown that high doses of some inorganic NPs (e.g., cadmium-based quantum dots) can be harmful to the liver and kidneys [17, 29]. Therefore, great emphasis is placed on surface functionalization of NPs to make them inert or "stealthy." PEGylation is a common method to reduce protein adsorption and avoid immune clearance, although PEGylated liposomes have still been shown to cause infusion-related reactions due to complement activation, also known as CARPA (complement activation-related pseudoallergy) [9, 32].

With lipid nanoparticles (LNPs) used in mRNA vaccines, reactogenicity (transient inflammation) in the majority of recipients is partly due to immune detection of both the nanoparticle and the RNA payload [25, 27, 33]. These reactions are typically acceptable (fever and soreness), and serious adverse events are rare.

Although long-term data on LNPs is still developing, as of 2025, no significant long-term safety issues have been reported across hundreds of millions of doses, which is reassuring [25, 33].

For metal NPs like gold, studies have demonstrated that AuNPs are inert and tend to accumulate in organs like the spleen and liver over extended durations without causing acute toxicity [10, 28]. However, the concern remains over potential long-term effects of persistence, and more studies are being conducted to determine whether gold NPs can be safely excreted or remain stable indefinitely [10, 29]. Research on ultrasmall gold NPs (<5 nm) shows they can be cleared renally, but this may compromise benefits like plasmonic behavior, which are size-dependent [10, 30].

Iron oxide NPs are more biodegradable, as iron is naturally stored in ferritin or hemoglobin. For instance, Ferumoxytol is taken up by macrophages, degraded, and contributes to the normal iron pool in the body, with minimal risk of iron overload or labile oxidative iron toxicity when used at therapeutic doses [13, 17]. These properties make SPIONs among the least toxic inorganic nanomaterials, and they were among the first NP-based therapies approved for clinical use [4, 13].

Ionic liquid (IL) toxicity has also been a longstanding concern, especially in the fields of green chemistry and biomedical sciences [1, 5]. Early ILs, such as imidazolium-based systems with PF_6^- or BF_4^- anions, are known to be toxic to aquatic organisms and cytotoxic to mammalian cells, sometimes comparable to benzalkonium chloride [1, 5, 6].

Toxicity is typically correlated with IL lipophilicity; those that penetrate lipid bilayers are more likely to disrupt cellular integrity. Research has shown that shortening the alkyl chain or incorporating hydrophilic functional groups reduces IL toxicity [5, 6]. Further, choosing biocompatible ions, such as choline, amino acids, or acetate, can produce milder ILs suitable for topical or injectable applications [1, 21].

For example, choline acetate consists of nutritional-grade choline and acetic acid, showing low cytotoxicity [1]. Similarly, choline geranate (CAGE) exhibits antibacterial activity and was shown to have only minor skin irritation in animal models, both of its components being food-grade compounds [21]. In such cases, the route and dose of exposure matter: topically applied ILs may be tolerated, even if their systemic use might raise toxicity concerns [21, 22].

Conversely, intravenous IL administration demands that the compound be non-toxic at effective concentrations or rapidly metabolized into safe products [5, 7, 20].

Regulators, including the FDA, treat ionic liquids as new chemical entities (NCEs). This classification requires extensive testing, including acute, subchronic, and chronic toxicity, as well as genotoxicity and pharmacokinetics [7, 22, 27].

There is a silver lining, however: some biomedical ILs are composed of building blocks already known to regulatory agencies. For example, cholinium salicylate, an API-IL, degrades into choline and salicylic acid, both of which are approved compounds with known safety profiles [1, 14]. Nonetheless, ILs must still be evaluated as distinct molecular entities, because ionic pairing can significantly alter distribution, toxicity, and bioactivity compared to their components in isolation [5, 14].

5.3 Environmental and Long-Term Considerations

If patients are exposed to or pass out these compounds, will they be an environmental threat? Ionic liquids, being present in waste streams, could be toxic to microbial life at wastewater treatment plants, especially since several ILs are poorly biodegradable [5, 6, 38]. Considerable research is focused on the development of biodegradable ionic liquids, typically by using enzyme-cleavable bonds or natural ionic components, which degrade into nontoxic, environmentally benign byproducts upon environmental release [1, 7, 22].

Eliminated nanoparticles may also find their way into the aquatic environment or bioaccumulate through food chains, although the risk depends on particle type and concentration. Most clinically used nanoparticles, such as liposomes, PLGA-based systems, or LNPs, are biodegradable, and those that are inorganic, like gold or iron, are generally used in low doses [4, 13, 29]. These inorganic NPs are likely to settle in sludge or become immobilized in soils, with low environmental mobility, resulting in a negligible ecological impact when compared to common metal-based environmental contaminants [7, 10].

Clinically, patient-specific considerations are equally important. Hypersensitivity to formulation components must be addressed, for example, anti-PEG antibodies have been identified in some individuals, likely due to prior exposure to PEG in drugs or cosmetics, and can lead to accelerated blood clearance or infusion reactions when PEGylated nanoparticles are administered [9, 32]. These rare but real effects highlight the need for rigorous preclinical and clinical testing to detect idiosyncratic responses, especially in diverse patient populations [32, 33].

It is also noteworthy that the industrial-scale manufacturing of such complex materials must adhere strictly to good manufacturing practices (GMP). Batch-to-batch reproducibility is critical; in the case of nanoparticles, even minor variations in size, charge, or surface chemistry can impact pharmacokinetics, toxicity, and clinical efficacy [7, 33]. Accordingly, the pharmaceutical industry has developed robust, validated processes, such as high-pressure homogenization for liposomes and microfluidic mixers for LNPs, to ensure consistent product quality [27, 29].

For ionic liquids, chemical purity is of paramount importance. Residual precursors, solvent impurities, or unreacted ionic species may pose toxicity risks, even if the IL itself is biocompatible [5, 20, 22]. Therefore, stringent analytical control and purification procedures are needed to ensure clinical-grade IL production aligns with regulatory expectations for biologically active excipients and therapeutic agents [7, 19].

5.4 Current Clinical Trials

Besides those products that have been licensed, there are several dozen more that are in trial. For example, magnetic nanoparticle hyperthermia is in trial for prostate and pancreatic cancer (some involving injections of magnetic fluids into tumors, others involving systemically delivered targeted magnetic particles for metastasis) [7, 17, 20]. Gold nanoshell therapy is being trialed for lung and brain cancer [10, 22]. Quantum dots (QDs) have been applied in human patients for sentinel lymph node mapping (as a QD imaging agent), but concerns about cadmium toxicity have prevented widespread adoption; however, investigations into indium-based QDs could help revive this platform [10, 23]. Mesoporous silica nanoparticles, loaded with therapeutic agents, are being tested in early phase human trials for anticancer applications. These porous NPs can carry large drug payloads and release them selectively within tumor cells [10, 29].

Ionic liquids may soon appear in clinical trials as liquid oral formulations for poorly soluble drugs if preclinical animal studies show enhanced bioavailability. Phase I trials may follow [1, 2, 5]. Also, IL-based electrolytes embedded in medical devices (e.g., implantable biosensors, glaucoma-sensing contact lenses) may undergo device-focused safety testing; in such cases, the IL is nonsystemic, but still must be proven nontoxic in case of leakage or exposure [22, 24].

Generally, the safety profile of each material should be assessed individually; however, some general principles can be reported [7, 9, 21].

Ionic Liquids

Target those composed of biologically well-characterized ions, screen extensively for cytotoxicity, hemocompatibility (e.g., impact on red and white blood cells), and organ toxicity in animal models [1, 5, 22]. Some ILs are irritating to local tissues, but if properly used (e.g., topically), they can be acceptable for biomedical applications [5, 21]. Biodegradability and metabolic fate are actively investigated, e.g., does an IL become metabolized by liver enzymes or excreted unchanged through renal or biliary pathways? [5, 20]. Quaternary ammonium-based ILs, for instance, can be eliminated via urine or bile, depending on their molecular charge, polarity, and size [5, 7, 20].

Nanoparticles

Should learn pharmacokinetics (ADME), absorption (or distribution if intravenous), distribution, metabolism, and excretion [9, 13, 29]. The vast majority of inorganic nanoparticles (NPs) are not metabolized enzymatically, but are instead physically sequestered in tissues or dissolve very slowly over time [10, 29]. Smaller is generally better for excretion; particles larger than renal or hepatic filtration thresholds may not be cleared efficiently and may persist in the body for extended periods [29, 30].

If they remain, extensive long-term toxicity studies are required to prove nonharmful accumulation and chronic safety [9, 10]. Additionally, nanoparticles can traverse biological barriers; for example, ultrasmall NPs (<10 nm) have been shown to cross the blood–brain barrier (BBB), which could be beneficial or hazardous depending on the application [9, 29]. One study reported that 5 nm gold nanoparticles successfully penetrated rodent brain tissue, while larger AuNPs were restricted [17]. Therefore, NP physicochemical parameters (e.g., size and surface charge) must be engineered based on therapeutic goals, including preventing BBB penetration unless targeting neurological diseases [10, 19].

Immunogenicity

ILs and NPs are both immunostimulants. Ionic liquids (ILs) possess structural motifs that can activate the complement system or trigger mast cell degranulation, while nanoparticles (NPs) are often recognized as foreign entities by phagocytes [5, 9, 14]. However, some nanoparticles can be engineered to deliberately engage the immune system beneficially, for example, as vaccine adjuvants. Most existing vaccines employ aluminum oxide nanoparticles, and newer studies are exploring PLGA-based and lipid-based nanoparticles as next-generation adjuvants [10, 23, 25].

So, immunogenicity is not inherently detrimental; it depends on context and design. However, undesired immune responses, such as anaphylaxis, hypersensitivity, or chronic inflammation, remain a major safety concern and must be rigorously assessed in preclinical and clinical trials [9, 32]. From a regulatory perspective, it is reassuring that multiple nanomedicines have already received FDA and EMA approval, demonstrating that these agencies have well-defined evaluation frameworks for nanotechnology-based therapeutics [4, 30]. In contrast, ionic liquids remain relatively novel to regulatory pipelines, but since they are chemical entities, they are typically evaluated via the small-molecule regulatory route, albeit with additional scrutiny for properties such as tissue persistence, toxicity, or organ-specific tropism [5, 7, 20].

Clinical Case Studies

One of the most notable clinical cases was the use of Magtrace, an iron oxide nanofluid for lymphatic mapping in breast cancer (approved in Europe) [4, 13]. It allowed surgeons to avoid radioactive tracers by injecting the magnetic nanofluid and sensing sentinel lymph nodes using a magnetometer (the particles both serve as the tracer and are afterwards cleared by macrophages with few side effects, aside from a brownish discoloration at the injection point for several months) [13, 17]. This shows how nanomaterials can provide safer or more convenient alternatives to older techniques (radioactive colloids in this case) [13, 32].

Briefly, biocompatibility can be designed for ionic liquids, nanofluids, and nanoparticles, but with careful engineering and rigorous testing [1, 5, 9]. The clinical record is strongest with nanoparticle technology, with many FDA- and EMA-approved nanotherapies leading to improved patient outcomes [4]. Ionic liquids are approaching clinical use, especially in specialty formulations and implantable device applications, though regulatory scrutiny will remain high [5, 7, 20]. Regulatory agencies will continue to evaluate on a risk–benefit basis, as is standard. If a new drug offers clear therapeutic or diagnostic benefits and the risks are minimal or manageable, it has a strong chance of approval [7, 10].

As an instance of benefit outweighing risk: the lipid nanoparticles (LNPs) used in mRNA vaccines caused greater local and systemic reactogenicity than traditional vaccines (most individuals experienced fever or chills), yet due to their high efficacy against a deadly virus, the benefit was enormous and the side effects tolerable for most; thus, these vaccines were approved in record time [25, 27, 33]. Similarly, a nanoparticle-based chemotherapy might introduce novel side effects, but if it greatly improves survival or reduces severe toxicity compared to standard chemotherapy, the benefit-to-risk ratio supports approval [10, 17]. Finally, ongoing postmarket surveillance of nanomedicines is essential. We keep learning from real-world experience, i.e., by monitoring long-term follow-up of patients treated with iron oxide or gold nanoparticles, we can detect any delayed adverse effects (none reported so far in short- to medium-term studies) [13, 21].

6 Future Prospects and Conclusions

Looking ahead, the convergence of ionic liquid technology, nanofluid engineering, and nanotechnology will provide new avenues in medicine. Several promising directions and trends can be anticipated.

6.1 Personalized Medicine and Smart Design

The possibility of tuning ILs and nanoparticles suggests that treatments can be customized according to the patient's needs. In the future, a patient's unique condition could dictate which ionic liquid (IL) is used as a solvent or carrier for their medication, or which surface ligands are attached to a nanoparticle formulation, for example, selecting a targeting ligand that is specific to the patient's tumor biomarkers.

As diagnostic and genomic technologies advance, it will become increasingly possible to develop personalized nanoparticle-based therapies, such as choosing an IL that best solubilizes a customized combination of drugs selected to match a cancer mutation profile, or using IL-assisted 3D printing to fabricate patient-specific scaffolds loaded with their growth factors.

6.2 Integration of NPs and ILs in Medical Devices

We can anticipate ionic liquids as long-lasting electrolytes in implantable biosensors monitoring glucose or electrolytes continuously inside the human body (ILs would not dry up or evaporate like water-based electrolytes). Nanoparticles can be integrated into closed-loop systems, for instance, nanoparticles that deliver insulin in response to elevated glucose levels, while also generating a detectable signal for wearable monitoring devices.

Other scientists are developing nanorobots or micro-nano machines: although these are still largely conceptual, they may use nanofluidic channels and magnetic nanoparticles to perform precise biomedical tasks such as targeted drug delivery, clot removal, or even microsurgery (e.g., magnetically controlled microdevices designed to dissolve thrombi).

6.3 Regenerative Medicine Progress

In tissue engineering, one of the main targets is 3D bioprinting of organs. Ionic liquids may stabilize printable bio-inks containing living cells, prevent ice-crystal formation in printable cryogels, or enable printing of otherwise nonprintable biomaterials. Nanoparticles will be incorporated into bio-inks to provide biochemical cues to the printed tissues or allow for external stimulation after printing, for example, printing a cardiac tissue patch with embedded magnetic NPs, which can then be conditioned using magnetic fields to promote tissue maturation.

In addition, gene delivery via nanoparticles could facilitate in vitro or in vivo gene editing, such as CRISPR/Cas9 systems delivered into stem cells to correct genetic disorders, as part of next-generation regenerative medicine. This concept is already advancing; lipid nanoparticles have been used experimentally to deliver CRISPR editors to the liver and correct genetic diseases in animal models, a technique that could become clinically available within the coming decade.

7 Overcoming Current Challenges

Some challenges must be overcome by future research despite the progress. For ionic liquids: improving our understanding of toxicological mechanisms (so we can design out toxicity systematically), and designing ILs that can be cleared or metabolized predictably. For nanomedicines: overcoming the heterogeneity of patient response, e.g., the EPR effect in tumors is not uniform; in some patients' tumors, nanoparticles will accumulate well, in others not at all. New strategies like active targeting or pretreatment of the tumor vasculature (potentially by IL-induced vaso-dilation?) would make nanoparticle treatment more reliably effective.

And even scale-up manufacturing and quality control are still issues for advanced nanoparticle systems, advances in microfluidics and automation technology may bring consistency and reduced costs.

Lastly, ionic liquids, nanofluids, and nanoparticles are emerging technologies that are making the vision of advanced, "smart" biomedical therapies a reality. The past two decades have witnessed these materials evolve from laboratory curiosities to clinically established remedies: from ILs enabling enzyme catalysis in biotech, to nanoparticle therapeutics saving lives in cancer therapy, to magnetic nanofluids offering new cancer therapy modalities. The synergy of their properties allows scientists and clinicians to address previously intractable challenges, like the delivery of insoluble drugs to where they need to have an effect, imaging small clusters of cancer cells, or the regeneration of complex tissues.

Finally, success will be gauged in terms of enhanced patient outcomes and public health. The advances described hold out the potential for better treatments with fewer side effects (with targeted delivery), earlier diagnosis of disease (with sensitive nanosensors and imaging), and even prevention or reversal of disease (with regenerative medicine). By cautious development, thorough safety evaluation, and innovative clinical trials, ionic liquids, nanofluids, and nanoparticles will necessarily be at the forefront of the future of medicine, one where disease can be precisely targeted and patients healed or restored to health with minimal collateral damage.

References

1. K.S. Egorova, E.G. Gordeev, V.P. Ananikov, Biological activity of ionic liquids and their application in pharmaceutics and medicine. Chem. Rev. **117**(10), 7132–7189 (2017). https://doi.org/10.1021/acs.chemrev.6b00562
2. M. Sheikhpour, M. Arabi, A. Kasaeian, A. Rokn Rabei, Z. Taherian, Role of nanofluids in drug delivery and biomedical technology: Methods and applications. Nanotechnol. Sci. Appl. **13**, 47–59 (2020). https://doi.org/10.2147/NSA.S260374
3. D. Bobo, K.J. Robinson, J. Islam, K.J. Thurecht, S.R. Corrie, Nanoparticle-based medicines: A review of FDA-approved materials and clinical trials to date. Pharm. Res. **33**(10), 2373–2387 (2016). https://doi.org/10.1007/s11095-016-1958-5
4. A.K. Bambam, A. Dhanola, K.K. Gajrani, A critical review on halogen-free ionic liquids as potential metalworking fluid additives. J. Mol. Liq. **380**, 121727 (2023). https://doi.org/10.1016/j.molliq.2023.121727
5. I.M. Marrucho, L.C. Branco, L.P.N. Rebelo, Ionic liquids in pharmaceutical applications. Annu. Rev. Chem. Biomol. Eng. **5**(1), 527–546 (2014). https://doi.org/10.1146/annurev-chembioeng-060713-040024
6. E.M. Hamad et al., Review of nanofluids and their biomedical applications. J. Nanofluids **10**(4), 463–477 (2021). https://doi.org/10.1166/jon.2021.1806
7. A.C. Anselmo, S. Mitragotri, Nanoparticles in the clinic: An update. Bioeng. Transl. Med. **4**(3) (2019). https://doi.org/10.1002/btm2.10143
8. W.L. Hough et al., The third evolution of ionic liquids: Active pharmaceutical ingredients. New J. Chem. **31**(8), 1429 (2007). https://doi.org/10.1039/b706677p
9. K. Maier-Hauff et al., Efficacy and safety of intratumoral thermotherapy using magnetic iron-oxide nanoparticles combined with external beam radiotherapy on patients with recurrent glioblastoma multiforme. J. Neuro-Oncol. **103**(2), 317–324 (2011). https://doi.org/10.1007/s11060-010-0389-0

10. M. Petkovic, K.R. Seddon, L.P.N. Rebelo, C. Silva Pereira, Ionic liquids: A pathway to environmental acceptability. Chem. Soc. Rev. **40**(3), 1383–1403 (2011). https://doi.org/10.1039/C004968A
11. K.S. Egorova, V.P. Ananikov, Toxicity of ionic liquids: Eco(cyto)activity as complicated, but unavoidable parameter for task-specific optimization. ChemSusChem **7**(2), 336–360 (2014). https://doi.org/10.1002/cssc.201300459
12. S.P.M. Ventura, C.S. Marques, A.A. Rosatella, C.A.M. Afonso, F. Gonçalves, J.A.P. Coutinho, Toxicity assessment of various ionic liquid families towards Vibrio fischeri marine bacteria. Ecotoxicol. Environ. Saf. **76**, 162–168 (2012). https://doi.org/10.1016/j.ecoenv.2011.10.006
13. K.S. Egorova, M.M. Seitkalieva, A.V. Posvyatenko, V.N. Khrustalev, V.P. Ananikov, Cytotoxic activity of salicylic acid-containing drug models with ionic and covalent binding. ACS Med. Chem. Lett. **6**(11), 1099–1104 (2015). https://doi.org/10.1021/acsmedchemlett.5b00258
14. E.J.A. Choi, Enhancing thermal conductivity of fluids with nanoparticles, in *Proceedings of the ASME International Mechanical Engineering Congress & Exposition*, (ASME, New York, 1995)
15. C.-J. Chiang, C.-H. Chang, Magnetic hyperthermia mediated by Escherichia coli for targeted cancer therapy. J. Taiwan Inst. Chem. Eng. **126**, 29–35 (2021). https://doi.org/10.1016/j.jtice.2021.07.011
16. X. Qu, Yao, Wang, Li, Zhang, Anti-CD30-targeted gold nanoparticles for photothermal therapy of L-428 Hodgkin's cell. Int. J. Nanomedicine **7**, 6095 (2012). https://doi.org/10.2147/IJN.S37212
17. H. Su, Y. Wang, Y. Gu, L. Bowman, J. Zhao, M. Ding, Potential applications and human biosafety of nanomaterials used in nanomedicine. J. Appl. Toxicol. **38**(1), 3–24 (2018). https:/ /doi.org/10.1002/jat.3476
18. K. Karikó, M. Buckstein, H. Ni, D. Weissman, Suppression of RNA recognition by toll-like receptors: The impact of nucleoside modification and the evolutionary origin of RNA. Immunity **23**(2), 165–175 (2005). https://doi.org/10.1016/j.immuni.2005.06.008
19. P.T. Heath et al., Safety and efficacy of NVX-CoV2373 Covid-19 vaccine. N. Engl. J. Med. **385**(13), 1172–1183 (2021). https://doi.org/10.1056/NEJMoa2107659
20. D.M. Correia et al., Ionic liquid-based materials for biomedical applications. Nano **11**(9), 2401 (2021). https://doi.org/10.3390/nano11092401
21. S.J. Bryant et al., Cryopreservation of mammalian cells using protic ionic liquid solutions. J. Colloid Interface Sci. **603**, 491–500 (2021). https://doi.org/10.1016/j.jcis.2021.06.096
22. J.F. Hainfeld, D.N. Slatkin, T.M. Focella, H.M. Smilowitz, Gold nanoparticles: A new X-ray contrast agent. Br. J. Radiol. **79**(939), 248–253 (2006). https://doi.org/10.1259/bjr/13169882
23. S. Bonvalot et al., First-in-human study testing a new radioenhancer using nanoparticles (NBTXR3) activated by radiation therapy in patients with locally advanced soft tissue sarcomas. Clin. Cancer Res. **23**(4), 908–917 (2017). https://doi.org/10.1158/1078-0432.CCR-16-1297
24. R. Bardhan, S. Lal, A. Joshi, N.J. Halas, Theranostic nanoshells: From probe design to imaging and treatment of cancer. Acc. Chem. Res. **44**(10), 936–946 (2011). https://doi.org/10.1021/ar200023x
25. J.V. Jokerst, S.S. Gambhir, Molecular imaging with theranostic nanoparticles. Acc. Chem. Res. **44**(10), 1050–1060 (2011). https://doi.org/10.1021/ar200106e
26. E. Pomerantseva, F. Bonaccorso, X. Feng, Y. Cui, Y. Gogotsi, Energy storage: The future enabled by nanomaterials. Science **366**(6468), 2019 (1979). https://doi.org/10.1126/science.aan8285
27. J.-W. Yoo, D.J. Irvine, D.E. Discher, S. Mitragotri, Bio-inspired, bioengineered and biomimetic drug delivery carriers. Nat. Rev. Drug Discov. **10**(7), 521–535 (2011). https://doi.org/10.1038/nrd3499
28. M. Du, J. Chen, K. Liu, H. Xing, C. Song, Recent advances in biomedical engineering of nano-hydroxyapatite including dentistry, cancer treatment and bone repair. Compos. Part B Eng. **215**, 108790 (2021). https://doi.org/10.1016/j.compositesb.2021.108790

29. B. Kang, M.A. Mackey, M.A. El-Sayed, Nuclear targeting of gold nanoparticles in cancer cells induces DNA damage, causing cytokinesis arrest and apoptosis. J. Am. Chem. Soc. **132**(5), 1517–1519 (2010). https://doi.org/10.1021/ja9102698
30. F.P. Polack et al., Safety and efficacy of the BNT162b2 mRNA Covid-19 vaccine. N. Engl. J. Med. **383**(27), 2603–2615 (2020). https://doi.org/10.1056/NEJMoa2034577
31. M. Delcea, H. Möhwald, A.G. Skirtach, Stimuli-responsive LbL capsules and nanoshells for drug delivery. Adv. Drug Deliv. Rev. **63**(9), 730–747 (2011). https://doi.org/10.1016/j.addr.2011.03.010
32. M. Fathi-Achachelouei et al., Use of nanoparticles in tissue engineering and regenerative medicine. Front. Bioeng. Biotechnol. **7**, 113 (2019). https://doi.org/10.3389/fbioe.2019.00113
33. I. Noshadi et al., Engineering biodegradable and biocompatible bio-ionic liquid conjugated hydrogels with tunable conductivity and mechanical properties. Sci. Rep. **7**(1), 4345 (2017). https://doi.org/10.1038/s41598-017-04280-w
34. S. Li et al., The multifunctional role of hydroxyapatite nanoparticles as an emerging tool in tumor therapy. Acta Biomater. (2025). https://doi.org/10.1016/j.actbio.2025.05.019
35. X. Qian et al., In vivo tumor targeting and spectroscopic detection with surface-enhanced Raman nanoparticle tags. Nat. Biotechnol. **26**(1), 83–90 (2008). https://doi.org/10.1038/nbt1377
36. A. Wolff et al., Direct gene transfer into mouse muscle in vivo. Science **247**(4949), 1465–1468 (1990). https://doi.org/10.1126/science.1690918
37. F.C.B.E. Vânia Vilas-Boas, Magnetic hyperthermia for cancer treatment: Main parameters affecting the outcome of in vitro and in vivo studies. Molecules **25**(12), 2874 (2020)
38. Z. Khan, D. Shahid, A.M. Khan, M.K. Khan, Nanoscale solutions for water treatment and purity enhancement, in *Breaking Boundaries: Pioneering Sustainable Solutions Through Materials and Technology*, (Springer, Singapore, 2025), pp. 193–211. https://doi.org/10.1007/978-981-97-9827-8_9

From Antiquity to Innovation: Biomedical Applications of Ionic Liquids, Nanofluids, and Nanoparticles

Ahmet Aykaç, Amirhossein Nassirzadehsaleh, Büşra Çakin, Salva Alizadeh Farahmand, and Sahra Dikmen

1 Introduction

Nanotechnology is generally perceived as a contemporary cutting-edge discipline, yet its origins unexpectedly extend back to ancient times. It is possible to trace early signs of nanoscience and nanotechnology in historical records. Throughout recorded history, the glitter of Mesopotamian temple windows or the fine craftsmanship in royal artifacts indicate the use of nano-sized structures. Over the recorded time, researchers have uncovered significant evidence of various applications. One of the best-known and most fascinating pieces of evidence for nanotechnology in the ancient world is the Lycurgus Cup, dating back to the fourth century AD. This remarkable dichroic glass artifact showcases the Romans' mastery of materials science. The cup's color-changing property is attributed to the presence of silver-gold alloy nanoparticles (NPs) embedded in the glass matrix, enabling the cup to display different colours in transmitted and reflected light. Following early examples like the Lycurgus Cup, the use of NPs and similar nanostructures continued over

A. Aykaç (✉)
Faculty of Engineering and Architecture, Department of Engineering Sciences, Izmir Katip Çelebi University, Izmir, Turkey

Graduate School of Natural and Applied Sciences, Department of Nanoscience and Nanotechnology, Izmir Katip Çelebi University, Izmir, Turkey
e-mail: ahmet.aykac@ikcu.edu.tr

A. Nassirzadehsaleh · B. Çakin · S. A. Farahmand
Graduate School of Natural and Applied Sciences, Department of Nanoscience and Nanotechnology, Izmir Katip Çelebi University, Izmir, Turkey

S. Dikmen
Graduate School of Natural and Applied Sciences, Material Science and Engineering Department, Izmir Katip Çelebi University, Izmir, Turkey

S. Kulkarni, S. Das (eds.), *Ionic Liquids, Nanofluids and Nanotechnology*,
Lecture Notes in Nanoscale Science and Technology 42,
https://doi.org/10.1007/978-3-032-04008-4_10

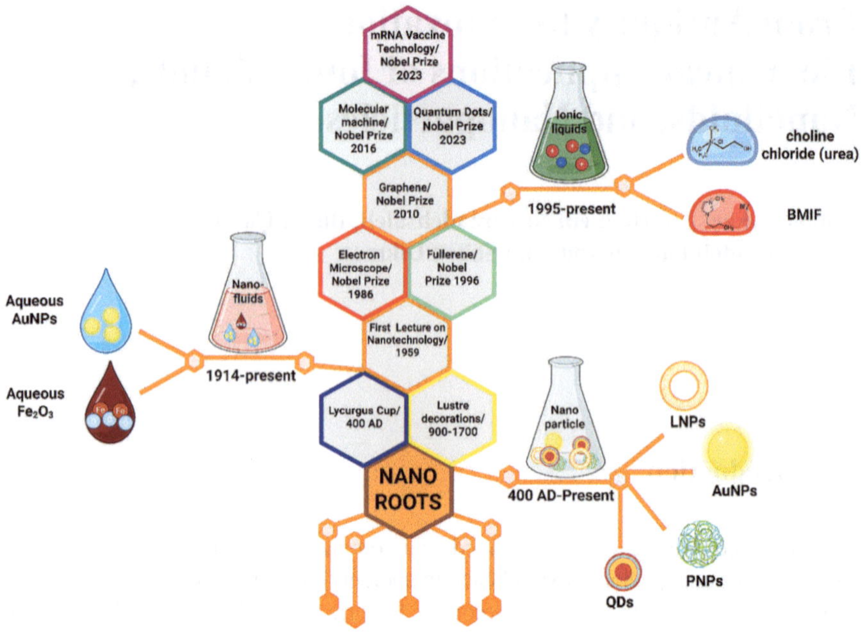

Fig. 1 A schematic illustration of the historical timeline of NPs, ILs, and NFs: Nobel Prize–winning discoveries

the centuries, either consciously or coincidentally. This is observed particularly in the Lustre decorations of medieval ceramics (900–1700), which developed in the medieval Islamic world and spread to Europe, as well as Italian antiques like Deruta pottery (1450–1600) and Damascus steel swords (1300–1800) [1, 2]. It is widely acknowledged that the modern scientific framework of nanotechnology took root with the seminal 1959 lecture by Nobel laureate physicist Richard P. Feynman at CALTECH. In this conversation, titled *There's Plenty of Room at the Bottom*, Feynman outlined the conceptual framework for nanotechnology by discussing the potential to individually manipulate atoms and molecules. He predicted that the ability to control individual atoms would offer revolutionary possibilities in various fields such as materials science, computer technology, and biology. Following Feynman's visionary ideas, the term "nanotechnology" was first used and defined in 1974 by Norio Taniguchi from the University of Tokyo. Taniguchi explained this term as the process of processing materials at the nanoscale by separating, combining, and reshaping them at the atomic or molecular level. Furthermore, he predicted that techniques achieving dimensional precision better than 100 nanometers (nm) would be developed by the late 1980s [3]. Some of the key Nobel Prize–related achievements are shown in Fig. 1 to visualize the progress of nanotechnology, thereby completing this historical background.

According to a widely accepted assumption in the scientific world, one of the most important factors that has contributed to the development of nanotechnology in this century is undoubtedly Feynman's famous speech at CALTECH. From that speech to the present day, in a little over half a century, which is not really a long time for the development of a scientific field, studies in nanoscience and nanotechnology have grown at remarkable momentum, and numerous significant studies have been initiated in medicine, pharmacy, chemistry, material science, electronics, agriculture, industry, and even art. It offers innovative approaches to a wide range of global problems, from increasing energy efficiency [4] to combating environmental pollution to providing smart solutions in agriculture [5]. In materials science, nanostructured coatings [6] stand out by offering superior performance in different industrial applications. Compared to the valuable advances in all these areas, the most striking effects of nanotechnology are probably emerging in the fields of biomedicine, biotechnology, and nano-biotechnology. In these fields, nanoscale materials and methods are significantly advancing early diagnosis and treatment [7, 8], enabling precise drug targeting [9], promoting the repair and regeneration of damaged tissues [10], facilitating the engineering of biomolecules, and promising revolutionary progress in advanced disease monitoring devices [11]. However, these significant advances also present new challenges and necessitate more novel approaches. It is clear that nanostructures alone have reached their limits, especially when it comes to the complexity of biological systems or harsh industrial conditions. Thus, different classes of nanomaterials can synergize to overcome certain limitations. This section provides an in-depth examination of important solid, liquid, and mixture forms relevant to materials such as ionic liquids (ILs), nanofluids (NFs), and NPs.

The development of next-generation fluids and solvents to overcome challenges in nanotechnology holds significant importance. Among these new materials, ILs are biocompatible solvents made up of bulky, asymmetric organic anions and cations. This purely ionic nature distinguishes them from molecular solvents and traditional ionic solutions. ILs were first synthesized in 1914 but attracted scientific interest in the 1990s with the discovery of more practical air- and moisture-stable types as the second and third generations were produced [12, 13]. The key feature of these liquids is their tunable nature [14]. By carefully selecting the cation and anion pair, their chemical and physical properties (hydrophilicity, hydrophobicity), biocompatibility, and production efficiency can be precisely tailored for specific purposes [12, 13, 15]. This unique tunability and other beneficial properties such as low melting point [15], solubility [16], negligible vapor pressure [14], high electrochemical stability, biocompatibility, and low volatility enable the investigation of ILs for various promising applications in biomedical fields [12]. ILs are used as functional excipients to address the issue of low drug solubility, and the integration of drugs into nanocarriers is a promising application, particularly in biosensor development [12, 17]. In addition, the thermal stability and flame-resistant properties of ILs ensure the safe production of electrochemical biosensors [18]. These liquids enhance the penetration of drugs by enabling their application to the barrier on the surface of the skin (ILs-Azone) or are used as polymeric ILs (PILs) to enhance antibacterial properties [19, 20]. According to the study by Borkowski, Jing, and Klähn, ILs exhibit strong antimicrobial activity by interacting with bacterial cell membranes [21–23], making them potential candidates for combating infections. Biocompatible and antibacterial ILs also play a significant role in tissue engineering. They are utilized for processing biopolymers like cellulose [18, 24], creating three-dimensional (3D) scaffolds [25] and supporting vital cell culture and regenerative medicine applications [19, 26, 27]. In addition to the unique properties and application areas offered by ILs, new functional systems are created by dispersing NPs, the fundamental building blocks of nanotechnology, within various liquid bases. NFs are engineered colloids with enhanced properties, formed by creating stable suspensions of nm-sized solid particles within a base fluid [28]. Although the initial theoretical foundations were laid by Maxwell [29], the modern concept of "NFs" utilizing nano-sized particles was introduced by Choi in 1995 [30]. These fluids are typically prepared using either one-step [31, 32] or the more common two-step [33, 34] methods. Their unique thermophysical and optical properties enable groundbreaking applications in biotechnology and medicine. In biomedical imaging, the targeted accumulation and enhanced signaling capabilities of NPs provide clearer and more sensitive diagnosis in techniques like magnetic resonance imaging (MRI) [35, 36]. In drug delivery, NFs hold the potential to increase treatment efficacy and reduce systemic side effects by facilitating the direct delivery of active substances to diseased tissues [9, 37]. Magnetic hyperthermia (MH), a key approach in cancer therapy, utilizes NFs containing magnetic NPs to generate heat within the targeted tumor region via an external magnetic field, thereby destroying or sensitizing cancer cells [38, 39]. Furthermore, the strong antibacterial activity exhibited by NPs within NFs through mechanisms such as damaging the cell membrane or producing reactive oxygen species (ROS) [40, 41], offers new strategies for infection control [41, 42].

The fundamental element determining the performance and functionality of NFs is the NPs within the suspension. NPs are materials specifically designed and produced at the atomic or molecular level, with sizes typically ranging from 1 to 100 nm [43]. Their unique physicochemical properties exhibited due to this nanoscale size, and their ability to interact with biological systems, grant them promising potential in various clinical applications. Specifically, capabilities such as reaching targeted tissues or cells, crossing biological barriers, and carrying and controllably releasing therapeutic or diagnostic agents enable the development of new approaches in the diagnosis and treatment of many diseases. Consequently, the clinical use of NPs is mainly directed toward cancer therapy, where they enhance chemotherapy efficacy and reduce side effects through targeted delivery, destroy tumors using photothermal therapy (PTT), and facilitate the safe delivery of genetic material in gene therapy [40, 44]. In regenerative medicine, NPs play vital roles in tracking the behavior of stem cells within the body and in repairing damaged tissues through tissue engineering approaches [45, 46].

This chapter aims to evaluate the historical background and recent advancements in ILs, NFs, and NPs in biomedical, biotechnology, and nano-biotechnology clinical applications. This chapter examines potential biomedical and clinical applications, respective regulatory and ethical issues, and safety and toxicity concerns of ILs (see Sect. 2), NFs (see Sect. 3), and NPs (see Sect. 4).

2 Ionic Liquids in Biomedical and Clinical Applications

2.1 Properties of Ionic Liquids

Ionic liquids are generally known as biocompatible solvents composed of bulky and asymmetric organic anions and cations that remain in liquid form below 100 °C [12]. The selection of cation and anion pairs enables the adjustment of specific properties, such as the materials' biocompatibility and production efficiency. The choice of cation largely determines the stability of the ILs, while the anion selection typically governs its functional role [12, 13]. When evaluated based on cation content, there are four categories: (1) dialkylimidazolium, (2) N-alkyl-pyridinium, (3) phosphonium, and (4) alkylammonium cations [17]; in this way these cations can combine with organic or inorganic anions, e.g., CH_3COO^-, Cl^-, Br^-, and I^- [24]. They can be either hydrophilic or hydrophobic depending on the type of ions they contain [15]. Historically speaking, Paul Walden succeeded in producing the first ILs, ethylammonium nitrate (EAN), in 1914. However, owing to its high sensitivity to air and moisture, EAN could not be widely used and the processing of EAN became difficult. In 1992, Wilkes and Zaworotko discovered the first set of ILs based on the imidazolium cation, which is not sensitive to air and moisture [12]. The second generation of ILs based on the 1-ethyl-3-methylimidazolium cation and different anions such as $[CH_3CO_2]^-$, $[NO_3]^-$, and $[BF_4]^-$ aimed to achieve stability against air and water. These ILs, with their unique properties and simple processing methods, have attracted attention in the scientific community. Subsequently, the third generation of ILs with tunable chemical and physical properties, depending on the desired processes, was discovered [13]. Since this discovery, ILs have become a popular research topic in fields such as engineering, chemistry, materials science, and biomedical applications. It has been achieved due to a host of beneficial properties such as low melting point [15], solubility [16], negligible vapor pressure [14], high electrochemical stability, tunability, biocompatibility, and low volatility [12].

One of the most important characteristics of ILs is their low volatility, which has led to their classification as "green solvents." This characteristic makes them promising alternatives to volatile organic compounds (VOCs), which are widely used in large quantities in the chemical and engineering industries, and contributes to significant environmental issues [47]. For instance, in the study conducted by Jain et al., thanks to the negligible volatility of ILs, which enabled the reaction to proceed without the need for volatile organic solvents (VOSs), thus aligning with the principles of green chemistry. Moreover, the study was conducted in an environmentally friendly manner [48]. Another fundamental characteristic of ILs is their tunable nature and low vapor pressure, which prevents the formation of ionic pairs due to weak Coulombic interactions between their ions, thereby hindering the volatilization of the substance [14]. This low vapor pressure combined with excellent thermal stability and tunable solubility properties. These features provide dissolved, regenerated, and functionalized large molecules [16, 24], including

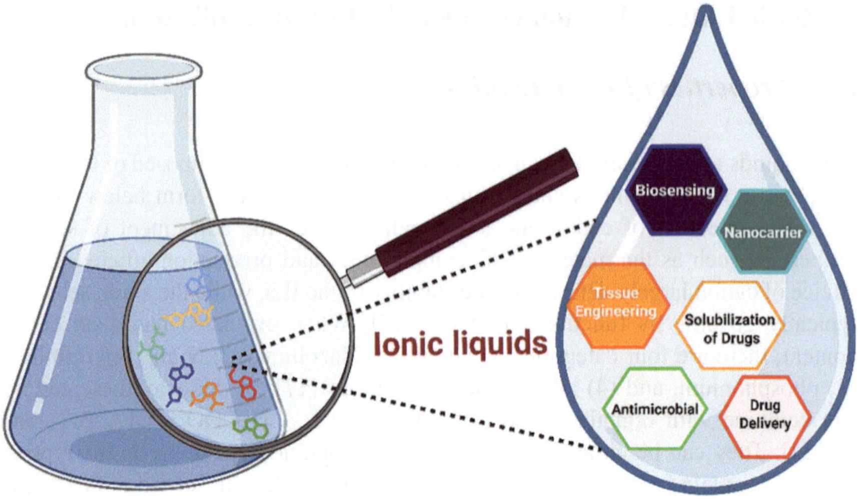

Fig. 2 Some applications of ILs in biomedicine

biopolymers, such as cellulose, wool keratin, and silk fibroin, which are difficult to dissolve. Utilizing ILs to dissolve polymers such as chitin presents a potential alternative method for improving their solubility and processability, avoiding the need for aggressive chemicals and solvents. For example, Brigham et al. focused on the dissolution of chitin using ILs to produce chitin membranes with different properties [24]. A further aspect of these liquids is that they have the properties that increase drug solubility and permeability and can be used as functional excipients and drug stabilizers. Employing functional excipients like ILs and incorporating drugs into nanocarriers is an effective approach to address the issue of low drug solubility [17]. Incorporating ILs into the creation or alteration of nanocarriers is a promising and emerging application for this class of solvents, particularly in biosensor development [12]. The thermal stability and flame-resistant properties of ILs are additional characteristics that make them highly safe for the production of electrochemical biosensors [18]. ILs are used not only as biosensors in nanocarriers, drug delivery, and drug solubilization processes but also in tissue engineering and antimicrobial activities in pharmaceutical applications. Figure 2 illustrates some of the pharmaceutical applications of ILs.

2.2 *Drug Delivery Systems of ILs*

Ionic liquids, owing to their tunable, biocompatible, and biodegradable properties, play a significant role in optimizing common issues encountered in medicine, such as poor drug solubility, low biological activity, and insufficient drug delivery efficiency. In drug delivery systems, drugs may not always be integrated into the carrier systems. In such cases, the drug concentration in the body rises rapidly and then falls below the optimal level due to degradation in the circulation. Then, the absorption of the drug at the target site becomes insufficient. While addressing these issues, ILs can be produced in a nontoxic manner using pharmaceutically acceptable organic cations and anions. These organic anions (acetate, phosphate, carboxylate, and amino acid), linear long-chain anions (hexanoate, octanoate, oleate), and cations (amino acid esters, choline, piperidinium, and pyrrolidinium) are easily biodegradable and nontoxic substances. Thanks to the properties such as biodegradability and nontoxicity, these serve as a guide in identifying suitable biocompatible ILs for pharmaceutical purposes [49, 50]. ILs are frequently used in pharmaceutical applications owing to their excellent properties as solvents for poorly soluble drugs, drug reservoirs, drug carriers, antimicrobial agents, antibiofilm agents, and ionogels [51]. As early as 2008, the Jaitely research group highlighted that imidazolium-based ILs possess solubilizing capabilities and can function as drug depots for both sustained and controlled drug release [52]. A study by Neugebauer et al. presented that choline, a water-soluble trimethylammonium salt with a chloride anion, is frequently used as a naturally occurring cationic component in biocompatible monomers. Also, it exhibits antibacterial properties and improves the pharmacodynamic and pharmacokinetic effects of the delivered drug [53].

ILs are particularly effective in enhancing skin penetration [12]. The transdermal administration of water-insoluble drugs is more difficult due to their low solubility in water and most organic solvents. To address this problem, ILs can serve as potential nanocarriers for certain drugs. Goto et al. conducted skin irritation studies using a human epidermis model with ILs. In vitro drug studies showed that ILs enhanced the amount of drug permeating into the skin. As a result of this study, it has been reported that systems containing biocompatible ILs may be promising for the transdermal delivery of insoluble or sparingly soluble drugs [50]. Another study by Oshizaka et al. also demonstrated the functionality of ILs in the pharmaceutical field. It was shown that an IL composed of choline bicarbonate and geranic acid enhanced the intestinal absorption of insulin. In this study, they discovered an IL structurally similar to azone, referred to as IL-azone, and defined it as an "anti-enhancer." ILs-azone enhances the penetration of drugs applied to the stratum corneum barrier of the skin and then reverts to its original components in the viable epidermis and dermis beneath the stratum corneum, causing minimal skin irritation. This results in a very low level of skin irritation [20].

The benefits of neat and aqueous solutions of ILs on the functionality of DNA and proteins have also been investigated through both experimental and simulation studies [54]. The study by Senapati et al. reported that the long-term structural stability of nucleic acids in aqueous solutions of ILs is supported by the site-specific interactions of their anions and cations with DNA [55]. ILs are still at an early stage of application in drug delivery. Nonetheless, the impressive preliminary outcomes seen in oral and transdermal delivery suggest that their potential influence could be substantial [12].

2.3 Antimicrobial Properties of ILs

It has been centuries since Fleming discovered penicillin, one of the first drugs effective against bacterial infections. However, nowadays, antimicrobials play a significant role in the fields of medicine and healthcare. Cationic surfactants, especially amphiphilic quaternary ammonium salts, with their superior antibacterial effects, are being extensively studied [56]. The primary basis of the antibacterial effect of ILs comes from the electrostatic attraction between the ILs and the bacterial cell membrane. Cationic antibacterial agents generally have a lipophilic component and a positively charged portion. The lipophilic part of these agents embeds within the lipid bilayer, while the cationic part attaches to the negatively charged surface of the bacterial membrane, leading to membrane disruption and bacterial cell death. In this process, hydrophobic alkyl chains penetrate the phospholipid bilayer, breaking down the membrane's hydrophobic structure [49]. The disruption caused by ILs in membrane diffusion leads to changes in membrane permeability [22]. For instance, in a study by Zhu et al., ILs demonstrated the ability to create differences in membrane lateral diffusion, which can result in changes in membrane permeability [23]. In another study by Klähn et al., the permeability and fluidity properties of the membrane changed with the integration of ILs into the lipid membrane [21]. These processes could lead to changes in lipid membrane composition or even the morphology of entire cells [22].

The potential biological activity of ILs and their possible interactions with living cells can be significant through both membrane and other cellular components. For example, in the study by Ghosh et al., the effects of imidazolium-based ILs, such as 1-butyl-3-methylimidazolium tetrafluoroborate ([BMIM][BF4]) and 1-ethyl-3-methylimidazolium tetrafluoroborate ([EMIM][BF4]), on *E. coli* bacteria were investigated. This study indicated ILs' antibacterial activity, and the survival rate of the bacteria decreased as a function of IL concentrations [57]. The antibacterial activity of ILs is widely recognized to be closely related to the cation/anion pairings and the lipophilic nature of the alkyl chains. The study by Borkowski et al. observed that ILs containing theophylline as an anion exhibited significant feeding deterrence and antifungal effects. Considering the theophylline ILs (TILs) mechanism, where cation insertion into the cell membrane occurs, it can be said that ILs play an important role in disrupting or altering bacterial membranes and reducing resistance to antibiotics [22]. In another study, Yan et al. synthesized imidazolium-type PILs by exchanging anions with L-proline or L-tryptophan. The PIL membranes exhibited high antimicrobial properties from the synergistic effects of imidazolium cations and anions [58]. Moreover, some antibacterial agents exhibit size-dependent effects. In ILs, the antibacterial activity may vary depending on the molecular size. However, the interaction between cationic small molecules and bacterial membranes requires more investigation, particularly with respect to Gram-negative bacteria. There is still a lack of comprehensive research at the molecular level that precisely controls the number of carbon atoms in the alkyl chain [26].

2.4 Tissue Engineering Applications

The central aim of research in tissue engineering is the creation of substances that integrate the mechanical strength, chemical resilience, and biological attributes vital for promoting tissue regeneration. Such biomaterials are required to be biologically compatible, capable of natural decomposition within the body, able to support cell multiplication, and their degradation residues must elicit only a negligible immunological response. There are some structures used in tissue engineering, such as bandages, scaffolds, and hydrogels that are commonly encountered in coating, encapsulation, or functionalization processes. ILs with biocompatibility features show antibacterial activities. They have widespread applications in tissue engineering and regenerative medicine [18]. The gelation and stereocomplex formation observed in ILs could provide a new alternative in tissue engineering for the use of polylactides as scaffolds with high thermal stability [25]. For instance, a study by Pauliukaitytė et al. demonstrated that combining ILs with BMIMAc and DMSO cosolvents in the electrospinning process can effectively enhance cellulose dissolution. This approach presents a promising method for improving the efficiency of this process. They hypothesized that matrices with superior nanofibrous morphology could be obtained. The resulting matrices would exhibit high biocompatibility and provide a suitable environment to support cell viability. As a result, it is expected that this will open up opportunities for further applications in tissue engineering and cell culture [26]. In another study conducted by Wahit et al., regenerated cellulose/β-cyclodextrin scaffolds were effectively created from a 1-butyl-3-methylimidazolium chloride (BMIMCl) solution, and in vitro cytotoxicity tests were performed to evaluate their potential use as tissue engineering scaffolds. The RC/β-CD tissue engineering scaffolds were produced using two particulate leaching methods with the eco-friendly ILs, namely 1-butyl-3-methylimidazolium chloride (BMIMCl). The scaffolds made through the water-insoluble leaching process demonstrated the highest water absorption capacity. The in vitro cytotoxicity tests confirmed that the RC/β-CD scaffolds are cytocompatible [59]. A compact table of some ILs commonly used in the literature and their areas of use is presented in Table 1. Poly ILs are versatile macromolecular compounds synthesized through the polymerization of IL-based monomers, either alone or in combination with other monomers, and are known to exhibit enhanced antibacterial performance. These characteristics make PILs promising candidates for bioengineering applications, particularly in wound healing research [30]. In the study carried out by Lee et al., ILs were used as a novel porogen to create biodegradable polymer scaffolds with three-dimensional (3D) porous structures for tissue regeneration. The biological properties of these ILs, including in vitro cell growth and differentiation as well as in vivo tissue compatibility, were evaluated. In this research, no adverse reactions were noted. Furthermore, the in vivo tissue compatibility of the scaffold was confirmed through the evaluation of cell growth, neo-vascularization, and the formation of connective tissues. As a result, the new product can be considered an alternative for use in tissue regeneration [31].

Table 1 Chemical formulas, IUPAC names, applications, and roles of some ILs

Chemical formula	IUPAC name	Application of ILs	References
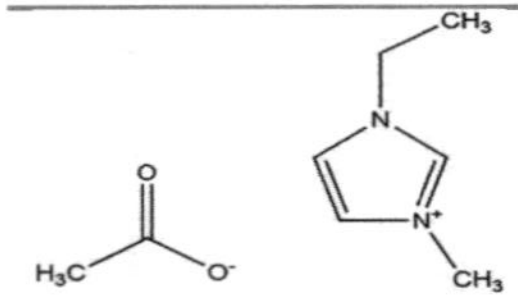	1-ethyl-3-methylimidazolium acetate	Green chemistry (as solvent)	[24]
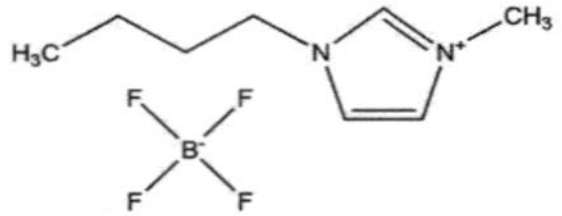	1-butyl-3-methylimidazolium tetrafluoroborate ([BMIM]BF4)	Anticancer and antibacterial activity,membrane disruptor	[48, 57]
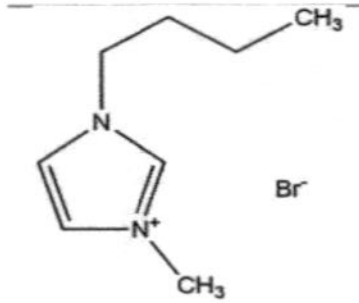	1-butyl-3-methylimidazolium bromide (BMIB)	Anticancer activity	[48]
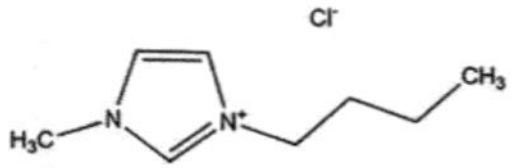	1-butyl-3-methylimidazolium chloride (BMIC)	Anticancer activity	[48]
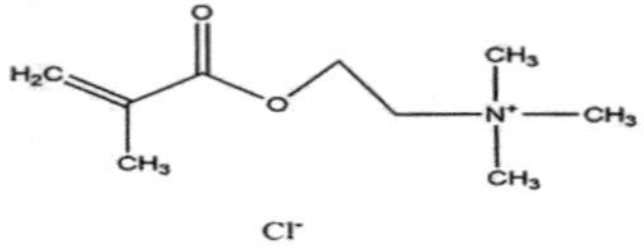	[2-(methacryloyloxy) ethyl]trimethylammonium chloride (TMAMA_Cl^{-})	Carrier in drug delivery systems	[53]

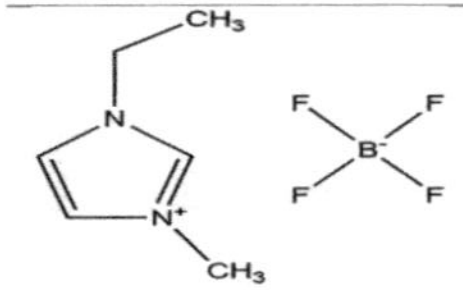	1-ethyl-3-methylimidazolium tetrafluroborate ([EMIM][BF4])	Antibacterial activity	[57]
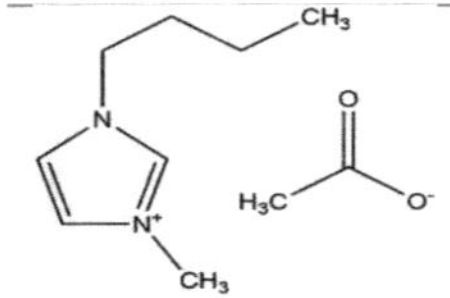	1-butyl-3-methylimidazolium acetate (BMIMAc)	Tissue engineering	[26]

2.5 Safety, Toxicity, and Future Perspectives of ILs

The applications of ILs are still under investigation, with new results being obtained and expanded. The discovery of potential antibiotics or the improvement of existing ones is essential owing to infections caused by antibiotic-resistant bacteria. IL-based therapeutics are a possible alternative to traditional antibiotics in the fight against multidrug-resistant infections. Strategies focused on these liquids are not yet ready for commercial use, and preclinical studies are still being conducted. It is believed that IL-based antibacterial agents will find a significant place in the future [18]. Although ILs hold great potential for diverse applications, their possible safety hazards to both ecosystems and human health should not be overlooked. There is no standard that can be considered appropriate for toxicity and safety studies of ILs, and many studies are still in the early stages. From the results of the many studies, it is understood that ILs exhibit low cytotoxicity; however, toxicity studies at the cellular level are not sufficient. Therefore, to ensure their reliability in the pharmaceutical field, animal studies should be more comprehensive and focused on biotoxicity. The most important aspect of overcoming these issues and improving toxicity is the selection of appropriate anions and cations. In the case of selecting nontoxic and biocompatible anions and cations, it is essential to carry out a thorough biotoxicity assessment [49]. Like most chemicals, ILs have limited applicability in in vitro systems for predicting in vivo outcomes. This problem is further amplified by detailed in vitro studies, such as the widespread use of only *E. coli* and *S. aureus* for antibacterial assessment or the use of single cell lines to evaluate toxicity. All safety-related studies should be conducted with ILs, as well as their final formulations to be applied. However, mitigating the potential toxicity risk can be made possible by utilizing ions that are natural metabolites or are approved for use in food or pharmaceutical materials [12].

3 Nanofluids in Biotechnology and Medicine

Nanofluids are suspensions formed by homogeneously dispersing nm-sized structures in an insoluble base liquid [28]. The foundations of the idea of homogeneous dispersion of solids in liquids were first formed by Maxwell's theoretical studies. Later, between 1975 and 1992, Ahuja, Liu, and researchers at Argonne National Laboratory conducted studies on dispersing millimeter and micrometer-sized particles in liquids by taking advantage of the high thermal conductivity of metals. However, micrometer-sized particles formed large clusters, causing blockages in systems with small passages, thus reducing their efficiency and even rendering them unusable [29]. These results led researchers to conduct studies on smaller particles. Stephen U.S. Choi, who is considered the "father" of NFs, introduced the term "nanofluid" in 1995 by adding nano-sized particles to conventional heat transfer fluids and conducted pioneering studies in this field [30]. The first step in the experimental studies of NFs starts with the preparation of NFs. There are two basic methods for preparing NFs: (1) one-step and (2) two-step preparation. The one-step NF preparation methods make NPs directly in the base liquid, which removes the need for drying, storing, and mixing them again. This method reduces agglomeration and creates more stable suspensions. Various techniques such as VEROS [31, 60], microwave-assisted chemical methods [32], and the vacuum arc synthesis technique [61] have been used to produce NFs from different materials such as copper, silver, and oxides. However, these techniques are generally limited to low-vapor-pressure liquids, which limits their application areas. The two-step NF preparation method is still the most preferred method for industrial-scale NF production because it is economically affordable. This method is widely used by mixing commercial NPs produced by methods such as milling, sol-gel, or vapor phase into base liquids. In this method, agglomeration of NPs is prevented by using ultrasonic mixers or high shear mixers. In some literature, it is seen that NFs from different materials such as alumina [33], TiO_2 [34], and copper [35] have been successfully synthesized with this method.

Building on these fundamental properties and the preparation methods shown in Fig. 3, the unique properties of NFs make them highly promising for various applications in biotechnology and medicine. Their nanoscale dimensions, enhanced thermal properties, large surface area, and the ability to functionalize NPs surfaces enable innovative solutions in diagnostics and therapeutics. This section will examine in detail how NFs are used in specific areas such as biomedical imaging, targeted drug delivery, hyperthermia therapy for cancer, and exploiting their natural antibacterial activity. Each subsection will elaborate on the mechanisms and potential of NFs in these different biomedical fields.

3.1 Biomedical Imaging

The unique structural properties of NFs make them increasingly important in the field of biomedical imaging. In particular, NPs are preferred in biological imaging because of their superior optical properties, customizable surface structures, and ability to carry therapeutic molecules with different detection. In addition, the selective accumulation of contrast agents in target areas and the clear imaging of the diseased area by these structures allow the development of more sensitive and effective diagnostic and treatment methods. These features have significant potential for the future of biomedical imaging techniques. In this way, in vivo imaging of pathological regions becomes possible at the molecular level, thus paving the way for the development of personalized and targeted therapeutic interventions [62]. For example, Shelat et al. investigated the modification of magnetic iron oxide (Fe_3O_4) NPs with biocompatible amino acids (arginine and histidine) in search of a reliable method for tracking mesenchymal stem cells (MSCs). The aim of this modification is to increase the stability and reduce the toxicity of the NPs. Stem cells readily absorbed modified nano-ferrofluid (AA@MNFs) with low toxicity, according to

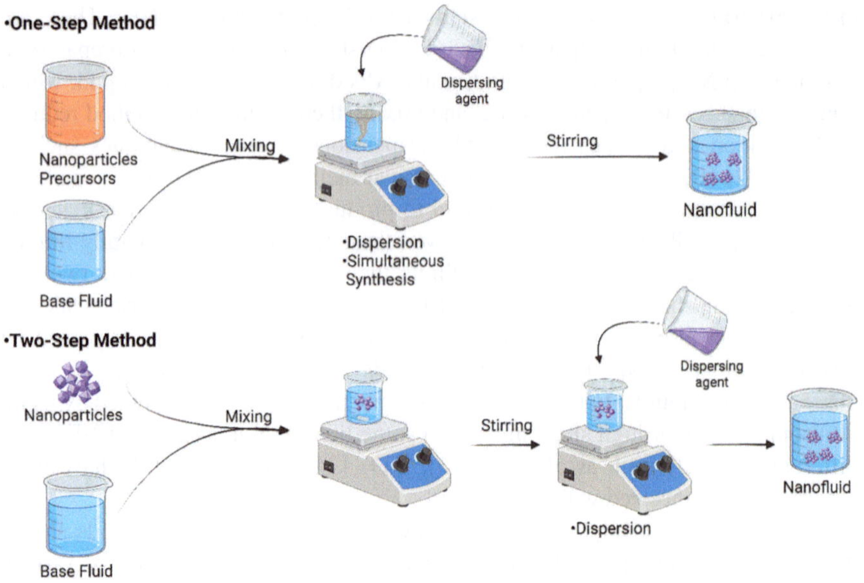

Fig. 3 Process flow diagram for nanofluids fabrication methods

in vitro and preclinical studies. These results further suggest that AA@MNFs could be effectively used as MRI contrast agents [63]. In another study, Umar Farooq and Tao Liu numerically investigated the flow of magnetized micropolar NFs on a curved surface. The heat transfer and flow dynamics of this special fluid containing blood silica NPs were analyzed using mathematical modeling methods. The partial differential equations (PDEs) describing the flow problem were reduced to ordinary differential equations (ODEs) with appropriate similarity transformations and solved by numerical methods. The obtained results indicated that the magnetic field, radiation, and NP concentration were effective in the flow and heat transfer. Also, increasing the magnetic field provided an improvement in heat transfer. These findings have potential applications in the fields of medical imaging and thermal therapy [64]. In a study by Sheng et al., a pH-sensitive MNF NFs was designed by combining poly (sulfadimethoxine acrylamide) polymers and approximately 8 nm Fe_3O_4 NPs. This was intended to improve MRI-based tumor detection and enable simultaneous treatment. The result of this study showed that MNF stays stable in the body but combines in the slightly acidic environment of tumor tissue, which boosts the MRI signal and may improve treatment methods like MH and the Fenton reaction (FR). In vitro experiments further revealed that these NFs inhibit tumor cell growth, highlighting its potential versatility for both imaging and treatment applications [36].

3.2 Drug Delivery

NFs are becoming a focus of increasing interest, especially in drug delivery systems, with the potential to offer innovative solutions to the challenges faced by modern medicine. The basis of this interest lies in the unique properties of the nanomaterials that make up NFs. The special ability of nanomaterials to get into damaged tissues and organs allows drugs to be sent straight to the right spot, which helps limit how much the drugs spread in the body and reduces the risk of harming healthy cells, unlike traditional methods. These remarkable properties of NFs offer significant potential for overcoming the limitations faced by traditional drug delivery methods such as uncontrolled distribution of drugs in the body and side effects at high doses [62, 65]. For instance, Kazemi-Beydokhti et al. investigated the design of special fluids containing cisplatin (cancer drug) formulated by coating carbon nanotube drug carriers with two biocompatible phospholipids (DSPE-PEG2000 and DSPG) using engineering principles. The main purpose of this approach is to increase the treatment efficacy and reduce systemic toxicity by directly targeting the drugs to the tumor cells. One of the unique aspects of the study is the experimental investigation of the thermal conductivity property of these medical NFs. The experiments indicated that the thermal conductivity of the NFs increased as the temperature increased, and drug release increased when the pH decreased. As a result, this research indicated that thermal conductivity can be an important parameter in the design of advanced drug delivery systems [9]. Another study by Shahzadi and Bilal explored the use of hybrid NFs to understand blood flow in a bifurcated stenosed artery. The researchers used Navier-Stokes equations to model blood flow and accounted for the properties of Cu-CuO/blood hybrid NFs. In this study, blood was considered a Newtonian fluid, and properties such as permeability and elasticity of arterial walls were stated as factors that could affect drug release processes. The findings revealed that permeability effectively mitigated hemodynamic effects in atherosclerotic arteries. Furthermore, a reduced bifurcation angle correlated with decreased shear stress in branch arteries, compliance parameters influenced the number of circulating boluses, and hybrid NFs demonstrated superior velocity profiles compared to Cu-blood and pure blood [37].

3.3 Hyperthermia Treatment

Hyperthermia is an important method in cancer treatment that aims to destroy tumor cells selectively or make them more sensitive to radiotherapy and chemotherapy. In magnetic hyperthermia (MH), a technique of significant interest, magnetic NPs are directly introduced into the tumor tissue. Subsequently, an alternating magnetic field is applied, leading to the generation of localized heat that triggers the demise or injury of malignant cells. The heating power of the NPs is measured by the specific absorption rate (SAR), and this value is an important measure of how effective the treatment method is. Compared to traditional treatments, hyperthermia offers the potential to cause less damage to healthy tissues. Therefore, the production, properties, and heating capacity of MNPs are being intensively investigated for better application of hyperthermia treatment. NFs also play an important role in this treatment because they can increase their effectiveness through different mechanisms. NFs provide better heat distribution and control in the tumor region by increasing the thermal conductivity of basic fluids. In MH, a technique of significant interest, MNPs are directly introduced into the tumor tissue. Subsequently, an alternating magnetic field is applied, leading to the generation of localized heat that triggers the demise or injury of malignant cells. In addition, the fluidity properties of NFs can affect their movement in blood vessels and tumor tissue, facilitating the transport of drugs or heat to the tumor site. The stability of NFs, their biocompatibility, and the controlled distribution of NPs are crucial factors in minimizing side effects and enhancing treatment safety, which is paramount for clinical translation [38, 66]. For instance, in this study by Kharat et al., oleic acid-coated cobalt ferrite NPs were successfully synthesized by the chemical precipitation method, and their structural, morphological, and magnetic properties were investigated by various techniques. X-ray diffraction (XRD), transmission electron microscopy (TEM), and Fourier transform infrared (FT-IR) spectroscopy analyses confirmed that the NPs formed in a cubic-spinel structure and single phase, coated with oleic acid, and were approximately 12 nm in size. Magnetic measurements revealed that the NPs behaved superparamagnetically and exhibited moderate saturation magnetization. The findings suggest that these NPs have significant potential in biomedical applications such as magnetic fluid hyperthermia, especially when their superparamagnetic properties and heating capacities are considered [67]. In a similar study, Yusefi et al. synthesized Fe_3O_4 NPs using Punica granatum fruit peel extract and then investigated the magnetic hyperthermia performance of the NPs, their MRI potential, and their efficacy as anticancer drug carriers. Magnetic test of this study indicates that the NPs can generate heat for magnetic hyperthermia treatments and may also work well as contrast agents in MRI. At the same time, the NPs were successfully loaded with 5-fluorouracil, and their cytotoxicity against cancer cells was evaluated, demonstrating the potential of the NPs for controlled drug delivery and cancer treatment. The findings of this research show that these NPs are promising in both cancer treatment and imaging fields [39].

3.4 Antibacterial Activity

Preventing bacterial growth in many areas, from medical devices to industrial cooling systems, is still an important issue. One of the approaches that has attracted attention recently is adding nanomaterials to liquids. Fluids prepared with this method, called "NFs," show promising results in suppressing or eliminating bacterial growth due to the effect of the nanostructures they contain. This antibacterial activity of NFs is based on the ability of the NPs to target bacteria through various mechanisms. The main ways they work include the NPs directly interacting with the bacteria's cell wall or membrane, which damages its structure. In addition, nanomaterials like metal oxides cause oxidative stress in the cell and damage to critical biomolecules by producing reactive oxygen species (ROS). MNPs (silver and copper) can inhibit bacterial enzymes and prevent vital functions through the release of toxic ions. Direct intervention in intracellular metabolic processes is another possible mechanism. Usually, a combination of these mechanisms is responsible for the observed strong antibacterial activity of NFs [40, 62]. For example, Fakhroueian et al., who addressed the antibacterial potential of zinc oxide (ZnO)-based NFs, obtained significant findings in vitro. Their team found that different formulations against various pathogens, especially the delivery setup with CNT-amine and a mixture supported by ZnO/PVP nanopolymer, showed a strong inhibitory effect (MIC, MBC, according to the agar diffusion test) on *Streptococcus pyogenes*, *Enterococcus faecalis*, and *Escherichia coli*. The study also indicated that one of the possible mechanisms of this effect is the weakening of the NFs by the bacterial cell membrane, increasing the production of ROS leading to cell death [41]. In another study by Pauzi et al., the antibacterial potential of ZnO-NFs stabilized using Gum arabic was investigated. The stability and antibacterial effect of NFs prepared with different concentrations of Gum arabic using a microwave-assisted green method were examined. This team found that NFs stabilized with 1.5% Gum arabic, exhibited stronger antibacterial activity against *Staphylococcus aureus* and *Escherichia coli* compared to unstabilized ones. This enhanced effect was suggested to be due to the smaller size of the stabilized NFs increasing their interaction with bacteria [42]. Table 2 summarizes various NFs compositions and their corresponding specific applications, aiming to illustrate the potential of NFs within the fields of biotechnology and medicine. The table lists examples of NFs created by dispersing different NPs types, such as ZnO, nickel ferrite ($NiFe_2O_4$), and silver (Ag), into various base fluids including deionized water, phosphate buffer, or blood. Each specific composition is linked to a particular application area, such as antimicrobial activity, drug delivery systems, cancer therapy via hyperthermia, or biomedical imaging. Furthermore, references to the relevant scientific studies supporting the information are provided for each entry.

3.5 *Safety, Toxicity, and Future Perspectives of Nanofluids*

NFs are special mixtures containing nano-sized materials suspended in a base fluid. Owing to these NPs, NFs attract attention by exhibiting superior thermophysical properties. These materials hold immense potential within the biomedical field. This is demonstrated by their utility in targeted drug delivery systems and their ability to enhance cancer treatments like hyperthermia. In addition, they serve as contrast agents in advanced imaging techniques such as MRI and offer promising next-generation solutions to address antibiotic resistance. The incorporation of magnetic NPs into magnetic NFs allows for the direct transport of drugs to target diseased

Table 2 Composition and applications of nanofluids in biotechnology and medicine

Nanoparticle type	Base fluid	Application	References
Zinc oxide (ZnO)	Deionized water	Antimicrobial activity	[68]
Nickel ferrite ($NiFe_2O_4$), cobalt ferrite ($CoFe_2O_4$), iron (II, III) oxide (Fe_3O_4)	Phosphate buffer	Drug delivery systems	[69]
Manganese ferrite ($MnFe_2O_4$)	Deionized water	Hyperthermia/cancer treatment	[70]
Titanium dioxide-silver (TiO_2-Ag)	Blood	Cancer therapy	[71]
Gold-silica (Au-SiO_2)	Blood	Drug delivery systems	[72]
Copper (Cu), alumina (Al_2O_3), titanium dioxide (TiO_2)	Water, oil, toluene, ethylene glycol	Vehicle cooling, power generation, microelectronics, biomedical devices	[73]
Silver (Ag)	Blood	Cancer therapy, biomedical imaging, hyperthermia, tumor therapy	[74]
Magnetite iron oxide	Deionized water	Hyperthermia, MRI, drug delivery	[39]
Zinc oxide (ZnO)	Deionized water	Antibacterial application	[42]
Copper oxide (CuO), titanium dioxide (TiO_2), aluminum oxide (Al_2O_3)	Blood	Blood filtration system, obstructed hemodynamics, nano-hemodynamics	[65]

areas using an external magnetic field. This capability offers the potential to enhance treatment efficacy while minimizing systemic toxicity. [75].

One of the most critical barriers to the widespread clinical application of NFs is the concern regarding their safety and potential toxicity. This is significantly influenced by the unique physicochemical properties of their constituent nanomaterials. While these characteristics, including their small size, high surface area-to-volume ratio, and increased surface reactivity, are advantageous, they also bring the potential for unwanted biological interactions and detrimental effects. The potential toxicity of NPs is a result of their intrinsic physicochemical qualities and the complex interactions they establish with biological systems. Factors such as the particle's size, shape, chemical structure, surface properties (charge, coating), and solubility in the biological medium significantly affect the toxicity potential. Furthermore, the route of NP uptake into the body, the administered dose, the duration of exposure, and the protein corona layer formed on their surface upon contact with biological fluids are other critical factors shaping this potential. As a result of these interactions, various toxic mechanisms can be activated. These include the triggering of oxidative stress in cells via the production of reactive oxygen species, and the occurrence of direct physical damage to cell membranes or internal structures. Consequently, damaged cells may then enter processes of

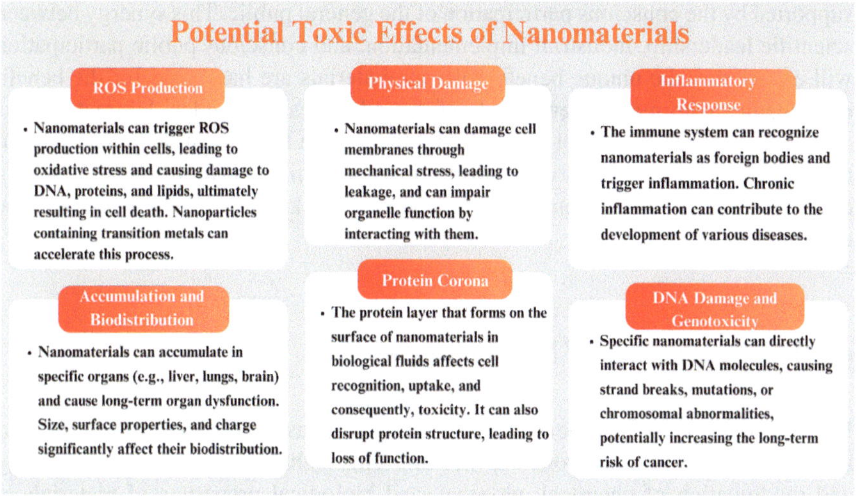

Fig. 4 Potential toxic effects of nanomaterials

programmed death (apoptosis) or uncontrolled destruction (necrosis). Figure 4 successfully summarizes these potential toxic effect mechanisms [76].

As a consequence of these mechanisms, the potential for NPs to accumulate in specific organs over time may carry the risk of leading to long-term toxic effects or chronic inflammation. Although, the potential of NFs and their constituent NPs to revolutionize the biomedical field is immense, the safe realization of this potential depends on the thorough understanding and control of the aforementioned toxicity mechanisms. Toxicity can be minimized through intelligent design of physicochemical properties, biocompatible coatings, and safer synthesis methods.

In the future, as the applications of NFs become more widespread across various sectors, concerns regarding safety and toxicity will continue to be central to scientific research and societal discussions. Considering the factors and mechanisms summarized in Fig. 4, academic studies will progress toward minimizing the multi-faceted safety and toxicity risks of these structures through strategies such as material selection, size/shape optimization, and modification of NP surfaces. Achieving the safe and effective translation of nanotechnology necessitates a multifaceted approach. This includes a thorough comprehension of the complex interactions of NPs with biological systems and their associated toxicity mechanisms, meticulous monitoring of environmental footprints, predictive modeling of long-term effects, the synthesis of well-characterized materials, and the implementation of robust safety protocols to protect public health, crucially. The integration of safe-by-design principles into industrial applications and the creation of regulatory frameworks will ensure that human health and environmental sustainability are balanced alongside the promotion of innovation. In this process, the future success of NFs depends on a collaboration led by scientists, implemented by industry representatives, and supported by the conscious participation of the general public. This synergy between scientific leadership, industrial implementation, and conscious public participation will ensure that the unique benefits of nanomaterials are harnessed for the benefit of humanity within a framework of safety and ethical principles. Ultimately, this comprehensive approach not only harness the unique benefits of nanomaterials for the benefit of humanity but also fulfill our responsibility to future generations by enabling the meticulous management of potential risks to their well-being and the environment.

4 The Emergence of Nanoparticles

Nanotechnology is the science of designing and using devices, materials, and systems at scales ranging from 1 to 100 nm. This technology is based on the study and modification of chemical, physical, and biological properties of materials at this scale. These properties can differ significantly from those at larger scales. The main focus of nanotechnology is the production of NPs with unique properties such as high reactivity and surface area [43]. Although nanotechnology is a new field, similar applications of nanoscale materials have been seen in the past. For example, the Lycurgus Cup from the fourth century changes color due to gold and silver NPs in the glass [3]. In addition, medieval stained glass was made with vibrant colors using gold and silver NPs. Damascus steel is renowned for its durability and sharpness and contains carbon nanotubes that contribute to these properties. With advances in nanotechnology, the use of NPs has become widespread in science and industry. Due to their small size, large surface area, chemical reactivity, and ability to interact with cells, NPs play an important role in modern technologies. In general, these nanoscale particles have opened up new possibilities in science and technology. In this section, four main types of NPs will be introduced and reviewed: Metallic NPs (MNPs), polymeric NPs (PNPs), liposomal NPs (LNPs), and quantum dots (QDs). These NPs are used in various clinical applications. For instance, MNPs are used in biomedical imaging, cancer therapy, and biosensor production due to their optical and electronic properties. PNPs are applied in drug delivery and targeted delivery systems. LNPs are preferred in drug delivery and gene therapy due to their biocompatibility. QDs are used in cellular imaging, disease diagnosis, and phototherapy due to their light-emitting properties [77]. Also, the classification of various types of NPs matches with examples for each group is presented in Fig. 5 and all of these NPs are explained in detail as follows.

MNPs are particles at the nanoscale and consist of clusters of metal atoms or ions. They have properties such as high surface area/volume ratio and chemical stability. These properties allow MNPs to interact with biological structures at the molecular and cellular levels. The foundation of developments in this field was laid in 1959 when Richard Feynman proposed the idea that reducing materials to the nanoscale would lead to technological advances [3]. During the 1960s and 1970s, the use of MNPs such as gold and silver increased in medicine and pharmacy.

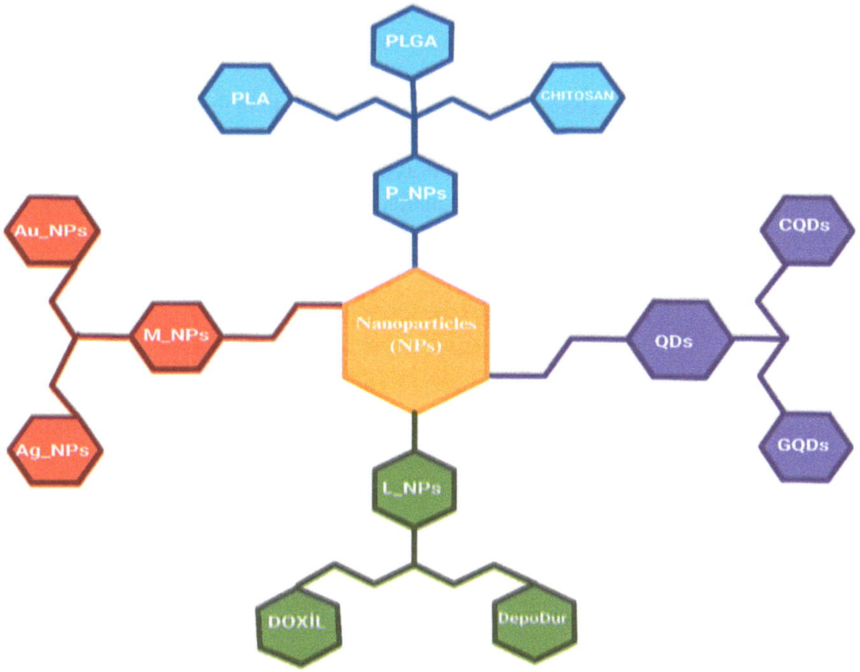

Fig. 5 Classification of various types of nanoparticles

In the 1990s, the research focus shifted to designing these NPs for drug delivery systems and therapeutic approaches [78]. Today, as the knowledge of the properties of MNPs increases, their areas of use have also expanded. These NPs play an important role in medical imaging, early disease diagnosis, and the development of new therapeutics due to their biological stability, ability to selectively absorb and emit light, and surface plasmon resonance (SPR) phenomenon. Various MNPs, such as gold NPs (Au NPs), silver (Ag NPs), Fe_3O_4 NPs, copper (Cu NPs), and zinc (Zn NPs), are used as photothermal therapy (PTT) to destroy cancer cells, contrast agents in fighting infections, and in magnetic resonance imaging (MRI), offering new possibilities in medical science [77]. Another important NP that is used clinical applications is PNPs. These NPs are derived from biocompatible polymers such as chitosan, gelatin, sodium alginate, polylactides (PLA), and poly(lactic-co-glycolic acid) (PLGA) as carriers at the nm scale. The use of these NPs in medicine dates back to the 1970s, when liposomes and micelles emerged as the first-generation drug carriers. These systems have increased the effectiveness of treatments by increasing drug solubility and prolonging their effects in the body. In the 1990s, NPs were coated with polyethylene glycol (PEG), which increased their biocompatibility and reduced their clearance by the immune system. These developments paved the way for the development of PNPs that can carry complex molecules such as proteins and nucleic acids. This ability is especially important for mRNA vaccines, such as those developed against COVID-19, which effectively deliver delicate genetic material. [79, 80]. As knowledge of nanomaterials increases, PNPs have become widely used in medical and pharmaceutical fields due to their versatile properties such as controlled drug delivery. These NPs play an important role in the design of advanced gene therapy and diagnostic tools, as they can carry drugs and genetic materials to target tissues with minimal side effects. Polymeric NPs are also used in biosensors and contrast agents to improve image quality in medical imaging techniques and facilitate disease diagnosis [44]. Among these NPs, LNPs are biocompatible and degradable phospholipid vesicles. The history of LNPs dates back to the 1960s when Alec Bangham discovered liposomes, which are vesicles composed of phospholipid bilayers [81]. The potential of these NPs as drug carriers was quickly recognized due to their ability to encapsulate both hydrophobic and hydrophilic compounds, protect drugs, and provide controlled release. A significant development was the approval by the US FDA of a liposomal formulation of doxorubicin (Doxil) in the 1990s, demonstrating the value of these NPs in targeted drug delivery. Since then, studies and applications of LNPs in the pharmaceutical and clinical fields have increased, leading to advances in disease diagnosis and treatment. As knowledge of their structural properties increases, LNPs have become more important in modern drug delivery systems. These NPs are widely used in drug delivery because they can effectively encapsulate both hydrophilic and hydrophobic drugs, protect drugs from degradation, reduce toxicity, and increase accumulation in target tissues. In addition, modification of their surfaces with substances such as chondroitin sulfate allows targeted delivery, increasing cellular uptake and therapeutic effects. These properties have made LNPs suitable for the treatment of diseases such as cancer, HIV, and fungal infections. For example, a liposomal formulation of amphotericin B, a potent but toxic antifungal drug, has been shown to reduce its toxicity and is more effective in treating fungal infections. Similarly, DaunoXome, a liposomal formulation of the chemotherapy drug daunorubicin, has reduced the side effects of chemotherapy and has been used to treat cancers such as Kaposi's sarcoma, leukemia, and lymphoma. In addition, Doxil, a liposomal formulation of another cancer drug, doxorubicin, has improved treatment outcomes in tumors and reduced side effects. In general, liposomal anthracyclines have been more successful in targeting cancer cells than free drugs. However, the high cost of these drugs has been a barrier to their widespread use in resource-limited settings [82, 83]. The last group of these NPs are QDs that are semiconductors compared to MNPs, LNPs, and PNPs. QDs are NPs with sizes ranging between about 1 and 10 nm. The first observations in this area were made by Katzschmann in 1977, who suggested that particles of Group II-VI elements with sizes of 1–5 nm exhibited an unusual luminescence and suggested that there might be a relationship between particle size and band gap. In the early 1980s, Elimov, Efros, and Brus conducted pioneering studies examining the effects of quantum size on semiconductor nanocrystals. In 1986, Reed named these nanocrystals with zero degrees of freedom, where quantum size effects were observed, "QDs." In the 1990s, interest in QDs increased significantly and the field experienced rapid development. In particular, Bawendi has succeeded in producing QDs with uniform size and shape, while Guyot-Sionnest obtained the first core-shell QDs, namely cadmium selenide/zinc sulfide (CdSe/ZnS) QDs by combining CdSe-QDs with ZnS. The production of high-quality QDs has significantly paved the way for the use of this new type of material in various applications. Also, they have great value in biomedical and clinical applications due to their unique optical and electronic properties such as high brightness, stability, and light emission that can be tuned according to their size. Thanks to these unique properties, QDs are effectively used in various biomedical fields, from cell labeling to the development of biosensors, and from imaging in living organisms to disease diagnosis. Among the different types, carbon and graphene QDs increase the variety of biomedical applications thanks to their ability to adjust the wavelength of light they emit. Being brighter and more stable than organic dyes, they allow cellular processes to be examined in much more detail. In addition to enabling precise measurement of molecular-level interactions in biosensors, especially the types that emit near-infrared light, they are promising as a guide for imaging tumors and surgical interventions. Moreover, their ability to detect more than one target at the same time facilitates comprehensive disease diagnosis and follow-up. The focus of current research is to improve their biocompatibility and reduce toxicity risks by modifying their surfaces and coating them with protective layers in order to increase their safety and effectiveness in clinical diagnosis and targeted treatments [84].

4.1 Cancer Treatment

Cancer is a significant and complex health problem encountered worldwide. This disease manifests itself with the uncontrolled growth of abnormal cells that disrupt the normal functioning of the body and their spread to other tissues and organs. The use of nanotechnology in cancer treatment offers an innovative and promising approach for researchers. NPs have the ability to selectively penetrate cancer cells thanks to their extremely small size. This characteristic feature provides a significant increase in treatment effectiveness while also allowing for a significant reduction in side effects. In addition, NPs play an active role in methods such as PTT, which locally destroys cancer cells by converting light into heat. In the fields of gene therapy and chemotherapy, these particles facilitate the safe and effective delivery of genetic material and chemotherapeutic drugs to target cells. Recent studies have shown that the use of NPs can also improve patients' response to immunotherapy and radiotherapy. In the following section, we will further examine the role of NPs in innovative cancer treatment approaches such as chemotherapy, PTT, and gene therapy [85].

Nanoparticles in Chemotherapy

Chemotherapy is a cancer treatment method that uses potent drugs to kill rapidly growing cells in the body. In chemotherapy, NPs are employed to deliver therapeutic agents to tumor regions in a targeted and controlled manner. This approach is specifically developed to address the limitations of traditional chemotherapy. A primary difficulty of conventional chemotherapy is its lack of specificity, resulting in insufficient targeting of cancer cells and detrimental effects on healthy tissues. To overcome this limitation, there is a growing need for the development of nanocarriers that can precisely deliver drugs to tumor sites while minimizing systemic toxicity. NPs, liposomes, micelles, and dendrimers are common nanocarrier systems used in this approach. Among them, particularly NPs are designed in different forms and can encapsulate or bind chemotherapeutic drugs to their surfaces, ensuring protection in circulation, penetration into the tumor microenvironment, and selective uptake by cancer cells. This strategic approach has a high potential to improve the efficacy of chemotherapy by reducing the side effects of the drug on healthy tissues and increasing its therapeutic concentration in the tumor area [86].

Nanoparticles in Photothermal Therapy

PTT offers an innovative approach to cancer treatment. In this method, light energy is converted into heat with the help of specially designed NPs, and tumor cells are selectively destroyed. In the process, NPs absorb light in the near infrared (NIR) wavelengths between 650 and 900 nm, whereas biological tissues absorb light at the lowest level. This leads to a sudden heating and a temperature increase only in the targeted area, allowing cancer cells to be effectively eliminated. AuNPs, especially gold nanorods (AuNRs), have attracted great interest in this field due to their controllable SPR properties and unique optical properties. When selecting the appropriate material for PTT, factors such as high light absorption ability in the NIR range, high uptake of NPs by target cells, and light-heat conversion efficiency that will increase the success of the treatment are of critical importance. An ideal PTT agent should be able to provide safe and selective tumor ablation even with low laser power and short application time. Research to improve photothermal conversion efficiency has focused on differentiating the shapes of NPs and developing functional strategies such as AuNRs coated with mesoporous silica, which integrate drug delivery with PTT. Furthermore, examining whether AuNRs containing two different metals (bimetallic AuNRs) exhibit better photothermal efficiency compared to conventional AuNRs provides valuable information for the advancement of NP-based PTT in cancer therapy [87].

Nanoparticles in Gene Therapy

Gene therapy is a groundbreaking strategy in the fight against cancer, aiming to provide healing by changing the genetic structure of tumor cells. In this innovative approach, methods such as transferring cancer-fighting genes to the organism, suppressing the activity of cancer-causing genes (oncogenes), or strengthening the immune system's response to cancer can be applied. Although viral or nonviral carrier systems (vectors) are mostly used in delivering genetic material to target cells, there are problems to be solved in terms of ensuring the safety and efficiency of these methods and ensuring sufficient gene expression. At this point, NPs play an important role in overcoming some limitations of traditional vectors by undertaking the gene delivery task. For example, liposome-based NPs can increase the success of the treatment process by carrying therapeutic genes to cancerous cells.

In a study conducted by Zhang and colleagues in 2022 [86], PEG-b-PLGA polymer-based NPs were developed. These specially designed NPs can deliver Cas9 DNA and guide RNA to the endothelial cells that form the inner surface of blood vessels with a high success rate. These NPs, which trigger gene editing in endothelial cells, cause a decrease in protein production by approximately 80%, resulting in a genetic "silencing" (gene knockout)-like result. This technique allows gene editing in endothelial cells or the addition of new genes to these cells at the same time, making it a promising tool for gene therapy applications in the clinical field [88].

4.2 *Regenerative Medicine: Stem Cell Tracking and Tissue Regeneration*

Regenerative medicine, as one of the important advances in medical science, is a highly innovative discipline that aims to repair, replace, or recreate damaged or dysfunctional tissues and organs. The main goal in this field is to restore the structural integrity of the body by triggering natural biological processes and supporting the self-healing potential of the organism. Stem cell monitoring and tissue regeneration are two basic components of regenerative medicine that play a critical role in the healing of damaged tissues and the restoration of normal bodily functions [46]. The concept of stem cell monitoring, which is discussed as follows, draws attention to its indispensable importance in the tissue regeneration process. Stem cells are special cells that have unique abilities such as differentiating into a wide variety of specialized cell types and repairing damaged tissues. NPs offer a wide range of uses in regenerative medicine applications and stem cell-based treatments. These nanostructures serve as drug carriers or imaging agents, allowing the precise monitoring and tracking of stem cell movements, viability, and differentiation processes within the organism. By means of NPs, growth factors, cytokines, and other therapeutic molecules can be delivered to damaged areas, supporting healing mechanisms and accelerating the tissue regeneration process. Cytokines, which play a central role in regulating immune and inflammatory responses, can trigger tissue repair and regeneration. For example, interleukin-6 (IL-6), an inflammatory cytokine, plays an active role in the body's response to tissue injury and can be directed directly to damaged areas via NPs. In addition, the integration of NPs with noninvasive imaging techniques such as ultrasound and photoacoustic imaging (US/PA) allows real-time and high-resolution observation of stem cell behavior, making significant contributions to the development of tissue regeneration treatments [45]. In this context, MSCs in particular offer considerable potential for cellular therapy approaches and tissue engineering strategies targeting tissue repair and functional recovery [46]. Our next focus will be on tissue regeneration. Tissue regeneration in regenerative medicine involves the use of innovative approaches for the repair and reconstruction of damaged tissues. Advanced nanomaterial-based systems, NPs, play a vital role in this complex process. These systems, such as biocompatible scaffolds and nanocomposite hydrogels, provide a suitable environment for cell proliferation and specialization. They accelerate functional tissue reconstruction through controlled release of growth factors and strengthening of cell-scaffold interactions, allowing the implementation of personalized treatment strategies. For example, a collagen model based on magnetic stimulation uses static magnetic fields and magnetic NPs to understand the effects of magnetic fields and magnetic NPs on bone formation and regeneration, and offers potential applications for the repair of different tissue types. In addition, magnetic NPs coated with biocompatible materials can be guided to the targeted tissue with the help of an external magnetic field, allowing precise delivery of drugs and growth factors to the injured area. This method improves both the speed and quality of wound healing and tissue regeneration by optimizing cell-scaffold interactions [10, 89].

4.3 Antibacterial and Antiviral Applications

NPs offer brand-new and promising solutions in the fight against antibacterial and antiviral diseases today. The unique mechanisms offered by NPs draw attention in the fight against infections where traditional methods are inadequate or resistance problems arise. The basis of this approach lies in the very large surface areas of NPs and their structures that can be designed for different purposes. Thanks to these features, NPs can directly interact with microorganisms and damage their vital structures or carry antimicrobial substances to the exact site of the infection with high sensitivity. This transport process has the potential to minimize the damage to healthy cells by ensuring that the drug focuses only on the target microorganisms [90]. A striking example of this application is the use of didodecyldimethylammonium bromide (DDAB). DDAB is a strong cationic (positively charged) surfactant developed specifically to combat bacteria and viruses. Thanks to its positive charge, it has the ability to easily bind to negatively charged microbial cell membranes and disrupt their integrity. Moreover, DDAB molecules can strongly adsorb onto the surface of silica NPs. In this way, the resulting NPs coatings provide a broad-spectrum and effective protection mechanism not only against bacteria but also against fungi, viruses, and yeasts. A particularly striking point is the non-release property of these coatings. This feature ensures that the antimicrobial agent on the NPs remains active for a long time and does not lose its effectiveness. As a result, this innovative approach makes a significant contribution to infection control by preventing the spread of infections and increasing the effectiveness of treatment in clinical applications [91].

4.4 Biosensors and Lab-on-a-Chip Devices

Biosensors are analytical devices that combine physical and chemical sensing techniques. These devices combine a biologically active component with a suitable physical transducer to produce a measurable signal proportional to the concentration of chemical species in a sample [92]. It is known that the first studies on biosensors began in the early 1900s. However, the first studies that turned into scientific publications in this field were carried out by Leland C. Clark—who is considered the father of biosensors—and his colleagues. Clark and his team first developed a chemical sensor for the detection of oxygen in 1956. Then, by modifying this chemical sensor with glucose oxidase enzyme, they published the first scientific article on the first enzyme-based glucose biosensor [93, 94]. Biosensors are extremely important analytical tools that offer a wide range of applications in various disciplines such as medical diagnosis and health monitoring (blood sugar measurement in diabetes or early pregnancy detection), environmental surveillance (determination of air and water pollutants), food and agriculture sectors (detection of disease-causing organisms and quality control), pharmaceutical industry (drug quality control), and security (detection of explosives). These devices play an important role in different areas by providing fast, precise, and often on-site diagnostics. For instance, glucose biosensors allow millions of individuals to easily monitor their blood sugar levels, while pathogen detection biosensors can quickly and reliably detect the presence of viruses or bacteria in clinical samples [95]. In addition, lab-on-a-chip devices are miniature laboratory systems that integrate various laboratory processes on a miniature chip. This integration has significantly transformed laboratory work by allowing experiments to be performed on a small scale, with high speed and efficiency. Lab-on-a-chip devices have revolutionized many areas, including medicine, drug development, biological and chemical analysis, environmental surveillance, food safety, and forensic science. By providing rapid and sensitive analysis with small samples, they are widely used in a wide range of applications, from disease diagnosis to drug production, environmental pollution monitoring to food safety [92, 96]. A summary of the classification, advantages, disadvantages, and biomedical applications of various types of most frequently encountered NPs is provided in Table 3.

4.5 Concerns About Safety and Toxicity of Nanoparticles

Although NPs have high potential for use in medicine, there are significant concerns about their safety and toxicity. The toxicity of NPs depends on factors such

Table 3 Classification, advantages, disadvantages, and biomedical applications of important types of nanoparticles

Classification	Advantages	Disadvantages	Applications	References
Graphene quantum dots	Photostability, water solubility	Toxicity, endosomal escape, bioconjugation problems	Cell labeling, biosensing, imaging, drug delivery tracking, cancer biomarker detection	[88]
Carbon quantum dots	High surface area, modifiability with various functional groups, water solubility, low toxicity, biocompatibility, chemical stability, fluorescent property	High cost (especially for activated carbon), difficult separation after purification process, low regeneration rate	Removal of synthetic dyes from aqueous solutions, use in adsorption processes due to high surface area and porosity; alternative to conventional treatment methods (ozonation), flocculation, membrane separation	[44, 77]
Gold nanoparticles	Non-toxic and inert, simple to synthesize, controllable diameter (1–150 nm), good drug carriers, enhanced X-ray absorption, nucleic acid protection, simple to alter	Various diameters can be harmful (from <2 nm to bigger); the cytotoxicity of different forms depends on morphology; aggregation/assembly increases toxicity and cytotoxicity via surface charge	Tumor imaging, radio sensitization, hyperthermia, and gene therapy for tumors	[77]
Silver nanoparticles	Microbicidal effect, low risk of bacteria resistance, high efficiency silver ion release, antiviral effect, wide application area, ecofriendly synthesis methods, durable antibacterial effect	Cytotoxicity risk, oxidative stress and lipid peroxidation, intracellular buildup and mitochondrial damage, limited knowledge on human toxicity and apoptosis induction, potential concerns with silver absorption	Medical device coatings, wound dressings, biosensors	[97]
Iron oxide nanoparticles	Magnetic resonance imaging, magnetic separation, photothermal therapy, magnetic hyperthermia benefits of ultrasmall size include effective leukemia therapy, enhanced chemotherapeutic efficacy, and efficient kidney clearance (high r_1)	Disadvantages of large-sized SPIONs include retention by RES and low image differentiation; ultrasmall iron oxide NPs have unknown bio safety and high toxicity potential; USPIONs generate ROS production, ferroptosis, cell aggregation, and efficient nuclei entry	MRI contrast agents, magnetic separation, photothermal therapy	[98]

as structure, size, shape, surface charge, and dose. Therefore, a comprehensive toxicological evaluation is necessary before clinical applications. For example, Wu et al. showed that high doses of Fe_3O_4 NPs caused serious side effects in animals. These studies reveal that Fe_3O_4 NPs smaller than 5 nm cause oxidative stress in the body due to their large surface area and ability to interact with biological molecules. Oxidative stress caused by NPs can damage cell membranes, proteins, and DNA by increasing the production of free radicals. This can lead to cell death and organ dysfunction. These findings indicate that NPs smaller than 5 nm may accumulate in organs such as the liver, spleen, and kidneys, posing greater risks. Similarly, Hassanen et al. reported that chitosan-coated Ag NPs, which are widely used as antibacterial and antiviral agents, have dose-dependent toxicity. These NPs can also damage various tissues in the body by increasing free radical levels. In addition, a study by de Lima et al. shows that NP-based chemotherapy has not yet provided significant benefits in the treatment of head and neck cancers. These results indicate that despite the therapeutic potential of NPs, there are issues regarding the efficacy and safety of NP-based treatments in some cancers, and further research is needed [97–99].

4.6 Future Research and Potential Developments

Nanoparticles hold promise for various medical applications, especially in areas such as cancer treatment and organ transplantation, as mentioned earlier. In particular, NPs smaller than 10 nm are increasingly used in clinical studies, and biological functionalization methods allow better access to tumors and uptake by cells. In addition, in organ transplantation, NP-based drug delivery systems aim to improve drug delivery and targeting, thereby increasing safety, efficacy, and patient compliance. Immunotherapies based on natural killer cells use NP interfaces that enable tracking and activation of cells in tumors. The rapid uptake of NPs by the mononuclear phagocyte system (MFS) reduces their efficacy. The temporary suppression of MFS has become known as a potential strategy to enhance particle circulation and distribution to tumors [100, 101]. Future work will focus on improving the design of NPs for specific purposes, improving drug delivery methods, and solving clinical issues such as long-term patient response and overall benefit from treatment.

5 Conclusion

In this chapter, we tried to focus on the use of nanotechnology and nanoscience from ancient times to the present day. In that context, a comprehensive literature review has been conducted on the growing field where nanotechnology intersects with biomedicine, highlighting the clinical applications of ILs, NFs, and NPs. These materials demonstrate significant promise across diverse therapeutic and diagnostic fields. ILs provide unique advantages in drug formulation and drug delivery systems because of their tunable properties, solvency, potential antimicrobial applications, and tissue engineering applications. NFs exploit the properties of suspended NPs to improve biomedical imaging, facilitate targeted drug delivery, and offer innovative antibacterial approaches. NPs (MNPs, PNPs, LNPs, and QDs) are essential for the advancement of targeted cancer therapies, regenerative medicine applications such as stem cell tracking and tissue scaffolds, infection control strategies, and highly sensitive biosensors for real-time disease monitoring. Despite the demonstrated transformative potential, there are significant obstacles to widespread clinical use, primarily related to safety, toxicity, and regulatory approval. We have a detailed discussion of all the carried research in the field including their "safe-by-design" principles applications related to the concerns about the biocompatibility, possible cytotoxicity, oxidative stress, bioaccumulation, and long-term biological effects. It is vital to successfully overcome these challenges through ongoing multidisciplinary research, comprehensive in vivo testing, and the development of precise regulatory guidelines.

References

1. M. Reibold, P. Paufler, A.A. Levin, W. Kochmann, N. Pätzke, D.C. Meyer, Carbon nanotubes in an ancient Damascus sabre. Nature **444**, 286–286 (2006)
2. P. Sciau, P. Sciau, *Nanoparticles in Ancient Materials: The Metallic Lustre Decorations of Medieval Ceramics* (IntechOpen, London, 2012). https://doi.org/10.5772/34080
3. S. Bayda, M. Adeel, T. Tuccinardi, M. Cordani, F. Rizzolio, The history of nanoscience and nanotechnology: From chemical–physical applications to nanomedicine. Molecules **25**, 112 (2019)
4. D.S. Markovic, D. Zivkovic, D. Cvetkovic, R. Popovic, Impact of nanotechnology advances in ICT on sustainability and energy efficiency. Renew. Sust. Energ. Rev. **16**, 2966–2972 (2012)
5. S. Garg, N.P. Rumjit, S. Roy, Smart agriculture and nanotechnology: Technology, challenges, and new perspective. Adv. Agrochem. **3**, 115–125 (2024)
6. D.R. Baer, P.E. Burrows, A.A. El-Azab, Enhancing coating functionality using nanoscience and nanotechnology. Prog. Org. Coat. **47**, 342–356 (2003)
7. A. Nasir, A. Khan, J. Li, M. Naeem, A.A.K. Khalil, K. Khan, et al., Nanotechnology, a tool for diagnostics and treatment of cancer. Curr. Top. Med. Chem. **21**, 1360–1376 (2021)
8. M. Azzawi, A. Seifalian, W. Ahmed, Nanotechnology for the diagnosis and treatment of diseases. Nanomedicine **11**, 2025–2027 (2016)
9. A. Kazemi-Beydokhti, S.Z. Heris, M.R. Jaafari, Experimental investigation of thermal conductivity of medical nanofluids based on functionalised single-wall carbon nanotube and conjugated cisplatin. Micro Nano Lett. **10**, 241–247 (2015)
10. B. Córdoba-Jover, A. Arce-Cerezo, J. Ribera, M. Pauta, D. Oró, G. Casals, et al., Cerium oxide nanoparticles improve liver regeneration after acetaminophen-induced liver injury and partial hepatectomy in rats. J. Nanobiotechnol. **17**, 1–12 (2019)
11. R. Savaliya, D. Shah, R. Singh, A. Kumar, R. Shanker, A. Dhawan, et al., Nanotechnology in disease diagnostic techniques. Curr. Drug Metab. **16**, 645–661 (2015)
12. A.M. Curreri, S. Mitragotri, E.E. Tanner, Recent advances in ionic liquids in biomedicine. Adv. Sci. **8**, 1–19 (2021)

13. C. Cruz, A. Ciach, Phase transitions and electrochemical properties of ionic liquids and ionic liquid-solvent mixtures. Molecules (2021). https://doi.org/10.3390/molecules26123668
14. M.G. Fuster, I. Moulefera, M.N. Muñoz, M.G. Montalbán, G. Víllora, Synthesis of cellulose nanoparticles from ionic liquid solutions for biomedical applications. Polymers (2023). https://doi.org/10.3390/polym15020382
15. A.K. Ziyada, C.D. Wilfred, M.A. Bustam, Z. Man, T. Murugesan, Thermophysical properties of 1-propyronitrile-3-alkylimidazolium bromide ionic liquids at temperatures from (293.15 to 353.15) K. J. Chem. Eng. Data **55**, 3886–3890 (2010)
16. J. Chakravarty, M.F. Rabbi, V. Chalivendra, T. Ferreira, C.J. Brigham, Mechanical and biological properties of chitin/polylactide (PLA)/hydroxyapatite (HAP) composites cast using ionic liquid solutions. Int. J. Biol. Macromol. **151**, 1213–1223 (2020)
17. A. Júlio, R. Caparica, S.A. Costa Lima, A.S. Fernandes, C. Rosado, D.M.F. Prazeres, et al., Ionic liquid-polymer nanoparticle hybrid systems as new tools to deliver poorly soluble drugs. Nanomaterials (2019). https://doi.org/10.3390/nano9081148
18. N. Nikfarjam, M. Ghomi, T. Agarwal, M. Hassanpour, E. Sharifi, D. Khorsandi, et al., Antimicrobial ionic liquid-based materials for biomedical applications. Adv. Funct. Mater. (2021). https://doi.org/10.1002/adfm.202104148
19. J. Xu, H. Fang, Y. Su, Y. Kang, D. Xu, Y.Y. Cheng, et al., A 3D bioprinted decellularized extracellular matrix/gelatin/quaternized chitosan scaffold assembling with poly(ionic liquid)s for skin tissue engineering. Int. J. Biol. Macromol. **220**, 1253–1266 (2022)
20. T. Oshizaka, S. Kodera, R. Kawakubo, I. Takeuchi, K. Mori, K. Sugibayashi, Enhanced drug skin permeation by a zone-mimicking ionic liquids: Effects of fatty acids forming ionic liquids. Pharmaceutics **17**, 1–12 (2025)
21. G.S. Lim, J. Zidar, D.W. Cheong, S. Jaenicke, M. Klahn, Impact of ionic liquids in aqueous solution on bacterial plasma membranes studied with molecular dynamics simulations. J. Phys. Chem. B **118**, 10444–10459 (2014)
22. A. Borkowski, Ł. Gutowski, M. Syczewski, T. Cłapa, G. Czerwonka, Adaptation of bacteria Escherichia coli in presence of quaternary ammonium ionic liquids. Ecotoxicol. Environ. Saf. **164**, 370–378 (2018)
23. B. Jing, N. Lan, J. Qiu, Y. Zhu, Interaction of ionic liquids with a lipid bilayer: A biophysical study of ionic liquid cytotoxicity. J. Phys. Chem. B **120**, 2781–2789 (2016)
24. J. Chakravarty, M.F. Rabbi, N. Bach, V. Chalivendra, C.L. Yang, C.J. Brigham, Fabrication of porous chitin membrane using ionic liquid and subsequent characterization and modelling studies. Carbohydr. Polym. **198**, 443–451 (2018)
25. J. Zhu, B. Na, R. Lv, C. Li, Enhanced stereocomplex formation of high-molecular-weight polylactides by gelation in an ionic liquid. Polym. Int. **63**, 1101–1104 (2014)
26. I. Pauliukaitytė, D. Čiužas, E. Krugly, O. Baniukaitienė, M. Bulota, V. Petrikaitė, et al., Regenerated nanofibrous cellulose electrospun from ionic liquid: Tuning properties toward tissue engineering. J. Biomed. Mater. Res. A (2025). https://doi.org/10.1002/JBM.A.37798
27. H.Y. Lee, G.Z. Jin, U.S. Shin, J.H. Kim, H.W. Kim, Novel porous scaffolds of poly(lactic acid) produced by phase-separation using room temperature ionic liquid and the assessments of biocompatibility. J. Mater. Sci. Mater. Med. **23**, 1271–1279 (2012)
28. T.T. Baby, S. Ramaprabhu, Synthesis and nanofluid application of silver nanoparticles decorated graphene. J. Mater. Chem. **21**, 9702–9709 (2011)
29. N. Ali, J.A. Teixeira, A. Addali, A review on nanofluids: Fabrication, stability, and thermophysical properties. J. Nanomater. (2018). https://doi.org/10.1155/2018/6978130
30. S.U.S. Choi, Enhancing thermal conductivity of fluids with nanoparticles. ASME Fluids Eng. Div. **231**, 99–105 (1995)
31. J.A. Eastman, U.S. Choi, S. Li, L.J. Thompson, S. Lee, Enhanced thermal conductivity through the development of nanofluids. MRS Online Proc. Libr. **457**, 3–11 (1997)
32. H.T. Zhu, Y.S. Lin, Y.S. Yin, A novel one-step chemical method for preparation of copper nanofluids. J. Colloid Interface Sci. **277**, 100–103 (2004)
33. S.U.S. Choi, S. Li, J.A. Eastman, Measuring thermal conductivity of fluids containing oxide nanoparticles. J. Heat Transf. **121**, 280–289 (1999)

34. S.M.S. Murshed, K.C. Leong, C. Yang, Enhanced thermal conductivity of TiO2 - water based nanofluids. Int J Therm Sci **44**, 367–373 (2005)
35. Y. Xuan, Q. Li, Heat transfer enhancement of nanofluids. Int. J. Heat Fluid Flow **21**, 58–64 (2000)
36. L. Sheng, X. Zhu, M. Sun, Z. Lan, Y. Yang, Y. Xin, et al., Tumor microenvironment-responsive magnetic nanofluid for enhanced tumor MRI and tumor multi-treatments. Pharmaceuticals (2023). https://doi.org/10.3390/ph16020166
37. I. Shahzadi, S. Bilal, A significant role of permeability on blood flow for hybrid nanofluid through bifurcated stenosed artery: Drug delivery application. Comput. Methods Prog. Biomed. **187**, 105248 (2020)
38. L. Benos, G. Ninos, N.D. Polychronopoulos, M.A. Exomanidou, I. Sarris, Natural convection of blood–magnetic iron oxide bio-nanofluid in the context of hyperthermia treatment. Computation (2022). https://doi.org/10.3390/computation10110190
39. M. Yusefi, K. Shameli, Z. Hedayatnasab, S.Y. Teow, U.N. Ismail, C.A. Azlan, et al., Green synthesis of Fe3O4 nanoparticles for hyperthermia, magnetic resonance imaging and 5-fluorouracil carrier in potential colorectal cancer treatment. Res. Chem. Intermed. **47**, 1789–1808 (2021)
40. H. Jannati, M. Sheikhpour, S.D. Siadat, P. Safarian, Antimicrobial activity and drug delivery ability of functionalized multi-walled carbon nanotubes nanofluid on *Staphylococcus aureus*. Nanomed. Res. J. **6**, 248–256 (2021)
41. Z. Fakhroueian, F. Katouzian, P. Esmaeilzadeh, S. Moradi Bidhendi, P. Esmaeilzadeh, Enhanced engineered ZnO nanostructures and their antibacterial activity against urinary, gastrointestinal, respiratory and dermal genital infections. Appl. Nanosci. **9**, 1759–1773 (2019)
42. N. Pauzi, N.M. Zain, N.A.A. Yusof, Gum Arabic as natural stabilizing agent in green synthesis of ZnO nanofluids for antibacterial application. J. Environ. Chem. Eng. **8**, 103331 (2020)
43. A. Aykaç, S. Elalaoui, S.H.M. Hosseini, A. Nassirzadehsaleh, Trash to treasure-revolutionizing bioengineering with waste-driven nanocomposites. Adv. Mater. Sci. Res. **69**, 149–192 (2024)
44. Q. Zhong, D.M.D. Chinta, S. Pamujula, H. Wang, X. Yao, T.K. Mandal, et al., Optimization of DNA delivery by three classes of hybrid nanoparticle/DNA complexes. J. Nanobiotechnol. **8**, 1–10 (2010)
45. K.P. Kubelick, E.J. Snider, C. Ross Ethier, S. Emelianov, Development of a stem cell tracking platform for ophthalmic applications using ultrasound and photoacoustic imaging. Theranostics **9**, 3812–3824 (2019)
46. P. Arora, A. Sindhu, N. Dilbaghi, A. Chaudhury, G. Rajakumar, A.A. Rahuman, Nano-regenerative medicine towards clinical outcome of stem cell and tissue engineering in humans. J. Cell. Mol. Med. **16**, 1991–2000 (2012)
47. Q. Yang, D.D. Dionysios, Photolytic degradation of chlorinated phenols in room temperature ionic liquids. J. Photochem. Photobiol. A Chem. **165**, 229–240 (2004)
48. R. Sangwan, M. Saini, R. Verma, S. Kumar, M. Banerjee, S. Jain, Synthesis of 1,8-dioxooctahydroxanthene derivatives using ionic liquids, quantum chemical studies and anticancer activity. J. Mol. Struct. **1208**, 127786 (2020)
49. Y. Zhuo, H.L. Cheng, Y.G. Zhao, H.R. Cui, Ionic liquids in pharmaceutical and biomedical applications: A review. Pharmaceutics **16**, 1–29 (2024)
50. M.R. Islam, M.R. Chowdhury, R. Wakabayashi, Y. Tahara, N. Kamiya, M. Moniruzzaman, et al., Choline and amino acid based biocompatible ionic liquid mediated transdermal delivery of the sparingly soluble drug acyclovir. Int. J. Pharm. (2020). https://doi.org/10.1016/j.ijpharm.2020.119335
51. S. Goindi, R. Kaur, R. Kaur, An ionic liquid-in-water microemulsion as a potential carrier for topical delivery of poorly water soluble drug: Development, ex-vivo and in-vivo evaluation. Int. J. Pharm. **495**, 913–923 (2015)
52. V. Jaitely, A. Karatas, A.T. Florence, Water-immiscible room temperature ionic liquids (RTILs) as drug reservoirs for controlled release. Int. J. Pharm. **354**, 168–173 (2008)

53. K. Niesyto, S. Keihankhadiv, A. Mazur, A. Mielańczyk, D. Neugebauer, Ionic liquid-based polymer matrices for single and dual drug delivery: Impact of structural topology on characteristics and in vitro delivery efficiency. Int. J. Mol. Sci. **25**, 1292 (2024)
54. F. Fadaei, M. Tortora, A. Gessini, C. Masciovecchio, S. Catalini, J. Vigna, et al., Structural specificity of groove binding mechanism between imidazolium-based ionic liquids and DNA revealed by synchrotron-UV resonance Raman spectroscopy and molecular dynamics simulations. J. Mol. Liq. **347**, 118350 (2022)
55. A. Chandran, D. Ghoshdastidar, S. Senapati, Groove binding mechanism of ionic liquids: A key factor in long-term stability of DNA in hydrated ionic liquids? J. Am. Chem. Soc. **134**, 20330–20339 (2012)
56. H. Wang, X. Shi, D. Yu, J. Zhang, G. Yang, Y. Cui, et al., Antibacterial activity of geminized amphiphilic cationic homopolymers. Langmuir **31**, 13469–13477 (2015)
57. G. Bhattacharya, R.P. Giri, A. Dubey, S. Mitra, R. Priyadarshini, A. Gupta, et al., Structural changes in cellular membranes induced by ionic liquids: From model to bacterial membranes. Chem. Phys. Lipids **215**, 1–10 (2018)
58. J. Qin, J. Guo, Q. Xu, Z. Zheng, H. Mao, F. Yan, Synthesis of pyrrolidinium-type poly(ionic liquid) membranes for antibacterial applications. ACS Appl. Mater. Interfaces **9**, 10504–10511 (2017)
59. M. Soheilmoghaddam, G. Sharifzadeh, R.H. Pour, M.U. Wahit, W.T. Whye, X.Y. Lee, Regenerated cellulose/β-cyclodextrin scaffold prepared using ionic liquid. Mater. Lett. **135**, 210–213 (2014)
60. A. Hiroshi, Y. Tsukasaki, YS and TA., Magnetic properties of ferromagnetic ultrafine particles prepared by vacuum. J. Cryst. Growth **45**, 495–500 (1978)
61. C.H. Lo, T.T. Tsung, L.C. Chen, C.H. Su, H.M. Lin, Fabrication of copper oxide nanofluid using submerged arc nanoparticle synthesis system. J. Nanopart. Res. **7**, 313–320 (2005)
62. E.M. Hamad, A. Khaffaf, O. Yasin, Z. Abu El-Rub, S. Al-Gharabli, W. Al-Kouz, et al., Review of nanofluids and their biomedical applications. J. Nanofluids **10**, 463–477 (2021)
63. R. Shelat, L.K. Bhatt, A. Khanna, S. Chandra, A comprehensive toxicity evaluation of novel amino acid-modifed magnetic ferrofuids for magnetic resonance imaging. Amino Acids **51**, 929–943 (2019)
64. U. Farooq, T. Liu, Numerical simulation of radiative flow with magnetized micropolar nanofluid along a curved stretched surface involving blood silica nanoparticles. Numer. Heat Transf. A Appl. **1**, 1–18 (2024)
65. A.S. Alnahdi, S. Nasir, T. Gul, Couple stress ternary hybrid nanofluid flow in a contraction channel by means of drug delivery function. Math. Comput. Simul. **210**, 103–119 (2023)
66. S.M. Fotukian, A. Barati, M. Soleymani, A.M. Alizadeh, Solvothermal synthesis of CuFe2O4 and Fe3O4 nanoparticles with high heating efficiency for magnetic hyperthermia application. J. Alloys Compd. **816**, 152548 (2020)
67. P.B. Kharat, S.B. Somvanshi, P.P. Khirade, K.M. Jadhav, Induction heating analysis of surface-functionalized nanoscale CoFe2O4for magnetic fluid hyperthermia toward noninvasive cancer treatment. ACS Omega **5**, 23378–23384 (2020)
68. Z. Yan, L. Zhang, Z. Zhao, H. Qi, Y. Li, D. Cang, Enhanced antimicrobial activity of ZnO nanofluids in sonophotocatalysis and its mechanism. Ultrason. Sonochem. **47**, 133–140 (2018)
69. R. Mannu, V. Karthikeyan, N. Velu, C. Arumugam, V.A.L. Roy, A.I. Gopalan, et al., Polyethylene glycol coated magnetic nanoparticles: Hybrid nanofluid formulation, properties and drug delivery prospects. Nano **11**, 1–18 (2021)
70. S.R. Patade, D.D. Andhare, S.B. Somvanshi, S.A. Jadhav, M.V. Khedkar, K.M. Jadhav, Self-heating evaluation of superparamagnetic MnFe2O4 nanoparticles for magnetic fluid hyperthermia application towards cancer treatment. Ceram. Int. **46**, 25576–25583 (2020)
71. N. Shanmugapriyan, S. Jakeer, Biomedical aspects of entropy generation on MHD flow of TiO2-Ag/blood hybrid nanofluid in a porous cylinder. Comput. Methods Biomech. Biomed. Engin. **27**, 1492–1509 (2024)

72. S.V.K. Varma, Magnetized electroosmotic drug delivery of Au-SiO2 nanocarriers through blood-based Jeffrey hybrid nanofluid in a microchannel: Caputo Fabrizio derivative modeling. Heliyon **10**, e36891 (2024)
73. N.A. Amirsom, M.J. Uddin, A.I. Ismail, MHD boundary layer bionanoconvective non-Newtonian flow past a needle with Stefan blowing. Heat Transf. Asian Res. **48**, 727–743 (2019)
74. A. Mathew, S. Areekara, A.S. Sabu, S. Saleem, Significance of multiple slip and nanoparticle shape on stagnation point flow of silver-blood nanofluid in the presence of induced magnetic field. Surf. Interfaces **25**, 101267 (2021)
75. M. Sheikhpour, M. Arabi, A. Kasaeian, A.R. Rabei, Z. Taherian, Role of nanofluids in drug delivery and biomedical technology: Methods and applications. Nanotechnol. Sci. Appl. **13**, 47–59 (2020)
76. X.Q. Zhang, X. Xu, N. Bertrand, E. Pridgen, A. Swami, O.C. Farokhzad, Interactions of nanomaterials and biological systems: Implications to personalized nanomedicine. Adv. Drug Deliv. Rev. **64**, 1363–1384 (2012)
77. S.J. Soenen, W.J. Parak, J. Rejman, B. Manshian, (intra)cellular stability of inorganic nanoparticles: Effects on cytotoxicity, particle functionality, and biomedical applications. Chem. Rev. **115**, 2109–2135 (2015)
78. A. Balfourier, A.P. Marty, F. Gazeau, Importance of metal biotransformation in cell response to metallic nanoparticles: A transcriptomic meta-analysis study. ACS Nanosci. **3**, 46–57 (2023)
79. M. Hauser, G. Li, B. Nowack, Environmental hazard assessment for polymeric and inorganic nanobiomaterials used in drug delivery. J. Nanobiotechnol. **17**, 1–10 (2019)
80. A. Suberi, M.K. Grun, T. Mao, B. Israelow, M. Reschke, J. Grundler, et al., *Polymer Nanoparticles Deliver mRNA to the Lung for Mucosal Vaccination* (American Association for the Advancement of Science, Washington, 2023)
81. A. Bangham, M.M. Standish, J.C. Watkins, The 1st description of liposomes - A citation classic commentary on diffusion of univalent ions across the lamellae of swollen phospholipids by bangham. J. Mol. Biol. **I**, 14 (1989)
82. M.C. Chung, H.R. Mendez-Gomez, D. Soni, R. McGinley, A. Zacharia, J. Ashbrook, et al., Multi-step assembly of an RNA-liposome nanoparticle formulation revealed by real-time, single-particle quantitative imaging. Adv. Sci. **12**, 2414305 (2025)
83. Y. Barenholz, Doxil® - The first FDA-approved nano-drug: Lessons learned. J. Control. Release **160**, 117–134 (2012)
84. J.T. Cao, Z.X. Chen, X.Y. Hao, P.H. Zhang, J.J. Zhu, Quantum dots-based immunofluorescent microfluidic chip for the analysis of glycan expression at single-cells. Anal. Chem. **84**, 10097–10104 (2012)
85. H. Barabadi, A. Alizadeh, M. Ovais, A. Ahmadi, Z.K. Shinwari, M. Saravanan, Efficacy of green nanoparticles against cancerous and normal cell lines: A systematic review and meta-analysis. IET Nanobiotechnol. **12**, 377–391 (2018)
86. S. Safwat, R.M. Hathout, R.A. Ishak, N.D. Mortada, Elaborated survey in the scope of nanocarriers engineering for boosting chemotherapy cytotoxicity: A meta-analysis study. Int. J. Pharm. **610**, 121268 (2021)
87. J. Tang, X. Jiang, L. Wang, H. Zhang, Z. Hu, Y. Liu, et al., Au@Pt nanostructures: A novel photothermal conversion agent for cancer therapy. Nanoscale **6**, 3670–3678 (2014)
88. T. Tashima, Non-invasive delivery of CRISPR/Cas9 ribonucleoproteins (Cas9 RNPs) into cells via nanoparticles for membrane transport. Pharmaceutics (2025). https://doi.org/10.3390/pharmaceutics17020201
89. Idulhaq, M.; Mudigdo, A.; Utomo, P.; Wasita, B.; Warman, FI. Platelet-rich fibrin as a tissue engineering material in accelerate bone healing in rat bone defects: A systematic review and meta-analysis. Ann. Med. Surg. 2022; 84: 104869
90. A. Aykaç, İ. Ok, E. Bakay, N. Topaloğlu, Chitosan-based hydrogel incorporating mesoporous silica/zinc oxide nanoparticles with controlled swelling and potent antibacterial activity. Int. J. Polym. Mater. Polym. Biomater. **2025**, 465–473 (2025)

91. D. Botequim, J. Maia, M.M.F. Lino, L.M.F. Lopes, P.N. Simões, L.M. Ilharco, et al., Nanoparticles and surfaces presenting antifungal, antibacterial and antiviral properties. Langmuir **28**, 7646–7656 (2012)
92. R.E. Madrid, R. Chehín, T.-H. Chen, A. Guiseppi-Elie, Biosensors and nanobiosensors, in *Nanomedicine*, (Seifalian University College, London, 2017), pp. 391–462
93. L.C. Clark, C. Lyons, Electrode systems for continuous monitoring in cardiovascular surgery. Ann. N. Y. Acad. Sci. **102**, 29–45 (1962)
94. L.C. Clark, R. Wolf, D. Granger, Z. Taylor, Continuous recording of blood oxygen tensions by polarography. J. Appl. Physiol. **6**, 189–193 (1953)
95. A. Aykaç, H. Gergeroglu, B. Beşli, E.Ö. Akkaş, A. Yavaş, S. Güler, et al., An overview on recent progress of metal oxide/graphene/CNTs-based nanobiosensors. Nanoscale Res. Lett. **16**, 1–19 (2021)
96. F. Güneş, A. Aykaç, M. Erol, Ç. Erdem, H. Hano, B. Uzunbayir, et al., Synthesis of hierarchical hetero-composite of graphene foam/α-Fe2O3 nanowires and its application on glucose biosensors. J. Alloys Compd. **895**, 162688 (2022)
97. S. Nakamura, M. Sato, Y. Sato, N. Ando, T. Takayama, M. Fujita, et al., Synthesis and application of silver nanoparticles (Ag NPS) for the prevention of infection in healthcare workers. Int. J. Mol. Sci. (2019). https://doi.org/10.3390/ijms20153620
98. L. Wu, W. Wen, X. Wang, D. Huang, J. Cao, X. Qi, et al., Ultrasmall iron oxide nanoparticles cause significant toxicity by specifically inducing acute oxidative stress to multiple organs. Part. Fibre Toxicol. **19**, 1–14 (2022)
99. J.M. de Lima, P.R. Bonan, D.E. Perez, M. Hier, M.A. Alaoui-Jamali, S.D. da Silva, Nanoparticle-based chemotherapy formulations for head and neck cancer: A systematic review and perspectives. Nano **10**, 1–16 (2020)
100. L. Carmès, M. Banerjee, P. Coliat, S. Harlepp, X. Pivot, O. Tillement, et al., Toward clinical transfer of tumor-targeted ultrasmall inorganic nanoparticles. Adv. Ther. **6**, 1–10 (2023)
101. I.V. Zelepukin, K.G. Shevchenko, S.M. Deyev, Rediscovery of mononuclear phagocyte system blockade for nanoparticle drug delivery. Nat. Commun. (2024). https://doi.org/10.1038/s41467-024-48838-5

Exploring the Biomedical and Biotechnological Potential of Ionic Liquids, Nanofluids, and Nanoparticles

Oluwabusayo O. Oluyide, George D. Ametefe, Babamotemi O. Itakorode, I. Muritala, Tolulope V. Ayo-Awe, Omotayo O. Oyedara, and Olumuyiwa P. Oluyide

1 Overview of Ionic Liquids, Nanofluids, and Nanoparticles

Ionic liquids, nanofluids, and nanoparticles represent three distinct yet interconnected fields of modern materials science, each offering unique properties and transformative applications. Ionic liquids, defined as salts melting below 100 °C, combine the ionic conductivity of salts with the variable solvent properties of organic compounds [1–4]. Nanofluids, colloidal suspensions of nanoparticles in base fluids, enhance thermal management systems through superior heat transfer capabilities [5, 6]. Nanoparticles, particles between 1 and 100 nm, exhibit size-dependent quantum effects that revolutionize fields from medicine to energy storage [7–10]. Together, these materials form the foundation of advanced technologies addressing global challenges in sustainability, energy efficiency, and healthcare innovation.

O. O. Oluyide (✉) · G. D. Ametefe · B. O. Itakorode · I. Muritala · O. O. Oyedara
Department of Biotechnology, Osun State University, Osogbo, Osun, Nigeria
e-mail: oluwabusayo.oluyide@uniosun.edu.ng

T. V. Ayo-Awe
Department of Basic Medical Science, College of Health Sciences and Technology, Ijero Ekiti, Ekiti, Nigeria

O. P. Oluyide
Department of Community Medicine, Osun State University Teaching Hospital, Osogbo, Osun, Nigeria

S. Kulkarni, S. Das (eds.), *Ionic Liquids, Nanofluids and Nanotechnology*, Lecture Notes in Nanoscale Science and Technology 42,
https://doi.org/10.1007/978-3-032-04008-4_11

2 Ionic Liquids: Structure and Fundamental Properties

2.1 Definition and Chemical Composition

Ionic liquids (ILs) are organic salts characterized by melting points below 100 °C, with room-temperature ionic liquids (RTILs) remaining liquid at ambient conditions [11]. Unlike conventional salts such as sodium chloride, ILs feature bulky, asymmetric organic cations paired with weakly coordinating anions. Common cations include imidazolium, pyrrolidinium, and ammonium derivatives. At the same time, anions range from halides (e.g., Cl^-) to complex species like bis(trifluoromethylsulfonyl)imide ($NTf2^-$) [12–14]. This structural asymmetry disrupts crystalline packing, suppressing melting points while maintaining ionic conductivity [15].

2.2 Physicochemical Characteristics

The dual nature of ILs—combining ionic character with liquid-state disorder—confers exceptional properties. Their negligible vapor pressure eliminates flammability risks, making them safer than volatile organic solvents 1. The physicochemical properties of ionic liquids commonly include low volatility and tunable melting points; others include high polarity, ionic conductivity, and thermal stability, enabling high-temperature applications [16].

2.3 Synthesis and Functionalization

IL synthesis typically involves metathesis reactions, where halide salts react with silver or ammonium salts to exchange anions [17]. Properties are tailored by modifying alkyl chains or selecting anions with specific coordinating strengths. Protic ionic liquids, synthesized via proton transfer from Brønsted acids to bases, exhibit hydrogen-bonding networks that enhance proton conductivity for fuel cell applications [18–21].

3 Nanofluids

3.1 Composition and Preparation Methods

Nanofluids consist of nanoparticles (1–100 nm) dispersed in base fluids like water, ethylene glycol, or oils [22]. Metallic (copper, silver), oxide (Al_2O_3, TiO_2), and carbon-based (nanotubes, graphene) nanoparticles are common. Preparation involves two approaches: single-step methods that synthesize nanoparticles directly within the fluid via chemical reduction (i.e., using techniques like direct evaporation, liquid chemical direct evaporation, and physical vapor deposition). Therefore, the single-step process synthesizes nanoparticles and suspends them in a base fluid, creating a nanofluid with excellent stability, homogeneity, regulated nanoparticle size, no aggregation, and no stabilizer usage, while two-step methods disperse pre-made nanoparticles using ultrasonication or surfactants [23, 24].

3.2 Enhanced Thermal Properties

The main benefit of nanofluids is their improved heat conductivity, which rises in direct proportion to the concentration of nanoparticles. The reason for this is that the nanoparticles, which are usually smaller than 100 nm, open up more channels for heat transfer in the fluid. So, the total thermal conductivity of the nanofluid rises with the concentration of these nanoparticles, improving the fluid's capacity for heat transmission [25, 26].

3.3 Rheological and Acoustical Behavior

Nanofluids display non-Newtonian shear-thinning behavior at high nanoparticle loadings (>5%), necessitating viscosity modifiers for pump-driven systems [27]. Acoustically, ultrasonic waves induce shear–wave reconversion in nanofluids, a phenomenon exploited in medical imaging and targeted drug delivery [28–31].

4 Nanoparticles

4.1 The Quantum Realm of Materials and Size-Dependent Phenomena

Nanoparticles exhibit quantum confinement effects when their size approaches the electron wavelength (1–10 nm for metals) [32]. Gold nanoparticles, for instance, transition from metallic conductivity to semiconducting behavior below 5 nm, enabling applications in plasmonic sensors [33, 34]. Surface-area-to-volume ratios exceeding 100 m^2/g enhance catalytic activity; platinum nanoparticles catalyze hydrogen fuel cell reactions 50 times faster than bulk platinum [32, 35].

Table 1 Synthesis methods [36]

Synthesis method	Process
Bottom-up	Chemical vapor deposition (CVD) assembles atoms into nanoparticles, as seen in carbon nanotube production
Top-down	Ball milling grinds bulk materials into nanoscale powders for ceramic reinforcements

4.2 Classification and Manufacturing

Joudeh and Linke and the Nanotechnologies—Plain Language Explanation of Selected Terms (from the ISO/IEC 80004 Series) stated that the International Organization for Standardization (ISO) defines nanoparticles as having all dimensions below 100 nm. The European Commission, however, includes materials with ≥50% particles in this range. Table 1 below shows the synthesis method.

4.3 The Following Are a Few Typical Physical Synthesis Techniques for Nanoparticles

Evaporation–condensation: In this process, a substance is heated or laser-treated to evaporate it, and the vapor that results is then condensed to create nanoparticles. With laser ablation, a target substance is ablated using a high-energy laser, creating a plasma that condenses to create nanoparticles. In sputtering, high-energy ions or atoms are fired at a target material, ejecting the atoms and depositing them as nanoparticles on a substrate [37, 38].

4.4 Environmental and Health Considerations

Despite their benefits, nanoparticles pose inhalation risks due to alveolar deposition. Through the induction of oxidative stress, inflammatory reactions, DNA damage, and cytotoxicity, nanoparticles can be harmful to human health whether consumed, inhaled, or applied topically. It is essential to look at drinking water quality and freshwater habitats. Nanoparticles are successfully eliminated from drinking water using a variety of mitigation and treatment techniques such as filtering systems, adsorption procedures, and sophisticated oxidation procedures; standardized safety testing and regulatory frameworks are necessary, nevertheless [39].

4.5 Interplay of Nanoparticles and Ionic Liquids: Hybrid Material Design

Integrating nanoparticles into ionic liquids creates multifunctional hybrids. Gold nanoparticles dispersed in 1-butyl-3-methylimidazolium hexafluorophosphate form conductive gels for flexible supercapacitors [40]. The ionic liquid's high dielectric constant stabilizes nanoparticles by electrostatic screening, preventing aggregation [40–42].

5 Importance of Convergence in Biomedical, Biotechnological, and Clinical Applications

The advancement of healthcare depends on convergence, the integration of several disciplines, including biomedicine, biotechnology, and clinical applications, as the creation of new technologies, personalized medicine, and improved diagnosis, consequently improving patient outcomes and disease management. For instance, smart implants that monitor physiological parameters and communicate wirelessly can facilitate remote patient monitoring and early intervention for chronic diseases [43].

Building a knowledge network for precision medicine calls for combining clinical, technological, and biomedical domains. The integration of genomic data with clinical information, facilitated by biomedical informatics, can help identify patients at risk for certain diseases and personalize their treatment plans. This network would allow doctors and scientists to find trends, create focused therapies, and enhance patient care by combining information on genetics, environment, medical history, and health outcomes. This development would need cooperation from several disciplines—including life sciences, medicine, behavioral science, and systems science—thereby improving treatment results and raising survival rates across different demographic and socioeconomic groups [44].

The uniquely multidisciplinary approach to problem-solving in industry makes it a major collaborator for convergence-driven innovation. The increasing complexity of healthcare issues—including chronic diseases, aging populations, evolving lifestyle variables, antimicrobial resistance, and climate change—calls for a cooperative approach outside conventional medicine. Medical personnel cannot manage these matters alone. We can hasten progress and discover new answers to urgent health problems by combining medical knowledge with non-medical inventions, including genomics, proteomics, artificial intelligence, and digital health technologies [44].

In a multidisciplinary field, population health gains momentum by merging disciplines like education, social services, economics, environmental science, and healthcare. Effectively combining these points of view will help us to create thorough solutions to enhance health results and general wellness [45]. In the end, we can turn data into knowledge and wisdom by using convergence science and leveraging the strength of cooperation and openness, therefore, guaranteeing more powerful and responsible uses [43].

5.1 Ionic Liquids, Nanofluids, and Nanoparticles in Biomedical Applications

Ionic liquids are molten salts; because of large, asymmetric cations and weakly coordinated anions distorting the crystal lattice, they are liquid at temperatures below 100 °C. Ionic liquids have numerous possible applications. They are potent solvents and may serve as electrolytes. In addition to their unique features and use as solvents, the adaptable characteristics of ionic liquids have facilitated the emergence of innovative and valuable subjects. Ionic liquids are regarded as the foundation of a significant transformation in materials science [46].

They demonstrate multiple advantageous characteristics, including minimal vapor pressure, excellent thermal stability, elevated ionic conductivity, extensive electrochemical potential ranges, favorable solubility, and significant synthetic versatility [47].

Ionic liquids offer some possible applications. They are potent solvents and can be used as electrolytes. In addition to their unique features and use as solvents, the adaptable characteristics of ionic liquids have facilitated the emergence of innovative and significant subjects. Ionic liquids are heralded as the foundation of a significant transformation in materials research [46].

They demonstrate numerous advantageous characteristics, including minimal vapor pressure, excellent thermal stability, elevated ionic conductivity, extensive electrochemical potential ranges, favorable solubility, and significant synthetic versatility [47]. Ionic liquids are utilized in numerous chemical processes, notably in the manufacture of medicinal compounds and medications. They have numerous benefits, including excellent thermal and chemical stability, high solubility, polarity, and hydrophilicity [48].

Bacteria possess inherent resistance to drugs, utilizing molecular pathways for self-protection. Antimicrobial resistance elevates morbidity and mortality rates and has emerged as a significant issue within healthcare organizations. Bacteria are more likely to acquire advanced resistance to the diverse antibiotics utilized globally. Bacterial resistance mechanisms render antibiotics ineffective against germs, resulting in increased antibiotic dosages or alterations in antibiotic classes within healthcare environments. Researchers are investigating novel and effective antimicrobial drugs to address the rising issue of antibiotic resistance [49].

Highly designable and adaptable ionic liquids possess numerous exceptional physicochemical features that render them promising in combating harmful germs. Electrostatic interaction spans between the cationic component of ionic liquids and the anionic bacterial cell wall, while the penetration of hydrophobic alkyl chains into the phospholipid bilayer leads to the disruption of the hydrophobic integrity of the bacterial cell membrane, culminating in bacterial cell death [50].

Research indicates that an increase in the length of the alkyl side chain in ionic liquids correlates with enhanced toxicity and antibacterial efficacy. Ionic liquids with 10–14 carbon atoms in their side chains have the most significant antibacterial activity [51]. Nanofluid is a compelling heat transfer fluid for enhanced heat transfer applications owing to its superior thermal conductivity and rheological characteristics [52].

Nanofluids are complex colloidal systems composed of nanoparticles, usually metal oxides or carbon-based compounds, suspended in standard base fluids such as water, ethylene glycol, or oil. The incorporation of nanoparticles enhances thermal conductivity, convective heat transfer rates, and fluid viscosity owing to their elevated surface area-to-volume ratio and superior thermal characteristics. This results in improved thermophysical properties, rendering nanofluids appropriate for high-performance cooling systems, heat exchangers, and industrial thermal applications [53].

Recent investigations have explored the biological applications of nanofluids for many purposes, including medication delivery and antibacterial therapies. The occurrence of excessive infections in skin lesions is prevalent during treatment, necessitating appropriate dressing. In 2013, Anghel et al. examined the application of iron nanofluid in wound dressings to inhibit the colonization of Candida albicans and the production of biofilms. The combination of ionic liquids with nanoparticles in biomedical applications has numerous advantages owing to their distinctive features, such as increased stability, greater drug transport, and the potential for tailored therapy.

Nanoparticles are substances having dimensions in the nanoscale, namely, less than 100 nm. Recently, these materials have become significant in contemporary medicine, utilized as contrast agents in medical imaging and as carriers for gene delivery into particular cells. Nanoparticles, encompassing nanocrystals, nanoclusters, nanometals, and nanotubes, demonstrate distinctive physical properties that facilitate numerous applications in catalysis, nanocomposites, electrochemistry, biotechnology, chemical analysis and sensing, diagnostics, drug delivery, and medicine [54].

The most remarkable applications of nanomaterials in diagnostic medicine are their utilization in magnetic resonance imaging (MRI). The incorporation has improved contrast and sensitivity in MRI. Nanomaterials are utilized in the healthcare sector for diagnosis, treatment, management, and illness prevention. The advancement of improved and safer pharmaceuticals, tissue-targeted interventions, and customized nanomedicines is propelled by nanoparticles. Nanoparticles can deliver medications precisely to the targeted location, significantly enhancing their effectiveness and minimizing harm to surrounding tissues. Nanoparticles may be engineered to facilitate medicine administration to the brain for challenging disorders by circumventing the blood–brain barrier [55]. Nanomedicine will undeniably play a crucial role in future personalized medicine, encompassing both prognosis and monitoring.

6 Clinical Applications of Ionic Liquids, Nanofluids, and Nanoparticles.

6.1 Nanoparticles

The discovery of nanotechnology and nanomedicine has led to diverse and impressive advancements in nanomedicine, which has increased the medication to a new level with noteworthy medical results. Findings on the substantial potency of nanotechnology in healthcare are necessary. Many studies are being conducted in the medical field to assess the optimal techniques and approaches, such as cancer treatment, cardiovascular disease therapeutic gene, and nephrology. Traditional treatment has advanced significantly, and nanotechnology and nanoparticle quality have both improved and produced promising outcomes [56]. Due to their nanoscale size, nanoparticles have a large surface-area-to-volume ratio that enables them to swiftly circulate through the bloodstream and absorb large amounts of drugs. They have unique properties due to their larger surface area, which enhances their mechanical, magnetic, optical, and catalytic properties and makes them suitable for a wider range of medicinal uses. Nanotechnology has the potential to significantly advance the diagnosis, treatment, and prevention of diseases in the healthcare industry [57]. The possible health applications of nanotechnology are attracting more and more attention from innovators, and this might lead to a new era of expansion for the sector. By delivering medications to the precise location needed, nanoparticles can significantly increase their effectiveness and limit harm to other bodily tissues [55, 58].

Advances in illness diagnosis and treatment through targeted medicines are made possible by innovations like gold nanoparticles and quantum dots, which improve antibacterial qualities. Drugs, diagnoses, cell treatment, and the production of biocompatible materials are all expected to be provided by medicines. Nanomedicine and the creation of nanoscale agents for the treatment of different diseases are included in the application of nanotechnology in therapeutics. By enhancing diagnostic methods, nanotechnology enables the early detection of diseases and the creation of regenerative medications, which advance tissue regeneration and prosthetic body parts. Early disease diagnosis is now possible due to the utilization of carbon nanotubes, gold nanorods, and rapid, low-cost detection. This method facilitates patients' overall quality of life by using tools and techniques of submicrometer dimensions to better identify, treat, and cure ailments [55].

The use of nanotechnology speeds up the development of regenerative medications. Patients with injuries or organ insufficiency can now take prosthetic skin, bone, cartilage, or other tissues as a result of the new techniques. Nanotechnology imitates the effects of natural tissues and organs by more accurately altering cellular function [59]. Nanotechnology enables the construction of tissues and organs that resemble natural ones, improving wound healing through engineered nanoparticles that target sick tissues and enhance treatment precision. For individuals who may have lost their organs and tissues owing to illness or damage, a person's cells can produce them from a holy grail for use in transplant and reconstructive surgery. Research on this topic aims to create tissues with structures and functions similar to those of real organisms, aided by an understanding of the topography of nanoscale tissues. Nanogenerators, polymer nanoparticles, and targeted solutions may improve the treatment of wounds [60]. Furthermore, several biological processes in the human body that enable overcoming natural barriers to reach new delivery sites and interact with DNA or tiny proteins at various levels in blood or within bodily tissues or cells are also included in the nanometric size [61]. Direct targeting of cancer cells is made possible by nanotechnology, which also improves precision in thermal therapies employing nanotubes and lessens damage to healthy cells during treatments like chemotherapy. By using nanotechnology in conjunction with chemotherapy, doctors may also provide heat treatment more precisely. In such a treatment, nanotubes attached to tumor-attracted antibodies, perhaps causing the tumor to burn [62].

Nanomedicine also makes considerable use of bio-biological, non-biological, biomimetic, or hybrid materials. Cell therapy and tissue engineering promote cell growth and repair by offering scaffolding for the development of biocompatible nanoparticle-based functional tissues. The biocompatibility of the cells and nanoparticles with the sample source is advantageous for the emerging tissues. Doctors can then control the processes of cell formation, development, and repair in the resulting tissue. Through improved medication delivery, more accurate disease targeting, and controlled therapeutic releases, nanomedicine maximizes therapy effectiveness and reduces side effects. By concentrating medications on the underlying cause of the disease, nanoparticles improve effectiveness and reduce side effects. Additionally, they offer fresh chances to control therapeutic releases. With a variety of nanodrugs, drug transport, and diagnostic uses, nanomedicine has greatly benefited from advances in equipment and pharmaceutical production [62].

6.2 *Nanofluids*

Nanofluids, which are emulsions or suspensions of nanoparticles (NPs) in fluids, have the potential to improve heat transfer directly in a variety of industrial applications, including heat exchangers, electronics, transportation, biomedicine, and the food sector. Since the application of nanofluids in nano biomedical technology is becoming more and more popular, their physicochemical characteristics alter significantly when compared to their solid counterparts. The potential advantages of using nanomaterials in drug delivery systems, including increased therapeutic qualities, decreased toxicity, improved biocompatibility, and precise administration to target cells, have drawn attention [63]. Long shelf-life, biocompatibility, and appropriate drug loading and release should all be guaranteed by efficient nanofluid formulations for drug delivery. The charge of nanoparticles (NPs) is a crucial component of these nanofluidic medications; therapeutic particles should be negatively charged to prevent aggregation, as blood particles and vessel walls are negatively charged [64].

The stability of magnetic iron oxide nanoparticles at the fluid interface is one of the prerequisites for their use in medical applications. In their 2015 study, Vazhnichaya et al. produced a nanofluid with sustained magnetic nanoparticles that exhibited anti-anemic effects. Mexidol (2-ethyl-6-methyl-3-hydroxy pyridine succinate) and PVP (polyvinyl pyrrolidine) were used to stabilize NPs. According to the study's findings, the nanofluid promoted the formation of hemoglobin and hematocrit in healthy animals as well as red blood cells (RBCs). A nanofluid helps restore blood loss and boost reticulocyte counts when severe anemia results from blood loss. In undamaged animals, this nanofluid promotes erythropoiesis; in anemia caused by blood loss, it initiates the erythron regenerative reaction [64].

Among other nanofluids, the antibacterial activity of ZnO nanofluids that are safe for human cells has been the subject of the most research [65]. Several facets of ZnO toxicity have been identified based on morphological analyses of the bacterial cell wall treated with ZnONPs using Scanning Electron Microscope (SEM). The stability of ZnO nanofluids has been increased by the use of dispersants such as polyethylene glycol (PEG), deflocculants made of sodium silicate, or sodium carbonate [66]. Additionally, because of their uses in the pharmaceutical industry, MgO and $Mg(OH)_2$ nanofluids have garnered a lot of attention throughout time.

It is typical for excessive infections to develop in skin lesions during therapy; for this reason, a suitable dressing is essential. In 2013, Anghel et al. looked at the application of iron nanofluid as a wound dressing to stop Candida albicans from colonizing and forming biofilms. The dressing patches were submerged in the nanofluid, which was a mixture of active magnetic nanoparticles, like iron oxide. The study's findings suggested that combining the special qualities of iron oxide nanoparticles with Satureja hortensis oil might provide a novel product that inhibits the growth of fungus and biofilms on a range of surfaces seen in medical and clinical equipment.

In the medical procedure known as radio frequency ablation (RFA), short-wavelength (350–500 kHz) radio waves produce a lot of heat in the targeted tissue. In actuality, this technique involves inserting an ablation electrode into the tumor's tissue, after which waves cause the temperature to rise. At that point, the surrounding tissue, typically the tumor tissue, would be heated, causing tissue damage. The tumor would vanish in a matter of minutes if the heat provided were adequate [66]. Wu et al. used the RFA approach to examine the use of iron nanoparticles (NPs) with carbon coating. According to the study's findings, using nanofluid increased thermal conductivity by 41%.

6.3 *Ionic Fluid*

A class of highly adjustable salts known as ionic liquids (ILs) preserves the characteristics of its parent ions. Because of the structural heterogeneity of sterically hindered asymmetric coordination complexes, the ionic interaction of ILs is weaker than the strong electrostatic forces between ions in normal ionic crystals. This disrupts the formation of an intensive crystal, which ultimately results in ILs' exceptionally low melting point. One of the characteristics of ILs that has been the subject of the most research is their antimicrobial activity. Imidazolium, pyridinium, phosphonium, and choline are among the ILs that have demonstrated antibacterial action in recent decades, either in their monomeric or polymeric forms [67, 68]. Antimicrobial ILs can be broadly classified as monomeric, dicationic, and polymeric based on their production technique and application form. Cationic moieties have greater bactericidal activity than anions among the antibacterial ILs. The extent of trying to influence the bactericidal activity has resulted in the onset of Dicationic ionic liquids (DILs). DILs are produced using the same concept of precursors as monomeric ILs, but an extra step of alkane substitution is added to the two cationic moieties. As counter ions, either dicarboxylate or two anions are used [61, 69]. PILs, or polymeric ionic liquids, have garnered a lot of attention.

Three components make up the monomer of PILs: A spacer or linker, a tail, and the head. The head structure improves the bactericidal efficacy of the ILs, solidifies them, and drastically lowers their toxicity [70, 71].

7 Prospects and Challenges

Viscosity, thermal capacity, and purity have been the major challenges in both the primary and advanced applications of ionic liquids. The challenges faced in the properties-based applications have been overcome to a large extent, and specific ILs with definite physico–chemical properties are constantly under development. Toward further development, their solubility has nanoscale and quantum-level material through attaining better stability and shelf-life from ILs that are liquid at room temperature. Among other challenges, ILs' use as chromatographic solvents and extractants is expected to rise to the extraction capabilities, including CO_2 and water, and the ongoing efforts toward increasing CO_2 solubility hold further promise.

Also, further study prospects should focus on the issue of additive selection, measurement, and performance of various types of surfactants for different nanofluids. Moreover, investigations should also be done on the corrosion and erosion behavior of surfactants as well as nanofluids, as this area of study is still shallow. So, overall economic aspects of important equipment and practical aspects may be an intriguing area to be addressed in future work.

8 Conclusion

In summary, three areas of materials science (ionic liquids, nanofluids, and nanoparticles) offer special qualities and vital uses. Nanofluids improve thermal management systems by having better heat transfer capacities, whereas liquids combine conductivity with flexible solvent characteristics. The size-dependent quantum effects of nanoparticles, which range from 1 to 10 nm, are revolutionizing industries including energy storage and health. By raising concentration, nanofluids enhance heat conductivity. Because of their quantum confinement properties, nanoparticles can be used in catalysis and plasmonic sensors. Standardized safety testing and regulatory systems are necessary, nonetheless, due to the associated inhalation hazards.

References

1. A. Banerjee, K. Ibsen, T. Brown, R. Chen, C. Agatemor, S. Mitragotri, Reply to Rogers and Gurau: Definitions of ionic liquids and deep eutectic solvents. Proc. Natl. Acad. Sci. USA **115**(47), E11000–E11001 (2018)
2. P. Domínguez de María, Ionic liquids: Definition, applications, and context for biotransformations and organocatalysis, in *Ionic Liquids in Biotransformations and Organocatalysis*, (Wiley, Hoboken, 2012), pp. 1–14
3. B.S. Jabes, L.D. Site, Nanoscale domains in ionic liquids: A statistical mechanics definition for molecular dynamics studies. ArXiv [Physics.Chem-Ph]. arXiv (2019), http://arxiv.org/abs/1903.05902
4. J.B. Shadrack, L.D. Site, Nanoscale domains in ionic liquids: A statistical mechanics definition for molecular dynamics studies. J. Chem. Phys. **149**(18), 184502 (2018)
5. Definition of Nanofluids, *In FlexPDE and Finite Element Method Applications in Thermal Energy Storage and Cavities* (IGI Global, London, 2023), pp. 52–66
6. M.I. Pryazhnikov, D.V. Guzei, A.V. Minakov, A.S. Lobasov, Development and testing of experimental methods: Definition of thermal conductivity of nanofluids. J. Siberian Fed. Univ. Eng. Technol. **8**(2), 153–165 (2015)
7. I. Ahamad, T. Fatma, Investigation of antifungal and antibiofilm activities of green-synthesized silver nanoparticles against Candida Glabrata. Biometals (2025). https://doi.org/10.1007/s10534-025-00680-y
8. K.R. Nagavenkatesh, P. Devendran, N. Nallamuthu, M. Murugesan, C. Sambathkumar, G. Ramadoss, Developing a high specific capacitance electrode design using lanthanum doped CuWO4 nanoparticles embedded on RGO sheets for cutting-edge energy storage devices. J Energy Storage **121**, 116542 (2025)
9. C. Saritha, V.K. Rajana, K. Choudhary, A. Vairagar, A. Mishra, S.M. Penumaka, S. Jain, et al., Highly selective ergosterol binding and impaired redox balance lead to improved antileishmanial efficacy for amphotericin B synthesized silver nanoparticles with reduced toxicity-in vitro and in vivo studies. Free Radic. Biol. Med. (2025). https://doi.org/10.1016/j.freeradbiomed.2025.03.033
10. X. Zheng, Y. Cui, J. Rong, S. Chen, X. Qu, X. Hu, GSH/PH-sensitive förster resonance energy transfer nanoparticles for synergistic chemotherapy and chemodynamic therapy. J. Mater. Chem. B Mater. Biol. Med. (2025). https://doi.org/10.1039/d5tb00243e
11. Z. He, P. Alexandridis, Nanoparticles in ionic liquids: Interactions and organization. Phys. Chem. Chem. Phys. **17**(28), 18238–18261 (2015)

12. N. Hassanshahi, G. Hu, J. Li, Application of ionic liquids for chemical demulsification: A review. Molecules **25**(21), 4915 (2020)
13. P.A. Hunt, I.R. Gould, B. Kirchner, The structure of imidazolium-based ionic liquids: Insights from ion-pair interactions. Aust. J. Chem. **60**(1), 9 (2007)
14. A. Sowińska, M. Maciejewska, A. Grajewska, Bis(trifluoromethylsulfonyl)imide ionic liquids were applied for fine-tuning the cure characteristics and performance of natural rubber composites. Int. J. Mol. Sci. **22**(7), 3678 (2021)
15. R. Hayes, G.G. Warr, R. Atkin, Structure and nanostructure in ionic liquids. Chem. Rev. **115**(13), 6357–6426 (2015)
16. H. Tokuda, K. Hayamizu, K. Ishii, M.A.B.H. Susan, M. Watanabe, Physicochemical properties and structures of room-temperature ionic liquids. 1. Variation of anionic species. J. Phys. Chem. B **108**(42), 16593–16600 (2004)
17. G.S. Sankaran, An overview of ring ring-closing metathesis reaction and its applications. Open Access J. Agric. Res. **3**(10), 203 (2018). https://doi.org/10.23880/oajar-16000203
18. Y. Ansari, K. Ueno, A. Angell, Protic ionic liquids can be both free proton conductors and benign superacids. ChemRxiv (2020). https://doi.org/10.26434/chemrxiv.13249814.v1
19. J. Bailey, E.L. Byrne, P. Goodrich, P. Kavanagh, M. Swadźba-Kwaśny, Protic ionic liquids for sustainable uses. Green Chem. **26**(3), 1092–1131 (2024)
20. S. Pradhan, P.K. Sahu, S. Priyadarshini, S. Puhan, Prospects and insights of protic ionic liquids: The new generation solvents used in fuel cells. Mater Today Proc **67**, 1046–1051 (2022)
21. D. Skarpalezos, A. Tzani, E. Avraam, C. Politidis, A. Kyritsis, A. Detsi, Synthesis, structure-properties relationship and biodegradability assessment of novel protic ionic liquids. J. Mol. Liq. **344**, 117754 (2021)
22. B. Robert, I. Robert, Magnetically triggered drug release from liposome embedded gel. J. Nanomed. Biotherap. Discov. **4**(3), 1000130 (2014). https://doi.org/10.4172/2155-983x.1000130
23. M. Afrand, R. Ranjbarzadeh, Hybrid nanofluids preparation method, in *Hybrid Nanofluids for Convection Heat Transfer*, (Elsevier, Amsterdam, 2020), pp. 49–99
24. J. Ji, W. Lu, C. Si, S. Zhang, X. Yao, W. Wang, D. Chu, Overview on the preparation and heat transfer enhancement of nanofluids. J. Phys. Conf. Ser. **1637**(1), 012003 (2020)
25. A. Arbak, A. Attar, M.A. Yapaoz, M. Armağan, Y. Bulbul, E.K. Demir, Y. Karagöz, Experimental investigation of bimetallic nanoparticles heat transfer characteristics in automotive radiators. Case Stud. Ther. Eng. **43**, 102763 (2023)
26. S. Koçak-Soylu, I. Atmaca, M. Asiltürk, A. Doğan, Improving heat transfer performance of an automobile radiator using cu and ag doped TiO2 based nanofluids. Appl. Therm. Eng. **157**(113743), 113743 (2019)
27. K. Yapici, O. Osturk, Y. Uludag, Dependency of nanofluid rheology on particle size and concentration of various metal oxide nanoparticles. Braz. J. Chem. Eng. **35**(2), 575–586 (2018)
28. R. Berkowitz, Making shear waves for ultrasonic imaging. Physics **10**, 57 (2017). https://doi.org/10.1103/physics.10.57
29. D.P. Darrow, Focused ultrasound for neuromodulation. Neurotherapeutics **16**(1), 88–99 (2019)
30. G. Pinton, J.F. Aubry, E. Bossy, M. Muller, M. Pernot, M. Tanter, Attenuation, scattering, and absorption of ultrasound in the skull bone. Med. Phys. **39**(1), 299–307 (2012)
31. C. Sierra, C. Acosta, C. Chen, S.Y. Wu, M.E. Karakatsani, M. Bernal, E.E. Konofagou, Lipid microbubbles as a vehicle for targeted drug delivery using focused ultrasound-induced blood-brain barrier opening. J. Cereb. Blood Flow Metab. **37**(4), 1236–1250 (2017)
32. N. Joudeh, D. Linke, Nanoparticle classification, physicochemical properties, characterization, and applications: A comprehensive review for biologists. J. Nanobiotechnol. **20**(1), 262 (2022)
33. V.E. Babicheva, Optical processes behind plasmonic applications. Nanomaterials **13**(7), 1270 (2023). https://doi.org/10.3390/nano13071270
34. I. Khan, K. Saeed, I. Khan, Nanoparticles: Properties, applications and toxicities. Arab. J. Chem. **12**(7), 908–931 (2019)

35. P. Nasr-Esfahani, A.A. Ensafi, Noble metals and nonnoble metal oxides based electrochemical sensors, in *Functionalized Nanomaterial-Based Electrochemical Sensors*, (Elsevier, Amsterdam, 2022), pp. 115–140
36. N. Patil, R. Bhaskar, V. Vyavhare, R. Dhadge, V. Khaire, Y. Patil, Overview on methods of synthesis of nanoparticles. Int. J. Curr. Pharm. Res. **13**, 11–16 (2021)
37. K.R. Jagdeo, Physical methods for synthesis of nanoparticles, in *Nanochemistry*, (CRC Press, Boca Raton, 2023), pp. 66–76
38. L.V. Srinivasan, S.S. Rana, A critical review of various synthesis methods of nanoparticles and their applications in biomedical, regenerative medicine, food packaging, and environment. Dis. Appl. Sci. **6**(7), 371 (2024). https://doi.org/10.1007/s42452-024-06040-8
39. G. Singh, N. Thakur, R. Kumar, Nanoparticles in drinking water: Assessing health risks and regulatory challenges. Sci. Total Environ. **949**, 174940 (2024)
40. J. Pereira, R. Souza, A. Moreira, A. Moita, An overview of the ionic liquids and their hybrids operating in electrochemical cells and capacitors. Ionics **30**(8), 4343–4385 (2024)
41. S.S. Mondal, D. Marquardt, C. Janiak, H.J. Holdt, Use of a 4,5-dicyanoimidazolate anion based ionic liquid for the synthesis of iron and silver nanoparticles. Dalton Trans. **45**(13), 5476–5483 (2016)
42. D. Van-Dao, T.T.D. Nguyen, H.Y. Song, J.K. Yang, T.W. Kim, Y.T. Yu, I.H. Lee, Ionic liquid-assisted preparation of ag-CeO2 nanocomposites and their improved photocatalytic activity. Mater. Des. **159**, 186–194 (2018)
43. H.A. Eyre, H. Lavretsky, M. Forbes, Convergence science arrives: How does it relate to psychiatry. Acad. Psychiatry **41**(1), 91–99 (2017)
44. NIH, Precision medicine initiative (2015), https://ghr.nlm.nih.gov/primer/precisionmedicine/initiative.
45. V.J. Dzao, C.A. Balatbat, *Precision Medicine: An Action Plan for California* (Elsevier, Amsterdam, 2018)., www.opr.ca.gov/
46. A.J. Greer, J. Jacquemin, C. Hardacre, Industrial applications of ionic liquids. Molecules **25**(21), 5207 (2020). https://doi.org/10.3390/molecules25215207
47. Z. He, P. Alexandridis, Ionic liquid and nanoparticle hybrid systems: Emerging applications. Adv. Colloid Interf. Sci. **244**, 54–70 (2017). https://doi.org/10.1016/j.cis.2016.08.004
48. M.K. Shukla, H. Tiwari, R. Verma, W.-L. Dong, S. Azizov, B. Kumar, S. Pandey, D. Kumar, Role and recent advancements of ionic liquids in drug delivery systems. Pharmaceutics **15**(2), 702 (2023). https://doi.org/10.3390/pharmaceutics15020702
49. J.M. Blair, M.A. Webber, A.J. Baylay, D.O. Ogbolu, L.J. Piddock, Molecular mechanisms of antibiotic resistance. Nat. Rev. Microbiol. **13**(1), 42–51 (2015)
50. M.C. Bubalo, K. Radošević, I.R. Redovniković, J. Halambek, V.G. Srček, A brief overview of the potential environmental hazards of ionic liquids. Ecotoxicol. Environ. Saf. **99**, 1–12 (2014)
51. S. Anvari, H. Hajfarajollah, B. Mokhtarani, M. Enayati, A. Sharifi, M. Mirzaei, Antibacterial and anti-adhesive properties of ionic liquids with various cationic and anionic heads toward pathogenic bacteria. J. Mol. Liq. **221**, 685–690 (2016)
52. V. Mikkola, S. Puupponen, H. Granbohm, K. Saari, T. Ala-Nissila, A. Seppälä, Influence of particle properties on convective heat transfer of nanofluids. Int J Therm Sci **124**, 187–195 (2018)
53. I. Razzaq, W. Xinhua, G. Rasool, T. Sun, A.S. Shflot, M.Y. Malik, A. Ali, Nanofluids for advanced applications: A comprehensive review on preparation methods, properties, and environmental impact. ACS Omega **10**(6), 5251–5282 (2025)
54. Z. He, P. Alexandridis, Micellization thermodynamics of Pluronic P123 (EO20PO70EO20) amphiphilic block copolymer in aqueous ethylammonium nitrate (EAN) solutions. Polymers **10**(1), 32 (2017)
55. E. Rastegari, Y.J. Hsiao, W.Y. Lai, et al., An update on mesoporous silica nanoparticle applications in nanomedicine. Pharmaceutics **13**(7), 1067 (2021)
56. K.H. Keskinbora, M.A. Jameel, Nanotechnology applications and approaches in medicine: A review. J. Nanosci. Nanotechnol. Res. **2**(2), 6 (2018)

57. C. Ren, Z. Wang, X. Zhang, et al., Construction of all-in-one peptide nanomedicine with photoacoustic imaging-guided mild hyperthermia for enhanced cancer chemotherapy. Chem. Eng. J. **405**, 127008 (2021)
58. A. Raina, M.I.U. Haq, A. Anand, J. Sudhanraj, Lubrication characteristics of oils containing nano additives: Influencing parameters, market scenario and advancements. J. Inst. Eng. India Ser. D **102**(2), 575–587 (2021)
59. W.K. Shafi, A. Raina, M.I.U. Haq, Friction and wear characteristics of vegetable oils usingnanoparticles for sustainable lubrication. Tribol. - Mater. Surf. Interfaces **12**(1), 27–43 (2018)
60. M.S. Amiri, V. Mohammadzadeh, M.E.T. Yazdi, M. Barani, A. Rahdar, G.Z. Kyzas, Plant-based gums and mucilages applications in pharmacology and nanomedicine: A review. Molecules **26**(6), 1770 (2021)
61. E. Mbunge, B. Muchemwa, J. Batani, Sensors and healthcare 5.0: Transformative shift in virtual care through emerging digital health technologies. Glob. Health J. **5**(4), 169–177 (2021)
62. M. Hofmann-Amtenbrink, H. Hofmann, A. Hool, F. Roubert, Nanotechnology in medicine: European research and its implications. Swiss Med. Wkly. **144**, 14044 (2014)
63. M. Bañobre-López, A. Teijeiro, J. Rivas, Magnetic nanoparticle-based hyperthermia for cancer treatment. Rep. Pract. Oncol. Radiother. **18**(6), 397–400 (2013)
64. W.H. De Jong, P.J.A. Borm, Drug delivery and nanoparticles: Applications and hazards. Int. J. Nanomedicine **3**(2), 133–149 (2008)
65. R.J. Vandebriel, W.H. De Jong, A review of mammalian toxicity of ZnO nanoparticles. Nanotechnol. Sci. Appl. **5**, 61–71 (2012)
66. Y. Wu, L. Wan, P. Li, Y. Zhang, M. Li, J. Gong, W. Zhang, Application of radiofrequency ablation for splenic preservation. J. Surg. Res. **193**(2), 781–787 (2015)
67. B.F. Gilmore, G.P. Andrews, G. Borberly, M.J. Earle, M.A. Gilea, S.P. Gorman, et al., Enhanced antimicrobial activities of 1-Alkyl-3-methyl imidazolium ionic liquids based on silver or copper containing anions. New J. Chem. **37**(4), 873–876 (2013)
68. K.N. Ibsen, H. Ma, A. Banerjee, E.E.L. Tanner, S. Nangia, S. Mitragotri, Mechanism of antibacterial activity of choline-based ionic liquids (CAGE). ACS Biomater Sci. Eng. **4**(7), 2370–2379 (2018)
69. B.L. Kuhn, B.F. Osmari, T.M. Heinen, H.G. Bonacorso, N. Zanatta, S.O. Nielsen, et al., Dicationic imidazolium-based dicarboxylate ionic liquids: Thermophysical properties and solubility. J. Mol. Liq. **308**, 112983 (2020)
70. J. Qin, J. Guo, Q. Xu, Z. Zheng, H. Mao, F. Yan, Synthesis of pyrrolidinium-type poly(ionic liquid) membranes for antibacterial applications. ACS Appl. Mater. Interfaces **9**(12), 10504–10511 (2017)
71. A.N. Vereshchagin, N.A. Frolov, K.S. Egorova, M.M. Seitkalieva, V.P. Ananikov, Quaternary ammonium compounds (QACs) and ionic liquids (ILs) as biocides: From simple antiseptics to tunable antimicrobials. Int. J. Mol. Sci. **22**(13), 6793 (2021)

Catalysis Applications of Ionic Liquid Nanoparticles

Ravula Rajasekhar, Ritul M. Chaudhari, S. N. Jain, and Kanchapogu Suresh

1 Introduction

The pursuit of efficient and sustainable catalytic systems is at the forefront of modern chemical research. Catalysis, the acceleration of chemical reactions by substances not consumed in the overall process, underpins a vast array of industrial processes, from the production of fuels and polymers to pharmaceuticals and fine chemicals [1]. Traditional catalytic systems often rely on volatile organic solvents, high reaction temperatures, and homogeneous or heterogeneous catalysts that can present challenges in terms of separation, recovery, and environmental impact. In this context, the emergence of ionic liquids (ILs) and their integration into nanoscale architectures, forming ionic liquid nanoparticles (ILNPs), offers a promising avenue for developing next-generation catalytic materials with enhanced performance and sustainability [2].

ILs are salts that are liquid at or below 100 °C. Unlike conventional molecular solvents, ILs are composed entirely of ions, typically a large, asymmetric organic cation and a smaller, inorganic or organic anion [2]. This unique ionic nature imparts them with a remarkable combination of properties, including negligible vapor pressure, high thermal and chemical stability, nonflammability, tunable polarity, and the ability to dissolve a wide range of organic, inorganic, and polymeric materials. The vast structural diversity achievable through various combinations of cations

R. Rajasekhar (✉) · R. M. Chaudhari · S. N. Jain
K.K. Wagh Institute of Engineering Education and Research, Nashik, India
e-mail: rajravula@kkwagh.edu.in

K. Suresh
Department of Chemical Engineering, Maulana Azad National Institute of Technology (MANIT), Bhopal, India

S. Kulkarni, S. Das (eds.), *Ionic Liquids, Nanofluids and Nanotechnology*,
Lecture Notes in Nanoscale Science and Technology 42,
https://doi.org/10.1007/978-3-032-04008-4_12

and anions allows for the design of task-specific ILs (TSILs) tailored for particular applications, including catalysis [3].

ILs have found widespread use as reaction media in various catalytic processes, often demonstrating enhanced reaction rates, selectivity, and catalyst stability. They can act as both solvents and co-catalysts, sometimes even participating directly in the catalytic cycle. The unique solvation environment provided by ILs can influence the activity and selectivity of homogeneous catalysts by affecting the stability of transition states and the solubility of reactants and products. Furthermore, the inherent nonvolatility of ILs mitigates the risk of solvent evaporation and allows for easier catalyst recovery and recycling in homogeneous catalytic systems [4].

While ILs themselves offer numerous advantages in catalysis, their immobilization onto solid supports or their organization into nanoscale structures has opened up new possibilities for further enhancing catalytic performance and facilitating catalyst separation and recovery. Ionic liquid nanoparticles (ILNPs) represent a fascinating class of materials where ILs are either encapsulated within nanoparticles, grafted onto their surface, or form the nanoparticle structure itself. These hybrid materials combine the unique properties of ILs with the high surface area, enhanced mass transfer, and potential for magnetic separation offered by nanoparticles [3].

Various strategies have been employed for the synthesis of ILNPs, including microemulsion techniques, miniemulsion polymerization, self-assembly, and surface modification of pre-formed nanoparticles. The resulting ILNPs can exhibit diverse morphologies, sizes, and compositions, allowing for the fine-tuning of their properties for specific catalytic applications. For instance, encapsulating a homogeneous catalyst within an ILNP can provide a pseudo-heterogeneous system that benefits from the high activity and selectivity of the homogeneous catalyst while enabling facile separation and recycling. Similarly, surface-functionalizing nanoparticles with catalytically active ILs can create highly efficient and robust heterogeneous catalysts [5].

The integration of ionic liquids into nanoscale architectures presents a compelling array of benefits for catalytic applications, effectively addressing some of the inherent limitations associated with both traditional homogeneous and heterogeneous catalysts, as well as bulk ionic liquid systems [6, 7]. The use of nanoparticles as supports for ILs and catalytic species leads to enhanced catalyst dispersion and stability. The high surface area provided by the nanoparticles prevents aggregation of the active components, thereby increasing the accessibility of catalytic sites. This improved dispersion also contributes to enhanced catalyst stability by mitigating leaching or decomposition of the active species within the ionic liquid environment [8].

Facile catalyst separation and recycling represent a significant advantage of ILNPs. Unlike homogeneous catalysts dissolved in bulk ILs, ILNPs can often be readily separated from the reaction mixture through straightforward physical methods such as centrifugation, filtration, or magnetic decantation when magnetic nanoparticles are employed. This ease of separation is crucial for developing sustainable catalytic processes, allowing for efficient recovery and reuse of the catalyst, which in turn reduces operational costs and minimizes waste generation.

Furthermore, the nanoscale dimensions of ILNPs can lead to improved mass transfer characteristics. The reduced diffusion distances for reactants to reach the active sites and for products to leave the catalytic environment can significantly enhance the overall reaction kinetics compared to bulk IL phases or larger supported catalyst particles. This improved mass transfer can translate to higher reaction rates and potentially allow for operation under milder conditions.

The ability to tune catalytic activity and selectivity is another key advantage offered by ILNPs. By carefully selecting the specific ionic liquid, the nature of the nanoparticle support, and the method of immobilization, the properties of the resulting ILNP catalyst can be precisely tailored for a particular chemical transformation. This includes fine-tuning parameters such as Lewis acidity or basicity, polarity of the microenvironment around the active site, and steric factors that can influence reaction pathways and product distribution.

Moreover, the combination of ionic liquids and nanoparticles can often result in synergistic effects, where the catalytic performance of the ILNP surpasses that of its individual components. For instance, the unique solvation properties of the ionic liquid can stabilize crucial reaction intermediates or favorably influence the electronic properties of supported metal nanoparticles, leading to enhanced activity or selectivity [9].

The high surface area and potentially enhanced intrinsic activity associated with nanoscale catalysts can often translate to the use of reduced catalyst loadings. This efficiency in catalyst utilization not only contributes to the economic viability of the process but also aligns with the principles of green chemistry by minimizing the amount of catalyst required to achieve the desired conversion [2].

1.1 Synthesis and Characterization of ILNPs

ILNPs are typically synthesized using a variety of methods, often involving the self-assembly of ionic liquids (ILs) in the presence of a stabilizing agent or a template [10]. One common approach is the microemulsion method, where ILs are dispersed as nanodroplets in a continuous phase, leading to the formation of stable nanoparticles. Another widely employed technique is the antisolvent precipitation method, which involves the rapid addition of a non-solvent to an IL solution, inducing nanoparticle nucleation and growth [11, 12]. Furthermore, ILs can be directly incorporated onto pre-formed nanomaterials, such as silica or metal oxides, through surface functionalization or impregnation techniques. These synthetic strategies allow for control over the size, morphology, and composition of the resulting ILNPs, which are crucial factors influencing their catalytic performance [13, 14].

1.2 Methods of ILNPs Synthesis

Several techniques have been developed to synthesize ILNPs, each offering specific advantages in terms of size control, morphology, and the nature of the resulting nanoparticles. Here is a more detailed look at some prominent methods.

Microemulsion Method

The microemulsion method is a widely used technique for the synthesis of nanoparticles, including ILNPs. Microemulsions are thermodynamically stable, isotropic dispersions of two immiscible liquids (typically oil and water) stabilized by a surfactant [15]. In the context of ILNP synthesis, the IL can be dispersed as nanodroplets within the microemulsion, acting as a confined reaction environment.

Mechanism: Typically, an IL is dissolved in one of the phases (e.g., the dispersed aqueous phase in a water-in-oil microemulsion). Upon mixing the microemulsion components, the IL molecules self-assemble within the nanodroplets. The size and shape of the resulting ILNPs are largely determined by the composition of the microemulsion (water-to-surfactant ratio, oil type). Subsequent steps might involve the addition of a co-surfactant or a change in conditions to induce further aggregation or stabilization of the nanoparticles. This method offers good control over particle size and can produce relatively uniform nanoparticles. It is also amenable to incorporating other functional materials within the ILNPs. The removal of the surfactant from the final product can be challenging. The choice of suitable microemulsion systems can also be specific to the IL being used.

Antisolvent Precipitation Method

The antisolvent precipitation method is a straightforward and often solvent-efficient approach for the synthesis of various nanoparticles, including ILNPs [16]. This technique relies on the principle of reducing the solubility of the ionic liquid by introducing a non-solvent, leading to supersaturation and subsequent nucleation and growth of nanoparticles.

Mechanism: Typically, the ionic liquid is first dissolved in a good solvent. This solution is then rapidly added to a miscible non-solvent (or a poor solvent for the IL) under controlled conditions, such as stirring rate and temperature. The sudden decrease in the IL's solubility in the mixed solvent system triggers the formation of a supersaturated state, resulting in the nucleation of IL molecules to form primary nanoparticles. These primary particles then grow through diffusion and aggregation, eventually forming a stable colloidal dispersion of ILNPs [17]. The size and morphology of the resulting nanoparticles can be influenced by factors, such as the concentration of the IL solution, the ratio of solvent to antisolvent, the rate of addition, and the presence of any stabilizing agents.

This method is often simple, cost-effective, and can be scaled up relatively easily. It can also be performed at room temperature and may avoid the use of surfactants in some cases, leading to potentially purer nanoparticles. Controlling the particle size and achieving narrow size distributions can sometimes be challenging. The choice of suitable solvent–antisolvent pairs is crucial and depends on the specific ionic liquid.

1.3 Structural and Physicochemical Properties

The structural and physicochemical properties of ILNPs are significantly influenced by the nature of the constituent ionic liquid, the synthesis method, and the presence of any supporting material [18, 19]. ILNPs often exhibit high surface areas and unique interfacial properties due to the amphiphilic nature of ILs, which can enhance their interaction with reactants and catalytic sites. The thermal stability and nonvolatility of many ILs are retained in their nanoparticle form, offering advantages over traditional organic solvents in catalytic applications. Moreover, the tunability of ILs through variations in their cation and anion combinations allows for the design of ILNPs with specific properties, such as tailored polarity and Lewis acidity/basicity, to optimize catalytic activity and selectivity [20, 21].

1.4 Characterization Techniques

Characterizing ionic liquid nanoparticles (ILNPs) is crucial for understanding their behavior and catalytic performance. Researchers employ several techniques to achieve this: transmission electron microscopy (TEM) and scanning electron microscopy (SEM) visualize their size, shape, and morphology. X-ray diffraction (XRD) reveals crystalline structure and phase purity. Brunauer–Emmett–Teller (BET) surface area analysis determines specific surface area and porosity. For chemical composition and surface functional groups, Fourier transform infrared spectroscopy (FTIR) and X-ray photoelectron spectroscopy (XPS) are utilized [22, 23]. Finally, thermogravimetric analysis (TGA) and differential scanning calorimetry (DSC) assess their thermal stability [24]. Collectively, these methods provide a comprehensive understanding of ILNP structure and properties, which is vital for correlating them with their catalytic activity.

2 Applications of ILNPs in Catalysis

2.1 *Homogeneous Vs. Heterogeneous Catalysis*

Ionic liquid nanoparticles (ILNPs) represent a fascinating class of catalytic materials that harness the unique properties of ionic liquids (ILs) at the nanoscale, often in conjunction with metal or metal oxide nanoparticles [25, 26]. This synergistic combination endows ILNPs with remarkable catalytic potential across a wide spectrum of chemical transformations, ranging from fundamental organic reactions to complex industrial processes [27]. Their ability to mediate reactions efficiently and selectively has garnered significant attention in the field of catalysis. To understand the diverse applications of ILNPs, it is crucial to first classify their catalytic action based on the physical state of the catalyst relative to the reactants, leading to the fundamental distinction between homogeneous and heterogeneous catalysis [28]. ILNPs exhibit versatility in operating under both regimes, often blurring the traditional boundaries and offering a unique set of advantages [2, 29].

Homogeneous catalysis: In homogeneous catalysis, the catalyst and the reactants exist in the same physical phase, typically a liquid solution [30]. Traditional homogeneous catalysts, such as soluble metal complexes or organocatalysts, are renowned for their high activity and selectivity [31]. This stems from the well-defined nature of their active sites and the unimpeded interaction between the catalyst and reactant molecules within the solution [32]. However, a significant drawback of homogeneous catalysis is the often-challenging separation of the catalyst from the reaction products, which can complicate downstream processing and limit catalyst reuse [33].

Heterogeneous catalysis: Conversely, heterogeneous catalysis involves a catalyst that is in a different phase from the reactants, most commonly a solid catalyst interacting with liquid or gaseous reactants [34]. Supported metal nanoparticles, metal oxides, and zeolites are common examples of heterogeneous catalysts [35]. Their primary advantage lies in the ease of separation from the reaction mixture, typically through simple filtration or decantation, facilitating catalyst recovery and reuse. However, heterogeneous catalysts can sometimes suffer from lower activity compared to their homogeneous counterparts due to mass transfer limitations (diffusion of reactants to the active sites on the solid surface) and the potential for a wider distribution of active sites with varying activity [36].

The role of ILNPs: Ionic liquid nanoparticles (ILNPs) provide a distinctive platform that bridges the gap between homogeneous and heterogeneous catalysis. Through precise design of their structure and optimization of reaction conditions, ILNPs can demonstrate the advantages of both catalytic systems, often resulting in improved performance and the emergence of new catalytic mechanisms [37]. The nanoscale ionic liquid phase plays a crucial role in enhancing the solubility and dispersion of catalytic particles, while also modifying the local reaction environment factors that significantly affect catalytic activity and ease of separation.

ILNPs in Homogeneous Catalysis

Soluble or dispersed ILNPs: In certain applications, ILNPs can be designed to be soluble or to form stable colloidal dispersions within the reaction medium. In such cases, the ILNPs effectively function as homogeneous catalysts, with the catalytic nanoparticles being finely dispersed and readily accessible to the reactants in solution. The ionic liquid shell surrounding the nanoparticles plays a crucial role in maintaining their stability and preventing aggregation in the chosen solvent [38].

Enhanced solubility and stability of homogeneous catalysts: The ionic liquid component of ILNPs can significantly enhance the solubility of catalytically active nanoparticles, particularly metal nanoparticles, in various organic solvents. This is often achieved through specific interactions between the ionic liquid's cation and anion and the surface of the nanoparticle, preventing their agglomeration and precipitation. Furthermore, the IL environment can provide a degree of stabilization to the nanoparticles under reaction conditions that might otherwise lead to their deactivation through sintering or leaching. For instance, ILNPs containing precious metal nanoparticles have shown remarkable stability in high-temperature reactions due to the protective effect of the surrounding ionic liquid layer.

ILNPs offer a highly tunable reaction environment by selecting ionic liquids with specific properties like polarity, viscosity, and hydrogen-bonding ability. This enables control over local reactant concentration and stabilization of transition states, enhancing catalytic activity and selectivity compared to traditional solvents. For example, certain ionic liquids can preferentially solvate one reactant, improving selectivity in competitive reactions. Regarding separation, ILNPs can be functionalized with thermo-responsive or magnetic components, enabling phase or magnetic separation. However, complete and efficient catalyst recovery remains challenging, making ILNP systems more complex to separate than conventional heterogeneous catalysts despite these advancements.

ILNPs in Heterogeneous Catalysis

Immobilized ILNPs on solid supports: A widely adopted strategy to harness the benefits of both ILNPs and heterogeneous catalysis involves immobilizing ILNPs onto various solid support materials. These supports can include high-surface-area materials like silica, alumina, carbon nanotubes, graphene oxide, and porous polymers. The immobilization prevents the ILNPs from dispersing in the reaction medium, allowing for easy separation of the catalyst by filtration or decantation after the reaction is complete, thus enabling catalyst reuse.

Enhanced dispersion and stability of nanoparticles: The ionic liquid layer surrounding the nanoparticles plays a crucial role in achieving uniform dispersion of the catalytic centers on the support surface and preventing their aggregation or sintering, even under harsh reaction conditions. The interaction between the ionic liquid and the support material also contributes to the overall stability of the catalyst. This enhanced dispersion leads to a higher number of accessible active sites and sustained catalytic activity over multiple reaction cycles, a significant advantage over traditional supported metal nanoparticles that can often suffer from nanoparticle agglomeration.

Tailored surface properties and reactivity: By carefully selecting the ionic liquid used to functionalize the nanoparticles, the surface properties of the resulting heterogeneous catalyst can be precisely tailored. The ionic liquid can influence the adsorption of reactants onto the catalytic sites, their activation, and the desorption of products. This tunability allows for the optimization of catalyst performance for specific reactions, often leading to higher activity and selectivity compared to conventional supported catalysts. For example, the ionic liquid can introduce specific interactions (e.g., hydrogen bonding, electrostatic interactions) that favor the adsorption and activation of certain reactants.

Biphasic catalysis with ILNPs: ILNPs can also be effectively employed as catalysts in biphasic reaction systems, where the reactants and products are distributed between two immiscible phases (e.g., an organic phase and an aqueous or ionic liquid phase). The ILNPs can be designed to reside preferentially in one of the phases, often the ionic liquid phase if one is used as a solvent. This approach can enhance mass transfer of reactants from one phase to the catalyst in the other and facilitate the separation of products into a different phase. The nanoscale nature of the ILNP catalyst can provide a high interfacial area for efficient reactions in such biphasic systems.

Bridging the Gap and Unique Advantages of ILNPs

ILNPs offer a unique opportunity to combine the desirable features of both homogeneous and heterogeneous catalysis. When dispersed, they can exhibit high activity and tunable selectivity reminiscent of homogeneous catalysts, while when immobilized, they offer the ease of separation and reusability characteristic of heterogeneous catalysts.

Enhanced stability and reduced leaching: The encapsulation of catalytic nanoparticles within the ionic liquid matrix often provides enhanced stability, protecting the nanoparticles from deactivation pathways such as aggregation and leaching of active metal species into the reaction medium, a common concern with both traditional homogeneous metal catalysts and some heterogeneous systems where metal nanoparticles can detach from the support.

Tunability at multiple levels: The design of ILNP catalysts allows for multi-level tunability. One can tailor the nature of the nanoparticle (metal, metal oxide, size, shape), and independently choose the ionic liquid based on its specific properties and functional groups to optimize the catalytic performance for a particular reaction. This dual tunability provides a powerful tool for catalyst design.

Nanoscale confinement effects: The nanoscale environment created by the ionic liquid surrounding the nanoparticles can sometimes lead to unique catalytic effects that are not observed with the bulk components. This confinement can influence reaction pathways, stabilize reactive intermediates, and enhance catalytic activity or selectivity in ways that are still being actively explored.

Ionic liquid nanoparticles represent a versatile class of catalysts that can function effectively in both homogeneous and heterogeneous catalytic regimes. Their unique structure allows them to bridge the traditional divide between these two approaches, often offering a synergistic combination of their advantages. The ability to tune the properties of both the nanoparticle and the ionic liquid component provides a powerful platform for designing highly efficient, selective, and potentially recyclable catalysts for a wide range of chemical transformations, contributing to the development of more sustainable and efficient catalytic processes. The choice between homogeneous-like or heterogeneous-like application of ILNPs depends on the specific reaction requirements and the desired balance between activity, selectivity, and ease of catalyst separation and reuse.

2.2 Organic Transformations

Ionic liquid nanoparticles (ILNPs) have emerged as a versatile and often highly effective class of catalysts for a wide array of organic transformations [39]. Their unique combination of nanoscale effects and the tunable properties of ionic liquids provides opportunities for enhanced reaction rates, improved selectivity, and potentially greener reaction protocols compared to traditional catalysts [2, 40]. This subsection will focus on the application of ILNPs in several fundamental organic transformations, including hydrogenation, oxidation, and coupling reactions, showcasing their catalytic prowess and the underlying principles contributing to their performance [41].

Organic transformations form the bedrock of chemical synthesis, enabling the construction of complex molecules from simpler building blocks. Catalysts play a pivotal role in these transformations by lowering activation energies and directing reaction pathways. ILNPs have demonstrated significant potential as catalysts in various key organic reactions, often exhibiting unique advantages due to the synergistic effects of the ionic liquid environment and the high surface area of the nanoparticles.

Hydrogenation Reactions

Hydrogenation, the addition of hydrogen across a multiple bond (e.g., C=C, C≡C, C=O, C=N), is a fundamental reaction in organic synthesis, crucial for the production of pharmaceuticals, fine chemicals, and bulk chemicals. Traditional hydrogenation often employs heterogeneous catalysts based on noble metals (Pd, Pt, Ru, Rh) supported on materials like carbon or alumina. ILNPs have shown promise as efficient catalysts for hydrogenation reactions, often exhibiting enhanced activity and selectivity [42].

Mechanisms and Advantages

1. Enhanced dispersion and stability of metal nanoparticles: The ionic liquid shell around the metal nanoparticles in ILNPs can prevent their aggregation and sintering under hydrogenation conditions, maintaining a high density of active sites. This is particularly beneficial for maintaining catalyst activity over extended periods and multiple reaction cycles [43].
2. Improved substrate solubility and mass transfer: The ionic liquid environment can enhance the solubility of organic substrates, particularly polar ones, in the vicinity of the catalytic metal nanoparticles, facilitating better contact and thus higher reaction rates. The ionic liquid can also influence the diffusion of hydrogen to the active sites [2].
3. Tunable selectivity: By carefully selecting the ionic liquid, the chemo-, regio-, and stereoselectivity of hydrogenation reactions can be influenced. The ionic liquid can interact with specific functional groups on the substrate, directing the hydrogen addition to the desired site or face of the molecule. For example, chiral ionic liquids incorporated into ILNPs have shown potential for enantioselective hydrogenation [44].

Examples

1. Hydrogenation of alkenes, alkynes, and nitro compounds using ILNPs containing Pd, Pt, or Ru nanoparticles. Studies have reported higher activity and selectivity compared to conventionally supported metal catalysts in some cases.
2. Selective hydrogenation of α, β-unsaturated aldehydes to unsaturated alcohols using ILNPs, where the ionic liquid can influence the adsorption mode of the substrate and promote the desired selectivity [45].
3. Hydrogenation reactions performed in ionic liquid solvents using ILNPs as catalysts, demonstrating the compatibility and potential synergy of combining ILNP catalysis with ionic liquid reaction media [46].

Oxidation Reactions

Importance: Oxidation reactions, involving the increase in the oxidation state of a molecule, are essential for the synthesis of various functional groups (e.g., alcohols, aldehydes, ketones, carboxylic acids, epoxides) and are crucial in many industrial processes. Traditional oxidation methods often involve stoichiometric oxidants or metal-based catalysts, which can have environmental drawbacks.

ILNPs in oxidation: ILNPs have emerged as effective catalysts for a range of oxidation reactions, often utilizing more environmentally benign oxidants like molecular oxygen or hydrogen peroxide.

Mechanisms and Advantages

Metal nanoparticles within ILNPs, such as gold, platinum, or manganese oxides, can effectively activate molecular oxygen for oxidation reactions. The ionic liquid environment can influence the adsorption and activation of oxygen on the nanoparticle surface [47]. The ionic liquid can help stabilize metal nanoparticles in high oxidation states, which are often the active species in oxidation catalysis, preventing their decomposition or reduction [48].

The ionic liquid can play a crucial role in controlling the chemoselectivity and regioselectivity of oxidation reactions. By influencing the interaction of the substrate with the active site, the desired oxidation product can be preferentially formed [49]. For example, selective oxidation of alcohols to aldehydes or ketones can be achieved with high selectivity using tailored ILNPs. Performing oxidation reactions with ILNPs in ionic liquid solvents can offer advantages such as enhanced solubility of oxygen and organic substrates, as well as stabilization of intermediates.

Examples

1. Selective oxidation of alcohols to aldehydes or ketones using ILNPs containing gold or supported metal oxides in ionic liquids or conventional solvents.
2. Epoxidation of alkenes using hydrogen peroxide as the oxidant, catalyzed by ILNPs containing transition metal complexes or metal oxides, often exhibiting high activity and selectivity.
3. Aerobic oxidation reactions catalyzed by ILNPs, offering a greener alternative to stoichiometric oxidants.
4. Oxidation of sulfides to sulfoxides or sulfones using ILNPs, with the ionic liquid influencing the degree of oxidation.

Coupling Reactions

Carbon-carbon and carbon-heteroatom bond formation reactions, known as coupling reactions (e.g., Suzuki–Miyaura, Heck, Sonogashira, Ullmann), are fundamental tools in organic synthesis for building molecular complexity. These reactions traditionally rely heavily on homogeneous catalysts based on palladium or other transition metals. ILNPs have shown significant potential as heterogeneous or quasi-homogeneous catalysts for various coupling reactions, offering the advantages of easier catalyst recovery and reuse while maintaining high activity and selectivity [50].

Mechanisms and Advantages

The ionic liquid shell can effectively stabilize the metal nanoparticles (e.g., Pd, Cu, Ni) that catalyze coupling reactions, preventing their aggregation and leaching, which are common deactivation pathways for traditional homogeneous catalysts. This allows for easier recovery and recycling of the catalyst [51]. The ionic liquid environment can enhance the solubility of organic substrates, including polar and non-polar ones, and can also play a role in the activation of reactants through specific interactions.

In some cases, ILNPs have demonstrated comparable or even superior catalytic activity to their homogeneous counterparts in coupling reactions, potentially due to the high surface area and the unique microenvironment provided by the ionic liquid at the nanoparticle surface. When ILNPs are supported on solid materials or designed for phase separation, the catalyst can be easily separated from the reaction mixture and reused for multiple cycles, making the process more sustainable and cost-effective [52].

Examples

1. Suzuki–Miyaura cross-coupling reactions catalyzed by Pd-containing ILNPs supported on various materials, showing good to excellent yields and ease of catalyst recovery.
2. Heck reactions using ILNPs as catalysts, demonstrating their effectiveness in forming carbon-carbon bonds between aryl halides and alkenes.
3. Sonogashira coupling reactions (formation of carbon-carbon triple bonds) catalyzed by Cu- or Pd-containing ILNPs.
4. Ullmann-type coupling reactions (formation of carbon-heteroatom or carbon-carbon bonds) using ILNPs containing copper nanoparticles.

Ionic liquid nanoparticles have proven to be highly versatile and effective catalysts for a wide range of crucial organic transformations, including hydrogenation, oxidation, and coupling reactions. Their unique structural features, combining the benefits of nanoscale materials and the tunable properties of ionic liquids, often lead to enhanced catalytic activity, improved selectivity, and the potential for easier catalyst recovery and reuse [51]. Ongoing research continues to explore and optimize ILNP catalysts for an even broader spectrum of organic reactions, paving the way for more sustainable and efficient chemical synthesis.

2.3 Biomass Conversion and Green Chemistry Applications

The extensive reliance on fossil fuels like coal, oil, and natural gas has propelled modern economic and societal advancement, improving living standards [53].

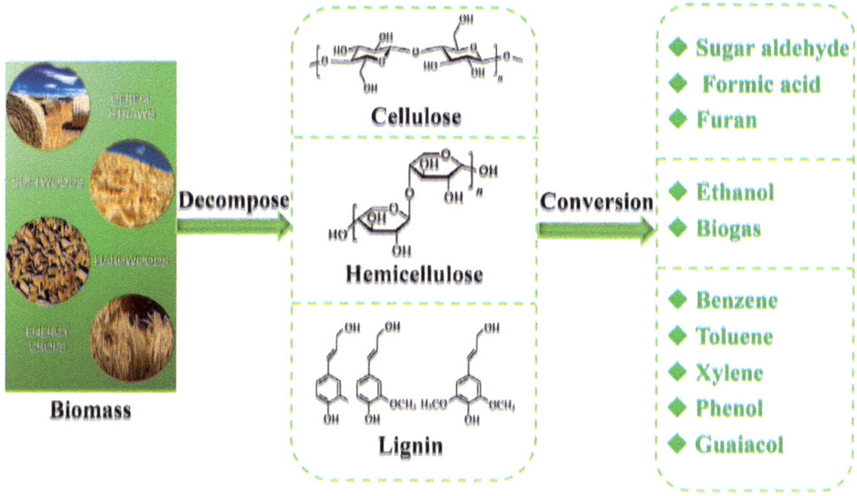

Fig. 1 The main components of biomass

However, increasing global population has led to a surge in energy demand, which finite fossil fuel reserves cannot sustainably meet [54, 55]. Centuries of fossil fuel use have severely harmed the environment, causing the greenhouse effect, global warming, and ozone depletion, hindering human sustainable development [56]. Geopolitical instabilities and uneven resource distribution have caused volatile energy prices, increasing living costs, and slowing global economic recovery [57].

Therefore, developing renewable resources to replace petrochemicals is critical. While wind, solar, and tidal energies are being explored, their intermittency due to weather conditions poses supply instability [58, 59]. In contrast, biomass stands out as a promising, carbon-containing, abundant, and widely distributed renewable resource with waste recycling potential [60].

Lignocellulose, the most abundant biomass, consists of cellulose (40–50 wt%), hemicellulose (15–30 wt%), and lignin (16–33 wt%) [61], as illustrated in Fig. 1. Cellulose breaks down into compounds like sugar aldehyde, formic acid, and furan [62]. Hemicellulose converts to ethanol and biogas [63]. Lignin, a complex network, produces aromatic compounds such as benzene, toluene, and xylene (BTX) [64].

Utilization Methods of Lignocellulosic Biomass

Lignocellulosic biomass is primarily utilized via three methods, as shown in Fig. 2:

(a) Direct combustion: Low efficiency and environmentally polluting [65, 66].
(b) Bio-enzymatic method: High production costs, long conversion times, and limited product diversity [67, 68].

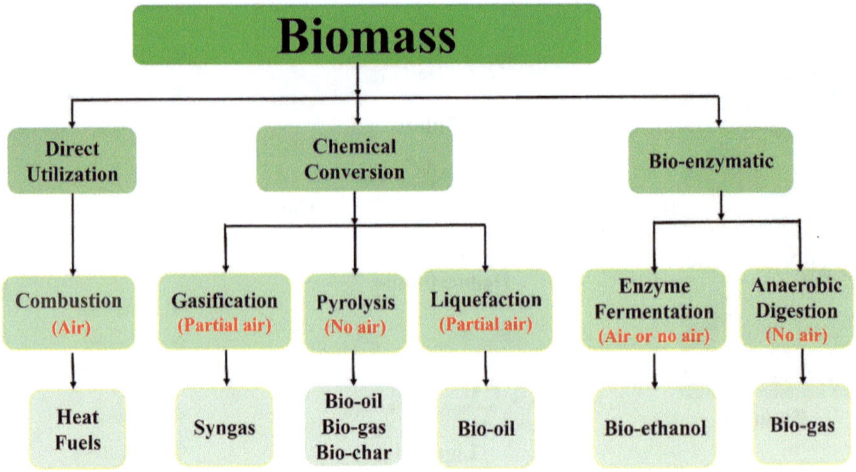

Fig. 2 The utilization pathways of biomass

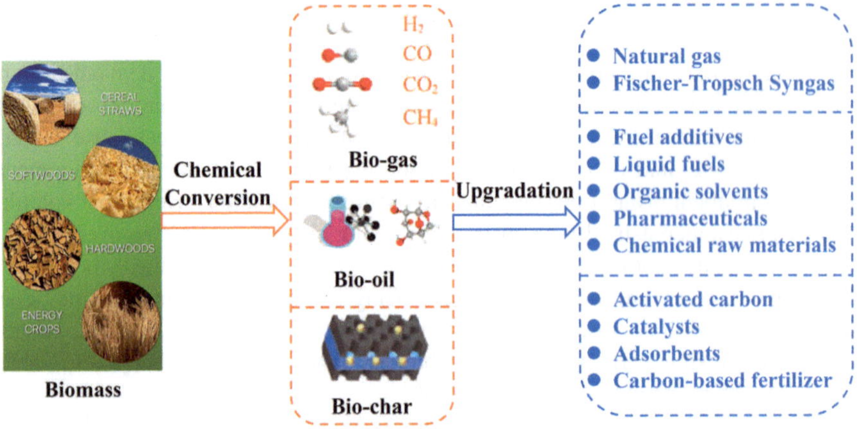

Fig. 3 The chemical conversion of biomass

(c) Chemical conversion: This method is preferred due to rapid production rates, diverse products, and high resource utilization [69–71]. It yields bio-char, bio-oil, and combustible gas. These intermediates can be further converted into high-value chemicals (e.g., 5-HMF, levulinic acid), biofuels (e.g., ethanol, hydrogen), and novel biomass-based materials (e.g., activated carbon, catalysts) [72, 73], as shown in Fig. 3.

Despite the potential of traditional chemical conversion of lignocellulose for green and sustainable biomass utilization, challenges such as insufficient conversion of desired products, difficulties in separation and purification, and poor selectivity significantly hinder its large-scale implementation [74]. The introduction of catalysts into the chemical conversion process offers an effective means to improve the selectivity, yield, and reaction rate of conversion products, while also reducing costs [75]. Therefore, this chapter provides a detailed review of the application of catalysts in the conversion of biomass and its derivatives, focusing on both the chemical conversion products of lignocellulosic biomass and the production of high-value-added chemicals. Furthermore, it offers an outlook on the challenges in future development, aiming to provide new insights for the efficient and green conversion of biomass into sustainable biomass energy and high-value-added chemical products.

Biomass Conversion and Green Chemistry Applications

Biomass, a renewable and abundant resource, holds immense potential as a feedstock for producing biofuels and valuable chemicals. However, the efficient and sustainable conversion of biomass, particularly its complex lignocellulosic components, often requires catalytic interventions. ILNPs, with their tunable properties and nanoscale characteristics, offer several advantages in this field, aligning well with the principles of green chemistry.

ILNPs in Biofuel Production

1. Catalysis of key conversion steps: ILNPs can act as catalysts in various crucial steps involved in biofuel production from different biomass sources (e.g., lignocellulosic biomass, vegetable oils, and algae). These steps include:
2. Esterification and transesterification for biodiesel production: ILNPs containing acidic or basic functional groups or metal nanoparticles can catalyze the conversion of fatty acids and triglycerides into biodiesel (fatty acid methyl esters or ethyl esters). The ionic liquid environment can enhance the solubility of reactants and promote the reaction. Supported ILNPs offer the advantage of catalyst recyclability in these processes [76].
3. Hydrodeoxygenation (HDO) for bio-oil upgrading: Bio-oil, produced from the pyrolysis of biomass, is a complex mixture requiring upgrading to improve its fuel properties. ILNPs containing metal nanoparticles (e.g., supported noble metals, transition metal phosphides) can catalyze HDO reactions, removing oxygen and improving the stability and energy density of the bio-oil. The ionic liquid can influence the reaction pathway and prevent catalyst deactivation [77].
4. Fischer–Tropsch synthesis: ILNPs supporting metal catalysts (e.g., Fe, Co) can be used in the Fischer–Tropsch process to convert syngas (CO and H_2), derived from biomass gasification, into liquid hydrocarbons for use as biofuels. The nanoscale environment and the ionic liquid can affect product selectivity.
5. Alcohol dehydration for biofuel production: ILNPs with acidic ionic liquid components or supported acid catalysts can catalyze the dehydration of bio-

alcohols (e.g., ethanol, butanol) derived from biomass fermentation to produce valuable biofuels like ethylene or butane [78].

Advantages of ILNPs

The high surface area of nanoparticles and the unique microenvironment provided by the ionic liquid can lead to enhanced reaction rates and conversions. The tunability of the ionic liquid can influence the reaction pathway and product distribution, leading to higher selectivity for desired biofuels. ILNPs can often catalyze reactions under milder conditions (lower temperatures and pressures) compared to traditional catalysts, reducing energy consumption and improving safety [79].

Supported ILNPs offer the potential for easy separation and reuse, aligning with green chemistry principles of catalyst recycling. IL can act as solvents for biomass and its derivatives, potentially enhancing the interaction between the catalyst and the feedstock.

ILNPs in Lignocellulose Degradation

The recalcitrant structure of lignocellulose (cellulose, hemicellulose, and lignin) poses a significant challenge for its efficient conversion. ILNPs can play a role in the pretreatment stage to disrupt this structure and enhance the accessibility of cellulose and hemicellulose to subsequent enzymatic or chemical hydrolysis. Ionic liquids themselves are known for their ability to dissolve cellulose, and ILNPs can potentially enhance this process or catalyze specific bond cleavages within the lignocellulosic matrix [80]. ILNPs with acidic or enzymatic functionalities can catalyze the hydrolysis of cellulose and hemicellulose into fermentable sugars (e.g., glucose, xylose).

Sulfonated ionic liquids or ILNPs supporting solid acid catalysts can effectively hydrolyze polysaccharides. The nanoscale dispersion can provide a high density of active sites. Immobilizing enzymes (cellulases, hemicellulases) onto nanoparticles stabilized by ionic liquids can enhance their stability and activity under certain reaction conditions. The ionic liquid environment can potentially improve enzyme solubility and substrate accessibility [81]. Lignin, the aromatic component of lignocellulose, is often underutilized. ILNPs can catalyze the depolymerization and oxidation of lignin into valuable aromatic chemicals. Metal nanoparticles supported on ILNPs can facilitate the cleavage of C-O and C-C bonds in lignin, yielding monomers and oligomers that can be used as platform chemicals.

Advantages of ILNPs

1. Enhanced reaction rates: Nanoscale catalysts can accelerate the degradation process.

2. Improved selectivity: Tailored ILNPs can potentially enhance the selective breakdown of different lignocellulosic components.
3. Catalyst stability in harsh conditions: The ionic liquid can provide a degree of stability to the catalytic nanoparticles under the often-harsh conditions required for lignocellulose degradation.
4. Integration with green solvents: ILNPs are inherently linked to ionic liquids, which are considered green solvents with low volatility and flammability.

Green Chemistry Aspects

1. Atom economy: ILNP-catalyzed reactions can often be designed to maximize atom economy, minimizing waste generation.
2. Less hazardous chemical syntheses: The use of ILs as part of the ILNP structure can reduce the need for volatile organic solvents, leading to safer and more environmentally friendly processes.
3. Safer solvents and auxiliaries: Ionic liquids are generally considered safer solvents compared to many traditional organic solvents.
4. Design for degradation: While the focus is on catalyst stability and reuse, research into biodegradable ionic liquids for ILNP synthesis aligns with the principle of designing for degradation.
5. Renewable feedstocks: ILNPs facilitate the utilization of renewable biomass resources, reducing dependence on fossil fuels.
6. Catalysis: ILNPs act as catalysts, reducing the energy input and often the amount of reagents required for chemical transformations.

Challenges and Future Directions

1. Scalability of ILNP synthesis: Developing cost-effective and scalable methods for ILNP production is crucial for industrial applications in biomass conversion.
2. Long-term stability under harsh biomass conversion conditions: Ensuring the stability and preventing the leaching of nanoparticles from ILNPs under the often-harsh conditions of biomass processing (high temperatures, acidic/basic environments) remains a challenge.
3. Understanding structure–activity relationships: Further research is needed to fully understand the intricate relationships between the ILNP structure (nanoparticle size, shape, composition, ionic liquid type, and functionalization) and their catalytic performance in biomass conversion reactions.
4. Integration with biorefinery concepts: Developing integrated biorefinery processes that effectively utilize ILNP catalysts for multiple stages of biomass conversion (pretreatment, hydrolysis, upgrading) is a key future direction.
5. Ionic liquid nanoparticles hold significant promise as green catalysts for various biomass conversion processes, including biofuel production and lignocellulose degradation. Their unique properties can enhance reaction efficiency, improve

selectivity, and promote the use of renewable resources under milder and more environmentally benign conditions. Continued research and development in the synthesis, characterization, and application of ILNPs will be crucial for realizing their full potential in advancing sustainable green chemistry practices in the field of biomass valorization.

2.4 *Environmental Applications*

Environmental pollution and the urgent need for sustainable remediation technologies have spurred the exploration of advanced materials for environmental applications. Ionic liquid nanoparticles (ILNPs), combining the unique properties of ionic liquids (ILs) with the enhanced surface area and reactivity of nanoparticles, have emerged as promising candidates for addressing critical environmental challenges such as CO_2 capture, wastewater treatment, and pollutant degradation. Their tunability and potential for synergistic effects make them attractive for developing efficient and environmentally benign solutions.

CO_2 Capture

The increasing concentration of CO_2 in the atmosphere, primarily resulting from fossil fuel combustion, is a leading cause of climate change, making the development of efficient and cost-effective CO_2 capture technologies critically important. Ionic liquids (ILs) have shown strong potential for CO_2 absorption, and incorporating nanoparticles into ILs to form ionic liquid nanoparticle (ILNP) systems can significantly enhance this capability. Nanoparticles dispersed within ILs increase the interfacial area between the gas and liquid phases, thereby improving CO_2 absorption rates. Additionally, the presence of nanoparticles can alter the viscosity and diffusion properties of ILs, facilitating better mass transfer. ILNPs can also be functionalized with specific chemical groups, such as amines, which show a strong affinity for CO_2, allowing for more selective capture from gas mixtures like flue gas. To further enhance practicality, ILNPs can be immobilized on porous solid supports such as silica, creating structured sorbents that simplify separation and regeneration processes. Studies have demonstrated that such supported and functionalized ILNP systems exhibit superior CO_2 absorption capacity and selectivity compared to bulk ILs, making them promising candidates for advanced carbon capture technologies.

Wastewater Treatment

Industrial and domestic wastewater often contains a complex mixture of pollutants, such as organic dyes, heavy metals, and pharmaceutical residues, which pose significant threats to both environmental and human health. Ionic liquid nanoparticle hybrids (ILNPs) have emerged as promising materials for wastewater treatment due to their unique catalytic and adsorptive properties. In catalytic degradation, ILNPs embedded with metal nanoparticles like Fe, Au, or Pt can accelerate the breakdown of organic pollutants through advanced oxidation processes (AOPs), utilizing oxidants, such as hydrogen peroxide or persulfate. The ionic liquid medium can stabilize the nanoparticles and influence the degradation mechanism. For adsorption applications, ILNPs with functionalized surfaces, such as chelating groups, exhibit high affinity for heavy metals and organic contaminants, with the ionic liquid enhancing nanoparticle dispersibility and pollutant interaction. Additionally, incorporating ILNPs into membrane filtration systems can improve membrane antifouling characteristics, permeability, and selectivity, as the ionic liquid component modifies surface properties and reduces organic fouling. ILNPs also show potential for catalytic reduction of inorganic pollutants like nitrates, converting them into benign products such as nitrogen gas. Notably, ILNPs supported on magnetic nanoparticles have enabled the efficient degradation of organic dyes with easy magnetic separation and recovery, while functionalization with carboxylate or amine groups has led to effective removal of heavy metal ions, demonstrating the versatility and efficiency of ILNPs in wastewater treatment applications.

Pollutant Degradation

Persistent organic pollutants (POPs), pharmaceuticals, and other emerging contaminants present a significant environmental threat due to their high toxicity and resistance to natural degradation processes. To address this challenge, ionic liquid nanoparticles (ILNPs) have emerged as promising materials capable of catalyzing the breakdown of such pollutants through multiple mechanisms. One approach involves photocatalysis, where ILNPs incorporating semiconductor nanoparticles like TiO_2 or ZnO utilize light energy to generate reactive species that degrade organic pollutants. The ionic liquid component of these systems can influence the dispersion and surface properties of the photocatalyst, thereby enhancing its efficiency. ILNPs also support enzymatic degradation by serving as immobilization platforms for enzymes, potentially improving their stability and activity even in non-aqueous or harsh environments. Additionally, ILNPs containing metal nanoparticles can facilitate catalytic oxidation or reduction reactions, detoxifying various organic and inorganic pollutants. The integration of multiple functionalities within a single ILNP system can yield synergistic effects, further boosting pollutant degradation performance. For instance, TiO_2 nanoparticles stabilized by ionic liquids have demonstrated enhanced photocatalytic activity in breaking down organic dyes and pharmaceuticals, while enzyme-functionalized ILNPs have proven effective in degrading pesticides in soil and water. The application of ILNPs in environmental remediation aligns with the principles of green chemistry. These include the use of catalysis to lower reagent and energy consumption, the incorporation of safer solvents by replacing volatile organics with ionic liquids, and the potential for designing ILNPs with biodegradable components. Moreover, ILNPs can contribute to waste prevention through their reusability and efficiency in pollutant degradation processes.

Challenges and Future Directions

While ionic liquid nanoparticles (ILNPs) show immense promise for environmental remediation, several critical factors must be addressed to ensure their safe and effective implementation. A thorough assessment of the long-term environmental fate and ecotoxicity of ILNPs is essential before they are widely applied, as their persistence or toxicity in ecosystems could pose unintended risks. Moreover, the development of scalable and cost-effective synthesis methods is crucial for transitioning ILNPs from laboratory research to practical, real-world applications. Evaluating the performance of ILNPs in complex environmental matrices, such as real wastewater or contaminated soil, is necessary to determine their actual efficacy under realistic conditions. Additionally, exploring how ILNP-based catalytic and adsorptive processes can be integrated with existing remediation technologies may lead to more robust, hybrid treatment systems. ILNPs represent a highly tunable and catalytically active platform that aligns with green chemistry principles, making them suitable for addressing a variety of environmental challenges, including CO_2 capture, wastewater treatment, and the degradation of persistent pollutants. Their compatibility with sustainable solvents and multifunctional design enhances their appeal for next-generation remediation technologies. However, continued interdisciplinary research is needed to ensure their environmental safety, optimize their large-scale deployment, and confirm their effectiveness in real-world scenarios, steps that are vital for unlocking their full potential in building a more sustainable and cleaner future.

2.5 *Stability and Recyclability*

Ionic liquid nanoparticles (ILNPs) have emerged as promising catalytic materials, combining the unique properties of ionic liquids (ILs) with the high surface area and tunable characteristics of nanoparticles. While they offer advantages in terms of activity and selectivity, significant challenges remain regarding their long-term stability and practical reusability, hindering their widespread industrial adoption. This section delves into the key factors affecting the stability and recyclability of ILNPs in catalytic applications.

Catalyst Deactivation Mechanisms

The catalytic efficiency of ionic liquid nanoparticles (ILNPs) can decline over time due to several deactivation mechanisms. A key factor is the leaching of the ionic liquid, which stabilizes metal nanoparticles and supports catalysis. In polar solvents or at high temperatures, ionic liquids may dissolve into the reaction medium, reducing active sites. Leaching depends on solvent polarity, temperature, interactions with reactants/products, support properties, and ionic liquid chemistry. Nanoparticle aggregation and sintering also lower activity by decreasing surface area; while ionic liquids can stabilize particles, they may sometimes promote aggregation. Additionally, ionic liquid decomposition under heat or reactive conditions can produce by-products that poison catalysts. Fouling and poisoning from adsorbed impurities or compounds like sulfur further deactivate ILNPs. Regeneration and reuse are critical but challenging due to nanoscale size; magnetic separation and supported ionic liquid phases help but have limitations. Enhancing ILNP durability requires stable supports, strong immobilization, and designing ionic liquids with low solubility, alongside optimized reaction conditions to ensure sustainable catalyst performance.

Scalability and Industrial Feasibility

The transition of ionic liquid nanoparticle (ILNP) catalysts from laboratory research to industrial application faces significant challenges related to scalability and cost-effectiveness. While ILNPs exhibit excellent catalytic performance in controlled environments, scaling up requires overcoming economic and technical barriers. Economically, the high cost of ionic liquids due to complex syntheses and expensive reagents, along with the intricate nanoparticle synthesis and functionalization processes, increases production expenses. To address this, researchers are investigating greener, more affordable synthetic routes using bio-based precursors, simplified methods, and continuous-flow systems to enhance scalability and reduce waste. Technically, large-scale production demands strict control over reaction conditions to maintain consistent particle size, morphology, and uniform ionic liquid coverage, which are vital for catalytic activity. Safety concerns also arise from handling dry nanoparticles, which tend to agglomerate and pose health risks; thus, developing stable dispersions or supported catalysts is essential. Industrial integration may require reactor design adaptations to optimize catalyst loading, contact efficiency, and separation, with continuous-flow reactors offering promising scale-up potential. Ensuring long-term stability and reusability under harsh industrial conditions, such as high temperatures and pressures, remains a critical research goal to realize the practical and economic feasibility of ILNP catalysts in real-world applications.

Toxicity and Environmental Concerns

Despite being promoted as "green" solvents due to their low volatility, ionic liquid nanoparticle (ILNP) systems raise significant concerns regarding environmental safety and toxicity. As ILNPs move toward industrial applications, understanding their ecological impact is essential for sustainable development.

A major issue is the biodegradability of ionic liquids, which varies with molecular structure. Some degrade easily, while others persist, creating long-term environmental risks. The fate of nanoparticles adds complexity, as they may aggregate, interact with natural matter, or bind pollutants, influencing their behavior and bioavailability in ecosystems.

The toxicological effects of ILNPs also require attention. Interactions with biological systems depend on properties such as particle size, surface charge, and ionic liquid composition, potentially leading to harmful effects on microbes, plants, animals, and humans. Synergistic toxicity from the combined components further complicates risk assessments.

Ecotoxicity data remain limited, but existing studies suggest possible harm to aquatic and terrestrial organisms, with concerns about oxidative stress and membrane disruption. Bioaccumulation potential is another emerging risk, with possible long-term consequences if ILNPs enter food chains.

Life cycle assessments (LCA) are crucial for evaluating the overall environmental impact of ILNP catalysts, from raw material sourcing to disposal. Meanwhile, research is progressing on greener alternatives, such as bio-based ionic liquids and low-toxicity nanoparticles.

3 Future Perspectives and Research Directions

The field of ionic liquid nanoparticle (ILNP)-based catalysis is rapidly advancing, offering exciting opportunities for innovation at the intersection of green chemistry, nanotechnology, and materials science. This section explores emerging trends, the integration of ILNPs with advanced materials, and their potential in sustainable industrial applications. Addressing current challenges and leveraging novel design strategies can unlock greater catalytic performance and broaden the applicability of ILNPs in real-world processes.

3.1 Emerging Trends in ILNP-Based Catalysis

Development of Multifunctional ILNPs

Future ILNP catalysts are being designed with multifunctionality at their core, enabling them to perform several roles in a single system enhancing efficiency, reducing steps, and increasing selectivity. Moving beyond single-purpose catalysts, multifunctional ILNPs combine two or more distinct capabilities, catalysis, separation, and sensing within a nanoscale system, either synergistically or independently.

Examples

- Catalysis and separation integration: ILNPs incorporating magnetic nanoparticles for magnetic separation post-reaction or product-selective surfaces for simplified recovery.
- Tandem catalysis: ILNPs with spatially separated active sites enabling consecutive reaction steps in a one-pot system, often supported by different metal nanoparticles or functional IL groups.
- Stimuli-responsive ILNPs: Catalytic behavior triggered or modulated by external stimuli such as temperature, pH, light, or electric fields. Thermo-responsive ILs, for instance, adjust solvation to control substrate accessibility.
- Enhanced mass transfer: ILNPs designed with hierarchical porosity or tailored surfaces to promote faster diffusion and efficient catalytic turnover.

Key Research Directions

- Precision synthesis of multifunctional ILNPs, including site-selective functionalization and tailored IL design.
- Understanding inter-functional synergy and optimizing multi-component interactions.
- Advanced characterization (e.g., TEM, XPS, FTIR) to elucidate structure–function relationships.
- Computational modeling to predict behavior and guide rational ILNP design.

3.2 Integration with Nanotechnology and Advanced Materials

Hybrid Nanostructures and Composite ILNPs

Merging ILNPs with other advanced materials such as Metal-Organic Frameworks (MOFs), carbon-based materials, or polymers allows for the creation of hybrid catalysts with improved performance and expanded functionality. Strategic assembly of ILNPs with support matrices or other nanostructures enhances stability, dispersion, and catalytic activity while adding new functionalities.

Examples

- Supported ILNPs (e.g., on MOFs, graphene oxide): Prevent aggregation, improve dispersion, and enhance thermal/mechanical stability. MOFs provide confinement effects, improving selectivity.
- Core-Shell ILNPs: Metal/metal oxide cores coated with IL shells allow decoupling of catalytic and solvation roles, with tunable IL thickness for fine control.

- ILNPs in polymer matrices: Embedding ILNPs into membranes or hydrogels yields reusable and stable catalysts with controllable permeability and separation characteristics.
- Plasmonic hybrid systems: Incorporating plasmonic nanoparticles (e.g., Au, Ag) creates photocatalysts with enhanced light absorption, leveraging ionic liquids to stabilize reactive interfaces.

Key Research Directions

- Scalable synthesis of well-defined composite ILNPs with controlled morphology and interfacial architecture.
- Probing nanoscale interactions and synergies between ILNPs and hybrid components.
- In situ/operando studies to connect structure to catalytic function.
- Testing in real reaction systems for industrial relevance and reusability.

3.3 *Potential for Sustainable Industrial Applications*

Energy Storage and Conversion

ILNPs show strong promise beyond traditional catalysis, particularly in electrochemical and photochemical energy systems due to their ionic conductivity, electrochemical stability, and versatile surface chemistry. ILNPs can be engineered to improve reaction kinetics, mass/charge transport, and catalyst lifetime in sustainable energy systems.

Examples

- Electrocatalysis (fuel cells, electrolyzers): ILNPs enhance oxygen reduction, hydrogen evolution, and methanol oxidation reactions by providing stable environments and facilitating ion transport.
- Photocatalysis for solar fuel generation: ILNPs can efficiently drive water splitting or CO_2 photoreduction. ILs assist in charge separation and solvation of reactants.
- Biomass conversion: ILNPs support dissolution and catalytic conversion of lignocellulosic biomass to fuels/chemicals, leveraging the dual roles of ILs in solvation and nanoparticle stabilization.
- CO_2 capture and utilization: ILNPs designed for CO_2 solubility and activation catalyze its conversion to value-added chemicals or fuels, contributing to carbon-neutral technologies.

Key Research Directions

- Designing ILNPs with optimized electrochemical, optical, or catalytic properties tailored to energy applications.
- Mechanistic studies of ILNP-mediated electrocatalytic and photocatalytic reactions.
- Developing scalable production methods and integrating ILNPs into real energy devices.
- Long-term performance evaluation under operational conditions to ensure stability and recyclability.

ILNP-based catalysts represent a dynamic and promising class of materials poised to revolutionize green chemistry and sustainable industrial processes. With continued advancements in multifunctionality, material integration, and eco-responsibility, ILNPs are likely to play a pivotal role in the transition to cleaner, more efficient catalytic technologies. Ongoing research should prioritize interdisciplinary approaches combining synthetic chemistry, materials science, computational modeling, and environmental assessment to fully realize their potential.

4 Conclusion

4.1 Summary and Perspectives

The field of ionic liquid nanoparticle (ILNP)-based catalysis has emerged as a transformative area of research, offering a wide array of advantages and versatile applications. Throughout this chapter, we have examined the fundamental principles, structural design, and various catalytic roles of ILNPs. The key benefits identified include:

- Enhanced catalytic performance: Due to the unique combination of ionic liquids (ILs) and nanoparticles (NPs), ILNPs exhibit high surface area, tunable physicochemical properties, and superior catalytic activity.
- Stabilization and reusability: ILs stabilize NPs and prevent agglomeration, leading to increased catalyst lifespan and recyclability.
- Green chemistry compatibility: The inherent properties of ILs—such as low volatility, thermal stability, and design flexibility—make ILNPs highly suitable for sustainable catalytic processes.
- Wide applicability: ILNPs have demonstrated efficacy in diverse fields, including organic transformations, electrocatalysis, photocatalysis, biomass conversion, and CO_2 utilization.
- Integration with advanced materials: Synergistic hybrid systems and composite materials have extended the performance and functionality of ILNPs for industrially relevant reactions.

These findings underscore the importance of ILNPs in advancing both academic and industrial catalysis, pushing the boundaries of reaction efficiency, selectivity, and environmental compatibility.

4.2 *Final Remarks on the Future of ILNPs in Catalysis*

Looking forward, ILNP-based catalysts hold tremendous promise for revolutionizing catalytic science and its application in sustainable industrial processes. Several future trajectories and considerations can be highlighted:

- Commercialization potential: With scalable synthesis methods and ongoing advancements in material stability and functionality, ILNPs are increasingly poised for industrial adoption. Their use in energy devices, chemical manufacturing, and environmental remediation presents attractive opportunities for commercialization.
- Adoption in green chemistry: The alignment of ILNP technology with the principles of green chemistry—such as waste reduction, energy efficiency, and the use of benign solvents—positions it as a frontrunner in the transition to environmentally responsible processes.
- Policy and regulatory support: As global focus intensifies on sustainability and climate goals, ILNPs may benefit from regulatory frameworks promoting green technologies and carbon reduction initiatives.
- Collaborative and interdisciplinary research: Realizing the full potential of ILNPs will require coordinated efforts across chemistry, materials science, chemical engineering, and environmental science, driving innovation in both fundamental and applied domains.

References

1. G. Ertl, Elementary steps in heterogeneous catalysis. Angew. Chemie Int. Ed. **29**(11), 1219–1227 (1990). https://doi.org/10.1002/ANIE.199012191
2. H.P. Steinrück, P. Wasserscheid, Ionic liquids in catalysis. Catal. Lett. **145**(1), 380–397 (2015). https://doi.org/10.1007/S10562-014-1435-X/FIGURES/9
3. X. Wang, Y. Chi, T. Mu, A review on the transport properties of ionic liquids. J. Mol. Liq. **193**, 262–266 (2014). https://doi.org/10.1016/J.MOLLIQ.2014.03.011
4. R.A. Sheldon, Green solvents for sustainable organic synthesis: State of the art. Green Chem. **7**(5), 267–278 (2005). https://doi.org/10.1039/B418069K
5. Z. He, P. Alexandridis, Ionic liquid and nanoparticle hybrid systems: Emerging applications. Adv. Colloid Interf. Sci. **244**, 54–70 (2017). https://doi.org/10.1016/J.CIS.2016.08.004
6. M. Hejazifar, O. Lanaridi, K. Bica-Schröder, Ionic liquid based microemulsions: A review. J. Mol. Liq. **303**, 112264 (2020). https://doi.org/10.1016/J.MOLLIQ.2019.112264
7. J. Zhang, Ionic liquid-based microemulsions, in *Ionic Liquid-Based Surfactant Science: Formulation, Characterization, and Applications*, (Elsevier, Amsterdam, 2015), pp. 325–341. https://doi.org/10.1002/9781118854501

8. Q. Zhang, S. Zhang, Y. Deng, Recent advances in ionic liquid catalysis. Green Chem. **13**(10), 2619–2637 (2011). https://doi.org/10.1039/C1GC15334J
9. O. Soleimani, Properties and applications of ionic liquids. J. Chem. Rev. **2**(3), 169–181 (2020). https://doi.org/10.33945/SAMI/JCR.2020.3.4
10. I. Jhanji, The role of ionic liquids in green solvent chemistry: Properties and applications. Int. Res. J. Eng. Appl. Sci. **11**(4), 57–67 (2023). https://doi.org/10.55083/IRJEAS.2023.V11I04009
11. G. Wang, X. Chang, J. Peng, K. Liu, K. Zhao, Y. Chunmeng, Y. Fang, Towards a new FRET system via combination of pyrene and perylene bisimide: Synthesis, self-assembly and fluorescence behavior. Liq. Interact. Organ. J. Phys. Chem. Chem. Phys. (2015). https://doi.org/10.1039/c0xx00000x
12. M. Zhen, J. Yu, S. Dai, Preparation of inorganic materials using ionic liquids. Adv. Mater. **22**(2), 261–285 (2010). https://doi.org/10.1002/ADMA.200900603
13. G. Rothenberg, *Catalysis: Concepts and Green Applications* (Wiley, Hoboken, 2017), p. 303
14. R.A. Sheldon, Green and sustainable manufacture of chemicals from biomass: State of the art. Green Chem. **16**(3), 950–963 (2014). https://doi.org/10.1039/C3GC41935E
15. L.L. Schramm, Emulsions, foams, and suspensions: Fundamentals and applications. Ind. Eng. Chem. **62**(1), 117–154 (2005)
16. A. Sainz Martinez, *Novel Strategies for Extraction and Synthesis via Combination of Supercritical Carbon Dioxide and Ionic Liquids* (Universität Wien, Wien, 2023). https://doi.org/10.34726/HSS.2023.113824
17. B. Richard, J.L. Lemyre, A.M. Ritcey, Nanoparticle size control in microemulsion synthesis. Langmuir **33**(19), 4748–4757 (2017). https://doi.org/10.1021/ACS.LANGMUIR.7B00773
18. A.J. Greer, J. Jacquemin, C. Hardacre, Industrial applications of ionic liquids. Molecules **25**(21), 5207 (2020). https://doi.org/10.3390/MOLECULES25215207
19. N.V. Plechkova, K.R. Seddon, Applications of ionic liquids in the chemical industry. Chem. Soc. Rev. **37**(1), 123–150 (2007). https://doi.org/10.1039/B006677J
20. M. Cvjetko Bubalo, K. Radošević, I. Radojčić Redovniković, J. Halambek, V. Gaurina Srček, A brief overview of the potential environmental hazards of ionic liquids. Ecotoxicol. Environ. Saf. **99**, 1–12 (2014). https://doi.org/10.1016/J.ECOENV.2013.10.019
21. V. Strehmel, Introduction to ionic liquids. Chem. Rev. **117**(10), 6633–6635 (2016). https://doi.org/10.1007/978-3-319-32489-0_1
22. M. Eddaoudi, Characterization of porous solids and powders: Surface area, pore size and density by S. Lowell (Quantachrome instruments, Boynton Beach), J. E. Shields (C. W. Post campus of Long Island university), M. A. Thomas, and M. Thommes (Quantachrome instruments). Kluwer academic publishers: Dordrecht, The Netherlands. 2004. Xiv + 348 pp. $159.00. ISBN 1-4020-2302-2. J. Am. Chem. Soc. **127**(40), 14117–14117 (2005). https://doi.org/10.1021/JA041016I
23. S. Lowell, J.E. Shields, M.A. Thomas, M. Thommes, *Characterization of Porous Solids and Powders: Surface Area, Pore Size and Density*, vol 16 (Springer, Cham, 2004). https://doi.org/10.1007/978-1-4020-2303-3
24. P.J. Haines et al., *Thermal Methods of Analysis – Principles, Applications and Problems* (Blackie Academic & Professional, London, 1995)
25. P. Wasserscheid, T. Welton, Ionic liquids in synthesis: Second edition. Ion. Liq. Synth. **1**, 1–721 (2008). https://doi.org/10.1002/9783527621194
26. G. Keglevich, N. Zsuzsa Kiss, Z. Radai, An overview of the applications of ionic liquids as catalysts and additives in organic chemical reactions. Curr. Org. Chem. **22**, 533–556 (2017). https://doi.org/10.2174/1385272822666171227152013
27. V.I. Pârvulescu, C. Hardacre, Catalysis in ionic liquids. Chem. Rev. **107**(6), 2615–2665 (2007). https://doi.org/10.1021/CR050948H
28. H.G.O. Alvim et al., Ionic liquid effect over the Biginelli reaction under homogeneous and heterogeneous catalysis. ACS Catal. **3**(7), 1420–1430 (2013). https://doi.org/10.1021/CS400291T

29. L. Graser, D. Betz, M. Cokoja, F.E. Kuhn, Ionic liquids as solvents for ionic transition-metal catalysts. Curr. Inorg. Chem. **1**(2), 166–181 (2011). https://doi.org/10.2174/1877944111101020166
30. B. Driessen-Hölscher, Multiphase homogeneous catalysis. Adv. Catal. **42**(C), 473–505 (1998). https://doi.org/10.1016/S0360-0564(08)60632-X
31. F. Huang, E.V. Anslyn, Introduction: Supramolecular chemistry. Chem. Rev. **115**(15), 6999 (2015). https://doi.org/10.1021/ACS.CHEMREV.5B00352
32. K. Götz, L. Hilterhaus, A. Liese, Industrial application of oxidoreductase catalyzed reduction of ketones and aldehydes. Enzym. Catal. Org. Synth. Third Ed. **2**, 1205–1223 (2012). https://doi.org/10.1002/9783527639861.CH29
33. D.J. Cole-Hamilton, Homogeneous catalysis - New approaches to catalyst separation, recovery, and recycling. Science **299**(5613), 1702–1706 (2003)
34. J.M. Thomas, W.J. Thomas, *Principles and Practice of Heterogeneous Catalysis* (Wiley, Hoboken, 2015), p. 767
35. J.C. Védrine, Acid-base characterization of heterogeneous catalysts: An up-to-date overview. Res. Chem. Intermed. **41**(12), 9387–9423 (2015). https://doi.org/10.1007/S11164-015-1982-9
36. M. Boudart, G. Djega-Mariadassou, *Kinetics of Heterogeneous Catalytic Reactions* (Princeton University Press, Princeton, 2014)
37. Y. Gu, G. Li, Ionic liquids-based catalysis with solids: State of the art. Adv. Synth. Catal. **351**(6), 817–847 (2009). https://doi.org/10.1002/ADSC.200900043
38. D. Woitassek et al., Tunable aryl alkyl ionic liquid supported synthesis of platinum nanoparticles and their catalytic activity in the hydrogen evolution reaction and in hydrosilylation. Molecules **28**(1), 405 (2023). https://doi.org/10.3390/MOLECULES28010405
39. W. Zhu, Z. Hou, Ionic liquid stabilized metal nanoparticles: Synthesis and catalytic hydrogenation. Curr. Inorg. Chem. **2**(2), 213–227 (2012). https://doi.org/10.2174/1877944111202020213
40. M.M. Seitkalieva, D.E. Samoylenko, K.A. Lotsman, K.S. Rodygin, V.P. Ananikov, Metal nanoparticles in ionic liquids: Synthesis and catalytic applications. Coord. Chem. Rev. **445**, 213982 (2021). https://doi.org/10.1016/J.CCR.2021.213982
41. J. Dupont, B.C. Leal, P. Lozano, A.L. Monteiro, P. Migowski, J.D. Scholten, Ionic liquids in metal, photo-, electro-, and (bio) catalysis. Chem. Rev. **124**(9), 5227–5420 (2024). https://doi.org/10.1021/ACS.CHEMREV.3C00379
42. H.U. Blaser, C. Malan, B. Pugin, F. Spindler, H. Steiner, M. Studer, Selective hydrogenation for fine chemicals: Recent trends and new developments. Adv. Synth. Catal. **345**(1-2), 103–151 (2003). https://doi.org/10.1002/ADSC.200390000
43. J. Dupont, M.R. Meneghetti, On the stabilisation and surface properties of soluble transition-metal nanoparticles in non-functionalised imidazolium-based ionic liquids. Curr. Opin. Colloid Interface Sci. **18**(1), 54–60 (2013). https://doi.org/10.1016/J.COCIS.2012.12.001
44. T. Sasaki, CO_2 hydrogenation in ionic liquids: Recent update. Curr. Opin. Green Sustain. Chem. **36**, 100633 (2022). https://doi.org/10.1016/J.COGSC.2022.100633
45. Y. Hu et al., The functionalized ionic liquid-stabilized palladium nanoparticles catalyzed selective hydrogenation in ionic liquid. Catal. Commun. **10**(14), 1903–1907 (2009). https://doi.org/10.1016/J.CATCOM.2009.06.025
46. K.J. Mohammed, S.K. Hadrawi, E. Kianfar, Synthesis and modification of nanoparticles with ionic liquids: A review. BioNanoScience **13**(2), 760–783 (2023). https://doi.org/10.1007/S12668-023-01075-4
47. N. Gunasekaran, Aerobic oxidation catalysis with air or molecular oxygen and ionic liquids. Adv. Synth. Catal. **357**(9), 1990–2010 (2015). https://doi.org/10.1002/ADSC.201400989
48. C. Vollmer, C. Janiak, Naked metal nanoparticles from metal carbonyls in ionic liquids: Easy synthesis and stabilization. Coord. Chem. Rev. **255**(17-18), 2039–2057 (2011). https://doi.org/10.1016/J.CCR.2011.03.005
49. Y. Zhao, B. Han, Z. Liu, Ionic-liquid-catalyzed approaches under metal-free conditions. Acc. Chem. Res. **54**(16), 3172–3190 (2021). https://doi.org/10.1021/ACS.ACCOUNTS.1C00251

50. M. Gholinejad, F. Zareh, H. Sheibani, C. Nájera, M. Yus, Magnetic ionic liquids as catalysts in organic reactions. J. Mol. Liq. **367**, 120395 (2022). https://doi.org/10.1016/J.MOLLIQ.2022.120395
51. F. Giacalone, M. Gruttadauria, Covalently supported ionic liquid phases: An advanced class of recyclable catalytic systems. ChemCatChem **8**(4), 664–684 (2016). https://doi.org/10.1002/CCTC.201501086
52. I. Geukens, D.E. De Vos, Organic transformations on metal nanoparticles: Controlling activity, stability, and recyclability by support and solvent interactions. Langmuir **29**(10), 3170–3178 (2013). https://doi.org/10.1021/LA304639Z
53. J. Theerthagiri et al., Electrochemical conversion of biomass-derived aldehydes into fine chemicals and hydrogen: A review. EnvCL **21**(3), 1555–1583 (2023). https://doi.org/10.1007/S10311-022-01543-5
54. M.S. Akhtar, M.T. Naseem, S. Ali, W. Zaman, Metal-based catalysts in biomass transformation: From plant feedstocks to renewable fuels and chemicals. Catalysts **15**(1), 40 (2025). https://doi.org/10.3390/CATAL15010040
55. S. Babu, L. Jojo, A. James, K. Melethil, B. Thomas, Metal-based catalysis for biomass and renewables valorization- Current status. Tetrahedron Green Chem. **2**, 100018 (2023). https://doi.org/10.1016/J.TGCHEM.2023.100018
56. Y.V. Samoylova, K.N. Sorokina, V.N. Parmon, Use of microalgae biomass to synthesize marketable products: 4. Production of biofuels from microalgae using bioengineering approaches. Catal. Ind. **16**(1), 89–101 (2024). https://doi.org/10.1134/S2070050424010069
57. G. Wu, C. Shen, S. Liu, Y. Huang, S. Zhang, H. Zhang, Research progress on the preparation and application of biomass derived methyl levulinate. Green Chem. **23**(23), 9254–9282 (2021). https://doi.org/10.1039/D1GC03474J
58. G.H. Delmas, J.H. Banoub, M. Delmas, Lignocellulosic biomass refining: A review promoting a method to produce sustainable hydrogen, fuels, and products. Waste Biomass Valoriz. **13**(5), 2477–2491 (2022). https://doi.org/10.1007/S12649-021-01624-6
59. A.D. Dwivedi, B. Priya, R. Chinthala, D.S. Pandey, S.K. Singh, Valorization of biomass-derived furans over molecular catalysts. Tetrahedron Green Chem. **1**, 100008 (2023). https://doi.org/10.1016/J.TGCHEM.2023.100008
60. K.N. Yogalakshmi et al., Lignocellulosic biomass-based pyrolysis: A comprehensive review. Chemosphere **286**, 131824 (2022). https://doi.org/10.1016/J.CHEMOSPHERE.2021.131824
61. S. Wang, Z. Li, X. Bai, W. Yi, P. Fu, Catalytic pyrolysis of lignin in a cascade dual-catalyst system of modified red mud and HZSM-5 for aromatic hydrocarbon production. Bioresour. Technol. **278**, 66–72 (2019). https://doi.org/10.1016/J.BIORTECH.2019.01.037
62. S. Nanda, J.A. Kozinski, A.K. Dalai, Lignocellulosic biomass: A review of conversion technologies and fuel products. Curr. Biochem. Eng. **3**(1), 24–36 (2015). https://doi.org/10.2174/2213385203666150219232000
63. K. Fan et al., Study on government subsidy in a two-level supply chain of direct-fired biomass power generation based on contract coordination. J. Ind. Manag. Optim. **19**(4), 2436–2451 (2023). https://doi.org/10.3934/JIMO.2022049
64. M.K. Meena, S. Anand, D.K. Ojha, Interdependency of pyrolysis and combustion: A case study for lignocellulosic biomass. J. Therm. Anal. Calorim. **148**(12), 5509–5519 (2023). https://doi.org/10.1007/S10973-023-12090-8/METRICS
65. S. Pati, S. De, R. Chowdhury, Exploring the hybrid route of bio-ethanol production via biomass co-gasification and syngas fermentation from wheat straw and sugarcane bagasse: Model development and multi-objective optimization. J. Clean. Prod. **395**, 136441 (2023). https://doi.org/10.1016/J.JCLEPRO.2023.136441
66. S. Manikandan et al., Critical review of biochemical pathways to transformation of waste and biomass into bioenergy. Bioresour. Technol. **372**, 128679 (2023). https://doi.org/10.1016/J.BIORTECH.2023.128679
67. W. Deng et al., Catalytic conversion of lignocellulosic biomass into chemicals and fuels. Green Energy Environ. **8**(1), 10–114 (2023). https://doi.org/10.1016/j.gee.2022.07.003

68. J.Y. Chen, Y. Xiao, F.S. Guo, K.M. Li, Y.B. Huang, Q. Lu, Single-atom metal catalysts for catalytic chemical conversion of biomass to chemicals and fuels. ACS Catal. **14**(7), 5198–5226 (2024). https://doi.org/10.1021/ACSCATAL.4C00208
69. S. Wang et al., Catalytic conversion network for lignocellulosic biomass valorization: A panoramic view. Ind. Chem. Mater. **1**(2), 188–206 (2023). https://doi.org/10.1039/D2IM00054G
70. E. Nzediegwu, M.J. Dumont, Chemo-catalytic transformation of cellulose and cellulosic-derived waste materials into platform chemicals. Waste Biomass Valoriz. **12**(6), 2825–2851 (2021). https://doi.org/10.1007/S12649-020-01179-Y
71. L. Peng, M. Wang, H. Li, J. Wang, J. Zhang, L. He, Tert-butanol intervention enables chemoselective conversion of xylose to furfuryl alcohol over heteropolyacids. Green Chem. **22**(17), 5656–5665 (2020). https://doi.org/10.1039/D0GC02001J
72. M.M. Rahman, R. Liu, J. Cai, Catalytic fast pyrolysis of biomass over zeolites for high quality bio-oil—A review. Fuel Process. Technol. **180**, 32–46 (2018). https://doi.org/10.1016/J.FUPROC.2018.08.002
73. A. Shetty, V. Molahalli, A. Sharma, G. Hegde, Biomass-derived carbon materials in heterogeneous catalysis: A step towards sustainable future. Catalysts **13**(1), 20 (2023). https://doi.org/10.3390/CATAL13010020
74. Q. Yan et al., Transition metals-catalyzed amination of biomass feedstocks for sustainable construction of N-heterocycles. Coord. Chem. Rev. **502**, 210622 (2024). https://doi.org/10.1016/J.CCR.2023.215622
75. A. Lodhi, K.C. Maheria, Zeolite-catalysed esterification of biomass-derived acids into high-value esters products: Towards sustainable chemistry. Catal. Commun. **187**, 106883 (2024). https://doi.org/10.1016/J.CATCOM.2024.106883
76. A. Bohre, A. Modak, V. Chourasia, P. Ram Jadhao, K. Sharma, K. Kishore Pant, Recent advances in supported ionic liquid catalysts for sustainable biomass valorisation to high-value chemicals and fuels. Chem. Eng. J. **450**, 138032 (2022). https://doi.org/10.1016/J.CEJ.2022.138032
77. M. Saidi, M. Safaripour, Ni–Mo nanoparticles stabilized by ether functionalized ionic polymer: A novel and efficient catalyst for hydrodeoxygenation of 4-methylanisole as a representative of lignin-derived pyrolysis bio-oils. Int. J. Hydrog. Energy **46**(2), 2191–2203 (2021). https://doi.org/10.1016/J.IJHYDENE.2020.10.274
78. M.L. Ibrahim et al., Ionic liquid functionalized nanomaterials for CO_2 capture and conversion. Nanomater. Carbon Dioxide Capture Convers. Technol. **12**, 361–383 (2023). https://doi.org/10.1016/B978-0-323-89851-5.00015-9
79. R. Villa, E. Alvarez, R. Porcar, E. Garcia-Verdugo, S.V. Luis, P. Lozano, Ionic liquids as an enabling tool to integrate reaction and separation processes. Green Chem. **21**(24), 6527–6544 (2019). https://doi.org/10.1039/C9GC02553G
80. A.A. Elgharbawy, M.Z. Alam, M. Moniruzzaman, M. Goto, Ionic liquid pretreatment as emerging approaches for enhanced enzymatic hydrolysis of lignocellulosic biomass. Biochem. Eng. J. **109**, 252–267 (2016). https://doi.org/10.1016/J.BEJ.2016.01.021
81. K. Zhuo, Q. Du, G. Bai, C. Wang, Y. Chen, J. Wang, Hydrolysis of cellulose catalyzed by novel acidic ionic liquids. Carbohydr. Polym. **115**, 49–53 (2015). https://doi.org/10.1016/J.CARBPOL.2014.08.078

Nanomaterials for Biomedical Sensing: A Focus on Refractive Index-Based Detection Inspired by Dual Self-Referenced Sensors

Uddipan Chowdhury, Samir Kumar, and Partha Sona Maji

1 Introduction

In the realm of global health challenges, malaria continues to demand urgent attention due to its widespread impact and high mortality rate. As highlighted by the World Health Organization, the year 2021 witnessed over 247 million reported cases and approximately 619,000 related deaths [1]. This persistent burden underscores the critical need for innovative diagnostic technologies. In this chapter, we explore the potential of nanomaterial-based biomedical sensors—particularly those leveraging refractive index changes through dual self-referenced detection—to offer enhanced sensitivity and accuracy in the early detection of malaria and similar infectious diseases. Human malaria is caused by several *Plasmodium* species, among which *P. falciparum* is the most lethal, accounting for the majority of malaria-related deaths [2]. Although *P. vivax* is the most widespread, over 200 species have been identified, each with unique host ranges and biological behaviors.

Conventional diagnostic methods—ranging from microscopic examination and rapid diagnostic tests to molecular techniques—have demonstrated varying levels of success [3–7]. Microscopy, though considered the gold standard, is time-intensive and heavily reliant on expert interpretation. Molecular methods offer precision but demand costly equipment and trained personnel, limiting their accessibility in endemic regions. There remains a strong need for diagnostic tools that are sensitive, specific, cost-effective, and capable of real-time detection.

U. Chowdhury · P. S. Maji (✉)
Department of Physics, Amity School of Applied Sciences, Amity University Kolkata, Kolkata, India
e-mail: psmaji@kol.amity.edu

S. Kumar
Department of Physics, Hotilal Ramnath College, Jai Prakash University, Chapra, Bihar, India

S. Kulkarni, S. Das (eds.), *Ionic Liquids, Nanofluids and Nanotechnology*,
Lecture Notes in Nanoscale Science and Technology 42,
https://doi.org/10.1007/978-3-032-04008-4_13

In recent years, photonic and plasmonic technologies have shown immense potential in biosensing applications due to their high sensitivity and adaptability. Refractive index (RI)-based plasmonic sensors, in particular, are gaining traction for biomedical diagnostics [8, 9]. Various sensor architectures—such as metal-insulator-metal (MIM) waveguides, ring resonators, and hybrid plasmonic devices—have been explored for their enhanced light-matter interaction capabilities and promising figures of merit [10, 11]. These technologies have been applied to a wide range of sensing tasks, including the detection of gases, magnetic fields, biomolecules, and environmental contaminants [12–18].

Building upon these advances, Tamm plasmon polaritons (TPPs)—localized optical modes at metal-dielectric interfaces—have emerged as a novel approach for sensitive RI sensing [19]. Recent studies indicate that combining TPPs with photonic quasicrystals (PQCs), particularly those based on Fibonacci sequences, enhances sensing performance due to their omnidirectional bandgaps and robust light confinement properties [20–22]. PQC-based configurations also allow for self-referenced sensing, minimizing noise and enhancing measurement accuracy [23].

In this work, we present a theoretical investigation into a dual self-referenced Tamm plasmon-based sensor using a PQC structure for malaria diagnosis [24]. The proposed sensor comprises dual-metal-PQC stacks separated by a cavity designed to hold blood samples. The quasi-periodic layering of SiO_2 and Ta_2O_5 follows a Fibonacci sequence and supports multiple resonance modes within the photonic bandgap—one of which remains index-insensitive, serving as a reference [25, 26]. We specifically target the detection of different stages of *Plasmodium* infection in red blood cells (RBCs), which exhibit distinct refractive indices as the parasite progresses through its life cycle.

Our simulations reveal consistent red shifts in resonant wavelengths correlating with stage-specific refractive index changes. Sensor performance metrics such as sensitivity (up to ~425 nm/RIU), detection accuracy, and quality factor demonstrate the system's potential for precise, early-stage malaria detection. The sensor design also benefits from fabrication feasibility using standard thin-film techniques like CVD, MBE, or electrochemical deposition, making it suitable for scalable production.

Despite its advantages, challenges such as fabrication complexity, cost constraints, and integration with existing diagnostic platforms must be addressed for real-world implementation. Nonetheless, this sensor configuration opens avenues for next-generation malaria diagnostics, offering high specificity, rapid detection, and adaptability to low-resource settings. Through detailed analysis of its optical properties and performance characteristics, this chapter aims to lay the groundwork for the development of practical and impactful plasmonic biosensors.

2 Theoretical Modeling

Figure 1 illustrates the structural framework of the proposed sensor configuration. In the pursuit of advanced biomedical sensing technologies, the integration of nanostructured materials has opened new avenues for precise and sensitive detection.

This work delves into a novel refractive index-based sensing approach, inspired by dual self-referenced mechanisms, utilizing a photonic quasi-crystal (PQC) architecture. The PQC is thoughtfully engineered with a sequence of five tantalum pentoxide (Ta_2O_5) layers, labeled as A, and seven silicon dioxide (SiO_2) layers, labeled as B. At the heart of this multilayered structure lies a central cavity, purposefully designed to house blood samples for refractive index analysis. To significantly enhance the interaction between light and the analyte, a plasmonically active metal layer is strategically embedded within the configuration, ensuring heightened sensitivity and efficient signal transduction. Figure 1 depicts the proposed sensor

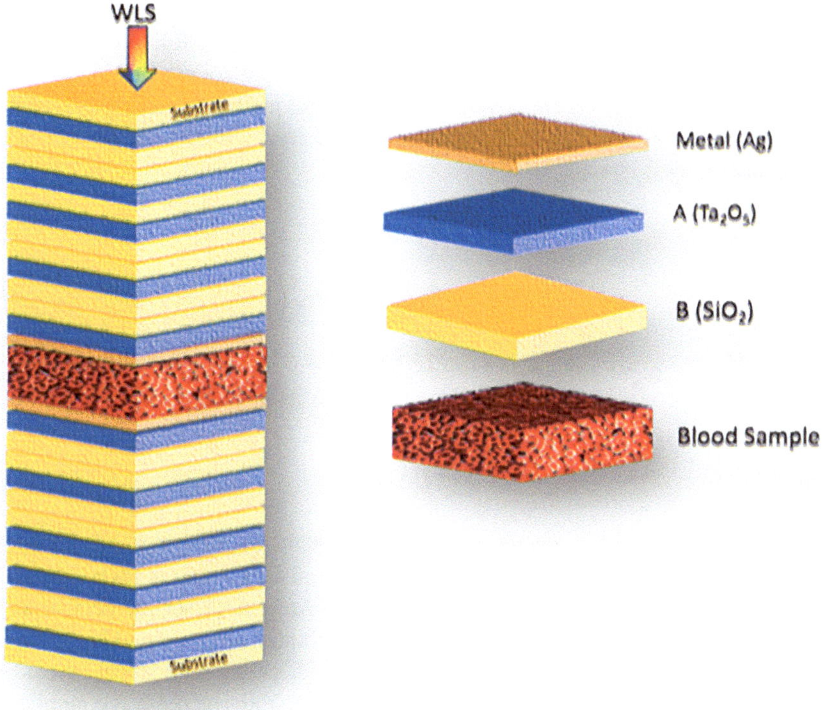

Fig. 1 Schematic representation of the proposed sensor architecture, incorporating dual silver–photonic quasi-crystal (Ag–PQC) structures arranged according to a Fibonacci sequence, with a centrally located layer designated for the introduction of blood samples

architecture incorporating a 30-nm-thick silver (Ag) layer. The photonic quasi-crystal (PQC) structure consists of alternating Ta_2O_5 and SiO_2 layers, where Ta_2O_5—positioned adjacent to the metal—exhibits a higher refractive index (RI), while the underlying SiO_2 possesses a comparatively lower RI. The RI values for Ta_2O_5 and SiO_2 are referenced from studies conducted by Bright et al. [27] and Ghosh et al. [28], respectively. The refractive index (RI) of silver is obtained using a wavelength-dependent dispersion model, as detailed in the work by Srivastava et al. [29].

$$\varepsilon_m(\lambda) = 1 - \frac{\lambda^2 \lambda_c}{\lambda_p^2 (\lambda_c + i\lambda)} \tag{1}$$

For experimental validation, the sensor setup is positioned beneath a white light source (WLS), which is vertically aligned to illuminate the substrate surface. The PQC layers are fabricated on a SiO_2 substrate for structural integrity. The thicknesses of SiO_2 and Ta_2O_5 layers are determined based on the quarter-wave condition, optimizing light confinement at a central wavelength of 600 nm in the visible spectrum. This calculation results in Ta_2O_5 layers being 70.4 nm thick and SiO_2 layers being 102 nm thick. The PQC arrangement follows a Fibonacci sequence, which can be succinctly described using a word count rule [30].

In our investigation, we employed the Fibonacci sequence defined by the relation $S_n = S_{n-1}S_{n-2}$, for $n \geq 2$, with initial conditions $S_0 = A$ and $S_1 = B$. The sequence unfolds as follows:

$$S_0 : A$$

$$S_1 : B$$

$$S_2 : \text{BA}$$

$$S_3 : \text{BAB}$$

$$S_4 : \text{BABBA}$$

For the purposes of this study, we focused on the fourth-order sequence, S_4, which results in a structure comprising 5 layers of Ta_2O_5 and 7 layers of SiO_2. This specific choice allowed us to maintain a manageable sensor design while still achieving the desired optical functionality. Although advancing to higher-order sequences could enable the formation of additional cavity resonances, including more Tamm plasmon modes, doing so would significantly increase the number of dielectric layers, thus complicating the fabrication process and enlarging the device footprint—undesirable traits for nanoscale sensing applications. By restricting the structure to the S_4 configuration, we were able to preserve the compactness of the sensor while effectively demonstrating its dual self-referencing capabilities. This configuration also offered an adequately broad spectral response to facilitate accurate identification of the different malaria parasite stages. Ultimately, this approach provides a well-balanced compromise between design simplicity and sensing performance, aligning with the practical requirements of miniaturized biomedical sensors.

The conventional transfer matrix method (TMM) was utilized to examine the dispersion characteristics of the multilayered geometry, including its reflection, transmission, and absorption spectra [31–33]. In this methodology, each layer within the multilayer structure is characterized by its own transfer matrix, constructed using essential parameters including phase shifts, internal angles of incidence, and propagation vectors [34, 35]. These individual matrices are then sequentially multiplied to formulate a comprehensive global transfer matrix that represents the entire optical system. Extensive insights into the calculation of reflectivity and the distribution of the electric field across multilayer configurations can be found in the relevant scientific literature [35, 36].

3 Reflection Characteristics and Mode Analysis of the Dual Ag-FS-PQC Sensor Configuration

Figure 2 presents the calculated reflection spectra for two configurations: the standalone FS-PQC structure and the proposed dual configuration denoted as "FS-PQC–Ag–da–Ag–FS-PQC," where d_a represents the thickness of the intervening blood sample layer. These spectra were obtained using the transfer matrix method (TMM). The sensor structure explored in this study is constructed following a fourth-order Fibonacci sequence, consisting of five layers of tantalum pentoxide (Ta_2O_5) and seven layers of silicon dioxide (SiO_2), with individual layer thicknesses of 70.4 nm and 102 nm, respectively. Silver layers of 30 nm thickness were incorporated on either side of the cavity to enable the excitation of hybrid plasmonic modes. To optimize the sensing performance, particularly sensitivity, the thickness of the analyte (blood) layer was varied, with a focus on achieving the maximum shift in resonance wavelength in response to changes in the refractive index. Sensitivity in this context is defined by the following relation:

$$S_n = \frac{\delta\lambda_r}{\delta n} \tag{2}$$

where $\delta\lambda_r$ represents the resonance wavelength shift, while δn corresponds to the variation in the refractive index of the analyte. The cavity thickness was further optimized by leveraging the dual self-referencing nature of the structure. Through detailed simulations, it was determined that a sample thickness of approximately 265 nm yielded optimal performance across all four malaria stages analyzed—namely *RBC* (red blood cells), *RPC* (ring parasite cells), *TPC* (trophozoite parasite cells), and *SPC* (schizont parasite cells) [37].

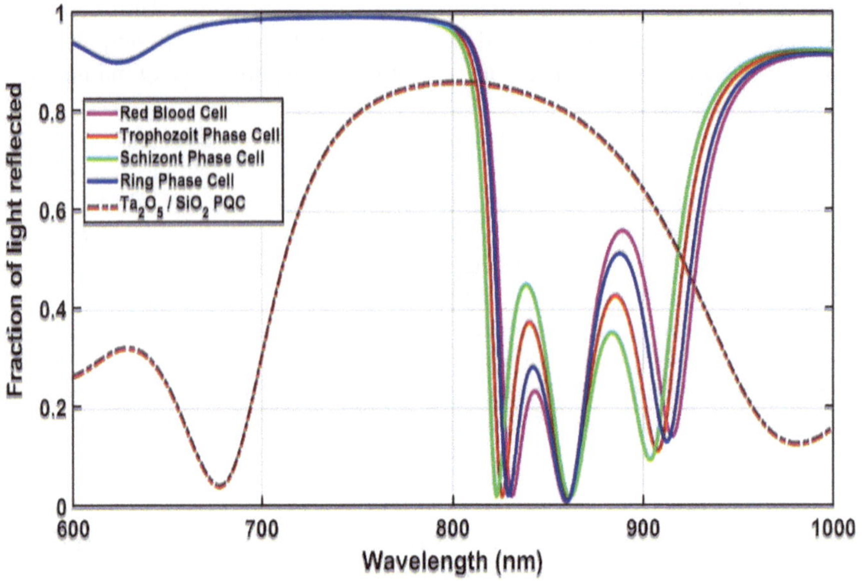

Fig. 2 Reflection spectra illustrating the comparison between the bare configuration and the dual "Ag-FS-PQC" setup, analyzed at various stages of malaria infection. The sample layer thickness (d_a) is maintained at 265 nm

The reflection spectra analyzed via TMM revealed three prominent dips within the photonic bandgap for all stages of analysis, as illustrated in Fig. 2. These resonance dips result from the coupling of two distinct optical modes: The Tamm plasmon mode, which originates at the interface between the metal layer and the photonic quasi-crystal, and the Fabry–Pérot cavity mode, which is established within the cavity formed by the two opposing metal layers. This interaction gives rise to distinct hybrid modes observable as dips in the reflection spectra. Among these, one dip consistently appeared around 862 nm across all refractive index variations, remaining largely unaffected. This robust spectral feature, which we refer to as *Mode-1*, serves as an invariant reference. In contrast, the remaining two dips exhibited notable redshifts as the refractive index of the analyte increased. The dip located at a shorter wavelength than Mode-1 is labeled *Mode-2*, while the one on the longer wavelength side is designated *Mode-3*. The presence of this invariant Mode-1, along with the dynamic shifts of Modes 2 and 3, establishes the dual self-referencing capability of the sensor. This characteristic enhances the reliability and accuracy of refractive index measurements, making the configuration especially suitable for biomedical sensing applications, such as detecting malaria stages through blood sample analysis.

Figure 2 demonstrates the reflection spectra characteristics observed across various stages of malaria infection. To analyze the reflectivity dip positions and their shifts due to variations in refractive indices (RIs), simulations were performed.

These simulations serve to visualize the electric field distributions at the resonance wavelengths corresponding to the dips. Figure 3a, b present a detailed analysis of the mode-field distributions in the "FS-PQC-Ag-da-Ag-FS-PQC" configuration, featuring a 265-nm-thick analyte layer with different refractive indices. Figure 3a presents a comparative analysis of the mode-field distribution between Mode-2 and Mode-1, whereas Fig. 3b delves into the differences in field distributions between Mode-3 and Mode-1. These visualizations facilitate a qualitative understanding of how hybrid modes undergo shifts in their resonance wavelength. The fluctuations in the field distribution of Mode-2 compared to Mode-1 (refer to Fig. 3a), alongside the variations in the field distribution of Mode-3 compared to Mode-1 (refer to Fig. 3b), provide critical insights into the functional behavior of the sensor at different stages of malaria infection. The mode-field distribution plots presented in Fig. 3a, b are not merely visual tools but serve multiple purposes in advancing the understanding of light-sample interactions. These include the visualization of field distributions, correlating resonance shifts with mode identification, and evaluating the influence of these shifts on sensing performance.

Further analysis of the parasitic modes enables a deeper understanding of their influence on the propagation of electromagnetic waves within the multilayer structure. This investigation helps clarify deviations from optimal sensing behavior and sheds light on potential areas for improvement. Moreover, comparing mode-field distributions across various parasitic stages enhances comprehension of the dynamic changes in sensing mechanisms introduced by these stages. Such an understanding is pivotal in decoding the intricate interplay between the sensor architecture, incident light, and the surrounding medium, such as blood samples.

These findings and visual representations collectively reinforce the significance of mode-field analysis in achieving accurate sensing results and offer a pathway to optimize sensor designs for high-precision applications in detecting malaria infection stages. The methodological approach adopted here, coupled with detailed simulation results, underscores the role of structural configurations and hybrid-mode interactions in tailoring the sensor's functionality for diverse analytical requirements.

The field distribution profiles provide critical insights into the interaction dynamics within the sensor structure. A particularly interesting observation emerged when analyzing the resonance dips associated with different modes. For *Mode-2*—located at shorter wavelengths relative to the invariant *Mode-1*—the corresponding electric field intensities exhibited more pronounced peaks when interacting with lower refractive index (RI) components, such as those characteristic of the schizont parasite cell (SPC) stage. In contrast, as the RI increased, particularly in the red blood cell (RBC) stage, the peak amplitudes of Mode-2 diminished noticeably. An inverse trend was observed for *Mode-3*, which appears at longer wavelengths than Mode-1. Here, the field intensities increased significantly with higher RI values. This opposing behavior between Mode-2 and Mode-3 reinforces the dual self-referencing nature of the structure, enhancing its diagnostic capability by providing distinct, mode-dependent responses to varying RI conditions.

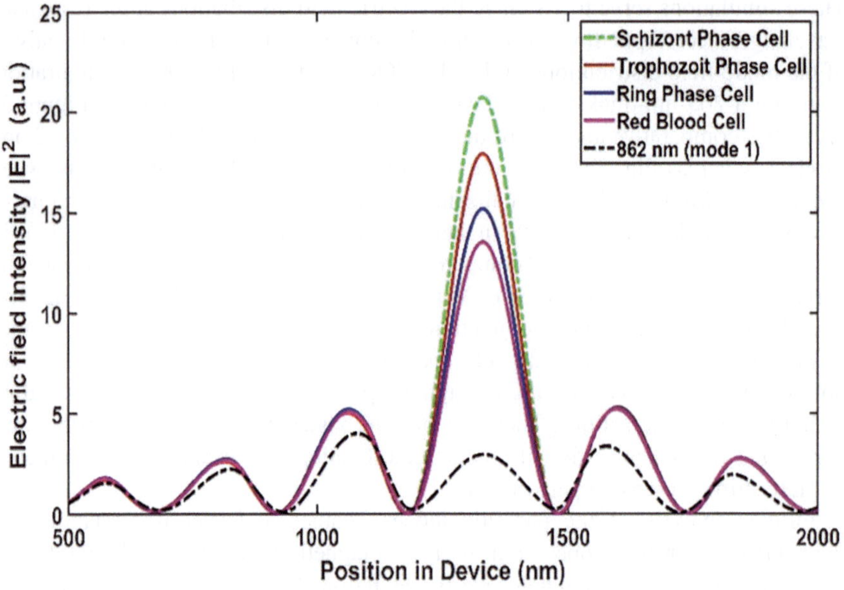

(a)

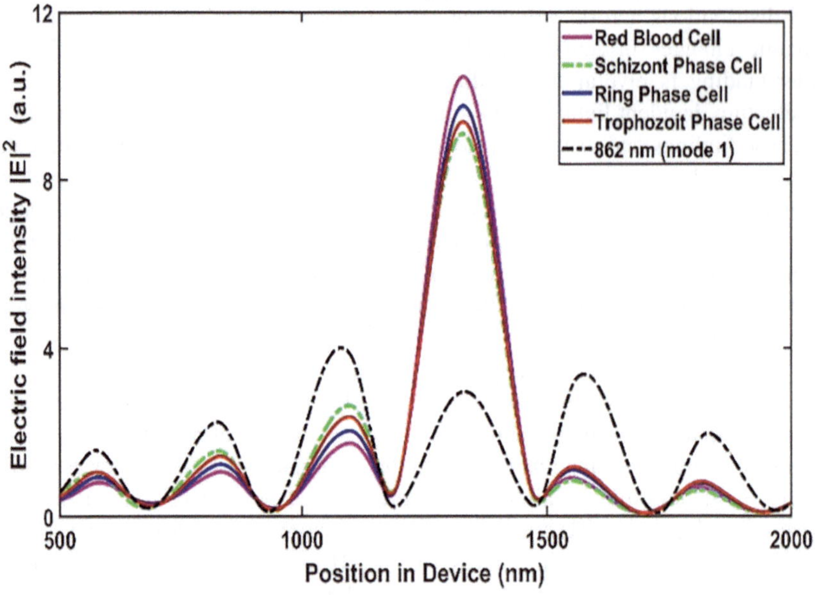

(b)

Fig. 3 (**a**) Mode-field distribution illustrating the characteristics of Mode-2 relative to Mode-1. (**b**) Mode-field distribution highlighting the differences between Mode-3 and Mode-1

Table 1 Variation of resonance wavelength as a function of refractive index changes for mode-2 and mode-3

Components	Resonance wavelength (nm) for mode-2	Resonance wavelength (nm) for mode-3
RBC (1.402)	832	915
RPC (1.395)	830	912
TPC (1.383)	827	907
SPC (1.373)	824	903

Moreover, it is worth highlighting that the maximum field intensities attained by Mode-3 across different parasitic stages consistently surpassed those observed for Mode-2. This enhanced localization and intensity suggest a stronger confinement and coupling efficiency for Mode-3, making it particularly advantageous for high-sensitivity detection and monitoring of RI variations associated with malaria progression. The resonance wavelengths associated with the distinct optical modes across various parasitic stages are systematically summarized in Table 1. As depicted in Fig. 3a, b, the mode-field distribution for *Mode-1*—characterized by its anti-symmetric profile—remains remarkably invariant despite changes in the refractive index (RI) of the sample. This invariance forms the cornerstone of our dual self-referencing sensor architecture. In this framework, *Mode-1* functions as a stable reference point, while the resonance positions of *Mode-2* and *Mode-3* shift corresponding to changes in the optical properties of the surrounding medium, specifically its refractive index (RI).

In the pursuit of high-precision optical sensing, the integration of self-referencing mechanisms has become increasingly vital. Traditional RI sensors often rely on a single resonance mode, rendering them susceptible to external perturbations that can obscure or distort meaningful data. The introduction of a reference mode—ideally one that remains insensitive to environmental fluctuations—provides a robust solution to this challenge. By serving as a stable baseline, the reference mode enables the isolation of analyte-induced changes from extrinsic noise, significantly improving the reliability and accuracy of the sensor. Our simulations confirm that *Mode-1* fulfills this role effectively, exhibiting consistent resonance behavior regardless of changes in the RI of the blood sample. In contrast, both *Mode-2* and *Mode-3* demonstrate pronounced shifts in response to RI variations, making them suitable for quantitative sensing. The differential sensitivity of these modes enhances the dynamic range and specificity of detection.

Ultimately, the dual self-referencing approach not only enables long-wavelength tunability and improved information extraction but also mitigates the likelihood of false positives. By leveraging the complementary behavior of multiple resonance modes, this design paradigm marks a significant advancement in the field of optical biosensing, particularly for applications demanding high sensitivity and precision, such as the detection of parasitic blood stages.

To systematically optimize the thickness of the sample layer for enhanced sensing performance, we carried out a comprehensive three-dimensional (3D) analysis across a range of thicknesses varying from 200 nm to 300 nm. This visual and quantitative approach provided greater insight into the influence of cavity thickness on the resonance behavior of the sensor. The optimization methodology involved incrementally adjusting the sample layer thickness and tracking the corresponding shifts in the resonance wavelengths as observed in the reflectivity spectra.

Through this parametric study, we identified the conditions under which the resonance features exhibited maximal sensitivity to refractive index variations—critical for the reliable detection of malaria-related biomarkers. The 3D visualization not only facilitated a clearer understanding of the resonance dynamics but also served as a powerful tool for pinpointing the optimal geometrical parameters that balance structural compactness with high sensor performance. These findings underscore the importance of precise cavity engineering in the development of compact, high-resolution optical biosensors for clinical diagnostics.

Our analysis of the mode-field distribution confirmed that Mode-1—referred to as the invariant mode—exhibits remarkable stability, showing negligible sensitivity to variations in the refractive index of the blood sample. This inherent stability underpins the implementation of a dual self-referencing optical sensing approach. In this configuration, Mode-1 serves as a robust internal reference, maintaining a nearly constant resonance wavelength regardless of changes in the surrounding medium. Meanwhile, the resonance wavelengths corresponding to Mode-2 and Mode-3 demonstrate significant shifts as the refractive index of the sample varies.

Figure 4a–d depict the reflectivity response maps as a function of analyte thickness (ranging from 200 to 300 nm) across different blood sample stages, each characterized by a distinct refractive index. These plots clearly reveal that changes in the sample layer thickness exert pronounced effects on Mode-2 and Mode-3, while Mode-1 remains largely unaffected. Notably, a sample layer thickness near 265 nm results in the simultaneous and consistent appearance of both Mode-2 and Mode-3 across all parasitic stages, a necessary condition for effective implementation of the dual self-referencing scheme. To assess the sensing performance of the proposed configuration, three critical performance metrics were systematically examined: sensitivity (Sn), quality factor (QF), and detection accuracy (DA). These parameters were evaluated using a fixed sample layer thickness of 265 nm, identified as the optimal value for enhancing resonance wavelength shifts linked to Tamm plasmon polariton (TPP) modes.

Sensitivity, as expressed in Eq. (2), quantifies the resonance wavelength shift per unit change in refractive index (RI). Detection accuracy is inversely related to the full width at half minimum (FWHM) of the resonance dip and offers insight into the sharpness and precision of the sensor's response. Quality factor (QF), calculated using Eq. (3), serves as a metric for the sensor's spectral selectivity and reliability in detecting small perturbations in the sample environment.

$$QF = \frac{\lambda_r}{FWHM} \tag{3}$$

Figure 5a–f summarize the variations of these performance metrics for both Mode-2 and Mode-3 across the different stages of malaria infection. Figure 5a illus-

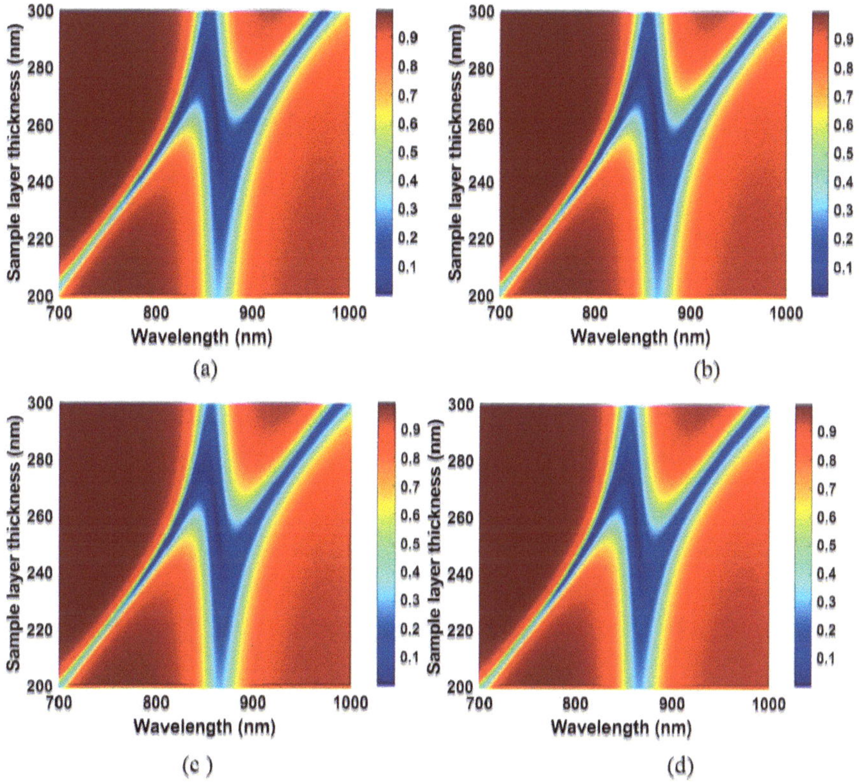

Fig. 4 Three-dimensional reflectivity plots depicting how reflectivity varies with sample layer thickness and wavelength across different blood sample types: (**a**) RBC, (**b**) RPC, (**c**) TPC, and (**d**) SPC. These plots highlight the resonance wavelength shifts across thicknesses ranging from 200 nm to 300 nm, aiding in the optimization of cavity thickness for enhanced sensitivity

trates that Mode-2's sensitivity declines with increasing sample RI, whereas Fig. 5b shows that Mode-3's sensitivity improves under similar conditions. Specifically, Mode-2 achieves a peak sensitivity of approximately 299 nm/RIU, while Mode-3 demonstrates an even higher maximum of around 425 nm/RIU. Detailed sensitivity values for Mode-3 across various parasitic stages are as follows: 425 nm/RIU (RBC), 415 nm/RIU (RPC), 398 nm/RIU (TPC), and 383 nm/RIU (SPC). For Mode-2, the corresponding values are 257 nm/RIU (RBC), 266 nm/RIU (RPC), 282 nm/RIU (TPC), and 299 nm/RIU (SPC).

The divergence in sensitivity trends between Mode-2 and Mode-3 is likely attributable to differences in the electric field distributions at different RI values, as suggested by the field plots in Fig. 4a, b. Figure 5c, d further show that detection accuracy follows a decreasing trend for Mode-2 and an increasing trend for Mode-3 as the sample RI increases. Mode-2's DA ranges from 0.0664 nm^{-1} (RBC) to 0.0792 nm^{-1} (SPC), with intermediate values for RPC (0.0694 nm^{-1}) and TPC

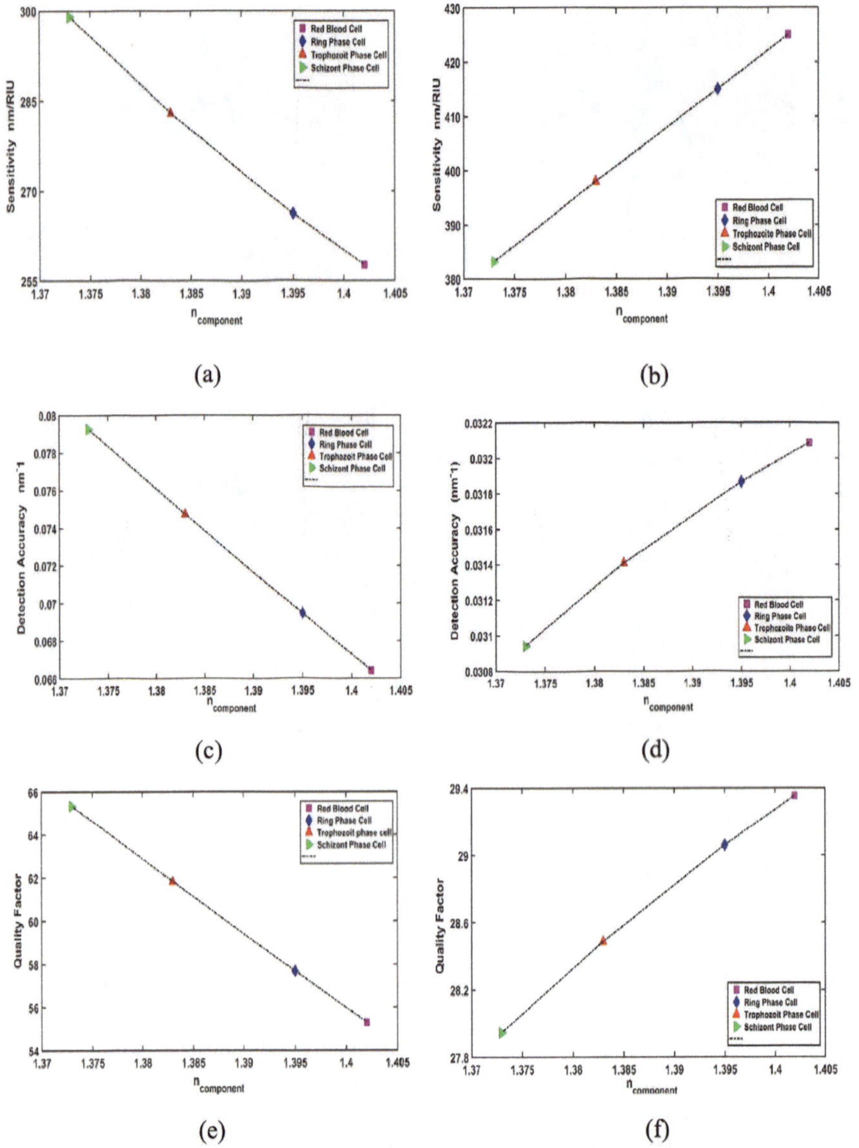

Fig. 5 Sensitivity variation for (**a**) mode-2 and (**b**) mode-3; detection accuracy for (**c**) mode-2 and (**d**) mode-3; and quality factor for (**e**) mode-2 and (**f**) mode-3. These parameters are presented for the proposed configuration as functions of the sample's refractive index (RI), considering a fixed sample thickness of 265 nm

(0.0747 nm^{-1}). In contrast, Mode-3 exhibits DA values spanning from 0.0309 nm^{-1} (SPC) to 0.0320 nm^{-1} (RBC), with RPC and TPC showing values of 0.0318 nm^{-1} and 0.0314 nm^{-1}, respectively.

Table 2 Sensitivity (Sn), detection accuracy (DA), and quality factor (QF) for Mode-2 and Mode-3 across various sample types

	Mode-2			Mode-3		
Components	S_n (nm/RIU)	DA (nm^{-1})	QF	S_n (nm/RIU)	DA (nm^{-1})	QF
RBC	257.72	0.0664	55.3	425.04	0.032	29.35
RPC	266.16	0.0694	57.68	415.11	0.0318	29.06
TPC	282.97	0.0747	61.83	397.98	0.0314	28.48
SPC	299.02	0.0792	65.34	383.31	0.0309	27.94

The quality factor results, shown in Fig. 5e, f, reinforce this trend. Mode-2 reaches a peak QF of 65.34 (SPC) and a minimum of 55.3 (RBC), with RPC and TPC registering values of 57.68 and 61.83, respectively. Conversely, Mode-3 exhibits a narrower QF range, peaking at 29.35 (RPC) and dipping to 27.94 (SPC), with other stages falling in between. Collectively, these findings underscore a fundamental contrast in the sensing behavior of Mode-2 and Mode-3. For Mode-2, sensitivity, detection accuracy, and QF all decrease with increasing RI. Conversely, Mode-3 shows a positive correlation with RI for these same metrics. This complementary behavior enhances the robustness of the dual self-referencing sensor, offering improved accuracy, extended tuning ranges, and reduced susceptibility to external disturbances. A consolidated summary of sensitivity, detection accuracy, and quality factor values for both modes across different parasitic stages is provided in Table 2.

The analysis of sensor performance parameters indicates that, at resonance wavelengths associated with mode-2—typically lower than those of mode-1—both detection accuracy (DA) and quality factor (QF) attain relatively higher values. These parameters demonstrate a decreasing trend with increasing refractive index (RI), whereby samples with lower RI yield superior DA and QF values compared to those with higher RI. Conversely, for mode-3, which corresponds to resonance wavelengths longer than mode-1, DA and QF are observed to be lower overall, and they follow an inverse trend to that of mode-2—higher RI materials exhibit enhanced DA and QF. These contrasting behaviors can be primarily attributed to differences in the full width at half maximum (FWHM) of the reflectance dips corresponding to various phases and their respective resonance wavelengths. The sharper resonance features in mode-2 contribute to enhanced DA and QF for low-RI materials, whereas broader dips in mode-3 influence the opposite trend. In terms of sensitivity (Sn), mode-3 demonstrates superior performance, with a direct correlation between RI and sensitivity—higher RI samples produce higher sensitivity values. In contrast, mode-2 exhibits lower overall sensitivity, with an inverse relationship between RI and Sn, where lower RI samples provide relatively higher sensitivity than higher RI counterparts. These observations underscore the distinct modal dependencies of sensor performance metrics and highlight the importance of tailoring resonance mode selection based on specific sensing requirements.

The data presented in Table 2 offer critical insights into the potential of the proposed photonic quasi-crystal (PQC)-based structure for sensing applications, particularly in biomedical diagnostics. This table summarizes key sensor performance metrics—namely, sensitivity (Sn), detection accuracy (DA), and quality factor (QF)—for both Mode-2 and Mode-3 across various blood-derived samples. As previously defined, sensitivity (Sn) reflects the sensor's responsiveness to variations in the refractive index (RI) of the target medium, while DA and QF serve as complementary indicators of sensing precision and spectral sharpness, respectively. Elevated sensitivity values signify an enhanced ability to detect minute RI changes, which is vital for identifying and characterizing biochemical substances based on their optical signatures. The observed trends emphasize the importance of a multi-parameter approach when evaluating sensor performance, as well as the nuanced dependence of sensor behavior on both modal characteristics and sample properties. These findings highlight the intricate interplay between structural parameters and sensing performance, thereby informing the rational design and optimization of optical sensors for targeted applications. Additionally, the values of sensitivity (Sn), quality factor (QF), and detection accuracy (DA) can be tailored by altering the number of dielectric layers in the PQC structure or by varying the thickness of individual layers, such as the cavity and sample layers. This design versatility allows for precise adjustment of the sensor's performance to suit targeted applications and functional demands.

Despite the promising numerical outcomes—such as high sensitivity, sharp resonances, and strong detection accuracy—the practical deployment of the proposed sensor, especially in clinical settings like malaria diagnostics, demands further investigation. A comprehensive evaluation encompassing factors such as cost-efficiency, manufacturability, and operational simplicity is imperative. Furthermore, experimental validation under real-world conditions, including controlled laboratory testing and field trials, is essential to verify the sensor's robustness, reproducibility, and comparative efficacy relative to existing diagnostic technologies. Such assessments will ultimately determine the feasibility of integrating the proposed PQC-based sensor into mainstream diagnostic workflows.

4 Conclusion

In this study, a novel and straightforward approach has been proposed for the development of a refractive index (RI)-based optical sensor aimed at malaria detection. The sensor design is based on a hybrid-mode configuration integrated into a dual-layer "Ag-FS-PQC" structure. These hybrid modes result from the interaction and coupling of distinct Tamm plasmon polariton (TPP) resonances occurring at the interface between the metal layer and the photonic quasi-crystal (PQC). The PQC structure is realized using a Fibonacci sequence arrangement of alternating SiO_2 and Ta_2O_5 dielectric layers, enabling a deterministic aperiodic configuration that enhances optical confinement. The sensing mechanism is centered around a cavity layer situated between the dual Ag-FS-PQC interfaces, into which the blood sample is introduced. The reflectance spectrum of the structure reveals the formation of three distinct resonant modes within the photonic bandgap. Of these, Mode-2 and Mode-3 demonstrate significant sensitivity to changes in the refractive index of the analyte, making them highly effective for biosensing applications. The operational principle of the proposed self-referenced sensor lies in monitoring the spectral shifts of the coupled TPP resonance dips, which correspond to variations in the RI of blood at different stages of malaria infection. By optimizing the thickness of the blood layer and operating at normal incidence of the incident electromagnetic wave, enhanced sensor performance is achieved. Under normal incidence conditions, Mode-2 exhibits a peak sensitivity of 299 nm/RIU, a detection accuracy (DA) of 0.0664 nm^{-1}, and a quality factor (QF) of 55.3. In contrast, Mode-3 offers even higher sensitivity at 425 nm/RIU, with a DA of 0.032 nm^{-1} and a QF of 29.35. These performance metrics underscore the sensor's ability to detect minute RI variations with high precision. The proposed dual-mode, self-referencing configuration presents a promising pathway for the advancement of compact, label-free biosensors with enhanced reliability and specificity. Beyond malaria detection, the underlying sensing strategy holds potential for broader biomedical applications, particularly in the diagnosis of other blood-related disorders characterized by RI fluctuations. Furthermore, this platform may serve as a foundation for the design of next-generation optoelectronic and biochemical sensing devices with tunable and application-specific functionalities.

References

1. World Malaria Report (2022), https://www.who.int/teams/global-malaria-programme/reports/world-malaria-report-2022
2. L.S. Garcia, Malaria. Clin. Lab. Med. **30**, 93–129 (2010)
3. T.E. Clendennen, G.W. Long, J.K. Baird, QBC and Giemsa-stained thick blood films: diagnostic performance of laboratory technologists. Trans. R. Soc. Trop. Med. Hyg. **89**, 183–184 (1995)
4. L.K. Erdman, K.C. Kain, Molecular diagnostic and surveillance tools for global malaria control. Travel Med. Infect. Dis. **6**, 82–99 (2008)
5. D. Bell, C. Wongsrichanalai, J.W. Barnwell, Ensuring quality and access for malaria diagnosis: how can it be achieved? Nat. Rev. Microbiol. **4**, 682–695 (2006)
6. R.C. She, M.L. Rawlins, R. Mohl, S.L. Perkins, H.R. Hill, C.M. Litwin, Comparison of immunofluorescence antibody testing and two enzyme immunoassays in the serologic diagnosis of malaria. J. Travel Med. **14**, 105–111 (2007)
7. Ankita B. Suthar, A. Bhargava, Biosensor application of 1D photonic crystal for malaria diagnosis. Plasmonics **16**, 59–63 (2021)
8. R. Rahad, A.K.M. Rakib, M.A. Haque, S.S. Sharar, R.H. Sagar, Plasmonic refractive index sensing in the early diagnosis of diabetes, anemia, and cancer: An exploration of biological biomarkers. Results Phys. **49**, 106478 (2023)
9. A.K.M. Rakib, R. Rahad, M.O. Faruque, R.H. Sagar, ZrN-based plasmonic sensor: A promising alternative to traditional noble metal-based sensors for CMOS-compatible and tunable optical properties. Opt. Express **31**, 25280–25297 (2023)
10. M.A. Haque, R. Rahad, M.O. Faruque, M.S. Mobassir, R.H. Sagar, Numerical analysis of a metal-insulator-metal waveguide-integrated magnetic field sensor operating at sub-wavelength scales. Sens. Bio-Sens. Res. **43**, 100618 (2024)

11. H. Bensalah, A. Hocini, H. Bahri, et al., A plasmonic refractive index sensor with high sensitivity and its application for temperature and detection of biomolecules. J. Opt. **52**, 1035–1046 (2023)
12. H. Li, J.T. Kim, J.-S. Kim, D.-Y. Choi, S.-S. Lee, Metasurface-incorporated optofluidic refractive index sensing for identification of liquid chemicals through vision intelligence. ACS Photonics **10**(3), 780–789 (2023)
13. R. Rahad, M.A. Haque, M.O. Faruque, A.S.M. Mohsin, M.S. Mobassir, R.H. Sagar, A novel plasmonic MIM sensor using integrated 1×2 demultiplexer for individual lab-on-chip detection of human blood group and diabetes level in the visible to near-infrared region. IEEE Sensors **24**, 12034–12041 (2024)
14. M.A. Haque, R. Rahad, A.K.M. Rakib, S.S. Sharar, R.H. Sagor, Plasmonic sensor for rapid detection of water adulteration in honey and quantitative measurement of lactose concentration in solution. Results Phys **51**, 106733 (2023)
15. R. Rahad, A.K.M. Rakib, M.K. Mahadi, O. Faruque, Fuel classification and adulteration detection using a highly sensitive plasmonic sensor. Sens. Bio-Sens. Res. **40**, 100560 (2023)
16. A.K.M. Rakib, M.R. Muttaqi, A.T.B. Siddique, A.B. Siddique Mohd, R.H. Sagor, Oxidation-resistant nanosensor: A label free, on-chip sensing approach of protein concentration, salinity, tissue classification. Results Phys. **53**, 106956 (2023)
17. S.S. Sharar, R. Rahad, M.A. Haque, R.H. Sagor, A numerical investigation of the opposing-face semi-circular refractive index sensor for detection of chemical pollutants and heavy metals in water. Opt. Commun. **549**, 129887 (2023)
18. R. Rahad, A.K.M. Rakib, M.A. Haque, S.S. Sharar, R.H. Sagor, Plasmonic refractive index sensing in the early diagnosis of diabetes, anemia, and cancer: An exploration of biological biomarkers. Results Phys. **49**, 106478 (2023)
19. S. Kumar, M.K. Shukla, P.S. Maji, R. Das, Self-referenced refractive index sensing with hybrid Tamm-Plasmon-Polariton modes in sub-wavelength analyte layers. J. Phys. D. Appl. Phys. **50**, 375106 (2017)
20. W. Gellermann, M. Kohmoto, B. Sutherland, P. Taylor, Localization of light waves in Fibonacci dielectric multilayers. Phys. Rev. Lett. **72**, 633–636 (1994)
21. Y. Trabelsi, N. Ben Ali, W. Belhadj, M. Kanzari, Photonic band gap properties of one-dimensional generalized Fibonacci photonic quasicrystal containing superconductor material. J. Supercond. Nov. Magn. **32**, 3541–3547 (2019)
22. Z.A. Zaky, M. Al-Dossari, E.I. Zohny, A.H. Aly, Refractive index sensor using Fibonacci sequence of gyroidal graphene and porous silicon based on Tamm plasmon polariton. Opt. Quant. Electron. **55**, 6 (2023)
23. U. Chowdhury, R. Mukherjee, A.R. Maity, S. Kumar, P.S. Maji, Refractive index sensing using Tamm plasmons in photonic quasicrystals. Sens. Imaging **24**, 33 (2023)
24. A. Shafkat, A.N.Z. Rashed, H.M. El-Hageen, A.M. Alatwi, Design and analysis of a single elliptical channel photonic crystal fiber sensor for potential malaria detection. J. Sol-Gel Sci. Technol. **98**, 202–211 (2021)
25. X. Hui, C. Yu, Photonic bandgap structure and long-range periodicity of a cumulative Fibonacci lattice. Photon. Res. **5**, 11 (2017)
26. L.D. Negro, C.J. Oton, Z. Gaburro, L. Pavesi, P. Johnson, A. Lagendijk, R. Righini, M. Colocci, D.S. Wiersma, Light transport through the band edge states of fibonacci quasicrystals. Phys. Rev. Lett. **90**, 055501 (2003)
27. T.J. Bright, J.I. Watjen, Z.M. Zhang, C. Muratore, A.A. Voevodin, D.I. Koukis, D.B. Tanner, D.J. Arenas, Infrared optical properties of amorphous and nanocrystalline Ta2O5 thin films. J. Appl. Phys. **114**, 083515 (2013)
28. G. Ghosh, M. Endo, T. Iwasaki, Temperature-dependent Sellmeier coefficients and chromatic dispersions for some optical fiber glasses. J. Light. Tech. **12**, 1338–1342 (1994)
29. T. Srivastava, R. Das, R. Jha, On the high performance of channel photonic crystal waveguide comprising different plasmonic active metals. Appl. Phys. B Lasers Opt. **108**, 629–634 (2012)
30. E.L. Albuquerque, M.G. Cottam, Theory of elementary excitations in quasiperiodic structures. Phys. Rep. **376**, 225–337 (2003)

31. S Kumar, Titanium nitride as a plasmonic material for excitation of Tamm plasmon states in visible and near-infrared region, Photonics Nanostruct. Fund. Appl. **46**, 100956 (2021)
32. P.S. Maji, M.K. Shukla, R. Das, Blood component detection based on miniaturized self-referenced hybrid Tamm-plasmon-polariton sensor. Sens. Act. B: Chem. **255**, 729–734 (2018)
33. P.S. Maji, R. Das, Hybrid-Tamm-plasmon-polariton based self-reference temperature sensor. J. Light. Tech. **35**, 2833–2839 (2017)
34. A.K. Ghatak, K. Thayagarajan, M.R. Shenoy, Numerical analysis of planar waveguide using matrix method. J. Lightw. Technol. **5**, 660–667 (1987)
35. L.A.A. Pettersson, L.S. Roman, O. Inganas, Modeling photocurrent action spectra of photovoltaic devices based on organic thin films. J. Appl. Phys. **86**, 487 (1999)
36. P. Peumans, A. Yakimov, S.R. Forrest, Small molecular weight organic thin-film photodetectors and solar cell. J. Appl. Phys. **93**, 3693–3723 (2003)
37. U. Chowdhury, P. Mandi, R. Mukherjee, et al., Dual self-referenced refractive index sensor utilizing Tamm Plasmons in photonic quasicrystal for multistage malaria parasite detection. Plasmonics **20**, 497–509 (2025)

Applications of Nanomaterials/Nanotechnology in Batteries, Energy Storage Devices, Supercapacitors, Fuel Cells, and Smart Materials

Oluwabusayo O. Oluyide, Babamotemi O. Itakorode, George D. Ametefe, Toyin F. Babalola, Ismaila Muritala, and Alaba T. Agbele

1 Introduction

The search for green energy alternatives, such as renewable energy sources and sustainable storage technologies, is necessary due to the depletion of fossil fuel supplies, the world's expanding population, and the need for energy. Maintaining grid stability and reliability has become increasingly difficult with the advent of intermittent renewable energy sources such as wind and solar. The development of next-generation energy storage systems with high efficiency has been made possible in large part by nanomaterials and nanotechnology. Short diffusion paths on the electrodes of the energy storage system are made possible by the high surface-to-volume ratio of various nanomaterials, which leads to the devices' desirable qualities, such as high power and energy densities. As a result, nanotechnology and nanomaterials are influencing all energy storage device businesses and technologies in a way never seen before. In response to the global energy crisis and the mounting environmental concerns resulting from the use of fossil fuels, experts have recently placed a greater emphasis on clean and renewable energy sources. Over the past few decades, there has been a lot of interest in novel energy storage technologies such as supercapacitors (SCs), batteries, and fuel cells. These technologies lessen the

O. O. Oluyide (✉) · B. O. Itakorode · G. D. Ametefe · I. Muritala
Department of Biotechnology, Osun State University, Osogbo, Osun, Nigeria
e-mail: oluwabusayo.oluyide@uniosun.edu.ng

T. F. Babalola
Department of Environmental Health Technology, College of Health Sciences and Technology, Ijero Ekiti, Ekiti, Nigeria

A. T. Agbele
Department of Physics, Bamidele Olumilua University of Education, Science and Technology, Ikere-Ekiti, Ekiti, Nigeria

S. Kulkarni, S. Das (eds.), *Ionic Liquids, Nanofluids and Nanotechnology*,
Lecture Notes in Nanoscale Science and Technology 42,
https://doi.org/10.1007/978-3-032-04008-4_14

harm that fossil fuels cause to our planet by enabling cleaner, more environmentally friendly methods of energy storage and consumption. They offer a feasible path to a future with more efficient and sustainable energy [1]. The amazing realm of nanoparticles has become a ray of hope and innovation in the search for effective and sustainable energy solutions [2]. With their special qualities and wide range of uses, the use of these tiny objects has created new opportunities in the fields of energy conversion and storage. Additionally, nanoparticles have become deeply embedded in the field of solar energy conversion, serving as catalysts to increase solar cell efficiency [3]. These little wonders have increased the light absorption and overall energy conversion efficiency of solar devices by permitting light trapping, arousing plasmonic phenomena, and enabling improved charge transfer. A new era of energy harvesting has begun with the incorporation of semiconductor nanoparticles, especially quantum dots, into these devices, leading to improved dissociation and increased energy yields [4]. Nanoparticles demonstrate their effectiveness as catalysts in energy conversion processes [5]. These particles have outstanding catalytic activity due to their extraordinary surface area and complex size-dependent electronic characteristics. Nanoparticles offer intriguing new avenues for a range of energy-related applications. The surface area of nanomaterials, including nanowires, nanotubes, nanoparticles, and lithium-ion battery electrodes, is greater than that of comparable materials. As a result, energy storage reactions have more active sites, leading to better energy densities and quicker charging and discharging rates [6]. In contemporary solar cells, nanoparticles enhance light absorption and conversion efficiency, hence augmenting energy transfer. The efficiency and performance of batteries and supercapacitors are increased by the special qualities of nanomaterials, which also enhance charge transport within energy storage devices [7]. Nanoparticles also contribute significantly to energy savings by improving the efficiency of light-emitting diode illumination, where quantum dots may precisely customize output light hues and reduce energy usage. In complex catalytic systems, nanoparticles also help optimize chemical reactions, reducing the energy requirements for numerous industrial applications. Nanoparticles are crucial for improving energy transmission, storage, and savings as well as for developing a more efficient and sustainable energy environment [2]. To fulfill the demands of the future and the environment, next-generation energy storage devices must be developed with low self-discharge rates, high energy densities, and low costs. We review the latest developments in nanotechnology and nanomaterials for energy storage applications, with an emphasis on batteries, energy storage devices, supercapacitors, fuel cells, and smart materials, given the extraordinary potential of nanoparticles in energy storage. Furthermore, we explore the fundamentals, mechanisms, and production methods of nanomaterials in battery technologies.

2 Principles and Mechanisms of Nanomaterials in Battery Technologies

Nanomaterials have emerged as transformative components in modern battery technologies, significantly enhancing performance, energy density, and longevity. These materials, with dimensions between 1 and 100 nm, exhibit unique physical, chemical, and electrical properties that address many limitations of conventional battery designs [8–10]. Energy storage capacities have advanced, thanks to the use of nanotechnology in battery manufacturing, especially for electric cars and renewable energy sources. This section looks at the fundamental processes of action of nanomaterials, how they work in battery systems, and how they may transform energy storage technology.

2.1 Fundamental Principles of Nanomaterials in Battery Applications

Enhanced Surface Area and Reaction Kinetics

Nanomaterials fundamentally transform battery performance through their exceptionally high surface-to-volume ratio. This increased surface area dramatically improves the interface between electrodes and electrolytes, creating more sites for electrochemical reactions [11]. When battery materials are engineered at the nanoscale, they provide significantly more active surfaces for lithium ions to interact with during charging and discharging processes [12]. Faster charging rates and better energy distribution capabilities are directly correlated with this increased response area. For example, nanostructured electrode materials facilitate more efficient lithium-ion insertion and extraction due to shortened diffusion paths for ions moving through the electrode material [13].

Improved Electrical Conductivity

Enhancing electrical conductivity across the battery structure is one of the key benefits of nanomaterials for battery applications. The remarkable electron transport capabilities of carbon nanotubes (CNTs) and graphene, in particular, greatly increase the total conductivity of battery components [12–15]. During charge and discharge cycles, these nanomaterials' integration into electrode designs produces highly conductive networks that speed up electron flow. Carbon nanomaterials excel in forming conductive pathways throughout the electrode matrix, enabling faster energy storage and release while maintaining structural integrity [8, 16]. This enhanced conductivity directly addresses one of the fundamental limitations of conventional battery materials, particularly for cathodes with traditionally poor electronic conductivity like lithium metal phosphates.

Volume Change Accommodation

Large volume fluctuations during charging and discharging cycles cause structural deterioration and limit battery life in many high-capacity battery materials, including lithium and sodium-ion batteries [17–19]. Nanomaterials offer elegant solutions to this challenge through their unique structural properties. For instance, silicon anodes, which can store significantly more lithium than traditional graphite, typically expand by up to 300% during lithium insertion [20]. However, when silicon is engineered as nanoparticles or nanotubes, these structures can better accommodate the volumetric changes while maintaining structural integrity through multiple cycles [21, 22]. The small dimensions of nanomaterials allow for elastic deformation rather than catastrophic fracturing, significantly extending battery lifespan and maintaining capacity over more charging cycles [23].

2.2 *Nanomaterials in Battery Anodes*

Carbon-Based Nanomaterials

Carbon nanotubes (CNTs) and graphene represent revolutionary advancements for battery anode materials. These nanomaterials provide exceptional electrical conductivity combined with mechanical strength, creating resilient anode structures with enhanced performance characteristics. Multi-walled carbon nanotubes, when applied as films to lithium metal anodes, have demonstrated a remarkable ability to suppress dendrite formation, a common failure mechanism in lithium metal batteries [24, 25]. So, the nanotube films act as controlled diffusion interfaces that regulate lithium-ion flux during charging and discharging, preventing the uneven deposition that leads to dendrite growth [24, 25]. Through this mechanism, the high-energy-density benefits of lithium metal anodes are maintained while battery safety and cycle life are greatly increased.

Nanostructured Silicon Anodes

Silicon represents an ideal theoretical anode material due to its extremely high capacity—approximately ten times that of conventional graphite anodes [26]. However, its practical application has been limited by massive volume expansion during lithium insertion. Through meticulously designed silicon nanostructures, nanotechnology provides an answer [27]. When silicon is reduced to nanoparticle dimensions (around 50 nm), it provides the highest surface area and lowest charge transfer resistance, leading to superior performance [28]. These nanostructured silicon anodes can withstand the mechanical stresses of repeated charging cycles while maintaining their high capacity. Additionally, silicon and carbon nanomaterials are frequently combined to create composite anodes, which improve conductivity across the electrode structure and act as a buffer against mechanical stresses [23].

2.3 *Nanomaterials in Battery Cathodes*

Nanostructured Lithium Metal Phosphates

Lithium metal phosphates ($LiMPO_4$, where M represents metals like iron, manganese, cobalt, or nickel) are among the most promising cathode materials for high-capacity lithium-ion batteries. However, their inherently low electronic conductivity presents significant performance limitations. To overcome this difficulty, nanoscale engineering optimizes shape, decorates particle surfaces with conductive phases, and reduces crystallite size to submicron levels [29]. By creating intergranular conductive phases on lithium metal phosphate nanoparticles, researchers have produced composite materials with conductivities ranging from 10^4 to 10^2 S/cm-orders of magnitude higher than unmodified materials [29]. These enhancements facilitate coupled electron and lithium-ion movement through the cathode material, significantly improving overall battery performance.

Yolk-Shell Nanostructures

Advanced cathode designs incorporate specialized nanoarchitectures like yolk-shell structures that combine high energy density with structural stability. For example, sulfur-titanium dioxide ($S\text{-}TiO_2$) yolk-shell constructs create nanoscale compartments that contain active materials while providing void space for expansion [27]. These complex nanostructures solve the volume change issues that usually restrict cathode performance while also increasing energy density. The outer shell maintains structural integrity while the internal void space accommodates expansion of the active material during cycling, representing an elegant nanoscale solution to a macroscale engineering challenge.

2.4 *Nanomaterials in Electrolytes and Interfaces*

Nanostructured Solid-State Electrolytes

The development of safe, high-performance solid-state electrolytes represents another frontier where nanomaterials are making significant contributions. Nanostructured solid electrolytes maximize the movement of lithium ions, increasing ionic conductivity, a crucial factor for fast-charging batteries [27]. The meticulously engineered nanoporous architectures solve the conventional trade-off between conductivity and structural stability in solid electrolytes by facilitating ion transport while preserving mechanical integrity. Additionally, these electrolytes based on nanomaterials help to inhibit dendritic development, which is necessary for the practical use of high-energy-density lithium metal anodes.

Engineered Interfaces

Nanomaterials excel at creating specialized interfaces that regulate the critical processes occurring between battery components. For example, the lithiated multi-walled carbon nanotube films developed at Rice University function as controlled lithium diffusion interfaces that regulate ion flux during cycling [24, 25]. This regulation prevents the uneven deposition of lithium that leads to dendrite formation. Similar nanoscale interface engineering approaches are being applied throughout battery architectures to control electrochemical reactions, prevent side reactions, and extend battery lifespan. By tackling basic issues at the molecular level, these tailored interfaces are among the most advanced uses of nanotechnology in battery systems.

2.5 *Challenges and Future Directions*

Manufacturing Scalability

Despite their promising performance characteristics, nanomaterials face significant challenges in scaling from laboratory demonstrations to commercial production. The precise synthesis conditions required to produce optimized nanomaterials with controlled size, morphology, and surface properties present manufacturing hurdles [27]. Future research must focus on developing scalable production methods that maintain the advantageous properties of nanomaterials while enabling cost-effective mass production. This includes innovations in continuous flow synthesis, quality control mechanisms, and integration with existing battery manufacturing infrastructure.

Cost Considerations and Material Selection

Economical material selection and processing are critical to the commercial feasibility of batteries augmented with nanomaterials. While nanomaterials offer impressive performance improvements, they must ultimately compete with established technologies on price as well as performance. Research into abundant, environmentally friendly nanomaterials represents a promising direction, as nanoscale engineering can potentially enhance the performance of common materials, reducing dependence on rare or harmful elements. This approach could simultaneously address cost concerns while improving the environmental profile of battery technologies.

Optimized Nanostructure Design

The future of nanomaterials in batteries lies in increasingly sophisticated and application-specific nanostructure designs. Advanced methods entail building hierarchical structures with carefully tailored characteristics at several scale levels, as opposed to merely decreasing particle size. Multiple performance factors, such as combining a large surface area with strong mechanical stability and ideal electron/ion transport channels, may be addressed simultaneously by these designs. Computational modeling coupled with advanced characterization techniques will play crucial roles in developing these next-generation nanostructured battery materials [27].

3 Types of Nanomaterials Used for Energy Storage in Battery Technologies

Materials for negative electrodes are crucial to the different parts of lithium-ion batteries (LiBs). The nanomaterials in the LiBs' anode enhance capacity and increase electron collection efficiency [30] and promote lithium intercalation and deintercalation. Currently, there are three different kinds of nanomaterial anodes for LiBs: alloy materials (usually silicon, germanium, and carbon-based), conversion materials (metals with oxides, sulfides, phosphides, and nitrides), and insertion materials (carbon-based and titanium-based).

3.1 $Li_4Ti_5O_{12}$ (LTO)

LiBs benefit from the intercalated titanium-based oxide anode's inherent qualities, which include increased safety while cycling and no volume change, among others. Nevertheless, a very small theoretical capacity (about 175 mAh/g) and negligible ionic conductivity are significant problems for industrial production on a wide scale. When used as electrodes directly, without the need for conductive binders or additives, titanium foils with nanowires showed outstanding rate performance since the theoretical capacity value is near the practical capacity (173 mAh/g at 0.2 °C rate) and has a better cycle life. After 100 cycles at 5 °C, the cycling performance reached a 95% capacity retention.

3.2 Carbon-Based (Carbon) Nanotubes (CNTs)

According to Desai et al., CNTs are rolled graphene sheets having one or more layers. Carbon nanotubes (CNTs) are produced by several techniques, such as chemical vapor deposition, ball milling, and acid oxidation, and are then employed as negative electrode materials for LiBs [30]. Because of their tiny diameter, CNTs may put strain on their hexagonal lattice structure, producing a robust graphitic sheet. They are therefore electronegative due to these characteristics. According to Senami et al., single-walled carbon nanotubes are single-walled sheets, and multi-walled carbon nanotubes are multi-layered sheets. Because of its greater capability for energy conversion and storage, a carbon nanotube electrode is superior to a conventional graphite electrode.

3.3 Oxide of Titanium (TiO_2)

Enhancing qualities, including capacity retention, structural stability, high oxidation ability, high chemical steadiness, structural diversity, safety, and low mass manufacturing costs, are all made possible by titanium oxide [31]. At 1.5 V versus Li/Li^+ (operating potential), it is also steady. These features increase the potential of LiB anode materials, especially for use in electric vehicle applications. Systems for inserting and removing lithium depend on the particle's size, shape, surface area, and crystallinity. Among the numerous allotropic forms of TiO_2 are rutile, anatase, and brookite. Materials with nanostructured anodes, such as graphene-assisted anatase TiO_2, porous TiO_2-C nanocomposite carapaces, and TiO_2-graphene nanocomposites, showed more than 90% capacity retention and more than 100 cycle stability [32].

3.4 Cobalt Oxides

Cobalt oxides have several exceptional redox characteristics, such as high theoretical capacity (Co_3O_4 is 890 mAh/g while CoO is 715 mAh/g) and capacity retention. Cobalt oxides (CoO and Co_3O_4) are utilized as negative electrodes for LiBs. Mass practical applications are hampered by their drawbacks, which include toxicity, volume change, and semiconducting nature when charging and discharging [33]. It is available in a variety of forms, including Co_3O_4 nanotubes, multiple shell-spheres, porous nanostructures (nanoplates), nanowires, and nanosheets. Guan et al. created CoO octahedral nanocages in a pure phase. The authors utilized template-free techniques to use NH_3. Thus, octahedral nanocages of uniform size were formed, with edge lengths ranging from 100 to 200 nm [34].

3.5 Iron Oxides

Nanostructured iron oxides are crucial for LiB anode materials. For instance, as anode materials, iron oxides (α-Fe_2O_3, γ-Fe_2O_3, and Fe_3O_4) are superior to semiconductor and carbon/graphite materials. These iron oxides are suitable for anode applications due to their inexpensive cost, high theoretical capacities (about 1000 mAh/g), non-toxicity, and reactivity with lithium conversion [35]. They do have certain drawbacks, though, including a short cycle life because of low ionic conductivity (10^{-7} S/gm at RT), limited diffusion of Li-ions, volume change, and iron accumulation during cycling.

3.6 Germanium

Compared to graphite anodes, germanium-based anodes are required for lithium-ion batteries due to their 10,000 times higher electrical conductivity, low energy/band gap (0.67 eV), high theoretical capacity (1623 mAh/g), and high Li-ion diffusion [36]. At normal temperature, Li-ions move into germanium (Ge) 15 times as fast as Si, and at 360 °C, they do so 400 times faster. However, due to the large volume difference (about 300%) brought on by mechanical stress during insertion and de-insertion, Ge is limited in practical applications when it comes to bulk structure [37].

3.7 Silicon

Due to its widespread use and environmental friendliness, silicon (Si) is a more suitable active material for an anode than graphite. Due to its high volumetric capacity (9786 mAh/cm^3), Si shortens the time of donor–acceptor interactions and boosts performance [38, 39]. It is evident why Si and its derivatives are imagined to be the most futuristic material for the production of LiBs. The production of intermetallic Li–Si binary composites has been referenced as the reason for the high specific capacity value. However, the significant volume changes that take place during the lithiation/delithiation process, which result in subpar cycling performance, are a significant obstacle to the widespread use of silicon-based anodes.

4 Synthetic Approaches for Nanomaterial Applications in Energy Storage

The influence of important fabrication processes on the structural, morphological, and electrochemical characteristics of nanomaterials is examined in this section. The techniques include sol-gel, hydrothermal synthesis, chemical vapor deposition (CVD), electrospinning, green synthesis, and microwave-assisted approaches. By analyzing recent advancements, it highlights how these methods address critical challenges in supercapacitors, lithium-ion batteries, and other energy storage devices, while also exploring future directions for sustainable and high-efficiency materials design. Energy storage systems have undergone a revolution, thanks to artificial intelligence, and synthetic techniques have been crucial in modifying nanomaterials for enhanced functionality.

4.1 Sol-Gel Synthesis for Tailored Nanostructures

A colloidal suspension (sol) is transformed into a gel-like network before being calcined to create nanostructured materials in sol-gel methods, a wet chemical technique that is frequently used to create metal oxide nanocomposites [40]. The ability to regulate particle size and composition at low temperatures is a significant benefit of this method, which makes it suited for producing homogeneous materials such as zinc oxide (ZnO) and titanium dioxide (TiO_2). The stability and high theoretical capacity of these oxides make them essential for lithium-ion battery anodes [40]. For instance, sol-gel-derived TiO_2 nanoparticles possess a porous structure that allows fast ion diffusion, attaining specific capacities surpassing 200 mAh/g [40]. Recent studies have optimized sol-gel parameters such as precursor concentration and pH to engineer materials with enhanced conductivity. By incorporating carbon matrices during gelation, researchers have developed hybrid electrodes that mitigate volume expansion during charge cycles, improving cycle life by up to 40% [40]. For industrial applications, the sol-gel method must be integrated with other methods, such as spray drying, due to its limitations in scalability and processing times, despite its versatility.

4.2 Hydrothermal Synthesis of Metal Ferrites

To precisely regulate shape and crystallinity, hydrothermal synthesis uses high-pressure aqueous conditions to crystallize nanomaterials. An investigation conducted in 2024 showed that this technique was effective in creating manganese ferrite ($MnFe_2O_4$) and nickel-manganese ferrite (Ni-$MnFe_2O_4$) nanoparticles, with average crystallite sizes of 17.88 nm and 23.75 nm, respectively [41]. These materials, which have spherical and semi-spherical shapes, were applied to nickel foam substances to test supercapacitors. The Ni-$MnFe_2O_4$ electrode exhibited a remarkable specific capacitance of 1255 F/g at 1 mA/cm^2, outperforming many conventional composites [41]. The success of hydrothermal synthesis lies in its ability to regulate reaction kinetics through temperature and precursor concentration. For example, using 2M NaOH as a mineralizer promotes the formation of phase-pure ferrites, while shorter reaction times yield smaller nanoparticles with higher surface areas [41]. The strong structural integrity and reduced electrode–electrolyte interfacial resistance of these materials allow them to exhibit remarkable cyclic stability, maintaining over 90% capacitance after 5000 cycles [41].

4.3 Chemical Vapor Deposition for 3D Nanostructures

Chemical vapor deposition (CVD) has emerged as a cornerstone for synthesizing graphene and transition metal dichalcogenides (TMDCs) with three-dimensional architectures [42]. By decomposing hydrocarbon precursors on structured substrates, CVD enables the growth of vertically aligned carbon nanotubes and crumpled graphene sheets. These 3D designs prevent layer restacking, increasing accessible surface areas by up to 300% compared to planar counterparts [42]. In electrocatalytic applications, CVD-grown molybdenum disulfide (MoS_2) nanosheets exhibit a low hydrogen evolution reaction (HER) overpotential of 140 mV, rivaling platinum-based catalysts [42]. Recent innovations involve hybridizing TMDCs with conductive scaffolds like nickel foam. A 2024 review highlighted that such hybrids, when used in sodium-ion batteries, achieve specific capacities of 523 mAh/g due to enhanced ion diffusion pathways [42]. Research on plasma-enhanced variants that lower synthesis temperatures by 40% is being driven by CVD's high energy consumption and substrate reliance, which continue to be obstacles [42].

4.4 Electrospinning of Nanofibrous Electrodes

Electrospinning transforms polymer solutions into continuous nanofibers through electrostatic forces, producing mats with high porosity and surface-to-volume ratios [43]. This method is particularly effective for fabricating lithium battery anodes, where carbon nanofibers embedded with silicon nanoparticles deliver capacities of 1226 mAh, three times higher than graphite [43]. The fibrous architecture accommodates silicon's volume expansion, maintaining 77% capacity retention after 1000 cycles [43]. In supercapacitors, polyaniline-coated electrospun fibers achieve specific capacitances of 501 F/g, with energy densities surpassing 30 Wh/kg [43]. Fiber diameter and alignment are fine-tuned by optimizing spinning parameters, such as voltage (15–25 kV) and collector distance (10–20 cm), which have a direct impact on electrochemical performance. For instance, reducing fiber diameter from

500 nm to 100 nm boosts charge–discharge rates by 50% due to shortened ion transport paths [43].

4.5 Green Synthesis for Sustainable Materials

An environmentally acceptable substitute for conventional techniques, green synthesis turns metal precursors into nanoparticles using plant extracts and biogenic waste [43]. For example, mango leaf extract-mediated synthesis produces silver-doped cobalt oxide (Ag-Co_3O_4) nanoparticles with sizes below 20 nm. These nanoparticles, when tested in symmetric supercapacitors, exhibit a capacitance of 422 F/g and 91.5% retention after 30,000 cycles [43].

Carbon quantum dots, which are conductive additions in zinc-ion batteries, are produced from agricultural waste, demonstrating the scalability of green synthesis. These dots enhance electrode kinetics, enabling a specific capacity of 423 mAh/g and 84.7% retention over 1000 cycles [43]. Despite these gains, batch-to-batch variability and sluggish reaction kinetics demand additional development for commercial application.

4.6 Microwave-Assisted Synthesis for Microstructural Control

Microwave irradiation enables rapid, uniform heating, facilitating the synthesis of nanomaterials with tailored microstructures [44]. A 2024 study showcased the fabrication of titanium niobium oxide ($TiNb_2O_7$) anode materials with a crystallite size of 23 nm, delivering a specific capacity of 299 mAh/g in lithium-ion batteries [44]. Microwave parameters such as power (800 W) and irradiation time (10 min) critically influence grain growth, with shorter durations yielding smaller particles and higher surface areas [44]. This technique excels in creating defect-engineered materials. For instance, nitrogen-doped ultrananocrystalline diamond (N-UNCD) films synthesized via microwave plasma CVD exhibit exceptional conductivity, enabling a lithium storage capacity of 350 mAh/g and 98% retention after 100 cycles [44]. In supercapacitors, microwave-synthesized $Mn_3O_4/CoWO_4$ composites achieve 529.8 mAh/g with 117.2% capacity retention over 13,000 cycles, attributed to synergistic ion storage mechanisms [44].

4.7 Comparative Analysis of Synthesis Techniques

Every synthetic method has unique benefits suited to particular energy storage uses. Sol-gel and hydrothermal methods excel in producing oxide-based nanomaterials with precise stoichiometry, while CVD and electrospinning are optimal for carbonaceous and fibrous architectures. Green synthesis prioritizes sustainability but lags in scalability, whereas microwave techniques provide rapid, energy-efficient microstructure control. Trade-offs are shown by performance metrics: ferrites produced hydrothermally have higher capacitance (>1200 F/g) but need longer reaction times (12–24 h) [42]. In contrast, microwave-derived materials achieve comparable performance in minutes but demand specialized equipment [44]. Electrospun fibers balance high capacity and flexibility, making them ideal for wearable energy devices [43].

5 Applications of Nanomaterial/Nanotechnology in Energy Storage Devices

Applications of nanomaterials in energy storage devices have transformed and improved the performance of energy storage technology, for example, lithium-ion batteries and supercapacitors. Also, enhanced charges, increased effectiveness, and longevity, which in turn are driving market growth. Nanomaterials' performance benefits are a necessary determinant in meeting the complete potential of energy storage innovations [11]. Nanotechnology and advanced nanomaterials have reformed energy storage technology industries and given access to new possibilities. General applications of nanomaterials have been practiced in microbial fuel cells (MFCs) and carbon nanofiber/PDMS nanocomposites and have been significantly used as a coating for copper anodes. Also, battery applications have worked with different performances of V4C3 MXene as an anode for Li-ion and Na-ion batteries, corresponding cathodes, and synthesis routes [45]. The application of nanomaterials and nanotechnology for energy storage devices is more than batteries, capacitors, and fuel cells.

6 Economic Impact of the Advancement of Nanomaterials in Energy-Related Technologies

Nanotechnology has had a profound impact on the result of greater efficiency and a next-generation energy storage system. Significant performance improvements in solar energy match traditional silicon-based photovoltaics. Nanomaterials have also made it possible for robust batteries and supercapacitors, tackling the energy storage issues built into renewable energy systems [46]. In the energy industry, nanotechnology-assisted breakthroughs have boosted the efficiency of photovoltaic cells and energy storage systems, thereby facilitating the shift to cleaner energy and minimizing pollution. Advanced materials involved in sustainable practices, such as lightweight materials in transportation, support lower fuel utilization and pollution, with efforts to combat climate change worldwide. Advancements in nanomaterials packaging have extended food product lifespans, reduced waste, and promoted sustainable supply chains [25].

7 Sustainability and Potential Environmental Implications

Nanotechnology is a fast-growing field and is seen as one of the areas with more potential to facilitate the sustainability of human systems by developing efficient production processes from an energy point of view [46]. The environment is one of the main concerns out of the three pillars of sustainability. Sustainability assesses nanotechnology products from the industry sector that require the evaluation of impacts along the supply chain, ranging from the raw materials to the final products. All nanoparticles have some similar characteristics but differ in their material properties. For example, some nanoparticles are effective catalysts, good sorbents, and others have perceiving capacity or are harmful to humans or the environment. Currently, implications among varied environmental sections of air, water, and soil contamination have emerged from fabrication systems and enhanced toxicological fouling of the ecosystem due to the compositions of some of the nanomaterials. Nanoparticles have higher coverage compared to the materials, which can cause more harm to the ecosystem. The environmental impact and hazards linked with nanotechnology are very finite and inconsistent. The potential environmental implications are

1. Nanoparticle synthesis demands high energy.
2. Release of hazardous nanosubstances producing environmental harm.
3. Reduced recovery and reuse rate.
4. Scarcity of trained or skilled engineers and workers is causing further concerns [46].

8 Conclusion

The application of nanotechnology and nanomaterials into energy storage technology denotes a remarkable advancement for sustainable energy storage systems. The unique properties of nanomaterials have led to significant improvements in their performance in devices. Innovations address the traditional energy storage technologies and open the door for cleaner and more efficient energy systems. Moreover, the economic implications of these are profound. The enhanced efficiency and performance of nanomaterials contribute to renewable energy sources. This shift reduces reliance on fossil fuels and also reduces pollution and combats climate change. This development can lead to significant cost savings in different sectors. As industries engage in these systems, it is likely to push toward sustainable practices, providing brand new opportunities and innovations. However, it is important to be observant concerning the synthesis and application of nanoparticles, which can put risks into ecosystems and human health. Therefore, there is a need for research and regulatory frameworks to ensure that the benefits of nanotechnology in energy storage are achieved by reducing potential environmental impacts.

References

1. I.I. Hussain, R. Shaheen, Ahmad, et al., Binder-free cupricion containing zinc sulfide nanoplates-like structure for flexible energy storage devices. Chemosphere **314**, 137660 (2023)
2. F. Jamil, F. Hassan, S. Shoeibi, M. Khiadani, Application of advanced energy storage materials in direct solar desalination: A state of art review. Renew. Sust. Energ. Rev. **186**, 113663 (2023)
3. H.R. Jiang, Y.K. Zeng, M.C. Wu, W. Shyy, T.S. Zhao, A uniformly distributed bismuth nanoparticle-modified carbon cloth electrode for vanadium redox flow batteries. Appl. Energy **240**, 226–235 (2019)
4. M.B. Gawande, A. Goswami, T. Asefa, et al., Core–shell nanoparticles: Synthesis and applications in catalysis and electrocatalysis. Chem. Soc. Rev. **44**(21), 7540–7590 (2015)
5. J. Zhao, A.F. Burke, Electrochemical capacitors: Materials, technologies and performance. Energy Storage Mater **36**, 31–55 (2021)
6. H. Tondan, A.K. Singh, Advances in energy harvesting and storage materials: Unlocking the potential of solid-state nanomaterials for renewable energy technologies, in *Futuristic Trends in Physical Sciences*, (Iterative International Publishers, Chikkamagaluru, 2024), pp. 21–32
7. S.M.M. Alonzo, J. Bentley, S. Desai, B.P. Bastakoti, Hydrothermal synthesis of hierarchical microstructure tungsten oxide/carbon nanocomposite for supercapacitor application. Sci. Rep. **13**, 21732 (2023)
8. M.Z.A. Mahmud, A concise review of nanoparticles utilized energy storage and conservation. J. Nanomater. **2023**(11), 1–14 (2023)
9. R. Sharma, L.K. Sharma, Nanotechnology's potential in advancing renewable energy solutions, in *2024 International Conference on Communication, Control, and Intelligent Systems (CCIS)*, (IEEE, New York, 2024), pp. 1–6
10. P. Simpa, N.O. Solomon, O.A. Adenekan, S.C. Obasi, Nanotechnology's potential in advancing renewable energy solutions. Eng. Sci. Technol. J. **5**(5), 1695–1710 (2024)
11. E. Pomerantseva, F. Bonaccorso, X. Feng, Y. Cui, Y. Gogotsi, Energy storage: The future enabled by nanomaterials. Science **366**(6468), 8285 (2019)
12. P. Kumar, H.K. Channi, A. Babbar, R. Kumar, J.K. Bhutto, T.M.Y. Khan, A. Bhowmik, A. Razak, A.W. Wodajo, A systematic review of nanotechnology for electric vehicles battery. Int. J. Low-Carbon Technol. **19**, 747–765 (2024)
13. L. Zhang, X. Qin, S. Zhao, A. Wang, J. Luo, Z.L. Wang, F. Kang, Z. Lin, B. Li, Advanced matrixes for binder-free nanostructured electrodes in Lithium-ion batteries. Adv. Mater. **32**(24), e1908445 (2020)
14. A.V. Bazanov, A.N. Prusov, S.M. Prusova, M.V. Radugin, Synthesis and properties of graphene-like porous carbon from modified cellulose. Full. Nanotubes Carbon Nanostruct. **33**(4), 415–429 (2025)
15. C. Qi, J. Wang, S. Li, Y. Cao, Y. Liu, L. Wang, Application of graphene in lithium-ion batteries, in *Nanotechnology and Nanomaterials*, (IntechOpen, London, 2024)
16. A.A.E. Abdou, Z.A.N. Khaled, N.M. Saman, Current insights and future prospects of graphene aerogel-enhanced supercapacitors: A systematic review. Heliyon **10**(17), e37071 (2024)
17. S.S. Madani, Y. Shabeer, F. Allard, M. Fowler, C. Ziebert, Z. Zuolu Wang, S. Panchal, et al., A comprehensive review on lithium-ion battery lifetime prediction and aging mechanism analysis. Batteries **11**(4), 127 (2025)

18. M.T. McDowell, S. Xia, T. Zhu, The mechanics of large-volume-change transformations in high-capacity battery materials. Extreme Mech. Lett. **9**, 480–494 (2016)
19. T. Wulandari, D. Fawcett, S.B. Majumder, G.E.J. Poinern, Lithium-based batteries, history, current status, challenges, and future perspectives. Batter. Energy **2**(6), 20230030 (2023). https://doi.org/10.1002/bte2.20230030
20. H. Seo, D. Kim, S. Park, E. Seo, P. Kim, J. Choi, J. Eeyoung Yoo, A comprehensive review of liquid electrolytes for silicon anodes in lithium-ion batteries. Adv. Ind. Eng. Chem. (2025). https://doi.org/10.1007/s44405-025-00004-1
21. X. Chen, W. Cheng, H. Liu, H. Chen, J. Ma, Y. Zhang, Z. Wu, et al., Research progresses on nano-structured silicon-based materials as anode for lithium-ion batteries. Materials **18**(4), 830 (2025). https://doi.org/10.3390/ma18040830
22. M. Khan, S. Yan, M. Ali, F. Mahmood, Y. Zheng, G. Li, J. Liu, X. Song, Y. Wang, Innovative solutions for high-performance silicon anodes in lithium-ion batteries: Overcoming challenges and real-world applications. Nano-Micro Lett. **16**(1), 179 (2024)
23. M.K. Aslam, Y. Niu, T. Hussain, H. Tabassum, W. Tang, M. Xu, R. Ahuja, How to avoid dendrite formation in metal batteries: Innovative strategies for dendrite suppression. Nano Energy **86**, 106142 (2021)
24. R.V. Salvatierra, G.A. López-Silva, A.S. Jalilov, J. Yoon, G. Wu, A.L. Tsai, J.M. Tour, Suppressing Li metal dendrites through a solid Li-ion backup layer. Adv. Mater. **30**(50), e1803869 (2018)
25. J. Pope, C. Ban, D. Rodriguez, D.P. Finegan, J. Pope, Title: Study of structural evolution of silicon-based anodes in lithium-ion batteries. Meet. Abstr. **1**(1), 174 (2024)
26. T.B. Taha, A.A. Barzinjy, F.H.S. Hussain, T. Nurtayeva, Nanotechnology and computer science: Trends and advances. Memories **2**, 100011 (2022)
27. D. Gyanprakash, Maurya, C.K. Rastogi, Investigation of silicon nanoparticle size on specific capacity of li-ion battery via electrochemical impedance spectroscopy. J. Electroanal. Chem. **931**, 117176 (2023)
28. B. Ellis, P.S. Herle, Y.H. Rho, L.F. Nazar, R. Dunlap, L.K. Perry, D.H. Ryan, Nanostructured materials for lithium-ion batteries: Surface conductivity vs. bulk ion/electron transport. Faraday Discuss. **134**, 119–141 (2007)
29. P. Roy, S.K. Srivastava, Nanostructured anode materials for lithium ion batteries. J Mater Chem A **3**, 2454–2484 (2015)
30. Y. Pan, S. Chen, Y. Jia, First-principles investigation of phonon dynamics and electrochemical performance of TiO2-x oxides lithium-ion batteries. Int. J. Hydrog. Energy **45**, 6207–6216 (2020)
31. W. Wang, Q. Sa, J. Chen, Y. Wang, H. Jung, Y. Yin, Porous TiO2/C nanocomposite shells as a high-performance anode material for lithium-ion batteries. ACS Appl. Mater. Interfaces **5**, 6478–6483 (2013)
32. Y. Li, Co3O4 nanoflake as high capacity anode materials for superior lithium storage performance. Int. J. Electrochem. Sci. **17**, 220454 (2022)
33. H. Guan, X. Wang, H. Li, C. Zhi, T. Zhai, Y. Bando, D. Golberg, CoO octahedral nanocages for high-performance lithium ion batteries. Chem. Commun. **48**, 4878 (2012)
34. I. Khan, A. Khalil, F. Khanday, A.M. Shemsi, A. Qurashi, K.S. Siddiqui, Synthesis, characterization and applications of magnetic iron oxide nanostructures. Arab. J. Sci. Eng. **43**, 43–61 (2018)
35. N.G. Rudawski, B.R. Yates, M.R. Holzworth, K.S. Jones, R.G. Elliman, A.A. Volinsky, Ion beam-mixed Ge electrodes for high capacity Li rechargeable batteries. J. Power Sources **223**, 336–340 (2013)
36. H. Zhang, P.V. Braun, Three-dimensional metal scaffold supported bicontinuous silicon battery anodes. Nano Lett. **12**, 2778–2783 (2012)
37. F. Wang, Y. Ma, P. Li, C. Peng, H. Yin, W. Li, D. Wang, Electrochemical conversion of silica nanoparticles to silicon nanotubes in molten salts: Implications for high-performance lithium-ion battery anode. ACS Appl. Nano Mater. **4**, 7028–7036 (2021)

38. H. Cheng, J.G. Shapter, Y. Li, G. Gao, Recent progress of advanced anode materials of lithium-ion batteries. J. Energy Chem. **57**, 451–468 (2021)
39. D. Bokov, A.T. Jalil, S. Chupradit, W. Suksatan, M.J. Ansari, I.H. Shewael, G.H. Valiev, E. Kianfar, Nanomaterial by sol-gel method: Synthesis and application. Adv. Mater. Sci. Eng. **2021**(1), 1–21 (2021)
40. P.P. Chavan, U.D. Babar, A.D. Chougale, D.J. Shetti, R.B. Patil, P.D. Kamble, M.H. Mahmoud, Hydrothermal synthesis of metal ferrite nanocomposites for energy storage applications. J. Nanoelectron. Optoelectron. **19**(7), 684–691 (2024)
41. Y. Peng, J. Hu, Y. Huan, Y. Zhang, Chemical vapor deposition growth of graphene and other nanomaterials with 3D architectures towards electrocatalysis and secondary battery-related applications. Nanoscale **16**(16), 7734–7751 (2024)
42. G. Sun, L. Sun, H. Xie, J. Liu, Electrospinning of nanofibers for energy applications. Nano **6**(7), 129 (2016)
43. H.M. Abuzeid, C.M. Julien, L. Zhu, A.M. Hashem, Green synthesis of nanoparticles and their energy storage, environmental, and biomedical applications. Crystals **13**(11), 1576 (2023)
44. P. Formoso, R. Muzzalupo, L. Tavano, G. De Filpo, F.P. Nicoletta, Nanotechnology for the environment and medicine. Mini Rev. Med. Chem. **16**(8), 668–675 (2016)
45. P. Padma, Nanotechnology and its applications in medicine, energy and transportation (2020), https://industrywired.com/nano-technology-and-itsapplications-in-medicine-energy-and-transportation/
46. M.S. Diallo, N.A. Fromer, M.S. Jhon, *Nanotechnology for Sustainable Development: Retrospective and Outlook* (Springer, Cham, 2014)

Applications of Nanofluids in Energy Storage Devices

Sunil Jayant Kulkarni

1 Introduction

Optimum energy utilization of energy resources has become the most crucial aspect of modern-day industrial and societal developments. The conventional energy resources are depleting. Also, the focus of people and society at large nowadays is on the use of green energy. The renewable energy resources are becoming the keyword in the energy sector. There are many types of non-conventional energy sources. Wind, tidal and solar energy are the forms of energy the experts are looking at to explore and utilize. Solar energy use has increased worldwide [1, 2]. Also, the use of hydrogen for vehicles is being explored. While one explores different energy types, three aspects of energy are very crucial: synthesis, storage and transportation. Storage is associated with the capacity and ability to transfer, convert or respond. Solar energy is collected with the help of solar collectors in various forms. It can be stored as latent heat (thermal), electrical energy (battery storage) or hydro or compressed gas (mechanical) energy. Energy storage in thermal form is commonly practised. In the use of thermal energy storage, charging and discharging of batteries need good thermal conductivity and heat transfer coefficient [1, 3]. The storage of the energy is facilitated by utilizing phase change materials (PCMs). Molten salts are used as electrolytes for ion transport. The nanofluids contain a base fluid enhanced with the nanoparticle dispersion. The nanofluids, as discussed in this chapter, have many advantages related to heat capacity, conductivity and heat transfer.

S. J. Kulkarni (✉)
Chemical Engineering Department, Gharda Institute of Technology, Lavel, Maharashtra, India

S. Kulkarni, S. Das (eds.), *Ionic Liquids, Nanofluids and Nanotechnology*,
Lecture Notes in Nanoscale Science and Technology 42,
https://doi.org/10.1007/978-3-032-04008-4_15

1.1 Nanotechnology

Nanotechnology includes the study of nanomaterials. Also, nanofabrication and nanometrology are important areas of nanotechnology. In nanotechnology, the atoms and molecules are manipulated to obtain customized materials with new, required properties and structures. Nanotechnology is becoming an important part of modern-day development in material engineering [1–3]. Nanotechnology is used in almost all applications involving materials science. Additionally, nanomaterials and nanoparticles are an inherent part of modern-day developments in the medical and pharmaceutical fields. In modern-day developments in sports gears, paint technologies, computer hardware, fertilizer and crop management and entertainment industry, nanotechnology has been a useful tool.

1.2 Nanomaterials

Many materials, when disintegrated to nanoscale, exhibit very different physical properties from the original material [4, 5]. Many physico-chemical and biological properties were also found to be significantly different. These properties, many times, are found to change in such a way that the transformation enhances certain properties, leading to novel applications. Classification of nanoparticles can be done on the basis of dimension, composition and morphology. Dimensionally, these can be one-dimensional like nanotubes, two-dimensional like nanofilms or three-dimensional like nanocomposites. Based on the composition, the nanomaterials can be organic (derived from natural origin), inorganic, composite and carbon-based. Based on the morphology, the nanomaterials can be classified according to their shape, size and structure.

1.3 Nanofluids

Water, ethylene glycol (EG), oil, etc. are used as base fluids to form nanofluids containing nanoparticles. These nanofluids possess enhanced thermal properties like improved thermal conductivity and heat transfer [6]. The stability of nanoparticles is important, and the application of surfactants can ensure it. Apart from the thermal conductivity and heat transfer, density and viscosity also get affected due to the addition of nanoparticles. Nanomaterials improve the thermal efficiency of solar heat collection systems and hence find wide application in solar energy systems. There are various types of nanofluids based on the composition and content [6]. Hybrid nanofluids contain more than one nanofluid, resulting in far better thermal properties. The use of PCM improves thermal properties considerably. Nanofluids find wider applications in heating and cooling applications in industry. As discussed, the solar energy systems are major beneficiaries of the nanofluid advantages over conventional materials.

1.4 Aim and Objectives

This chapter aims at providing insight into enhancing energy storage with the use of nanofluids in energy storage systems, related investigations and future perspectives. This chapter sheds light on the nanofluid, its general applications, classification and properties. In the consecutive sections, the synthesis of nanofluids, properties, applications and recent developments are discussed, citing the significant research and literature in the field. This chapter can serve as a guide for the researchers willing to work on this topic.

1.5 Overview

Heat transfer fluids are essential in thermal energy systems for increasing and decreasing the temperatures for various applications. The fluids used for energy transfer are water, EG, oil and various thermic fluids. The low thermal conductivity of these materials is a disadvantage. The nanofluids have provided a solution to the problem [7].

Nanoparticles are explored for increasing the heat transfer coefficient of the fluids. In solar collectors, nanofluids can be used as absorbing materials. Nanofluids based on aluminium, carbon nanotubes (CNTs), Al_2O_3, etc. have been used by investigators to find out the enhancement in heat transfer properties of the nanofluids than a conventional medium [8–12]. These studies have reported significant improvement in heat transfer properties. Evacuation-type solar collectors are superior to the conventional collectors because of advantages such as high efficiencies, low cost and lower heat losses. The use of nanofluids can increase the thermal properties of such systems [12]. By using nanofluids instead of water, enhanced thermal efficiency can be obtained [13]. In solar collectors, the absorbed solar radiations are converted into heat. The fluid is used for sending this heat to various applications. The ratio of useful gain to available energy is collector efficiency, which can be increased by the use of nanofluids.

There are many more advantages of using nanofluids in solar energy systems. Nanofluids have a small particle size and a large surface area. Also, the optical properties of nanofluids are much better than those of conventional fluids. High thermal conductivity is another advantage. Nanofluids are an excellent absorbing medium due to their high absorption capacity and high thermal stability [14, 15]. Figure 1 depicts the properties that are fundamental for their applications in thermal energy systems. Figure 2 depicts the application areas of nanofluids in thermal energy storage systems.

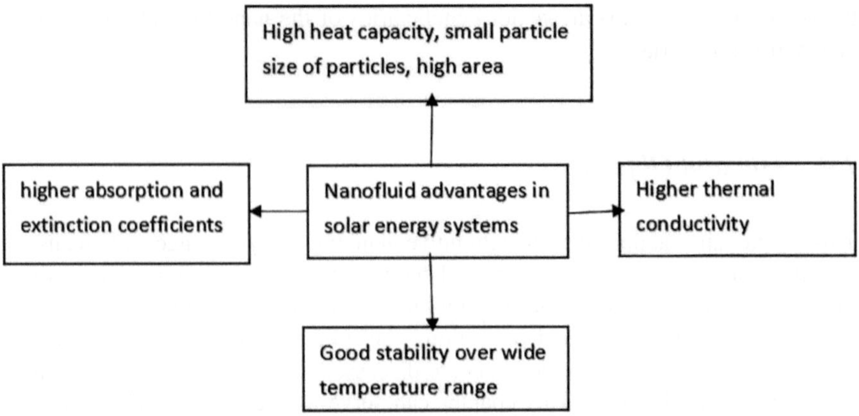

Fig. 1 Nanofluid advantage in solar energy systems [14]

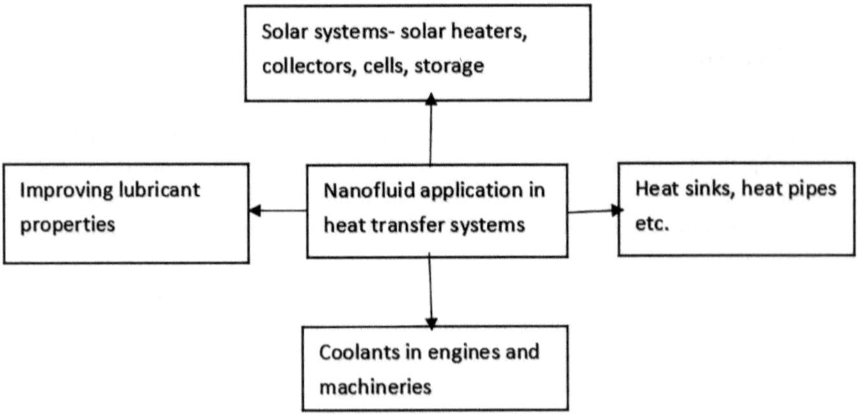

Fig. 2 Nanofluid application in solar energy systems [14, 15]

2 Nanofluids in Energy Storage

The thermal energy can be stored in the form of sensible heat, latent heat and thermochemical energy. Accordingly, the storage systems are classified into these three types of storage as indicated in Fig. 3 [16].

Sensible heat storage systems are simple and efficient and capable of recovering about 70–90% of the stored energy [17]. The disadvantages of this system are less heat storage per unit volume, heat losses and temperature distribution. Latent heat-based storage systems utilize organic or inorganic PCMs [18–21]. Paraffins, fatty acids, esters and fatty alcohols are examples of organic phase materials. Water and molten salts are commonly used inorganic PCMs. Thermochemical heat storage

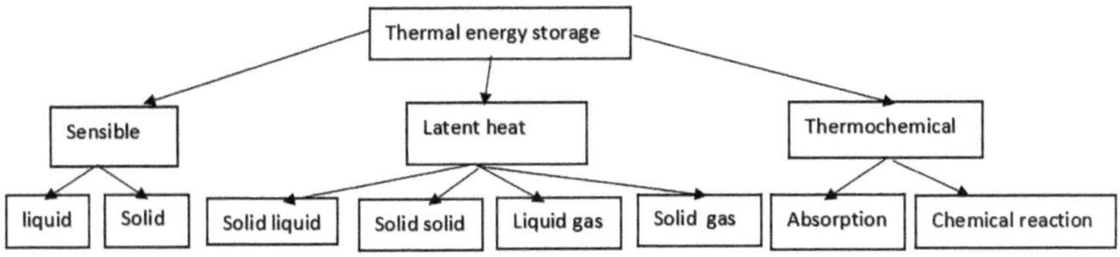

Fig. 3 Thermal energy storage systems [16]

includes the utilization of reversible chemical reactions [20]. Slow kinetics and low energy density are limitations of thermochemical storage systems.

2.1 *Nanofluid Enhanced Energy Storage*

The nanofluid application for energy storage has a number of advantages, namely, improved heat transfer rate and better heat distribution, resulting in lesser heat losses and improved heat transfer efficiency. The heat transfer enhancement due to better thermal conductivity is an obvious advantage. High retrieval and storage efficiency due to better thermal properties can be attributed to better conductivity, higher heat transfer and improved energy density [21–25]. Also, the degree of supercooling is reduced by the addition of nanoparticles. Another advantage of using nanofluids is that it helps to prevent the sedimentation, clogging and fouling of pipes and pumps [25]. This is due to the micro or millimetre-sized particles. Application of nanofluids is not limited to their heat transfer properties but also their optical properties. By means of selective coatings, nanomaterials can improve the optical properties of the receivers [26]. The nanofluids are capable of absorbing 100% solar energy, while water can absorb 80% [27]. The increase in viscosity can result in a rise in pumping power. In passive systems, nanofluids have better efficiency [27].

2.2 *Nanofluid Synthesis [28, 29]*

Nanoparticles are dispersed in the base fluid to form nanofluids. Proper mixing is crucial for the formation of a uniform concentration. Stability of nanofluids is also important from their application point of view. The nanofluids can be prepared by using a one-step or two-step process.

In a two-step process, dry powder is produced by using a physical or chemical method. Then, ball milling, ultrasound agitation, sonication and homogenization are used for proper dispersion. Due to high surface area and activity, nanoparticles have a tendency to aggregate. Surface-active species can be used to stabilize them. Ceramic compounds such as oxides of aluminium, copper, silica, zinc, caesium and titanium can be used as nanoparticles [28–34]. Glycerol, EG, ethanol, acetone, oil, water, deionized water and distilled water are often used as base fluids for nanofluid synthesis [35]. Surfactants are used for increasing the stability of nanofluids [36]. The surfactant layer avoids the agglomeration of nanoparticles. Sodium dodecyl benzene sulfonate, cetyltrimethylammonium bromide (CTAB), oleic acid, anionic/non-ionic surfactants, Laurate salt, etc. are used as surface agents [37–43]. The one-step method for nanofluid synthesis avoids transportation, storage, drying and dispersion of nanoparticles. Nanofabrication techniques such as laser ablation, vapour deposition, sputtering, plasma arc and electric deposition methods can be used for single-step nanofluid synthesis [44]. Retention of nanoparticle species, varying concentrations of nanoparticles in various batches and structural fluctuations are a few limitations of the one-step process [35, 44]. In a single-step method, preparation and dispersion of nanoparticles are carried out in one step. In a two-step process, first, nanoparticles are prepared, and then in the second step, they are dispersed and stabilized in a base fluid. Nanofluids are used in solar collectors and batteries to reduce the cycle time and store more energy per unit volume or space [45–51]. Most of the literature has reported enhancement in heat transfer after the addition of nanoparticles in base fluids [14, 52–56]. Figure 4 depicts the one-step and two-step synthesis of nanofluids. Some significant investigations on nanofluid synthesis are listed in Table 1. Figure 5 shows important base fluids and nanoparticles for nanofluid synthesis.

3 Application of Nanofluids in Energy Storage

3.1 Solar Collectors

Solar energy is clean energy with the potential to provide solutions to energy problems across industries. The energy storage in solar energy is always a topic of discussion and research. Various energy storage devices and batteries are explored for efficient storage of the energy. An investigation was carried out by Krishna et al. [45] to study the thermal energy storage system. They used alumina and titanium dioxide nanoparticles for heat transfer. Their investigation indicated that only a 0.2% addition of nanoparticles for working fluid indicated a reduction in charging time, 19% for TiO_2 and 28% for Al_2O_3. The decrease in the charging time was affected by the flow rate of working fluids and the nanoparticle content. For the same flow rate, an increase in the nanoparticle from 0.2 to 0.4% resulted in a reduction in charging time from 19% to 28% for TiO_2 and 28% to 38% for Al_2O_3.

Obaid et al. carried out an investigation to study the effect of the addition of nanoparticles to water in the building heating system by solar energy [46]. They observed that the addition of 0.4–0.5% of MgO nanoparticles resulted in higher temperature gradients than pure water. Reddy et al. carried out experimentation using a packed-bed thermal energy system in which latent and sensible heat storage

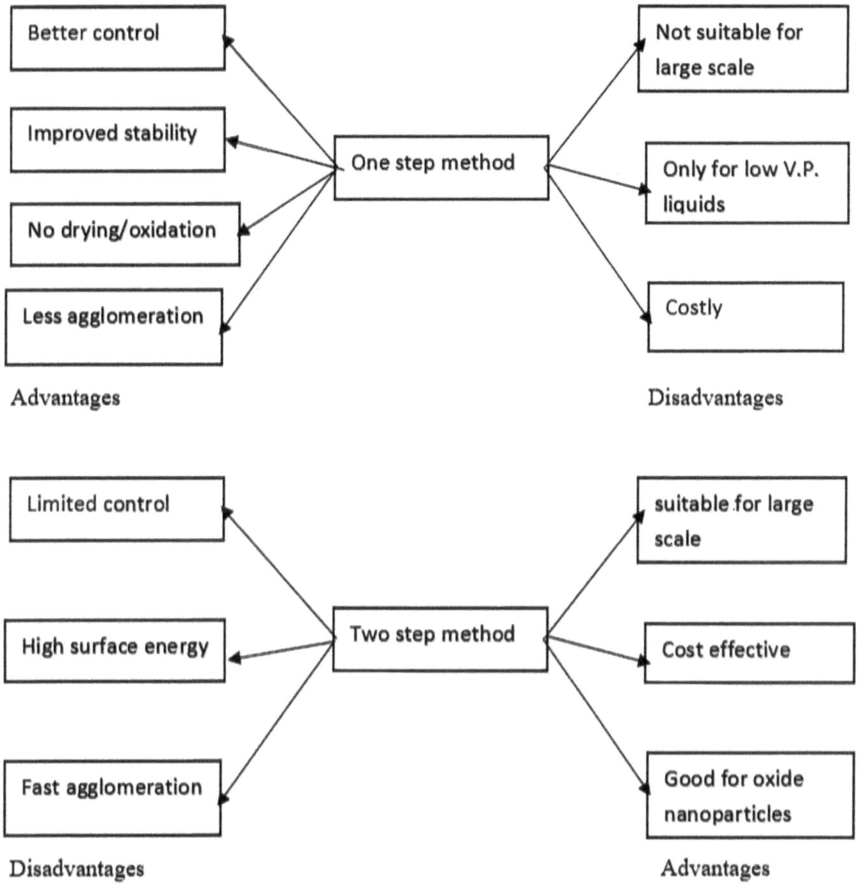

Fig. 4 One-step and two-step methods for nanofluid synthesis [20]

was integrated [47]. The heat source used for this experimentation was solar energy. They observed that at higher flow rates, the reduction in charging time was greater than at low flow rates [47]. The charging time was reduced by 15% by increasing the nanoparticle Al_2O_3 composition from 0.2% to 0.8%. The reduction in charging time was 33% for Al_2O_3 compared to 51% for MgO. Pérez De Albéniz Azqueta, in his investigations, used Al_2O_3 and multiwalled nanotubes in nanofluids with distilled water and EG as base liquids [48]. There was an enhancement of efficiency by 48% for 0.08% nanofluid in distilled water. For multiwalled carbon nanotubes, an enhancement of up to 51% was obtained.

Table 1 Investigations on nanofluid synthesis

References	Base fluid	Nanoparticles	Highlight
[57]	Gum Arabic, water	Multiwalled carbon nanotube (MWCNT)	MWCNT shortened by ultrasound
[58]	Diesel oil	Single-walled carbon nanotubes	Enhanced lubricity and heat transfer
[59]	Water	Carbon black	Better absorption of solar energy
[60]	Water	Titanium dioxide(TiO_2)	30–33% enhancement in base fluid
[61]	Transformer oil	Al_2O_3	For 8% by volume, 20% enhancement in the heat transfer coefficient
[62]	Water	Silver	Increase in heat transfer—27% at 0.3 vol% and 80% at 0.9 vol%
[63]	Ethylene glycol	Silver	In the early stage of preparation, aggregation of nanoparticles was observed
[64]	Ethylene glycol	ZnO	26.5% rise in thermal conductivity
[65, 66]	Water	CuO	22.4% rise in thermal conductivity
[67]	Ethylene glycol	CuO	One-step synthesis, no agglomeration for 20 days
[68]	Water	SiO_2	Two-step synthesis, stable for 0–8 days
[69]	Water	MgO, ZnO	550% enhancement in thermal conductivity.
[70]	Ethylene glycol	MgO–SWCNT	35% maximum rise in thermal conductivity
[71]	Ethylene glycol	Al_2O_3/MWCNT	At 50 °C, for 0.8% concentration, 17% increase in thermal conductivity
[72]	Water	Ag–MWCNT	3% AG, 0.05% MWCNT, 14.5% maximum enhancement at optimum temperature
[73]	Water	Au–CNT	Stability was affected by nanoparticle characteristics
[74]	Water	Graphene–Al_2O_3	Increase in the cooling capacity and COP by 72%
[75]	Water	γ-Al_2O_3/MWCNT	At 55 °C, maximum enhancements of 442.9% and 26.3%
[76]	Water/ethylene glycol	Al_2Cu/Ag_2Al	50–150% more thermal conductivity by using nanofluid
[77]	Lubricant oil	Carbon quantum dots	The wear decreases by 52% with the application of a new material
[78]	Water	Copper ferrite nanoparticles ($CuFe_2O_4$ NPs)	4.8% improvement in thermal conductivity
[79]	1:1 mixture of water and ethylene glycol	Cuprous oxide nanofluid	Thermal conductivity increased by 3.59 W m^{-1} K^{-1}
[80]	Water	Silver nanofluid	0.01 wt%, 65.66% rise in critical heat flux, 49.31% increase in heat transfer coefficient
[81]	Water	Sugarcane juice-mediated ZnO	Heat transfer coefficients increase by 37.20% by using the nanoparticles in the base fluid

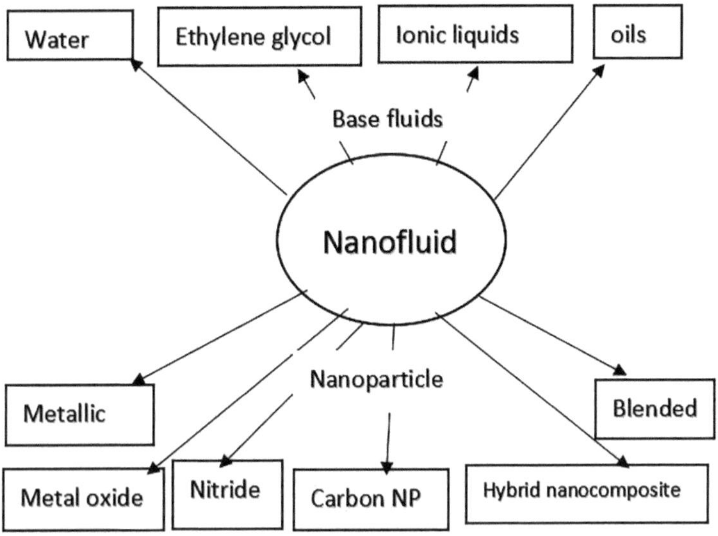

Fig. 5 Base fluids and nanoparticles for nanofluids [82]

3.2 Improving Heat Transfer Properties

The engine oils used are not efficient as a heat transfer medium in automotive coolant applications. The addition of nanoparticles has the potential to reduce the size of the coolant system due to improved heating properties of the fluid [49]. Nanoparticle-doped EG can serve as an efficient coolant material. Nikkam, in his investigation, explored various nanoparticles for nanofluids in order to increase the heat transfer coefficient [50]. In his work, SiC NPs with alpha-crystal structure in water/EG were dispersed with 9% concentration by weight. At 20 °C, they obtained a 20% increase in the thermal conductivity and a 14% rise in viscosity.

3.3 Improving Electrical Properties

Boiko et al. investigated the critical factors that are responsible for improving the electrical properties of nanofluids [51]. The addition of TiO_2 filler increased the AC breakdown voltage (ACBV) by 22% for 0.02% doping [52]. AC breakdown voltage can be improved with the addition of multi-layered graphene sheets (MGS) to cottonseed oil. Also, there is an increase in the ACBV with an increase in the sonication [53].

3.4 Phase Change Materials

Water-based EG was used as a PCM for performance enhancement of cold thermal energy systems by Kee et al. [54]. Cooling thermal energy systems for large buildings for maintaining comfort are becoming a necessity in society. These systems have energy usage. EG-based PCM was used in this investigation. The CuO nanoparticles were used for heat transfer enhancement. They obtained a 15% enhancement in heat transfer for a 0.025 volume fraction of nanofluids. Hasan et al., in their investigation, used nanofluids as PCMs in melting paraffin [14]. Simple co-precipitation route was employed by them for the preparation of $ZnO/\alpha\text{-}Fe_2O_3$ nanocrystals. These nanoparticles were ultrasonically dispersed into the paraffin. X-ray diffraction analysis was used by them to study the behaviour of the nanocrystals. Transmission electron microscopy was used for morphology. As an industrial ecology concept, they utilized aluminium-based waste derived from water-works plants' alum sludge (AS). They augmented it with $ZnO/\alpha\text{-}Fe_2O_3$ nanocrystals. The nanofluid with AS -$ZnO/\alpha\text{-}Fe_2O_3$ nanocrystals exhibited very good stability in melting paraffin. While various proportions of the nanocrystals were tried, 1% nanocrystal composition showed a maximum 93% performance enhancement with a heat gain of 47 kJ [14].

Navarrete et al., in their investigation, explored a novel nanomaterial composed of aluminium copper alloy that was encapsulated with Al-oxide layer [55]. The storage medium was solar salt containing 60% sodium nitrate and 40% potassium nitrate. The energy storage capacity was increased up to 18% with the use of nanomaterials. In their investigation, Reddy and Reddy used copper oxide nanoparticles in PCMs [56]. They observed a 15–20% drop in the charging time by using nanoparticles.

4 Recent Investigations on Application of Nanofluids in Energy Storage

Abir et al. explained the nanoparticle applications in molten salts [83]. Molten salts are basically nitrate, chloride and carbonate-based mixtures that are promising thermal energy system (TES) fluids for high-temperature plants. Investigations have revealed that molten salts with nanoparticles exhibit enhanced heat transfer and high thermal conductivity. Mirahmad et al. examined the effect of nanoparticles on the properties of base fluids. They considered the base fluids like molecular, molten salts and ionic liquids [84]. The addition of nanoparticles can increase the thermal conductivity and heat transfer, but it may also result in a reduction in specific heat capacity. Also, this study emphasizes the importance of proper mixing and dispersion of nanoparticles in base fluids. Nanofluids hold great potential in sustainable green energy solutions if the balancing between heat capacity, thermal conductivity and viscosity is managed appropriately.

Sathish Kumar et al. carried out investigations on the use of aloe vera for PCM in cooling thermal energy systems [85]. They prepared the PCM by using graphene nanoplatelets in deionized water along with aloe vera. The stability of the PCM was found to increase with the addition of aloe vera. About 40% rise in thermal conductivity was observed with aloe vera gel. Also, a 25% reduction in charging time was observed with aloe vera gel. According to Svobodova-Sedlackova et al., 60:40 mixture of $NaNO_3$ and KNO_3 has emerged as the primary thermal energy storage medium [86]. The corrosive nature of the salts at high temperature limits the use of these salts at high temperatures. The investigation to study the corrosive nature of the salts with addition of silica dioxide nanoparticles was carried out by the investigators [86]. The study concluded that the addition of nanoparticles reduced the corrosive products and hence helped to reduce the rate of corrosion. Kalbande et al., in their review, highlighted the use of hybrid nanofluids [87]. Multiwalled carbon nanotube (MWCNTs) and alumina nanoparticles dispersed in oil exhibit excellent heat transfer properties. Gao et al., in their investigations, observed that with the addition of nanoparticles, there was a decrease in the corrosion current density [88]. SiO_2/(LiCl–KCl) nanofluids were doped with the chloride salts by Tian et al. [89]. They observed that there was about a 17% increase in viscosity, more than a 6% increase in the thermal conductivity and a 1.22% increase in the density. Nanoparticles increase the collision frequency of ions and enhance the heat transfer. However, the diffusion ability of the salt gets affected adversely.

In their investigation, Pineda et al. examined the effect of multilayer Ti_3C_2Tx MXenes at different concentrations on the thermal properties of nanofluids consisting of solar salt [90]. They observed a rise in decomposition temperature and heat capacity due to structural and chemical effects of added particles. Roy et al. obtained a 19% enhancement in heat transfer with a 0.06% addition of Al_2O_3 in water as a base fluid [91]. Basha et al. performed a detailed numerical simulation considering volume fraction, Rayleigh number, bottom slit position and Hartmann number [92]. A 2.5% addition of magnetite nanoparticles to GO fluid resulted in a maximum 4% increase in heat transfer rate. Colak used deep learning methods for predicting heat transfer in nanoencapsulated PCM [93]. Wavy enclosures containing encapsulated PCMs indicated better heat transfer and energy properties due to a larger heat transfer area and also distinctive shape [93]. The research carried out by Emara indicated that the addition of 0.08% nanoparticles (TiO_2) increases the heat transfer by 21% and 37.5% for Al_2O_3 nanoparticles [94]. In Table 2, important investigations on nanofluid application in energy storage systems in the last 3 years are briefed.

5 Ionanofluids (INFs)

When ionic liquids and nanoparticles are mixed, they form ionanofluids (INFs). Cavieres et al. carried out an investigation on INF formed by using ionic liquid as a base fluid and MWCNTs as nanoparticles. With the use of

Table 2 Recent investigations on nanofluids for energy storage

References	Aim and objective	Methodology	Major findings
[95]	Studying the use of hybrid nanofluid to enhance liquid cooling capacity	Water as a base fluid and copper and aluminium oxide	3.26% lower battery temperature with the addition of Cu nanoparticles to Al_2O_3/H_2O NFs
[96]	Improving the heat-related properties of the hybrid nanofluid	Cu and MWCNTs with water-based fluid	The increase in the value of Eckert number and thermophoretic diffusion parameter. The fast rise in Brownian motion parameter
[97]	Studying the melting process of PCM	SiO_2, olive leaf extracts and TiO_2 with water as base fluid	More than 66% drop in the melting time of solid PCM
[98]	Study of battery thermal management systems (BTMSs)	Study of nanofluids and PCM-based BTMS	Metal foams with PCM in BTMS enhance the performance related to heat transfer
[99]	To increase the thermal storage and electrical energy of solar cells in a photovoltaic thermal system	Paraffin wax and water for storage, CuO water nanofluid as a working fluid	For the nanofluid-PVT/PCM, 37% more electric efficiency and energy conversion efficiency of 82%
[100]	Increasing the heat transfer-related properties of binary alkali chloride	Use of Al_2O_3 nanoparticles	Enhancement in heat capacity and thermal conductivity by 13–15%
[101]	Investigating the effect of hybrid nanofluids on the cooling performance of batteries	Al_2O_3, CuO and Al_2O_3–CuO nanofluids with water as base fluid	Hybrid nanofluids show better conductivity than mono nanofluids
[102]	Enhancement of the thermo-physical properties of erythritol and xylitol	Copper (Cu), aluminium (Al) and zinc (Zn) nanoparticles, heat transfer fluid—Therminol-66	Thermal conductivity, diffusivity, density and specific heat capacity increased significantly due to nanoparticle doping
[103]	To study the properties of PEG200/ZnO semiconducting nanofluids	ZnO nanoparticles and ethylene glycol-base fluid	Efficient medium for heat transfer, especially in flow-type devices
[104]	Comparison of different fluids for Li battery cooling	Comparison of nano-enhanced fluids when added to base fluids	Nanofluids as coolants show cooling performance 10%–15% higher than conventional fluids

nanoparticles in ionic liquid, the heat capacity increased by 3.7 times [105]. In their investigation, Hosseinghorbani et al. used 1-butyl-3-methylimidazolium-bis(trifluoromethylsulfonyl)imide ([Bmim][NTf2]) as an ionic solvent and dispersed graphene oxide nanoparticles to obtain ionanofluid [106]. The investigation confirmed the enhancement in the thermal conductivity (6.5%) and heat capacity (27%) upon addition of nanoparticles in the ionic liquids. Duarte et al. used carbon dots as nanoparticles in the ionic liquid [107]. Their investigation indicated that INFs can convert light energy into heat energy with a conversion efficiency of 28%. Moulefera et al., in their work, dispersed the graphene oxide nanoparticles into four different ionic liquids [108]. They obtained more than a 37% increase in efficiency by replacing water with ionic liquid under suitable conditions. Kanti et al., in their investigation, used oxide in ionic liquid for the enhancement of thermal properties [109]. At 60 °C, a maximum of 27% enhancement in thermal conductivity was obtained by them. Also, there was a significant rise in the heat capacity due to nanoparticle addition. Jozwiak et al., in their investigation, highlighted the role of morphological properties in the enhancement of thermal conductivity of the ionic liquids and ionanofluids [110].

6 Conclusion

Nanofluids are explored for increasing the heat transfer coefficient of the fluids. In solar collectors, nanofluids can be used as absorbing materials. Faster melting and solidification can be facilitated in the case of the material that changes phase (PCM) with the application of nanofluids. Improved energy transfer, energy storage and reduced size of the equipment are important aspects of the application of nanofluids in the energy storage system. There are many advantages of using nanofluids in solar energy systems. Nanofluids have a small particle size and a large surface area. Also, the optical properties of nanofluids are much better than conventional fluids. Significantly high thermal conductivity is another advantage. Nanofluids are excellent absorbing mediums due to their high absorption and high thermal stability. Application of nanofluid is limited not only to its heat transfer properties but also its optical properties. Using selective coatings, nanomaterials can improve the optical properties of the receiver. Nanoparticles when dispersed into ionic liquids form ionanofluids. These ionanofluids indicate significant improvement over the ionic liquids.

References

1. J.A. Ceron, E. Gomez-Luna, J.C. Vasquez, Driving the energy transition in Colombia for off-grid regions: Microgrids and non-conventional renewable energy sources. Energies **18**(4), 1010 (2025). https://doi.org/10.3390/en18041010

2. W. Apollon, I. Rusyn, N.E. Paucar, M. Hibbert, S.K. Kamaraj, C. Sato, Energy recovery from organic wastes using microbial fuel cells: Traditional and nonconventional organic substrates. Resources **14**(3), 47 (2025). https://doi.org/10.3390/resources14030047
3. A.M. Omer, A review of non-conventional energy systems and environmental pollution control. J. Soil Sci. Environ. Manage. **1**(7), 127–154 (2010) https://academicjournals.org/article/article1380029583_Abdeen%20pdf.pdf
4. R. Ragadhita, S. Sriyati, N. Nahadi, N. Rustaman, I. Kaniawati, Context evaluation in nanotechnology courses: A context, input, process, and product (CIPP) model perspective. Indonesian J. Teach. Sci. **5**(1), 73–92 (2025) https://ejournal.upi.edu/index.php/IJoTis/article/view/83191
5. K. Anbarasu, Y. Devarajan, Nanomaterials-based wastewater treatment: Addressing challenges and advancing sustainable solutions. BioNanoScience **15**(1), 1–14 (2025). https://doi.org/10.1007/s12668-024-01780-8
6. S.A. Kadhim, Q.J. Abdul-Ghafoor, K.A. Hammoodi, H.A. Abdul Wahhab, O.A.A.M. Ibrahim, M.K. Al-Ghezi, Enhancement of the heat transfer performance of a double-pipe heat exchangers in various configurations using Al_2O_3 nanofluid: A focused review. J. Therm. Anal. Calorim. (2025). https://doi.org/10.1007/s10973-025-14042-w
7. A. Kasaeian, A.T. Eshghi, M. Sameti, A review on the applications of nanofluids in solar energy systems. Renew. Sust. Energ. Rev. **43**, 584–598 (2015). https://doi.org/10.1016/j.rser.2014.11.020
8. H. Tyagi, P. Phelan, R. Prasher, Predicted efficiency of a low-temperature nanofluid-based direct absorption solar collector. J. Solar Energy **131**(4), 041004 (2009). https://doi.org/10.1115/1.3197562
9. E. Sani, L. Mercatelli, S. Barison, C. Pagura, F. Agresti, L. Colla, P. Sansoni, Potential of carbon nanohorn-based suspensions for solar thermal collectors. Sol. Energy Mater. Sol. Cells **95**, 2994–3000 (2011). https://doi.org/10.1016/j.solmat.2011.06.011
10. V. Khullar, H. Tyagi, P.E. Phelan, T.P. Otanicar, H. Singh, R.A. Taylor, in Solar energy harvesting using nanofluids-based concentrating solar collector. Proceedings of MNHMT2012 3rd Micro/Nanoscale Heat & Mass Transfer International Conference on March 3–6, Atlanta, Georgia, USA (2012)
11. A.K. Tiwari, P. Ghosh, J. Sarkar, in performance analysis of a flat plate solar collectors using Al_2O_3/water nanofluid. An International Conference on Energy Security, Global Warming and Sustainable Climate, Banaras Hindu University (BHU), Varanasi, India (2012), https://www.researchgate.net/publication/235785074_Performance_analysis_of_a_flat_plate_solar_collectors_using_aluminawater_nanofluid
12. R.A. Taylor, P.E. Phelan, T.P. Otanicar, C.A. Walker, M. Nguyen, S. Trimble, R. Prasher, Applicability of nanofluids in high flux solar collectors. J. Renew. Sustain. Energy **3**, 023104 (2011). https://doi.org/10.1063/1.3571565
13. Y. Tong, J. Kim, H. Cho, Effects of thermal performance of enclosed-type evacuated U-tube solar collector with multi-walled carbon nanotube/water nanofluid. Renew. Energy **83**, 463–473 (2015). https://doi.org/10.1016/j.renene.2015.04.042
14. E.A. Hassan, M.A. Tony, M.M. Awad, Thermal energy storage using hybrid nanofluid phase change material (PCM) based on waste sludge Incorp rated ZnO/α-Fe2O3. Nano **14**(7), 604 (2024). https://doi.org/10.3390/nano14070604
15. E.O. Atofarati, M. Sharifpur, Z. Huan, Nanofluids for heat transfer enhancement: A holistic analysis of research advances, technological progress and regulations for health and safety. Cogent Eng. **11**(1), 2434623 (2024). https://doi.org/10.1080/23311916.2024.2434623
16. M.S. Peer, M. Cascetta, L. Migliari, M. Petrollese, Nanofluids in thermal energy storage systems: A comprehensive review. Energies **18**(3), 707 (2025). https://doi.org/10.3390/en18030707
17. M.H. Shojaeefard, M. Jourabian, A. Ali Rabienataj Darzi, A. Bayat, Inward melting inside a horizontal multilobed capsule with conductive wall affected by Ag-MgO/water hybrid and MgO/water nanofluids. J. Heat Mass Transf. Res. **8**, 205–223 (2021). https://doi.org/10.22075/jhmtr.2021.22194.1326

18. W. Hua, X. Lv, X. Zhang, Z. Ji, J. Zhu, Research progress of seasonal thermal energy storage technology based on supercooled phase change materials. J Energy Storage **67**, 107378 (2023). https://doi.org/10.1016/j.est.2023.107378
19. Q. Ren, H. Xu, Z. Luo, PCM charging process accelerated with combination of optimized triangle fins and nanoparticles. Int J Therm Sci **140**, 466–479 (2019). https://doi.org/10.1016/j.ijthermalsci.2019.03.005
20. D. Aydin, S.P. Casey, S. Riffat, The latest advancements on thermochemical heat storage systems. Renew. Sust. Energ. Rev. **41**, 356–367 (2015). https://doi.org/10.1016/j.rser.2014.08.054
21. B. Zalba, J.M. Marín, L.F. Cabeza, H. Mehling, Review on thermal energy storage with phase change: Materials, heat transfer analysis and applications. Appl. Therm. Eng. **23**, 251–283 (2003). https://doi.org/10.1016/S1359-4311(02)00192-8
22. A. Mukherjee, R. Majumdar, S.K. Saha, L. Kumar, C. Subramaniam, Assessment of open thermochemical energy storage system performance for low temperature heating (2 applications). Appl. Therm. Eng. **156**, 453–470 (2019). https://doi.org/10.1016/j.applthermaleng.2019.04.096
23. X.F. Long, L. Dai, B. Lou, J. Wu, The kinetics research of thermochemical energy storage system ca$(OH)_2$/CaO. Int. J. Energy Res. **41**, 1004–1013 (2017). https://doi.org/10.1002/er.3688
24. S. Wu, C. Zhou, E. Doroodchi, R. Nellore, B. Moghtaderi, A review on high-temperature thermochemical energy storage based on metal oxides redox cycle. Energy Convers. Manag. **168**, 421–453 (2018)
25. A.H. Elsheikh, S.W. Sharshir, M.E. Mostafa, F.A. Essa, Applications of nanofluids in solar energy: A review of recent advances. Renew. Sust. Energ. Rev. **82**(3), 3483–3502 (2017). https://doi.org/10.1016/j.rser.2017.10.108
26. K.S. Reddy, N.R. Kamnapure, S. Srivastava, Nanofluid and nanocomposite applications in solar energy conversion systems for performance enhancement: A review. Int. J. Low-Carbon Technol. **12**(1), 1–23 (2017). https://doi.org/10.1093/ijlct/ctw007
27. N. Çobanoglu, Z.H. Karadeniz, A. Turgut, Carbon-based nanofluid applications in solar thermal energy. E3S Web Conf. **111**, 01056 (2019). https://doi.org/10.1051/e3sconf/201911101056
28. A. Angayarkanni, J. Philip, Fundamentals and transport properties of nanofluids. R. Soc. Chem. Lond. **2022**, 1–40 (2022). https://doi.org/10.1039/9781839166457
29. R.K. Mudidana, V. Miditana, V. Rambabu, Synthesis of nanofluids preparation—A review. Mater Today Proc **115**, 265–273 (2024). https://doi.org/10.1016/j.matpr.2023.09.086
30. X. Li, Y. Chen, S. Mo, L. Jia, X. Shao, Effect of surface modification on the stability and thermal conductivity of water-based SiO2-coated graphene nanofluid. Thermochim. Acta **595**, 6–10 (2014). https://doi.org/10.1016/j.tca.2014.09.006
31. K. Wongcharee, S. Eiamsa-Ard, Enhancement of heat transfer using CuO/water nanofluid and twisted tape with alternate axis. Int. Commun. Heat Mass Trans. **38**(6), 742–748 (2011). https://doi.org/10.1016/j.icheatmasstransfer.2011.03.011
32. P.V. Durga Prasad, A.V.S.S.K.S. Gupta, M. Sreeramulu, L. Syam Sundar, M.K. Singh, C.M.S. Antonio, Experimental study of heat transfer and friction factor of Al_2O_3 nanofuid in U-tube heat exchanger with helical tape inserts. Exp. Thermal Fluid Sci. **62**, 141–150 (2015). https://doi.org/10.1016/j.expthermflusci.2014.12.006
33. L.S. Sundar, M.K. Singh, C.M. Sousa, Enhanced heat transfer and friction factor of MWCNT–Fe3O4/water hybrid nanofluids. Int. Commun. Heat Mass Trans. **52**, 73–83 (2014). https://doi.org/10.1016/j.icheatmasstransfer.2014.01.012
34. M.C.S. Reddy, V.V. Rao, Experimental investigation of heat transfer coefficient and friction factor of ethylene glycol water based TiO2 nanofuid in double pipe heat exchanger with and without helical coil inserts. Int. Commun. Heat Mass Trans. **50**, 68–76 (2014). https://doi.org/10.1016/j.icheatmasstransfer.2013.11.002

35. K.V. Modi, P.R. Patel, S.K. Patel, Applicability of mono-nanofluid and hybrid-nanofluid as a technique to improve the performance of solar still: A critical review. J. Clean. Prod. **387**, 135875 (2023). https://doi.org/10.1016/j.jclepro.2023.135875
36. K.A. Jehhef, M.A.A.A. Siba, Effect of surfactant addition on the nanofluids properties: A review. Acta Mech. Malaysia **2**(2), 1–19 (2019). https://doi.org/10.26480/amm.02.2019.01.19
37. H.K. Gupta, G.D. Agrawal, J. Mathur, An overview of nanofluids: A new media towards green environment. Int. J. Environ. Sci. **3**(1), 433–441 (2012) https://citeseerx.ist.psu.edu/document?repid=rep1&type=pdf&doi=e679cb6f6e39e88b77aa75393ba6d9e4a7c570d2
38. M. Drzazga, A. Gierczycki, G. Dzido, M. Lemanowicz, Influence of nonionic surfactant addition on drag reduction of water based nanofluid in a small diameter pipe. Chin. J. Chem. Eng. **21**(1), 104–108 (2013). https://doi.org/10.1016/S1004-9541(13)60447-4
39. S.E.B. Maiga, S.J. Palm, C.T. Nguyen, G. Roy, N. Galanis, Heat transfer enhancement by using nanofluids in forced convection flows. Int. J. Heat Fluid Flow **26**(4), 530–546 (2005). https://doi.org/10.1016/j.ijheatfluidflow.2005.02.004
40. D. Wongwises, W. Duangthongsuk, S. Wongwises, Heat transfer enhancement and pressure drop characteristics of TiO_2–water nanofluid in a double-tube counter flow heat exchanger. Int. J. Heat Mass Transf. **52**, 2059–2067 (2009). https://doi.org/10.1016/j.ijheatmasstransfer.2008.10.023
41. J.A. Lopez, F. González, F.A. Bonilla, G. Zambrano, E. Maria, Gómez, Synthesis and characterization of Fe3O4 magnetic nanofluid. Materials **30**(1), 60–66 (2010) https://ve.scielo.org/scielo.php?script=sci_arttext&pid=S0255-69522010000100007
42. A.R.I. Ali, B.A. Salam, Review on nanofluid: Preparation, stability, thermophysical properties, heat transfer characteristics and application. SN Appl. Sci. **2**, 1636 (2020). https://doi.org/10.1007/s42452-020-03427-1
43. H.B. Bacha, N. Ullah, A. Hamid, N.A. Shah, A comprehensive review on nanofluids: Synthesis, cutting-edge applications, and future prospects. Int. J. Thermofluids **22**, 100595 (2024). https://doi.org/10.1016/j.ijft.2024.100595
44. B. Wang, F. Qiu, G.C. Barber, Q. Zou, J. Wang, S. Guo, Y. Yuan, Q. Jiang, Role of nano-sized materials as lubricant additives in friction and wear reduction: A review. Wear **490**, 204206 (2022). https://doi.org/10.1016/j.wear.2021.204206
45. K. Krishna Reddy, R. Meenakshi Reddy, B. Durga Prasad, Experimental investigation of a packed bed latent heat thermal energy storage system with nanofluids as heat transfer fluid (HTF) for solar heating applications. Int. J. Innov. Eng. Technol. **14**(3), 11–19 (2019) https://ijiet.com/wp-content/uploads/2019/11/3.pdf
46. H.N. Obaid, M.A. Habeeb, F.L. Rashid, A. Hashim, Thermal energy storage by nanofluids. J. Energy Technol. Policy **3**(5), 34–36 (2013) https://iiste.org/Journals/index.php/JETP/article/view/5595/5707
47. K.K. Reddy, R.M. Reddy, K.M. Reddy, Y.M. Reddy, K.P. Reddy, G.G. Dattatreya, Performance of thermal energy storage system using water-based nanofluids for solar heating applications, in *International Conference on Advanced Materials, Manufacturing and Sustainable Development (ICAMMSD 2024)*, (Atlantis Press, Dordrecht, 2025), pp. 592–610. https://doi.org/10.2991/978-94-6463-662-8_47
48. P. De Albéniz, S. Azqueta, *Application of Nanofluids in a Flat Plate Solar Collector* (Universitat Politècnica de Catalunya, Barcelona, 2022) http://hdl.handle.net/2117/376935
49. N. Rao, L. Gahane, S. Ranganayakulu, Synthesis, applications and challenges of nanofluids-review. IOSR J. Appl. Phys. **6**, 21–28 (2014) https://www.iosrjournals.org/iosr-jap/papers/ICAET-2014/volume-1/6.pdf
50. N. Nikkam, *Engineering Nanofluids for Heat Transfer Applications* (KTH Royal Institute of Technology, Stockholm, 2014) https://www.diva-portal.org/smash/get/diva2:712511/FULLTEXT01.pdf
51. O. Boiko, H.D. Stryczewska, G.K. Komarzyniec, K. Ebihara, S.I. Aoqui, M. Yamazato, M. Zagirnyak, Key factors enhancing the electrical properties of nanofluids. A mini-review of the applications in the energy-related sectors. Arch. Electr. Eng. **73**(4), 152115 (2024). https:/ /doi.org/10.24425/aee.2024.152115

52. K.N. Koutras, I.A. Naxakis, A.E. Antonelou, V.P. Charalampakos, E.C. Pyrgioti, S.N. Yannopoulos, Dielectric strength and stability of natural ester oil based TiO_2 nanofluids. J. Mol. Liq. **316**, 113901 (2020). https://doi.org/10.1016/j.molliq.2020.113901
53. R.A. Farade, N.I.A. Wahab, D.E.A. Mansour, N.B. Azis, J.B.T. Jasni, V. Veerasamy, M. Thirumeni, A.X.R. Irudayaraj, A.S. Murthy, Investigation of the effect of sonication time on dispersion stability, dielectric properties, and heat transfer of graphene based green nanofluids. IEEE Access **9**, 50607–50623 (2021). https://doi.org/10.1109/ACCESS.2021.3069282
54. C.H. Kee, N.A.C. Sidik, S.N.A. Yusof, M. Beriache, T.A. Mohamad, Performance enhancement of cold thermal energy storage system using nanofluid phase change materials. J. Adv. Res. Appl. Mech. **62**(1), 16–32 (2019) https://www.akademiabaru.com/submit/index.php/aram/article/view/1850
55. N. Navarrete, R. Mondragón, D. Wen, M.E. Navarro, Y. Ding, J.E. Juliá, Thermal energy storage of molten salt–based nanofluid containing nano-encapsulated metal alloy phase change materials. Energy **167**, 912–920 (2019). https://doi.org/10.1016/j.energy.2018.11.037
56. N. Sudheer Kumar Reddy, R. Meenakshi Reddy, Analysis of thermal energy storage systems by using nanoparticles. Int. J. Appl. Eng. Res. **14**(14), 3304–3310 (2019) https://www.ripublication.com/ijaer19/ijaerv14n14_26.pdf
57. A. Indhuja, K.S. Suganthi, S. Manikandan, K.S. Rajan, Viscosity and thermal conductivity of dispersions of gum Arabic capped MWCNT in water: Influence of MWCNT concentration and temperature. J. Taiwan Inst. Chem. Eng. **44**(3), 474–479 (2013). https://doi.org/10.1016/j.jtice.2012.11.015
58. F. Marquis, L. Chibante, Improving the heat transfer of nanofluids and nanolubricants with carbon nanotubes. JOM J. Miner. Met. Mater. Soc. **57**(12), 32–43 (2005). https://doi.org/10.1007/s11837-005-0180-4
59. D. Han, Z. Meng, D. Wu, C. Zhang, H. Zhu, Thermal properties of carbon black aqueous nanofluids for solar absorption. Nanoscale Res. Lett. **6**, 457 (2011). https://doi.org/10.1186/1556-276X-6-457
60. S.M.S. Murshed, K.C. Leong, C. Yang, Enhanced thermal conductivity of TiO2—Water based nanofluids. Int J Therm Sci **44**(4), 367–373 (2005). https://doi.org/10.1016/j.ijthermalsci.2004.12.005
61. C. Choi, H.S. Yoo, J.M. Oh, Preparation and heat transfer properties of nanoparticle in transformer oil dispersions as advanced energy-efficient coolants. Curr. Appl. Phys. **8**, 710–712 (2008). https://doi.org/10.1016/j.cap.2007.04.060
62. L. Godson, B. Raja, D.M. Lal, S.J.E.H.T. Wongwises, Experimental investigation on the thermal conductivity and viscosity of silver-deionized water nanofluid. Exp. Heat Transf. **23**(4), 317–332 (2010). https://doi.org/10.1080/08916150903564796
63. P. Sharma, I.H. Baek, T. Cho, S. Park, K.B. Lee, Enhancement of thermal conductivity of ethylene glycol based silver nanofluids. Powder Technol. **208**(1), 7–19 (2011). https://doi.org/10.1016/j.powtec.2010.11.016
64. W. Yu, H. Xie, L. Chen, Y. Li, Investigation of thermal conductivity and viscosity of ethylene glycol based ZnO nanofluid. Thermochim. Acta **491**(1-2), 92–96 (2009). https://doi.org/10.1016/j.tca.2009.03.007
65. M.S. Liu, M.C. Lin, I.T. Huang, C.C. Wang, Enhancement of thermal conductivity with CuO for nanofluids. Chem. Eng. Technol. **29**(1), 72–77 (2006). https://doi.org/10.1002/ceat.200500184
66. W. Chamsa-Ard, S. Brundavanam, C.C. Fung, D. Fawcett, G. Poinern, Nanofluid types, their synthesis, properties and incorporation in direct solar thermal collectors: A review. Nano **7**(6), 131 (2017). https://doi.org/10.3390/nano7060131
67. S.A. Kumar, K.S. Meenakshi, B.R.V. Narashimhan, S. Srikanth, G.J.M.C. Arthanareeswaran, Synthesis and characterization of copper nanofluid by a novel one-step method. Mater. Chem. Phys. **113**(1), 57–62 (2009). https://doi.org/10.1016/j.matchemphys.2008.07.027
68. T. Zhang, Q. Zou, Z. Cheng, Z. Chen, Y. Liu, Z. Jiang, Effect of particle concentration on the stability of water-based SiO_2 nanofluid. Powder Technol. **379**, 457–465 (2021). https://doi.org/10.1016/j.powtec.2020.10.073

69. S.O. Giwa, M. Momin, C.N. Nwaokocha, M. Sharifpur, J.P. Meyer, Influence of nanoparticles size, percent mass ratio, and temperature on the thermal properties of water-based MgO–ZnO nanofluid: An experimental approach. J. Therm. Anal. Calorim. **143**, 1063–1079 (2021). https://doi.org/10.1007/s10973-020-09870-x
70. M.H. Esfe, S. Esfandeh, M. Afrand, M.K. Amiri, A novel applicable experimental study on the thermal behavior of SWCNTs (60%) MgO (40%)/EG hybrid nanofluid by focusing on the thermal conductivity. Powder Technol. **342**, 998–1007 (2019). https://doi.org/10.1016/J.POWTEC.2018.10.008
71. A.A.A. Arani, F. Pourmoghadam, Experimental investigation of thermal conductivity behavior of MWCNTS-Al_2O_3/ethylene glycol hybrid nanofluid: Providing new thermal conductivity correlation. Heat Mass Transf. **55**(1), 2329–2339 (2019). https://doi.org/10.1007/s00231-019-02572-7
72. B. Munkhbayar, M.R. Tanshen, J. Jeoun, H. Chung, H. Jeong, Surfactant-free dispersion of silver nanoparticles into MWCNT aqueous nanofluids prepared by one-step technique and their thermal characteristics. Ceram. Int. **39**(6), 6415–6425 (2013). https://doi.org/10.1016/j.ceramint.2013.01.069
73. S. Jana, A.S. Khojin, W.H. Zhong, Enhancement of fluid thermal conductivity by the addition of single and hybrid nano-additives. Thermochim. Acta **462**, 45–55 (2007). https://doi.org/10.1016/j.tca.2007.06.009
74. N. Ahammed, L.G. Asirvatham, S. Wongwises, Entropy generation analysis of graphene–alumina hybrid nanofluid in multiport minichannel heat exchanger coupled with thermoelectric cooler. Int. J. Heat Mass Transf. **103**, 1084–1097 (2016). https://doi.org/10.1016/j.ijheatmasstransfer.2016.07.070
75. S.O. Giwa, M.J.P. Sharifpur, S.W. Meyer, O. Mahian, S. Wongwises, Experimental measurement of viscosity and electrical conductivity of water based γ-Al_2O_3/MWCNT hybrid nanofluids. J. Therm. Anal. Calorim. **143**(4), 1037–1050 (2020). https://doi.org/10.1007/s10973-020-10041-1
76. M. Chopkar, S. Kumar, D. Bhandari, P.K. Das, I. Manna, Development and characterization of Al2Cu and Ag2Al nanoparticle dispersed water and ethylene glycol based nanofluid. Mater. Sci. Eng. B **139**, 141–148 (2017). https://doi.org/10.1016/j.mseb.2007.01.048
77. H. Yu, X. Zuo, X. Zhang, X. Wang, F. Zhou, Bioinspired carbon quantum dots nanofluid fabricated by a facile two-step synthesis toward high performance lubrication. Chem. Eng. J. **510**, 161810 (2025). https://doi.org/10.1016/j.cej.2025.161810
78. B. Cardoso, G. Nobrega, M. Machado, R.A. Lima, Green synthesis of copper ferrite-based nanofluids using Chlorella vulgaris for heat transfer enhancement. J. Mol. Liq. **428**, 127498 (2025). https://doi.org/10.1016/j.molliq.2025.127498
79. D.K. Bhat, S.P. Kumar, U.S. Shenoy, Enhancing the thermal conductivity and stability of cuprous oxide nanofluids: Ribose-mediated single step chemical synthesis for solar energy applications. Nano Trends **9**, 100071 (2025). https://doi.org/10.1016/j.nwnano.2024.100071
80. L. Huang, J. Hou, L. Ma, Z. Ding, Y. Yu, J. Wu, X. Yu, W. Zhou, N. Hao, Rational design of 3D-printed microfluidic chips for one-step production of silver nanofluids in thermal management of power electronic chips. ACS Appl. Mater. Interfaces (2025). https://doi.org/10.1021/acsami.5c00015.s001
81. G. Sai, V.R. Anna, K. Swapna, M.S. Prasad, R.P. Ch, B.C. Krishna, V.N. Mandhala, B.T.P. Madhav, S. Pravinkumar, S. Saleh, M.C. Rao, S. Shanmugan, Evaluating the effects of sugarcane juice-mediated ZnO nanofluids on solar light activation for enhancing double-slope solar still performance. Appl. Mater. Today **42**, 102542 (2025). https://doi.org/10.1016/j.apmt.2024.102542
82. S. Sreekumar, N. Shah, J. Mondol, N. Hewitt, S. Chakrabarti, Broadband absorbing mono, blended and hybrid nanofluids for direct absorption solar collector: A comprehensive review. Nano Futures **6**, 022002 (2022). https://doi.org/10.1088/2399-1984/ac57f7
83. F.M. Abir, Q. Altwarah, M.T. Rana, D. Shin, Recent advances in molten salt-based nanofluids as thermal energy storage in concentrated solar power: A comprehensive review. Materials **17**(4), 955 (2024). https://doi.org/10.3390/ma17040955

84. A. Mirahmad, R.S. Kumar, B.P. Doldán, C.P. Rios, J. Díez-Sierra, Beyond thermal conductivity: A review of nanofluids for enhanced energy storage and heat transfer. Nano **15**(4), 302 (2025). https://doi.org/10.3390/nano15040302
85. A. Sathishkumar, P. Sundaram, S.P. Ranga, P.G. Kumar, M. Cheralathan, R. Velraj, Role of aloe vera based nanofluids for cool thermal energy storage system: A comparative study. J Energy Storage **90**, 111710 (2024). https://doi.org/10.1016/j.est.2024.111710
86. A. Svobodova-Sedlackova, A. Palacios, Z. Jiang, A.I.F. Renna, Y. Ding, H. Navarro, C. Barreneche, Thermal stability and durability of solar salt-based nanofluids in concentrated solar power thermal energy storage: An approach from the effect of diverse metal alloys corrosion. J Energy Storage **75**, 109715 (2024). https://doi.org/10.1016/j.est.2023.109715
87. V.P. Kalbande, M.S. Choudhari, Y.N. Nandanwar, Hybrid nano-fluid for solar collector based thermal energy storage and heat transmission systems: A review. J Energy Storage **86**, 111243 (2024). https://doi.org/10.1016/j.est.2024.111243
88. Q. Gao, Y. Lu, Y. Wang, Y. Wu, C. Zhang, Y. Wang, Electrochemical study on the corrosion behaviour of 316L stainless steel in quaternary nitrate molten salt nanofluids for thermal energy storage applications. J Energy Storage **83**, 110491 (2024). https://doi.org/10.1016/j.est.2024.110491
89. H. Tian, W. Zhang, Z. Kou, Molecular dynamics simulations on the structure and thermal property of SiO2/(LiCl-KCl) nanofluids for high temperature thermal energy storage. Ceram. Int. **51**(4), 5125–5134 (2025) https://linkinghub.elsevier.com/retrieve/pii/S0272884224056062
90. F. Pineda, D.F. Zambrano, M.I. Lasanta, D. Guzmán, A. Angel, F. Palay, P. Rios, R. Gonzalez, M. Ramirez, J. Rogan, J. Valdivia, F. Perez, A. Rosenkranz, MXene-enhanced nanofluids for superior thermal energy storage in concentrated solar power plants. Sol. Energy Mater. Sol. Cells **283**, 113461 (2025). https://doi.org/10.1016/j.solmat.2025.113461
91. A. Roy, K.P. Venkitaraj, P. Vigneshwaran, S. Saboor, E. Cuce, K.K. Saxena, Enhanced convective heat transfer with Al_2O_3-water nanofluid in a PCM-based thermal energy storage system. J Energy Storage **97**, 112853 (2024). https://doi.org/10.1016/j.est.2024.112853
92. H.T. Basha, H. Kim, B. Jang, Buoyancy-driven heat transfer and entropy analysis of a hydromagnetic GO-Fe3O4/H_2O hybrid nanofluid in an energy storage enclosure partially filled with non-Darcy porous medium under an oblique magnetic field. Int. J. Numer. Methods Heat Fluid Flow **35**(2), 491–523 (2025). https://doi.org/10.1108/HFF-03-2024-0193
93. A.B. Çolak, Numerical investigation of thermal energy storage in wavy enclosures with nanoencapsulated phase change materials using deep learning. Energy **320**, 135272 (2025). https://doi.org/10.1016/j.energy.2025.135272
94. K. Emara, A.M.M. Abd-Elgawad, A. Emara, Effect of nanofluids-PCM heat exchanging on engine downsizing and heat transfer enhancement via the heat engine's cooling system: A novel saving tactic. J Energy Storage **117**, 115815 (2025). https://doi.org/10.1016/j.est.2025.115815
95. B. Venkateswarlu, S. Chavan, S.W. Joo, S.C. Kim, Impact of hybrid nanofluids on thermal management of cylindrical battery modules: A numerical study. J Energy Storage **99**, 113266 (2024). https://doi.org/10.1016/j.est.2024.113266
96. A. Kumar, B.K. Sharma, T. Muhammad, L.M. Pérez, Optimization of thermal performance in hybrid nanofluids for parabolic trough solar collectors using Adams–Bashforth–Moulton method. Ain Shams Eng. J. **15**(12), 103106 (2024). https://doi.org/10.1016/j.asej.2024.103106
97. A. Hosseinpour, M. Pourfallah, M. Gholinia, Analysis of phase change material (PCM) melting utilizing environmentally friendly nanofluids in a double tube with spiral fins: A numerical study. Int. J. Thermofluids **22**, 100620 (2024). https://doi.org/10.1016/j.ijft.2024.100620
98. Z. Haddad, D. Belkadi, A. Mourad, A. Aissa, Z. Said, O. Younis, A. Alazzam, E. Abu-Nada, Advancements and comprehensive overview of thermal management systems for lithium-ion batteries: Nanofluids and phase change materials approaches. J. Power Sources **603**, 234382 (2024). https://doi.org/10.1016/j.jpowsour.2024.234382

99. M.A. Almeshaal, A.A. Altohamy, Experimental analysis of a photovoltaic thermal collector using phase change materials and copper oxide nanofluid. J Energy Storage **93**, 112265 (2024). https://doi.org/10.1016/j.est.2024.112265
100. Z. Huang, Q. Li, Y. Qiu, Enhancements in thermal properties of binary alkali chloride salt by Al_2O_3 nanoparticles for thermal energy storage. Energy **301**, 131584 (2024). https://doi.org/10.1016/j.energy.2024.131584
101. P.K. Kanti, E.S.J. Yang, V.V. Wanatasanappan, P. Sharma, N.M. Said, Impact of hybrid and mono nanofluids on the cooling performance of lithium-ion batteries: Experimental and machine learning insights. J Energy Storage **101**, 113613 (2024). https://doi.org/10.1016/j.est.2024.113613
102. V. Jayaprakash, S. Ganesan, N. Beemkumar, M. Sunil Kumar, K. Kamakshi Priya, Enhancing thermal energy storage efficiency: Synthesis and analysis of hybrid Nano-PCMs. Res. Eng. Des. **26**, 104899 (2025). https://doi.org/10.1016/j.rineng.2025.104899
103. M. Saraswat, R.J. Sengwa, Investigations on the optical, dielectric, electrical, and rheological properties of PEG200/ZnO semiconducting nanofluids for soft matter based technological innovations. Hybrid Adv. **5**, 100140 (2024). https://doi.org/10.1016/j.hybadv.2024.100140
104. P. Thorat, S. Sanap, S. Gawade, A comparative study of thermo-physical properties of different nanofluids for effective heat transfer leading to Li-ion battery pack cooling. Energy Storage **6**(1), e566 (2024). https://doi.org/10.1002/est2.566
105. J. Cavieres, M.J. Inestrosa-Izurieta, D.A. Vasco, J.I. Urzúa, Ionanofluids based on ionic liquid mixtures, a new approach as an alternative material for solar energy storage. J. Mol. Liq. **351**, 118677 (2022). https://doi.org/10.1016/j.molliq.2022.118677
106. A. Hosseinghorbani, M. Mozaffarian, G. Pazuki, Application of graphene oxide IoNanofluid as a superior heat transfer fluid in concentrated solar power plants. Int. Commun. Heat Mass Transf. **111**, 104450 (2020). https://doi.org/10.1016/j.icheatmasstransfer.2019.104450
107. T.A. Duarte, R.F. Pereira, B. Medronho, E.S. Maltseva, E.F. Krivoshapkina, A. Varela-Dopico, P. Taboada, L. Fu, R. Ferreira, B.V. de Zea, A glance at novel ion anofluids incorporating silk-derived carbon dots. Chem. Mater. **36**(3), 1136–1152 (2024). https://doi.org/10.1021/acs.chemmater.3c01370
108. I. Moulefera, A.R. Pastor, M.G. Fuster, J.J. Delgado-Marín, M.G. Montalbán, I. Rodríguez-Pastor, A. Lopez Perez, I. Martin Gullon, R. Gonzalez, M. Alarcon, G. Víllora, Novel application for graphene oxide-based ionanofluids in flat plate solar thermal collectors. Sci. Rep. **14**(1), 17610 (2024). https://doi.org/10.1038/s41598-024-67874-1
109. P. Kanti, A.A. Minea, K.V. Sharma, M. Revanasiddappa, Improved thermophysical properties of graphene Ionanofluid as heat transfer fluids for thermal applications. J. Ionic Liq. **2**(2), 100038 (2022). https://doi.org/10.1016/j.jil.2022.100038
110. B. Jóźwiak, G. Dzido, E. Zorebski, A. Kolanowska, R. Jedrysiak, J. Dziadosz, M. Libera, S. Boncel, M. Dzida, Remarkable thermal conductivity enhancement in carbon-based ionanofluids: Effect of nanoparticle morphology. ACS Appl. Mater. Interfaces **12**(34), 38113–38123 (2020). https://doi.org/10.1021/acsami.0c09752

Effect of Magnetically Controlled Viscosity on Double-Diffusive Convection in Magnetic Nanofluids

Mahesh Kumar Sharma and Monika Arora

1 Introduction

Nanofluids, comprising engineered suspensions of nanoparticles dispersed within conventional fluids, have emerged as promising media for advanced thermal applications. Since the seminal work of Choi [1], substantial research has focused on their potential in improving heat transfer efficiency in fields such as electronics thermal management, biomedical cooling systems, and industrial heat exchangers. A significant theoretical contribution was made by Buongiorno [2], who introduced a two-component transport model involving Brownian diffusion and thermophoresis, which together account for the enhanced thermal behavior of these nanofluids.

Among nanofluid categories, $M_g\mathcal{N}_f$, composed of ferromagnetic nanoparticles suspended in nonmagnetic liquids, exhibit unique properties under magnetic fields. Their viscosity and thermal conductivity can be tuned externally, making them highly adaptable for targeted applications such as drug transport, magnetic sealing technologies, and responsive heat transfer systems [3].

$\mathcal{D}_{convec}$, driven by both thermal and solutal gradients, plays a critical role in numerous natural and engineering processes [4]. Previous studies [5–8] have shown that this dual-gradient mechanism can either stabilize or destabilize flow depending on which buoyant force dominates. The introduction of field-dependent viscosity into such systems adds further complexity, altering stability thresholds and flow dynamics in significant ways.

M. K. Sharma
Department of Mathematics, Maharaja Agrasen University, Baddi, Himachal Pradesh, India

M. Arora (✉)
Department of Mathematics, School of Chemical Engineering and Physical Sciences, Lovely Professional University, Phagwara, Punjab, India

S. Kulkarni, S. Das (eds.), *Ionic Liquids, Nanofluids and Nanotechnology*,
Lecture Notes in Nanoscale Science and Technology 42,
https://doi.org/10.1007/978-3-032-04008-4_16

Within this context, $\mathcal{M}_f\mathcal{V}_{cos}$ is used to represent the modulation of viscosity through magnetic field interaction. Investigations by Sunil and colleagues [9], as well as Vaidyanathan et al. [10], emphasized its relevance in shifting convective onset conditions. Sekar et al. [11] showed that $\mathcal{M}_f\mathcal{V}_{cos}$ enhances flow resistance in rotating porous domains, while Nanjundappa and Gopalakrishna [12] demonstrated similar stabilizing trends in Bénard-Marangoni convection settings.

Building on this foundation, Arora et al. [13] explored $\mathcal{M}_f\mathcal{V}_{cos}$ in rotating systems of $\mathcal{W}_b - M_g\mathcal{N}_f$ and $\mathcal{E}_b - M_g\mathcal{N}_f$ nanofluids, concluding that rotation and magnetic tuning jointly delay the initiation of convection. Despite such progress, a comprehensive assessment of how $\mathcal{M}_f\mathcal{V}_{cos}$ affects $\mathcal{D}_{convec}$ in $M_g\mathcal{N}_f$ remains lacking.

This work aims to fill that research void by evaluating the interplay between $\mathcal{M}_f\mathcal{V}_{cos}$ and $\mathcal{D}_{convec}$ in horizontally layered $M_g\mathcal{N}_f$ systems. Specifically, it investigates how external magnetic fields modulate the relative effects of thermal and solutal buoyancy. The analysis framework incorporates microscale mechanisms, such as Brownian motion, thermophoresis, and magnetophoresis, and varies critical parameters—including the Langevin parameter, Lewis number, and Rayleigh numbers—to systematically assess their impact on convection behavior in both $\mathcal{W}_b - M_g\mathcal{N}_f$ and $\mathcal{E}_b - M_g\mathcal{N}_f$ configurations.

The subsequent sections of this chapter are organized as follows: Sect. 2 introduces the physical model and assumptions. Section 3 formulates the governing equations and presents their non-dimensionalization. Section 4 discusses the steady-state conditions, while Sect. 5 develops the linear stability analysis. Section 6 presents the numerical simulations and physical interpretations. Section 7 concludes the paper with a summary of key findings.

2 Physical Model of the Flow

We analyze a system consisting of an unbounded horizontal layer of incompressible $M_g\mathcal{N}_f$, which is influenced by thermal and solutal gradients imposed from below. The dynamic viscosity in this configuration is considered to be sensitive to magnetic fields, and it follows the functional relationship $\mu_1 = \mu(1 + \boldsymbol{\delta} \cdot \boldsymbol{B})$. Here, μ denotes the fluid's reference viscosity in the absence of magnetic forces, $\boldsymbol{B}$ signifies the magnetic induction vector, and $\boldsymbol{\delta}$ quantifies the isotropic variation characteristic of $\mathcal{M}_f\mathcal{V}_{cos}$.

The fluid occupies a slab defined by the region $z \in [0, d]$, with gravitational acceleration $\boldsymbol{g}$ oriented vertically downward along the negative z-axis. An external magnetic field $\boldsymbol{H}$ is applied transversely across the fluid layer. Additionally, the thermal boundary conditions and the nanoparticle volume concentrations are maintained uniformly at the upper and lower surfaces of the domain (Fig. 1).

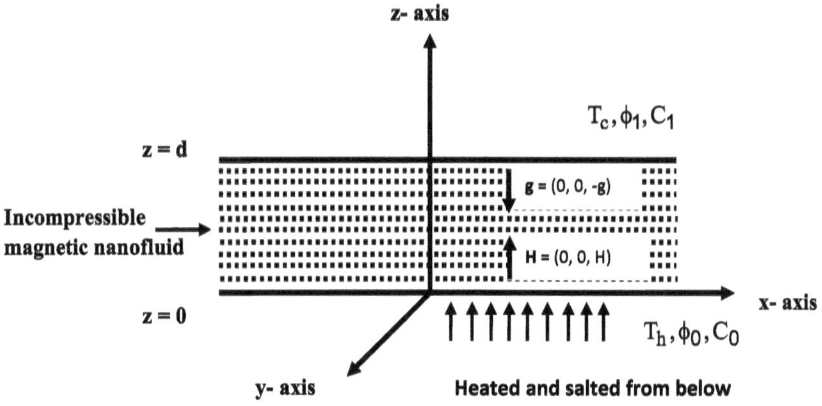

Fig. 1 Schematic illustration of the modeled configuration

3 Analytical Framework and Mathematical Setup

In alignment with the formulations presented in [14, 15], and [16], the present study considers a comprehensive system of fundamental equations governing the dynamics of the model. These include the continuity equation, momentum equation, nanoparticle transport equation, energy conservation equation, Maxwell's equations, magnetization relation, and solute concentration equation, each capturing essential aspects of the underlying physics. The system dynamics are described through a set of coupled differential equations, derived based on established thermofluidic principles in [14, 15], and [16]. These include:

To describe the physical behavior of the system, we consider a set of fundamental equations derived from fluid mechanics and electromagnetic theory. These equations capture the transport of mass, momentum, energy, nanoparticle concentration, and the effects of magnetic field-induced forces within the nanofluid domain.

The equation governing the conservation of mass, based on the incompressibility assumption, is given by

$$\nabla \cdot \boldsymbol{u} = 0. \tag{1}$$

The momentum balance incorporates viscous, pressure, magnetic, and gravitational contributions:

$$\rho_f \left(\frac{\partial \boldsymbol{u}}{\partial t} + \boldsymbol{u} \cdot \nabla \boldsymbol{u} \right) = -\nabla p + \mu_1 \nabla^2 \boldsymbol{u} + \mu_0 (\boldsymbol{M} \cdot \nabla) \boldsymbol{H} - \rho g \boldsymbol{k}. \tag{2}$$

The nanoparticle concentration ϕ evolves as per a modified advection-diffusion equation including thermophoretic and magnetophoretic influences:

$$\frac{\partial \phi}{\partial t} + \boldsymbol{u} \cdot \nabla \phi = \nabla \cdot \left(D_B \nabla \phi + D_T \frac{\nabla T}{T_c} - D_H \nabla H \right). \tag{3}$$

The energy conservation equation accounts for conduction, nanoparticle interactions, and magnetic heating effects:

$$(\rho c)_f \left(\frac{\partial T}{\partial t} + \boldsymbol{u} \cdot \nabla T \right) = \nabla \cdot (k_1 \nabla T) + \rho_p c_p \left(D_B \nabla T \cdot \nabla \phi + D_T \frac{\nabla T \cdot \nabla T}{T_c} - D_H \nabla T \cdot \nabla H \right) + (\rho c)_f D_{TC} \nabla^2 C. \tag{4}$$

The electromagnetic field components are governed by Maxwell's quasi-static relations:

$$\nabla \cdot \boldsymbol{B} = 0, \quad \nabla \times \boldsymbol{H} = 0, \quad \boldsymbol{B} = \mu_0 (\boldsymbol{M} + \boldsymbol{H}). \tag{5}$$

The equilibrium magnetization $\boldsymbol{M}_{eq}$ of the fluid aligns with the field direction, modeled as

$$\boldsymbol{M}_{eq} = \frac{\boldsymbol{H}}{H} M_s \phi L(\alpha_L) = \frac{\boldsymbol{H}}{H} M_{eq}(H\phi, T, C). \tag{6}$$

The solute concentration C satisfies a diffusion equation with thermal cross-diffusion (Dufour) effect:

$$\frac{\partial C}{\partial t} + \boldsymbol{u} \cdot \nabla C = D_S \nabla^2 C + D_{CT} \nabla^2 T. \tag{7}$$

The system is subject to thermal and concentration constraints at the horizontal boundaries. The corresponding boundary conditions are

$$\left. \begin{aligned} w = 0, \quad T = T_h, \quad \phi = \phi_0, \quad C = C_0 \quad \text{at} \quad z = 0, \\ w = 0, \quad T = T_c, \quad \phi = \phi_1, \quad C = C_1 \quad \text{at} \quad z = d. \end{aligned} \right\} \tag{8}$$

Additionally, we impose a no-shear condition on the lower rigid boundary and a stress-free condition on the upper surface:

$$\left. \frac{\partial w}{\partial z} \right|_{z=0} = 0, \quad \left. \frac{\partial^2 w}{\partial z^2} \right|_{z=d} = 0.$$

Dimensionless System Derived from Eqs. (1)–(7) To facilitate analytical treatment and uncover key scaling behaviors, the governing equations are reformulated using a set of suitable nondimensional variables. This normalization preserves the physical essence of each process, while abstracting the problem into a parameter-dependent structure.

The continuity equation remains unaltered in form:

$$\nabla \cdot \boldsymbol{u} = 0. \tag{9}$$

The dimensionless momentum equation now includes the impact of magnetic induction, field-sensitive viscosity, and composite buoyancy effects:

$$\begin{aligned}\frac{1}{Pr}\left(\frac{\partial \boldsymbol{u}}{\partial t} + \boldsymbol{u} \cdot \nabla \boldsymbol{u}\right) &= -\nabla p + \nabla^2 \boldsymbol{u} + \{c_1(\boldsymbol{\delta} \cdot \boldsymbol{M}) + c_2(\boldsymbol{\delta} \cdot \boldsymbol{H})\} \nabla^2 \boldsymbol{u} \\ &\quad + \lambda_1(\boldsymbol{M} \cdot \nabla)\boldsymbol{H} - \Big(Rn\phi - RaT + \frac{Rs}{Le_s}C \\ &\quad + Ra_n N_\phi T\phi - Rs_n N_\phi C\phi + \rho_1 - \rho_2\phi\Big)\boldsymbol{k}. \end{aligned} \tag{10}$$

The equation governing nanoparticle volume fraction accounts for diffusion, thermophoresis, and magnetophoresis:

$$\frac{\partial \phi}{\partial t} + \boldsymbol{u} \cdot \nabla \phi = \frac{1}{Le}\nabla^2 \phi + \frac{N_a}{Le}\nabla^2 T - \frac{N_a'}{Le}\nabla^2 H. \tag{11}$$

The temperature field evolves according to a heat balance equation that incorporates classical conduction as well as cross-gradient and particle-coupled effects:

$$\begin{aligned}\frac{\partial T}{\partial t} + \boldsymbol{u} \cdot \nabla T &= \nabla^2 T + \frac{N_b}{Le}(\nabla \phi \cdot \nabla T) + \frac{N_a N_b}{Le}(\nabla T \cdot \nabla T) \\ &\quad - \frac{N_a' N_b}{Le}(\nabla H \cdot \nabla T) + N_{ct}\nabla^2 C. \end{aligned} \tag{12}$$

The coupling between the magnetic field and magnetization is governed by

$$\chi_2 \nabla \cdot \boldsymbol{M} + \nabla \cdot \boldsymbol{H} = 0. \tag{13}$$

The dimensionless expression for $\boldsymbol{M}$ reflects its alignment with the applied magnetic field and incorporates the cumulative influence of temperature, concentration, and particle fraction variations:

$$\boldsymbol{M} = \frac{\boldsymbol{H}}{H} \cdot \frac{1+\chi}{\chi_2}\left\{\frac{\chi}{1+\chi}H - \frac{M_{p1}}{M_{p3}}T + \frac{M_{p1}'}{M_{p3}'}\phi + \frac{M_{p1}''}{M_{p3}''}C + \frac{\chi_2 - \chi}{1+\chi}\right\}. \tag{14}$$

The concentration field C satisfies the following advection-diffusion equation with cross-coupling:

$$\frac{\partial C}{\partial t} + \boldsymbol{u} \cdot \nabla C = \frac{1}{Le_s} \nabla^2 C + N_{ct} \nabla^2 T. \tag{15}$$

The auxiliary parameters used in the above expressions are given by

$$\rho_1 = \frac{d^3 \rho_f}{\kappa \mu}(1 + \alpha T_c - \alpha' C_1) g, \quad \rho_2 = \frac{d^3 \rho_f}{\kappa \mu}(\phi_0 - \phi_1)(\alpha T_c - \alpha' C_1) g,$$

$$\lambda_1 = \frac{\mu_0 M_0 H_0 d^2}{\kappa \mu}, \quad N_\phi = \frac{\phi_0 - \phi_1}{1 - \phi_0}, \quad c_1 = \mu_0 M_0, \quad c_2 = \mu_0 H_0.$$

The nondimensional boundary conditions associated with this formulation, as adapted from the physical domain, are

$$\left.\begin{array}{lllll} w = 0, & T = \dfrac{T_h}{T_h - T_c}, & \phi = \dfrac{\phi_0}{\phi_0 - \phi_1}, & C = \dfrac{C_0}{C_0 - C_1} & \text{at} \quad z = 0, \\ w = 0, & T = \dfrac{T_c}{T_h - T_c}, & \phi = \dfrac{\phi_1}{\phi_0 - \phi_1}, & C = \dfrac{C_1}{C_0 - C_1} & \text{at} \quad z = 1. \end{array}\right\} \tag{16}$$

Nondimensional Parameters The behavior of the system is governed by several dimensionless quantities that represent thermal, solutal, magnetic, and transport effects.

$$Ra = \frac{\rho_f g \alpha d^3 (T_h - T_c)}{\mu \kappa} \text{ (thermal Rayleigh number)},$$

$$Le = \frac{\kappa}{D_B} \text{ (Lewis number)},$$

$$Pr = \frac{\mu}{\rho_f \kappa} \quad \text{(Prandtl number)},$$

$$Rn = \frac{(\rho_p - \rho_f)(\phi_0 - \phi_1) g d^3}{\mu \kappa} \quad \text{(nanoparticle Rayleigh number)},$$

$$N_a = \frac{D_T (T_h - T_c)}{D_B T_c (\phi_0 - \phi_1)} \quad \text{(thermophoretic diffusivity ratio)},$$

$$N_a' = \frac{D_H H_0}{D_B (\phi_0 - \phi_1)} \quad \text{(magnetophoretic diffusivity ratio)},$$

$$Rs = \frac{\rho_f g \alpha' d^3 (C_0 - C_1)}{\mu D_S} \quad \text{(solutal-based Rayleigh number)},$$

$$Le_s = \frac{\kappa}{D_S} \quad \text{(solute-based Lewis number)},$$

$$N_{ct} = \frac{D_{TC}(C_0 - C_1)}{k(T_h - T_c)} \quad \text{(Soret parameter)},$$

$$N_{tc} = \frac{D_{CT}(T_h - T_c)}{k(C_0 - C_1)} \quad \text{(Dufour parameter)},$$

$$M_{p1} = \frac{\mu_0 \chi^2 H_0^2 (T_h - T_c)}{\rho_f g \alpha d (1 + \chi) T_h^2} \quad \text{(magnetization coefficient for temperature field)},$$

$$M_{p3} = \frac{\mu_0 \chi H_0^2}{\rho_f g \alpha d T_h} \quad \text{(reference magnetic scaling for temperature)},$$

$$M'_{p1} = \frac{\mu_0 \chi^2 H_0^2 (\phi_0 - \phi_1)}{\rho_f g \alpha d (1 + \chi) \phi_0^2} \quad \text{(magnetization coefficient for nanoparticle volume)},$$

$$M'_{p3} = \frac{\mu_0 \chi H_0^2}{\rho_f g \alpha d \phi_0} \quad \text{(reference magnetic scaling for nanoparticle field)},$$

$$M''_{p1} = \frac{\mu_0 \chi^2 H_0^2 (C_0 - C_1)}{\rho_f g \alpha' d (1 + \chi) C_0^2} \quad \text{(magnetization coefficient for solute concentration)},$$

$$M''_{p3} = \frac{\mu_0 \chi H_0^2}{\rho_f g \alpha' d C_0} \quad \text{(reference magnetic scaling for concentration field)},$$

$$N_b = \frac{(\rho C)_p (\phi_0 - \phi_1)}{(\rho C)_f} \quad \text{(normalized heat capacity of particle phase)}.$$

The kinematic viscosity is denoted by $\nu = \mu/\rho_f$. The adjusted Rayleigh number accounting for base nanoparticle concentration is $Ra_N = (1 - \phi_0)Ra$, while the normalized solutal Rayleigh number is $Rs_n = (1 - \phi_0)Rs/Le_s$.

4 Steady-State Configuration

To determine the base state, we assume the system reaches a stationary profile with no fluid motion:

$$\boldsymbol{u}_b = \boldsymbol{0},$$

and the scalar fields p_b, T_b, ϕ_b, C_b as well as the vector fields $\boldsymbol{H}_b, \boldsymbol{M}_b$ are functions of the vertical coordinate z only.

Substituting these steady-state assumptions into the nondimensional governing equations (10)–(15), we obtain the following set of ordinary differential equations:

The pressure gradient in the vertical direction satisfies

$$-\frac{dp_b}{dz} + \lambda_1 M_b \frac{dH_b}{dz} - Rn\phi_b + RaT_b - \frac{Rs}{Le_s} C_b - Ra_n N_\phi T_b \phi_b + Rs_n N_\phi C_b \phi_b - \rho_1 + \rho_2 \phi_b = 0. \tag{17}$$

The steady-state equation for nanoparticle concentration becomes

$$\frac{d^2\phi_b}{dz^2} + N_a \frac{d^2 T_b}{dz^2} - N_a' \frac{d^2 H_b}{dz^2} = 0. \tag{18}$$

The temperature field in steady condition follows:

$$\frac{d^2 T_b}{dz^2} + \frac{dT_b}{dz}\left(\frac{N_b}{Le}\frac{d\phi_b}{dz} + \frac{N_a N_b}{Le}\frac{dT_b}{dz} - \frac{N_a' N_b}{Le}\frac{dH_b}{dz}\right) + N_{ct}\frac{d^2 C_b}{dz^2} = 0. \tag{19}$$

The divergence condition for the magnetic field yields

$$\chi_2 \frac{dM_b}{dz} + \frac{dH_b}{dz} = 0. \tag{20}$$

The magnetization $\boldsymbol{M}_b$ at steady state relates to local fields via

$$M_b = \frac{1+\chi}{\chi_2}\left(\frac{\chi}{1+\chi} H_b - \frac{M_{p1}}{M_{p3}} T_b + \frac{M_{p1}'}{M_{p3}'}\phi_b + \frac{M_{p1}''}{M_{p3}''} C_b + \frac{\chi_2 - 2\chi}{1+\chi}\right). \tag{21}$$

Solute concentration in the absence of time dependence evolves as

$$\frac{1}{Le_s}\frac{d^2 C_b}{dz^2} + N_{ct}\frac{d^2 T_b}{dz^2} = 0. \tag{22}$$

The above system is solved subject to the boundary conditions given in (16). Adopting the approach from prior works [2, 16], and [17], the exact solutions for the base state fields are determined to be

$$u_b = \mathbf{0}, \quad p = p_b(z), \quad T_b = \frac{T_h}{T_h - T_c} - z,$$

$$\phi_b = \frac{\phi_0}{\phi_0 - \phi_1} - z, \quad C_b = \frac{C_0}{C_0 - C_1} - z,$$

$$H_b = 1 - \frac{M_{p1}}{M_{p3}} z + \frac{M'_{p1}}{M'_{p3}} z + \frac{M''_{p1}}{M''_{p3}} z,$$

$$M_b = 1 + \frac{1}{\chi_2} \left(\frac{M_{p1}}{M_{p3}} - \frac{M'_{p1}}{M'_{p3}} - \frac{M''_{p1}}{M''_{p3}} \right) z. \tag{23}$$

5 Linearized Perturbation Analysis

To examine the stability of the base state, we introduce infinitesimal perturbations to all dependent variables:

$$\boldsymbol{u}, \quad p, \quad T, \quad C, \quad \boldsymbol{M}, \quad \phi, \quad \boldsymbol{H}.$$

Inserting these perturbed variables into the nondimensional equations (9)–(15) and performing linearization around the steady state result in the following set of linearized equations.

The vertical component of the linearized momentum equation, obtained by applying the curl of a curl operator and extracting its z-component, becomes

$$\begin{aligned}
\frac{1}{Pr}\frac{\partial \nabla^2 w}{\partial t} &= \nabla^4 w - \left(RaM_{p3} - Ra_s M'_{p3} - \frac{Rs}{Le_s} M''_{p3} \right) \frac{\partial \nabla_H^2 \psi}{\partial z} \\
&+ \left(Ng - Ra\frac{M_{p3}M'_{p1}}{M'_{p3}} + Ra_n(1 + N_\phi z) - \frac{Rs}{Le_s}\frac{M''_{p3}M_{p1}}{M_{p3}} \right) \nabla_H^2 \theta \\
&- \left(Ra\frac{M_{p3}M'_{p1}}{M'_{p3}} - Ra_s M'_{p1} - \frac{Rs}{Le_s}\frac{M''_{p3}M'_{p1}}{M'_{p3}} + Rn + Ra_n N_\phi (1 - z) \right. \\
&\left. - Rs_n N_\phi (1 - z) \right) \nabla_H^2 \phi \\
&+ \left(\frac{Rs}{Le_s}\frac{M''_{p3}M'_{p1}}{M'_{p3}} + \frac{Rs}{Le_s} M''_{p1} - \frac{Rs}{Le_s}\frac{M''_{p3}M_{p1}}{M_{p3}} - Rs_n(1 + N_\phi z) \right) \nabla_H^2 C.
\end{aligned} \tag{24}$$

The equation for nanoparticle concentration fluctuation is

$$\frac{\partial \phi}{\partial t} = w + \frac{1}{Le}\nabla^2\phi + \frac{N_a}{Le}\nabla^2\theta - \frac{N_a'}{Le}\frac{\partial \nabla^2\psi}{\partial z}. \tag{25}$$

The temperature perturbation evolves as

$$\frac{\partial \theta}{\partial t} = \nabla^2\theta + w - \frac{N_b}{Le}\frac{\partial \phi}{\partial z} + \frac{N_b N_a'}{Le}\frac{\partial^2\psi}{\partial z^2} - \left(\frac{N_b}{Le} + \frac{2N_a N_b}{Le} - \frac{N_b N_a' M_{p1}}{Le M_{p3}} + \frac{N_b N_a' M_{p1}'}{Le M_{p3}'} + \frac{N_b N_a' M_{p1}''}{Le M_{p3}''}\right)\frac{\partial \theta}{\partial z} + N_{ct}\nabla^2 C. \tag{26}$$

The perturbed magnetic potential satisfies

$$\frac{\partial^2\psi}{\partial z^2} = -\frac{(1+\chi_2)}{(1+\chi)}\nabla_H^2\psi + \frac{M_{p1}}{M_{p3}}\frac{\partial \theta}{\partial z} - \frac{M_{p1}'}{M_{p3}'}\frac{\partial \phi}{\partial z} - \frac{M_{p1}''}{M_{p3}''}\frac{\partial C}{\partial z}. \tag{27}$$

The solute concentration perturbation evolves according to

$$\frac{\partial C}{\partial t} = w + \frac{1}{Le_s}\nabla^2 C + N_{ct}\nabla^2\theta. \tag{28}$$

Here, the horizontal Laplacian operator is $\nabla_H^2 = \frac{\partial^2}{\partial x^2} + \frac{\partial^2}{\partial y^2}$, the magnetic buoyancy factor is $Ng = RaM_{p1}$, and the nanoparticle Rayleigh number is defined as $Ra_s = \frac{\rho_f g \alpha d^3(\phi_0 - \phi_1)}{\mu\kappa}$.

To facilitate normal mode analysis, we express all perturbation fields using the following normal form:

$$[w, \phi, \theta, \psi, C] = [w(z), \phi(z), \theta(z), \psi(z), C(z)]\exp(\sigma t + i(k_x x + k_y y)), \tag{29}$$

where k_x, k_y are the horizontal wavenumbers and $k = \sqrt{k_x^2 + k_y^2}$ is the total horizontal wavenumber.

Substituting this form into Eqs. (24)–(28), we obtain the reduced eigenvalue system:

$$\frac{\sigma}{Pr}(4D^2 - k^2)w = (4D^2 - k^2)^2 w - k^2\left(Ng - Ra\frac{M_{p3}M_{p1}'}{M_{p3}'} - \frac{Rs}{Le_s}\frac{M_{p3}''M_{p1}}{M_{p3}} + Ra_n(1 + N_\phi \tfrac{z+1}{2})\right)\theta$$

$$+k^2\left(Ra\frac{M_{p3}M'_{p1}}{M'_{p3}}-Ra_s M'_{p1}+Rn-\frac{Rs}{Le_s}\frac{M''_{p3}M'_{p1}}{M'_{p3}}+Ra_n N_\phi \tfrac{1-z}{2}-Rs_n N_\phi \tfrac{1-z}{2}\right)\phi$$

$$+\,2k^2\left(RaM_{p3}-Ra_s M'_{p3}-\frac{Rs}{Le_s}M''_{p3}\right)D\psi$$

$$+\,k^2\left(\frac{Rs}{Le_s}\frac{M''_{p3}M_{p1}}{M_{p3}}-\frac{Rs}{Le_s}\frac{M''_{p3}M'_{p1}}{M'_{p3}}+Rs_n(1+N_\phi \tfrac{z+1}{2})-\frac{Rs}{Le_s}M''_{p1}\right)C. \tag{30}$$

$$\sigma\phi = w+\frac{1}{Le}(4D^2-k^2)\phi+\frac{N_a}{Le}(4D^2-k^2)\theta-\frac{2N'_a}{Le}(4D^2-k^2)D\psi, \tag{31}$$

$$\sigma\theta = w+(4D^2-k^2)\theta$$
$$-\,2\left(\frac{N_b}{Le}+\frac{2N_aN_b}{Le}-\frac{N_bN'_aM_{p1}}{LeM_{p3}}+\frac{N_bN'_aM'_{p1}}{LeM'_{p3}}+\frac{N_bN'_aM''_{p1}}{LeM''_{p3}}\right)D\theta$$
$$-\,\frac{2N_b}{Le}D\phi+\frac{4N_bN'_a}{Le}D^2\psi+N_{ct}(4D^2-k^2)C, \tag{32}$$

$$\left(4D^2-\frac{k^2(1+\chi_2)}{1+\chi}\right)\psi-\frac{2M_{p1}}{M_{p3}}D\theta+\frac{2M'_{p1}}{M'_{p3}}D\phi+\frac{2M''_{p1}}{M''_{p3}}DC=0, \tag{33}$$

$$\sigma C = w+\frac{1}{Le_s}(4D^2-k^2)C+N_{ct}(4D^2-k^2)\theta. \tag{34}$$

The linear system is closed with the following boundary conditions:

$$\left.\begin{array}{lll} w=0,\quad \theta=0,\quad \phi=0, & C=0 & \text{at } z=\pm 1,\\ Dw=0, & 2(1+\chi)D\psi-k\psi=0 & \text{at } z=-1,\\ D^2w=0, & 2(1+\chi)D\psi+k\psi=0 & \text{at } z=+1. \end{array}\right\} \tag{35}$$

This eigenvalue problem, composed of Eqs. (30)–(34) with conditions (35), is numerically solved using the Chebyshev pseudospectral method as described in [18].

6 Results and Discussion

This section presents the findings derived from the linear stability analysis of $M_g\mathcal{N}_f$ under various physical conditions. The objective is to determine the circumstances leading to the initiation of convective motion, assessed via critical Rayleigh number (Ra_c). The analysis encompasses both $\mathcal{W}_b - M_g\mathcal{N}_f$ and $\mathcal{E}_b - M_g\mathcal{N}_f$ formulations.

The influence of layer thickness d and Langevin parameter α_L on Ra_c is demonstrated in Fig. 2. It becomes evident that as d increases, the corresponding Ra_c also rises, implying that a thicker fluid domain resists convection more effectively. Simultaneously, greater values of α_L promote the alignment of magnetic particles, thereby enhancing the stabilizing effect of the magnetic field.

Figure 3 explores the relationship between Ra_c and the diffusivity ratio δ for varying values of $\Delta\phi$. The findings suggest that a higher δ corresponds to increased Ra_c, primarily due to intensified solutal diffusion suppressing thermal perturbations. This stabilizing effect becomes more noticeable as $\Delta\phi$ rises, indicating the interactive role of concentration gradients in convective dynamics.

In Fig. 4, the variation of Rayleigh number with respect to wavenumber k is plotted for different δ values. A trend of increasing Ra with higher k is identified,

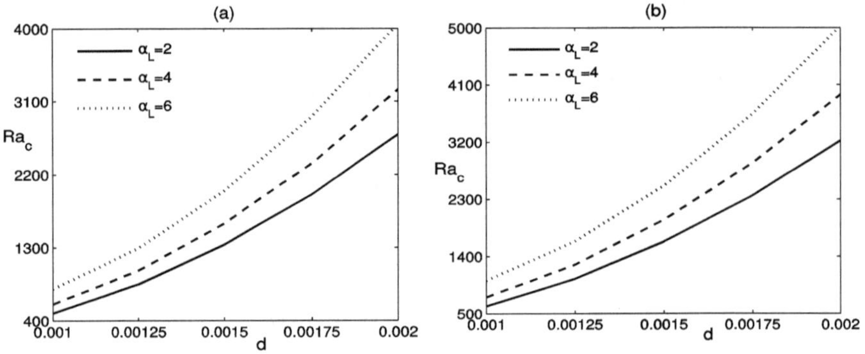

Fig. 2 C_sC_{urves} depicting the dependence of Ra_c on d across varying values of α_L, for (**a**) $\mathcal{W}_b - M_g\mathcal{N}_f$ and (**b**) $\mathcal{E}_b - M_g\mathcal{N}_f$

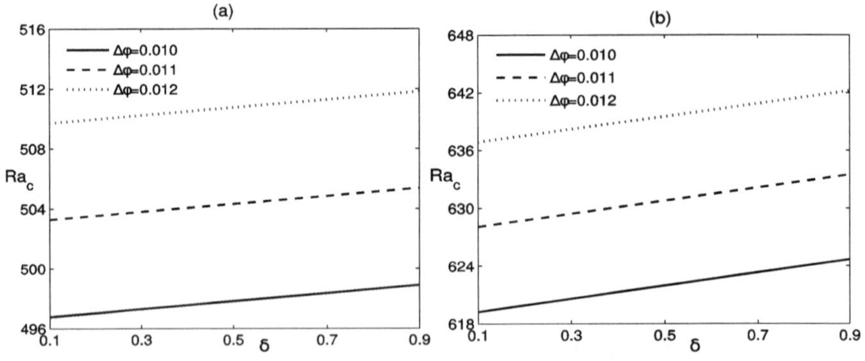

Fig. 3 C_sC_{urves} illustrating the dependence of Ra_c on δ for various values of $\Delta\phi$, shown for (**a**) $\mathcal{W}_b - M_g\mathcal{N}_f$ and (**b**) $\mathcal{E}_b - M_g\mathcal{N}_f$

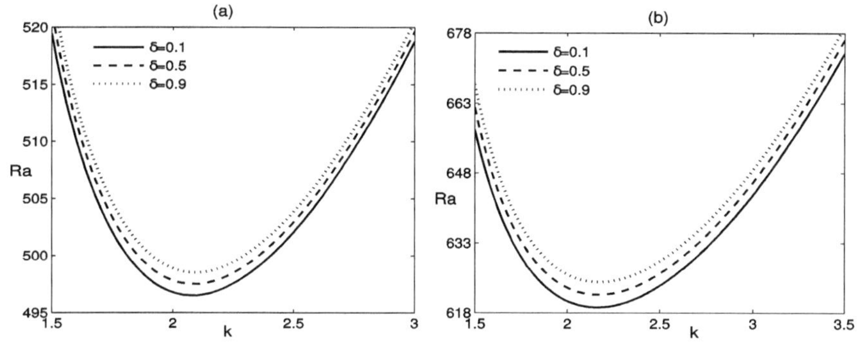

Fig. 4 $\mathcal{N}_s\mathcal{C}_{urves}$ showing the influence of δ on Ra for varying wavenumber k, corresponding to (**a**) $\mathcal{W}_b - M_g\mathcal{N}_f$ and (**b**) $\mathcal{E}_b - M_g\mathcal{N}_f$

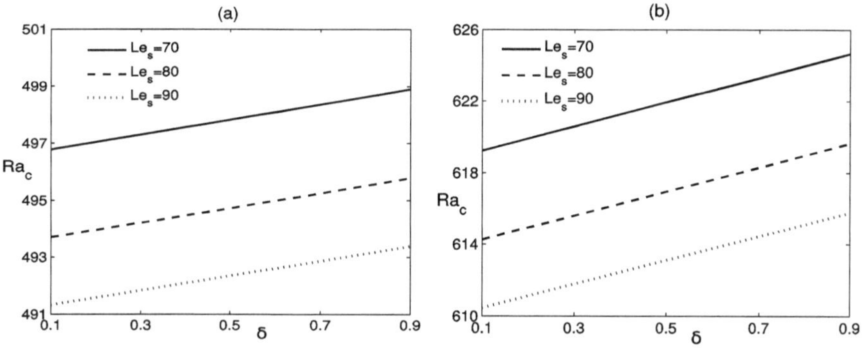

Fig. 5 $\mathcal{C}_s\mathcal{C}_{urves}$ illustrating how Ra_c varies with δ for different values of Le_s, for (**a**) $\mathcal{W}_b - M_g\mathcal{N}_f$ and (**b**) $\mathcal{E}_b - M_g\mathcal{N}_f$

indicating that initiating smaller convection cells requires more energy due to increased resistance from diffusive processes.

Figure 5 presents the response of Ra_c to variations in δ for multiple values of the solutal Lewis number (Le_s). As Le_s increases, Ra_c also rises, demonstrating that dominant solute diffusion over thermal conduction enhances system stability by resisting the onset of convection.

Figure 6 examines how the solutal Rayleigh number Rs influences Ra_c. The results confirm that increasing Rs elevates Ra_c, showcasing the inhibitory role of solute-driven buoyancy forces in triggering convection.

The interplay between Lewis number Le and Rayleigh number for different k values is illustrated in Fig. 7. With larger Le, Ra increases, underscoring that when thermal diffusivity dominates, a higher thermal gradient is needed to induce instability.

Figure 8 investigates the impact of the magnetic parameter M_3. A direct correlation is observed: Higher M_3 values lead to increased Ra, confirming that

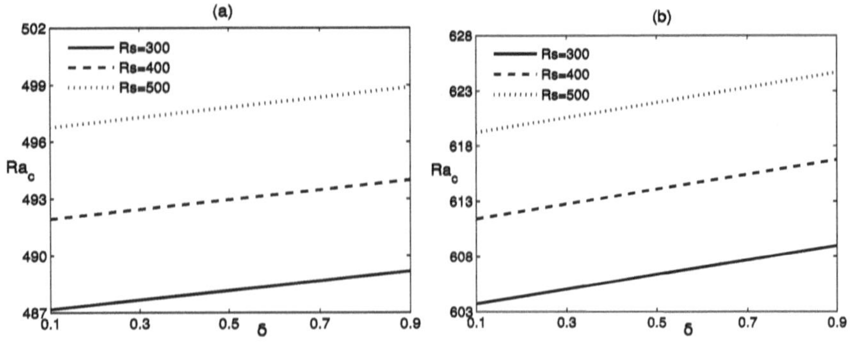

Fig. 6 $\mathcal{C}_s\mathcal{C}_{urves}$ representing variation of Ra_c as a function of δ for different values of Rs for (**a**) $\mathcal{W}_b - M_g\mathcal{N}_f$ and (**b**) $\mathcal{E}_b - M_g\mathcal{N}_f$

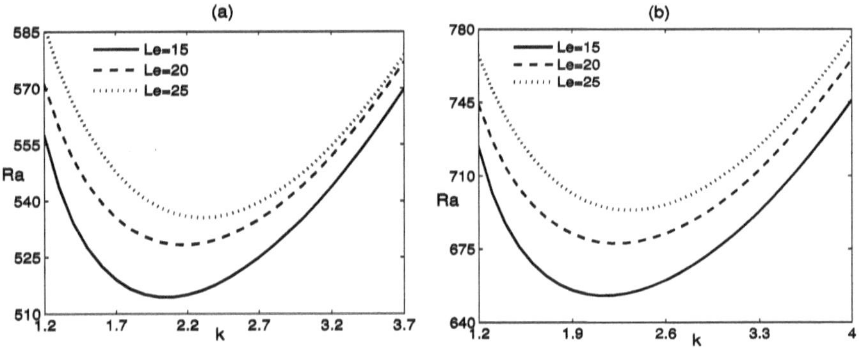

Fig. 7 $\mathcal{N}_s\mathcal{C}_{urves}$ for different values of Le for (**a**) $\mathcal{W}_b - M_g\mathcal{N}_f$ and (**b**) $\mathcal{E}_b - M_g\mathcal{N}_f$

magnetic alignment of nanoparticles contributes to stabilizing the fluid by reducing convective vigor.

Figure 9 displays the variation in Ra with the concentration Rayleigh number Rn. Unlike other parameters, an increase in Rn leads to a decrease in Ra, suggesting a destabilizing influence. This behavior is attributed to enhanced Brownian motion and solute-induced buoyancy forces facilitating convection.

Table 1 summarizes numerical values of Ra_c and the corresponding critical wavenumber k_c for different boundary configurations—rigid-rigid ($\mathcal{RR}$), rigid-free ($\mathcal{RF}$), and free-free ($\mathcal{FF}$). Both $\mathcal{W}_b - M_g\mathcal{N}_f$ and $\mathcal{E}_b - M_g\mathcal{N}_f$ cases are considered across diverse δ and Rs scenarios. The tabulated data validate that rigid boundary setups impose greater restrictions on convection compared to those with free surfaces.

Overall, the study reveals a nuanced interaction among magnetic, solutal, and thermal influences. While parameters such as α_L, M_3, and Rs tend to stabilize the fluid, others including Rn and $\Delta\phi$ exhibit destabilizing tendencies depending on the magnitude and interaction of governing forces.

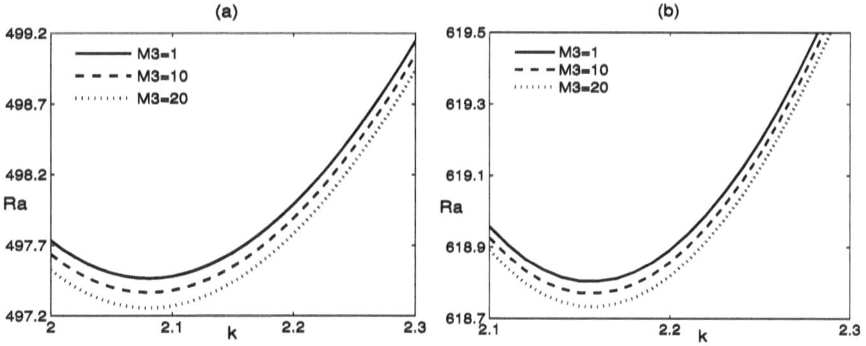

Fig. 8 $\mathcal{N}_s\mathcal{C}_{urves}$ for different values of M_3 for (**a**) $\mathcal{W}_b - M_g\mathcal{N}_f$ and (**b**) $\mathcal{E}_b - M_g\mathcal{N}_f$

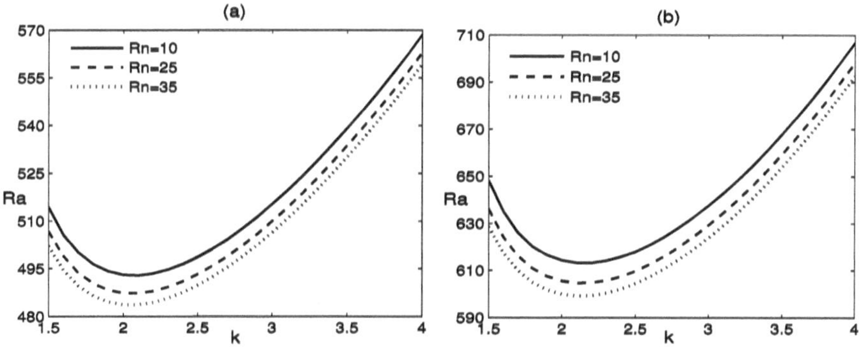

Fig. 9 $\mathcal{N}_s\mathcal{C}_{urves}$ for different values of Rn for (**a**) $\mathcal{W}_b - M_g\mathcal{N}_f$ and (**b**) $\mathcal{E}_b - M_g\mathcal{N}_f$

7 Conclusions

In this chapter, a linear stability analysis of $\mathcal{D}_{convec}$ in $M_g\mathcal{N}_f$ was carried out. The investigation focused on both $\mathcal{W}_b - M_g\mathcal{N}_f$ and $\mathcal{E}_b - M_g\mathcal{N}_f$ incorporating the influence of $\mathcal{M}_f\mathcal{V}_{cos}$ in the model. The analysis revealed that $\mathcal{M}_f\mathcal{V}_{cos}$, in conjunction with thermophoresis, Brownian motion, and magnetophoresis, significantly influences the onset of convective motion in these fluids.

The parametric results indicate that increasing certain dimensionless parameters enhances the stability of the nanofluid layer. For example, raising δ, α_L, or M_3 increases the value of Ra_c required for convection onset, thereby stabilizing the system. Likewise, larger values of Rs and Le_s further delay the onset of convective motion, underscoring the stabilizing roles of solute diffusion and magnetic field effects.

By contrast, some parameter variations were observed to have a destabilizing effect under specific conditions. For example, Rn or $\Delta\phi$ was found to precipitate the onset of instability in the layer. Additionally, the results confirm that rigid boundary

Table 1 The values of Ra_c and k_c for $\mathcal{W}_b - M_g\mathcal{N}_f$ and $\mathcal{E}_b - M_g\mathcal{N}_f$

		$\mathcal{RR}$				$\mathcal{RF}$				$\mathcal{FF}$			
		$\mathcal{W}_b$		$\mathcal{E}_b$		$\mathcal{W}_b$		$\mathcal{E}_b$		$\mathcal{W}_b$		$\mathcal{E}_b$	
Rs	δ	k_c	Ra_c	k_c	Ra_c	k_c	Ra_c	k_c	Ra_c	k_c	Ra_c	k_c	Ra_c
500	0.01	2.6	545.0	2.9	726.7	2.1	496.3	2.2	618.6	1.9	466.4	1.8	544.5
	0.11	2.6	545.6	2.9	727.9	2.1	496.5	2.2	619.3	1.9	466.4	1.8	545.0
	0.21	2.6	546.3	2.9	729.2	2.1	496.8	2.2	620.0	1.9	466.5	1.8	545.4
	0.41	2.7	547.6	2.9	731.7	2.1	497.3	2.2	621.3	1.9	466.5	1.8	546.4
	0.51	2.7	548.2	2.9	732.9	2.1	497.5	2.2	622.0	1.9	466.6	1.8	546.8
	0.71	2.7	549.5	2.9	735.5	2.1	498.0	2.2	623.4	1.9	466.6	1.8	547.8
	0.91	2.7	550.7	2.9	738.0	2.1	498.5	2.2	624.7	1.9	466.7	1.8	548.7
1000	0.01	2.8	576.0	3.1	784.1	2.1	521.2	2.2	660.1	1.9	489.1	1.8	583.8
	0.11	2.8	576.7	3.1	785.5	2.1	521.5	2.2	660.9	1.9	489.2	1.8	584.3
	0.21	2.8	577.4	3.1	787.0	2.1	521.8	2.2	661.6	1.9	489.2	1.8	584.8
	0.41	2.8	578.9	3.1	789.9	2.1	522.4	2.2	663.2	1.9	489.3	1.8	585.8
	0.51	2.8	579.6	3.1	791.3	2.1	522.7	2.2	663.9	1.9	489.4	1.8	586.3
	0.71	2.8	581.0	3.1	794.3	2.2	523.2	2.3	665.5	1.9	489.5	1.8	587.3
	0.91	2.8	582.4	3.1	797.3	2.2	523.8	2.3	667.0	1.9	489.6	1.8	588.3
1500	0.01	2.9	610.1	3.3	850.7	2.2	548.0	2.3	706.4	1.9	513.5	1.9	626.5
	0.11	2.9	610.9	3.3	852.5	2.2	548.3	2.3	707.3	1.9	513.6	1.9	627.0
	0.21	2.9	611.7	3.3	854.2	2.2	548.7	2.4	708.2	1.9	513.7	1.9	627.6
	0.41	2.9	613.3	3.3	857.8	2.2	549.3	2.4	709.9	1.9	513.8	1.9	628.7
	0.51	2.9	614.1	3.3	859.6	2.2	549.6	2.4	710.7	1.9	513.9	1.9	629.3
	0.71	2.9	615.8	3.3	863.2	2.2	550.3	2.4	712.4	1.9	514.1	1.9	630.4
	0.91	2.9	617.4	3.3	866.9	2.2	551.0	2.4	714.1	1.9	514.3	1.9	631.5

conditions provide greater resistance to convective instability than free boundary conditions, highlighting the influence of boundary constraints on system stability.

Overall, the combined influence of magnetic fields and solutal concentration gradients provides a means to finely tune the convective behavior of nanofluid systems. These insights carry significant implications for the design of efficient thermal management strategies and magnetically responsive flow systems. Future studies could extend the present analysis to nonlinear regimes and pursue experimental validation to further strengthen and verify the theoretical predictions.

References

1. S. U. S. Choi, . Enhancing thermal conductivity of fluids with nanoparticles, in *Proceedings of the ASME International Mechanical Engineering Congress and Exposition*, vol. 66 (1995), pp. 99–105
2. J. Buongiorno, Convective transport in nanofluids. J. Heat Transfer **128**, 240–250 (2006)
3. R.E. Rosensweig, *Ferrohydrodynamics* (Dover Publications, New York, 1985)

4. D.A. Nield, A. Bejan, *Convection in Porous Media*, 3rd edn. (Springer, New York, 2006)
5. S. P. Sharma, A. Mahajan, A nonlinear stability analysis of a rotating double-diffusive magnetized ferrofluid. Appl. Math. Comput. **218**, 2785–2899 (2011)
6. M.K. Sharma, R. Singh, Linear stability analysis of double-diffusive convection in magnetic nanofluids in porous media. J. Porous Media **17**, 883–900 (2014)
7. D. Yadav, D. Lee, H.H. Cho, J. Lee, The onset of double-diffusive nanofluid convection in a rotating porous medium layer with thermal conductivity and viscosity variation: a revised model. J. Porous Media **19**, 105–21 (2016)
8. J.C. Umavathi, M.A. Sheremet, O. Ojjela, C.J. Reddy, The onset of double-diffusive convection in a nanofluid saturated porous layer: cross-diffusion effects. Eur. J. Mech. B/Fluids **65**, 70–87 (2017)
9. S. A. Sharma, D. Sharma, P. Kumar, Effect of magnetic field-dependent viscosity on thermal convection in a ferromagnetic fluid. Chem. Eng. Commun. **195**, 571–583 (2008)
10. G. Vaidyanathan, R. Sekar, R. Vasanthakumari, A. Ramanathan, The effect of magnetic field dependent viscosity on ferroconvection in a rotating sparsely distributed porous medium. J. Magn. Magn. Mater. **250**, 65–76 (2002)
11. R. Sekar, D. Murugan, Stability analysis of ferrothermohaline convection in a Darcy porous medium with Soret and MFD viscosity effects. Int. J. Appl. Mech. Eng. **23**(4), 911–928 (2018)
12. C. E. Nanjundappa, I. S. Shivakumara, K. Srikumar, Effect of MFD viscosity on the onset of ferromagnetic fluid layer heated from below and cooled from above with constant heat flux. Meas. Sci. Rev. **9**(3), 75–80 (2009)
13. M. Arora, M. Danesh, M. K. Sharma, K. Kaur, Effect of rotation on the onset of convection in magnetic nanofluids with magnetic field dependent viscosity. AIP Conf. Proc. **3081**(1), 050001 (2024)
14. A. Mahajan, M.K. Sharma, Double-diffusive convection in a magnetic nanofluid layer with cross diffusion effects. J. Eng. Math. **115**, 67–87 (2019)
15. D.A. Nield, A.V. Kuznetsov, The onset of double-diffusive convection in a nanofluid layer. Int. J. Heat Fluid Flow **32**, 771–6 (2011)
16. M. Arora, R. Singh, M.K. Panda, Effects of magnetic-field-dependent viscosity at onset of convection in magnetic nanofluids. J. Eng. Math. **101**, 201–217 (2016)
17. D.A. Nield, A.V. Kuznetsov, The onset of convection in a horizontal nanofluid layer of finite depth. Eur. J. Mech. B/Fluids **29**, 217–23 (2010)
18. C. Canuto, M.Y. Hussaini, A. Quateroni, T. Zang, *Spectral Methods in Fluid Dynamics* (Springer, New York, 1998)

Synthesis and Antimicrobial Applications of Silver Nanoparticle-Based Nanofluids

Renat R. Khaydarov

1 Introduction

Silver nanoparticle-based nanofluids have attracted wide attention owing to their enhanced thermal and catalytic properties, which make them suitable for various industrial applications. These nanofluids are created by dispersing silver nanoparticles (AgNPs) in a base fluid, which significantly improves the thermal conductivity and other thermophysical properties of the fluids. The synthesis, characterization, and applications of these nanofluids have been extensively studied, revealing their potential for heat transfer, catalysis, and environmental remediation [1–5].

AgNPs are known for their high thermal conductivity, which significantly enhances the thermal properties of base fluids. For instance, silver nanorods in a base fluid showed a thermal conductivity enhancement of 78.4% at 323 K, outperforming other forms of silver nanoparticles [6]. The addition of silver nanoparticles to graphene-based nanofluids resulted in a 102.85% increase in thermal conductivity at 0.1 vol% concentration, demonstrating the synergistic effect of combining silver with other high-conductivity materials [7]. Smaller NPs and higher concentrations generally lead to greater enhancements, with a 54% increase observed at a 0.1% volume fraction and 20 nm particle size [8].

AgNPs can be used to enhance the catalytic performance of nanofluids. For example, silver-loaded alumina/silica nanofluids demonstrated superior catalytic efficiency in the reduction of 4-nitrophenol, with a rate constant significantly higher than those of other metal-based catalysts [9]. These nanofluids also show promise in environmental remediation, as they can be used to degrade toxic chemicals, thereby reducing water pollution and its impact on biodiversity and human health [9].

R. R. Khaydarov (✉)
Institute of Nuclear Physics, Academy of Sciences of Republic of Uzbekistan, Tashkent, Uzbekistan

S. Kulkarni, S. Das (eds.), *Ionic Liquids, Nanofluids and Nanotechnology*,
Lecture Notes in Nanoscale Science and Technology 42,
https://doi.org/10.1007/978-3-032-04008-4_17

Silver-nanoparticle-based nanofluids can also improve the conversion efficiency of solar energy equipment by enhancing both their thermal and electrical components [10]. The inclusion of luminescent organic molecules with Ag nanoparticles in nanofluids can further optimize their optical properties for solar energy applications, making them suitable candidates for large-scale renewable energy systems [10].

Stability is a critical factor for the practical application of nanofluids. Silver nanoparticle-decorated graphene-like porous carbon nanofluids have shown excellent stability, remaining dispersed for over 6 months without surfactants [11]. The viscosity of silver nanofluids is generally higher than that of the base fluid, which can affect their flow characteristics. For instance, a 19.84% increase in viscosity was observed for a hybrid nanofluid containing Ag and graphene [7]. However, an increase in viscosity is often a trade-off for enhancing the thermal properties.

While silver nanoparticle-based nanofluids offer numerous advantages, challenges such as increased viscosity and the potential environmental impacts of silver nanoparticles must be addressed. The development of green synthetic methods and the use of environmentally friendly stabilizing agents can mitigate some of these concerns [12]. In addition, optimizing the nanoparticle concentration and surface modification can enhance the performance and stability of these nanofluids for specific industrial applications [13].

In the current work, we elucidate a straightforward and effective methodology that facilitates the production of AgNP-infused nanofluids using cellulose fibers, which are abundant and renewable materials. The proposed synthetic pathway holds significant potential for large-scale applications owing to its simplicity in preparation and absence of stringent operational requirements, thus making it an attractive option for industrial implementation. This research also endeavors to contribute to the expanding body of knowledge on environmentally sustainable and efficient methodologies for nanomaterial synthesis, with the potential to enhance the applicability of AgNPs in biomedical and technological domains.

2 Methodology

In this study, we have used silver nitrate ($AgNO_3$, analytical grade), polyvinylpyrrolidone (PVP), cellulose fibers, Hawach BIO-40 quantitative filter paper (8 μm), and polytetrafluoroethylene Gelman syringe filters (0.2 μm). All materials were used as received without any additional purification. The methodology proposed for the synthesis of AgNPs is predicated on the utilization of a cost-effective device, as depicted in Fig. 1. The $AgNO_3$ concentration in distilled water within the vessel was evaluated over a range of 40–1000 mg/L. The temperature varied from 18 °C to 70 °C. The contact duration of cellulose with the $AgNO_3$ dispersion ranged from 40 s to 120 s. The equipment used for the generation of Ag^+ ions has been described in our previous publications [14–16]. The quantification of AgNPs and ionic species in aqueous solutions was accomplished using mass and optical emission spectroscopy methods. The dimensions and morphology of AgNPs in the solution were assessed by transmission electron microscopy (TEM).

3 Results and Discussion

3.1 Morphology of Obtained AgNPs

In the course of the device operation (refer to Fig. 1), $AgNO_3$ solution, maintained at temperatures ranging from 18 °C to 70 °C, is transferred from (1) through cellulose (4) and filter paper into (3), subsequently collecting in (5). The silver ions undergo reduction to Ag as the aqueous solution containing Ag^+ ions traverses cellulose (4). The reduction of Ag^+ ions is an intricate multistep phenomenon (cf. Figs. 2 and 3). The synthesized AgNPs exhibited a positive charge and displayed the structural characteristics illustrated in Fig. 4. The phenomenon of silver nanoparticles' growth, influenced by the duration of exposure, was meticulously examined for various initial temperatures of the $AgNO_3$ aqueous solution, utilizing transmission electron microscopy (TEM) as a key measurement technique. As illustrated in Fig. 5, this intricate relationship is clearly delineated for two distinct initial solution temperatures, specifically, 25 °C and 60 °C. Upon analysis, the mean sizes of the $Ag(0)_n$ particles, which were measured at the exit point of container (3) following their traversal through the cellulose layers (4), were found to be 1.5 ± 0.4 nm and 3.5 ± 1.0 nm, respectively, indicating a notable variation in particle size relative to the temperature conditions. A typical TEM image capturing the AgNPs, which were synthesized after a continuous exposure period of 5 days, is presented in Fig. 6. In this image, the particles exhibit a nearly spherical morphology, with

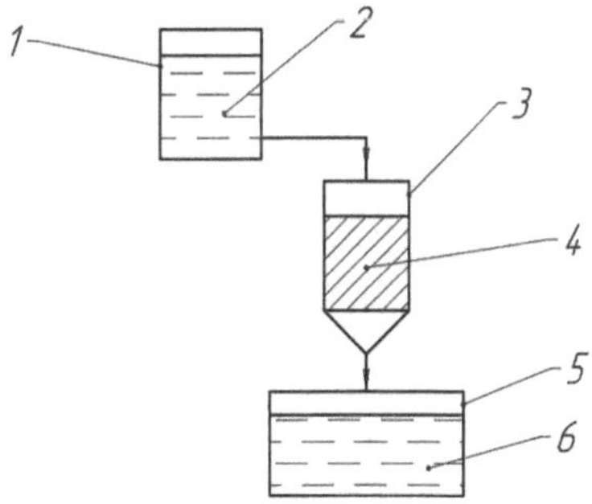

Fig. 1 The schematic representation illustrating the apparatus designed for the continuous synthesis of AgNPs solutions: 1—a vessel that serves as the primary reaction chamber; 2—a solution of silver nitrate that acts as the precursor material for AgNPs formation; 3—a container that holds and facilitates the mixing of reactants; 4—cellulose; 5—an additional vessel that is utilized for the subsequent processing or manipulation of materials; and 6—an aqueous dispersion of AgNPs, which is the final product resulting from the aforementioned chemical processes

Fig. 2 Synthesis of silver nanoparticles

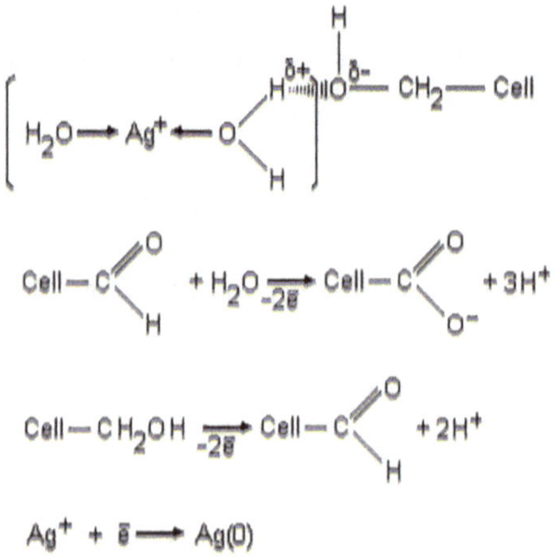

Fig. 3 $Ag(0)_n$ clusters formation

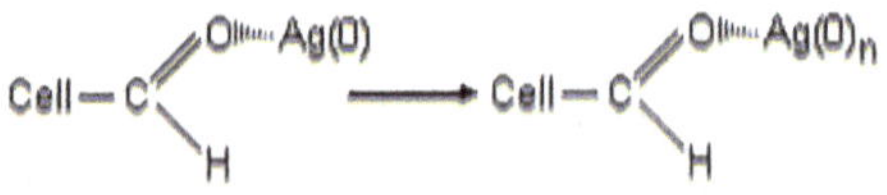

Fig. 4 Synthesized structure of nanosilver

$$\{[Ag]_n \cdot mAg^+ \cdot (m\text{-}x)NO_3^- \}^{x+} \cdot xNO_3^-$$

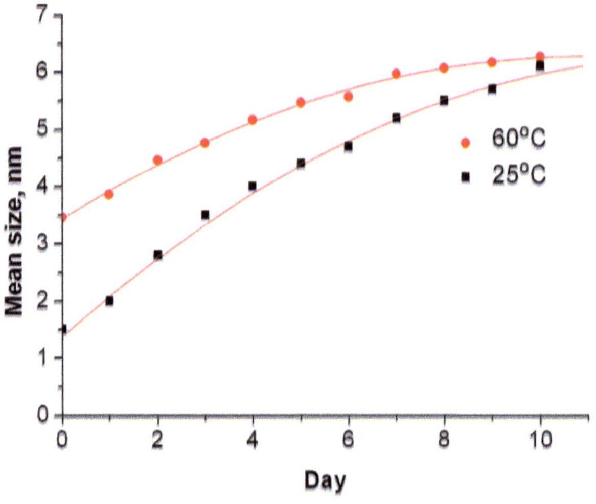

Fig. 5 Average dimensions (diameter) of AgNPs in relation to duration of exposure to $AgNO_3$ solution (100 mg/l)

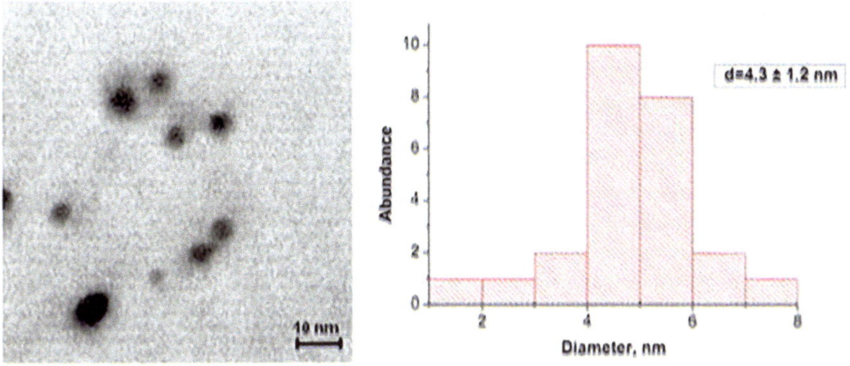

Fig. 6 A conventional transmission electron microscopy (TEM) image depicting AgNPs synthesized from an aqueous $AgNO_3$ solution (0.1 g/l, 25 °C)

Fig. 7 XRD patterns of AgNPs

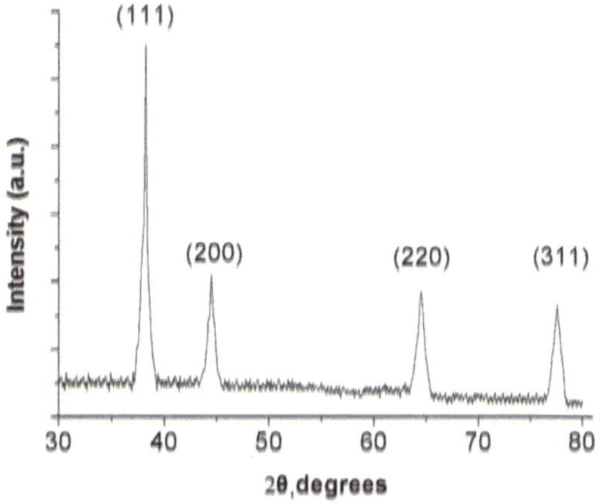

an average diameter calculated to be 4.3 ± 1.2 nm. In addition to morphological analysis, the structural properties of the silver colloids were rigorously investigated through X-ray powder diffraction (XRD) analysis, a technique renowned for its ability to elucidate crystalline structures. A representative XRD pattern of the obtained AgNPs is shown in Fig. 7. In this analysis, four distinct diffraction peaks corresponded to the crystalline planes of cubic silver. This correlation between the observed peaks and the known crystallographic planes reinforces the conclusion that the synthesized AgNPs exhibit crystalline characteristics that are consistent with established structural data. The AgNP solutions exhibited a transition from colorless to light yellow as the concentration of aqueous $AgNO_3$ increased from 40 to 1000 mg/L, accompanied by a concurrent increase in AgNPs concentration within 25–600 mg/L. An increase in the concentration of $AgNO_3$ leads to the growth of the mean diameter of NPs as well as the concentration of AgNPs fluid. The size distribution (13 ± 5 nm) was also broadened (cf. Fig. 8).

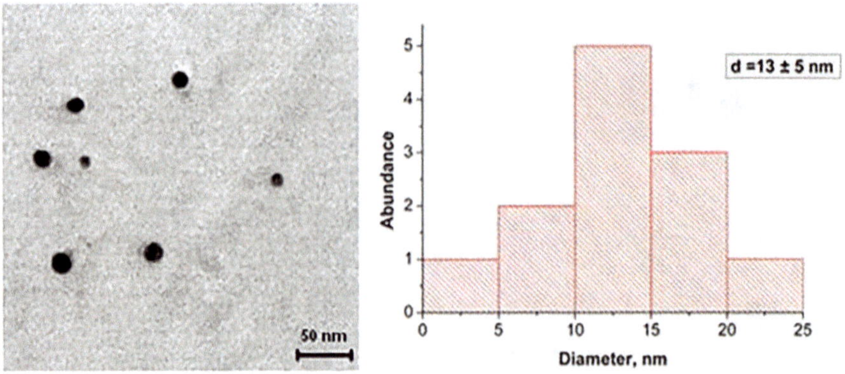

Fig. 8 Transmission electron microscopy (TEM) image depicting AgNPs solutions synthesized from an aqueous solution (800 mg/l) of $AgNO_3$ (25 °C), acquired following a duration of 5 days of exposure

Table 1 Average dimensions of silver nanoparticles (in nanometers) synthesized from an aqueous silver nitrate solution (25 °C)

	Silver nitrate solution (0.1 g/l)		Silver nitrate solution (0.8 g/l)	
Days	Without PVP	PVP-stabilized	Without PVP	PVP-stabilized
1	2.0 ± 0.7	14 ± 6	8 ± 3	16 ± 7
10	6.2 ± 1.9	16 ± 6	20 ± 7	19 ± 7
20	8 ± 2	17 ± 6	26 ± 9	21 ± 7
100	23 ± 8	18 ± 7	59 ± 16	24 ± 8
200	34 ± 10	23 ± 9	76 ± 25	30 ± 9

3.2 *AgNPs Stability*

Average dimensions of silver nanoparticles (in nanometers) synthesized from an aqueous silver nitrate solution (25 °C) are demonstrated in Table 1.

In our research, polyvinylpyrrolidone (PVP) was utilized as a stabilizing agent. PVP impacts the molecular dynamics of reduced silver, effectively preventing nanoparticle agglomeration. PVP was incorporated into the samples within the vessel (5). The stability of the post-synthesis nanoparticle solution was evaluated over a period of 200 days. As shown in Table 1, the nanoparticles synthesized with the inclusion of PVP in the solution of vessel (5) exhibited rapid growth during the initial day, reaching an average size of 14–16 nm, and demonstrated considerable stability throughout the testing period, specifically for a minimum of 200 days. In contrast, the nanoparticles produced without the addition of PVP showed a significantly smaller size. Although these nanoparticles also displayed a reasonable level of stability over the 200-day evaluation period, as indicated in Table 1, their average size increased at a faster rate than that of the nanoparticles synthesized with PVP. A freshly prepared nanosilver solution devoid of PVP is advantageous for scenarios requiring smaller AgNPs, which typically require the

AgNP-solution to be utilized within a timeframe of 1–2 days. Although PVP is an effective stabilizer for silver nanoparticles, its performance can vary based on environmental conditions and specific applications. In some cases, other stabilizers may offer better stability, particularly in high-ionic-strength environments or when long-term chemical stability is required. Nonetheless, PVP remains a popular choice because of its ability to provide steric stabilization and compatibility with a wide range of applications.

3.3 *Antimicrobial Activity and Applications of AgNPs*

Nanofluids containing AgNPs have exhibited substantial antimicrobial efficacy against a broad spectrum of microorganisms [17–21]. This efficacy is primarily attributed to the distinctive physicochemical properties of AgNPs. The antimicrobial activity of silver nanofluids is modulated by factors such as nanoparticle size, concentration, and the presence of stabilizing agents. AgNPs can adhere to bacterial cell membranes, inducing structural alterations and increased permeability, ultimately resulting in cell death [22]. Furthermore, AgNPs can penetrate microbial cells and disrupt essential cellular processes, including enzyme function and DNA replication [22]. Chitosan/silver nanofluids prepared with plant extracts effectively inhibited bacteria such as *Bacillus cereus* and *Escherichia coli*, with smaller nanoparticles exhibiting higher efficacy [23]. In addition, AgNPs have proven effective against multidrug-resistant bacteria, with a more pronounced inhibitory effect on gram-negative bacteria [24]. AgNPs also display strong antifungal activity. For instance, AgNPs coated with silica exhibited significant antifungal effects against *Candida* and *Aspergillus* species [25]. To elucidate the bactericidal potency of silver nanoparticles synthesized via the current methodology, minimum inhibitory concentration (MIC) assays were conducted on the gram-negative bacterium *Escherichia coli*, as well as thc gram-positive bacteria *Staphylococcus aureus* and *Bacillus subtilis*. The resultant data were compared with those derived from AgNPs samples produced using our proposed electrochemical technique [14–16, 26–30] and electrically generated silver ions [26] (refer to Table 2).

Table 2 The findings of the minimum inhibitory concentration (MIC) evaluation pertaining to AgNPs and silver ions

Bacterium	MIC (mg (Ag)/l) (~4 nm)[a]	MIC (mg (Ag)/l) (~13 nm)[a]	MIC (mg (Ag)/l) (~70 nm)	MIC (mg (Ag)/l) for electrically generated silver ions
E. coli	3	9	34	1
S. aureus	2	5	25	0.6
B. subtilis	22	40	68	2.4

[a] The samples used in the MIC assay are representative of those illustrated in Figs. 5 and 6

The observed discrepancies in MIC among the nanofluid samples (see Table 2) can be attributed to (a) the variation in AgNP size among the tested samples and (b) the observation that a larger mass concentration ratio of silver ions to silver nanoparticles results in heightened bactericidal effectiveness. We posit that the proposed straightforward and economically viable technique for the fabrication of aqueous dispersions of silver nanoparticles holds significant potential for application in both scientific research and industrial settings, where nanosilver is employed as a bactericidal agent.

In medical contexts, AgNPs are seamlessly integrated into wound coatings for surgical implants, effectively preventing infections and promoting healing by disrupting microbial proliferation [17, 31]. This antimicrobial efficacy extends to biomedical devices, where AgNP coatings inhibit biofilm formation, significantly reducing infection risks associated with implants and instrumentation [32, 33]. In veterinary medicine, AgNPs serve as feed supplements, enhancing immune responses and optimizing gut microbiota, thus offering a sustainable alternative to antibiotics [17]. Beyond healthcare, AgNPs enhance personal protective equipment, such as masks, by improving antibacterial and filtration efficiency, effectively filtering bacteria and particulate matter over extended periods [34]. In environmental applications, AgNPs are pivotal in water and air purification, leveraging their ability to target and destroy microbial cell membranes, as demonstrated in polyvinyl alcohol/chitosan membranes and gelatin sponge filtration systems, which achieve near-complete bacterial removal, including pathogens like *E. coli* [35–37]. Furthermore, AgNPs enhance solar-driven desalination systems by improving photothermal conversion efficiency, facilitating water distillation and the removal of heavy metals and dyes [38]. In the food industry, AgNPs are incorporated into packaging materials to suppress spoilage and pathogenic microorganisms, thereby extending shelf life and ensuring safety. Despite their promise, the potential toxicity of AgNPs necessitates further research to modify synthesis methods and explore combinational therapies, ensuring both safety and efficacy in their widespread application.

4 Conclusion

The development of silver nanoparticle (AgNP)-based nanofluids through a cellulose-mediated synthesis approach represents a significant advancement in the field of nanotechnology, offering a scalable and environmentally friendly method for producing stable high-performance nanomaterials. The synthesized AgNPs, with controlled sizes and crystalline structures, exhibited remarkable antimicrobial activity, making them highly effective for biomedical applications, including infection prevention in wound care, biomedical device coatings, and veterinary health. Their environmental applications, such as water and air purification, solar-thermal systems, and food packaging, highlight their multifaceted utility in addressing global health and sustainability challenges. The stability of these nanofluids, enhanced by polyvinylpyrrolidone (PVP) and electrostatic interactions with nitrate ions, ensures their practical applicability over extended periods. However, the observed increase in viscosity and potential environmental concerns underscore the need for continued research on green synthesis methods, surface modifications, and combinational therapies to mitigate risks while maximizing efficacy. By addressing these challenges, AgNP-based nanofluids have immense potential to revolutionize industries, ranging from healthcare to renewable energy, contributing to a sustainable and technologically advanced future.

Acknowledgments The present research work has been funded by the State budget of Uzbekistan within the fundamental research topic of the Laboratory of Interdisciplinary Technologies of the Institute of Nuclear Physics of the Uzbekistan Academy of Sciences.

References

1. G.S. Reddy, V. Sumalatha, Nanofluids: Bridging nanotechnology and fluid dynamics for enhanced thermal performance. J. Phys. Conf. Ser. **2837**, 012053 (2024). https://doi.org/10.1088/1742-6596/2837/1/012053
2. M. Tawalbeh, I. Shomope, A. Al-Othman, Comprehensive review on non-Newtonian nanofluids, preparation, characterization, and applications. Int. J. Thermofluids **22**, 100705 (2024). https://doi.org/10.1016/j.ijft.2024.100705
3. S. Shaw, A review of nanofluids synthesis for oil and gas applications. Nano/Micro Eng. Mol. Syst. (2020). https://doi.org/10.1109/NEMS50311.2020.9265611
4. M. Schneider, K. Cesca, S. Maria, D. de Amorim, E.R.-C. Hotza, F.P.M. Regina, Moreira, Synthesis and characterization of silica-based nanofluids for enhanced oil recovery. J. Mater. Res. Technol. **24**, 4143–4152 (2023). https://doi.org/10.1016/b978-0-323-90540-4.00002-8
5. J.M. Silva, T. Moreno, G. Chango, Recent progress on preparation, properties and applications of nanofluids. Lat. Am. Appl. Res. **53**, 620 (2023). https://doi.org/10.52292/j.laar.2023.620
6. S.M. Nyamgoudar, V.P. Silaparasetti, M. Shilpa, et al., Analysis of shape dependency of thermal conductivity of silver-based nanofluids. J. Therm. Anal. Calorim. (2022). https://doi.org/10.1007/s10973-022-11604-0
7. A.N. Zaaba, A.S. Muhsan, M.S. Nasif, M.U. Shahid, Thermophysical properties of silver/graphene hybrid nanoparticles reinforced ethylene glycol–glycerol mixture based nanofluids. Res. Sq. (2024). https://doi.org/10.21203/rs.3.rs-4476760/v1
8. S. Iyahraja, J.S. Rajadurai, Study of thermal conductivity enhancement of aqueous suspensions containing silver nanoparticles. AIP Adv. **5**, 057134 (2015). https://doi.org/10.1063/1.4919808
9. R. Mannu, V. Karthikeyan, M.M. Veerappa, et al., Facile use of silver nanoparticles-loaded alumina/silica in nanofluid formulations for enhanced catalytic performance toward 4-nitrophenol reduction. Int. J. Environ. Res. Public Health **18**, 2994 (2021). https://doi.org/10.3390/ijerph18062994
10. J. Walshe, P.M. Carron, C. McLoughlin, et al., Nanofluid development using silver nanoparticles and organic-luminescent molecules for solar-thermal and hybrid photovoltaic-thermal applications. Nano **10**, 1201 (2020). https://doi.org/10.3390/nano10061201
11. C. Jin, Q. Wu, H. Zhang, et al., Study on preparation, stability, thermal conductivity, and viscosity of silver nanoparticles-decorated three-dimensional graphene-like porous carbon hybrid nanofluids. Nanotechnology **32**, 135704 (2021). https://doi.org/10.1088/1361-6528/abed77

12. M.M.R. Mollick, B. Bhowmick, D. Maity, et al., Green synthesis of silver nanoparticles-based nanofluids and investigation of their antimicrobial activities. Microfluid. Nanofluid. **17**, 345–355 (2014). https://doi.org/10.1007/s10404-013-1252-3
13. W. Bunpheng, R. Dhairiyasamy, Enhancement in turbulent convective heat transfer using consider using silver nanofluids: Impact of citrate, lipoic acid, and silica coatings. Chem. Eng. **8**, 65 (2024). https://doi.org/10.3390/chemengineering8040065
14. R.R. Khaydarov, R.A. Khaydarov, O. Gapurova, et al., Antimicrobial effects of silver nanoparticles synthesized by an electrochemical method, in *Nanostructured Materials for Advanced Technological Applications*, (Springer, Cham, 2009), pp. 215–218
15. R.A. Khaydarov, R.R. Khaydarov, O. Gapurova, Y. Estrin, Silver nanoparticles: Environmental and human health impact, in *Nanomaterials: Risk and Benefits*, (Springer, Cham, 2009), pp. 287–299
16. R.R. Khaydarov, R.A. Khaydarov, O. Gapurova, et al., Environmental and human health issues of silver nanoparticles applications, in *Environmental Security and Ecoterrorism*, (Springer, Cham, 2011), pp. 117–127
17. T. Frippiat, T. Art, C. Delguste, Silver nanoparticles as antimicrobial agents in veterinary medicine: Current applications and future perspectives. Nano **15**, 202 (2025). https://doi.org/10.3390/nano15030202
18. A.S. Rodrigues, J.G.S. Batista, M.A.V. Rodrigues, et al., Advances in silver nanoparticles: A comprehensive review on their potential as antimicrobial agents and their mechanisms of action elucidated by proteomics. Front. Microbiol. **14**, 1440065 (2024)
19. R.M. Pattabi, M. Pattabi, Antibacterial applications of silver nanoparticles. Mater. Sci. Forum **754**, 131–140 (2013). https://doi.org/10.4028/www.scientific.net/MSF.754.131
20. A.M. Díez-Pascual, Antibacterial activity of nanomaterials. Nanomaterials **8**, 359 (2018). https://doi.org/10.3390/nano8060359
21. V.M. Holubnycha, Y. Husak, V. Korniienko, et al., Antimicrobial activity of two different types of silver nanoparticles against wide range of pathogenic bacteria. Nano **14**, 137 (2024). https:/ /doi.org/10.3390/nano14020137
22. D.S. Shafi, A. Kazi, et al., Silver nanoparticle as a modern era in antimicrobial activity. Asian J. Pharm. Res. Dev. **11**, 1407 (2024). https://doi.org/10.22270/ajprd.v11i3.1407
23. S.A. Umoren, M.M. Solomon, A. Nzila, I.B. Obot, Preparation of silver/chitosan nanofluids using selected plant extracts: Characterization and antimicrobial studies against gram-positive and gram-negative bacteria. Materials **13**, 1629 (2020). https://doi.org/10.3390/ma13071629
24. X. Na, Synthesis and antibacterial effects of silver nanoparticles (AgNPs) against multi-drug resistant bacteria. Biomed. Mater. Eng. (2024). https://doi.org/10.3233/bme-240034
25. G.J. Mohammed, The antimicrobial effect of silver nanoparticles coated with silica against human pathogenic bacteria and fungi. Int. J. Health Sci. **6**, 5652 (2022). https://doi.org/10.53730/ijhs.v6ns1.5652
26. R.A. Khaydarov, R.R. Khaydarov, O. Gapurova, Y. Estrin, A novel method of continuous fabrication of aqueous dispersions of silver nanoparticles. Int. J. Nanoparticles **3**, 77–91 (2010)
27. R.A. Khaydarov, R.R. Khaydarov, O.U. Gapurova, Nano-photocatalysts for the destruction of chloro-organic compounds and bacteria in water. J. Colloid Interface Sci. **406**, 105–110 (2013)
28. I.T. Garipov, R.R. Khaydarov, O.U. Gapurova, et al., Silver nanoparticles as a new generation of antimicrobial prophylaxis. J. Sib. Fed. Univ. Biol. **12**, 266–276 (2019)
29. R. Khaydarov, O. Gapurova, M. Abdukhakimov, et al., Antimicrobial properties of nanofiltration membranes modified with silver nanoparticles. Emerg. Mater. **5**, 1477–1483 (2022)
30. J.R. Morones, J.L. Elechiguerra, A. Camacho, et al., The bactericidal effect of silver nanoparticles. Nanotechnology **16**, 2346–2353 (2005)
31. H. Jangid, S. Singh, P.K. Kashyap, A. Singh, Advancing biomedical applications: An in-depth analysis of silver nanoparticles in antimicrobial, anticancer, and wound healing roles. Front. Pharmacol. **15**, 1438227 (2024). https://doi.org/10.3389/fphar.2024.1438227
32. J. Yadav, H. Tare, Silver nanoparticles as antimicrobial agents: Mechanisms, challenges, and applications. Int. J. Pharm. Qual. Assur. **15**, 82 (2024). https://doi.org/10.25258/ijpqa.15.1.82

33. Y. Ren, Y. Zhang, X. Li, Application of AgNPs in biomedicine: An overview and current trends. Nanotechnol. Rev. **13**, 0030 (2024). https://doi.org/10.1515/ntrev-2024-0030
34. Y. Kang, J. Park, D. Kim, et al., Silver nanoparticles for enhanced antibacterial and filtration efficiency in personal protective equipment. Materials **15**(14), 4742 (2024)
35. A. Taha, E. Ahmed, A. Hassanin, et al., Polyvinyl alcohol/chitosan nanofibers incorporating silver nanoparticles for water purification. J. Environ. Sci. **83**, 67–78 (2019)
36. L. Wei, J. Zhang, Q. Wang, et al., Gelatin sponge functionalized with silver nanoparticles for antibacterial air filtration. ACS Appl. Mater. Interfaces **10**, 22866–22874 (2018)
37. Y. Liu, X. Zhang, J. Wang, et al., Silver nanoparticle-enhanced filtration systems for water purification. Water Res. **190**, 116678 (2021)
38. L. Xu, Y. Wang, H. Zhang, et al., Silver nanoparticles in solar-driven desalination: Enhanced photothermal efficiency and pollutant removal. Appl. Catal. B Environ. **310**, 121345 (2023)

Ionic Liquids in the Pretreatment of Plant Biomass: Sustainable Pathways for Biorefinery Applications

Sharmistha Sharma, Anjali Roy, and Himanshu Kishore Prasad

1 Introduction

Global demand for energy and materials, coupled with environmental concerns associated with fossil fuel dependency, necessitate the transition to renewable and sustainable sources [1]. The introduction of biodiesel and bioethanol into the market has replenished the need for nonrenewable petroleum products. Biodiesel and bioethanol are easily produced using renewable resources such as lignocellulosic biomass (LCB) [2, 3]. Agricultural residues, forest byproducts, and organic waste are some of the most notable LCB resources available globally. Harnessing LCB is pivotal for developing a sustainable bioeconomy, providing feedstocks for biofuels (e.g., ethanol and butanol) [4], biochemicals (e.g., platform chemicals and organic acids) [5], and advanced biomaterials (e.g., composites and fibers) [6]. Circular economies aim to reduce waste and optimize resource utilization, which is consistent with this principle [4, 7].

However, efficient conversion of LCB into valuable intermediates, particularly fermentable sugars, is significantly hindered by its inherent recalcitrance [6, 8]. The intricately structured plant cell wall confers resistance because of its deeply integrated cellulosic microfibrils in the hemicellulose and lignin matrix, which are frequently strengthened by lignin-carbohydrate complexes (LCCs) [9]. Enzymatic and chemical degradation are obstructed by factors such as the high crystallinity of cellulose, the defensive roles of lignin [10], and the restricted accessible surface area. Therefore, a strategic pretreatment method may be implemented for the effective degradation of lignocellulosic biomass and to access crucial polysaccha-

S. Sharma · A. Roy · H. K. Prasad (✉)
Laboratory of Microbial Biotechnology, Department of Life Science and Bioinformatics, Assam University, Silchar, Assam, India
e-mail: sharmistha.sharma1@aus.ac.in; anjali.roy@aus.ac.in; himanshu.k.prasad@aus.ac.in

S. Kulkarni, S. Das (eds.), *Ionic Liquids, Nanofluids and Nanotechnology*,
Lecture Notes in Nanoscale Science and Technology 42,
https://doi.org/10.1007/978-3-032-04008-4_18

rides (cellulose, hemicellulose, and lignin), thereby ensuring their hydrolysis into fermentable monosaccharides [9, 11].

Although sound and productive, conventional pretreatment methods are unduly burdened with major drawbacks, such as excessive energy consumption, reliance on strong chemicals, production of inhibitory secondary products, and ecological concerns [4]. This has led to the search for suitable, sustainable, and effective alternatives. Their relatively low melting points indicate that ILs are promising solvents for LCB pretreatment (Fig. 1) [15, 16]. High thermal tolerance, low vapor pressure, and strong biopolymer solvation capacities are some of the remarkable physicochemical attributes that aid biomass deconstruction [5], component fraction-

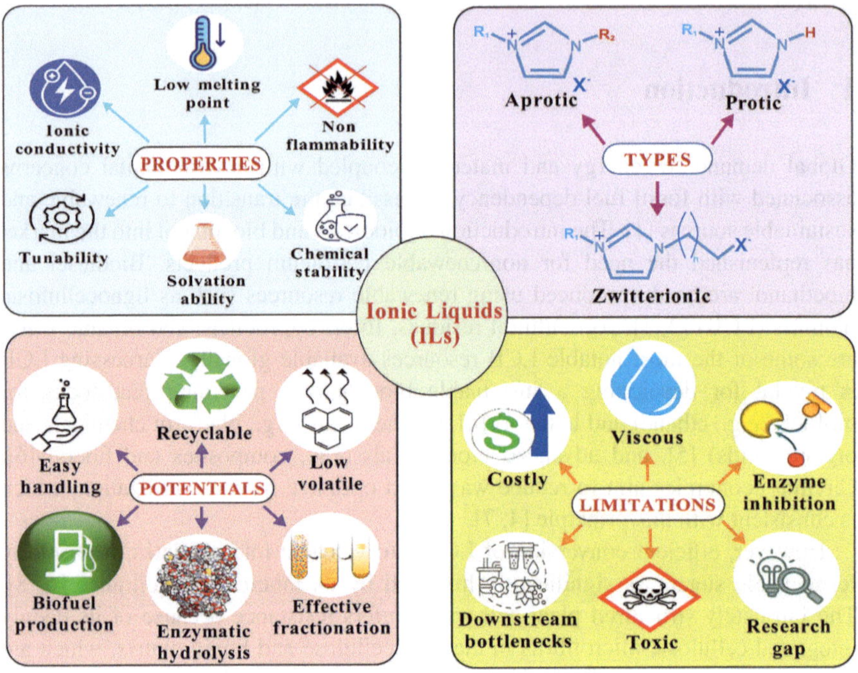

Fig. 1 An illustrative representation displaying the distinct features of ionic liquids (ILs). (Illustration created with BioGDP). This includes their types, properties, potentials, and limitations. ILs are generally classified as protic, aprotic, or zwitterionic ILs. Protic ILs are capable of proton transfer, whereas aprotic ILs are not. Zwitterionic ILs are another unique type, in which both charges lie within the same molecule. Among the various properties, the key distinct properties of IL include higher conductivity of ions, lower melting point due to their structural asymmetry, higher thermal stability leading to nonflammability, higher chemical stability, wider solvation capacity allowing the solvation of both nonpolar and polar solvents and a unique property called "tunability" allowing the modification of cations and anions. These alternative green solvents offer various potential applications such as ease of handling, recyclability, and low volatility, allowing effective fractionation, enzymatic hydrolysis, and higher biofuel production. Although ILs offer numerous applications, certain limitations persist. This includes their higher cost, elevated viscosity, toxicity, enzyme inhibitory properties, downstream processing limitations, and limited information [12–14]

ation [17], and saccharification [18]. Although ILs are often regarded as "green solvents" owing to their minimal volatility and sustainability, biodegradability, recyclability, and toxicity require detailed analysis [15, 16, 19].

An analysis of the use of ionic liquids (ILs) in the pretreatment of plant biomass for use in biorefineries is presented in this chapter. Detailed explanations of the structure and resilience of lignocellulose along with its dissolution mechanisms are provided. Furthermore, in subsequent sections, we assess the relative comparison between traditional and IL-based pretreatment methods. Additionally, the chapter appraises sustainable parameters in both economic and ecological aspects, explores challenges related to IL recovery and recycling, and highlights new prospects in IL-based biorefineries.

2 Structure and Recalcitrance of Lignocellulosic Biomass

Lignocellulose forms the core structural framework of plants and is characterized by its complex composition, which mainly includes cellulose, hemicellulose, and lignin [20, 21]. Depending on the biomass sources, such as agricultural residues, hardwood, and softwood, the relative abundances of cellulose, hemicellulose, and lignin range between 30–60%, 14–45%, and 5–35% by dry weight, respectively [4, 20].

Cellulose is the most abundant component, constituting 40–50% of the dry mass in many plants. It is a linear homopolymer consisting of D-glucose units linked by β-1,4-glycosidic bonds [10]. The repeating unit is cellobiose, a disaccharide [10]. Due to extensive intramolecular and intermolecular hydrogen-bonding networks, cellulose chains pack tightly into highly ordered crystalline structures interspersed with amorphous regions [4, 10]. At the same time, cellobiose is a disaccharide that comprises intramolecular and intermolecular hydrogen bonds, organized in a tightly packed crystalline domain [22]. The resilience of cellulose to enzymatic and chemical hydrolysis is mainly due to its polymerized and crystalline form [4, 10]. Cellulose II, a thermodynamically more stable form, can be produced via regeneration from solution or mercerization [5, 23].

Hemicellulose is a heterogeneous branched polysaccharide and the second most prevalent biopolymer [10, 24]. In contrast to cellulose, it comprises diverse sugar moieties, including pentoses, such as xylose and arabinose, and hexoses, such as mannose, glucose, and galactose [4, 25]. Its lower degree of polymerization, more amorphous, and relatively disorganized structure make it easily hydrolyzable [4]. The acetyl groups present on the backbone contribute to its properties and enzymatic hydrolysis proficiencies [22].

Lignin is an amorphous complex aromatic polymer [5, 26] composed of three phenylpropane units: coniferyl alcohol (G-unit), p-coumaryl alcohol (H-unit), and sinapyl alcohol (S-unit). There are considerable variations in the distribution of these three units among various plant species [26]. Lignin acts as a natural additive, occupying the voids between cellulose and hemicellulose, improving the cell walls' tensile strength, resilience, and hydrophobicity [4, 22].

The three biopolymers are intricately arranged within the plant cell wall, with cellulose clustering into bundles forming microfibrils enveloped by hemicelluloses and embedded within the lignin matrix [4]. Furthermore, covalent bonds, known as lignin-carbohydrate complexes (LCCs), exist between lignin and hemicellulose primarily through ester and ether linkages [27].

The dense and coherent structure of lignocellulose provides exceptional endurance and resilience to deterioration, which is known as biomass recalcitrance. Lignin's distribution and presence contribute to this recalcitrance [11]. However, other factors that contribute to this resistance include cellulose crystallinity, degree of polymerization (DP) [4], LCC linkages, accessible surface area [27], presence of acetyl groups on hemicellulose [28], and hemicellulose sheathing [22]. Therefore, it is necessary to have thorough knowledge of biomass recalcitrance, which is an intricate interplay of numerous structural and chemical variables, rather than the result of a single cause [11, 22]. To significantly enhance the accessibility and dissolution of polysaccharides, efficient pretreatment must address several facets of this resistance. Furthermore, biomass composition and structural attributes vary significantly among different plant types. For instance, the lignin content in water hyacinth is 3–20%, whereas it is 44% in rapeseed straw [29]. Thus, a pretreatment approach administered to one feedstock may not be appropriate for another.

3 Mechanisms and Drawbacks of Conventional Biomass Pretreatment

Enzymatic hydrolysis of LCB broadly includes four categories of conventional pretreatment techniques (Table 1). These include physical, physicochemical, chemical, and biological methods (Fig. 2) [4, 34]. Chipping, grinding, and milling are physical methods that reduce cellulose crystallinity and increase surface area [15]. Other methods of this process involve irradiation and ultrasonication, which distort the structural integrity of the LCB [35]. The primary disadvantage of this process is the energy-consuming steps [15], which often limit the lignin removal mechanisms [20].

Chemical and physical methods are effectively combined in the physicochemical approach [15]. Similar to the physical disruption technique, biomass is first subjected to high-temperature steam explosions, which cause hemicellulose hydrolysis and lignin transformation [4]. In liquid hot water (LHW) or autohydrolysis treatment, hot water hydrolyzes hemicellulose at high pressure. It provides a comparably pure hemicellulose stream but produces inhibitors at greater severities [15, 36]. The ammonia fiber expansion (AFEX) mechanism involves biomass treatment with liquid ammonia at modest temperatures and high pressures, leading

Table 1 Comparison of different traditional biomass pretreatment methods

Pretreatment method	Mode of treatment	Significances	Limitations	References
Milling/grinding	Physical	Reduces particle size, increases surface area, and decreases crystallinity of LCB	High energy consumption, costly, and does not remove lignin	[30]
Deep eutectic solvents (DESs)	Physicochemical	Removes high lignin content (18–86%) at 150 °C, yielding a saccharification of 40–90%	The technique for recovering the solubilized hemicellulose remains unknown	[31]
Organosolv	Chemical	Produces high-purity cellulose, recovers potentially high-value lignin, and the solvent can be recycled	High solvent cost, requires efficient solvent recovery, and potential flammability or toxicity of solvents	[30]
Biological	Biological	Low energy input, high selectivity (esp. lignin removal), mild conditions, minimal inhibitor formation	Very slow process rate, low efficiency, difficult to scale up and control, and may consume some carbohydrates	[32]

to swelling, decrystallization of cellulose, and partial delignification. It generates a few inhibitors, but ammonia recovery is expensive [15, 37, 38].

In the chemical treatment method, the agents used change the composition and structure of biomass [15]. Dilute H_2SO_4 in the acid pretreatment process, hydrolyzes hemicellulose into pentoses, changing lignin structure, and making cellulose more accessible (Table 1) [39]. The formation of inhibitory compounds such as furfural and 5-hydroxymethylfurfural (5-HMF) has been reported [4]. However, NaOH, lime, ammonia, and other alkaline agents help saponify the ester bonds of lignin, generating fewer inhibitors than acid treatment, but it is quite a slow process [8]. Therefore, different chemical treatments have different advantages and limitations. Microorganisms or their enzymes may also be employed to degrade lignin and hemicellulose under mild conditions biologically. The advantages include low energy requirements and minimal inhibitor formation. However, these methods are typically very slow and difficult to control on an industrial scale, making them commercially challenging [40].

The limitations inherent in these traditional methods highlight the critical need for alternative pretreatment strategies. Many conventional approaches involve a difficult trade-off: achieving high biomass deconstruction efficiency often comes at the cost of high-energy input, harsh chemical usage, or significant inhibitor generation [4]. Conversely, milder methods, such as biological pretreatment, are environmentally benign but too slow for practical industrial applications. Combined pretreatment approaches attempt to leverage the strengths of different methods but

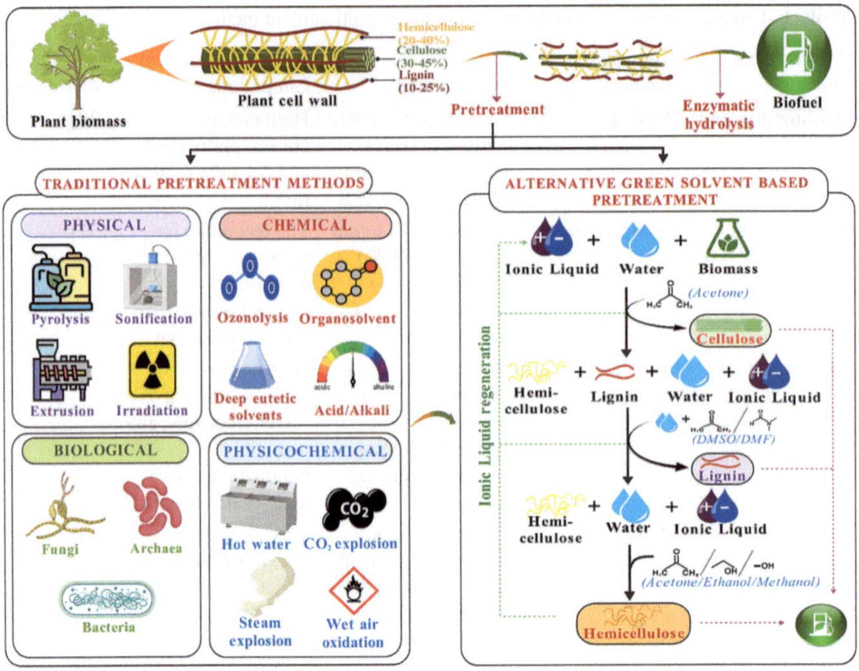

Fig. 2 A schematic representation of plant biomass pretreatment comparing Ionic Liquids (ILs) with traditional pretreatment methods. (Illustration created with BioGDP). Plant cell walls contain cellulose, hemicellulose, and lignin, which are decomposed during the pretreatment process of lignocellulosic biomass (LCB). In the following step, the breakdown products are hydrolyzed enzymatically to facilitate fermentation and make it possible to make biofuels and other value-added bioproducts. Physical, chemical, physicochemical, and biological pretreatments, however, can be energy-intensive, produce toxic byproducts, and be very expensive. The use of ionic liquids (ILs) has subsequently emerged as a unique approach to biomass dissolution, enabling the production of biofuels from biomass. The addition of molecular cosolvents, such as acetone, DMSO (dimethyl sulfoxide), DMF (dimethyl formamide), methanol, and ethanol, further enhances the efficiency of ILs [4, 14, 33]

can add complexity and cost [39]. This persistent challenge drives the exploration of novel solvents and processes, such as those involving ionic liquids, which promise high efficiency under potentially milder conditions, better selectivity, and improved sustainability profiles, in alignment with the principles of green chemistry [25].

4 Ionic Liquids: Properties and Potential as Pretreatment Solvents

Ionic liquids (ILs) represent a unique class of solvents defined as salts composed entirely of ions, typically a large, asymmetric organic cation and an organic or inorganic anion, with melting points at or below 100 °C [41]. Common cation families employed in biomass research include imidazolium, pyridinium, carboxylates, sulfonates, and phosphonates. ILs containing imidazolium have been widely studied because of their good solvation attributes and stability [16]. Phosphonium-containing ILs are known for their high thermal stability but are less commonly consumed in biomass pretreatment [42], whereas pyridinium, a heterocyclic cation, has been extensively explored for biomass applications [43]. Amino acid anions also offer potential as bio-based ILs and are highly effective for biomass dissolution [41, 44].

ILs are classified into different subclasses based on their specific functions (Fig. 1). Aprotic ILs (APILs) lack readily available protons, whereas protic ILs (PILs) are formed by proton transfer between Brønsted acids and bases [44]. The structural parameters of PILs, especially the cation/anion pairs, play a critical role in solubilizing lignin and carbohydrates [45]. Zwitterionic ILs (ZILs) incorporate cationic and anionic units that are covalently attached to the same molecule. It provides IL-like characteristics regarding volatility and melting points, with modest-to-high heat stability [46]. ILs are also recyclable and prepared from precursors such as amino acids, carbohydrates, lignin derivatives (furfural, vanillin), or choline [44]. A major advantage of ILs is their "designer solvent" nature, which allows them to be tailored for specific applications by carefully selecting their cation-anion pair. Plant biomass, especially lignocellulosic biomass, is a readily available renewable resource, but its intricate structure renders it recalcitrant to effective conversion [47]. Ionic liquids (ILs) have been considered as potential solvents for plant biomass pretreatment in biorefineries, with sustainable routes to produce biofuels and other valuable chemicals (Table 2) [54]. ILs can disrupt the complex structure of biomass and improve the accessibility of cellulose and hemicellulose for subsequent processing [55]. Several attributes (Fig. 2) make ILs very effective lignocellulosic biomass disruptors. ILs have diverse solvation capabilities owing to their ability to form varied intermolecular interactions, such as Coulombic forces, hydrogen bonding, van der Waals forces, and π-π stacking [44]. This allows them to dissolve insoluble polymers such as cellulose and lignin in many standard solvents [56]. The solvation ability, especially for cellulose, is closely correlated with the anion's hydrogen-bond basicity (β), a Kamlet-Taft parameter [28]. High thermal stability is also common in many ILs, with the onset of decomposition above 200–300 °C, making them applicable to processes involving higher temperatures [16]. High stability is a characteristic of phosphonium ILs [57]. However, their stability differs, and certain ILs degrade at pretreatment temperatures [58]. Other characteristic features of ILs include the minimization of solvent loss through evaporation [16] and nonflammability [12]. The ionic nature of ILs confers polarity, which is adjustable and affects the solvation nature and interactions with solutes [44]. The viscosity of ILs can be troublesome during mixing, mass transfer, and pumping. Viscosity is very sensitive to the ion structure (e.g., length of the alkyl chain and symmetry) and temperature. It gets reduced by the addition of cosolvents [59].

Table 2 Comparative efficacy of different IL classes in biomass pretreatment

IL class	Representative ILs	Typical feedstock (s)	Typical delignification (%)	Typical glucose yield (%)	Notable features	References
Metal-based	[BMIM][$AlCl_4$]	Bagasse	72.58	70.1	Bioconversion of complex cellulose into levulinic acid. Regenerated lignins exhibit potent antioxidant and antibacterial attributes	[48]
Imidazolium- and organic acid-based	[Bmim]Cl-OA	Bagasse	90.26	–	Ultra high pressure homogenization (UHPH), along with IL-OA enhanced the process of bioconversion and delignification	[49]
Imidazolium aprotic	[Emim][OAc], [Cl]	Switchgrass, corn Stover, wood, bagasse	15–50	55–96	Benchmark performance, good cellulose dissolution, often toxic, costly, and petroleum-derived	[50]
Choline amino acid	[Ch][Lys], [Ch][OAc]	Switchgrass, wood, sorghum, agave bagasse	70–80 [Ch][Lys]	78–96 [Ch][Lys]	Bio-derived, lower toxicity, biodegradable, [Ch][Lys] excellent for delignification, can be viscous	[51]
Bio-derived aromatic	Vanillin-IL, furfural-IL	Switchgrass	20–43	90–95	Renewable source, comparable sugar yields to [Emim][OAc], "closed-loop" potential	[52]
Zwitterionic (ZILs)	[C_2mim][OAc], histidine-ZIL	Cellulose, wood	Variable (lower than APILs)	Variable (can be comparable)	Potential for low toxicity/biocompatibility, the mechanism may differ (crystallinity reduction without complete dissolution)	[53]

5 Mechanisms of Biomass Deconstruction by Ionic Liquids

Ionic liquids facilitate the decomposition of lignocellulosic biomass primarily by disrupting the intricate assemblage of non-covalent interactions, predominantly hydrogen bonds, which uphold the three-dimensional structure of the plant cell wall [60]. Breaking this network results in solvation of the biomass components, distortion of the ultrastructure, and increased polysaccharide accessibility for successive enzymatic or chemical conversions.

Applying ILs to biomass pretreatment can lead to more environmentally friendly biorefinery processes [61]. The solubilization of cellulose, the most crystalline component, measures the effectiveness of the ILs. The primary mechanism involves IL ions interfering with the extensive hydrogen bonds (intra- and intermolecular) that hold cellulose chains together in the crystalline fibrils [60]. Anions characterized by strong hydrogen-bond basicity, such as chloride, acetate, formate, carboxylates, and phosphates, function as potent hydrogen-bond acceptors and hold significant importance [62]. As hydrogen-bond acceptors, these anions compete with cellulose hydroxyl groups for hydrogen bonding with the hydroxyl protons of cellulose, particularly at the C3-OH and C6-OH positions, thereby disrupting the preexisting hydrogen-bonding network [50]. Research has demonstrated a correlation between the hydrogen-bond basicity of anions (β value) and the ability of an ionic liquid (IL) to dissolve cellulose [28]. Some researchers have suggested that anions may intercalate between cellulose chains, thereby interrupting bonding [63]. Furthermore, cations, particularly imidazolium cations, are implicated in the dissolution process [50]. These cations may interact with cellulose hydroxyl groups at oxygen atoms through either slightly weaker hydrogen bonds or alternative interaction modes [63]. Conversely, some studies propose that anions play a more significant role than cations in the dissolution process [23]. However, molecular dynamics simulations indicate that both ions collaboratively act to sequentially disrupt intermolecular hydrogen bonds between cellulose chains [64]. The structure of the cation, such as variations in the length of alkyl side chains on imidazolium rings, may influence the properties of ILs and contribute to steric hindrance, thereby enhancing cellulose solubility [63]. Upon the introduction of an anti-solvent, such as water or ethanol, the precipitation of dissolved cellulose occurs, often resulting in the regeneration of a less crystalline and more reactive cellulose II allomorph [5]. Water competes with IL ions and cellulose hydroxyl groups for hydrogen bonding, disrupting IL-cellulose interactions and causing cellulose chains to re-aggregate [65].

Ionic liquids (ILs) also demonstrate efficacy in interacting with and dissolving lignin, a complex phenolic polymer found in biomass [44]. The dissolution process involves interactions between IL ions and the various functional groups and aromatic rings of lignin. For instance, imidazolium cations can engage in π–π interactions with the aromatic rings of lignin. These anions can form hydrogen bonds with the hydroxyl groups present in lignin [66]. The overall polarity, encompassing hydrophilicity and hydrophobicity, along with specific interactions, determines lignin solubility in ILs [52]. Modeling lignin dissolution is complicated due to the heterogeneity of its structure [67].

Ionic liquids (ILs) can disrupt the covalent bonds between lignin and carbohydrate molecules, specifically within lignin-carbohydrate complexes (LCCs), allowing fractionation. ILs with basic anions, such as acetate or amino acid anions, are especially effective in disrupting ester bonds that link hemicellulose to lignin (e.g., arabinoxylans via ferulic acid bridges in grasses) [68]. This saponification reaction breaks down the cross-links. The observed deacetylation of hemicellulose during pretreatment with acetate ILs provides further evidence of ester bond cleavage [5]. Specific ILs and process conditions can be chosen to preferentially dissolve lignin while leaving the cellulose fraction largely intact as a solid residue. This "ionoSolv" approach is particularly relevant for PILs and certain APILs [52]. The selectivity appears to be linked to the specific interactions between the IL ions and different biopolymers [66].

In addition to dissolution and chemical bond cleavage, the IL treatment induces significant physical changes in the biomass structure. The penetration of IL ions into the biomass matrix significantly causes swelling, disrupts the cell wall structure, and increases porosity [62]. The disruption of hydrogen bonding within cellulose reduces crystallinity, leading to amorphization and making cellulose chains more susceptible to enzymatic degradation [60]. This contributes to enhanced saccharification yield. Certain protic ionic liquids (PILs), designed for delignification, may have less impact on cellulose crystallinity [69]. The synergistic effects of swelling, removal of lignin and hemicellulose, and structural disruption increase enzyme-accessible surface area [60]. The ionic liquid (IL) mechanism combines physical disruption (swelling, decrystallization, increased porosity) and chemical interactions (hydrogen-bond disruption, solvation, lignin-carbohydrate complex cleavage). Efficacy varies with IL cation and anion structure and biomass composition. ILs with basic anions effectively disrupt cellulose hydrogen bonds and hydrolyze ester bonds [5], while PILs function through acid-catalyzed hydrolysis and lignin solvation [69].

Data suggest distinct interaction modes for cellulose and lignin in IL systems [66]. Anions that are strong hydrogen-bond acceptors primarily interact with cellulose hydroxyl groups [64]. Aromatic cations like imidazolium engage in π-π stacking with lignin's aromatic units. This differential interaction enables the design of ILs with enhanced selectivity for delignification or cellulose modification [66].

6 Evaluating the Efficiency of IL Pretreatment

IL pretreatment has been consistently demonstrated to offer substantial advantages over traditional methods. Numerous studies have shown that IL pretreatment significantly enhances the yields of fermentable sugars, such as glucose from cellulose and xylose from hemicellulose, following enzymatic hydrolysis, in comparison to untreated biomass or treatments like dilute acid pretreatment [60]. Certain

ILs, particularly those used for fractionation (e.g., PILs), exhibit a high capacity for lignin removal (70–85% or higher) [62], akin to conventional delignification processes such as alkali or Organosolv, but typically under milder conditions [60]. This effective delignification markedly improves the digestibility of the residual cellulose [42]. Compared to severe thermochemical processes like acid hydrolysis or steam explosion, IL pretreatment generally results in lower levels of sugar degradation products (furfural, HMF) and other inhibitory compounds that affect downstream fermentation enzymes and microbes [15]. This reduction may facilitate process integration and enhance biofuel yields. Nevertheless, it is important to mention that the remaining IL can be inhibitory, requiring effective removal or applying IL-tolerant biological systems [70].

Various IL classes have different strengths and limitations in biomass pretreatment (Fig. 1). The most widely researched ILs for biomass treatment are imidazolium-based APILs. They are very efficient at dissolving cellulose and destroying its crystallinity, resulting in high sugar recoveries after hydrolysis [16]. Nevertheless, they are usually based on petroleum feedstocks, may be costly, and tend to show substantial toxicity against enzymes and microbes, requiring extensive washing procedures [71]. Choline-derived ILs, which are frequently derived from renewable feedstocks (choline, amino acids), tend to exhibit lower toxicity and enhanced biodegradability compared to imidazolium analogues [41]. Viscosity is a disadvantage for certain choline-derived ILs [18]. Protic ILs (PILs) are very efficient for delignification (ionoSolv process), usually preserving cellulose crystallinity intact [44]. Certain PILs exhibit good water tolerance and are highlighted for their low cost (potentially ~\$1.24/kg) [62]. Phosphonium-based APILs are characterized by high thermal stability [57]. Their use in biomass pretreatment is less common than that of imidazolium or ammonium ILs. Some studies suggest they can influence thermal decomposition behavior [72]. Toxicity and cost can be concerns [73]. ILs synthesized from lignin or hemicellulose derivatives (e.g., vanillin, furfural) have demonstrated pretreatment efficacy, achieving 90–95% glucose and 70–75% xylose yields [44]. This offers a pathway toward closed-loop biorefineries. Explored for their potential low toxicity, some ZILs can reduce cellulose crystallinity or pretreat biomass, potentially comparable to conventional ILs under specific conditions (e.g., solid ZILs with water removal) [53]. However, their overall pretreatment efficacy compared to leading APILs or PILs may be lower or require further optimization [74].

Thus, the effectiveness of any given IL pretreatment is significantly affected by the type of biomass being processed [60]. For example, ILs highly effective on herbaceous materials like switchgrass or corn stover might be less effective on more recalcitrant woody biomass like pine due to differences in lignin content and structure [75]. Studies comparing the pretreatment of multiple feedstocks (e.g., corn stover, switchgrass, miscanthus, pine, sorghum, eucalyptus, bagasse) using the same IL often show wide variations in sugar yields and delignification rates [60]. This highlights the need for feedstock-specific optimization. Blending different feedstocks has been explored to create a more uniform input for biorefineries and mitigate supply chain variability, with some studies suggesting IL pretreatment can effectively handle such mixtures [76].

A significant challenge in comparing the efficacy of different ILs arises from the wide variation in experimental conditions reported in the literature [62]. Parameters like temperature, time, biomass loading, particle size, and water content drastically influence the outcome [21]. Consequently, direct comparisons between studies are often difficult, highlighting the need for standardized testing protocols or comprehensive parametric studies to assess the relative performance of different IL systems reliably.

7 Process Optimization and Intensification Strategies

Maximizing the efficiency and economic viability of IL-based biomass pretreatment necessitates careful optimization of process parameters and exploring intensification strategies. Increasing temperature generally accelerates dissolution and reaction rates [21]. However, elevated temperatures may result in the degradation of sugars and potentially the IL itself and may cause undesirable lignin condensation [69]. An optimal temperature balances reaction speed with degradation, often falling between 90 and 160 °C, depending on the IL and desired outcome [77]. An optimum duration of time is essential for efficient biomass modification and downstream processing [78]. Usually, longer pretreatment times allow for more extensive dissolution and deconstruction but increase energy costs and the risk of degradation reactions [21]. Increasing biomass concentration (solids loading) is also crucial for process economics, as it reduces the volume of solvent needed per unit of biomass, leading to smaller reactor sizes and lower energy requirements for solvent handling and recovery [15]. Typical lab-scale loadings are 5–15% [78], but achieving higher loadings (e.g., >20–30%) [62] is a key scale-up challenge. Other optimization techniques include reducing biomass particle size (e.g., through milling or grinding), which increases the surface area available for interaction with the IL, generally accelerating dissolution [21]. However, extensive size reduction is energy-intensive and expensive [35]. The presence of water significantly impacts IL pretreatment. For many conventional APILs effective at cellulose dissolution (e.g., Cl, [Emim][OAc]), water acts as an anti-solvent, drastically reducing their ability to dissolve cellulose and biomass [21]. However, for certain PILs used in fractionation processes (ionoSolv), controlled amounts of water (e.g., 10–40 wt%) are often beneficial or even necessary, potentially aiding hydrolysis reactions, reducing viscosity, and lowering solvent cost [44, 62]. The effect of water is highly specific to the IL system. Washing pretreated biomass with water is also critical for removing residual IL before enzymatic hydrolysis. However, it adds cost and complexity related to IL recovery from dilute aqueous solutions [51].

To overcome limitations like long reaction times or high-energy input, intensification techniques are being explored in IL-mediated pretreatments. In microwave-assisted pretreatment, microwave irradiations are used to heat the IL-biomass mixture. The mechanism involves dielectric heating, where polar molecules (like water within biomass and the IL ions) rapidly oscillate in the electromagnetic field, generating heat volumetrically and efficiently [79]. This leads to faster heating rates, reduced processing times, potentially lower overall energy consumption, and enhanced disruption of the biomass structure compared to conventional heating [50]. Ultrasound-assisted (US) pretreatment is another technique for applying high-frequency sound waves to the IL-biomass slurry [80]. The primary mechanism is acoustic cavitation, where microbubbles' formation, growth, and implosive collapse generate localized high temperatures and pressures, shockwaves, and micro-jets. These physical effects enhance mass transfer, increase porosity, erode surfaces, and physically disrupt the biomass structure. Chemical effects from radical generation (•OH, •H) can also contribute to lignin degradation. Ultrasound (US) may facilitate synergies in ionic liquid (IL) pretreatment by enhancing solvent penetration, reducing viscosity constraints, and influencing reaction durations. Promising outcomes have been achieved by integrating the US with cost-effective protic ionic liquids (PILs) [62, 80]. Additionally, ILs combined with other co-solvent systems present further opportunities for process enhancement [39]. Molecular solvents, such as dimethyl sulfoxide (DMSO), dimethylacetamide (DMAc), ethanolamine, or even water, can be incorporated to alter the properties of the IL system [28]. Potential benefits include mitigating the high viscosity typically associated with ILs and biomass slurries, decreasing the solvent cost by minimizing the quantity of expensive IL required, and potentially enhancing pretreatment selectivity or efficacy for specific components [28]. For example, mixtures of [Emim][OAc] with DMSO or ethanolamine have demonstrated pretreatment capabilities [56]. However, introducing a co-solvent may lead to increased complexities in solvent interactions, optimization, and recovery processes.

By integrating pretreatment and enzymatic saccharification into a single vessel, "One-Pot" eliminates the need for intermediate steps such as separation, washing, and transfer [81]. Traditional ionic liquid (IL) pretreatment often requires extensive washing to eliminate residual IL from the pretreated biomass, as these residues can inhibit subsequent enzymatic and microbial activities [70]. This washing process is water-intensive, produces dilute IL streams that are costly to recover, and leads to the loss of soluble sugars [81]. The "One-pot" approach effectively mitigates these issues. For successful one-pot operations, it is essential to either employ ILs that are naturally biocompatible, exhibiting low toxicity to enzymes and microbes, or to develop or identify enzymes and microbial strains that can withstand specific ILs at relevant concentrations [70]. Several studies have successfully demonstrated one-pot or separation-free processes by integrating IL pretreatment with saccharification and fermentation. Examples include the use of [Ch][Lys] pretreatment followed by fermentation with *Rhodosporidium toruloides* [45] and [Ch][OAc] pretreatment with IL-tolerant bacterial cellulases [70].

Consequently, intensification and integration strategies are pivotal in overcoming the primary challenges associated with IL pretreatment, such as cost, energy consumption, water usage, and process complexity [23]. The adoption of one-pot systems that incorporate biocompatible ILs, along with the integration of effective intensification techniques like MW or US, offers substantial potential for the development of sustainable and economically viable IL-based biorefineries [62, 70]. Additionally, cosolvents serve as another optimization variable, as they can alter the solvent environment to achieve synergistic reductions in viscosity, costs, and selective fractionation, thereby enhancing the adjustments possible through IL ion selection alone.

8 Roles of IL-Based Pretreatment in Sustainability

ILs facilitate the enzymatic hydrolysis of pretreated biomass, thereby enhancing sugar yields from biomass [47]. These pretreatments also enable the separation of lignin from lignocellulosic biomass, leading to its valorization into various products [67]. Despite the potential advantages in biomass deconstruction, a comprehensive evaluation of sustainability is necessary, considering the complete life cycle, including synthesis, use-phase effects (toxicity, energy consumption), and end-of-life impacts (biodegradability, recovery/recycling) [82]. The notably low vapor pressure of most ILs offers significant advantages over VOCs, reducing air pollution, explosion risk, and solvent loss during processing [16]. This attribute initially led to ILs being widely categorized as "green solvents"[41]. Most typical ILs, especially imidazolium-based ones, are prepared from petroleum-based precursors in multi-step syntheses involving toxic reagents or waste generation, undermining their green nature [41]. With their stable aromatic core, they exhibit the phenomenon of poor biodegradability and persistence in the environment. Unlike initial expectations, most ILs have been found to exhibit significant toxicity to a broad spectrum of organisms, including bacteria, fungi, algae, and aquatic invertebrates, and even mammalian cell lines [12]. Toxicity is strongly structure-dependent, frequently rising with hydrophobicity and chain length of alkyl groups on the cation and being dependent on the anion. Although antimicrobial activity is helpful for specific medical purposes, ecotoxicity on discharge into the environment is a serious issue [41]. Hence, it is insufficient to label an IL "green" based on low volatility alone [83]. Attention is being diverted to the development of "greener" ILs through the utilization of renewable starting materials (bio-based ILs) and integration of structural motifs for improved biodegradability and lowered toxicity (e.g., ester groups, ether groups, application of amino acid or choline groups) [41]. Cholinium amino acid ILs, for instance, have proven promising results toward decreased toxicity and facile biodegradability [12, 41].

Life cycle assessment (LCA) provides a systematic methodology for quantifying the environmental impacts of a process or product throughout its entire life span [84]. Using LCA on IL-based biomass pretreatment identifies environmental hotspots and contrasts the ecological impact of IL processes with conventional methods that help design more sustainable ILs and processes [85]. Some challenges to using LCA on ILs involve the lack of thorough inventory data for many ILs and their precursors and the difficulty in characterizing impacts such as toxicity and biodegradability within conventional LCA frameworks [82]. Simplified or predictive LCA techniques are frequently used for initial-stage evaluations [86]. Existing LCA studies show that the environmental footprint of ILs can be considerable, especially concerning their synthesis routes and the energy needed for recovery and recycling [87]. For instance, the cost of production and environmental impact may be significantly different based on a particular IL. PILs are more effective when compared to certain imidazolium-based ILs based on cost, environment, and synthesis factors [85].

LCA results often highlight the critical importance of achieving very high IL recovery rates (>99%) to minimize environmental impacts associated with solvent loss and replacement [88]. Comparing IL pretreatment to traditional methods via LCA reveals complex trade-offs; at the same time, ILs might reduce specific impacts (e.g., related to harsh acid use), but they can introduce others (e.g., related to IL synthesis or toxicity) [89]. The overall environmental benefit depends heavily on the specific IL, the process design (especially recovery efficiency), and the system boundaries considered. Thus, it may be concluded that the sustainability of IL-based biomass pretreatment is not guaranteed simply by the nature of the solvent. A critical evaluation considering synthesis routes, toxicity, biodegradability, energy requirements for use and recovery, and overall process integration, ideally through LCA and techno-economic analysis, is essential to identify sustainable pathways. The development of bio-derived, biodegradable, and low-toxicity ILs, coupled with efficient recovery and integrated processing schemes, represents the most promising direction for realizing the environmental benefits of IL technology in biorefineries.

9 Recycling and Recovery of Ionic Liquids

IL-based biomass pretreatment processes' economic viability and environmental sustainability arc critically dependent on the efficient recuperation and reutilization of the ionic liquid solvent [15]. Given the relatively high cost of many ILs compared to conventional solvents [21], even small losses during recycling can significantly impact process economics. Techno-economic analyses indicate that IL recovery rates exceeding 99%, and ideally much higher, are necessary for IL pretreatment to be competitive with traditional methods [90]. After pretreatment, the IL typically contains dissolved lignin, hemicellulose-derived sugars or degradation products, and residual moisture. Furthermore, washing the pretreated biomass solids (necessary for many ILs due to toxicity concerns) generates dilute aqueous IL solutions [51]. Recovering the IL in a pure and reusable form from these complex mixtures presents several challenges.

Several methods have been explored to recycle, reuse, and recover the ILs. Water and other volatile components (like cosolvents or volatile products like furfural) can be removed by evaporation or distillation, leaving the IL behind [91]. This is particularly relevant for dehydrating ILs after water washing. However, removing large amounts of water is energy-intensive [91]. Distillation can also recover certain distillable PILs directly [92]. Challenges include potential thermal degradation of the IL or dissolved components at high temperatures and the high vacuum often required [91]. Introducing anti-solvents, such as water or ethanol, causes the dissolved cellulose and/or lignin to precipitate out of the IL solution, allowing their separation as solids [60]. The IL remains in the liquid phase, now mixed with the anti-solvent, which requires a subsequent separation step (e.g., evaporation) to recover the pure IL.

Techniques such as pervaporation (PV), nanofiltration (NF), and electrodialysis (ED) offer energy-efficient alternatives for separating ILs from water or other components [91]. PV is particularly effective for IL dehydration. A selective membrane allows preferential passage of water vapor, leaving a concentrated IL solution [91]. High separation factors and IL recovery (>99.9%) have been demonstrated, making it a promising, scalable technology [88]. NF can potentially separate ILs from smaller molecules like sugars or salts based on size exclusion [91]. In electrodialysis, ions are separated using membranes that exchange ions and an electric field [88]. It has been investigated for IL recovery but has lower efficiencies and limitations on the levels of IL concentration that can be achieved [88]. The liquid-liquid extraction process with conventional organic solvents can also selectively extract products (like lignin or sugars) and IL from process streams [91]. However, this method introduces another solvent to recover and incurs the risk of contamination with other organic solvents [91]. Some ILs may be capable, with some salts (like phosphates) or polymers, of producing two nonmiscible aqueous phases in water [54]. This method could also facilitate separations by preferentially collecting sugars in the IL-lean phase and retaining the IL in the other, potentially allowing recovery [93]. Phase separation has also been shown using an alkali solution after acidolysis in imidazolium ILs to collect sugars into the alkaline phase, which can isolate IL in the original phase [93]. Solid adsorbents can also provide a means to selectively bind either IL or impurities like lignin or sugars and then elute from the solution [91]. The solvents that have been adsorbed are finally recovered through desorption.

Although there are multiple methods, many obstacles remain in IL recovery and recycling. Achieving the very high recovery rates (>99%) needed for economic viability while maintaining IL purity for effective reuse might be difficult [17]. Impurities like residual lignin, sugars, degradation products, or accumulated water can build up over recycling loops, potentially reducing pretreatment efficacy [94]. Some processes exhibit "self-cleaning" properties where impurities are removed with product streams [17]. Methods involving evaporation of large amounts of water (from washing or anti-solvent addition) are energy-intensive, adding significant operating costs [91]. Membrane processes are typically lower in energy consumption but may face issues with flux or stability of the membrane [91]. Some ILs may also be damaged during pretreatment or recovery conditions (e.g., high temperatures, acids/bases), leading to solvent loss or unwanted impurities [94]. The capital and operating costs of the recovery system contribute significantly to the process costs [51]. Some recovery methods demonstrated at the lab scale do not easily scale to industrial levels due to several factors [91]. It will be vital to start developing effective and performance-efficient recovery technologies for ILs, either in a stand-alone or integrated fashion, to enable successful industrial uptake of IL-based biorefineries. Integrated processes that reduce water use and limit the usage of washing steps (e.g., one-pot operations) will significantly aid in reducing the stressors placed on the recovery section of the process [81].

10 Innovations and Future Directions

Research into IL-based biomass pretreatment continues to evolve, focusing on enhancing efficiency, sustainability, and economic viability through innovative approaches. Since general-purpose ILs might not be the best for particular tasks such as selective lignin removal or targeted cellulose modification, scientists are developing task-specific ILs (TSILs) with functional groups optimized for particular interactions or catalytic functions [12]. ILs can be engineered to selectively interact with and dissolve lignin while leaving polysaccharides relatively intact. This principle is behind fractionation processes such as ionoSolv, commonly using PILs [95]. The design principle typically involves tailoring the IL's properties (e.g., acidity/basicity, H-bonding ability) to complement lignin's characteristics with minimal interaction with cellulose [66]. This enables the regeneration of a high-purity cellulose pulp for enzymatic hydrolysis and an independent lignin stream for valorization [95]. Integrating catalytic groups (e.g., acidic or basic functional groups) into the IL backbone can facilitate concurrent pretreatment and partial hydrolysis or lignin depolymerization of biomass constituents [44]. Acidic ILs, for instance, can catalyze hemicellulose hydrolysis [96], while others may enhance lignin depolymerization [33]. To reduce cost and environmental issues related to conventional ILs, considerable work focuses on producing ILs from renewable feedstocks and their design for improved biodegradability [68].

Amino acids [41], sugars [13], choline [41], and lignin/hemicellulose degradation products (e.g., vanillin, furfural) [44] are being employed as cationic and anionic building blocks. Various bio-derived ILs, including cholinium lysinate [77] and those based on aromatic aldehydes [52], have shown pretreatment performance equal to or even better than traditional ILs such as [Emim][OAc]. Introducing moieties subject to microbial attack (i.e., ester or ether bonds, amino acid configurations) can increase biodegradability [41]. Cholinium amino acid ILs have also been found to be very biodegradable [97]. Producing ILs from lignin and hemicellulose derivatives promises a "closed-loop" process, enhancing resource efficiency and economics [52]. Sophisticated methods facilitate IL integration in conversion processes, reducing separation steps, water use, and costs [81]. Developing IL pretreatment processes yields accessible carbohydrates and lignin fractions for conversion to high-value products [44], requiring ILs that minimize lignin degradation during pretreatment [98]. Combining IL pretreatment with other methods can reduce IL use and enhance selectivity [39]. COSMO-RS, DFT, and MD simulations offer insights into IL properties, predicting parameters like viscosity and solubility, aiding in IL tailoring for specific applications [28]. These simulations clarify IL and biomass interactions, optimizing IL-based processes and guiding the design of TSILs with targeted functions [28]. Modeling these interactions at the molecular level fosters biomass pretreatment innovations, improving efficiency and sustainability. Integrating predictive methods with experimental data advances IL development, meeting biorefinery needs by maximizing recovery, minimizing toxicity, and enabling integration with downstream processes [28]. ML algorithms quickly screen IL candidates, predicting performance for biomass applications [67]. Computational simulations are pivotal in evolving IL technologies toward a sustainable bioeconomy.

11 Challenges and Bottlenecks

Despite the significant potential and advancements in IL-based biomass pretreatment, several major challenges and bottlenecks hinder its large-scale industrial implementation. The high cost of ILs remains arguably the most significant barrier [15]. While costs have decreased and low-cost PILs are being developed [62], many high-performance ILs (especially imidazolium-based APILs) are still considerably more expensive than conventional solvents or reagents [56]. Techno-economic analyses suggest IL costs need to drop significantly (e.g., to ~$1/kg) to be competitive [90]. Due to the high cost, near-quantitative recovery (>99%, ideally >99.5%) and efficient recycling of the IL are essential for achieving economic viability [44]. Developing energy-efficient and scalable recovery processes remains difficult [91]. Impurities like lignin, sugars, and water can degrade IL performance over cycles [94]. Many ILs and dissolved biomass are highly viscous, complicating mixing, pumping, heat, and mass transfer, especially with high solids [54]. Higher temperatures or cosolvents might be needed, raising complexity and cost [56]. Water, integral to biomass and often an anti-solvent or co-solvent, majorly affects IL performance in cellulose dissolution, posing a significant energy demand for its removal [62]. Progress requires water-tolerant ILs or effective dehydration techniques. Most ILs' toxicity to microbial and enzymatic processes, potential eco-toxicity, and lack of biodegradability question their status as "green" technologies, demanding cautious handling and more eco-friendly alternatives [41]. Integrating IL pretreatment with downstream hydrolysis and fermentation, especially in one-pot methods, requires compatible enzymes/microbes and precise process control [70]. Scaling processes from lab to industry faces engineering challenges like handling viscous slurries, reactor design, energy integration, and cost-effective IL recovery [44]. Some ILs, particularly acidic PILs, may cause equipment corrosion [87]. Optimal IL and pretreatment conditions depend considerably on biomass origin, requiring flexible processes or specific optimization [98]. Mixed feedstocks add complexity [76].

While ionic liquids (ILs) are applicable for lignin extraction, the resulting lignin may undergo alteration or condensation, especially under rigorous conditions, which could hinder its subsequent valorization into high-value products [99]. Formulating an ionic liquid (IL) process capable of generating high-quality, reactive lignin is crucial for improving the economic viability of biorefineries. Addressing the associated challenges necessitates ongoing research and advancement in IL synthesis (cost-effective, bio-derived, and biodegradable), process engineering (effective recovery, high-capacity solids handling, and process intensification), development of biocatalysts (enzymes/microbes with IL tolerance), and integrated system design. These endeavors are critical for harnessing the full potential of ionic liquids in the sustainable conversion of biomass.

12 Conclusions

Ionic liquids represent a compelling and versatile class of solvents for the pretreatment of lignocellulosic biomass, offering significant potential to overcome the limitations of traditional methods and enable efficient biorefinery processes. ILs are specifically designed to overcome limitations associated with high cellulose disruption (e.g., [Emim][OAc]) or lignin fractionation ([Ch][Lys]). Additionally, the "one-pot" pretreatment process enables the combined effect of saccharification and fermentation using biocompatible ILs, which offers sustainability. Therefore, though the IL-based pretreatment method often faces obstacles, it holds a promising contribution for developing sustainable and competitive economic biorefineries. Thus, moving a step closer toward achieving a circular bioeconomy.

References

1. S. Ethaib, R. Omar, M.S. Mazlina, A.D. Radiah, S.L. Zubaidi, Bioresources **15**(4), 10063 (2020). https://doi.org/10.15376/biores.15.4.10063-10088
2. A. Patel, D.K. Sindhu, N. Arora, R.P. Singh, V. Pruthi, P.A. Pruthi, Bioresour. Technol. **197**, 91–98 (2015). https://doi.org/10.1016/j.biortech.2015.08.039
3. A. Yousuf, Waste Manag. **32**(11), 2061–2067 (2012). https://doi.org/10.1016/j.wasman.2012.03.008
4. A. Woźniak, K. Kuligowski, L. Świerczek, A. Cenian, Sustainability **17**(1), 287 (2024). https://doi.org/10.3390/su17010287
5. A. Brandt, J. Gräsvik, J.P. Hallett, T. Welton, Green Chem. **15**(3), 550–583 (2013). https://doi.org/10.1039/C2GC36364J
6. A.P. Ingle, S. Saxena, M.P. Moharil, J.D. Rivaldi, L. Ramos, A.K. Chandel, Biotechnol. Sustainable Mater. **2**(1), 3 (2025). https://doi.org/10.1186/s44316-025-00025-2
7. D.Y. Limeneh, M.T. Abate, K.T. Yilma, *Lignocellulosic Biomass and Enzymes* (Springer, Cham, 2025), pp. 3–31. https://doi.org/10.1007/978-981-96-3037-0_1
8. A.L. Woiciechowski, C.J.D. Neto, L.P. De Souza Vandenberghe, D.P. De Carvalho Neto, A.C.N. Sydney, L.J. Letti, S.G. Karp, L.A.Z. Torres, C.R. Soccol, Bioresour. Technol. **304**, 122848 (2020). https://doi.org/10.1016/j.biortech.2020.122848
9. J. Ma, N.L. Ma, W. Zhang, Y. Wu, Y. Ma, G. Chen, C. Sun, Ind. Crop. Prod. **230**, 121091 (2025). https://doi.org/10.1016/j.indcrop.2025.121091

10. A. Etale, A.J. Onyianta, S.R. Turner, S.J. Eichhorn, Chem. Rev. **123**(5), 2016–2048 (2023). https://doi.org/10.1021/acs.chemrev.2c00477
11. M. Siddique, A.N. Mengal, S. Khan, L.A. Khan, E.K. Kakar, MOJ Biol. Med. **8**(1), 39–43 (2023). https://doi.org/10.15406/mojbm.2023.08.00181
12. K.S. Egorova, E.G. Gordeev, V.P. Ananikov, Chem. Rev. **117**(10), 7132–7189 (2017). https://doi.org/10.1021/acs.chemrev.6b00562
13. J. Flieger, M. Flieger, Int. J. Mol. Sci. **21**(17), 6267 (2019). https://doi.org/10.3390/ijms21176267
14. S. Jiang et al., Nucleic Acids Res. **53**(1), D1670–D1676 (2025). https://doi.org/10.1093/nar/gkae973
15. G. Brodeur, E. Yau, K. Badal, J. Collier, K.B. Ramachandran, S. Ramakrishnan, Enzyme Res. **2011**, 787532 (2011). https://doi.org/10.4061/2011/787532
16. S. Peleteiro, S. Rivas, J.L. Alonso, V. Santos, J.C. Parajó, J. Agric. Food Chem. **63**(37), 8093–8102 (2015). https://doi.org/10.1021/acs.jafc.5b03461
17. A. Brandt-Talbot, F.J.V. Gschwend, P.S. Fennell, T.M. Lammens, B. Tan, J. Weale, J.P. Hallett, Green Chem. **19**(13), 3078–3102 (2017). https://doi.org/10.1039/c7gc00705a
18. M. Guncheva, B. Yakimova, Appl. Sci. **15**(3), 1515 (2024). https://doi.org/10.3390/app15031515
19. E. Pereira, D.T.V. Pereira, S.C. Rabelo, R. Ceriani, A.C. da Costa, J. Environ. Chem. Eng. **2025**, 115303 (2025). https://doi.org/10.1016/j.jece.2024.115303
20. W. Den, V.K. Sharma, M. Lee, G. Nadadur, R.S. Varma, Front. Chem. **6**, 337170 (2018). https://doi.org/10.3389/fchem.2018.00141
21. G. Gogoi, S. Hazarika, Int J Environ Sci Nat Resour **20**(2), 48–56 (2019). https://doi.org/10.19080/IJESNR.2019.20.556033
22. F.H. Isikgor, C.R. Becer, Polym. Chem. **6**(25), 4497–4559 (2015). https://doi.org/10.1039/c5py00263j
23. Y. Li, J. Wang, X. Liu, S. Zhang, Chem. Sci. **9**(17), 4027 (2018). https://doi.org/10.1039/c7sc05392d
24. G. Albornoz-Palma, D. Ching, O. Valerio, R.T. Mendonça, M. Pereira, Cellulose **27**(18), 10631–10647 (2020). https://doi.org/10.1007/s10570-020-03304-5
25. S. Roy, S.P.S. Chundawat, Bioenergy Res. **16**(1), 263–278 (2023). https://doi.org/10.1007/s12155-022-10425-1
26. W. Boerjan, J. Ralph, M. Baucher, Annu. Rev. Plant Biol. **54**(1), 519–546 (2003). https://doi.org/10.1146/annurev.arplant.54.031902.134938
27. H.V. Lee, S.B. Hamid, S.K. Zain, Sci. World J. **2014**, 631013 (2014). https://doi.org/10.1155/2014/631013
28. P. Moyer, M.D. Smith, N. Abdoulmoumine, S.C. Chmely, J.C. Smith, L. Petridis, N. Labbé, Phys. Chem. Chem. Phys. **20**(4), 2508–2516 (2018). https://doi.org/10.1039/c7cp07195g
29. S. Rezania, M.F.M. Din, S.E. Mohamad, J. Sohaili, S.M. Taib, M.B.M. Yusof, A. Ahsan, Bioresources **12**(1), 2108–2124 (2017)
30. D.P. Maurya, A. Singla, S. Negi, 3 Biotech **5**(5), 597 (2015). https://doi.org/10.1007/s13205-015-0279-4
31. S.B. Jamaldheen, M.B. Kurade, B. Basak, C.G. Yoo, K.K. Oh, B. Jeon, T.H. Kim, Bioresour. Technol. **346**, 126591 (2022). https://doi.org/10.1016/j.biortech.2021.126591
32. S. Baksi, D. Saha, S. Saha, U. Sarkar, D. Basu, J.C. Kuniyal, Int. J. Environ. Sci. Technol. **20**(12), 13895–13922 (2023). https://doi.org/10.1007/s13762-023-04838-4
33. C.G. Yoo, Y. Pu, A.J. Ragauskas, Curr. Opin. Green Sustain. Chem. **5**, 5–11 (2017). https://doi.org/10.1016/j.cogsc.2017.03.003
34. J. Chen, X. Ma, M. Liang, Z. Guo, Y. Cai, C. Zhu, Z. Wang, S. Wang, J. Xu, H. Ying, Waste **2**(4), 451–473 (2024). https://doi.org/10.3390/waste2040024
35. C. Huang, J. Liu, W. Geng, W. Tang, Q. Yong, Paper Biomater. **6**(3), 7 (2021). https://doi.org/10.1213/j.issn.2096-2355.2021.03.007
36. H. Lu, S. Liu, M. Zhang, F. Meng, X. Shi, L. Yan, Energy Fuel (2016). https://doi.org/10.1021/acs.energyfuels.5b02658

37. O. Awogbemi, D.V. Von Kallon, Case Stud. Chem. Environ. Eng. **6**, 100229 (2022). https://doi.org/10.1016/j.cscee.2022.100229
38. B. Kumar, N. Bhardwaj, K. Agrawal, V. Chaturvedi, P. Verma, Fuel Process. Technol. **199**, 106244 (2020). https://doi.org/10.1016/j.fuproc.2019.106244
39. X. Yin, L. Wei, X. Pan, C. Liu, J. Jiang, K. Wang, Front. Plant Sci. **12**, 670061 (2021). https://doi.org/10.3389/fpls.2021.670061
40. S.B. Ummalyma, R.D. Supriya, R. Sindhu, P. Binod, R.B. Nair, A. Pandey, E. Gnansounou, *Second and Third Generation of Feedstocks* (Elsevier, Amsterdam, 2019), pp. 197–212. https://doi.org/10.1016/b978-0-12-815162-4.00007-0
41. A.R. Gonçalves, X. Paredes, A.F. Cristino, F.J. Santos, C.S. Queirós, Int. J. Mol. Sci. **22**(11), 5612 (2021). https://doi.org/10.3390/ijms22115612
42. F. Wang, S. Li, Y. Sun, H. Han, B. Zhang, B. Hu, Y. Gao, X. Hu, RSC Adv. **7**(76), 47990–47998 (2017). https://doi.org/10.1039/c7ra08110c
43. A. Atashnezhad, J. Scott, M.F.A. Dushaishi, ACS Sustain. Chem. Eng. **12**(40), 14684–14693 (2024). https://doi.org/10.1021/acssuschemeng.4c04606
44. P.Y. Nakasu, C. Hopson, A. Ovejero-Pérez, J.M. Costa, J.P. Hallett, *Ionic Liquids and their Role in a Lignocellulosic Biorefinery* (Elsevier, Amsterdam, 2024). https://doi.org/10.1016/b978-0-443-15742-4.00082-x
45. V. Rigual, G. Papa, A. Rodriguez, M. Wehrs, K.H. Kim, M. Oliet, M.V. Alonso, J.M. Gladden, A. Mukhopadhyay, B.A. Simmons, S. Singh, ACS Sustain. Chem. Eng. **8**(2), 782–791 (2019). https://doi.org/10.1021/acssuschemeng.9b04451
46. V.R. Zeger, B. Thapa, D. Shamsaei, J.F. DeLair, T.L. Taylor, J.L. Anderson, Anal. Chem. **97**(9), 4793 (2025). https://doi.org/10.1021/acs.analchem.5c00264
47. S. Naz, M. Uroos, *Biofibers and Biopolymers for Biocomposites* (Springer, Cham, 2020), pp. 181–207. https://doi.org/10.1007/978-3-030-40301-0_9
48. Y. Du, M. Li, X. Zheng, Y. Chai, Y. Zheng, Ind. Crop. Prod. **223**, 120281 (2025). https://doi.org/10.1016/j.indcrop.2024.120281
49. Y. Du, J. Guo, Z. Wang, Y. Zhang, Y. Zheng, Int. J. Biol. Macromol. **286**, 138318 (2025). https://doi.org/10.1016/j.ijbiomac.2024.138318
50. N. Mohd, S.F.S. Draman, M.S.N. Salleh, N.B. Yusof, AIP Conf. Proc. **1809**(1), 4975450 (2017). https://doi.org/10.1063/1.4975450
51. A. Ovejero-Pérez, M. Ayuso, V. Rigual, J.C. Domínguez, J. García, M.V. Alonso, M. Oliet, F. Rodriguez, ACS Sustain. Chem. Eng. **9**(25), 8467–8476 (2021). https://doi.org/10.1021/acssuschemeng.1c01361
52. A.M. Socha, R. Parthasarathi, J. Shi, S. Pattathil, D. Whyte, M. Bergeron, A. George, K. Tran, V. Stavila, Venkatachalam, et al., Proc. Natl. Acad. Sci. **111**(35), E3587–E3595 (2014). https://doi.org/10.1073/pnas.1405685111
53. G. Huet, M. Araya-Farias, R. Alayoubi, S. Laclef, B. Bouvier, I. Gosselin, C. Cézard, R. Roulard, M. Courty, C. Hadad, et al., Green Chem. **22**(9), 2935–2946 (2020). https://doi.org/10.1039/d0gc01002b
54. P. Reddy, S. Afr. J. Sci. **111**(11), 9 (2015). https://doi.org/10.17159/sajs.2015/20150083
55. S. Swami, S. Suthar, R. Singh, A.K. Thakur, L.R. Gupta, V.S. Sikarwar, Environ. Sci. Pollut. Res. **31**(9), 12871–12891 (2024). https://doi.org/10.1007/s11356-024-32100-y
56. M. Ungurean, Z. Csanádi, L. Gubicza, F. Péter, Bioresources **9**(4), 443–447 (2014)
57. K. Ghandi, Green Sustain. Chem. **4**(1), 44–53 (2014). https://doi.org/10.4236/gsc.2014.41008
58. J. Hulsbosch, D.E. De Vos, K. Binnemans, R. Ameloot, ACS Sustain. Chem. Eng. **4**(6), 2917–2931 (2016). https://doi.org/10.1021/acssuschemeng.6b00553
59. M. Javvad Ali, M. Praveen, A. Ghouse, F. Fatima, Int. J. Biomed Res. **1**(2), 11 (2021). https://doi.org/10.31579/IJBR-2021/011
60. C. Li, B. Knierim, C. Manisseri, R. Arora, H.V. Scheller, M. Auer, K.P. Vogel, B.A. Simmons, S. Singh, Bioresour. Technol. **101**(13), 4900–4906 (2010). https://doi.org/10.1016/j.biortech.2009.10.066
61. A.M. Asim, M. Uroos, S. Naz, M. Sultan, G. Griffin, N. Muhammad, A.S. Khan, J. Mol. Liq. **287**, 110943 (2019). https://doi.org/10.1016/j.molliq.2019.110943

62. Z. Ziaei-Rad, M. Pazouki, J. Fooladi, M. Azin, S.N. Gummadi, A. Allahverdi, Sci. Rep. **13**, 446 (2023). https://doi.org/10.1038/s41598-022-27258-9
63. L. Yang, H. Peng, H. He, L. Liu, G. Fu, Y. Liu, Y. Wan, Bioresources **18**(1), 1590–1601 (2023). https://doi.org/10.15376/biores.18.1.1590-1601
64. T. Uto, K. Yamamoto, J. Kadokawa, J. Phys. Chem. B **122**(1), 258–266 (2017). https://doi.org/10.1021/acs.jpcb.7b09525
65. L. Fu, Z. Ju, M. Yu, H. Luo, C. Zhang, X. Zhang, H. Cheng, M. Zheng, L. Jin, C. Ge, ACS Omega **7**(46), 42170–42180 (2022). https://doi.org/10.1021/acsomega.2c04915
66. Q. Hou, M. Ju, W. Li, L. Liu, Y. Chen, Q. Yang, Molecules **22**(3), 490 (2017). https://doi.org/10.3390/molecules22030490
67. B.R. Taylor, N. Kumar, D.K. Mishra, B.A. Simmons, H. Choudhary, K.L. Sale, Molecules **29**(21), 5073 (2023). https://doi.org/10.3390/molecules29215073
68. Z. Usmani, M. Sharma, P. Gupta, Y. Karpichev, N. Gathergood, R. Bhat, V.K. Gupta, Bioresour. Technol. **304**, 123003 (2020). https://doi.org/10.1016/j.biortech.2020.123003
69. P.Y. Nakasu, P.V. Barbará, A.E. Firth, J.P. Hallett, Trends Chem. **4**(3), 175–178 (2022). https://doi.org/10.1016/j.trechm.2021.12.001
70. M. Sriariyanun, N. Kitiborwornkul, P. Tantayotai, K. Rattanaporn, L. Show, Bioengineering **9**(1), 17 (2022). https://doi.org/10.3390/bioengineering9010017
71. H. Zhu, X. He, Z. Xu, L. Dai, Green Chem. (2024). https://doi.org/10.1039/d4gc05662k
72. S. Saeed, M. Saleem, Iran. J. Chem. Chem. Eng. **37**(5), 225–234 (2018)
73. Q. Zhang, J. Hu, D. Lee, Renew. Energy **111**, 77–84 (2017). https://doi.org/10.1016/j.renene.2017.03.093
74. A. Hachisu, K. Takahashi, K. Kuroda, New J. Chem. (2024). https://doi.org/10.1039/d4nj03379e
75. C.L. Williams, C. Li, H. Hu, J.C. Allen, B.J. Thomas, Front. Energy Res. **6**, 374816 (2018). https://doi.org/10.3389/fenrg.2018.00067
76. J. Shi, V.S. Thompson, N.A. Yancey, V. Stavila, B.A. Simmons, S. Singh, Biofuels **4**(1), 63–72 (2012). https://doi.org/10.4155/bfs.12.82
77. N. Sun, R. Parthasarathi, A.M. Socha, J. Shi, S. Zhang, V. Stavila, S. Singh, Green Chem. **16**(5), 2546–2557 (2014). https://doi.org/10.1039/C3GC42401D
78. S. Chuetor, E.J. Panakkal, T. Ruensodsai, K. Cheenkachorn, S. Kirdponpattara, S. Cheng, M. Sriariyanun, Bioengineering **9**(3), 115 (2022). https://doi.org/10.3390/bioengineering9030115
79. A. Aguilar-Reynosa, A. Romaní, R.M. Rodríguez-Jasso, C.N. Aguilar, G. Garrote, H.A. Ruiz, Energy Convers. Manag. **136**, 50–65 (2017). https://doi.org/10.1016/j.enconman.2017.01.004
80. P. Chakraborty, R. Kumar, A. Banerjee, S. Chakrabortty, M. Pal, A. Upadhyaya, S. Chowdhury, M.A. Khan, B. Jeon, S.K. Tripathy, A.K. Ghosh, Biomass Bioenergy **192**, 107495 (2024). https://doi.org/10.1016/j.biombioe.2024.107495
81. M. Zhang, L. Jia, M. Li, H. Peng, Y. Tan, S. Arvelli, Y. Huang, A.C. Neves, E.J. Oh, J. Zhao, ACS Sustain. Chem. Eng. (2025). https://doi.org/10.1021/acssuschemeng.4c10101
82. V.G. Maciel, D.J. Wales, M. Seferin, C.M.L. Ugaya, V. Sans, J. Clean. Prod. **217**, 844–858 (2019)
83. T. Welton, R. Soc. **471**(2183), 20150502 (2015). https://doi.org/10.1098/rspa.2015.0502
84. B. Zheng, S. Yu, Z. Chen, Y. Huo, Front. Microbiol. **13**, 933882 (2022). https://doi.org/10.3389/fmicb.2022.933882
85. H. Baaqel, I. Díaz, V. Tulus, B. Chachuat, G. Guillén-Gosálbez, J.P. Hallett, Green Chem. **22**(10), 3132–3140 (2020). https://doi.org/10.1039/d0gc00058b
86. H.A. Baaqel, A. Bernardi, J.P. Hallett, G. Guillén-Gosálbez, B. Chachuat, ACS Sustain. Chem. Eng. **11**(18), 7157–7169 (2023). https://doi.org/10.1021/acssuschemeng.3c00547
87. A.S. Putra, Y. Anita, E3S Web Conf. **503**, 04001 (2024). https://doi.org/10.1051/e3sconf/202450304001
88. J. Sun, J. Shi, N. Murthy Konda, D. Campos, D. Liu, S. Nemser, J. Shamshina, T. Dutta, P. Berton, G. Gurau, et al., Biotechnol. Biofuels **10**, 154 (2017). https://doi.org/10.1186/s13068-017-0842-9

89. A.J. David, S. Abinandan, V.K. Vaidyanathan, C.C. Xu, T. Krishnamurthi, 3 Biotech **13**(7), 233 (2023). https://doi.org/10.1007/s13205-023-03657-1
90. N.R. Baral, A. Shah, Biofuels. Bioprod. Biorefin. **10**(1), 70–88 (2016). https://doi.org/10.1002/bbb.1622
91. C. Regmi, C. Thamaraiselvan, Z. Zhu, X. Qian, S.R. Wickramasinghe, PRO **12**(1), 55 (2023). https://doi.org/10.3390/pr12010055
92. E.C. Achinivu, B.W. Blankenship, N.R. Baral, H. Choudhary, R. Kakumanu, M. Mohan, E.E. Baidoo, C.D. Scown, A. George, B.A. Simmons, J. Gladden, Chem. Eng. J. **479**, 147824 (2023). https://doi.org/10.1016/j.cej.2023.147824
93. N. Sun, H. Liu, N. Sathitsuksanoh, V. Stavila, M. Sawant, A. Bonito, K. Tran, A. George, K.L. Sale, S. Singh, B.A. Simmons, B.M. Holmes, Biotechnol. Biofuels **6**(1), 39 (2013). https://doi.org/10.1186/1754-6834-6-39
94. P. Weerachanchai, J.M. Lee, Bioresour. Technol. **169**, 336–343 (2014). https://doi.org/10.1016/j.biortech.2014.06.072
95. J.L. Espinoza-Acosta, P.I. Torres-Chávez, E. Carvajal-Millán, B. Ramírez-Wong, L.A. Bello-Pérez, B. Montaño-Leyva, Bioresources **9**(2), 3660–3687 (2014)
96. J.R. Bernardo, F.M. Gírio, R.M. Łukasik, Molecules **24**(4), 808 (2018). https://doi.org/10.3390/molecules24040808
97. D. Hou, P. Liu, T.J. Smith, N. Li, H. Zong, PLoS One **8**(3), e59145 (2013). https://doi.org/10.1371/journal.pone.0059145
98. C. Li, D. Tanjore, W. He, J. Wong, J.L. Gardner, K.L. Sale, B.A. Simmons, S. Singh, Biotechnol. Biofuels **6**, 154 (2013). https://doi.org/10.1186/1754-6834-6-154
99. T.C. Pin, S.C. Rabelo, Y. Pu, A.J. Ragauskas, A.C. Costa, ACS Sustain. Chem. Eng. **9**(50), 16965–16976 (2021). https://doi.org/10.1021/acssuschemeng.1c05353

Toxicity of Ionic Liquids, Nanofluids, and Nanomaterials/Nanotechnology

Prajakta Kapadnis

1 Introduction

1.1 Ionic Liquids

Ionic liquids (ILs) are low-melting-point salts that can exist as liquids at temperatures under 100 °C. They are (usually) entirely composed of ions, usually a large, asymmetrical organic anion and a small inorganic or organic cation, which accounts for their low melting points and interesting physico-chemical properties [1].

Ionic liquids have been of great interest because of their special characteristics, such as negligible vapor pressure, high thermal stability, modifiable solubility, and outstanding ionic conductivity [2]. These properties render them attractive as greener replacements for traditional organic solvents in a number of applications such as green chemistry, catalysis, electrochemistry, and separation studies [3]. The environmental advantages of reduced volatility and potential recyclability also make them attractive in green industrial scenarios.

1.2 Nanofluids

Nanoparticles can be described as the colloidal suspension of nanoparticles (typically <100 nm) in a base fluid (water, ethylene glycol, or oils) called a nanofluids [4]. Nanomaterials made up of metals, oxides, carbides or carbon nanotubes (CNT)

P. Kapadnis (✉)
School of Pharmacy, Vishwakarma University, Pune, Maharashtra, India
e-mail: prajakta.kapadnis@vupune.ac.in

S. Kulkarni, S. Das (eds.), *Ionic Liquids, Nanofluids and Nanotechnology*,
Lecture Notes in Nanoscale Science and Technology 42,
https://doi.org/10.1007/978-3-032-04008-4_19

in nature, applicable for the cooling systems, in electronics, for the automotive industries, for nuclear reactors, nanofluids exhibit much higher thermal conductivity and heat transfer performance than the conventional fluids. They have a conspicuous potential role to play in new technology and renewable energy systems including thermal systems, with regard to improving the energy efficiency of heat transfer processes [5].

1.3 Nanosystems

Nanomaterials are materials in which at least one dimension of the said material is within the nanometer range (1–100 nm). They are made up of metal, ceramic, polymers, and composites and are composed of nanoparticles, nanotubes, nanowires, and quantum dots [6].

As a result of such quantum effects and their large surface area-to-volume ratios, nanomaterials display unique mechanical, electrical, optical, and catalytic properties that are distinctly different from bulk materials. They are used by electronics, medicine (diagnostics, drug delivery), environmental remediation, and energy storage [7] among other industries. Their dynamic power can be used as a driving force for the development of next-generation materials and advances in nanotechnology.

Rising Concerns About Toxicity and Environmental Impact of Ionic Liquids, Nanofluids, and Nanomaterials

Examples of such groundbreaking advances in chemical engineering and nanotechnology are ionic liquids, nanofluids, and nanomaterials with excellent properties such as enhanced conductivity, thermal stability, and novel capabilities. However, as their uses in many fields have been expanded, concerns about their toxicological effects as well as their environmental fate and bioaccumulation potential have intensified. They are pioneer, but they are not without risk. A reasonable discussion of therapeutic benefits and risks is warranted as these treatments come under greater scrutiny regarding their tolerability and potential to do harm to the ecosystem and human health.

Ionic Liquids

Toxicity in Water

Despite the fact that these were denoted as "green solvents" because of their low vapor pressure, most of the ionic liquids (ILs) were found to be highly toxic to the aquatic organisms like Vibrio fischeri and Daphnia magna [8]. Their degree of toxicity is largely determined by their structural and/or compositional factors (viz., alkyl chain length).

Bioaccumulation and Persistence

A few ILs are also resistant to biodegradation, and they can be persistent in the environment. Typical ILs, including imidazolium-based salts, have been found to bioaccumulate in organisms and sediments, with the potential to pose long-term ecological problems [9].

Cellular Cytotoxicity

Accidental exposure and occupational safety concerns arise from the fact that ILs can also be cytotoxic toward human cell lines. According to literature, ILs can be able to disrupt metabolic pathways and cell membranes [10].

Nanofluids

Release into the Environment

Nanoparticles in nanofluids may be released into the environment when synthesized, applied, or undeployed. Interactions of these particles in the water and soil environments may lead to unknown ecological consequences [11].

Skin and Respiratory Risks

Inhaling and dermal contact with nanofluid aerosols can result in health problems. Reactive oxygen species and pulmonary inflammation have been linked with various nanofluids, such as titanium dioxide and other metal oxide nanoparticles [12].

Absence of Rules for Disposal

The risk of accidental environmental contamination is also high as there are no specific regulations on the safe end-of-life management of nanofluids or draining to the environment [13].

Nanomaterials

The Area of Nanotoxicity Research

Due to their small size and high reactivity, nanomaterials will likely induce toxicity at the cellular, genetic, and neurological levels, and they are able to cross biological barriers [14]. The risk of lung disease from exposure to carbon nanotubes, for example, has been likened to that of asbestos.

Ecotoxicological Impact

Nanomaterials can change the physicochemical properties of terrestrial and aquatic ecosystems. For example, silver nanoparticles kill friendly soil microorganisms and insects with their toxic activity [15].

Ethical Issues on Regulation

Balanced transmission and public skepticism are consequences of the present lack of international consensus on regulating engineered nanomaterials. Besides, ethical questions regarding long-term exposure and health impacts on generations to come can also arise [16].

2 Toxicity of Ionic Liquids

2.1 Physicochemical Properties and Chemical Structure

Large organic cations and a range of anions constitute ionic liquids (ILs), which are responsible for their low melting points and unique solvent properties.

Their structural components namely, the alkyl chain length, the type of cation, and the anion—play an important role in determining their toxicity.

Key structural features and toxicological effects of some of the commonly used ionic liquids are mentioned in Table 1.

The bioaccumulation potential of cationic liquids is strongly influenced by lipophilicity. The chemical stability of the ionic liquids determines their persistence in the environment.

Table 1 Toxicity of ionic liquids

Structural component	Toxicological effect	Example
Imidazolium cations with long alkyl chains	Enhances membrane rupture and hydrophobicity	[C2mim][Cl] More toxic than [C8mim][Cl] [17]
Aromatic cations (such as imidazolium)	They intercalate into membranes and cause oxidative stress	More toxic than non-aromatic analogues [18]
Fluorinated anions (PF_6^-, BF_4^-, etc.)	Release Hydrogen Fluoride (HF) and fluorinated acids through hydrolyzing	Ecotoxicity and environmental persistence [19]

Mechanism of Toxicity in Biological and Environmental Systems

Disruption of Cellular Membranes

Since imidazolium-based ILs are amphiphilic, they can intercalate between membrane lipids and disrupt the bilayer structure. Zhao et al. reported that the presence of the long alkylside group increased fluidity and permeability when it was embedded within the membrane.

Mitochondrial Dysfunction and ROS Generation

Reactive oxygen species (ROS) are generated during exposure to the ILs, which injure the cells. This study has been proven on human and microbial cell lines [20]. The exposure may also lead to mitochondrial dysfunction in some instances.

Inhibition of Enzymes

ILs may act as neurotoxic in some cases by inhibiting enzymes like acetylcholinesterases. This mode of inhibition presents problems for regulatory issues since it is the same mode through which organophosphates operate [21].

Aquatic Systems Ecotoxicity

ILs have exhibited acute and chronic toxicity to fish, algae, and crustaceans, among other aquatic organisms. Their effect on the environment is further aggravated by their low biodegradability [22].

Regulatory Views and Case Studies

Aquatic Toxicity of [C8mim][Cl]: In comparing the toxicity of imidazolium ILs of different alkyl chain lengths toward Daphnia magna, García-Lorenzo et al. found mortality to increase exponentially with increasing chain length. For 180 mg/L ([C4mim][Cl]) to 30 mg/L ([C8mim][Cl]), LC_{50} values decreased.

Human Hepatocyte Cytotoxicity: HepG2 cells treated with 1-butyl-3-methylimidazolium tetrafluoroborate exhibited increased oxidative stress and cell death, suggesting mitochondrial involvement in cytotoxicity [20].

Inhibition of Soil Microbes: ILs have been reported by Stepnowski et al. to cause such a dramatic decrease in soil microbial activity that nitrogen cycling could be impaired. This poses concern for their runoff and disposal in industrial or agricultural settings.

Regulatory Overview

Ionic liquids are yet to be governed by an overarching international regulatory body. However, because of their persistence, bioaccumulation, and toxicity (PBT) nature, the European Chemicals Agency (ECHA) has taken certain ILs up for assessment under REACH [23, 24]. Moreover, the Green Chemistry Institute is in fact promoting research for safer toxicological and biodegradable IL alternatives [25].

Environmental Persistence and Biodegradation Fate of Ionic Liquids

The low biodegradability of ionic liquids (ILs) and their ability to persist in aquatic and terrestrial environments have generated growing interest in their persistence in natural systems. Even though ILs are often promoted as green solvents, their environmental fate is generally controlled by structural features, such as the chain length of the alkyl and the cation (e.g., imidazolium or pyridinium) and anion (e.g., PF_6^- or BF_4^-) [26]. Most common ILs, especially those with fluorinated anions, are poorly biodegradable and persist in soil and water environments, where they can negatively affect trophic levels and microbial communities.

Based on microbial degradation studies, ILs with functional groups such as hydroxyl or ether moieties and shorter alkyl chains are more susceptible to enzymatic cleavage and biodegradation. In addition, the toxicity of IL biotransformation products can vary from or even exceed that of the parent substance. Incomplete degradation, for example, can lead to intermediate metabolites that interfere with microbial respiration or aquatic photosynthesis [27]. These findings highlight that in order to fully understand the entire ecological footprint of ILs beyond their direct chemical footprint, lifecycle-based environmental analysis is required.

3 Toxicity of Nanofluids

Wide use of nanofluids in medical diagnostics, nuclear reactors, and renewable energy systems increases growing concern regarding the toxicological and environmental profile of nanofluids, mostly when the nanofluids are mishandled, released into the environment, or disposed of without sufficient regulatory concern.

3.1 Compositions of Nanofluids and Their Application Areas

Nanofluids may consist of the following:

Nanoparticles—Examples: Carbon-based compounds like graphene or CNTs, metal bases like Ag, Cu, metal oxides like ZnO, TiO_2, Al_2O_3, Base fluids like

Table 2 Toxicity of nanofluids

Nanofluid type	Primary use	Toxicity concern
ZnO in ethylene glycol	Solar collectors	Cytotoxic to lung cells and skin [28]
Al_2O_3 in water	Electronics cooling	Pulmonary inflammation [29]
Ag nanoparticle in saline	Antimicrobial coating and textiles	Cytotoxicity and microbial resistance [30]
TiO_2 in water	Biomedical imaging studies	ROS generation and DNA damage [31]

water, oil, glycols, and last but not least surfactants or dispersants to stabilize the suspension. Table 2 depicts commonly used nanofluids in different areas along with their toxicity issues in specific organs/systems.

Use of nanofluids for these applications starts releasing into the environment through manufacturing, usage, and disposal making it a serious concern for environmental and occupational safety.

3.2 *Relationships with Biological Systems and Potential Health Risks*

The combination of small particle size and extensive surface area enables nanoparticles to breach biological barriers. When people consume or breathe or their skin contact with these particles, it causes:

Nanofluids are biologically reactive since they can cross biological membrane barriers due to their nanoparticulates. These reactions with human and animal cells have raised many toxicological concerns.

Some mechanisms of toxicity cellular uptake and mitochondrial dysfunction: Nanoparticles in nanofluids can be taken up into cells through endocytosis and accumulated within intracellular organelles, thus disrupting mitochondrial function and triggering apoptotic pathways [32].

Immunotoxicity and inflammation: Chronic exposure may elicit inflammatory responses that interfere with immune, cutaneous, or pulmonary functions [14].

Genotoxicity: By way of oxidative stress, a few nanoparticles cause DNA fragmentation and chromosomal aberrations [33]. Table 3 describes the effect of various nanoparticles and their effects on the biological systems at the cellular level.

3.3 *Assessment Methods and Mitigation Strategies*

Assessment methods for toxicity can be categorized into three types as follows:

Table 3 Effect of nanofluids on biological systems

Nanoparticles	Nanofluid matrix	Biological system	Effect observed	References
Fe_3O_4	Water	Human lung epithelial cell (A549)	Induced oxidative stress and mitochondrial dysfunction	[34]
CeO_2 (cerium oxide)	Ethylene glycol	Zebrafish embryos (Danio rerio)	Pericardial edema, hatching delay	[35]
Carbon nanotubes (MWCNTs)	Oil-based coolant	Mouse	Pulmonary tissue granuloma formation, inflammation	[36]
Copper nanoparticles (CuNPs)	Water	Human liver HepG2 cells	DNA damage, enhanced ROS generation	[37]

- In vitro experiments using liver, kidney, and lung epithelial cells.
- In vivo animal testing to determine histopathology and organ accumulation.
- Environmental fate studies employing aquatic models (e.g., Danio rerio, Daphnia).

Mechanism of Interaction

Reactive Oxygen Species (ROS) generation: Nanoparticles like copper (CuNPs) and iron oxide (FeO_4) can produce ROS inside the cell. The ROS can damage DNA by causing breaks in the DNA, oxidization of proteins, and lipid peroxidation, which damages the cell organelles and may lead to cell death [37]. Magnetic fluids containing magnetite may also disrupt normal mitochondrial function, alter cell membrane potential, which may lead to cell death, which is also known as apoptosis [34].

When it comes to development, cerium oxide nanoparticles have been found to gather in the yolk sacs of zebrafish embryos, causing developmental delays and abnormalities [35]. On the other hand, multi-walled carbon nanotubes (MWCNTs) can build up in lung tissue because their shape resembles asbestos fibers. This can trigger fibrotic responses, which means scar tissue forms in the lungs, potentially leading to lung problems [36].

Mitigation Strategies

- Nanoparticles may be surface-coated with silica or Polyethylene glycol (PEG), for instance, to reduce toxicity and reactivity.
- Use of cellulose nanocrystals and other biodegradable nanomaterials.

Table 4 Mitigation strategies for reducing nanoparticle toxicity issues

Mitigation strategy	Effect	References
PEG coating on silver nanoparticles	Reduced cytotoxicity in human cells	[38]
Green synthesis of ZnO nanoparticles	Improved biocompatibility	[39]
Use of High Efficiency Particulate Air (HEPA) filters in manufacturing	Reduction in airborne particle exposure	[40]

- Green synthesis and lifecycle assessment are employed for evaluating long-term impacts.
- Principles on handling and workplace personal protection equipment (PPE). Table 4 describes different mitigation strategies to reduce the extent of nanoparticle toxicity.

3.4 *Effect of Nanofluid Agglomeration and Stability on Toxicity*

Colloidal stability and agglomeration tendency in biological and environmental systems influence nanofluid toxicity alongside nanoparticle composition. Agglomerated particles that often sediment, coalesce, and induce mechanical or concentration-dependent toxicities possess varied interaction patterns with cells compared to stable nanofluids with uniformly dispersed particles [41].

Agglomeration, especially around mucosal and epithelial tissues, raises the effective size of the particles, reducing cellular uptake but enhancing surface reactivity. Under physiological conditions, for example, TiO_2 nanofluids aggregate into micron clusters that, upon inhalation, induce localized inflammation within lung tissue [42].

In addition, the base fluid and surfactant choice drastically affect stability. Despite ethylene glycol-based nanofluids often showing greater stability, they also create fresh concerns regarding chemical toxicity. On the contrary, water-based nanofluids are highly biocompatible but require stringent dispersion procedures. So, evaluation of nanotoxicity needs a real-time determination of zeta potential, rate of sedimentation, and hydrodynamic diameters to assess real-time exposure hazards.

4 Toxicity of Nanomaterials and Nanotechnology

Nanomaterials and nanotechnology offer revolutionary changes in medicine, electronics, and energy. The nanoscale features make several qualities undesirable. The foremost among these are toxicity, bioavailability, and high surface reactivity. They can also injure cells, induce inflammation, or even cause ecological damage due to their unanticipated interaction within biological systems, thus subjecting them to strict safety evaluation and regulatory control [14].

4.1 Classification of Nanomaterials

Nanomaterials research classifies them into morphologies, dimensionalities, and chemistry, all of which define their physicochemical behavior and interaction with biological systems. A more advanced understanding would thus make it easier to predict the likely toxicological behavior.

By Chemical Composition

1. Metal nanoparticles include Ag, Au, and Pt, which find application in antimicrobial coatings and diagnostics. Their toxicity and effect are dependent on their ion-releasing capabilities.
2. Metal oxide nanoparticles: These nanoparticles such as cerium oxide (CeO_2), zinc oxide (ZnO), and titanium dioxide (TiO_2) can perform photocatalytic activities but may also instigate the formation of reactive oxygen species (ROS) within biological systems [33].
3. Carbon nanomaterials include graphene, fullerene, and carbon nanotubes (CNTs). These nanomaterials of high aspect ratios are usable for a host of applications, including drug delivery and electronics, depending on surface functionalization.
4. Polymeric nanoparticles such as poly lactic-co-glycolic acid (PLGA) and PEGylated systems are biodegradable. The logic goes that their breakdown products are less harmful; however, in some instances, these breakdown products can cause an immunological response [43].
5. Hybrid nanomaterials: Being core-shell nanoparticle blends, $FeO_4@SiO_2$ is meant to either enhance functionality or stability, but on the downside due to so many interfaces, they have very complex risk profiles [32].

Classification on the Basis of Nanomaterial Dimensions

1. 0D is the spherical nanoparticles of nanoscale size to 100 nm in any given dimension.
2. 1D are the formulations having a very high aspect ratio, which are believed to permeate the cell membrane e.g., nanorods and nanotubes.
3. 2D formulations can create protein coronas due to their high surface areas, e.g., graphene and nanosheets.

4. 3D formulations have branching or hollow shapes, which can encapsulate drugs but are also capable of bioaccumulation in tissues. e.g., dendrimers and nanocages.

4.2 Biocompatibility and Bioaccumulation Concern

When considering safety for nanomaterials for biomedical or environmental purposes, bioaccumulation and biocompatibility are significant concerns. Bioaccumulation is the accumulation and retention of nanoparticles in tissues or in an ecological system, while biocompatibility is the ability of a material to function in a biological system without eliciting deleterious effects.

Factors Affecting Biocompatibility

Size and Shape of Particles: Particles having a size smaller than 50 nm would easily pass through the cell membrane efficiently, increasing their uptake and hence cytotoxicity [44].

Surface Charge: Nanoparticles with a positive charge can cause greater cytotoxicity because of increased interactions with the negatively charged cell membrane.

Surface Functionalization: Surface coating of nanomaterials with PEG-like polymers can increase circulation by minimizing immune system recognition, but the surface coating degradation may lead to toxic side effects. Table 5 explains nanomaterial toxicity in various organisms and their sites of accumulation.

These cases highlight the complexity of designed nanomaterials. Even biodegradable materials may persist under certain biological conditions or interfere with normal cellular metabolism.

Table 5 Examples of bioaccumulation of nanomaterial residues in living organs/cells

Nanomaterials	Model organism	Site of accumulation	Toxicity	References
TiO_2 nanoparticles	Mice	Liver and kidney	Enzyme elevation and histopathological changes	[45]
CdSe/ZnS quantum dots	Zebrafish embryos	Yolk sac, gill tissues	Developmental delay, oxidative stress	[46]
Polymeric PLGA nanoparticles	Rats	Brain and spleen	Mild inflammation, persistent retention	[47]
Graphene oxide	In vitro human lung cells	Cellular membrane binding	ROS generation, cellular integrity disruption	[48]

4.3 Long-Term Health Effects and Environmental Persistence

There are important considerations about the long-term effects of being exposed to nanomaterials, whether through work, medical treatments, or environmental release. These include disrupting hormone functions, building up toxicity over time, possibly causing cancer, and leaving persistent residues in the environment.

Long-Term Health Effects

Research shows that some polymer-based nanoparticles can interfere with hormone balance by affecting estrogen receptors [49]. Neurotoxicity: Some metal-based nanoparticles, like manganese oxide and aluminum oxide, can pass through the blood-brain barrier. This allows them to build up in the nervous system and potentially affect brain function [50]. Lung Damage: Inhalation of fibrous nanomaterials, such as nanowires and carbon nanotubes, over long periods has been linked to lung scarring and impacts similar to mesothelioma.

Environmental Sustainability

Nanomaterials often resist degradation within wastewater treatment systems and persist in aquatic environments, where they can interact with various pollutants and transfer through food webs. Their longevity depends on factors such as their tendency to aggregate and their solubility profiles, their capacity to bind with humic substances, and their stability within different environmental redox conditions. A study conducted by Mitrano et al. revealed that silver nanoparticles present in sewage sludge remained stable for several months before eventually leaching into groundwater and agricultural soils [51].

Environmental Impact

In a relevant environmental simulation, Wang et al. observed that titanium dioxide (TiO_2) nanoparticles present in sunscreens tend to accumulate in coastal sediments. These particles were shown to cause oxidative stress and impair reproductive functions in the benthic invertebrate *Lumbriculus variegatus*. Such findings raise important concerns about the potential long-term ecological effects of consumer nanoproducts [52].

5 Comparative Analysis of Toxicity Assessment Methods

To maintain health and environmental equilibrium, an exact study of nanomaterial toxicity should be undertaken. Varied physicochemical properties of nanomaterials, including size, shape, surface charge, and reactivity, constitute peculiar challenges to characterize toxicity potential. Since testing methods for chemicals are not always suitable or exist, several testing platforms have been developed. This section compares different methodologies—in vitro, in vivo, in silico, and advanced ones—for assessing the safety profile of nanomaterials.

5.1 In Vitro Toxicity Assessment Methods

When single cells or tissues are separated from organisms for screening outside the organisms, it is in vitro. These methods are considered better for several reasons: they are cheaper, more reproducible, allow multiple chemicals to be screened in a short time, and have ethical implications.

Exposure Model and Cell Lines

Various animal and human cellular lines like liver cells (HepG2), lung (A549), kidney (HK-2), and immune cells (THP-1 macrophages) can mimic target organs. These cellular models are helpful to ensure the organ-specific nanomaterial toxicity pathways [53].

Key Assays

1. Cytotoxicity assays like MTT, LDH release assays can be used for cytotoxicity testing, monitoring cell membrane integrity, and altering cellular metabolism.
2. Oxidative stress assays can detect intracellular ROS, which generally increases in the presence of nanoparticles [54].
3. Genotoxicity assays like the comet assay and micronucleus tests can determine the extent of DNA damage induced by nanomaterials.
4. Necrosis/apoptosis assay can study cell death mechanism by caspase activation or flow cytometry.

Limitations of In Vitro Assessments

Even though in vitro models are quite useful, they often fall short when it comes to truly mimicking how the body works. Plus, nanoparticles can sometimes interact with test reagents in ways that lead to false or misleading results [55]. On top of that, when nanoparticles are in a biological environment, they tend to develop a layer of proteins, called a protein corona, which can change how they behave in lab tests.

In vitro models are highly useful, but they often fall short in exactly mimicking how the interaction will happen in the actual body. In addition, nanoparticles can also interact with test reagents, which may lead to false or misleading results [55]. When nanoparticles are in a biological environment, they tend to develop a protein coating known as a protein corona, which may alter their behavior in lab tests.

5.2 *In Vivo Toxicity Assessment Methods*

The best overall insight into nanomaterial toxicity is gained through in vivo methods, which utilize whole animal systems. In vivo models contribute to the identification of biodistribution, metabolism, clearance profiles, and organ-level, long-term effects.

Models of Animals

For high-throughput testing with ethical adaptability, popular animal models are rodents (mice and rats) and alternative models such as zebrafish embryos (Danio rerio) [56].

Endpoint Analysis

1. Loss of weight, organ damage, and rate of survival are markers of acute and chronic toxicity.
2. Bioaccumulation: Metal nanoparticles can be traced by ICP-MS or histopathology studies.
3. Immunotoxicity: Immune cell infiltration and release of cytokines in tissues can be measured to study the toxicity profile of nanomaterials.
4. Nervous System and Behavior Evaluation: Nervous system behavioral alterations when interacting with silver nanoparticles or carbon nanotubes.

Advantages and Limitations

In vivo studies provide systemic analysis and precise risk assessment for human health. However, ethical considerations, high cost, and variation in species complicate translation to humans [57].

5.3 *In Silico Toxicity Assessment Methods*

In silico methods rely on computational simulations and modeling for the prediction of nanomaterial toxicity. These methods are getting increasing importance due to the fact that they reduce and replace the use of animal models.

Types of In Silico Methods

1. Quantitative structure–Activity Relationship models (QSAR) are in silico models that aim to correlate the physical-chemical properties of nanomaterials to biological activity.
2. Molecular docking and dynamics: Biological macromolecule-toxicant interaction modeling.
3. Machine learning models: Algorithms are trained by available datasets to predict the toxic effects [58].

Data repositories eNanoMapper: Summarizes data about nanomaterials for QSAR simulation.

NanoCommons is a European initiative that promotes the standardization of nanotoxicological databases.

Advantages and Limitations

Being inexpensive and quick, an in silico approach is great for early screening and regulatory assessment. Yet, such methods do have limitations stemming from a needed model validation and, most importantly, a lack of data on nanomaterial variations [59]. Table 6 compares different in vivo, in vitro, and in silico studies for toxicity assessment studies.

Table 6 Summary of in vitro, in vivo, and in silico methods

Parameter	In vivo studies	In vitro studies	In silico studies
Biological relevance of the method	Moderate	High	Low-moderate (depends on model)
Ethical considerations	Low	High	None
Time for analysis	Short analysis time	Long analysis time	Very short analysis time
Cost	Low cost	Expensive	Low cost
Systemic effects	Absent	Present	Predictive study only
Mechanistic insights	High for the cellular level	Moderate to high	High if based on accurate models

An experiment conducted to explain pulmonary toxicity by Wang et al. combined an in vitro assay with in silico modeling to study pulmonary cell toxicity of cerium oxide nanoparticles [60]. The in vitro studies on ROS generation and cytokine expression were the markers used to establish the QSAR model, which accurately predicted the inflammatory response in the mouse lung model, indicating the importance of these testing models.

The increasing use of nanomaterials across industries like medicine, cosmetics, and environmental sciences asks for robust studies of their toxicological impacts. Toxicological studies like in vitro, in vivo and in silico studies act as a basis for the evaluation of environmental and biological risks posed by these materials. An integrated strategy combining all three of these methods is increasingly advocated for toxicological studies and risk management. Future prospective shall focus on protocol standardization, data interoperability, and promoting the use of a predictive modeling framework. The multidisciplinary approach will be important for the development of safe and effective applications of nanomaterials, confirming that the benefits of the materials outweigh the environmental risk and health harms.

5.4 Computational Models and Predictive Toxicology

Because of their fast-growing applications, nanomaterials require rapid toxicology predictions that are cost-effective and ethical. Although they can be informative, methods such as animal testing are increasingly costly, time-consuming, and banned by global legislation. Through the integration of data science, cheminformatics, bioinformatics, and systems biology, computational toxicology brings a paradigm shift by modeling and predicting adverse effects from chemical and nanomaterial exposure [61].

Due to the extreme diversity in nanoparticle composition, size, and surface modification, which renders fully conventional testing impossible, prediction plays a central role in nanotoxicology. Computational modeling for precise risk estimates of nanomaterials properly accommodates this diversity.

Types of Computational Models in Toxicology

Quantitative Structure–Activity Relationship (QSAR) Model

Predictive toxicology is highly dependent on QSAR models. QSAR models connect biological effects with quantifiable characteristics of chemical or nanomaterial structures, including hydrophobicity, electronic distribution, or surface charge [62].

For instance, Gajewicz et al. utilized a QSAR model to predict that metal oxide nanoparticles such as ZnO, TiO_2, and CuO would be toxic to human keratinocytes. The model was more than 85% correct, based on factors such as particle size and band gap energy. Such models minimize unnecessary exposure risks by enabling new nanoparticles to be screened prior to being manufactured [63].

Read-Across Methods

When experimental data are unavailable, read-across methods are used. They assume that unstudied nanoparticles behave in the same way as ones that have been well studied with regard to physical and biological characteristics so that information can be extrapolated from known examples.

Read-Across Risk Assessment

Read-across is defined as the prediction of properties for a substance by comparing it to substances that are structurally or mechanistically alike and for which data is available. In nanotoxicology, read-across is the increasingly employed methodology to fill data gaps [64]. By utilizing data from ZnO and SiO_2, Hristozov et al. used read-across to assess the pulmonary toxicity of TiO_2 nanoparticles. In doing so, it kept safe risk assessment parameters while sharply curtailing the requirement for further animal testing [65].

Physiologically Based Pharmacokinetics (PBPK) Models

PBPK models mimic the way nanomaterials are absorbed, distributed, metabolized, and excreted by the body. They are made up of compartments that mimic major tissues, regulated by parameters like blood flow, organ volume, and the nanoparticles' affinity.

Bachler et al. developed a PBPK model in rats for oral exposure to silver nanoparticles. The model was verified subsequently by in vivo data and predicted tissue-specific organ uptake, like spleen and liver [66].

AI-Based Techniques and Machine Learning

AI and machine learning transform model construction since dangerous datasets multiply. They can scan big, intricate, and nonlinear data and identify hidden relationships.

For instance, Zhao et al. utilized a random forest model to estimate oxidative stress potential for more than 500 nanomaterials. The two most important factors were surface area and elemental composition. The model was incorporated into a cloud-based system for regulatory scientists to utilize [67].

Omics and Systems Biology Integration

Systems toxicology integrates genomic, proteomic, and metabolomic information to model the way cells react to nanoparticle exposure. It can predict long-term consequences such as endocrine disruption or cancer formation.

Table 7 Current limitations and future directions in nanomaterial toxicology studies

Challenge	Effect	Suggested solution
Nanomaterial heterogeneity	Variable physical and chemical properties	Creating standardized descriptors like nanoinformatics
Imbalanced and less information	Model training is limited for rare endpoints	Synthetic data generation using AI
Black box nature of the AI model	Difficult to interpret	Focusing on explainable AI models
Regulatory gaps	Limited acceptance among regulatory agencies	Collaboration via OECD, EU nanosafety cluster

Example: A biological model mimicking the liver toxicity model for carbon nanotubes by utilizing gene expression profile shows disturbances in lipid metabolism and inflammatory pathways [68].

5.5 Challenges and Opportunities

There are various challenges in nanomaterial toxicity studies, like regulatory gaps, nanomaterial heterogeneity, etc. Table 7 suggests suitable solutions for the challenges.

6 Regulatory Framework and Risk Management

Nanomaterials and ionic liquids have emerged faster than a conventional regulatory framework, which requires rules particularly to address public health, occupational safety, and environmental security. The materials though have various industrial and technological advantages; they also have toxicity issues, bioaccumulation potential in living tissues, and persistence as distinctive challenges. Safety standards, global regulations, and sustainable policies must be collaborated to manage the risk of nanomaterial toxicity effectively.

6.1 International Regulations of Ionic Liquids and Nanomaterials

Unlike conventional organic solvents, ionic liquids (ILs) are usually not considered in current solvent regulatory guidelines due to their non-volatile nature. Still, issues related to their toxicity in water sources, bioaccumulation, and lesser biodegradation have triggered some countries to revise their safety regulations. Although said to be green solvents, they may pose long-term ecological effects and health hazards [9].

Certain ionic liquids are now required to submit toxicological data under the REACH regulation (Registration, Evaluation, Authorization, and Restriction of Chemicals), especially for compounds produced in large quantities. A pyridinium-based ionic liquid used in catalysis was found to bioaccumulate in aquatic life in rivers across the European Union. As a result, REACH has enforced stricter, substance-specific risk assessments and reporting [69].

Nanoparticles due to their unique behavior at the nanoscale, nanomaterials receive closer scrutiny. The OECD's Working Party on Manufactured Nanomaterials (WPMN) has identified 14 representative nanomaterials for safety testing and supports standardized international testing methods. The U.S. FDA regulates nanomaterials under the Federal Food, Drug, and Cosmetic Act and other existing laws. Since 2014, the FDA has issued guidance documents specifically focused on nanotechnology. India's Nano Mission encourages nanotechnology research and development considering safety through the Institutional Biosafety Committee under the initiative of the Department of Science and Technology. Table 8 discusses the regulatory authorities of various countries and their policies for assuring the safety of nanomaterials.

Table 8 Various regulatory authorities and policies for the safety of nanomaterials

Region/country	Regulatory authority	Policy	Notes
Europe	European Chemical Agency (ECHA)	REACH (Registration, Evaluation, Authorization, and Restriction of 502 Chemicals)	Includes physicochemical and toxicology data
USA	FDA	Federal Food, Drug and Cosmetics Act (FFDCA), Toxic Substances Control Act (TSCA)	Case review of nanoformulations
Japan	New Energy and Industrial Technology Development Organization (NEDO)	Guidelines for risk assessment	Focuses on workplace exposure
India	Department of Science and Technology (DST)	Nano mission, biosafety committee	Mostly focuses on policy
Australia	The National Industrial Chemicals Notification and Assessment Scheme (NICNAS) now replaced by Australian Industrial Chemicals Introduction Scheme (AICIS)	For new chemicals, it gives nano-notification	Uses OECD guidelines

6.2 Sustainable Development Strategies and Safety Guidelines

Ionic Liquids: Green Design and Safety

There is increased emphasis on “Designing Safer Ionic Liquids,” which seeks to develop compounds that are biodegradable, nontoxic, and do not accumulate in the environment. This method attempts to strike a balance between efficacy and safety. For example, researchers have developed cholinium-based ionic liquids with functional groups that help microbes to degrade them. One such ionic liquid was found to be more than 70% biodegradable in 28 days [22]. The Safer Choice Program of the U.S. EPA is also testing some ionic liquids in its green chemistry programs.

Sustainable and Safe-by-Design (SbD) Nanomaterials

Safety is factored in early in nanomaterial design using the Safe-by-Design (SbD) approach. USA-EU Collaboration on the NanoRelease Project: The project created test procedures for engineered nanomaterials applied to food packaging. It also investigated the possible leaching and health effects of nanosilver and titanium dioxide. ZnO and TiO_2 nanoparticles are widely used in sunscreens as they are effective UV blockers. Yet, owing to environmental issues such as coral reef destruction, Hawaii and Australia have imposed limitations on the application of nanoformulations in sunscreens [70].

Institutional and Corporate Guidelines

- The ISO/TR 13121:2011 offers a guide to the assessment of the risks of nanomaterials.
- NIOSH issues occupational exposure limits, for example, 1 $\mu g/m^3$ for nanofibers and carbon nanotubes.
- The DuPont–Environmental Defense Fund Framework provides industrial risk assessment techniques that integrate hazard identification, exposure estimation, and life cycle analysis.

6.3 Global Risk Management Strategies

1. In order to facilitate global conformity and trade, every country should implement the OECD’s nano-specific guidelines and have harmonized testing procedures.
2. Cosmetic and pharmaceutical industries should mandatorily submit data related to nanoformulations to the regulatory authorities.

3. Public-private collaborations like NanoReg2 can generate safety data through the experiments conducted in collaborations with industry, government, and academia in less time.
4. Predictive model-based data sharing and decision-making shall be facilitated through the digital databases and tools like eNanoMapper and NanoCommons.
5. European Chemical Agency (ECHA) recommends using futuristic outlooks like the QSAR toolbox for nanomaterials to maintain procedure validity and transparency. The US-EPA's ToxCast and Tox21 programs use high-throughput screening and modeling to predict chemical safety. The EU NanoReg2 project uses modeling tools in the REACH guidelines for nanomaterials.
6. Connecting with big public database platforms such as eNanoMapper, NanoCommons, and caNanoLab will improve the reliability of experimental models. Online models in open science and cloud computing, like Jaqpot, CompTox, and Knime, enable real-time toxicology studies.

7 Future Perspectives

7.1 Recommendations for the Use of Advanced Materials and Safer Design Parameters

Several scientific and policy recommendations are needed for safe integration of these new materials into commercial, clinical, and environmental disciplines:

Embedding Safe-by-Design (SbD) Principles into Practice

Since nanomaterials and ionic fluids will always be a subject of design, they must follow the Safe-by-Design idea, which amalgamates issues of low toxicity, environmental sustainability, and functional performance at a molecular level. For example, in this regard, anions should be chosen with biodegradability in mind, and ILs with shorter alkyl chains should be preferred over persistent ones [71].

Improving AI and Predictive Modeling Integration

In the early stages of development pipelines, state-of-the-art computational toxicology methods, such as QSARs, PBPK models, and AI-based simulations, should be utilized to forecast cytotoxicity, accumulation in organs, and persistence so as to minimize animal tests [72]. A unified global test protocol for nanotoxicology is needed. OECD Test Guidelines for Nanomaterials need to be made commonly accepted for comparative data collection between industries and countries.

Develop Regulatory Databases for Nanotechnology

Extension of real-time open-access databases such as eNanoMapper and NanoSafety Cluster, inculcating safety profiles of new formulations, openness, and compliance by the industry can help include safety profiles of new formulations, openness, and industry compliance. This would be preferable.

Foster Lifecycle Risk Evaluations

It is vital to look at the entire life cycle of a material, from its production through its use and disposal (with regard to environmental accumulation potential). A thorough environmental impact assessment can be obtained by using Life Cycle Assessment (LCA) methodologies combined with toxicity prediction techniques [73].

Improve Government-Private Sector Relationship

Industry, academia, and regulatory agency collaborations will speed the development of new "safe" materials, employee training, information sharing, etc.

7.2 Need for Data Sharing and Global Collaborations

Innovative materials such as ionic liquids and nanomaterials must be used sustainably due to their sophistication and developmental rate. Global and concrete solid efforts shall be made in the development of nanomaterials. Research data duplication, data silos, and variability in safety standards are the major issues induced due to differences in the regulatory environments, safety, and toxicity data and safety protocols in different countries [74]. Global collaboration and open-access sharing of data are essential to bridging these gaps. Some of the main examples of initiatives promoting standardization in testing and terminology are the EU NanoSafety Cluster and the OECD WPMN. In addition, integration of the physicochemical and toxicological data from many different sources has been facilitated through international platforms including eNanoMapper, NanoCommons, and the US-EPA CompTox Dashboard [75].

Limited database interoperability, varying reporting quality, and non-standard metadata are a few of the primary barriers. Transparency would be enhanced and Safe-by-Design innovations would be fostered by developing minimum information standards (e.g., MIAN—Minimum Information about Nanomaterials) and encouraging industry and academic data submission. In order to harmonize safety evaluations, facilitate reproducibility, and fill the gap of knowledge between innovation and regulation, research funding cooperation, trans-boundary policy negotiations, and public-private collaborations are essential.

8 Conclusion

Actually, the practical risk assessment of human exposure in toxicological studies is rarely given, so there is an urgent need to fix these drugs, although their patent has been extended. The tunable physicochemical properties, thermal stability, and nanofunctional aspects make these materials a great fulfillment in catalysis, energy, drug delivery, electronics, and environmental remediation. High surface areas, ionic mobility, and peculiar interfacial behaviors have been the source of possible interactions with biological and environmental systems, and their existence is unpredictable.

Ionic liquids do impart toxicity related to their counterions, type of cation, and length of the alkyl chain, notwithstanding their claims as green substitutes. Some ILs resist degradation and thus raise concerns about bioaccumulation and aquatic toxicity [76]. By way of aggregation of nanoparticles, inhalation hazards, and cytotoxicity of nanofluids, these nanofluids pose safety threats, most especially those with metal-based nanostructures such as CuO or AlO_3 in heat-exchange systems [77]. On top of that, in their ability to go through tissue and cellular barriers, nanomaterials—especially carbon-based and metallic nanoparticles—have also been implicated in chronic inflammation, mitochondrial dysfunction, and oxidative stress [78].

At present, regulatory regimes remain fragmented, disjointed, and far from consistent at the international level, even as the uses of these substances continue to develop. Many other countries regulate these materials under traditional chemical laws due to the absence of nano-specific regulations, thereby inadequately managing risks and delaying safety measures.

A peculiar element should be enlightening the Safe-by-Design (SbD) type principle with respect to designing newer formulations to try to steer away from such adverse consequences. It consists of evaluating materials for better biodegradability, reduced persistence, and intrinsic safety. Also, computational toxicology in assistance with in silico models can predict biological effects of nanomaterials by limiting the use of cumbersome in vivo tests [79]. AI assistance can further simplify the studies. Interdisciplinary databases need to be developed to facilitate risk management and data openness, which strengthens occupational safety standards and fosters public-private collaborations.

In conclusion, technological development along with environmental and health safety considerations is the pillar for the sustainability of ionic fluids, nanofluids, and nanomaterials. Effective regulation, well-developed risk assessment models, and worldwide acceptable characteristics are the main components for the development of safe and ethically acceptable nanomaterials.

References

1. N.V. Plechkova, K.R. Seddon, Applications of ionic liquids in the chemical industry. Chem. Soc. Rev. **37**, 123–150 (2008). https://doi.org/10.1039/B006677J

2. D. Zhao, M. Wu, Y. Kou, E. Min, Ionic liquids: Applications in catalysis. Catal. Today **74**(1-2), 157–189 (2007). https://doi.org/10.1016/S0920-5861(01)00541-7
3. P. Wasserscheid, T. Welton, *Ionic Liquids in Synthesis*, 2nd edn. (Wiley, Hoboken, 2008). https://doi.org/10.1002/9783527621199
4. S.U.S. Choi, A. Eastman, Enhancing thermal conductivity of fluids with nanoparticles. ASME Int. Mech. Eng. Congr. Expo. **66**, 99–105 (1995)
5. J.A. Eastman, S.U.S. Choi, S. Li, W. Yu, L.J. Thompson, Anomalously increased effective thermal conductivities of ethylene glycol-based nanofluids containing copper nanoparticles. Appl. Phys. Lett. **78**(6), 718–720 (2001). https://doi.org/10.1063/1.1341218
6. B. Bhushan, *Springer Handbook of Nanotechnology*, 3rd edn. (Springer, Cham, 2010). https://doi.org/10.1007/978-3-642-02525-9
7. M.C. Roco, Nanotechnology: Convergence with modern biology and medicine. Curr. Opin. Biotechnol. **14**(3), 337–346 (2003). https://doi.org/10.1016/S0958-1669(03)00068-5
8. C. Pretti, C. Chiappe, I. Baldetti, S. Brunini, G. Monni, L. Intorre, Acute toxicity of ionic liquids for three freshwater organisms: Pseudokirchneriella subcapitata, Daphnia magna and Danio rerio. Ecotoxicol. Environ. Saf. **72**(4), 1170–1176 (2006). https://doi.org/10.1016/j.ecoenv.2008.02.003
9. T.P.T. Pham, C.W. Cho, Y.S. Yun, Environmental fate and toxicity of ionic liquids: A review. Water Res. **44**(2), 352–372 (2010). https://doi.org/10.1016/j.watres.2009.09.030
10. J. Jodynis-Liebert, M. Nowicki, M. Murias, T. Adamska, M. Kujawska, Cytotoxicity of selected ionic liquids to human hepatoma HepG2 cells. Toxicol. In Vitro **24**(3), 789–799 (2010). https://doi.org/10.1016/j.tiv.2009.12.006
11. S.J. Klaine, P.J. Alvarez, G.E. Batley, T.F. Fernandes, R.D. Handy, D.Y. Lyon, J.R. Lead, Nanomaterials in the environment: Behavior, fate, bioavailability, and effects. Environ. Toxicol. Chem. **27**(9), 1825–1851 (2008). https://doi.org/10.1897/08-090.1
12. G. Oberdörster, E. Oberdörster, J. Oberdörster, Nanotoxicology: An emerging discipline evolving from studies of ultrafine particles. Environ. Health Perspect. **113**(7), 823–839 (2005). https://doi.org/10.1289/ehp.7339
13. G. Grasso, F. Frusteri, G. Cacciola, Waste disposal issues of nanofluids. Renew. Sust. Energ. Rev. **29**, 556–563 (2014). https://doi.org/10.1016/j.rser.2013.08.078
14. A. Nel, T. Xia, L. Mädler, N. Li, Toxic potential of materials at the nanolevel. Science **311**(5761), 622–627 (2006). https://doi.org/10.1126/science.1114397
15. J. Fabrega, J.C. Renshaw, J.R. Lead, Interactions of silver nanoparticles with the environment and human health. Sci. Total Environ. **408**(20), 3965–3973 (2011). https://doi.org/10.1016/j.scitotenv.2010.06.001
16. A.D. Maynard, R.J. Aitken, T. Butz, V.L. Colvin, K. Donaldson, G. Oberdörster, Safe Work Australia, Safe handling of nanotechnology. Nature **444**(7117), 267–269 (2006). https://doi.org/10.1038/444267a
17. A. García-Lorenzo, E. Tojo, J. Tojo, M. Teijeira, A. Rodríguez, J. Rodríguez-Fernández, Acute toxicity of imidazolium ionic liquids to Daphnia magna. Chemosphere **73**(1), 101–106 (2008). https://doi.org/10.1016/j.chemosphere.2008.05.014
18. R.F.M. Frade, A. Matias, L.C. Branco, C.A.M. Afonso, C.M.M. Duarte, Effect of ionic liquids on human colon carcinoma HT-29 and CaCo-2 cell lines. Green Chem. **9**(7), 873–877 (2007). https://doi.org/10.1039/b618220f
19. P. Stepnowski, A.C. Skladanowski, A. Ludwiczak, E. Laczyńska, Evaluating the cytotoxicity of ionic liquids using human cell line HeLa. Hum. Exp. Toxicol. **23**(11), 513–517 (2004). https://doi.org/10.1191/0960327104ht490oa
20. H. Chen, S. Zhang, X. Zhang, J. Wang, Cytotoxicity evaluation of ionic liquids on human hepatocytes. Ecotoxicol. Environ. Saf. **104**, 155–161 (2014). https://doi.org/10.1016/j.ecoenv.2014.02.004
21. S.P.M. Ventura, A.M.M. Gonçalves, T. Sintra, J.L. Pereira, F. Gonçalves, J.A.P. Coutinho, Designing ionic liquids: The chemical structure role in their toxicity. Ecotoxicology **21**, 1411–1422 (2012). https://doi.org/10.1007/s10646-012-0891-0
22. N. Gathergood, M.T. García, P.J. Scammells, Biodegradable ionic liquids: Part II. Effect of the anion and toxicology. Green Chem. **8**(2), 156–160 (2006). https://doi.org/10.1039/B512632K

23. S.E. Hale, H.P.H. Arp, I. Schliebner, M. Neumann, Persistence of ionic liquids in the environment. Environ. Sci. Technol. **44**(4), 1290–1295 (2010). https://doi.org/10.1021/es903256g
24. ECHA, *Read-Across Assessment Framework (RAAF)* (European Chemicals Agency, Helsinki, 2021) https://echa.europa.eu/documents/10162/19126370/raaf_nano_en.pdf
25. R.D. Rogers, J. Zhang, Green chemistry: A niche for ionic liquids? Chem. Commun. **8**, 992–993 (2003). https://doi.org/10.1039/B301034N
26. T.P.T. Pham, C.W. Cho, Y.S. Yun, Recent insights into environmental fate and toxicity of ionic liquids. Environ. Res. **192**, 110322 (2021). https://doi.org/10.1016/j.envres.2020.110322
27. J. Arning, P. Behrend, J. Müller, S. Stolte, Environmental biodegradability and aquatic toxicity of ionic liquids: A review. Chemosphere **263**, 128204 (2021). https://doi.org/10.1016/j.chemosphere.2020.128204
28. J. Chen, X. Dong, J. Zhao, G. Tang, In vivo acute toxicity of zinc oxide nanoparticles to mice after oral administration. J. Nanomater. **2016**, 1–9 (2016). https://doi.org/10.1155/2016/5635379
29. V. Sarma, N. Kumar, S. Ghosh, Effect of nanoparticle concentration on thermophysical properties and cytotoxicity of Al_2O_3 nanofluids. J. Mol. Liq. **249**, 641–648 (2018). https://doi.org/10.1016/j.molliq.2017.11.099
30. P.O. Rujitanaroj, N. Pimpha, P. Supaphol, Wound-dressing materials with antibacterial activity from electrospun gelatin fiber mats containing silver nanoparticles. Polymer **49**(21), 4723–4732 (2008). https://doi.org/10.1016/j.polymer.2008.08.052
31. M. Baek, H.E. Chung, J. Yu, J.A. Lee, T.H. Kim, J.M. Oh, J.K. Lee, Pharmacokinetics, tissue distribution, and excretion of zinc oxide nanoparticles. Int. J. Nanomed. **6**, 539–548 (2011). https://doi.org/10.2147/IJN.S16549
32. J. Park, S. Lee, D. Lee, Multifunctional hybrid nanoparticles: A new paradigm for targeted therapy and imaging. Nanomedicine **20**, 102013 (2019). https://doi.org/10.1016/j.nano.2019.102013
33. A.V. Singh, P. Laux, A. Luch, C. Sudrik, S. Wiehr, A.M. Wild, R. Patil, Review of emerging concepts in nanotoxicology: Opportunities and challenges for safer nanomaterial design. ACS Nano **13**(8), 9541–9570 (2018)
34. M. Mahmoudi, A. Simchi, A.S. Milani, P. Stroeve, Cell toxicity of superparamagnetic iron oxide nanoparticles. J. Colloid Interface Sci. **336**(2), 510–518 (2012). https://doi.org/10.1016/j.jcis.2009.04.046
35. G.C. Nallani, T.T.V. Phan, K.L. Chen, Influence of cerium oxide nanoparticle shape on their adsorption behavior and toxicity to aquatic organisms. Environ. Sci. Technol. **45**(17), 7311–7317 (2011). https://doi.org/10.1021/es200426f
36. K. Donaldson, C.A. Poland, R.P. Schins, C.L. Tran, Potential toxicity of carbon nanotubes: Mechanisms and safety assessment. Part. Fibre Toxicol. **7**(1), 32 (2010). https://doi.org/10.1186/1743-8977-7-32
37. R. Foldbjerg, D.A. Dang, H. Autrup, Cytotoxicity and genotoxicity of silver nanoparticles in the human lung cancer cell line, A549. Arch. Toxicol. **85**(7), 743–750 (2015). https://doi.org/10.1007/s00204-010-0545-5
38. X.D. Zhang, D. Wu, X. Shen, P.X. Liu, N. Yang, B. Zhao, F.Y. Fan, Size-dependent in vivo toxicity of PEG-coated gold nanoparticles. Int. J. Nanomed. **6**, 2071–2081 (2017). https://doi.org/10.2147/IJN.S21657
39. T. Ahmad, I.A. Wani, N. Manzoor, J. Ahmed, A.M. Asiri, Biosynthesis, structural characterization and antimicrobial activity of gold and silver nanoparticles. Colloids Surf. B: Biointerfaces **107**, 227–234 (2015). https://doi.org/10.1016/j.colsurfb.2013.02.014
40. A. Gogos, K. Knauer, T.D. Bucheli, Nanomaterials in plant protection and fertilization: Current state, foreseen applications, and research priorities. J. Agric. Food Chem. **60**(39), 9781–9792 (2012). https://doi.org/10.1021/jf302154y
41. A. Amiri, S. Haghighat, A. Khodakarami, M. Dadashzadeh, Influence of nanoparticle aggregation on the toxicity of nanofluids in biological systems: A comprehensive review. Environ. Sci. Pollut. Res. **28**(35), 47963–47980 (2021). https://doi.org/10.1007/s11356-021-14988-8

42. A.M. Nasir, P. Dufour, M. Lalancette-Hébert, P. Pouponneau, M.A. Fortin, Agglomeration of titanium dioxide nanoparticles increases their inflammatory effects on lung tissue in vivo. Toxicol. Lett. **333**, 27–35 (2020). https://doi.org/10.1016/j.toxlet.2020.08.002
43. D. Bobo, K.J. Robinson, J. Islam, K.J. Thurecht, S.R. Corrie, Nanoparticle-based medicines: A review of FDA-approved materials and clinical trials to date. Pharm. Res. **33**(10), 2373–2387 (2016). https://doi.org/10.1007/s11095-016-1958-5
44. M. Czajka, E. Sawosz, M. Grodzik, Nanoparticles—Interactions with living cells. J. Nanomater. **2015**, 783761 (2015). https://doi.org/10.1155/2015/783761
45. Y. Song, X. Li, X. Du, Hepatotoxicity of titanium dioxide nanoparticles in mice and its possible mechanisms. Environ. Res. **195**, 110810 (2021). https://doi.org/10.1016/j.envres.2021.110810
46. T. Wang, Y. Zhang, X. Lin, Cadmium-based quantum dots disrupt embryonic development of zebrafish by oxidative stress. Ecotoxicol. Environ. Saf. **195**, 110437 (2020). https://doi.org/10.1016/j.ecoenv.2020.110437
47. S. Jain, K. Thakur, A. Sharma, Biodistribution and biocompatibility of PLGA nanoparticles in Sprague Dawley rats: A 90-day study. J. Biomed. Nanotechnol. **18**(7), 1392–1401 (2022). https://doi.org/10.1166/jbn.2022.3260
48. Y. Luo, X. Yang, M. Li, Toxicity of graphene oxide to human lung epithelial cells: Mechanisms and mitigation strategies. Environ. Sci. Technol. Lett. **6**(9), 567–572 (2019). https://doi.org/10.1021/acs.estlett.9b00413
49. Q. Zhang, L. Wei, J. Shen, Endocrine-disrupting effects of nanoplastics on estrogen signaling pathways in breast cancer cells. Sci. Total Environ. **891**, 164313 (2023). https://doi.org/10.1016/j.scitotenv.2023.164313
50. T. Ahmed, M. Nagae, Y. Kimura, Neurotoxic effects of manganese oxide nanoparticles in cultured rat neurons. Toxicol. In Vitro **65**, 104804 (2020). https://doi.org/10.1016/j.tiv.2020.104804
51. D.M. Mitrano, B. Nowack, N. von Goetz, Tracking the transformation and fate of silver nanoparticles in sewage sludge-amended soil. Environ. Sci. Nano **8**(2), 539–550 (2021). https://doi.org/10.1039/D0EN01049K
52. Y. Wang, X. Xie, Z. Zhang, R. Li, Environmental exposure and ecotoxicity of titanium dioxide nanoparticles in coastal ecosystems. Mar. Pollut. Bull. **163**, 111957 (2021). https://doi.org/10.1016/j.marpolbul.2021.111957
53. K. Luyts, D. Napierska, D. Dinsdale, P.H.M. Hoet, A critical review of the effects of nanomaterials on the nervous system. Crit. Rev. Toxicol. **43**(5), 393–424 (2013). https://doi.org/10.3109/10408444.2013.770385
54. X. Wang, T. Xia, S.A. Ntim, Z. Ji, S. George, H. Meng, A.E. Nel, Quantitative techniques for assessing nanoparticle toxicity in vitro. Wiley Interdiscip. Rev. Nanomed. Nanobiotechnol. **6**(4), 347–362 (2014). https://doi.org/10.1002/wnan.1262
55. E. Fröhlich, The role of surface charge in cellular uptake and cytotoxicity of medical nanoparticles. Int. J. Nanomed. **7**, 5577–5591 (2012). https://doi.org/10.2147/IJN.S36111
56. O. Bar-Ilan, K.M. Louis, S.P. Yang, J.A. Pedersen, R.J. Hamers, R.E. Peterson, W. Heideman, Titanium dioxide nanoparticles produce phototoxicity in the developing zebrafish. Nanotoxicology **6**(6), 670–679 (2012). https://doi.org/10.3109/17435390.2011.604438
57. G. Oberdörster, A. Elder, A. Rinderknecht, Nanoparticles and the brain: Cause for concern? J. Nanosci. Nanotechnol. **9**(8), 4996–5007 (2015). https://doi.org/10.1166/jnn.2009.1397
58. A.B. Raies, V.B. Bajic, In silico toxicology: Computational methods for the prediction of chemical toxicity. Wiley Interdiscip. Rev. Nanomed. Nanobiotechnol. **6**(2), 147–172 (2016). https://doi.org/10.1002/wcms.1240
59. S. Halappanavar, S. van den Brule, P. Nymark, L. Gaté, C. Seidel, S. Valentino, V. Stone, Adverse outcome pathways as a tool for the design of testing strategies to support the safety assessment of nanomaterials. Part. Fibre Toxicol. **18**(1), 52 (2021). https://doi.org/10.1186/s12989-021-00424-2
60. R. Wang, M. Chen, Y. Liu, Predictive modeling of nanoparticle-induced pulmonary inflammation by integrating in vitro and in vivo data. Toxicol. Lett. **266**, 1–9 (2017). https://doi.org/10.1016/j.toxlet.2016.11.008

61. L.G. Valerio, The role of in silico modeling in the chemical risk assessment of nanomaterials. Regul. Toxicol. Pharmacol. **117**, 104776 (2020). https://doi.org/10.1016/j.yrtph.2020.104776
62. K. Roy, S. Kar, J. Leszczynski, *In Silico Toxicology: Principles and Applications* (CRC Press, Boca Raton, 2022)
63. A. Gajewicz, B. Rasulev, D. Leszczynska, J. Leszczynski, Toward understanding mechanisms governing cytotoxicity of metal oxides nanoparticles: In silico study of cellular responses. Nanotoxicology **11**(3), 299–310 (2017). https://doi.org/10.1080/17435390.2017.1292371
64. H. Raunio, In silico toxicology—Non-testing methods. Front. Pharmacol. **10**, 1641 (2020). https://doi.org/10.3389/fphar.2019.01641
65. D. Hristozov, A. Zabeo, C.M. Foran, P. Isigonis, A. Critto, A. Marcomini, A weight of evidence approach for the environmental risk assessment of engineered nanomaterials. Nanotoxicology **8**(1), 72–87 (2014). https://doi.org/10.3109/17435390.2012.750695
66. G. Bachler, N. von Goetz, K. Hungerbühler, Using physiologically based pharmacokinetic (PBPK) modeling for cross-species extrapolation of the systemic exposure to nanoparticles. Regul. Toxicol. Pharmacol. **65**(3), 393–403 (2013)
67. F. Zhao, Y. Wang, Y. Gao, H. Liu, Predicting nanotoxicity using machine learning models based on physicochemical properties. Comput. Toxicol. **21**, 100226 (2022). https://doi.org/10.1016/j.comtox.2022.100226
68. O. Schmid, T. Stoeger, R.V. Diaz, M. Semmler-Behnke, W.G. Kreyling, Systems biology-based insights into nanomaterial-induced hepatotoxicity. Nanotoxicology **15**(4), 412–430 (2021). https://doi.org/10.1080/17435390.2021.1881390
69. K. Kümmerer, R. Scholz, W. Scholz, Persistence, bioaccumulation, and toxicity of ionic liquids: A review of the environmental hazards of pyridinium-based ionic liquids. Green Chem. **13**(2), 312–322 (2011). https://doi.org/10.1039/C0GC00346J
70. A. Tovar-Sánchez, D. Sánchez-Quiles, G. Basterretxea, J.L. Benede, A. Chisvert, A. Salvador, I. Moreno-Garrido, Sunscreen products as emerging pollutants to coastal waters. PLoS One **8**(6), e65451 (2013)
71. K. Binnemans, P.T. Jones, Rare earths and the balance problem: How to deal with changing markets? J. Sustain. Metall. **6**(3), 593–610 (2020). https://doi.org/10.1007/s40831-020-00264-2
72. R.D. Walters, E.J. Weber, M.A. Stevens, Artificial intelligence in toxicology: A review of applications and limitations. Comput. Toxicol. **21**, 100206 (2022). https://doi.org/10.1016/j.comtox.2021.100206
73. T.Y. Sun, F. Gottschalk, K. Hungerbühler, B. Nowack, Comprehensive probabilistic modeling of environmental emissions of engineered nanomaterials. Environ. Pollut. **213**, 838–845 (2016). https://doi.org/10.1016/j.envpol.2016.03.042
74. H. Rauscher, K. Rasmussen, B. Sokull-Klüttgen, Regulatory challenges of nanomaterials and their risk assessment in the EU. Regul. Toxicol. Pharmacol. **91**, 257–267 (2017). https://doi.org/10.1016/j.yrtph.2017.01.013
75. A. Haase, A. Baun, P. Sayre, Research strategies for safety evaluation of nanomaterials. Nat. Nanotechnol. **16**(6), 614–621 (2021). https://doi.org/10.1038/s41565-021-00908-1
76. U. Schröder, J.L. Dorne, S.J. More, Environmental risk assessment of ionic liquids: A case study from the perspective of chemical mixtures. Sci. Total Environ. **713**, 136575 (2020). https://doi.org/10.1016/j.scitotenv.2020.136575
77. H.M. Ali, M.M. Janjua, M.S. Khan, A comprehensive review on nanofluids: Challenges and future directions. Appl. Therm. Eng. **185**, 116386 (2021). https://doi.org/10.1016/j.applthermaleng.2020.116386
78. Y. Liu, Y. Liu, L. Wang, C. Liu, Y. Ma, Toxicity of carbon-based nanomaterials to human health and the environment. Environ. Res. **204**, 112057 (2022). https://doi.org/10.1016/j.envres.2021.112057
79. C.T. Mahapatra, S. Nayak, S. Giri, Role of artificial intelligence in predicting toxicity and drug discovery: A modern toxicological approach. Comput. Biol. Chem. **100**, 107821 (2023). https://doi.org/10.1016/j.compbiolchem.2022.107821

Advancements in Nanotechnology: The Role of Machine Learning and AI in the Development of Ionic Liquids, Nanofluids, and Bionanofluids

Jisha Joseph, Jyoti Sahu, Jayesh Baldota, Vishnu Alias Vipul Parab, Tanishka Prakash Sharma, and Srushti Walsinge

Abbreviations

AFM	Atomic force microscopy
AI	Artificial Intelligence
ANN	Artificial Neural Network
CNN	Convolutional Neural Network
DL	Deep Learning
HAADF-STEM	High-angle annular dark-field scanning transmission electron microscopy
IL	Ionic fluids
ML	Machine learning
XRD	X-ray diffraction

1 Introduction

Nanotechnology, often described as "technology working at a nanoscale," typically refers to manipulating materials sized between 1 and 100 nanometers (nm). A nanometer is one-billionth of a meter, and this tiny scale allows for unique physical, chemical, and biological properties. Nanoscience, the study of materials at this scale, emphasizes structures smaller than 10 nm, precise control at the molecular level, and integration of nanoscale components into complex systems [1]. Nanoparticles (NPs) are of immense importance in many areas of application in

J. Joseph (✉) · J. Sahu · J. Baldota · V. A. V. Parab · T. P. Sharma · S. Walsinge
School of Pharmacy, Vishwakarma University, Pune, India
e-mail: jisha.joseph@vupune.ac.in

S. Kulkarni, S. Das (eds.), *Ionic Liquids, Nanofluids and Nanotechnology*, Lecture Notes in Nanoscale Science and Technology 42,
https://doi.org/10.1007/978-3-032-04008-4_20

agriculture, medicine, and environmental science. Nanoparticles in agriculture, with accurate chemical delivery, increase crop yield and environmental coping success. Nanotheranostics is revolutionizing both diagnostic and therapeutic sublime of medicine. The production of NPs from plant-derived materials in a green synthesis process enhances biocompatibility and lessens toxicity and might open up genetic engineering, as a response to genome interactions [2].

Ionic liquids (ILs) contain cations and anions, have a low melting point (below 100 °C), are non-hazardous to the environment, and have property-tunable characteristics and low vapor pressure. ILs become very important in the nanoparticle synthesis process because they increase stability, prevent the formation of aggregated clusters, and allow for particle size and shape control. Exposure to IL–NPs results in superior drug delivery due to increased penetration and chemical stability and decreased toxicity [3]. The exceptional thermal qualities of nanofluids, composed of nanoparticles in a colloid form, propel developments in energy technology and biomedical cooling. They are suitable for drug delivery and antibacterial use due to the functional modification and thermal change adaptability [4].

Bionanofluids are obtained when nanosized particles interact with biological solvents or biosurfactants. They have better thermal conductivity, stability, and dispersion capacity [5–9]. Demonstrative examples include improved oil recovery [5], functional nanotube-based anti-cancer strategies [10], and investigation of biovection in the presence of fluids under magnetic fields and Brownian motion [6]. Their use in the cross-cutting fields, such as medicine, energy, and environmental management, is promising.

1.1 Role of Machine Learning (ML) and Artificial Intelligence (AI) in Nanotechnology

The intersection of machine learning (ML) and artificial intelligence (AI) is rapidly changing the manner in which nanotechnology is functioning. Machine learning and artificial intelligence are providing growth in material discovery, progressing safety assessments, and enhancing the development of intelligent device technology. Following overview gives an insight of how ML and AI are used in nanotechnology with their primary functions, typical applications, and major challenges that were identified on recent studies.

Accelerating Discovery and Design: Using AI and ML in nanotechnology allows for analysis of big data, forecast of the material properties, as well as optimization of the new nanomaterial design considerably simplifying the research and development lifecycle. Examples of AI/ML techniques applied in this context are summarized in Table 1.

Enhancing Safety and Sustainability: AI and ML are used for hazard identification, risk management, and assessment of the sustainability of nanomaterials, thus facilitating a safe and responsible integration [13].

Table 1 Examples of AI/ML techniques used in nanotechnology

AI/ML technique	Application area
Support vector machine (SVM)	Material characterization, property prediction [11]
Artificial neural network (ANN)	Computational modeling, property prediction [11, 12]
Deep learning (DL)	Molecular discovery, generative models [11, 12]

Enabling Intelligent Devices: The integration of AI/ML in nanogenerators results in the creation of highly competent, automatically replenished items that perform activities such as sound monitoring and energy creation [14].

2 Fundamentals of Machine Learning and AI in Nanotechnology

The fields of machine learning (ML) and artificial intelligence (AI) are closely related and are transforming society and technology. AI addresses the vision of drawing up machines that can facilitate the performance of activities normally requiring human intelligence, while machine learning is one of the main aspects of AI through which systems learn how to perform better and improve its performance without being explicitly arranged.

2.1 Key Concepts and Definitions

Artificial intelligence (AI) involves designing intelligent agents that learn, reason, and find solutions to problems. Such agents generally emulate the cognitive techniques employed by humans during the thought and action processes [13].

Machine learning (ML) is a branch of artificial intelligence (AI) that makes use of statistical models and algorithms to let computers learn from data, spot trends, and make judgments with little help from humans [13].

Deep Learning: A subfield of machine learning that models intricate patterns using multilayered artificial neural networks, especially useful for interpreting images and language.

2.2 Relationship Between AI and ML

ML is the primary method by which AI systems are built and improved, serving as the "building blocks" for intelligent agents [13, 14]. The terms are sometimes used interchangeably, but ML is specifically about learning from data, while AI includes broader reasoning and problem-solving capabilities [13]. Table 2 summarizes the fundamental methods and approaches of machine learning which has contributed significantly in the development of AI systems.

Table 2 Fundamental methods and approaches in machine learning

Method/concept	Description
Supervised learning	Learning from labeled data to predict outcomes [14]
Unsupervised learning	Discovering patterns in unlabeled data [14]
Reinforcement learning	Learning optimal actions through trial and error [15]
Neural networks	Computing systems inspired by the human brain, core to deep learning [15]
Support vector machines	Algorithms for classification and regression tasks [15]
Bayesian networks	Probabilistic models for reasoning under uncertainty [15]

2.3 Applications of AI/ML in Nanomaterial Design and Discovery

Artificial intelligence (AI) and machine learning (ML) are transforming nanomaterial design and discovery by enabling faster, more efficient, and more targeted development of new materials. AI/ML approaches accelerate the prediction, optimization, and synthesis of nanomaterials, supporting applications from medicine to clean energy and sustainability.

2.4 Key Applications in Nanomaterial Design

1. Property Prediction and Structure Optimization: ML models efficiently predict nanomaterial properties and optimize structures, reducing reliance on trial-and-error and enabling rational design of materials with desired functionalities [16].
2. Synthesis Design and High-Throughput Screening: AI/ML facilitate the design of synthesis routes and automate high-throughput experimentation, speeding up the discovery and development of novel nanomaterials [16].
3. Characterization and Image Analysis: ML algorithms analyze complex characterization data and images, improving the understanding of nanoscale phenomena and supporting quality control [16].
4. Inverse Design: AI/ML enable the inverse design of nanomaterials, where desired properties are specified first and the optimal material structure is then identified [17]. Sector-specific applications of AI/ML are summarized in Table 3.

2.5 AI-Driven Optimization in Nanomanufacturing Processes

AI-driven optimization is transforming nanomanufacturing by enabling smarter process control, improved material design, and real-time resource allocation. The integration of artificial intelligence (AI), particularly machine learning (ML) and

Table 3 Sector-specific applications of AI/ML

Application area	AI/ML role and impact
Nanomedicine	Design of drug delivery systems, smart materials, and personalized therapies [17]
Clean energy and CCUS	Discovery of materials for solar cells, batteries, hydrogen energy, and CO2 capture [18]
Nanozymes	ML-guided design and optimization of enzyme-mimicking nanomaterials [19]
Sustainability and safety	Risk assessment, hazard identification, and sustainable nanomaterial development [20]

Table 4 Key techniques and strategies [21–23]

AI technique	Application area	Benefits
Machine learning	Process control, metrology, yield prediction	Real-time adjustments, improved quality
Deep learning	Image analysis, quality assurance	High-fidelity nanoscale characterization
Evolution strategies	Inverse design of nanostructures	Fast, scalable optimization
Predictive modeling	Nanomaterial synthesis, drug formulation	Tailored properties, reduced trial and error
IoT integration	Resource allocation, machine availability	Dynamic, decentralized manufacturing

deep learning, is addressing challenges in process efficiency, product quality, and scalability across various nanomanufacturing domains.

2.6 AI Applications in Nanomanufacturing

1. Process control and optimization: AI and ML algorithms are used for real-time process monitoring, high-speed metrology, and dynamic adjustment of process parameters to maximize yield and quality in continuous nanomanufacturing, such as roll-to-roll (R2R) systems [21].
2. Resource allocation: Smart agent systems powered by artificial neural networks (ANNs) and IoT enable real-time simulation and allocation of manufacturing resources, improving machine availability and process efficiency across distributed operations [22].
3. Material and product design: AI-driven inverse design and optimization strategies accelerate the development of multifunctional nanostructures and nanomaterials, allowing for tailored properties and rapid prototyping, especially in fields like nanophotonics and drug delivery [23]. The key AI techniques and strategies are summarized in Table 4.

2.7 Role of AI in Enhancing Nanoscale Characterization Techniques

Artificial intelligence (AI) is rapidly transforming nanoscale characterization by automating data analysis, improving accuracy, and enabling new insights from complex datasets. AI enhances image processing, predictive modeling, and real-time analysis, making nanoscale characterization faster, more reliable, and more informative.

Key Applications and Advances

1. Image Processing and Analysis: AI, especially deep learning, significantly improves the analysis of images from techniques like scanning electron microscopy (SEM), transmission electron microscopy (TEM), atomic force microscopy (AFM), and X-ray microscopy. AI algorithms enhance image quality, automate feature detection, and enable high-throughput analysis, outperforming traditional methods in accuracy and speed [24].
2. In super-resolution microscopy, deep learning models such as deformable DETR excel at identifying and differentiating nanoscale vesicles, revealing population heterogeneity, and reducing misidentification and computational demands [24].
3. Predictive modeling and structure–property relationships: AI and machine learning models are used to predict material properties, analyze structure–property relationships, and model complex phenomena like nanoscale friction with high accuracy. Symbolic regression and genetic programming approaches provide interpretable models for understanding multidimensional dependencies [25]. In battery research, AI-driven analysis of nano-CT images enables predictive modeling of electrode microstructure impacts on performance, supporting the development of advanced batteries [25].
4. Automation and efficiency: AI automates routine experimental processes, optimizes sample selection, and accelerates data interpretation, increasing the efficiency and reproducibility of nanoscale characterization workflows [26]. AI-powered tomographic reconstruction allows for faster, higher-quality 3D imaging, providing unprecedented detail and insight into internal microstructures [27].

2.8 AI/ML in Nanotechnology Applications and Breakthroughs

1. Accelerated nanomaterial discovery and design: AI and ML have revolutionized the discovery and design of novel nanomaterials with tailored properties, expediting the development of advanced materials for medicine, electronics, and energy applications. Predictive modeling and data-driven optimization enable

Table 5 AI/ML in nanotechnology breakthroughs

Application area	AI/ML role/breakthrough	Notable impact/example
Nanomaterial discovery	Predictive modeling, design optimization	Rapid synthesis of novel materials [28]
Drug delivery	Data-driven nanoformulation, personalized dosing	Enhanced efficacy, reduced side effects [29]
Diagnostics	ML-based data analysis, nanosensors	Early cancer detection, improved prognosis [29]
DNA sequencing	AI-powered nanopores, ML classification	High-precision, high-throughput sequencing [30]
Sustainability	Risk assessment, real-world monitoring	Safer, more sustainable nanomaterial use [31]
Nanofluids	AI prediction of thermophysical properties	Improved heat transfer applications [32]

rapid identification and synthesis of nanomaterials with desired functionalities, surpassing traditional trial-and-error approaches [28].

2. Precision medicine and drug delivery: AI-driven nanotechnology has enabled the creation of highly efficient, targeted drug delivery systems, improving therapeutic efficacy and minimizing side effects. In cancer management, AI optimizes nanomedicine design, enhances molecular profiling, and supports personalized dosing, leading to more effective and safer treatments [29, 30].
3. Diagnostics and early disease detection: The integration of AI with nanotechnology has led to the development of sensitive diagnostic tools and nanoscale devices for early disease detection, particularly in oncology. Machine learning algorithms analyze large healthcare datasets, improving the accuracy of early-stage cancer diagnosis and prognosis prediction [29, 30].
4. High-throughput DNA sequencing: AI-powered nanopores, combined with machine learning and quantum tunneling, have achieved high-precision, high-throughput DNA sequencing. These systems can classify nucleotides with over 96% accuracy, paving the way for rapid genomic analysis [30].
5. Sustainable nanotechnology and risk management: AI and ML enhance the sustainability of nanomaterials by improving simulations, refining safety assessments, and monitoring real-world behavior. These technologies support hazard identification and risk minimization, promoting the safe and responsible use of nanotechnology [31].
6. Advanced nanofluid applications: AI models, such as artificial neural networks and genetic algorithms, outperform traditional methods in predicting the thermophysical properties of nanofluids, advancing their use in heat transfer and energy systems [32]. Key AI/ML breakthroughs in nanotechnology is summarized in Table 5.

3 Ionic Liquids in Nanofluid Research

Ionic liquids are a type of salt that do not freeze at low temperatures and consist only of ions. These unique features make them useful in both research and industry, mostly as eco-friendly alternative solutions to classic solvents.

3.1 *Key Properties*

Tunable Structure: ILs are created by bringing together specific cations and anions and as a result, almost endless structure varieties and tailored uses are possible [33–35].

Thermal and Chemical Stability: They are stable at high temperatures and in chemical reactions, less likely to evaporate and are not easily set on fire which makes them safer and more eco-friendly than many common solvents [33–35].

Ionic Conductivity: Because ILs have both high electrical conductivity and do not evaporate, they are well suited for electrochemical uses [34, 35].

Eco-Friendly Potential: Because ILs have a low vapor pressure and could be more environmentally friendly, they are often advertised as green solvents, despite the fact that questions about their toxicity and use on a large scale are still unresolved [33, 35]. Major application of Ionic fluids is shown in Table 6.

3.2 *Synthesis and Stabilization of Nanofluids Using Ionic Liquids*

Nanofluids are colloidal suspensions comprising nanoparticles dispersed in a base fluid. The synthesis and stabilization of nanofluids are critical for their functionality and long-term performance. Two principal methodologies are employed for

Table 6 Major applications of ionic fluids

Application area	Description and benefits
Electrochemistry and energy	Electrolytes in batteries, supercapacitors, fuel cells, and as heat transfer fluids [34, 35]
Solvents and catalysts	Designer solvents for chemical synthesis, separation, and catalysis [33–35]
Biomedicine	Drug solubilization, delivery, protein stabilization, biosensors, antibacterial agents [33]
Environmental applications	CO_2 capture, biomass processing, and green chemistry processes [36]
Industrial processes	Enhanced oil recovery, lubrication, and other specialized industrial uses [36]

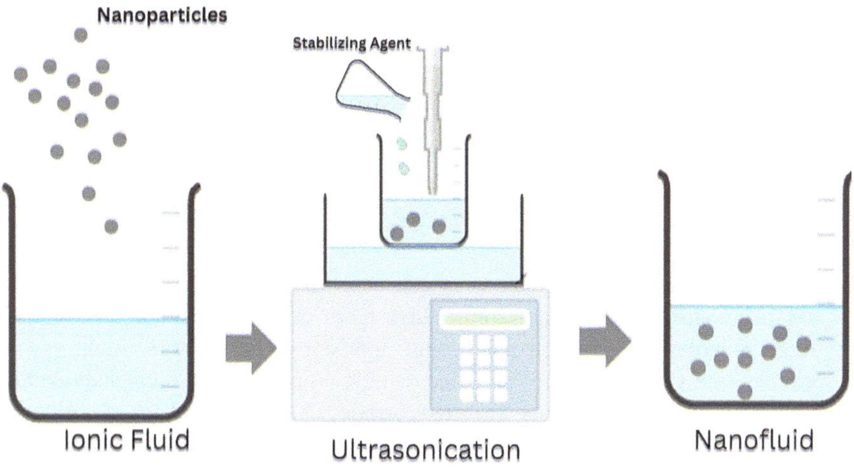

Fig. 1 Synthesis of nanofluid using ionic fluid

nanofluid preparation: the one-step method and the two-step method as illustrated in Fig. 1.

1. One-step method: The one-step synthesis technique involves the simultaneous formation and dispersion of nanoparticles within the base fluid. This method circumvents the need for drying, storage, and transportation of nanoparticles, thus significantly reducing the risk of agglomeration. The key advantages include in situ nanoparticle production, improved morphological control, and enhanced colloidal stability due to reduced particle clustering.
2. Two-step method: The vast majority of large-scale nanofluid is made using the two-step method. The initial step is to fabricate nanoparticles using either chemical or physical means and afterwards, disperse them in the base liquid. The best way to reduce agglomeration in dispersion is by using ultrasound and including stabilizing agents such as anionic, cationic, or non-ionic surfactants and polymers in the mix. Despite being affordable and able to work at scale, this method struggles with stability more than the one-step process [37].

Stability Mechanism

The Derjaguin–Landau–Verwey–Overbeek (DLVO) theory governs colloidal stability by balancing van der Waals attraction and electrostatic repulsion. A higher Hamaker constant amplifies attractive forces, while the Debye length determines the effective range of repulsion [37].

Stabilization Techniques

Both mechanical and chemical strategies are adopted to stabilize nanofluids [37].

1. Mechanical stabilization: Ultrasonication (bath or probe type), ball milling, and high-shear homogenization disrupt aggregates, enhancing dispersion. Probe sonicators typically offer superior outcomes [38]. demonstrated effective stabilization using a combination of planetary ball milling (PBM), ultrasonication, and magnetic stirring.
2. Chemical stabilization: Stabilization can be electrostatic (via ionic surfactants and pH control), steric (non-ionic surfactants or polymers), or electrosteric (ionic polymers). Polyelectrolytes with oppositely charged functional groups provide both steric hindrance and electrostatic repulsion. Chemical stabilization is classified into three types: (a) electrostatic, (b) steric, and (c) electrosteric stabilization. Electrostatic stabilization is achieved using ionic surfactants, pH adjustment, and covalent surface functionalization. Steric stabilization is attained by non-ionic surfactants and polymer addition, while electrosteric uses ionic polymers or liquids.

   (a) Electrostatic stabilization: Electrostatic stabilization balances van der Waals attraction with electrostatic repulsion using surface charge from ionization or surfactant adsorption. The electrical double layer (EDL) consists of surface charge, stern layer, and diffuse layer, with the zeta potential determining stability. Higher potential barrier leads to superior nanofluid stability.
   (b) Steric stabilization: The steric stabilization of nanofluid can also be termed a polymeric stabilization technique, which involves the addition of polymer molecules in the suspension of nanoparticles to prevent particle aggregation and eventual sedimentation. Polymeric molecules attached to the nanoparticle surface prevent the two nanoparticles from coming close to each other. This stabilization technique falls under the category of noncovalent functionalization.
   (c) Electrosteric stabilization: Electrosteric stabilization combines electrostatic and steric stabilization, using ionic polymers for both steric hindrance and electrostatic repulsion. Ionic polymers create a protective polymeric barrier and electric potential barrier between nanoparticles, enhancing stability. Polyelectrolytes, made of repeating polymer units with ionizable groups, are used, requiring opposite charges between nanoparticle and polyelectrolyte for good polymer adsorption [37].

Stability Evaluation Methods of Nanofluids Using Ionic Liquids

Nanofluid stability is a paramount criterion for their practical deployment in heat transfer, lubrication, and biomedical applications. Several techniques are utilized to assess colloidal stability [37] as illustrated in Fig. 2.

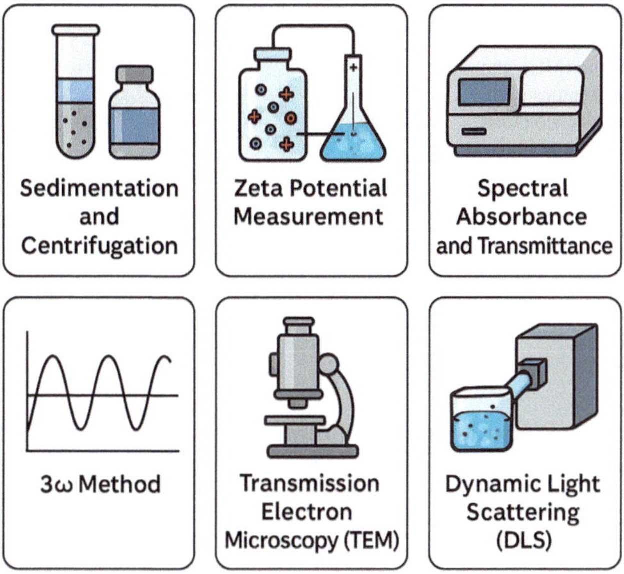

Fig. 2 Stability evaluation of nanofluids

1. Sedimentation and centrifugation: Sedimentation analysis, where prolonged precipitation time suggests better stability, is the most frequently employed technique. Centrifugation accelerates this process by applying centrifugal force, with sedimentation rate influenced by particle size, fluid viscosity, and density differences.
2. Zeta potential measurement: Zeta potential quantifies the electrostatic repulsion between particles. High absolute zeta potential values signify strong repulsive forces and improved colloidal stability.
3. Spectral absorbance and transmittance: Spectrophotometric methods analyze nanoparticle dispersion through absorbance and transmittance values. A decline in absorbance over time indicates sedimentation and instability. Monitoring transmittance periodically helps assess the dynamic stability of nanofluids.
4. 3ω method: The 3ω method correlates thermal conductivity with particle dispersion [39] observed increased thermal conductivity in Al_2O_3–water and Al_2O_3–ethylene glycol nanofluids due to clustering, which was prominent within the first hour post-preparation. A reduction in conductivity is associated with particle agglomeration, which reduces effective surface area for heat transfer.
5. Transmission electron microscopy (TEM): TEM enables visualization of particle morphology and dispersion quality employed TEM to investigate the impact of surfactants and polymers on the dispersion behavior of TiO_2 and Cu-Zn-Al LDH nanoparticles in aqueous media [37, 40, 41].
6. Dynamic light scattering (DLS): DLS is utilized to evaluate particle size distribution within colloidal suspensions. The method measures fluctuations in

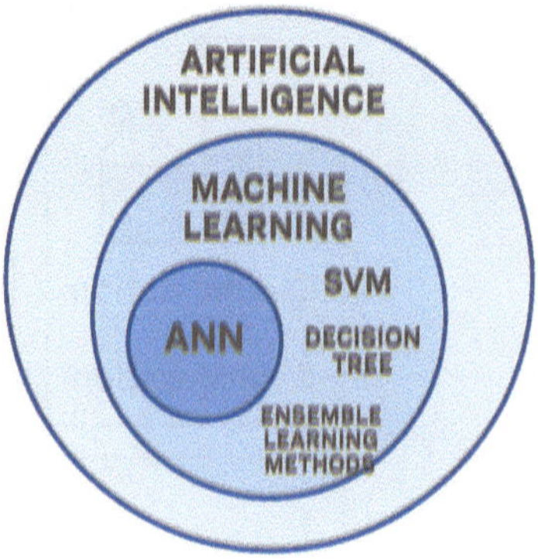

Fig. 3 Hierarchical relationship among AI, Machine Learning (ML), and specific ML techniques such as ANN, SVM, decision tree, and ensemble learning [42]

light scattering intensity to infer diffusion coefficients, from which instability due to aggregation can be inferred [37].

3.3 *Machine Learning Techniques in Ionic Liquid Research*

Many studies on ionic liquids (ILs) have significantly changed due to the fast growth of machine learning (ML). Applying these methods makes it possible to predict important properties, invent innovative ILs, and adjust IL-based systems for different manufacturing needs. Among all the ML algorithms in use, some have proven to be very effective and are frequently picked in IL work. An overview of the most popular techniques such as ANNs, SVMs, and DTs, is included, together with their advanced forms that use many classifiers [42]. Figure 3 illustrates the hierarchical relationship among AI, Machine Learning (ML), and specific ML techniques such as ANN, SVM, decision tree, and ensemble learning [42].

Artificial Neural Networks (ANNs) and Deep Learning (DL)

Artificial Neural Networks (ANNs) follow the structure and way the human brain works. They have many nodes (neurons) that work together to change input data into outcomes we can use to make predictions. In such research, ANNs are usually used to discover the relationships between molecular descriptors and target properties that are not exactly linear, for example, viscosity, conductivity, melting point, and thermal stability.

Especially, the use of Deep Learning (DL) and Deep Neural Networks (DNNs) has increased ANNs' abilities. These models have hidden layers, so they can pull out useful data features straight from the raw input. With this feature, DNNs are able to notice hidden patterns and connections that standard models could ignore.

Both ANNs and DL models have weaknesses in addition to all their benefits:

Large, good-quality datasets are usually required by these models and sometimes that data is not accessible in IL research. DL models are challenging to train as they need a lot of processing power and time. Since they have many parameters and can become very detailed, it's easy for them to adjust to the training data too closely, especially if the data is small or contains a lot of noise. How well deep learning works depends largely on correctly adjusting many hyperparameters. Superior modeling is possible with DNNs, but using them in IL research means finding a good balance between accuracy and computational expenses [42, 43].

Radial Basis Function (RBF) Neural Networks

Radial basis function (RBF) neural networks are powerful tools in machine learning, particularly suitable for classification and regression tasks. Introduced by Broomhead and Lowe in 1988, RBF networks are feedforward neural networks with two layers: a hidden layer using radial basis functions and an output layer with linear activation. These networks are highly effective for non-linear problems due to their localized and symmetric response properties. Their reliance on the distance from a central point allows them to efficiently capture spatial patterns, even with relatively small datasets [44].

GMDH-Type Neural Networks

GMDH-Type Neural Networks developed by Ivakhnenko in the 1960s, represent a self-organizing approach in machine learning. This method builds multilayered models by selecting and combining neurons based on their correlation with the target variable [138]. The selection is typically performed using criteria like the least squares method, which ensures that only the most relevant neurons contribute to the model, enabling the capture of complex, non-linear relationships [44].

Support Vector Machines (SVMs)

SVMs are a popular tool used in ML, especially when classifying or regressing IL problems. SVMs use a process to help separate data into separate categories or position data points near an average, given an allowance of errors.

Kernel functions allow SVMs to reshape input data into a new dimension so that linear separation is now possible, in contrast to the original data. Because of this property, SVMs are well suited to modeling situations in which relationships are not straight lines, an issue found in IL applications.

This method solves the optimization issue in a simpler way by converting it to a sequence of linear equations. Since LSSVM trains models more quickly and provides better numerical performance, it has become popular among researchers who handle few or interfered datasets.

Support Vector Machines (SVMs) offer robust performance in ionic liquid (IL) research due to their effectiveness with small, noisy datasets. They reduce overfitting risk compared to neural networks and deliver high precision in both classification and regression. Their ability to model complex relationships makes them ideal for predicting IL properties accurately. These strengths make SVMs an appealing alternative, especially when experimental data is scarce or computational resources are limited [42].

Decision Trees (DTs) and Ensemble Learning Methods

DTs use a straightforward yet effective way to study data and make predictions. In ILs, to relate molecular features with target characteristics, DTs apply a process that repeats and splits the data based on specific values. By producing a tree structure, the model makes its reasoning behind each prediction very easy for researchers to understand [42].

Decision trees are easy to understand and visualize, making them highly interpretable compared to complex models like neural networks. However, they are prone to overfitting with large or intricate datasets. Over time, several enhancements have been developed to improve the decision tree framework [44].

(a) Decision tree regression: It adapts decision trees for continuous outputs by assigning mean or median values at the leaf nodes, effectively capturing non-linear data patterns [45].
(b) Alternating decision trees (ADTrees): It enhances accuracy by combining multiple weak learners. They alternate between decision and prediction nodes, aggregating prediction scores to form final outputs [46].
(c) M5 trees: They are tailored for regression, integrating linear regression models at the leaves. This hybrid approach captures both hierarchical and linear feature relationships, improving predictive accuracy [47].
(d) Random forests (RFs): It uses an ensemble of decision trees trained on bootstrapped subsets. The final prediction is made by averaging the outputs, significantly reducing overfitting and improving model robustness [48].
(e) Extra trees regression (ETR): It adds randomness by selecting split points arbitrarily during training. This introduces greater diversity and further enhances generalization [49].
(f) AdaBoost: It builds strong classifiers by sequentially combining weak learners, typically decision stumps. It increases the focus on previously misclassified instances, refining accuracy over iterations [50].
(g) Boosted regression trees: It sequentially fit new trees to the residuals of prior models, enhancing performance by focusing on hard-to-predict data points.
(h) Gradient boosting machines (GBMs): It also build models iteratively, correcting previous errors at each stage. They offer strong performance on both classification and regression tasks [51].

(i) XGBoost is an optimized, scalable implementation of gradient boosting that delivers high speed and accuracy, making it a top choice in practical ML applications [52].

3.4 Prediction of Thermophysical Properties of Ionic Liquid Using ML

DTs provide an easy yet effective way to study data and forecast future events. For ILs, DTs let you relate the values of molecular characteristics to properties by going through the data set by feature and cutting it. Such a tree structure makes it easy for researchers to identify the reasons behind every prediction.

In the field of IL, researchers commonly divide data using the CART algorithm which goes on splitting until each leaf node is made up of identical cases. Even so, there are several issues with standard DTs.

IL nanofluids are being acknowledged for their advanced properties in heat transfer because their thermal conductivity, heat capacity, and strength at high temperatures are better than those of most heat transfer fluids. Because of these improvements, they can be applied in solar energy systems, storing heat and other application for efficient heat transfer [42].

Melting Point

To design ionic liquids (ILs) well for different areas such as electrochemistry and green chemistry, their melting points must be predicted accurately. Older approaches including group contribution models and QSPR tools have afforded some ability to forecast, but ML tools are now seen as stronger, able to address hard-to-predict, curved relationships, and boost accuracy.

Multilayer perceptron neural network (MLP-NN) was used to predict the melting points of 62 ILs from many chemical families. To investigate non-linear effects, there were studies of constitutional, topological, and electrostatic features from classical molecules, and this approach improved on previous methods that used regression. The computational demands of deep neural networks may hold back their use in situations that require high processing speeds [42].

Another promising approach is the use of Recurrent Neural Networks (RNNs) combined with graph convolutional theory to directly work with IL molecular structures. Unlike with descriptors, these methods use algorithms to extract molecular graph structure, so the process of generating descriptors is not needed [53]. Their model also worked well with the same pyridinium data, though it required far less time to get the data ready [54]. This paper demonstrated that deep learning can help create and speed up the process of molecular representation.

After gathering a dataset of 2000 ILs with diverse chemistry and testing SVM, decision tree, and neural network models, this study achieved a significant advancement. For input, semi-empirical quantum mechanical methods were used through PM6, giving good quality without overloading the system. Even though the models predicting the absolute values were moderate, they accurately showed the ranking of melting points—a crucial feature needed for high-throughput work [55].

In 2019, the researchers followed this up by releasing an extensive IL property prediction library including more than eight million hypothetical examples. With this database, multiple IL properties—melting point, viscosity, and density—were predicted at the same time by the selected best ML models. Since then, the resource has sealed its role in IL informatics by giving IL design a stable and data-based base [42].

To deal with data differences, there was a study with imidazolium-based ionic liquids that proposed a regression framework using clustering. ML models were created and refined within the groups of structurally similar ILs. Aligning the model structure with molecular patterns led to much better prediction results [56]. This research led to the creation which predicts many illuminants characteristics and was first used to predict their melting points using artificial neural networks [57].

Researchers [58] conducted a study to evaluate how different types of molecular descriptors affect the accuracy of predicting the melting points of ionic liquids (ILs). They used a dataset compiled by Venkatraman, which contains melting point data for 2200 ILs, and applied quantum mechanical descriptors to develop predictive models [58]. The most accurate model showed an average deviation of approximately 30 °C between the predicted and experimental melting points. While this deviation is significant, it still represents an improvement over traditional predictive method. Interestingly, the study found that since the selected descriptors were not structure-specific, the model could potentially be applied to any IL family—provided the training set is appropriate. This suggests that with sufficient high-quality data, the model could be generalized across different IL structures without needing unique structural descriptors.

Density

Machine learning has demonstrated it can accurately predict the density of ionic liquids (ILs) which is needed for engineering work. In 2009, Valderrama et al. published a study using group contribution (GC) theory with artificial neural networks (ANNs). Even though they looked at just 25 functional groups, their method worked well at predicting ILs, even for those they didn't look at in training.

To build on this approach, [59] used two ANN architectures to predict the densities of imidazolium-based ILs and IL-water mixtures. The team found that it is valuable to model mixtures, since ILs are often used in solutions.

In 2019, further studies were carried out in this field by accumulating a dataset of more than 2000 ILs and using MLP, FFANN, and LSSVM algorithms together.

Using this strategy, the team was able to make more accurate predictions than had been done before using previous models [60].

All of these studies together suggest that ML is advancing in predicting IL property values and future improvements could result from using more extensive data and creating hybrid models.

Viscosity

Nanoparticles typically raise the viscosity of the base fluid [58]. If more nanoparticles are added to a solution, the viscosity increases as well. The prediction of viscosity for ionic liquids (ILs) by machine learning (ML) has become much more precise. In 2011, there was introduction of new approach by combining ANNs with molecular connectivity indices (MCI) to predict the viscosity of 26 ionic liquids, and the results were only 5% different from those obtained from experiments [61]. Another study used Artificial Neural Network (ANN) to predict viscosity without structural features. Instead, they used the logarithm of viscosity at 323.15 K and the inverse of the reduced reference temperature as inputs. Their model outperformed empirical viscosity equations like Vogel–Tammann–Fulcher and the Arrhenius model, showing lower overall residual errors. Additionally, the ANN provided reliable results across different ionic liquid families, making it a versatile tool for viscosity prediction [62].

The combination of Group Contribution (GC) theory with Artificial Neural Networks (ANNs) for predicting the viscosities of ionic liquids (ILs) was used by researchers [63]. They used a database of 1484 ILs but found that predictions for ammonium and dicationic ILs were less accurate due to insufficient training data. But in 2019, Paduszyński expanded his model to over 2000 ILs, integrating Feedforward Artificial Neural Networks (FFAAN) and Least Squares Support Vector Machines (LSSVM), enhancing prediction accuracy compared to traditional Quantitative Structure-Property Relationship (QSPR) methods.

Fatehi and coworkers designed an ANN that includes the impact of pressure on material properties [64]. The authors used ELMs and thermodynamic descriptors proposed [65] and were able to accurately predict properties over a wide range of temperatures and pressures.

LSSVM was used by Baghban and colleagues with GC descriptors, where only the temperature parameter was necessary. Such research proves that, over time, ML is being used to predict IL viscosity more reliably and accurately than before, with emphasis on data rather than fixed rules [42].

Specific Heat Capacity

Specific heat capacity is defined as the number of units of energy required to raise the temperature of a substance's one unit mass by 1 °C (or 1 K) [43]. Nanofluids of silicon dioxide, zinc oxide, and alumina particles in a mixture of water and ethylene glycol showed a lower specific heat capacity than plain water. Besides, experiments showed that the specific heat capacity also became lower as the amount of nanoparticles was raised. Many studies have examined the use of ML methods to forecast the SHC of nanofluids as an alternative. To do this, [66] used Gaussian Process Regression, Support Vector Machines, and Extreme Gradient Boosting (XGBoost) to evaluate the SHC of oxide-MWCNT/water-based hybrid nanofluids. In their study, the researchers created and evaluated ML models using 432 samples taken from experiments. Of the algorithms used, XGBoost performed best, recording a mean absolute percentage error (MAPE) of only 0.01%.

They also use similar methods to calculate the SHC of molten salt-based nanofluids, finding that the approach provides accurate results. In addition, [67] designed enhanced ANN and SVMR models by using temperature, nanoparticle concentration, and ratio of microcapsules as input parameters. Existing models were able to accurately predict how quickly ternary hybrid oxide/water nanofluids would heat in the presence of a heat source at a rate of 99.67% [68, 69].

Thermal Conductivity

Thermal conductivity refers to a substance's ability to transfer heat through conduction. In this process, thermal energy moves via the kinetic energy of molecules, atoms, ions, or other particles in motion. Unlike convection, conduction does not involve the bulk movement of the substance; instead, heat transfer occurs at the microscopic level through direct particle interactions. Several methods are used to measure TC, including the steady-state parallel plate method, the three-omega technique, the transient hot-wire method, and the thermal constants analyzer [61]. In general, adding nanoparticles to a base fluid enhances its thermal conductivity, making nanofluids effective heat-removal agents in applications with high thermal loads. Studies have found that TC increases more significantly at higher temperatures when particle fractions are greater [70, 71].

Researchers have examined various factors influencing thermal conductivity, including volume fraction, particle size and shape, base fluid conductivity, temperature, and particle aggregation. Additionally, there have been multiple theoretical attempts to understand and predict this enhancement [72].

The study developed an Artificial Neural Network (ANN) model to predict the thermal conductivity ratio (knf/kf) of nanofluids based on experimental data. A total of 776 data points were collected from 21 different sources, covering various materials and conditions. The chosen input variables included the nanoparticle diameter (dnp), volume fraction of nanoparticles (φ), temperature (T), and specific properties of the base fluid and nanoparticles. To train the model efficiently, the dataset was randomly split into training (70%), validation (15%), and testing (15%) sets, and all inputs were normalized between -1 and 1 to enhance training performance. For the network architecture, a single hidden layer feedforward Multilayer Perceptron (MLP) ANN was employed, using the Levenberg–Marquardt backpropagation (LM-BP) algorithm, which is known for its rapid convergence and effectiveness in training. The tangentsigmoid (tansig) function was applied in the hidden layer, while the linear (purelin) function was used in the output layer. Since the number of neurons in the hidden layer plays a critical role in prediction accuracy, it was optimized through trial and error. To find the best topology, different configurations ranging from 1 to 23 neurons were tested, and each was run 50 times to determine the most effective setup. The model's performance was assessed using Mean Squared Error (MSE), Mean Absolute Percentage Error (MAPE), and the Correlation Coefficient (R), which measured the accuracy of predictions compared to actual values. The results demonstrated that the ANN model could effectively predict the thermal conductivity enhancement of nanofluids, making it a reliable tool for understanding thermal properties across various materials [72].

Machine learning (ML), especially convolutional neural networks (CNNs), has demonstrated potential in forecasting the effective thermal conductivity of composite materials and porous media [71] achieved impressive accuracy ($R^2 = 0.986$, RMSE = 1.9%) with a CNN trained on 1400 samples, surpassing traditional models like Maxwell–Eucken. In a similar vein, created a CNN model with $R^2 = 0.93$ and a mean absolute error of 5.2 W/(m K). However, earlier models fell short in optimization and lacked generalizability beyond their training datasets. To overcome these limitations, a recent study introduced a CNN model inspired by VGG architectures, trained on over 130,000 images with simulated conductivity values derived from a parallelized numerical heat transfer (NHT) model. This model exhibits high prediction accuracy, quick processing (milliseconds per prediction), and robust generalization across various 2D structures.

Importantly, the CNN was validated against unseen data and analytical benchmarks, showcasing its durability. It was further utilized in iterative topology optimization tasks, where swift property prediction is crucial due to continuous structural modifications. This model markedly decreases computational time compared to traditional NHT or random-walk simulations, facilitating optimization over larger domains. This development establishes CNNs as formidable tools for both predictive modeling and real-time material design [73].

3.5 Challenges in Experimental Characterization of Ionic Liquid Nanofluids

(a) High viscosity of ionic liquids: Many ILs exhibit significantly higher viscosity than conventional fluids, making accurate measurement of rheological, thermal, and flow properties difficult. This affects nanoparticle dispersion and hinders techniques like dynamic light scattering (DLS) and viscometry [74].

(b) Nanoparticle aggregation and sedimentation: Ensuring stable dispersion of nanoparticles in ILs is challenging due to strong van der Waals interactions. Aggregation affects consistency and reliability of measurements like thermal conductivity, zeta potential, and specific heat capacity [63].

(c) Complex interfacial interactions: Ionic liquids possess unique ionic structures that interact with nanoparticles via electrostatic, hydrogen bonding, and steric mechanisms. These complex interactions complicate the interpretation of spectroscopic and microscopic data (e.g., FTIR, TEM, or SEM) [64].
(d) Lack of standardization: Characterization methods for IL nanofluids are not standardized, making comparisons across studies difficult. Variations in synthesis routes, nanoparticle types, concentrations, and measurement protocols lead to inconsistencies in reported results [64].
(e) Thermal property measurement limitations: Accurate determination of thermal conductivity and specific heat capacity is difficult due to the intrinsic thermal instability and sensitivity of ILs to temperature. Conventional techniques like transient hot wire or laser flash analysis often require modification [64].
(f) Sensitivity to moisture and impurities: ILs are hygroscopic and can absorb atmospheric moisture, altering their physicochemical properties and affecting the reproducibility of characterization results [65].
(g) Instrumentation compatibility: Some characterization instruments and sensors may degrade or become contaminated when exposed to ILs due to their ionic nature and corrosive tendencies, limiting long-term measurements [64].

3.6 Potential of AI/ML in Addressing these Challenges

Artificial intelligence (AI) and machine learning (ML) offer transformative solutions to many challenges faced in the experimental characterization of ionic liquid (IL) nanofluids. These technologies can enhance accuracy, reduce experimental overhead, and provide predictive insights, especially in areas where traditional approaches struggle with complexity, non-linearity, and high-dimensional datasets [66–69].

(a) Predictive modeling: ML algorithms, such as artificial neural networks (ANNs), support vector machines (SVMs), and decision trees, can be trained on experimental data to predict thermophysical properties (e.g., thermal conductivity, viscosity, specific heat) of IL nanofluids across different compositions, temperatures, and particle concentrations. This lessens the need for thorough testing.
(b) Data-driven optimization: By pinpointing critical variables that affect thermal performance, dispersion quality, and nanofluid stability, AI can optimize experimental settings. Methods such as Bayesian optimization and genetic algorithms aid in adjusting the size, type, and IL composition of nanoparticles for best outcomes.
(c) Signal processing and noise reduction: Artificial intelligence (AI) technologies can be used to preprocess and denoise experimental data, particularly when observations are impacted by phase instability, high viscosity, or instrument restrictions. Regression filters based on machine learning, for example, are capable of identifying trustworthy patterns in noisy datasets [66–69].

(d) Image and spectral analysis: Compared to manual methods, deep learning models like convolutional neural networks (CNNs) can analyze microscopic and spectroscopic images (such as TEM, SEM, and FTIR) more quickly and accurately. Using this approach, they can recall the interaction between nanoparticles and ILs (surface ones, aggregation, overall dispersion).
(e) Machine learning methods can easily detect strange outcomes or mistakes that can be prompted by inconsistent samples or procedural difficulties. Moreover, such models as Gaussian processes provide the opportunity to estimate uncertainty in predictions and make researchers able to make more efficient conclusions about further experiments.
(f) Creation of digital twins: Employment of AI in parallel with the data of experimental nature ensures the possibility of making digital twins of the IL nanofluid systems to simulate all kinds of reactions without physical testing, therefore, simplify and facilitate the process.
(g) Standardization and reproducibility: By learning from large datasets and identifying the most dependable techniques across labs and setups, AI-based frameworks can help establish standardized characterization protocols [66–69].

4 Nanofluids: Thermal, Rheological, and AI-Optimized Performance

Nanofluids are specially formulated colloidal suspensions containing nanoparticles (typically less than 100 nm) dispersed in base fluids such as oils, ethylene glycol, or water. These fluids exhibit enhanced thermophysical properties compared to conventional fluids, demonstrating increased convective heat transfer coefficients and thermal conductivity. The enhancements in thermal performance are primarily attributed to mechanisms like Brownian motion, nanoparticle clustering, and the development of nanolayers at the solid–liquid interface performance [70, 71].

The synthesis of nanofluids involves either a one-step or two-step method. The one-step method simultaneously synthesizes and disperses nanoparticles into the base fluid, reducing agglomeration. In contrast, the two-step method first produces nanoparticles, which are then dispersed into the base fluid using techniques like ultrasonication. Stability is a critical concern; thus, surfactants or pH adjustments are employed to prevent sedimentation and ensure uniform dispersion.

Nanofluids find applications across various sectors. In heat exchangers, they enhance heat transfer rates, leading to more compact and efficient systems. Electronics benefit from improved cooling solutions, addressing the challenges of miniaturization and high heat fluxes. In renewable energy, nanofluids are utilized in solar collectors and photovoltaic/thermal systems to boost energy conversion efficiency. Additionally, their role in biomedical applications, such as targeted drug delivery and hyperthermia treatment, is under exploration due to their unique thermal properties [70, 71].

Despite these advantages, challenges persist. Issues like nanoparticle agglomeration, increased viscosity at higher concentrations, and potential environmental impacts necessitate ongoing research. Future directions focus on developing hybrid nanofluids, integrating phase change materials, and ensuring sustainable production and disposal methods [70, 71].

4.1 Role of Nanoparticles in Enhancing Heat Transfer and Stability

Nanoparticles enhance the way heat is transferred in base fluids, making them very steady and strong due to their excellent physical features. Scattering nanoparticles in a fluid makes thermal conductivity stronger, increases convective heat transfer rates, and for some cases, changes viscosity which is useful in heat exchangers, solar collectors, and cooling systems for electronics [74, 75]. The way nanoparticles are made, including their size, shape, material, and concentration, affects their influence on heat transfer. When size is taken into account, smaller nanoparticles tend to be better at helping energy transfer [76]. To maintain the stability of nanoparticles, it is essential to either use surfactants or modify the surface of the particles. As stated by [77], hybrid nanofluids mixed with varied nanoparticles transfer heat more effectively and remain steady in suspension. Putting bio- or polymer-based stabilizers onto the surface and using zeta potential helps the suspension to remain stable over a longer period. Therefore, properly modifying the particles and how they react with the main fluid is important to maximize heat transfer and maintain good fluid performance in a system for a long time.

4.2 Machine Learning Models for Predicting Thermophysical Properties

Thanks to latest advances in high-end machine learning, we are now able to measure the thermophysical properties of many materials such as fluids, nanofluids, alloys, and biological systems more precisely than before. Traditional methods such as empirical or regression factors tend to fail at precisely describing the complicated and non-straight nature of e-materials. Meanwhile, applying techniques such as artificial neural networks (ANNs), support vector machines (SVMs), random forests (RFs), and deep learning leads to better results compared to conventional models when it comes to forecasting thermal conductivity, viscosity, and specific heat and how an agent undergoes phase changes [78]. Models, when combined with automated features and large libraries, are able to predict new properties quickly, so fewer experiments are required. In agreement with [79], both ANNs and SVMs have the ability to reliably predict how concentrated, hot and what type of nanofluid will change its thermal conductivity [80], also explain how transfer learning and ensembles allow the prediction of properties in both fluid and material categories more accurately, using less data. Greater use of machine learning thanks to stronger computing will boost research in areas such as energy, health care, and aerospace.

4.3 AI-Driven Optimization of Nanofluid Formulation and Performance

Thanks to AI and ML, the development of nanofluids has grown and improved. Computer algorithms can be used to estimate the key characteristics of nanofluids by accounting for the size, type, and number of nanoparticles as well as the temperature and fluid [79, 81]. With technologies such as genetic algorithm, particle swarm optimization, and neural networks, scientists now have a way to find the best nanofluids mixture for various applications. The application of AI in combination with experimental data accelerates nanofluids development with the help of reliable predictive modeling in variable or dynamic environments [82]. These benefits find special relevance for cases when such applications are used in the areas of energy systems, solar thermal collectors, electronics cooling, and biomedical devices; the fast and trustable optimization is critical in these cases. By intelligent data analysis and automatic learning, AI enables economical and environment friendly production of high performance nanofluids.

4.4 Real-World Applications: Energy, Cooling Systems, and Manufacturing

Nanofluids, cutting-edge cooling technologies, and embedded devices have gained importance as enablers in enhancing energy efficiency, sustainability, and system reliability for worldwide industries. Cold thermal energy storage (CTES), solar-assisted cooling, and thermoelectric cooling systems are becoming increasingly common in energy systems because of their capacity to reduce fossil fuel consumption and manage peak loads [83, 84]. CTES and solar collectors directly benefit from increased conductivity and heat capacity in nanofluids and can be supported by enhanced thermal performance that allows for effective energy capture and conservation in these technologies. The high temperatures characteristic of electronic and data centers find application in technologies like spray cooling, radiative cooling, and hybrid nanofluids providing superior thermal management. Additionally, manufacturing areas such as mining, food and beverage processing, and building management can also enjoy combined cooling, heating, and power while utilizing nanofluids to facilitate reduced energy consumption and increased operational stability [85]. Building of these systems has a fundamental role in manufacturing and construction processes attained through emission reduction and increased equipment reliability and lifecycle. It is, therefore, no longer a secret that the use of nanofluids in cooling systems is rapidly hailed as an integral part of sustainable development in industry.

5 Bionanofluids: Biocompatibility, Stability, and AI-Enhanced Insights

Bionanofluids are a combination of biological substances such as blood and interstitial fluids and nanoparticles, size 1–100 nm, designed to enhance their thermal, mechanical, and transport properties [86]. The selection of nanoparticles metallic, metallic oxides, carbon-based such as nanotubes and graphene, or polymers provides a large surface area which interacts with the biological fluid, enhancing the general effectiveness of the bionanofluid system [87].

Biological media such as blood and saline are prevalent in bionanofluids, and the nanoparticles affect behavior through better thermal conductivity. Biocompatibility which is paramount in medical and pharmaceutical fields is essentially based on this interaction [88]. In addition, the stimuli-responsive nature of nanoparticles in bionanofluids demonstrates their underutilization potential in dynamic treatments like drug delivery and biosensing [89].

Medicine applications of bionanofluids include targeted drug delivery, whereby nanoparticles can be used to deliver therapeutic materials at pinpoint locations [90] and thermal cancer therapy where nanoparticles convert electromagnetic energy to heat to kill cancer cells (Gao & Nie). Apart from the imaging and wound healing, they are also used in artificial organs and biomedical systems [91].

5.1 Role of Biomolecules and Nanoparticles in Bionanofluid Design

The advancement of bionanofluids is critically dependent on the synergistic interaction between nanoparticles and biomolecules, which collectively govern their physicochemical properties and biomedical functions. The conjugation of biomolecules—such as proteins, lipids, nucleic acids, and peptides—with nanoparticles imparts functional specificity and improves biocompatibility, facilitating their application in targeted drug delivery, diagnostics, and therapeutics.

Biomolecular components are selected based on their compatibility with biological systems, targeting capability, and ability to enhance therapeutic efficacy. Functionalization with proteins or peptides enables cellular uptake and site-specific targeting, particularly in cancer therapy by binding to overexpressed receptors [92]. Antibody- or aptamer-conjugated nanoparticles significantly improve site-specific delivery, reducing off-target effects, and enhancing therapeutic outcomes [90]. Moreover, biomolecule-coated nanoparticles mimic endogenous surfaces, thereby evading immune clearance and enhancing in vivo stability [93].

Nanoparticles contribute to bionanofluid engineering by providing a high surface area and tunable size, facilitating efficient transport and thermal conduction. Metallic nanoparticles like gold and silver offer superior thermal conductivity for hyperthermia-based cancer therapy [94]. Additionally, biodegradable polymeric nanoparticles enable safe degradation post-treatment, while stimuli-responsive systems permit controlled drug release [87]. The integration of nanoparticles and biomolecules yields a synergistic effect wherein nanoparticles determine the physical properties, and biomolecules confer biological specificity. For instance, antibody-coated nanoparticles ensure selective action on tumorous tissues while sparing healthy cells [95]. This strategic integration enables the rational design of bionanofluids with enhanced efficacy and safety for biomedical applications.

5.2 Machine Learning Approaches for Biocompatibility Prediction

Machine learning methods, such as ensemble methods, and deep neural network models have been very effective in predicting biocompatibility, as well as other material properties on biology and material science's fronts. Optimizing synthesis parameters, improving process efficiency, and streamlining experimental protocols, these methods contribute to the acceleration of generation of biocompatible materials such as nanoparticles employed in the biomedical environment.

Ensemble Methods in Biocompatibility Prediction

Random Forest (RF), Gradient Boosting Machines (GBM), and AdaBoost are methods where their predictions are reliable on account of the aggregation of results of several different models. Due to their ability to manage intricate, high-dimensional information from chemical structure, physicochemical features, and bioassay results, these methods make good tools for biocompatibility studies.

(a) Random Forest (RF): This method works by combining several decision trees and consolidating their results to improve general prediction reliability. Since RF can handle big and variated datasets, it is commonly used to forecast factors such as toxicity, biocompatibility, and biological aspect of materials [96]. These models via RF can emphasize the critical features that impact biocompatibility, especially in relation to the chemical morphology and size of nanoparticles.
(b) Gradient boosting machines (GBM): GBM creates strong predictive models by means of an iterative process of having weaker models deal with error on top of the previous models. Owing to its ability to predict nanoparticle-

biological system interactions, this approach is suitable for the assessment of the biocompatibility of new materials (Joubert, McGuire, & Hend).

(c) AdaBoost: Mislabeled data are weighted more in the AdaBoost algorithm causing progressively better estimates of the predictions. It shines for datasets with an unevenly distributed class, which is representative of the standard case in biocompatibility study where most samples are non-toxic.

Deep Neural Networks (DNNs) for Biocompatibility

Deep neural networks (DNNs), with their layers of neurons that emulate the processing of the human brain, have become indispensable to predicting complex non-linear correlations in biocompatibility study. DNNs are capable of handling large and disparate datasets which may range from molecular genetics to clinical imaging and assay responses and it is on this basis that they provide reliable biocompatibility assessment.

(a) Convolutional neural networks (CNNs): While not originally used in image recognition, CNNs have also been valuable in determining interaction between nanoparticles and biological systems by evaluating structural or 3D molecular information. They can identify the subtle relationships including how particle geometry and surface affect general biocompatibility [97].
(b) Recurrent neural networks (RNNs): LSTM networks, belonging to the RNN family, do an outstanding job in handling long and short data cuts and are suited for investigating biological data related to time-series interaction of nanoparticles between cells [98].

5.3 Advantages of Using ML in Predicting Biocompatibility

ML gives the field of biocompatibility prediction with some important advantages.

(a) Reduced experimental workload: Thanks to machine learning, scientists can accurately predict whether materials are suitable for the body which means they need fewer experiments, saving both time and resources. Using machine learning can speed up the entire phase of developing and testing materials.
(b) Optimization of synthesis parameters: Machine learning-based prediction modeling helps identify the optimal synthesis parameters for nanoparticles, including their size, shape, and surface coating, directly impacting their biocompatibility and potential clinical applications [99].
(c) Enhanced accuracy and efficiency: Researchers can handle huge amounts of various data and get precise results for their predictions by using ensemble methods and deep learning architectures in ML. This technique is particularly useful for the prediction of biocompatibility of complex, experimental materials for which there is little existing data.

(d) Personalized medicine: Machine learning allows the manufacturing of customized materials that help satisfy specific needs of each patient, such as the production of biocompatible drug delivery systems targeted for the specific biological characteristics of each patient.

5.4 AI-Driven Optimization of Drug Delivery and Therapeutic Applications

Artificial Intelligence (AI) is also changing drug delivery and therapeutic applications because it enhances precision, efficiency, and personalization in the mode of treatment approach. All these developments have left their mark in areas such as nanomedicine, adapted dosing strategies and complex delivery systems.

Smart Nanocarriers and Controlled Drug Release

The use of target drug delivery was facilitated by AI-assisted development of smart multifunctional nanocarriers (SMNs). AI has been utilized by researchers in the context of nanomaterials such as polymeric, lipidic, and inorganic nanoparticles to enhance cargo loading, release modulation, and target delivery in managing oncology and hematologic disorders. Using machine learning, scientists are in a position to play with nanomaterials manufacturing to optimize sizes, surface characteristics, and controlled drug release that meets the requirements for targeted therapies.

Personalized Dosing and Adaptive Therapies

With the real-time dose adjustments facilitated by AI, clinicians are able to make small changes in allocation of treatment according to patient's prior reactions, thus making that treatment more effective and minimizing side effects. Having MIPD, the interaction of pharmacometric models with AI technologies provides personalized dosing specific to individuals, leading to elevated treatment success.

AI in Drug Discovery and Development

AI speeds up drug discovery by examining extensive datasets to pinpoint potential drug candidates and anticipate their interactions. Firms such as insilico medicine leverage AI to create innovative drugs, with several advancing to mid-stage human trials. Likewise, insitro uses machine learning to evaluate chemical and biological markers, striving to reduce the duration of drug development [100].

AI-Enabled Nanorobotics for Targeted Therapy

Recent advancements in AI have resulted in the creation of nanorobots that can navigate precisely and deliver drugs directly to their targets. These nanorobots detect and engage with cancer cells by analyzing biomarkers, opening up possibilities for personalized medicine and less invasive treatment options [101].

5.5 *Case Studies: AI-Powered Innovations in Bionanofluids for Medicine*

Artificial Intelligence (AI) is progressively integrated with bionanofluid technologies to improve multiple medical applications. Here are several case studies that highlight the influence of AI-driven innovations in this area:

AI-Optimized Nanoparticles for Targeted Drug Delivery

Scientists are utilizing AI algorithms to create nanoparticles that can efficiently transport therapeutic molecules such as mRNA directly to cancer cells. This groundbreaking method greatly enhances the accuracy and efficacy of treatments while substantially reducing side effects. The effectiveness of these AI-crafted nanoparticles exceeds that of conventional models, highlighting the significant potential of AI in the field of nanotherapeutics.

AI-Aided Geometric Design of Anti-Infection Catheters

A recent investigation applied artificial intelligence-based methods like Fourier neural operators, powered by AI, to update the geometry of urinary catheters. It works by breaking up the movements that push bacteria up the filter, so there is less bacterial contamination. Those catheters improved by using artificial intelligence had lower infection rates, highlighting the key role AI plays in designing biomedical equipment.

AI-Driven Nanoparticle Drug Discovery

Artificial Intelligence (AI) is transforming drug discovery and delivery by accelerating research, optimizing treatment regimens, and improving patient outcomes. AI-driven techniques, including machine learning and deep learning, enable pharmaceutical companies to identify drug targets, predict molecular interactions, and streamline clinical trials. AI also enhances personalized medicine by analyzing patient data to tailor treatments based on genetic and lifestyle factors. Additionally, AI facilitates drug repurposing, uncovering new therapeutic uses for existing drugs. Despite its benefits, challenges such as regulatory concerns, ethical considerations, and data privacy remain. Continued collaboration among researchers, clinicians, and policymakers is essential for AI's successful integration into healthcare [101].

AI in Personalized Medicine with Nanodiamonds

Dr. Dean Ho is investigating the combination of artificial intelligence and nanodiamond particles for the delivery of drugs to different cancer patients. CURATE.AI, an AI system, adjusts drug combinations and doses for patients, helping improve the results of their treatments. It postulates that AI and nanotechnology are working together to improve precision medicine [102].

AI-Enhanced Nanofluids for Biomedical Applications

Artificial intelligence is now being used by researchers to improve and understand how nanofluids function in biomedical applications. Using neural networks, experts can enhance the thermal and fluid properties of nanofluids for use in drug delivery and other health fields.

AI and ML for Predicting and Optimizing Nanomaterial Properties

Today, AI helps predict the features of complex fluids, for instance, ionic liquids (ILs), nanofluids and bionanofluids. By depending on machine learning, scientists can understand why tiny changes in molecules may cause significant modifications in major properties with just a few laboratory tests.

AI in Predicting Properties of Ionic Liquids (ILs)

Thanks to their range of properties, ionic liquids are difficult to model with normal methods since they come in various shapes and forms. Approaches using machine learning such as ANNs, SVMs, and ensemble methods have been useful in forecasting various characteristics of ionic liquids which include their viscosity, conductivity, and toxicity. For example, research has found that machine learning models can estimate the viscosity of mixtures involving ionic liquids and water and when these models are combined with the UNIFAC model, they provide an extremely accurate result [103].

AI in Nanofluid Property Prediction

Nanofluids have higher thermal capabilities, but it is difficult to model them due to the effect of their nanoparticles such as size and shape. Both ANNs and genetic algorithms have proven to be more effective than traditional methods in simply predicting thermal conductivity and specific heat capacity. It has been found that AI models outperform older methods in predicting the thermal properties of nanofluids.

Applications in Bionanofluids

More and more, bionanofluids that include both biological parts and nanomaterials are being used in the field of biomedical engineering. Techniques in AI are useful for predicting important characteristics like whether materials are safe for use in the body and how the blood will react to them. For instance, machine learning helps to predict how hybrid nanofluids transfer heat, helping to design better drug delivery systems.

6 Deep Learning and Neural Networks for Material Design

DL and neural networks have helped the field of material design to predict complicated material features and led to the development of new mixtures. They are very effective at understanding complex patterns found in material data that have many dimensions. For example, deep learning techniques have been used to research material design spaces and address challenges in predicting the behavior of new domains by means of active learning and increasing training data. Moreover, DL techniques have been used in atomistic simulations, imaging of materials, and analysis of spectra, further showcasing their usefulness in materials science [63].

6.1 Hybrid AI Approaches: Combining Computational and Experimental Data

Mixing simulations with experimental observations in AI increases both the effectiveness and accuracy of finding new materials. Experimental methods combined with machine learning algorithms help researchers to predict material properties faster. Using AI together with Integrated Computational Materials Engineering (ICME) has helped make discoveries in materials science and accelerate its digital transformation. They help scientists smoothly progress from predictions made by computers to experiments in the lab.

6.2 ML-Enabled High-Throughput Screening for Novel Nanofluids

Advances in machine learning (ML) have greatly improved high-throughput screening (HTS) for seeking and refining new nanofluids. Quickly processing extensive information, ML models can tell us the thermal conductivity and viscosity of various nanofluids. Using machine learning, HTS has boosted the development of next-generation fuels, proving ML's ability to improve material discovery. Moreover, using microfluidic technology, high-content imaging and active ML at each stage have helped to streamline the creation of nanoparticles.

6.3 Limitations and Challenges in AI-Driven Nanotechnology

Despite the importance of AI in nanotechnology, there are many problems to be addressed. Issues that need to be handled are working with excellent data, crafting consistent methods, and linking up scientists working in various fields. Issues related to ethics and putting the right regulations in place are preventing nanotechnology tools connected to AI from being approved. Studies from recent years discovered challenges linked to cooperation between different teams, doubts about data accuracy, issues with ethics, and difficulties applying the involved regulations. To apply AI in nanotechnology without risk, these situations must be managed well.

7 Experimental and Computational Synergies in AI-Driven Nanotechnology

XRD, AFM, and HAADF-STEM tools use AI to improve their performance. As an illustration, by training AI-STEM on simulated images with Bayesian convolutional neural networks, scientists can discover crystal structures and interfaces in both simulated and real images. Thanks to artificial intelligence, XRD, AFM, and HAADF-STEM become more accurate, have less bias, and discover results more quickly. Therefore, AI-STEM uses Bayesian convolutional neural networks on simulated pictures, allowing it to recognize both crystal structures and interfaces in both natural and created images which makes materials analysis both more accurate and efficient [104]. Using machine learning in electron microscopy makes it simpler to analyze data and fastens the process of making new discoveries [105].

7.1 Role of Quantum Computing and AI in Nanoscience

If AI is added to quantum computing, progress can be made in the field of nanoscience. Quantum computers enable scientists to manufacture new nanomaterials with more exact precision. As an illustration, exploring quantum theories help develop novel materials and equipment in nanotechnology [106]. Jointly, AI and quantum experiments prepare the ground for research to progress in nanoscience at a faster rate [106].

7.2 AI in Molecular Dynamics and Computational Fluid Dynamics (CFD)

Artificial intelligence has made molecular dynamics (MD) and computational fluid dynamics (CFD) more effective and efficient. MD applications using AI, for example AI^2BMD, can precisely model large biomolecules with the same level of accuracy as ab initio calculations, without using a lot of computing time [107]. Using AI, CFD can now improve flow prediction, boost optimization, and reduce computing costs which aids in designing and examining complex fluid systems [108].

7.3 Data Acquisition, Curation, and Interpretation for AI Modeling

For AI models to function well in nanoscience, accurate collection and management of data are key. Data that is correctly annotated gives machines and programs the ability to provide accurate insights. When data curation is paired with machine learning, it allows for more effective predictions of nanoparticle behaviors as well as better formulation adjustments, compared to traditional methods [109]. Reliable and repeatable AI research depends a lot on strong data curation skills [110].

8 Industrial and Technological Applications of AI in Nanotechnology

The use of artificial intelligence is improving nanotechnology through better material designs, more efficient processes and increased scalability in industry. AI is speeding up the linking of structure with function and aiding in developing new kinds of nanostructures. AI is being used in manufacturing to assist in controlling quality and detecting faults right away. It is important for developing main innovations in nanoelectronics, for instance, neuromorphic systems and flexible electronics. AI technology helps smart nanosensors monitor healthcare and the environment, so data can be analyzed while events happen. Advantages of AI in nanomedicine involve better delivery of nanoparticle medications and better outcomes for treatments. Additionally, AI assists in studying materials by handling image analysis in very detailed microscopy. It advances green nanotechnology by finding sustainable methods and lowering its effect on the environment. In the end, AI helps maintain quality and predict when maintenance is needed in large-scale nanomaterial production. The meeting of AI and nanotechnology is bringing new ideas to various sectors and offering solutions that are more effective, correct, and good for the environment.

8.1 AI-Driven Nanofluid Applications in Heat Transfer and Energy Storage

Nanofluids are produced by combining nanoparticles within base fluids which leads to much higher thermal conductivity, allowing them to see wide use in heat transfer and energy storage. Research on nanofluids using AI has allowed for swift improvement in optimization by predicting properties such as temperature, adjusting the particle levels, and boosting heat transfers. AI and especially ML are used to predict the flow of fluids and the heat transferred depending on various operating factors [44]. They have proven successful in predicting what happens with nanofluids in different systems which improves the work of both heat exchangers and thermal storage devices [111].

Currently, research shows that AI can evaluate the role of PCMs and nanofluids in creating more sustainable ways to store energy [112]. Support vector machines and neural networks are used to aid in creating energy storage models, reducing the amount of trial-and-error testing, and helping with better selection of materials.

8.2 AI in Drug Delivery and Biomedical Nanofluid Applications

Thanks to drug delivery systems, cancer treatment and therapy for other persistent diseases can now be focused on specific areas in the body. Thanks to bionanofluids, drugs in biological fluids are more easily soluble, absorbed in the body and guided to the proper parts of the body. Developing and improving these vehicles depend a lot on AI [113] suggest using deep learning to estimate release of the drug, control nanoparticle size, and confirm biocompatibility.

Furthermore, AI can copy nanoparticle's behaviors toward biological materials which helps determine how they interact with cells, tissues, and organs. By studying numerous datasets collected in labs and experiments with animals, AI can discover the best routes to administer drugs and evaluate how effective they might be. AI assists in producing medications that fit a patient's needs well and things up health problems.

8.3 AI for Sustainable and Eco-Friendly Nanomaterials

Since more people want environmentally friendly materials, AI is now vital for producing nanomaterials. AI helps to identify sustainable materials by forecasting their properties and how they influence the environment. Machine learning allows scientists to design nanomaterials useful for various tasks that also support the environment.

Using AI, researchers are able to predict the breakdown of nanomaterials over time, ensure their safety, and suggest alternative materials to replace those often used in industries. Moreover, AI is being exploited to keep waste and energy low in nanomaterials, helping to preserve their sustainability [114]. As a result, this is in line with the global shift to eco-friendly chemicals and ecological ways of making products.

8.4 AI-Enabled Quality Control and Process Optimization in Nanomanufacturing

Nanomanufacturing describes the crafting of objects with very large attention to details. With the help of AI and mainly machine learning, the manufacturing of nanomaterials has gotten better and they now meet higher standards. With AI, manufacturing is possible to watch live and it highlights problems or differences in goods before they negatively impact the final quality [115].

As a result, AI manages things like temperature, pressure, and ingredients to produce more and eliminate waste. It is simpler to stop machinery from malfunctioning and maintain manufacturing process effectiveness with the use of predictive maintenance supported by AI [116]. Thanks to AI, companies may achieve more for less and have a lighter impact on nature.

8.5 Emerging Trends in AI-Driven Industrial Nanotechnology

The industrial application of nanotechnology is accelerating due to AI which is now bringing many new trends and changes to how nanomaterials are used. More researchers are working with generative AI-driven algorithms to invent fresh and exciting nanomaterials that fit users' needs [117]. Using this technique, it becomes easier to find relevant materials needed in energy, electronics, and biomedicine.

AI is now helping by ensuring that nanomanufactured products are monitored and maintained to keep them meeting strict quality rules at all times. AI is now used in nanotechnology to predict the outcomes of nanomaterials in different uses, making it simpler to design and produce prototypes [118]. There is a notable trend in using generative design algorithms that allow AI to produce imaginative nanomaterials based on the user's requirements [119]. It makes it possible to introduce innovative new products for energy, electronics, and medical purposes more swiftly.

The use of AI to continuously track and look after nanomanufacturing is now also becoming popular, to keep the quality of items at a high level. Using AI in models is now helping predict the results of nanomaterials used in everyday applications and saves time by shortening the prototyping and testing phases.

9 Challenges, Ethical Considerations, and Future Directions

A main difficulty in using AI and ML in nanotechnology is dealing with ethics related to AI-assisted observations of materials, for example, by collecting impartial data and considering any public consequences. Additionally, issues related to data protection, openness, and understanding how an AI system works are crucial to people's trust in it.

Generative models, reinforcement learning, and AI working with experimental approaches look set to impact the future of AI-driven nanotechnology. Improvements and proper attention to ethics will help to change the way ionic liquids, nanofluids, and bionanofluids are explored and made. Following ethical practices, using AI combined with experiments will form the basis for significant and lasting development.

1. Enhancing generative models and reinforcement learning techniques: Deep learning and neural networks are expected to make generative models important in discovering new materials. The models by themselves can develop fresh molecular or nanomaterial designs that meet the conditions of stability, suitability for biological solutions, or special functions. With the help of generative models in nanotechnology, predicting synthesis of new nanoparticles and fluids becomes easier, thus significantly decreasing the use of traditional methods in experiments.

   Reinforcement Learning (RL) is a highly praised technique because it looks for the best solution by making updates as it goes. Changes in temperature, pressure, and concentration during nanomaterial synthesis can be adjusted through RL to improve their performance. With real experimental environments, RL algorithms learn and help identify materials that conduct or transfer heat effectively in nanofluids and those that are suitable for use in bionanofluids [120].

In addition, reinforcement learning allows for the automatic finding of new nanofluids used in energy systems by repeatedly improving the material's structural properties, thermal conductivity, and capacity to remain stable at different temperature and pressure values. As the algorithm gains insights from its interactions with the material characteristics, it can devise a set of experimental parameters that are most likely to yield an ideal fluid formulation. This method would significantly decrease the time and resources generally needed for experimental testing.

2. Integrating AI with experimental methods: When AI is used with experiments, it will lead to faster and better discoveries of new materials. Running experiments through classic methods is key, but they regularly face difficulties with the difficult and large amounts of data involved in nanotechnology. Analyzing data from experiments with AI can provide valuable insights into material performance, improving theory confirmation and ways to design better experiments.

   Teaming up high-throughput experimental systems with AI will allow for the collection of vast datasets needed to train machine learning models. Commonly, AFM, XRD, and HAADF-STEM are used to check the structure of nanomaterials. With these fast algorithms, AI is able to identify trends and relationships that experts might miss in all the gathered data.

   Also, by combining AI and molecular dynamics simulations, we can guess how nanoparticles interact at the atomic level. This information allows us to design better models, which results in improved predictions of the characteristics of ionic liquids, nanofluids, and bionanofluids.

   If AI is used to review and improve experimental designs based on past results, it will help to speed up the development of materials and encourage new ideas in the field [121].

3. Addressing ethical considerations in AI-driven nanotechnology: Since AI impacts nanotechnology by developing new materials, we must pay attention to the ethical matters that come with these developments. There are many ethical issues to manage in AI-based discoveries and manufacturing, which need to be closely examined to ensure irresponsibility does not creep into the development.

   The need to store and provide many datasets for AI systems raises the issue of protecting people's personal and health data from being accessed. Since AI is sometimes used in drug delivery in bionanofluids, it is necessary to put solid data privacy rules in place to ensure no patient's information gets misused. AI systems, particularly those based on deep learning, frequently function as "black boxes," making decisions without offering clear reasoning. For AI applications in nanotechnology, especially in the production of materials with specific biological or chemical properties, it is crucial to enhance the interpretability of these systems. It is necessary for AI researchers and developers to be able to back up and explain the AI model's decisions, especially for gaining approvals and keeping products safe. Developing XAI techniques can help make AI models more understandable.

   Bias in AI models: Biases present in the dataset can be passed to the AI and cause unfair outcomes in the development of educational materials. Because of biased AI, the products created by nanotechnology could be suitably effective for some situations but not everywhere. To explain, AI-generated nanomaterials may not consider every group but might intentionally target some. It is important to deal with bias in AI so that everyone is represented fairly in the development process.

The use of AI in search for new materials should always consider the ethical consequences regarding potential impacts on nature and society. When rare and dangerous substances are used in nanotechnology, they can harm the environment for an extended period. AI helps reduce the damage caused by production by detecting natural resources and recycling waste. The topic of AI-assisted manufacturing and the outcomes for job holders and new distribution of benefits should be addressed properly [122].

4. Fostering responsible innovation in nanotechnology: For AI-driven nanotechnology to develop successfully, experts from diverse areas need to collaborate. Researchers, policymakers, and ethicists should all contribute to how we should ethically develop AI-enhanced materials. With advancements in AI and nanotechnology, new rules should be created to verify new discoveries in science and ethics.

Moreover, working closely with residents and informing them can help support responsible innovation. Those involved in science, industry, and the public should understand both the achievements and the problems faced by AI and nanotechnology. Therefore, nanomaterials will help to achieve what is wanted and needed in healthcare, energy, and the environment [123].

10 Conclusion

The integration of ML and AI within the realm of nanotechnology has led to significant progress in the development and enhancement of advanced materials, including ionic liquids, nanofluids, and bionanofluids. These materials hold great promise for a range of applications, such as energy storage, heat transfer, drug delivery, and biomedical diagnostics.

Machine learning approaches, especially ensemble techniques and deep learning models, have enabled the prediction of material properties, optimized synthesis processes, and accelerated the discovery of new materials. Because of AI, it is now possible to make materials with fewer experimental tests which save both time and money. Thanks to artificial intelligence, the improvement of nanofluids has become easier by predicting the best features of nanomaterials (such as size, shape, and surface charge) that boost their results in thermal functions.

Through machine learning, researchers have found ionic compounds that are reliable, less volatile, and effective for uses in carbon capture and energy storage. For this reason, artificial intelligence has assisted in making nanofluids suited to living bodies and effective for targeting specific areas. Machine learning has allowed scientists to forecast the effects of nanoparticles on biological systems, so they can be used more effectively in drug delivery. Furthermore, AI combines information from different types of experiments and studies to support the creation of nanomaterials that are stronger and more practical. AI is also being used in reinforcement learning to develop systems capable of generating, testing, and improving materials, helping the world of material science progress at a quicker rate.

Even so, data quality, explaining how models work and the ability to use these methods on large scales remain issues. For artificial intelligence in nanotechnology to succeed, it must consider and solve these issues for broader use, clear understanding and ethical oversight when establishing new materials.

Basically, the integration of machine learning, artificial intelligence, and nanotechnology points to a significant transformation in our approach to developing and improving advanced materials. They can help improve many areas such as energy and healthcare, leading to achievements that people once believed could not be done. Future studies are expected to help merge AI with nanotechnology which will bring about smarter and eco-friendly materials that might resolve some of the significant issues facing society today.

References

1. M. Nasrollahzadeh, S.M. Sajadi, M. Sajjadi, Z. Issaabadi, An introduction to nanotechnology, in *An Introduction to Green Nanotechnology*, ed. by M. Nasrollahzadeh, S.M. Sajadi, M. Sajjadi, Z. Issaabadi, (Elsevier, Amsterdam, 2019), pp. 1–27. https://doi.org/10.1016/B978-0-12-813586-0.00001-8
2. M.S. Javed, T. Batool, S. Ahmad, I. Noreen, S. Javed, Z.A. Bajwa, Rimsha, K. Nazir, Haya Saudi J. Life Sci. (2024). https://doi.org/10.36348/sjls.2024.v09i12.004
3. P. Ray, M. Pilania, The application and influence of ionic liquids in nanotechnology. Mater. Today Proc. (2021). https://doi.org/10.1016/j.matpr.2021.03.602
4. H. Ben Bacha, N. Ullah, A. Hamid, A. Shah, A comprehensive review on nanofluids: Synthesis, cutting-edge applications, and future prospects. Int. J. Fluid Thermal Eng. **10**, 100595 (2024). https://doi.org/10.1016/j.ijft.2024.100595
5. D. Wang, S. Sun, T. Sha, T. Liu, H. Dong, K. Cui, H. Li, Y. Gong, J. Hou, Z. Zhang, P. Fu, Synergistic effect of silica nanoparticles and rhamnolipid on wettability alteration of low permeability sandstone rocks. Energy Fuel **32**(12), 12412–12420 (2018). https://doi.org/10.1021/acs.energyfuels.8b01207
6. K. Mishra, K. Rahman, S. Eldin, M. Bani-Fwaz, Investigation of blood flow characteristics saturated by graphene/CuO hybrid nanoparticles under quadratic radiation using VIM: Study for expanding/contracting channel. Sci. Rep. **13**, 35695 (2023). https://doi.org/10.1038/s41598-023-35695-3
7. J. Hou, H. Sui, J. Du, L. Sun, Synergistic effect of silica nanofluid and biosurfactant on bitumen recovery from unconventional oil. J. Dispers. Sci. Technol. **43**(5), 704–715 (2020). https://doi.org/10.1080/01932691.2020.1844741
8. S. Amirthalingam, B. Thangavel, On the thermal conductivity and viscosity of bionanofluid with neem (Azadirachta indica) assisted zinc oxide nanoparticles. J. Therm. Sci. Technol. **15**(1), JTST0023 (2020). https://doi.org/10.1299/jtst.2020jtst0023

9. J. Jiménez-Pérez, G. López-Gamboa, Z. Correa-Pacheco, J. Sánchez-Ramírez, M. Salazar-Villanueva, Preparation and enhancement of thermal diffusivity of biodiesel from Nutrioli oil filled with ag nanoparticles. J. Renew. Sustain. Energy **7**(6), 063104 (2015)
10. I. Dotan, P. Roche, M. Paliouras, E. Mitmaker, M. Trifiro, Engineering multi-walled carbon nanotube therapeutic bionanofluids to selectively target papillary thyroid cancer cells. PLoS One **11**(3), e0149723 (2016). https://doi.org/10.1371/journal.pone.0149723
11. Y. Cao, A. Nakhjiri, M. Ghadiri, Different applications of machine learning approaches in materials science and engineering: Comprehensive review. Eng. Appl. Artif. Intell. **135**, 108783 (2024). https://doi.org/10.1016/j.engappai.2024.108783
12. Y. Hong, B. Hou, H. Jiang, J. Zhang, Machine learning and artificial neural network accelerated computational discoveries in materials science. Wiley Interdiscip. Rev. Comput. Mol. Sci. **10**(5), e1450 (2019). https://doi.org/10.1002/wcms.1450
13. N. Kühl, M. Schemmer, M. Goutier, G. Satzger, Artificial intelligence and machine learning. Electron. Mark. **32**, 2235–2244 (2022). https://doi.org/10.1007/s12525-022-00598-0
14. S. Badillo, B. Bánfai, F. Birzele, I. Davydov, L. Hutchinson, T. Kam-Thong, J. Siebourg-Polster, B. Steiert, J. Zhang, An introduction to machine learning. Clin. Pharmacol. Therap. **107**(4), 871–885 (2020). https://doi.org/10.1002/cpt.1796
15. F. Ernst, A. Schweikard, *Fundamentals of Machine Learning* (Elsevier, Amsterdam, 2020)
16. L. Yang, H. Wang, D. Leng, S. Fang, Y. Yang, Y. Du, Machine learning applications in nanomaterials: Recent advances and future perspectives. Chem. Eng. J. **468**, 156687 (2024). https://doi.org/10.1016/j.cej.2024.156687
17. J. Li, K. Lim, H. Yang, Z. Ren, S. Raghavan, P. Chen, T. Buonassisi, X. Wang, AI applications through the whole life cycle of material discovery. Matter **3**(3), 892–907 (2020). https://doi.org/10.1016/j.matt.2020.06.011
18. H. Chen, Y. Zheng, J. Li, L. Li, X. Wang, AI for nanomaterials development in clean energy and carbon capture, utilization and storage (CCUS). ACS Nano **17**(5), 4567–4581 (2023). https://doi.org/10.1021/acsnano.3c01062
19. Y. Gao, Z. Zhu, Z. Chen, M. Guo, Y. Zhang, L. Wang, Z. Zhu, Machine learning in nanozymes: From design to application. Biomater. Sci. **12**(4), 1234–1245 (2024). https://doi.org/10.1039/d4bm00169a
20. K. Kasture, P. Shende, Amalgamation of artificial intelligence with nanoscience for biomedical applications. Arch. Comput. Methods Eng. **30**(3), 4667–4685 (2023). https://doi.org/10.1007/s11831-023-09948-3
21. S. Venkatesan, M. Cullinan, M. Baldea, Recent advances in continuous nanomanufacturing: Focus on machine learning-driven process control. Rev. Chem. Eng. **40**(1), 123–145 (2024). https://doi.org/10.1515/revce-2024-0029
22. N. Almakayeel, S. Desai, S. Alghamdi, M. Qureshi, Smart agent system for cyber nanomanufacturing in industry 4.0. Appl. Sci. **12**(12), 6143 (2022). https://doi.org/10.3390/app12126143
23. D. Gulevich, I. Nabiev, P. Samokhvalov, Machine learning–assisted colloidal synthesis: A review. Mater. Today Chem. **25**, 101837 (2024). https://doi.org/10.1016/j.mtchem.2023.101837
24. H. Lim, G. Kim, G. Heo, U. Jeong, M. Kim, D. Jeong, Y. Hyun, D. Kim, Nanoscale single-vesicle analysis: High-throughput approaches through AI-enhanced super-resolution image analysis. Biosens. Bioelectron. **263**, 116629 (2024). https://doi.org/10.1016/j.bios.2024.116629
25. J. Scharf, M. Chouchane, D. Finegan, B. Lu, C. Redquest, M. Kim, W. Yao, A. Franco, D. Gostovic, Z. Liu, M. Riccio, F. Zelenka, J. Doux, Y. Meng, Bridging nano- and microscale X-ray tomography for battery research by leveraging artificial intelligence. Nat. Nanotechnol. **17**(5), 446–459 (2021). https://doi.org/10.1038/s41565-022-01081-9
26. M. Lau, A. Burleigh, J. Terry, M. Long, Materials characterization: Can artificial intelligence be used to address reproducibility challenges? J. Vac. Sci. Technol. A **41**(2), 021001 (2023). https://doi.org/10.1116/6.0002809

27. S. Kelly, Y. Trenikhina, W. Harris, AI-powered 3D image reconstruction for analyzing lithium-ion battery microstructures at the nanoscale. Meet. Abstr. **1**, 360 (2024)
28. S. Zafar, N. Rana, The convergence of nanotechnology and artificial intelligence: Unlocking future innovations. Recent Innov. Chem. Eng. **14**(2), 125–156 (2025). https://doi.org/10.2174/0124055204359215250127071256
29. P. Tan, X. Chen, H. Zhang, Q. Wei, K. Luo, Artificial intelligence aids in development of nanomedicines for cancer management. Semin. Cancer Biol. **86**, 1–12 (2023). https://doi.org/10.1016/j.semcancer.2023.01.005
30. M. Jena, B. Pathak, Development of an artificially intelligent nanopore for high-throughput DNA sequencing with a machine-learning-aided quantum-tunneling approach. Nano Lett. **23**(5), 1234–1245 (2023). https://doi.org/10.1021/acs.nanolett.2c04062
31. A. Edriss, S. Yarra, J. Vomo, K. Ismael, Y. Elshiekh, AI-powered nano formulation: Revolutionizing drug development and delivery. Int. J. Sci. Res. Arch. **14**(2), 49–60 (2025). https://doi.org/10.30574/ijsra.2025.14.2.0491
32. I. Rehman, I. Ullah, H. Khan, M. Guellil, J. Koo, J. Min, S. Habib, M. Islam, M. Lee, A comprehensive systematic literature review of ML in nanotechnology for sustainable development. Nanotechnol. Rev. **13**(1), 112–179 (2024). https://doi.org/10.1515/ntrev-2024-0069
33. S.K. Singh, A.W. Savoy, Ionic liquids synthesis and applications: An overview. J. Mol. Liq. **297**, 112038 (2020). https://doi.org/10.1016/j.molliq.2019.112038
34. H. Liu, H. Yu, Ionic liquids for electrochemical energy storage devices applications. J. Mater. Sci. Technol. **35**(4), 674–686 (2019). https://doi.org/10.1016/j.jmst.2018.10.007
35. M. Watanabe, M.L. Thomas, S. Zhang, K. Ueno, T. Yasuda, K. Dokko, Application of ionic liquids to energy storage and conversion materials and devices. Chem. Rev. **117**(10), 7190–7239 (2017). https://doi.org/10.1021/acs.chemrev.6b00504
36. N. Nasirpour, M. Mohammadpourfard, S.Z. Heris, Ionic liquids: Promising compounds for sustainable chemical processes and applications. Chem. Eng. Res. Des. **160**, 264–300 (2020). https://doi.org/10.1016/j.cherd.2020.06.006
37. S. Chakraborty, P.K. Panigrahi, Stability of nanofluid: A review. Appl. Therm. Eng. **174**, 115259 (2020). https://doi.org/10.1016/j.applthermaleng.2020.115259
38. J. Boopathy, R. Pari, M. Kavitha, P.C. Angelo, Preparation of nano fluids by mechanical method. AIP Conf. Proc. **1461**(1), 218–221 (2012)
39. D.W. Oh, A. Jain, J.K. Eaton, K.E. Goodson, J.S. Lee, Thermal conductivity measurement and sedimentation detection of aluminum oxide nanofluids by using the 3ω method. Int. J. Heat Fluid Flow **29**(5), 1456–1461 (2008). https://doi.org/10.1016/j.ijheatfluidflow.2008.04.007
40. S. Chakraborty, I. Sarkar, D.K. Behera, S.K. Pal, S. Chakraborty, Experimental investigation on the effect of dispersant addition on thermal and rheological characteristics of TiO_2 nanofluid. Powder Technol. **307**, 10–24 (2017). https://doi.org/10.1016/j.powtec.2016.11.016
41. S. Chakraborty, I. Sengupta, I. Sarkar, S.K. Pal, S. Chakraborty, Effect of surfactant on thermo-physical properties and spray cooling heat transfer performance of CuZn-Al LDH nanofluid. Appl. Clay Sci. **168**, 43–55 (2019). https://doi.org/10.1016/j.clay.2018.10.034
42. S. Koutsoukos, F. Philippi, F. Malaret, T. Welton, A review on machine learning algorithms for the ionic liquid chemical space. Chem. Sci. **12**(23), 6820–6843 (2021). https://doi.org/10.1039/d1sc01000j
43. Y. Chen, C. Shen, L. Yang, Deep learning-enabled rapid prediction and optimization of nanofluid thermal properties. Energy Rep. **9**, 1221–1235 (2023). https://doi.org/10.1016/j.egyr.2022.12.085
44. A. Basu, A. Saha, S. Banerjee, P.C. Roy, B. Kundu, A review of artificial intelligence methods in predicting thermophysical properties of nanofluids for heat transfer applications. Energies **17**(6), 1351 (2024). https://doi.org/10.3390/en17061351
45. N. Halalsheh, O. Alshboul, A. Shehadeh, R.E. Al Mamlook, A. Al-Othman, M. Tawalbeh, A.S. Almuflih, C. Papelis, Breakthrough curves prediction of selenite adsorption on chemically modified zeolite using boosted decision tree algorithms for water treatment applications. Water **14**(16), 2519 (2022). https://doi.org/10.3390/w14162519

46. Y. Freund, L. Mason, The alternating decision tree learning algorithm, in *Proceedings of the 16th International Conference on Machine Learning*, (Morgan Kaufmann, San Francisco, 1999), pp. 124–133. https://cseweb.ucsd.edu/~yfreund/papers/atrees.pdf
47. I.A. Nepomuceno-Chamorro, J.S. Aguilar-Ruiz, J.C. Riquelme, Inferring gene regression networks with model trees. BMC Bioinf. **11**, 517 (2010). https://doi.org/10.1186/1471-2105-11-517
48. P.T. Prasetyaningrum, I. Pratama, A.Y. Chandra, Implementation of machine learning to determine the best employees using random forest method. Int. J. Comput. Netw. Secur. Inf. Syst. **2**(1), 53–59 (2021)
49. B. Padmaja, V. Prasa, K. Sunitha, A novel random split point procedure using extremely randomized (extra) trees ensemble method for human activity recognition. EAI Endors. Trans. Pervasive Health Technol. **6**(20), e5 (2020). https://doi.org/10.4108/eai.13-7-2018.163994
50. S. Chen, B. Shen, X. Wang, S.-J. Yoo, A strong machine learning classifier and decision stumps based hybrid AdaBoost classification algorithm for cognitive radios. Sensors **19**(22), 5077 (2019). https://doi.org/10.3390/s19225077
51. J.H. Friedman, Greedy function approximation: A gradient boosting machine. Ann. Stat. **29**(5), 1189–1232 (2001). https://doi.org/10.1214/aos/1013203451
52. T. Chen, C. Guestrin, XGBoost: A scalable tree boosting system, in *Proceedings of the 22nd ACM SIGKDD International Conference on Knowledge Discovery and Data Mining*, (ACM, New York, 2016), pp. 785–794. https://doi.org/10.1145/2939672.2939785
53. M.H. Fatemi, P. Izadian, Quantitative structure–property relationship study of specific heat capacity of ionic liquids using molecular descriptors. J. Theor. Comput. Chem. **11**(1), 127–141 (2012). https://doi.org/10.1142/S0219633612500086
54. R. Bini, C. Chiappe, C. Duce, A. Micheli, R. Solaro, A. Starita, M.R. Tinè, Thermal properties of ionic liquids: A calorimetric study. Green Chem. **10**(3), 306–309 (2008). https://doi.org/10.1039/B718524F
55. G. Carrera, J. Aires-de-Sousa, Prediction of melting points of ionic liquids using machine learning methods. Green Chem. **7**(1), 20–27 (2005). https://doi.org/10.1039/B412882A
56. V. Venkatraman, S. Evjen, H.K. Knuutila, A. Fiksdahl, B.K. Alsberg, Predicting CO_2 solubility in ionic liquids using machine learning. J. Mol. Liq. **264**, 318–326 (2018). https://doi.org/10.1016/j.molliq.2018.05.003
57. J.A. Cerecedo-Cordoba, J.J. González Barbosa, J. Frausto-Solís, N.V. Gallardo-Rivas, Predicting thermophysical properties of ionic liquids using machine learning. J. Chem. Inf. Model. **59**(7), 3144–3153 (2019). https://doi.org/10.1021/acs.jcim.9b00345
58. J.A. Cerecedo-Cordoba, J. Frausto-Solís, J.J. González Barbosa, ILThermoML: A software for the prediction of thermophysical properties of ionic liquids. SoftwareX **11**, 100448 (2020). https://doi.org/10.1016/j.softx.2020.100448
59. K. Low, R. Kobayashi, E.I. Izgorodina, Predicting the heat capacity of ionic liquids using quantum chemical descriptors. J. Chem. Phys. **153**(10), 104101 (2020). https://doi.org/10.1063/5.0019021
60. A. Najafi-Marghmaleki, M.R. Khosravi-Nikou, A. Barati Harooni, Prediction of heat capacity of ionic liquids using group contribution methods. J. Mol. Liq. **220**, 232–237 (2016). https://doi.org/10.1016/j.molliq.2016.04.006
61. K. Paduszynski, Comprehensive thermodynamic modeling of ionic liquids: A group contribution approach. Ind. Eng. Chem. Res. **58**(42), 17049–17066 (2019). https://doi.org/10.1021/acs.iecr.9b03436
62. J.O. Valderrama, Modeling thermophysical properties of ionic liquids: A review. Ind. Eng. Chem. Res. **53**(3), 1004–1014 (2014). https://doi.org/10.1021/ie402193d
63. I. Goodfellow, Y. Bengio, A. Courville, *Deep Learning* (MIT Press, Cambridge, 2016)
64. N. Dutt, Y. Ravikumar, K.Y. Rani, Artificial neural network modeling of thermophysical properties of ionic liquids. Chem. Eng. Commun. **200**(11), 1600–1622 (2013). https://doi.org/10.1080/00986445.2012.752701
65. K. Paduszynski, U. Domanska, Predicting thermophysical properties of ionic liquids using group contribution methods. J. Chem. Inf. Model. **54**(5), 1311–1324 (2014). https://doi.org/10.1021/ci5000973

66. M.-R. Fatehi, S. Raeissi, D. Mowla, Prediction of heat capacity of ionic liquids using artificial neural networks. J. Mol. Liq. **227**, 309–317 (2017). https://doi.org/10.1016/j.molliq.2016.12.091
67. X. Kang, J. Qian, J. Deng, U. Latif, Y. Zhao, Prediction of thermophysical properties of ionic liquids using machine learning techniques. J. Mol. Liq. **265**, 756–764 (2018). https://doi.org/10.1016/j.molliq.2018.06.001
68. Z. Said, P. Sharma, R.M. Elavarasan, A.K. Tiwari, M.K. Rathod, Exploring the specific heat capacity of water-based hybrid nanofluids for solar energy applications: A comparative evaluation of modern ensemble machine learning techniques. J. Energy Storage **54**, 105230 (2022). https://doi.org/10.1016/j.est.2022.105230
69. Y. Oh, Z. Guo, Applicability of machine learning techniques in predicting specific heat capacity of complex nanofluids. Heat Transf. Res. **55**(3), 39–60 (2024)
70. H. Adun, D. Kavaz, I. Wole-Osho, M. Dagbasi, Synthesis of Fe_3O_4-Al_2O_3-ZnO/water ternary hybrid nanofluid: Investigating the effects of temperature, volume concentration and mixture ratio on specific heat capacity, and development of hybrid machine learning for prediction. J. Energy Storage **41**, 102947 (2021). https://doi.org/10.1016/j.est.2021.102947
71. H. Wei, S. Zhao, Q. Rong, H. Bao, Predicting the effective thermal conductivities of composite materials and porous media by machine learning methods. Int. J. Heat Mass Transf. **127**, 908–916 (2018). https://doi.org/10.1016/j.ijheatmasstransfer.2018.08.081
72. E. Ahmadloo, S. Azizi, Prediction of thermal conductivity of various nanofluids using artificial neural network. Int. Commun. Heat Mass Transf. (2016). https://doi.org/10.1016/j.icheatmasstransfer.2016.03.008
73. A. Adam, H. Fang, X. Li, Effective thermal conductivity estimation using a convolutional neural network and its application in topology optimization. Energy AI (2023). https://doi.org/10.1016/j.egyai.2023.100310
74. S.K. Das, S.U.S. Choi, W. Yu, T. Pradeep, *Nanofluids: Science and Technology* (Wiley, Hoboken, 2007)
75. R. Saidur, K.Y. Leong, H.A. Mohammad, A review on applications and challenges of nanofluids. Renew. Sust. Energ. Rev. **15**(3), 1646–1668 (2011). https://doi.org/10.1016/j.rser.2010.11.035
76. W. Yu, H. Xie, A review on nanofluids: Preparation, stability mechanisms, and applications. J. Nanomater. **2012**, 435873 (2012). https://doi.org/10.1155/2012/435873
77. S.M.S. Murshed, K.C. Leong, C. Yang, Enhanced thermal conductivity of TiO_2–water based nanofluids. Int. J. Therm. Sci. **44**(4), 367–373 (2005). https://doi.org/10.1016/j.ijthermalsci.2004.12.005
78. G. Pilania, C. Wang, X. Jiang, S. Rajasekaran, R. Ramprasad, Accelerating materials property predictions using machine learning. Sci. Rep. **3**, 2810 (2013). https://doi.org/10.1038/srep02810
79. A. Asadi, M. Asadi, A. Rezania, W. Chen, Artificial neural network modeling for thermal conductivity of nanofluids: A review. J. Mol. Liq. **301**, 112417 (2020). https://doi.org/10.1016/j.molliq.2019.112417
80. M.A. Bessa, R. Bostanabad, Z. Liu, A. Hu, D.W. Apley, L.C. Brinson, W. Chen, W.K. Liu, A framework for data-driven analysis of materials under uncertainty: Countering the curse of dimensionality. Comput. Methods Appl. Mech. Eng. **320**, 633–667 (2017). https://doi.org/10.1016/j.cma.2017.03.037
81. A.P. Teixeira, J. Ribeiro, B. Neto, C.G. Soares, C. Guedes Soares, Machine learning techniques for thermophysical properties prediction of fluids: A systematic review. Appl. Energy **306**, 118097 (2022). https://doi.org/10.1016/j.apenergy.2021.118097
82. W. Chen, C. Zou, X. Li, An investigation into the thermophysical and optical properties of SiC/ionic liquid nanofluid for direct absorption solar collector. Sol. Energy Mater. Sol. Cells **163**, 157–163 (2017). https://doi.org/10.1016/j.solmat.2017.01.029
83. R.K. Chinnam, A. Sharma, N.L. Panwar, Recent advancements in solar-assisted cooling technologies with nanofluids. Renew. Energy **184**, 848–866 (2022). https://doi.org/10.1016/j.renene.2021.12.077

84. R.S. Babu, J.S. Rajadurai, T. Sundararajan, Performance enhancement of cold thermal energy storage system using nanofluids: A review. Renew. Sust. Energ. Rev. **139**, 110705 (2021). https://doi.org/10.1016/j.rser.2020.110705
85. J.Y. Jang, M.Y. Tsai, C.C. Wang, Integration of nanofluid-based CCHP systems in smart manufacturing and building applications. Energy Convers. Manag. **285**, 116924 (2023). https://doi.org/10.1016/j.enconman.2023.116924
86. S.U.S. Choi, J.A. Eastman, Enhancing thermal conductivity of fluids with nanoparticles. ASME Int. Mech. Eng. Congr. Expo. **66**(1), 99–106 (2001)
87. S.K. Das, H. Patel, *Nanofluids: Science and Technology* (CRC Press, Boca Raton, 2016)
88. S.M. Moghimi, A.C. Hunter, Nanomedicine: Current status and future prospects. Eur. J. Pharm. Biopharm. **57**(3), 243–256 (2001)
89. K. Saha, S.S. Agasti, Gold nanoparticles in chemical and biological sensing. Chem. Rev. **112**(5), 2739–2779 (2012)
90. M. Faraji, P. Wipf, Nanoparticles in cellular drug delivery. Bioorg. Med. Chem. **17**(8), 2950–2962 (2009). https://doi.org/10.1016/j.bmc.2009.02.034
91. S.-Y. Qin, A.-Q. Zhang, S.-X. Cheng, L. Rong, X.-Z. Zhang, Drug self-delivery systems for cancer therapy. Biomaterials (2016). https://doi.org/10.1016/j.biomaterials.2016.10.016
92. S.M. Moghimi, A.C. Hunter, Nanomedicine: Current status and future prospects. Eur. J. Pharm. Biopharm. **57**(3), 243–256 (2001). https://doi.org/10.1096/fj.04-2747rev
93. B. Bhushan, Z. Zhang, *Nanotribology and Nanomechanics: An Introduction* (Springer, Cham, 2007)
94. E. Petryayeva, W.R. Algar, I.L. Medintz, Quantum dots in bioanalysis: A review of applications across various platforms for fluorescence spectroscopy and imaging. Appl. Spectrosc. **67**(3), 215–252 (2013). https://doi.org/10.1366/12-06948
95. S.K. Jain, A. Jain, Development of liposomal formulation of doxorubicin for intravenous use in cancer therapy. Drug Dev. Ind. Pharm. **36**(6), 711–718 (2010)
96. F. Liu, D. Xu, M. Wang, Random forest modeling for predicting the toxicity of nanoparticles: An application in biocompatibility prediction. J. Nanopart. Res. **19**(12), 452 (2017). https://doi.org/10.1007/s11051-017-0862-3
97. Y. Zhou, J. Li, Q. Zhang, Application of convolutional neural networks in predicting the interactions between nanoparticles and biological systems. Nano **9**(6), 784 (2019). https://doi.org/10.3390/nano9060784
98. X. Li, Y. Zhang, Y. Liu, Long short-term memory networks for predicting nanoparticle interaction with biological systems. Comp. Toxicol. **14**, 100118 (2020). https://doi.org/10.1016/j.comtox.2020.100118
99. H. Tao, T. Wu, M. Aldeghi, et al., Nanoparticle synthesis assisted by machine learning. Nat. Rev. Mater. **6**, 701–716 (2021). https://doi.org/10.1038/s41578-021-00337-5
100. D.R. Serrano, F.C. Luciano, B.J. Anaya, B. Ongoren, A. Kara, G. Molina, B.I. Ramirez, S.A. Sánchez-Guirales, J.A. Simon, G. Tomietto, C. Rapti, H.K. Ruiz, S. Rawat, D. Kumar, A. Lalatsa, Artificial intelligence (AI) applications in drug discovery and drug delivery: Revolutionizing personalized medicine. Pharmaceutics **16**(10), 1328 (2024). https://doi.org/10.3390/pharmaceutics16101328
101. Author, Simulation of nanorobots with artificial intelligence and reinforcement learning for advanced cancer cell detection and tracking. arXiv (n.d.), https://arxiv.org/abs/2411.02345
102. L. Wang, D. Ho, Personalized dose selection for the first Waldenström macroglobulinemia patient using CURATE.AI: A case report. NPJ Dig. Med. **7**(1), 195 (2024). https://doi.org/10.1038/s41746-024-01195-5
103. M. Huang, J. Deng, G. Jia, Predicting viscosity of ionic liquids–water mixtures by bridging UNIFAC modeling with interpretable machine learning. J. Mol. Liq. **370**, 122095 (2023). https://doi.org/10.1016/j.molliq.2023.122095
104. M. Ziatdinov, O. Dyck, X. Li, et al., Automatic identification of crystal structures and interfaces via Bayesian deep learning. Npj Comput. Mater. **9**(1), 1–9 (2023). https://doi.org/10.1038/s41524-023-01133-1

105. S.V. Kalinin, M. Ziatdinov, S. Jesse, Machine learning in electron microscopy for advanced materials analysis. Nat. Rev. Mater. **7**(5), 357–372 (2022). https://doi.org/10.1038/s41578-021-00331-9
106. M. Krenn, J. Landgraf, T. Foesel, F. Marquardt, Artificial intelligence and machine learning for quantum technologies. arXiv preprint (2022), https://arxiv.org/abs/2208.03836
107. Y. Zhang, H. Wang, Y. Liu, Ab initio characterization of protein molecular dynamics with AI^2BMD. Nature **615**(7951), 123–130 (2024). https://doi.org/10.1038/s41586-024-08127-z
108. Y. Zhang, T. Zhao, Y. Wang, Application of artificial intelligence in computational fluid dynamics: A review. Ind. Eng. Chem. Res. **59**(37), 16759–16773 (2020). https://doi.org/10.1021/acs.iecr.0c05045
109. Y. Zhang, H. Wang, Y. Liu, Merging data curation and machine learning to improve nanomedicine development. Adv. Drug Deliv. Rev. **180**, 113–125 (2022). https://doi.org/10.1016/j.addr.2022.113125
110. J. Corso, Data curation: The beast behind every AI model. Medium (2024). https://medium.com/@jasoncorso/data-curation-the-beast-behind-every-ai-model-d136eac4da6b
111. S. Kakaç, H. Liu, *Heat Transfer Enhancement with Nanofluids* (CRC Press, Boca Raton, 2021)
112. T. Zhou, X. Wan, D.Z. Huang, Z. Li, Z. Peng, A. Anandkumar, J.F. Brady, P.W. Sternberg, C. Daraio, AI-aided geometric design of anti-infection catheters. Sci. Adv. **9**(1), 1741 (2023). https://doi.org/10.1126/sciadv.adj1741
113. M. Zhang, J. Liu, Z. Xu, Deep learning for predicting drug release from nanocarriers: Applications in biomedical nanofluids. J. Control. Release **329**, 99–109 (2021). https://doi.org/10.1016/j.jconrel.2021.01.037
114. J. Wang, X. Zhang, X. Li, Machine learning for green nanomaterial synthesis: Strategies and challenges. Nat. Sustain. **3**(5), 361–374 (2020). https://doi.org/10.1038/s41893-020-0474-3
115. A. Kumar, R. Sharma, R. Kaur, AI-based quality control and optimization in nanomanufacturing: A review. Mater. Today Commun. **26**, 101835 (2021). https://doi.org/10.1016/j.mtcomm.2020.101835
116. F. Pintar, M. Zupan, B. Simić, Machine learning for process optimization and quality control in nanomanufacturing. Comput. Ind. **115**, 103174 (2020). https://doi.org/10.1016/j.compind.2020.103174
117. J.H. Lee, K.S. Hwang, S.P. Jang, S.U.S. Choi, Effective viscosities and thermal conductivities of aqueous nanofluids containing low volume concentrations of Al_2O_3 nanoparticles. Int. J. Heat Mass Transf. **51**(11–12), 2651–2656 (2008). https://doi.org/10.1016/j.ijheatmasstransfer.2007.10.017
118. Y. Zhao, L. Wang, J. Li, Real-time process optimization in nanomanufacturing through AI-based predictive modeling. Nat. Mater. **22**(3), 456–468 (2023). https://doi.org/10.1038/s41563-023-01614-5
119. K. Lee, H. Park, Y. Cho, Generative design of nanomaterials using artificial intelligence. Nat. Nanotechnol. **17**(2), 124–134 (2022). https://doi.org/10.1038/s41565-021-00949-6
120. A.K. Shakya, G. Pillai, S. Chakrabarty, Reinforcement learning algorithms: A brief survey. Expert Syst. Appl. **231**, 120495 (2023). https://doi.org/10.1016/j.eswa.2023.120495
121. K.T. Butler, D.W. Davies, H. Cartwright, et al., Machine learning for molecular and materials science. Nature **559**, 547–555 (2018). https://doi.org/10.1038/s41586-018-0337-2
122. P. Goktas, A. Grzybowski, Shaping the future of healthcare: Ethical clinical challenges and pathways to trustworthy AI. J. Clin. Med. **14**, 1605 (2025). https://doi.org/10.3390/jcm14051605
123. G. Mishra, A comprehensive review of smart healthcare systems: Architecture, applications, challenges, and future directions. Int. J. Innov. Res. Technol. Sci. **12**(2), 210–218 (2024)

Machine Learning and AI in Nanotechnology with Ionic Liquids, Nanofluids and Bio-Nanofluids

Sundram Singh, Swati Sangolgi, and Tapas Ranjan Das

1 Introduction

Nanotechnology refers to the analysis and manipulation of materials at the nanometer scale, i.e., a scale of 1–100 nm where unique physical and chemical properties emerge. Materials at this scale display distinct chemical, physical and biological properties that can be radically different from their bulk counterparts [1]. Moreover, at this scale, researchers can manipulate the elements of matter for a wide variety of applications, such as agriculture, electronics, healthcare and manufacturing.

In agriculture, innovative approaches such as nano-engineered formulations of fertilisers and pesticides are used to improve the performance of traditional fertilisers. They can boost performance and mitigate environmental effects. For example, nano fertilisers enhance the absorption of nutrients by plants while reducing leaching into groundwater [2]. In the field of electronic and information technology, nanotechnology has emerged as a transformative force. Nanotechnology-based electronics operate at significantly higher speeds and are more compact. Today, state-of-the-art electronic devices are being engineered through the application of nanoscale technologies. This approach enhances the performance of gadgets, increases the density of memory chips and contributes to reducing both energy consumption and the size of the components of electronic devices [3]. Healthcare has also greatly benefited from nanotechnology, particularly in diagnostics. Devices equipped with nanotechnology are being used for rapid and early disease recognition at the molecular level by enhancing the precision and sensitivity of

S. Singh · S. Sangolgi
Animal Biochemistry Division, National Dairy Research Institute, Karnal, India

T. R. Das (✉)
Division of Genetics, ICAR-Indian Agricultural Research Institute, New Delhi, India

S. Kulkarni, S. Das (eds.), *Ionic Liquids, Nanofluids and Nanotechnology*,
Lecture Notes in Nanoscale Science and Technology 42,
https://doi.org/10.1007/978-3-032-04008-4_21

tests, leading to the development of a quick understanding of treatment options [4, 5]. Biosensors, capable of detecting biomolecules at extremely low concentrations in body fluids such as blood, saliva, sweat and urine, represent a significant advancement in nanotechnology applications [6, 7]. Furthermore, nanotechnology's versatility allows it to be applied across all aspects of life, making it capable of being applied in every facet of life. Therefore, it is easy to predict that this technology will become an essential field of study in the future [8].

Artificial intelligence (AI) and machine learning (ML) emerged in the mid-twentieth century. Alan Turing, often regarded as the father of AI, proposed the concept of a 'learning machine' in the 1950s, establishing the foundation for machines capable of self-improvement over time.

ML focuses on the exploration and development of diverse algorithms to process extensive datasets, recognise trends and generate independent predictions [9, 10]. The algorithms can later be optimised for greater predictive accuracy through iterative refinement [11].

Nanotechnology and machine learning (ML) are a powerful combination, driving innovation across various fields. The synergy between them is transforming industries and paving the way for groundbreaking advancements.

Researchers investigated the integration of nanotechnology through scanning electron microscopy (SEM) and machine learning for classifying wheat species [12]. Two convolutional neural network (CNN) architectures, EffectiveNet-B4 and EffectiveNetV2-M, were employed for image analysis tasks alongside the SEM imaging technique, which can capture intricate surface details at the nanoscale. This study demonstrates the ability of deep learning models to extract meaningful features from SEM images, which could be valuable for various applications beyond wheat classification, such as in material science and other agricultural domains. Moreover, nanochips can potentially be used to enhance the performance and efficiency of computers, significantly lowering the time and expense associated with training advanced machine learning models [13].

A nanofluid can be defined as a colloidal mixture consisting of nanoparticles suspended in a base fluid, primarily recognised for its ability to enhance the fluid's thermal conductivity and heat transfer properties. The properties of nanofluids vary significantly based on the type of nanoparticles they contain. For instance, nanofluids infused with metallic nanoparticles like copper or silver exhibit enhanced thermal conductivity, making them ideal for heat transfer applications. Similarly, nanofluids containing oxide nanoparticles, such as aluminium oxide or titanium dioxide, are often used to improve the fluid's stability and viscosity. Carbon-based nanoparticles like graphene or carbon nanotubes provide exceptional electrical conductivity, making them suitable for electronics and energy systems. These diverse properties enable nanofluids to cater to a wide range of applications across industries and emphasise their potential in future energy applications, particularly in thermal energy recovery and aerospace [14]. Other than electroactive nanofluids, having integration of nanoparticles with electrical attributes in a base fluid, ionic solutions can also be stimulated under electric fields. The preparation of nanofluids, ionic liquids (ILs) and bio-nanofluids is illustrated in Fig. 1.

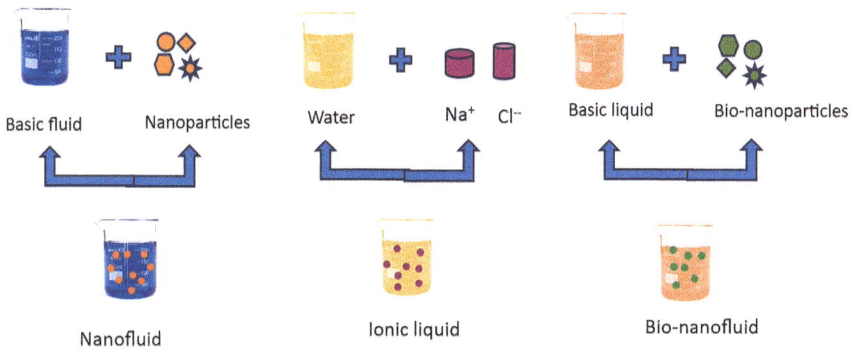

Fig. 1 Preparation of Nanofluids, Ionic liquids and Bio-nanofluids

This chapter explores the intricate relationship between nanotechnology and machine learning (ML), highlighting how their integration drives innovation across various fields, revealing new opportunities for innovation within fields of ionic liquids, nanofluids and bio-nanofluids.

2 Machine Learning Approaches for Nanoparticle Synthesis

In the context of nanoparticle synthesis, ML can be used in two ways: in predicting synthetic outcomes and optimising experimental planning. Prediction models focus on establishing mathematical relationships between key nanoparticle properties such as shape, size and behaviour, along with other input parameters. These models require experimental datasets for training. In contrast, experimental planning models do not predict outcomes directly but instead focus on optimising the data collection process to refine experimental efficiency and achieve specific objectives. By leveraging these ML techniques, researchers can enhance the precision and effectiveness of nanoparticle synthesis while reducing trial-and-error approaches.

2.1 Predictive ML Models

One of the most basic ML models used for prediction is linear regression, which identifies a direct relationship between experimental inputs and nanoparticle characteristics. Due to its simplicity, linear regression enables clear and quick predictions; however, it is limited to capturing linear patterns in the data only. As a result, non-linear ML models are often preferred for higher accuracy in predictive tasks.

Non-linear models include instance-based and model-based methods. Instance-based models predict outcomes by comparing new inputs to existing examples from

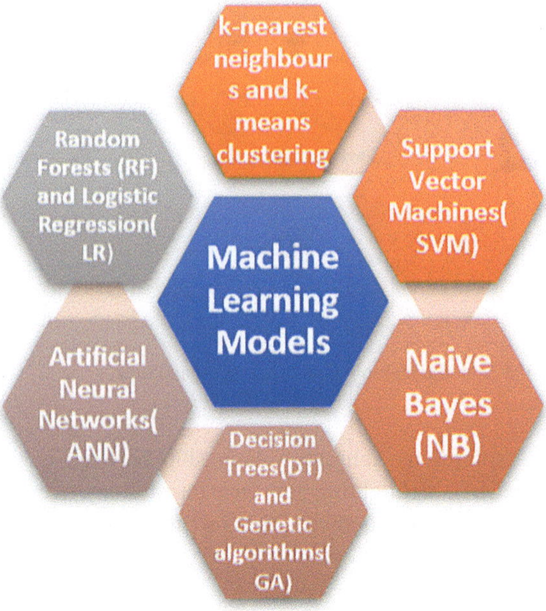

Fig. 2 Various machine learning models

the training data. In contrast, model-based approaches rely on a set of parameters learned during training to generalise predictions. Among model-based approaches, tree-based algorithms and deep learning methods have gained prominence in nanoparticle synthesis predictions, as they can effectively capture complex relationships between experimental conditions and nanoparticle properties, relationships that cannot be represented by simple linear equations. Different types of machine learning models used in nanotechnology are illustrated in Fig. 2.

Instance-Based Learning

Instance-based algorithms vary in their approach to defining similarity between data points. The simplest among them is k-nearest neighbour (k-NN), which predicts outcomes by averaging the most similar samples from the training dataset. More advanced instance-based methods, such as Gaussian processes [15] and support vector machines (SVMs) [16], incorporate sophisticated similarity measures to enhance predictive accuracy. Gaussian processes not only estimate experimental outcomes but also provide a confidence level for predictions. However, their major limitation lies in computational complexity; training costs increase exponentially with the data set size due to extensive linear algebra operations. This makes them more suitable for data sets with fewer than 1000 samples. Conversely, support vector machines offer greater computational efficiency compared to Gaussian processes but cannot assign confidence levels to their predictions. Each algorithm presents a unique strength and trade-off depending on dataset size, computational resources and prediction requirements.

Tree-Based Algorithms

In tree-based algorithms, the core element of tree-based algorithms, such as random forests (RFs) [17] and boosted trees [18, 19], is the decision tree (DT), which is a flow chart linking potential experimental outcomes to a series of decisions regarding experimental conditions. This diagram is built, for instance, from previous data about the synthesis of nanoparticles.

Deep Learning Method

Deep learning method, on the other hand, hinges on the combination of multiple non-linear functions [20]. Mapping of these functions is termed as an artificial neuron, which processes identical inputs and is grouped into a neuron layer to capture more complex relationships. The layered stacking of neurons in this mirrors the structure of brain-inspired neurons. These networks can be arranged in a variety of configurations, leading to numerous deep learning architectures optimised for various kinds of prediction tasks, for example convolutional neural networks for processing images and recurrent neural networks for sequential data like text. These specialised techniques are yet to be fully applied in nanoparticle synthesis [21].

3 ML Algorithms for Experimental Planning

A wide variety of algorithms exist to design experiments in nanoparticle synthesis. Traditional approaches, such as design of experiments (DOEs), systematically investigate a range of experimental conditions to understand their effects [22]. In contrast, more advanced heuristic and Bayesian optimisation methods exploit information gathered during the optimisation process to reach specific goals such as tailoring nanoparticle properties.

Heuristic strategies include evolutionary algorithms such as genetic algorithms (GAs) and the covariance matrix adaptation evolution strategy [23], which mimics the principles of natural selection. By adapting experimental conditions used in earlier trials, these methods iteratively improve a user-defined fitness metric. Although these evolutionary algorithms are highly effective for handling simultaneous and large-scale experimental processes, they generally demand a larger number of trials to achieve optimal results compared to Bayesian techniques.

One notable example of a heuristic method frequently used in nanoparticle synthesis is the Stable Noisy Optimisation by Branch and Fit (SNOBFIT) algorithm

[24, 25], which uses linear and quadratic models to link parameters with nanoparticle properties, guiding new experiments to explore unknown conditions or refining new ones.

Bayesian optimisation uses ML techniques to model the interdependencies between experimental conditions and outputs, providing uncertainty estimates that guide the search for optimal experiments. The most commonly used in this algorithm is Gaussian processes [26]; however, other methods, such as random forests [27] and kernel density [28], have also been employed. Bayesian optimisation focuses not on prediction but on selecting the next best experiment by balancing exploration and exploitation, an efficient strategy for costly, time-consuming experiments that minimises trials and conserves resources. One problem with this method is that it only excels in low-dimensional tasks, typically of less than 10 parameters, with limited experiments. For high-dimensional problems, heuristic methods like genetic algorithms or covariance matrix adaptation are preferred due to their parallelism, making them suitable for high-throughput set-ups. Ultimately, these strategies are judged by how effectively they find optimal conditions within time and budget constraints [29].

3.1 Other ML Algorithms

Active learning algorithms can identify the most informative experiments to enhance model accuracy. When predicting nanoparticle properties based on experimental conditions, these algorithms help generate datasets that, in turn, train predictive ML models. In contrast, Bayesian optimisation focuses on tuning specific parameters, such as nanoparticle size, to achieve optimal properties [21]. Both approaches fall within the scope of reinforcement learning, a comprehensive framework that enables AI systems to make decisions based on rewards and penalties, guiding them towards achieving a user-defined objective.

While applying ML to scientific challenges like nanoparticle synthesis requires some coding know-how, accessibility has improved, thanks to user-friendly Python libraries. Tools like scikit-learn [30, 31], MXNet, Microsoft Cognitive Toolkit (CNTK) and TensorFlow enable efficient model building, while libraries such as scikit-optimise [32], Olympus, Optuna, SigOpt and Hyperopt [33] offer an intuitive command-line interface for experimental planning.

4 Ionic Liquid Modelling Using Neural Networks

ILs have a unique nature beneficial for applications like catalytic reactions, lubricant additives [34] and carbon capture [35]. However, selecting the ideal IL is challenging, as key physical properties depend on cation and anion structures. The lack

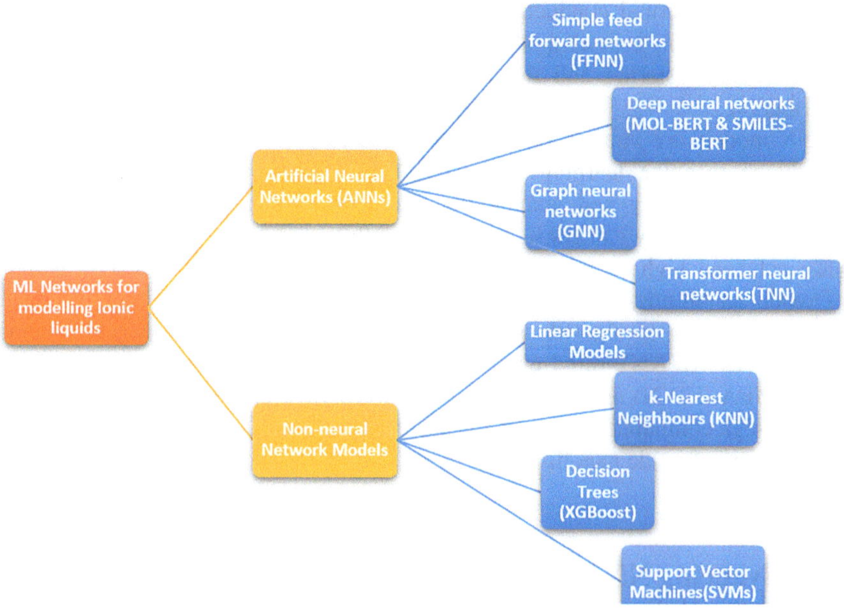

Fig. 3 Machine learning networks for modelling Ionic liquids

of comprehensive experimental data further complicates optimisation, necessitating reliable predictive models.

To overcome this problem, ML helps establish quantitative structure–property relationships (QSPRs) [36–38] with artificial neural networks (ANNs), which are particularly effective [39, 40], and using non-neural network models for reliable predictions [41, 42]. Given ILs' complex charge distributions and electrostatic interactions [37], addressing data limitations is crucial for accurate predictions. Figure 3 enlists various predictive models used in the modelling of ionic liquids.

The work examines recent advances in applying artificial neural networks (ANNs) to model the properties of ionic liquids [43]. The study highlights the effectiveness of ANNs over conventional machine learning methods for predicting IL thermodynamic and physical properties. Moreover, techniques like SHapley Additive exPlanations (SHAP) are also used to assess feature significance.

5 Need for NN Models for Predicting Nanofluid Thermal Conductivity

Li et al. [14] reviewed various models to predict the thermal conductivity of nanofluids. Their study discusses the limitations of classical models like Maxwell's and Hamilton-Crosser's, which do not adequately capture the mechanism behind the enhanced thermal conductivity of nanofluids. While useful, these theoretical models do not fully capture modern nanofluid behaviours such as Brownian motion, nanoparticle clustering and thermophoresis and often assume uniform particle distribution.

NN models are vital for overcoming these limitations of classical models by leveraging machine learning techniques to improve prediction accuracy, computational efficiency and adaptability. The key advantages of neural network models are that they can consider multiple influencing factors simultaneously and have achieved high accuracy, especially for predicting the thermal conductivity of oxide nanofluids, compared to traditional analytical models. Moreover, with the optimisation of algorithms like genetic algorithm-back propagation neural network (GA-BPNN), mind evolutionary algorithm-back propagation neural network (MEA-BPNN) and self-organising map (SOM), neural network models can achieve high computational speed and accuracy, especially in predicting the thermal conductivity of oxide nanofluids.

6 Advancing NN Models for Predicting Nanofluid Thermal Conductivity

The development of neural network prediction models now accounts for key factors such as particle size and average nanoparticle cluster size [44]. Similarly, Wang et al. [45] employed a backpropagation (BP) algorithm to construct an NN model that evaluates thermal conductivity of base fluid along with other physical parameters.

Komeilibirjandi et al. [46] developed an NN model to estimate the thermal conductivity of copper oxide nanofluids using group method of data handling (GMDH) algorithm, which handles mathematical models and enhances the overall prediction accuracy. Meanwhile, Jiang et al. integrated the backpropagation algorithm with both a genetic algorithm (GA and a minimax evolutionary algorithm (MEA), significantly refining the computational precision of BP-based models [47].

Elsheikh et al. utilised the self-organizing maps (SOM) algorithm, known for its high proficiency in managing intricate input–output relationships, to construct an NN model for predicting nanofluid thermal conductivity [48]. Sadeghzadeh et al. extended this approach by implementing both the SOM algorithm and the backpropagation–Levenberg–Marquardt (BP-LM) algorithm to develop predictive models specifically for TiO_2-Al_2O_3/H_2O nanofluids [49].

Ahmadi et al. (2018) conducted a comparative analysis of three distinct neural network models for estimating the thermal conductivity of Al_2O_3/H_2O nano fluids [50]. Their findings highlighted the effectiveness of genetic algorithm-least squares support vector machine (GA-LSSVM) in generating precise thermal conductivity predictions across varying conditions.

7 ML Models for Nanofluid Viscosity Assessment

Water and ethylene glycol, which possess industrial application, often exhibit limited efficiency in heat transfer due to inefficient thermal conductivity. Leveraging advancements in nanotechnology, nanofluids comprising nanoparticles suspended in base fluids were developed [51, 52]. The dispersed solid particles enhance heat transfer capabilities compared to conventional fluids; nano fluids offer superior thermophysical properties, influencing both flow behaviour and heat transfer efficiency [53, 54]. Key characteristics, including viscosity, must be assessed before their application in industrial processes.

Various methods have been adopted in the last decade to estimate the viscosity of nanofluids. Einstein pioneered viscosity prediction for fluids with suspended small particles, laying the foundation for later models. Researchers such as Brinkman [55], Lundgren [56], Frankel [57] and others proposed refinements or entirely new methodologies.

Recent advancements in artificial intelligence have introduced data-driven models, such as artificial neural networks (ANNs) and radial basis function neural networks (RBF-NN), which excel at capturing complex non-linear relationships with efficiency [58, 59]. Karimi et al. [60] pioneered an ANN integrated with a genetic algorithm (GA), to predict viscosity with high accuracy. Following this, multiple AI-based modelling techniques emerged, including fuzzy C-means-adaptive neuro-fuzzy inference system (FCM-ANFIS) and GMDH-based polynomial networks [61–65].

Researchers developed the most comprehensive model, a committee machine intelligent system (CMIS) by combining multiple MLP-NNs, RBF-NNs and LSSVM models optimised using various algorithms, demonstrating high accuracy and broad applicability across parameter ranges [51]. Moreover, multilayer perceptron (MLP) networks, decision trees (DTs), and random forests (RF) further refine viscosity prediction accuracy and computational efficiency, experimental data points from nano fluid systems to develop ML models utilised for predicting relative viscosity [66]. The models tested comprised multilayer perceptron (MLP) networks with Nadam and Adam exoptimisers, support vector regression (SVR) within RBF-NN, DT, RF and deep MLP architectures. All eight proposed models outperformed existing baseline models, highlighting the effectiveness of deeper MLP structures and ensemble tree models in improving viscosity predictions.

8 ML-Driven Advances in Bio-Nanofluids

Magnetised nanoparticles significantly enhance thermal transport in bio-nanofluids, presenting promising applications in medical science. Their suspension in biological fluids, such as blood, improves thermal conductivity, aiding biomedical fields like targeted drug delivery and bio imaging [67–71]. Ongoing research continues to refine their potential for healthcare innovations.

Elsaid et al. examined the irregular peristaltic flow of nanoparticles under magnetic instability, ensuring smooth heat transport [72], while Ali et al. reported hybrid magnetised nanofluid dynamics in blood [73], and Alnahdi et al. focused on tri-hybrid nanofluid interactions in perforated capillaries for drug delivery [74]. Further investigations analysed the Lorentz effect, nanoparticle aggregation and size-dependent thermophysical properties [75–78].

Quadratic convection in bio-nanofluid significantly influences nanoparticle mobility, affecting heat and mass transport in biological systems. Understanding these mechanisms could enhance biomedical treatment and technologies. Various studies have explored how biological fluids move with microscopic particles, magnetic forces and hybrid nanoparticles in different medical applications [79–81]. Additional research explored linear and quadratic convection processes within thermal management and magnetised environments [82–87].

Recent advances in artificial intelligence have enabled sophisticated fluid modelling. The Levenberg–Marquardt neural network (LMNN) dynamically adjusts parameters for real-time optimisation, improving accuracy across engineering, healthcare and computational fields. Amini and Mohaghegh utilised AI to model fluid dynamics in porous media [88], while Alizadeh et al. applied machine learning to mixed biothermal convection [89]. These studies highlight AI's growing role in refining nanofluid applications and optimising thermal transport.

The recent study by Ayub et al. provides valuable insights into the complex interplay of quadratic convection, thermal radiation and magnetic effects in blood-based nanofluids [90]. The study's integration of multiple nanoparticles and the LMNN technique enhances its depth and scientific impact. Findings could significantly impact biomedical engineering, aiding advancements in targeted drug delivery and bioimaging.

9 ML Boosts Design and Discovery of Nanomaterials

Nano materials have revolutionised various fields, including medicine, agriculture, electronics and heat transfer due to their unique properties absent at larger scales [91, 92]. However, traditional research methods often demand significant time to predict structure and properties. Machine learning, a powerful AI-driven tool, efficiently processes vast experimental data to uncover governing patterns, thereby accelerating chemical discoveries through insights extracted from large datasets [93–95].

Common ML models used in the design and discovery of nanomaterials are mentioned in Table 1.

Table 1 Common ML models used in the design of nanomaterials

ML model	Description	Application in nanomaterials
Naïve Bayes (NB)	Probabilistic classifier based on Bayes' theorem	Used for material classification and predicting nanomaterial properties based on probability distributions
Support vector machine (SVM)	Separates data into classes using a hyperplane	Helps identify nanomaterial phases and optimise synthesis conditions
Decision tree (DT)	A hierarchical model for classification and regression	Determines key features influencing nanomaterial performance
Random forest (RF)	Ensemble of decision trees to improve accuracy	Predicts stability and functional properties of hybrid nanomaterials
Artificial neural network (ANN)	Mimics brain neurons to learn complex patterns	Models structure–property relationships and optimises material formulations
Logistic regression (LR)	Statistical method for binary classification problems	Determines material feasibility for specific applications
Genetic algorithm (GA)	Optimisation technique inspired by natural selection	Finds optimal nanomaterial compositions and designs new structures
K-nearest neighbour (KNN)	Classifies data points based on nearest neighbours	Identifies similar nanomaterials and predicts their behaviour
K-means clustering	Groups data points into clusters based on similarity	Categorises nanomaterials based on functional attributes for targeted applications

10 Conclusions

The integration ofmachine learning (ML) and nanotechnology marks a groundbreaking shift in technological advancement. Machine learning's ability to process complex data complements nanotechnology's precision at the atomic scale, fostering breakthroughs in material science, healthcare, energy and environmental monitoring. Machine learning has evolved significantly due to advancements in big data, computational power and algorithm development [96–98].

This chapter highlights the successful application of various AI techniques in nanofluid, ionic liquid, and bio-nanofluid research. Machine learning models simplify formulation, adapt efficiently and reduce lengthy processing times without requiring complete system knowledge. By bridging these technologies, researchers can unlock new frontiers in intelligent material design, efficient energy storage and advanced biomedical applications.

For advancing particularly ionic liquid (IL) research, future efforts should focus on integrating artificial neural networks (ANNs) with optimisation algorithms and self-supervised learning to design ILs with precise, application-specific properties. A key priority is improving data preparation and ensuring model interpretability, enabling researchers to derive meaningful insights from predictive models. Additionally, leveraging advanced computational approaches will significantly enhance IL development, streamlining industrial applications and fostering innovations in the field [43].

Beyond computational advancements, ML also enhances healthcare and nanomedicine, offering insights into biological interaction with nanomaterials (NMs), a critical aspect for the biomedical applications [99]. It aids in discovering functional NMs and predicting biological responses, though deep ML models often lack interpretability. Compared to chemical and drug molecule databases, ML databases remain underdeveloped. Strengthening ML's role in material science requires expanding databases and creating interpretable models. Additionally, nanopattern image recognition offers further opportunities for ML to propel nanotechnology innovations [100].

The convergence of machine learning and nanotechnology offers immense potential but also presents pleasant challenges, including ethical concerns, complex manufacturing, data security, regulatory frameworks and interdisciplinary collaboration [101]. ML enhances quantum computing by optimising algorithms, while nanoscale components improve quantum processes, pushing computational capabilities to new levels [102, 103].

As machine learning (ML) accelerates material discovery, nanotechnology is poised to revolutionize healthcare and sustainable energy solutions by enabling precise drug delivery, improving environmental monitoring, and enhancing energy efficiency through advanced materials. Its transformative applications will drive innovation across multiple industries, addressing global challenges with unprecedented precision and effectiveness. By addressing ethical considerations, ensuring the responsible use of machine learning, and fostering collaborative research, this synergy can redefine technological progress and unlock new scientific frontiers.

References

1. J. Hu, T.W. Odom, C.M. Lieber, Chemistry and physics in one dimension: Synthesis and properties of nanowires and nanotubes. Acc. Chem. Res. **32**(5), 435–445 (1999)
2. M. Ghosh, C.R. Patra, Nanobiosensors and nanoformulations in agriculture: New advances and challenges for sustainable agriculture. J. Nanobiotechnol. **18**(1), 1–21 (2020)
3. M.K. Paras, Nanotechnology and its applications: An overview. Int. J. Phys. Appl. **6**(2), 179–185 (2024)
4. M. Dessale, G. Mengistu, H.M. Mengist, Nanotechnology: A promising approach for cancer diagnosis, therapeutics, and theragnosis. Int. J. Nanomedicine **17**, 3735–3749 (2022)
5. K.E. Fox, N.L. Tran, T.A. Nguyen, T.T. Nguyen, P.A. Tran, Surface modification of medical devices at nanoscale—Recent development and translational perspectives, in *Biomaterials in Translational Medicine*, (Academic Press, New York, 2019), pp. 163–189
6. Q.Y. Xu, Z. Tan, X.W. Liao, C. Wang, Recent advances in nanoscale metal-organic frameworks biosensors for detection of biomarkers. Chin. Chem. Lett. **33**, 22–32 (2022)
7. M. Sadeghi, S. Sadeghi, S.M. Naghib, H.R. Garshasbi, A comprehensive review on electrochemical nano biosensors for precise detection of blood-based oncomarkers in breast cancer. Biosensors **13**(4), 481 (2023)

8. N.A. Singh, Nanotechnology innovations, industrial applications and patents. Environ. Chem. Lett. **15**(1), 185–191 (2017)
9. I.H. Sarker, Machine learning: Algorithms, real-world applications and research directions. SN Comput. Sci. **2**, 160 (2021)
10. K.D. Foote, *A Brief History of Machine Learning* (Dataversity, Studio City, 2019)
11. L. Zhou, A.M. Yao, Y. Wu, Z. Hu, Y. Huang, Z. Hong, Machine learning assisted prediction of cathode materials for Zn-ion batteries. Adv. Theory Simul. **4**(6), 2100196 (2021)
12. Y. Anagun, S. Isik, M. Olgun, O. Sezer, Z.B. Basciftci, N.G.A. Arpacioglu, The classification of wheat species based on deep convolutional neural networks using scanning electron microscope (SEM) imaging. Eur. Food Res. Technol. **249**(6), 1023–1034 (2023)
13. M. Verhelst, B. Murmann, Machine learning at the edge, in *Nano-Chips 2030: On-Chip AI for an Efficient Data-Driven World*, (Springer, Berlin, 2020), pp. 293–322
14. J. Li, X. Zhang, B. Xu, M. Yuan, Nanofluid research and applications: A review, in *International Communications in Heat and Mass Transfer*, vol. 127, (Elsevier, Amsterdam, 2021), p. 105543
15. C.E. Rasmussen, C.K.I. Williams, *Gaussian Processes for Machine Learning* (MIT Press, Cambridge, 2006)
16. C. Cortes, V. Vapnik, Support-vector networks. Mach. Learn. **20**(3), 273–297 (1995)
17. L. Breiman, Random forests. Mach. Learn. **45**(1), 5–32 (2001)
18. J.H. Friedman, Greedy function approximation: A gradient boosting machine. Ann. Stat. **29**(5), 1189–1232 (2001)
19. J.H. Friedman, Stochastic gradient boosting. Comput. Stat. Data Anal. **38**, 367–378 (2002)
20. Y. Lecun, Y. Bengio, G. Hinton, Deep learning. Nature **521**(7553), 436–444 (2015)
21. H. Tao, T. Wu, M. Aldeghi, T.C. Wu, A. Aspuru-Guzik, E. Kumacheva, Nanoparticle synthesis assisted by machine learning. Nat. Rev. Mater. **6**(8), 701–716 (2021)
22. G.E.P. Box, J.S. Hunter, W.G. Hunter, *Statistics for Experimenters: Design, Innovation, and Discovery*, 2nd edn. (Wiley, Hoboken, 2005)
23. P.A. Vikhar, in Evolutionary algorithms: A critical review and its future prospects. 2016 International Conference on Global Trends in Signal Processing, Information Computing and Communication (ICGTSPICC). Presented at the 2016 International Conference on Global Trends in Signal Processing, Information Computing and Communication (ICGTSPICC), Jalgaon, India (2016)
24. W. Huyer, A. Neumaier, SNOBFIT: Stable noisy optimization by branch and fit. ACM Trans. Math. Softw. **35**, 1–25 (2008)
25. S. Krishnadasan, R.J.C. Brown, A.J. DeMello, J.C. DeMello, Intelligent routes to the controlled synthesis of nanoparticles. Lab Chip **7**, 1434–1441 (2007)
26. B. Shahriari, K. Swersky, Z. Wang, R.P. Adams, N. De Freitas, Taking the human out of the loop: A review of Bayesian optimization. Proc. IEEE **104**, 148–175 (2016)
27. F. Hutter, H.H. Hoos, K. Leyton-Brown, Sequential model-based optimization for general algorithm configuration, in *Lecture Notes in Computer Science*, (Springer, Cham, 2011), pp. 507–523. https://doi.org/10.1007/978-3-642-25566-3_40
28. F. Häse, L.M. Roch, A. Aspuru-Guzik, Gryffin: An algorithm for Bayesian optimization of categorical variables informed by expert knowledge. arXiv (2020), https://arxiv.org/abs/2003.12127
29. F. Häse et al., Olympus: A benchmarking framework for noisy optimization and experiment planning. Mach. Learn. Sci. Technol. (2021). https://doi.org/10.1088/2632-2153
30. F. Pedregosa, G. Varoquaux, A. Gramfort, V. Michel, B. Thirion, O. Grisel, M. Blondel, P. Prettenhofer, R. Weiss, V. Dubourg, J. Vanderplas, A. Passos, D. Cournapeau, M. Brucher, M. Perrot, E. Duchesnay, Scikit-learn: Machine learning in python. J. Mach. Learn. Res. **12**, 2825–2830 (2011)
31. F. Chollet, Keras [computer software]. GitHub (2015), https://github.com/fchollet/keras
32. T. Head et al., *scikit-optimize/scikit-optimize (Version 0.5.2) [Computer Software]* (Zenodo, Geneva, 2018). https://doi.org/10.5281/zenodo.1207017

33. J. Bergstra, B. Komer, C. Eliasmith, D. Yamins, D.D. Cox, Hyperopt: A python library for model selection and hyperparameter optimization. Comput. Sci. Discov. **8**, 014008 (2015)
34. Y. Zhou, J. Qu, Ionic liquids as lubricant additives: A review. ACS Appl. Mater. Interfaces **9**, 3209–3222 (2017)
35. M. Aghaie, N. Rezaei, S. Zendehboudi, A systematic review on CO_2 capture with ionic liquids: Current status and future prospects. Renew. Sust. Energ. Rev. **96**, 502–525 (2018)
36. S.M. Hosseini, A. Mulero, M.M. Alavianmehr, Predictive methods and semi-classical equations of state for pure ionic liquids: A review. J. Chem. Thermodyn. **130**, 47–94 (2019)
37. S. Koutsoukos, F. Philippi, F. Malaret, T. Welton, A review on machine learning algorithms for the ionic liquid chemical space. Chem. Sci. **12**, 6820–6843 (2021)
38. F. Yusuf, T. Olayiwola, C. Afagwu, Application of artificial intelligence-based predictive methods in ionic liquid studies: A review. Fluid Phase Equilib. **531**, 112898 (2021)
39. D. Dhall, R. Kaur, M. Juneja, Machine learning: A review of the algorithms and its applications. Proc. ICRIC **597**, 47–63 (2020)
40. R. Choudhary, H.K. Gianey, Comprehensive review on supervised machine learning algorithms, in *2017 International Conference on Machine Learning & Data Science (MLDS)*, (IEEE, New York, 2017), pp. 37–43
41. M. Nait Amar, M. Ghriga, M. Ben Seghier, H. Ouaer, Predicting solubility of nitrous oxide in ionic liquids using machine learning techniques and gene expression programming. J. Taiwan Inst. Chem. Eng. **128**, 156–168 (2021)
42. A. Shojaeian, M. Asadizadeh, Prediction of surface tension of the binary mixtures containing ionic liquid using heuristic approaches: An input parameters investigation. J. Mol. Liq. **298**, 111976 (2020)
43. A. Racki, K. Paduszyński, Recent advances in the modeling of ionic liquids using artificial neural networks. J. Chem. Inf. Model. **65**(7), 3161–3175 (2025). https://doi.org/10.1021/acs.jcim.4c02364
44. G.A. Longo et al., Application of artificial neural network (ANN) for the prediction of thermal conductivity of oxide–water nanofluids. Nano Energy **1**(2), 290–296 (2012)
45. X.H. Wang, X.N. Yan, N. Gao, G.M. Chen, Prediction of thermal conductivity of various nanofluids with ethylene glycol using artificial neural network. J. Therm. Sci. **3**(29), 1504–1512 (2019)
46. A.H. Komeilibirjandi, M.A. Raffiee, M.S.S. Nazari, Thermal conductivity prediction of nanofluids containing CuO nanoparticles by using correlation and artificial neural network. J. Therm. Anal. Calorim. **139**(4), 2679–2689 (2020)
47. Q. Jiang, R. Huang, Y. Huang, S. Chen, Y. He, L. Lan, C. Liu, Application of BP neural network based on genetic algorithm optimization in evaluation of power grid investment risk. IEEE Access Pract. Innov. Open Solut. **7**, 154827–154835 (2019). https://doi.org/10.1109/access.2019.2944609
48. A.H. Elsheikh, S.W. Sharshir, A.S. Ismai, R. Sathyamurthy, H. Zhang, An artificial neural network-based approach for prediction the thermal conductivity of nanofluids. SN Appl. Sci. **2**, 235 (2020)
49. M. Sadeghzadeh, H. Maddah, M.H. Ahmadi, A. Khadang, M. Ghazvini, A. Mosavi, N. Nabipour, Prediction of thermo-physical properties of TiO2-Al_2O_3/water nanoparticles by using artificial neural network. Nano **10**, 4 (2020)
50. M.H. Ahmadi, M.A. Nazari, R. Ghasempour, H. Madah, M.B. Shafii, M.A. Ahmadi, Thermal conductivity ratio prediction of Al_2O_3/water nanofluid by applying connectionist methods. Colloids Surf. A Physicochem. Eng. Asp. **541**, 154–164 (2018)
51. A. Hemmati-Sarapardeh, A. Varamesh, M.M. Husein, K. Karan, On the evaluation of the viscosity of nanofluid systems: Modeling and data assessment. Renew. Sust. Energ. Rev. **81**, 313–329 (2018)
52. L. Yang, J. Xu, K. Du, X. Zhang, Recent developments on viscosity and thermal conductivity of nanofluids. Powder Technol. **317**, 348–369 (2017)
53. H. Divandari, A. Hemmati-Sarapardeh, M. Schaffie, M. Ranjbar, Integrating functionalized magnetite nanoparticles with low salinity water and surfactant solution: Interfacial tension study. Fuel **281**, 118641 (2020)

54. A. Rezaei, H. Abdollahi, Z. Derikvand, A. Hemmati-Sarapardeh, A. Mosavi, N. Nabipour, Insights into the effects of pore size distribution on the flowing behavior of carbonate rocks: Linking a nano-based enhanced oil recovery method to rock typing. Nano **10**, 972 (2020)
55. H. Brinkman, The viscosity of concentrated suspensions and solutions. J. Chem. Phys. **20**, 571 (1952)
56. T.S. Lundgren, Slow flow through stationary random beds and suspensions of spheres. J. Fluid Mech. **51**, 273–299 (1972)
57. N. Frankel, A. Acrivos, On the viscosity of a concentrated suspension of solid spheres. Chem. Eng. Sci. **22**, 847–853 (1967)
58. R. Moghadasi, A. Rostami, A. Hemmati-Sarapardeh, M. Motie, Application of nanosilica for inhibition of fines migration during low salinity water injection: Experimental study, mechanistic understanding, and model development. Fuel **242**, 846–862 (2019)
59. M.S. Mazloom, F. Rezaei, A. Hemmati-Sarapardeh, M.M. Husein, S. Zendehboudi, A. Bemani, Artificial intelligence-based methods for asphaltenes adsorption by nanocomposites: Application of group method of data handling, least squares support vector machine, and artificial neural networks. Nano **10**, 890 (2020)
60. H. Karimi, F. Yousefi, M.R.J.H. Rahimi, M. Transfer, Correlation of viscosity in nanofluids using genetic algorithm-neural network (GA-NN). Heat Mass Transf. **47**, 1417–1425 (2011)
61. E. Gholami, B. Vaferi, M.A. Ariana, Prediction of viscosity of several alumina-based nanofluids using various artificial intelligence paradigms—Comparison with experimental data and empirical correlations. Powder Technol. **323**, 495–506 (2018)
62. S.A. Adio, M. Mehrabi, M. Sharifpur, J.P. Meyer, Experimental investigation and model development for effective viscosity of MgO–ethylene glycol nanofluids by using dimensional analysis, FCM-ANFIS, and GA-PNN techniques. Int. Commun. Heat Mass Transf. **72**, 71–83 (2016)
63. S. Atashrouz, G. Pazuki, Y. Alimoradi, Estimation of the viscosity of nine nanofluids using a hybrid GMDH-type neural network system. Fluid Phase Equilib. **372**, 43–48 (2014)
64. M.K. Meybodi, S. Naseri, A. Shokrollahi, A. Daryasafar, Prediction of viscosity of water-based Al_2O_3, TiO_2, SiO_2, and CuO nanofluids using a reliable approach. Chemom. Intell. Lab. Syst. **149**, 60–69 (2015)
65. E. Heidari, M.A. Sobati, S. Movahedirad, Accurate prediction of nanofluid viscosity using a multilayer perceptron artificial neural network (MLP-ANN). Chemom. Intell. Lab. Syst. **155**, 73–85 (2016)
66. M. Shateri, Z. Sobhanigavgani, A. Alinasab, A. Varamesh, A. Hemmati-Sarapardeh, A. Mosavi, Shahab., Comparative analysis of machine learning models for nanofluids viscosity assessment. Nano **10**(9), 1767 (2020). https://doi.org/10.3390/nano10091767
67. S. Li, M.I. Khan, F. Ali, S.S. Abdullaev, S. Saadaoui, Habibullah., Mathematical modeling of mixed convective MHD Falkner-Skan squeezed Sutterby multiphase flow with non-Fourier heat flux theory and porosity. Appl. Math. Mech. **44**(11), 2005–2018 (2023)
68. A. Wakif, M. Zaydan, R. Sehaqui, Further insights into steady three-dimensional MHD Sakiadis flows of radiating-reacting viscoelastic nanofluids via Wakif's-Buongiorno and Maxwell's models. J. Umm Al-Qura Univ. Appl. Sci. **2024**, 1–13 (2024)
69. A. Wakif, A. Alshehri, T. Muhammad, Influences of blowing and internal heating processes on steady MHD mixed convective boundary layer flows of radiating titanium dioxide-ethylene glycol nanofluids. ZAMM J. Appl. Math. Mech. **2024**, e202300536 (2024)
70. M. Alghamdi, A. Wakif, T. Muhammad, Efficient passive GDQLL scrutinization of an advanced steady EMHD mixed convective nanofluid flow problem via Wakif-Buongiorno approach and generalized transport laws. Int. J. Mod. Phys. B **2024**, 2450418 (2024)
71. A. El Harfouf, A. Wakif, S. Hayani Mounir, New insights into MHD squeezing flows of reacting-radiating Maxwell nanofluids via Wakif's–Buongiorno point of view. J. Umm Al-Qura Univ. Appl. Sci. **10**, 718–732 (2024)
72. E.M. Elsaid, A.A.M. Sayed, M.S. Abdel-wahed, Electromagnetohydrodynamic unsteady blood flow with ternary nanoparticles in a vertical irregular peristaltic flow: An exact treatment. J. Therm. Anal. Calorim. **148**(24), 14163–14181 (2023)

73. A. Ali, S. Das, T. Muhammad, Dynamics of blood conveying copper, gold, and titania nanoparticles through the diverging/converging ciliary micro-vessel: Further analysis of ternary-hybrid nanofluid. J. Mol. Liq. **390**, 122959 (2023)
74. A.S. Alnahdi, S. Nasir, T. Gul, Blood-based ternary hybrid nanofluid flow-through perforated capillary for the applications of drug delivery. Waves Random Complex Media **2022**, 1–19 (2022)
75. N.A. Ahammad, S. Majeed, F.A. Gamaoun, A.U. Awan, B. Ali, Significance of hybrid nanoparticles, Lorentz and Coriolis forces on the dynamics of water-based flow. Int. Commun. Heat Mass Transf. **135**, 106084 (2022)
76. B. Ali, N.A. Ahammad, A.U. Awan, A.S. Oke, A.U. Awan, E.M. Tag-ElDin, F.A. Shah, S. Majeed, The dynamics of water-based nanofluid subject to the nanoparticle's radius with a significant magnetic field: The case of rotating micropolar fluid. Sustainability **14**(17), 10474 (2022)
77. M.H. Shahzad, S. Akhtar, A.U. Awan, S. Nadeem, Entropy and stability analysis on blood flow with nanoparticles through a stenosed artery having permeable walls. Sci. Prog. **105**(2), 00368504221096000 (2022)
78. S. Majeed, B. Ali, L.A. Ali, A.U. Awan, Significance of nanoparticles aggregation and Coriolis force on the dynamics of Prandtl nanofluid: The case of rotating flow. Chin. J. Phys. **79**, 264–274 (2022)
79. C. Caro, M. Dalmases, A. Figuerola, M.L. García-Martín, M.P. Leal, Highly water-stable rare ternary Ag–Au–Se nanocomposites as long blood circulation time X-ray computed tomography contrast agents. Nanoscale **9**(21), 7242–7251 (2017)
80. S. Islam, B.M.J. Rana, M.S. Parvez, M.S. Hossain, M.M. Rahman, Electroosmotic flow in ternary (TiO_2-SiO_2-Al_2O_3) blood-based Sutterby nanomaterials with bio-active mixers. Int. J. Thermofluids **18**, 100363 (2023)
81. M.N. Khan, S. Ahmad, S. Nadeem, Flow and heat transfer investigation of bio–convective hybrid nanofluid with triple stratification effects. Phys Scr **96**(6), 065210 (2021)
82. S. Li, M.I. Khan, S. Ali, S. Ullah Khan, S.A. Althobaiti, I. Khan, et al., Influence of variable fluid properties on mixed convective Darcy-Forchheimer flow relation over a surface with Soret and Dufour spectacle. Open Phys. **22**(1), 20240010 (2024)
83. R. Zhang, M. Zaydan, M. Alshehri, C.S.K. Raju, A. Wakif, N.A. Shah, Further insights into mixed convective boundary layer flows of internally heating Jeffery nanofluids: Stefan's blowing case study with convective heating and thermal radiation impressions. Case Stud. Thermal Eng. **2024**, 104121 (2024)
84. B. Ali, N. Ahammad, K. Guedri, A. Tag-ElDin, A.U.E.M. Awan, S. Majeed, Dynamics of rotating micropolar fluid over a stretch surface: The case of linear and quadratic convection significance in thermal management. Nano **12**(18), 3100 (2022)
85. S.A.A. Shah, N.A. Ahammad, B. Ali, K. Guedri, A.U. Awan, F. Gamaoun, et al., Significance of bio-convection, MHD, thermal radiation, and activation energy across Prandtl nanofluid flow: A case of stretching cylinder. Int. Commun. Heat Mass Transf. **137**, 106299 (2022)
86. Q. Ali, S. Riaz, N.A. Shah, A.U. Awan, J.D. Chung, A thermal optimization through an innovative mechanism of free convection flow of Jeffrey fluid using non-local kernel. Case Stud. Thermal Eng. **24**, 100851 (2021)
87. A.A. Akbar, M.Z. Bani-Fwaz, E.M. Tag-ElDin, K. Guedri, M.F. Yassen, A.U. Awan, et al., Linear and quadratic convection significance on the dynamics of MHD Maxwell fluid subject to stretched surface. Front. Phys. **10**, 974681 (2022)
88. S. Amini, S. Mohaghegh, Application of machine learning and artificial intelligence in proxy modeling for fluid flow in porous media. Fluids **4**(3), 126 (2019)
89. R. Alizadeh, M.N. Abad, A. Fattahi, M. Mesgarpour, M.H. Doranehgard, Q. Xiong, Machine-learning enhanced analysis of mixed biothermal convection of single particle and hybrid nanofluids within a complex configuration. Ind. Eng. Chem. Res. **61**(24), 8478–8494 (2021)
90. A. Ayub, A. Darvesh, S.Z.H. Shah, M.-U.-D. Junjua, E.D. Santos, S. Ahmad, A.E. Ragab, Magnetized and quadratic convection based thermal transport in ternary radiative bio-nanofluid via intelligent neural networks: Two hidden layers mechanism. Results Phys **65**, 107973 (2024)

91. J. Carrete, N. Mingo, S. Wang, S. Curtarolo, Nanograined half-Heusler semiconductors as advanced thermoelectrics: An ab initio high-throughput statistical study. Adv. Funct. Mater. **24**(47), 7427–7432 (2014)
92. G. Takasao, T. Wada, A. Thakur, P. Chammingkwan, M. Terano, T. Taniike, Machine learning-aided structure determination for $TiCl_4$-capped $MgCl_2$ nanoplate of heterogeneous Ziegler-Natta catalyst. ACS Catal. **9**(3), 2599–2609 (2019)
93. A. Tkatchenko, Machine learning for chemical discovery. Nat. Commun. **11**(1), 4125 (2020)
94. O.A. von Lilienfeld, K.-R. Mueller, A. Tkatchenko, Exploring chemical compound space with quantum-based machine learning. Nat. Rev. Chem. **4**(7), 347–358 (2020)
95. A.P. Bartók, S. De, C. Poelking, N. Bernstein, J.R. Kermode, G. Csányi, M. Ceriotti, Machine learning unifies the modeling of materials and molecules. Sci. Adv. **3**(12), e1701816 (2017)
96. S. Natale, A. Ballatore, Imagining the thinking machine: Technological myths and the rise of artificial intelligence. Convergence **26**(1), 3–18 (2020)
97. Y. Duan, J.S. Edwards, Y.K. Dwivedi, Artificial intelligence for decision making in the era of big data: Evolution, challenges, and research agenda. Int. J. Inf. Manag. **48**, 63–71 (2019)
98. Y.K. Dwivedi, L. Hughes, E. Ismagilova, G. Aarts, C. Coombs, T. Crick, Y. Duan, R. Dwivedi, J. Edwards, A. Eirug, V. Galanos, Artificial intelligence (AI): Multidisciplinary perspectives on emerging challenges, opportunities, and agenda for research, practice, and policy. Int. J. Inf. Manag. **57**, 101994 (2021). https://doi.org/10.1016/j.ijinfomgt.2019.08.002
99. A. Mikolajczyk, N. Sizochenko, E. Mulkiewicz, A. Malankowska, B. Rasulev, T. Puzyn, A chemoinformatics approach for the characterization of hybrid nanomaterials: Safer and efficient design perspective. Nanoscale **11**(24), 11808–11818 (2019)
100. Y. Jia, X. Hou, Z. Wang, X. Hu, Machine learning boosts the design and discovery of nanomaterials. ACS Sustain. Chem. Eng. **9**(18), 6130–6147 (2021)
101. S. Mitchell, R. Qin, N. Zheng, J. Pérez-Ramírez, Nanoscale engineering of catalytic materials for sustainable technologies. Nat. Nanotechnol. **16**(2), 129–139 (2021)
102. S. Shivalkar, P.K. Gautam, S. Chaudhary, S.K. Samanta, A.K. Sahoo, Recent development of autonomously driven micro/nanobots for efficient treatment of polluted water. J. Environ. Manag. **281**, 111750 (2021). https://doi.org/10.1016/j.jenvman.2020.111750
103. A.V. Singh, M.H. Ansari, P. Laux, A. Luch, Micro-nanorobots: Important considerations when developing novel drug delivery platforms. Expert Opin. Drug Deliv. **16**(11), 1259–1275 (2019)

AI and Machine Learning Applications in Nanotechnology: Advances in Nanofluids, Bionanofluids, and Ionic Liquids

Ajay Verma, Arvind Singh Parmar, and Ayushi Godiya

1 Introduction

Machine learning (ML) and artificial intelligence (AI) are two closely linked topics that are transforming our interactions with technology. The technique of training computers to learn from data without having to be specifically programmed to do so is known as machine learning. This entails utilizing algorithms and statistical models to discover patterns in data and then using these patterns to make predictions or choices [1].

On the other hand, machine learning and other methods of creating intelligent systems are included in the larger area of AI. The goal of AI is to build robots that are capable of activities that would typically need human intellect, such as voice recognition, natural language comprehension, and complicated data-driven decision-making [2]. AI and machine learning strive to build computers that can learn and adjust to novel circumstances without explicit programming. By giving computers the ability to learn from data and make judgments based on that data, we may build systems that are more precise and more effective at a variety of jobs [3].

Self-driving vehicles, virtual assistants, identifying fraud, and medical testing are just a few of the many uses for machine learning and AI. We may anticipate seeing even more creative uses of machine learning and AI in the future as the technology develops [4]. Globally, machine learning and AI are generating a lot of interest. Artificial intelligence (AI) has transformed technology with its vast applications. While the terms AI and machine learning (ML) are often used interchangeably, they

A. Verma · A. S. Parmar
Parul Institute of Pharmacy and Research [PIPR], Parul University, Vadodara, Gujarat, India

A. Godiya (✉)
ADY Patil University, Pune, India

S. Kulkarni, S. Das (eds.), *Ionic Liquids, Nanofluids and Nanotechnology*,
Lecture Notes in Nanoscale Science and Technology 42,
https://doi.org/10.1007/978-3-032-04008-4_22

are not the same—a distinction that even some industry professionals overlook. AI refers to the broader concept of machines performing tasks that typically require human intelligence, such as reasoning, problem-solving, and decision-making. ML, on the other hand, is a subset of AI focused on training algorithms to learn patterns from data and improve over time without explicit programming [5].

First, we'll look at a typical example: virtual personal assistants. AI and ML are two distinct fields, despite their connections. A subfield of AI called ML is concerned with developing statistical models and algorithms that enable computers to learn and make decisions without the need for explicit programming [6]. Data may be used to train ML algorithms to find trends and forecast future occurrences.

On the other hand, ML and other methods for developing intelligent systems are included in the larger subject of AI. AI is the creation of computer systems that are capable of interpreting natural language, identifying pictures, and making decisions—tasks that normally require cognitive ability [7].

1.1 There Are Several Types of Machine Learning, Including

Supervised learning: A labeled dataset, where the intended output is known, is used to train the algorithm.

Unsupervised learning: Since the algorithm is not provided with any labeled data, it must independently identify the data's underlying structure.

Reinforcement learning: The algorithm gains knowledge from the input it gets from its activities in a given environment [8].

1.2 Overview of AI and ML in Nanotechnology

The fusion of AI and ML with nanotechnology marks a transformative era in materials science. Nanotechnology, concerned with manipulating matter at the nanoscale (1–100 nm), has been revolutionized by computational intelligence, which accelerates discovery, enhances material properties prediction, and optimizes experimental designs [9]. AI and ML enable researchers to quickly process vast amounts of complex data from nanotechnology studies, uncovering hidden patterns and valuable insights that conventional analytical approaches might miss. These approaches are now indispensable in materials synthesis, characterization, and applications, significantly reducing time and cost [10].

Importance of Nanofluids, Bionanofluids, and Ionic Liquids

Nanofluids (nanoparticle suspensions in a base fluid) and bionanofluids (nanofluids including biological materials) have emerged as critical in energy transfer, biomedical, and electrical cooling applications. Ionic liquids have significantly broadened the scope of green chemistry and nanomaterial stabilization due to their adjustable characteristics and low vapor pressure. Combining AI with these specialty fluids allows for precise customization of characteristics like stability, viscosity, and heat conductivity, promoting advancements in both industry and healthcare [11].

Role of Computational Intelligence in Material Science

Advanced methods like AI, ML, deep learning (DL), and evolutionary algorithms are all part of computational intelligence, which is changing the field of material science [12]. In the past, intensive, expensive, and time-consuming laboratory research was needed to find and optimize novel materials. Nevertheless, computational intelligence provides a potent substitute by making it possible to accurately forecast material behaviors using data-driven models and simulations, greatly speeding up the research and development process [13].

The application of AI simulations in material science may predict a range of material properties, including thermal conductivity, mechanical strength, chemical reactivity, electrical conductivity, and stability, with a small amount of input data, such as molecule structure, composition, and process parameters. These models can generalize to forecast the behavior of novel materials that have not yet undergone physical investigation by learning from recent experimental or simulation data [14].

For example, an ML model may be trained on a limited dataset and then predict the behavior over a wider range of circumstances, saving dozens of tests to determine how varying concentrations of nanoparticles impact a nanofluid's viscosity and thermal conductivity. This predictive capability not only saves time and costs but also allows researchers to explore large design domains that would be impossible to cover empirically [15]. Additionally, computational intelligence facilitates inverse design, in which scientists first specify the ideal material characteristics (such as a goal thermal conductivity), then the AI model recommends the best possible composition and structure to get those attributes. Reinforcement learning and genetic algorithms are increasingly being utilized to iteratively tune materials for complicated goals while balancing trade-offs such as performance and cost [16].

In the context of nanofluids, bionanofluids, and ionic liquids, computational intelligence plays a critical role in:

- Determining the best kinds and amounts of nanoparticles for improved thermal control.
- Forecasting bionanofluids' functional behavior and biocompatibility for use in medicine.
- Optimizing ionic liquids for energy storage, catalysis, and solvent applications.

Beyond making predictions, AI also helps find novel phenomena by revealing patterns in big, multifaceted datasets that are hard for humans to decipher. Combined with automated lab equipment and high-throughput computer simulations (like density functional theory and molecular dynamics), computational intelligence opens the door to a new paradigm in material discovery: Research on autonomous materials [17].

In the end, incorporating computational intelligence into material science promotes the creation of next-generation nanofluids and systems based on ionic liquids that are more economical, sustainable, efficient, and customized for particular energy, industrial, and medical uses. Important global issues including energy efficiency, environmental sustainability, and better healthcare delivery might be addressed by this revolution [18].

2 Fundamentals of Nanofluids and Ionic Liquids

2.1 Definition of Nanofluid

Over the past two decades, advances in nanotechnology and thermal engineering have led to the development of an innovative class of heat transfer fluids known as nanofluids. These fluids contain tiny suspended nanoparticles—typically between 1 and 100 nm in size—dispersed within a base liquid like water, oils, or organic compounds (such as ethylene glycol or refrigerants). Keeping the particles below 100 nm is crucial, as it enhances the fluid's stability and performance. Researchers have explored various nanoparticle materials and preparation methods, uncovering new insights into how nanofluids improve heat transfer efficiency. From industrial cooling systems to advanced biomedical applications, these engineered fluids are unlocking new possibilities in thermal management—making them a key focus in modern engineering and energy research [19].

2.2 Methods of Preparation

The thermophysical properties of nanofluids, particularly their stability and thermal conductivity, are significantly influenced by the preparation method used. These techniques not only determine whether the resulting suspension contains nanoparticles (1–100 nm) or larger microparticles but also directly affect the fluid's overall stability. A crucial aspect of the preparation process is sonication [20]—studies show that longer ultrasonication times and higher power levels enhance thermal conductivity, boost heat transfer efficiency, reduce viscosity, and minimize pressure drops. Researchers primarily employ two established preparation approaches: the one-step and two-step methods (detailed in Table 1). While these techniques differ in their methodology, both can effectively produce nanofluids with uniform particle distribution and optimal stability for various applications [22].

Table 1 Method of preparation by the one- and two-step methods

Particular	One-step method	Two-step method
Synthesis process	Simultaneous production of nanoparticles and nanofluid	Manufacturing of nanoparticles by chemical or mechanical means, then releasing them into the base fluid [21]
Production scale	Small-scale production	Large-scale production
Cost of production	High cost	Low cost
Control on NPs size	Difficult and limited control over the nanoparticle size during the preparation stage	Able to control the nanoparticle size during the preparation stage [21]
Particle oxidation	Oxidation of particles does not occur due to elimination of drying, transportation, and storage processes	No such problem
Advantages	Reduced chances of particle agglomeration; more stable nanofluid	Cheaper and more applicable in industry
Disadvantages	Residual reactants remain in nanofluids, causing potential problems during application; they can only be produced in batch	Prone to agglomeration; constant stabilizing process needed for long-term stability [21]

2.3 Nanofluids Classification

Nanofluids are categorized based on the type of nanoparticles used in their formulation. These advanced heat transfer fluids typically fall into four main groups: metal-based, metal oxide-based, carbon-based, and hybrid/mixed metal nanofluids. Manufacturers create these specialized fluids by dispersing the selected nanoparticles into common base fluids such as water, ethylene glycol, or various oils. The performance of these nanofluids heavily depends on their stability—any particle agglomeration can significantly alter their thermophysical properties. This makes stability a critical factor alongside enhanced physical properties when selecting nanofluids for specific applications. The following sections provide a comprehensive overview of the various nanofluid types documented in research literature, highlighting their unique characteristics and potential uses [23].

2.4 Metal-Based Nanofluids

Metal-based nanofluids are formulated by dispersing metallic nanoparticles—such as gold, silver, or aluminum—within a base fluid. A notable study by Beicker et al. demonstrated this process by developing a gold/water nanofluid through a two-step preparation method. Their research focused on examining the nanofluid's photothermal conversion capabilities, revealing impressive results. Remarkably, even at an extremely low concentration of just 0.004% by volume, the gold nanofluid exhibited exceptional performance. The study also confirmed the nanofluid maintained stability for an extended period of up to 120 h, highlighting its potential for practical applications [24].

2.5 *Metal Oxide-Based Nanofluids*

Metal oxide-based nanofluids have become indispensable across various industries due to their exceptional stability, excellent thermal properties, and cost-effectiveness. These advantages make them particularly attractive for large-scale industrial applications where both performance and budget matter. A key benefit lies in their economical synthesis, offering industries an affordable yet high-performance alternative to more expensive options. This cost advantage, combined with their reliable thermal conductivity, has made metal oxide nanofluids the go-to cooling solution in numerous engineering systems. Among the most popular choices are titanium dioxide (TiO_2) and aluminum oxide (Al_2O_3) nanoparticles, which strike an ideal balance between performance and practicality. These metal oxides consistently deliver results while remaining accessible for widespread industrial use, explaining their dominance in nanofluid applications [24].

2.6 *Carbon-Based Nanofluids*

Carbon-based nanofluids demonstrate superior thermal and physical properties compared to conventional base fluids, as evidenced by numerous studies. However, their high production costs currently limit large-scale commercial adoption. Research by Beicker et al. reveals an interesting cost-performance dynamic: while MWCNT/water nanofluids prove more economical than gold/water alternatives, they also offer better stability and slower quality degradation.

The stability potential of carbon nanofluids is particularly impressive. Studies document remarkable longevity:

Sarafraz et al. achieved 504 h of stability for COOH-CNT/water nanofluid using nonylphenol ethoxylate as a stabilizer.

Xie et al. reported exceptional 1440-h (2 months) stability for treated CNTs in both water and ethylene glycol—without any surfactants.

Interestingly, nonpolar base fluids like decene require minimal surfactant (oleylamine) to maintain stable TCNT suspensions for extended periods. These findings highlight both the promise and practical considerations of carbon-based nanofluids for industrial applications [25].

2.7 Hybrid Nanofluids

As nanofluids continue to demonstrate remarkable potential across various applications, researchers are pioneering an innovative approach: hybrid nanofluids. These advanced fluids incorporate multiple nanoparticle types within a single base fluid, creating materials that combine the best properties of their constituent parts at the nanoscale. The fundamental advantage of hybrid nanofluids lies in their ability to overcome the limitations of single-component systems. Since no single material possesses all the ideal characteristics for every application, these hybrid solutions strategically blend complementary properties from different nanoparticles. This synergistic approach enables engineers to tailor nanofluids with precisely optimized thermal, chemical, and physical properties for specific industrial needs [25]. Hybrid nanofluids represent a major advancement, demonstrating superior hydrodynamic, thermophysical, and heat transfer capabilities compared to conventional single-nanoparticle fluids. A compelling study by Sun et al. illustrates this enhancement through their work with a 50:50 Ag-MWCNT/water hybrid nanofluid in jet impingement cooling systems. Their findings revealed that the hybrid formulation substantially outperformed its MWCNT/water counterpart in thermal conductivity measurements, showcasing the tangible benefits of combining different nanoparticle types [26]. Experimental results demonstrated that the Ag-MWCNT/water hybrid nanofluid achieved significantly higher thermal conductivity compared to the standard MWCNT/water nanofluid. Researchers typically prepare these advanced hybrid nanofluids using conventional methods—particularly the well-established one-step and two-step preparation techniques. The next section details these standard nanofluid production methods and their specific applications in hybrid nanofluid synthesis [27].

2.8 Nanofluid Stabilization Methods

Achieving stable nanofluids is crucial for maintaining consistent thermophysical properties in practical applications. The key to stability lies in overcoming nanoparticle aggregation—a phenomenon governed by competing forces at the nanoscale. When the electrical double layer repulsive force (EDLRF) dominates over Van der Waals attractive forces, nanoparticles remain uniformly dispersed. However, if Van der Waals forces prevail, particles begin clustering through agglomeration and aggregation, ultimately leading to sedimentation. This delicate balance determines whether a nanofluid maintains its enhanced properties or degrades over time [28]. For any nanofluid—especially those prepared via the two-step method—undergoing stability enhancement is essential before real-world application. Research shows carbon-based nanofluids typically demonstrate superior stability compared to other types. Even metal-based nanofluids can achieve excellent stability when prepared properly through the two-step method with appropriate surfactants.

The stability of a nanofluid depends on two key factors that researchers must document:

The observed stability duration.
The specific characteristics of the nanoparticles used.
This section details proven stabilization techniques used in the two-step preparation process to prevent particle aggregation and maintain homogeneous dispersion—critical for reliable performance in engineering systems [29].

2.9 *Ionic Liquids*

A novel family of completely ionic, salt-like substances, liquid at very low temperatures, is known as an "ionic liquid." At the moment, its "official" definition refers to the boiling point of water: "These are ionic compounds which are liquids below 100 °C." These are more often found to have melting points lower than room temperature; some even have melting points lower than 0 °C. These novel materials are liquid between these substances' melting and decomposition temperatures, so they cover a wide temperature range from 300 to 400 °C [30].

A comparison between a normal inorganic salt, such as table salt (NaCl, m.p. 801 °C), and a typical ionic liquid, such as 1-ethyl-3-methylimidazolium ethylsulfate (m.p. <-20 °C), makes the distinction clear. The symmetry of the ionic liquid is much smaller! Additionally, the cation and anion charges are dispersed across the molecule's greater volume via resonance. As a result, the ionic liquid will solidify at lower temperatures. Sometimes a glass transition is seen in place of a point of melting, especially if longer aliphatic flank chains are present [31].

The strong ionic (Coulomb-) interaction of these substances produces a non-flammable material, a high mechanical, thermal, and electrochemically stable product, or a minimal pressure of vapor (until breakdown happens). Apart from this intriguing mix of qualities, they contain several benefits, like highly desired solvent qualities and immiscibility with natural solvents and water, which creates biphasic systems [32].

The cation selection will frequently determine its stability and has a significant impact on a cationic liquid's characteristics. The choice of an anion generally controls the chemistry and functioning of the ionic liquid. It is theoretically conceivable to have ionic liquids when an array of anions and cations are combined. An actual figure, however, will be magnitudes lower. Currently, research describes roughly 1000 of these compounds, and about 300 of them are sold commercially. Typical configurations include either inorganic or organic anions alongside organic cations as given in Table 2 [34].

Table 2 Challenges in optimizing nanofluid performance

Challenge	Description	Implications
Nanoparticle agglomeration	Nanoparticles tend to cluster together due to Van der Waals forces and high surface energy	Reduces effective surface area, deteriorates thermal conductivity, and leads to clogging issues [33]
Stability of suspension	Maintaining a stable and homogeneous suspension of nanoparticles in the base fluid over time is difficult	Affects heat transfer performance and causes sedimentation in systems [33]
High viscosity	Addition of nanoparticles often increases the viscosity of the base fluid	Increases pumping power requirement and energy consumption [33]
Compatibility with base fluids	Not all nanoparticles are compatible with all base fluids; chemical interactions may occur	Can lead to degradation of nanofluids or production of unwanted byproducts [33]
Thermal conductivity enhancement limits	Despite improvements, thermal conductivity enhancement may not always justify the costs or complexities involved	Limits practical applicability in certain systems [33]
Lack of standardization	No universally accepted standards for preparation, characterization, or testing of nanofluids	Makes comparison of experimental results and performance data difficult across studies [33]
Cost of nanoparticles	High-quality nanoparticles and their surface modification processes are often expensive	Restricts large-scale application and industrial use [33]
Environmental and health concerns	Long-term environmental and biological impacts of nanoparticles are not fully understood	Raises safety concerns for production, handling, and disposal [33]
Heat transfer mechanism complexity	Complex mechanisms including Brownian motion, thermophoresis, and particle-fluid interactions are not fully understood	Limits predictive modeling and optimization strategies [33]
Erosion and corrosion	Nanoparticles may cause erosion or corrosion in metal components of heat transfer systems	Shortens equipment life and increases maintenance costs [33]

3 AI and ML Techniques for Nanotechnology

An artificial neural network (ANN) is a group of neurons or perceptrons at each layer. Because inputs are only taken into account once, an ANN is also referred to as a feedforward neural network. These neural networks are some of the most fundamental types. Up until it reaches the output node, they send data in a single direction across several input nodes. Hidden node layers, which would make it simpler to comprehend how the network functions, may or may not be present [35].

Feedforward neural network: The feedforward neural network is one of the simplest ANNs. The input or data in this artificial neural network moves only in one direction. The layer of inputs is where it enters the ANN, and the output layer is where it leaves, with hidden layers perhaps present or absent. The feedforward neural network, thus, consists of a wave that is propagated forward; backpropagation is often absent [36].

Convolutional neural network: A convolutional neural network is similar to a feedforward neural network because the connections between units are weighted to identify the effect of one unit on another. However, a convolutional neural network (CNN) contains one or more convolutional layers that apply a convolution operation to an input and then transmit the result to the subsequent layer. CNNs can be used for image and audio processing, which is very helpful in computer vision [37].

Modular neural network: Modular neural networks employ multiple independent neural networks that work in parallel to solve complex problems efficiently. Unlike traditional architectures, each module specializes in processing distinct inputs and handling specific subtasks. This decentralized approach offers key advantages:

Task decomposition: Breaks down computationally intensive processes into manageable components.

Specialized processing: Enables dedicated networks to focus on specific aspects of the problem.

Reduced complexity: Maintains system performance while lowering overall computational demands.

By dividing labor among specialized modules, this architecture achieves target outcomes more efficiently than monolithic networks, particularly for multifaceted problems requiring diverse processing capabilities [38].

Radial basis function neural network: Radial basis functions are those that take into account a point's distance from the center. There are two levels in RBF functions. In the first layer, the input is mapped into all of the Radial basis functions in the hidden layer, and the output layer computes the output in the next step. Data that indicates any underlying trend or function is often modeled using radial basis function nets [39].

Recurrent neural network: The Recurrent Neural Network retains a layer's output and feeds it back into the input to improve the layer's predictability. In the recurrent neural network (RNN), the first layer is very similar to the feedforward neural network, so its recurrent phase begins after the first layer's output has been calculated. Following this layer, each unit will retain some data from the preceding stage, enabling it to function as a memory cell during calculations [40].

Advantages

- Maintaining data for the whole network
- The capacity to operate with little information
- Being tolerant of errors
- Having a memory that is spread

Disadvantages

- Hardware reliance
- Inexplicable network behavior
- Establishing the appropriate network structure [40]

3.1 Convolutional Neural Network (CNN)

One of the most popular models nowadays is a CNN. This computational model of neural networks uses a variation of multilayer perceptrons with one or more convolutional layers that can be fully linked or pooled accordingly. Following the development of feature maps, these convolutional layers divide the image into rectangular segments and forward the data to nonlinear processing [41].

CNNs are composed of numerous layers, including the input layer, convolutional layer, pooling layer, and fully connected layers. Let us examine CNNs in more detail with the help of Fig. 1 [42].

Convolution Neural Networks all share the same set of parameters.

Assume you have a photo. Given its length, width, and height (the image's dimensions and the channel, since most images have red, green, and blue channels), it could be depicted as a cuboid [43].

A new neural network with different depths, widths, and heights can then be created by moving that network over the entire image. Compared to the R, G, and B channels, the new channels we have are smaller in both width and height. They call this process convolution. A regular neural network is one in which the patch size is equal to the image size. We are lighter because of this small patch [44].

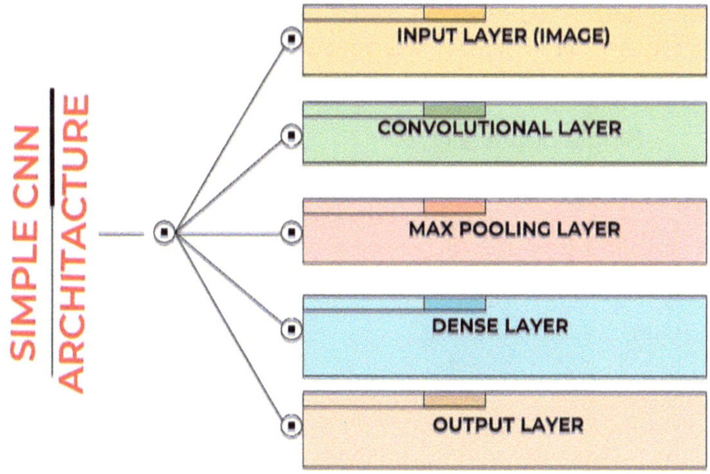

Fig. 1 Simple CNN architecture

3.2 Mathematical Overview of Convolution

- Learnable filters (also known as kernels) with modest widths and heights and the same depth as the input volume (three for image input) comprise convolution layers.
- Assume, for example, that we must perform a convolution on a 34 × 34 × 3 image. With "a" being any value between 3 and 7, but less than the image's dimensions, filter sizes can be axax3.
- Strides, which can vary from two to four for high-dimensional images, are the steps used to slide the filters over the input volume during the forward pass. Next, the dot product of the input volume's patch and kernel weights is calculated.
- Sliding each filter will produce a 2D output, and stacking them together will result in an output volume with a depth equal to the number of filters. The network will learn all of the filters [45].

3.3 Layers Used to Build ConvNets

ConvNet is another term for a complete CNN architecture. A network is made up of layers, each of which uses a differentiable function to change one volume to another. To demonstrate, let us run a CNN on a 32 × 32 × 3 image display in Fig. 2 [46].

- Input layers: This layer is where we feed data into our model. Typically, an image or series of images will be used as the input in a CNN. The raw input of the image with dimensions of 32 × 32 × 3 is stored in this layer [47].
- Convolutional layers: This layer is used to extract features from the input dataset. It processes the incoming images using a set of learnable filters known as kernels. The filters/kernels are smaller matrices, usually 2 × 2, 3 × 3, or 5 × 5. It computes the dot product of the kernel weight and the corresponding input image

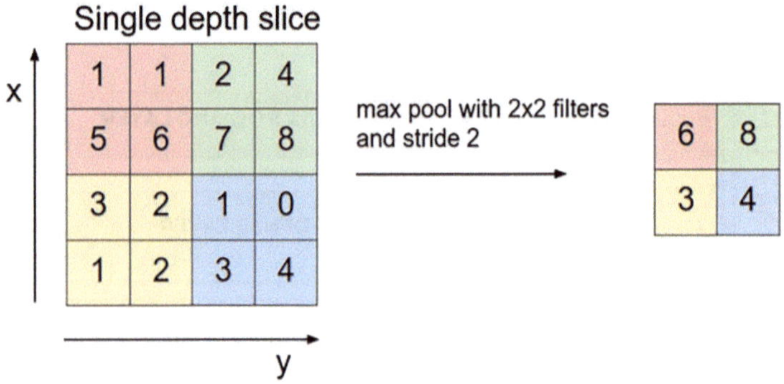

Fig. 2 Pooling layers [3231n.stanford.edu]

patch as it moves across the input image data. A feature map is the result of this layer. Using a total of 12 filters for this layer will produce an output volume with dimensions of 32 × 32 × 12 [48].

- Activation layer: In order to add nonlinearity to the network, activation layers apply an activation function to the output of the preceding layer. There will be an element-wise activation function applied to the convolution layer's output. Examples of common activation functions are RELU: max[0, x], Tanh, and Leaky RELU. Due to the constant volume, the output volume will be 32 × 32 × 12 [49].
- Pooling layer: The main function of this layer, which is frequently added to the covnets, is to reduce the volume, which expedites computation, reduces memory usage, and prevents overfitting. Maximum pooling and average pooling are the two most widely used types of pooling layers. Using a max pool with 2 × 2 filters and stride 2 will result in a volume that is 16 × 16 × 12 [50].
- Flattening: The resulting feature maps are flattened into a one-dimensional vector after the convolution and pooling layers, allowing them to be fed into a fully connected layer for classification or regression.
- Fully connected layers: The resulting feature maps are flattened into a one-dimensional vector after the convolution and pooling layers so that they can be fed into a fully connected layer for classification or prediction [51].

Output Layer: The output from the fully connected layers is then fed into a logistic function that converts each class's output into its likelihood score for classification tasks like sigmoid or softmax [51].

Advantages

- High accuracy in picture recognition
- Automatically recognizes significant aspects without human intervention
- The sharing of weight [52]

Disadvantages

- CNN does not encode an object's orientation or location
- Being spatially invariant to the incoming data is not possible
- A large amount of training data is needed

3.4 Recurrent Neural Network (RNN)

RNNs are more complex. The information was not transmitted in a single direction; instead, they store the output of processing nodes and feed it back into the model. In this way, the model is said to learn to predict the outcome of a layer. Each node in the RNN model continues to perform calculations and operations as a memory cell. The system learns on its own and continues attempting to make the correct prediction via backpropagation if the network's forecast is incorrect [5].

By feeding the output from one phase into the next, RNNs enable the network to "remember" previous information. This aids the network in comprehending the background of past events and improving its forecasts accordingly. For instance, the

RNN utilizes the preceding words to determine which word is most likely to appear next in a phrase [53].

3.5 RNN Work

At each time step, RNNs process units with a fixed activation function. These units have an internal hidden state that acts as memory that retains information from previous time steps. This memory allows the network to store past knowledge and adapt based on new inputs [53].

3.6 Updating the Hidden State in RNNs

The current hidden state ht. is calculated from the prior state h_{t-1} and the current input x_t using the following relations [54].

1. State update:

$$h_t = f\ [h_{t-1}, x_t]$$

   where h_t is the current state, h_{t-1} is the previous state, x_t is the input at the current time step.
2. Activation function application:

$$h_t = \tanh\ [W_{hh} \cdot h_{t-1} + W_{xh} \cdot x_t]$$

   Here W_{hh} is the weight matrix for the recurrent neuron, and W_{xh} is the weight matrix for the input neuron.
3. Output calculation:

$$yt = W_{hy}\ ht$$

   where yt is the output and W_{hy} is the weight at the output layer.

To update these parameters, backpropagation is used. But because RNN operates on sequential data, we employ an updated backpropagation technique called backpropagation via time in this case refer to Table 3 [56].

Advantages

- Over time, an RNN retains all of the information. Its ability to recall prior inputs is the only reason it is helpful for time series prediction. We refer to this as long short-term memory.
- To increase the effective pixel neighborhood, RNNs are even utilized in conjunction with convolutional layers.

Table 3 Summation of all three networks in a single table

Type of data	ANN	CNN	RNN
Type of data	Tabular data, text data	Image data	Sequence data
Parameter sharing	No	Yes	Yes
Fixed length input	Yes	Yes	No
Recurrent connections	No	No	Yes
Vanishing and exploding gradient	Yes	Yes	Yes
Spatial relationship	No	Yes	No
Performance	ANN is considered to be less powerful than CNN and RNN	CNN is considered to be more powerful than ANN and RNN	RNN includes less feature compatibility when compared to CNN [55]
Application	Facial recognition and computer vision	Facial recognition, text digitization, and natural language processing	Text-to-speech conversions [55]
Main advantages	Having fault tolerance, ability to work with incomplete knowledge	High accuracy in image recognition problems, such as weight sharing	Remembers each and every information, time series prediction [55]
Disadvantages	Hardware dependence, unexplained behavior of the network	Large training data needed, as it does not encode the position and orientation of object	Gradient vanishing, exploding gradient

Disadvantages

- Issues with gradient vanishing and exploding.
- RNN training is a very challenging process.
- If tanh or relu are used as activation functions, it is unable to process very long sequences [57].

4 AI-Enabled Property Prediction and Optimization

4.1 Predicting Thermal Conductivity and Viscosity of Nanofluids

SVM-Based Models

SVM-based techniques have shown notable efficacy in predicting the output of various thermal systems, such as solar power plants and exchangers for heat. Numerous research demonstrate how SVM-based techniques may be used to predict the thermophysical characteristics of nanofluids. Several databases and search engines, including Google Scholar, Scopus, and the websites of scientific publishers, were utilized to locate the relevant publications for the present review. “SVM,” “Support Vector Machine,” “Nanofluid,” “Dynamic Viscosity,” and “Thermal Conductivity” were the search terms utilized. Applications of SVM-based techniques for TC and DV of NFs are given in the ensuing subsections [58].

Thermal Conductivity Modeling by SVM

When it comes to the NFs’ capacity and rate of heat transmission, TC is one of the most crucial characteristics. In general, fluids with higher TCs are favored for heat transfer because they transport thermal energy more effectively. An increase in nanofluids’ TC results in an increase in the rate of convective heat transfer; as a result, these fluids are more desirable for thermal media that rely on convection heat transfer processes. As stated in the introduction, augmentation in fluid TC due to nanostructure dispersion is determined by a variety of parameters. These relevant aspects must be incorporated in the inputs to create models with the maximum potential accuracy [59].

Some of the basic and broad equations for the TC of NFs include just VF to calculate the TC ratio of the NFs; however, the bulk of the models presented for the TC of NFs include additional components. VF and temperature are two of the most often employed factors in most investigations that use clever methodologies [60].

Several studies have looked at how various components impact TC in NFs. Due to the intricacies of microscopic interactions between solid and liquid phases and the participation of several elements, sophisticated instruments should be used to achieve high precision. Among the most promising methods for this are SVM-based models, whose performance is evaluated against that of other clever methods. Metal oxides have been widely used as the solid phase of NFs because of their ease of manufacture, chemical stability, and comparatively low density [61].

Various scholars have used SVM-based methods to simulate the features of certain types of NFs. The TC of CuO/EG NF was predicted in a research using the genetic algorithm (GA)-least square support vector machine (LSSVM) and GMDH. The models’ inputs in their study were temperature, VF, and the diameter of the solid particles. With corresponding R2 values of 0.991 and 0.994, both models have shown a high degree of accuracy. Another research estimated the TC of Al_2O_3/EG NF using the same inputs. Using GA-LSSVM, the model’s R2 was 0.9902, and its MSE was 8.64×10^{-4}.

In another study, the TC ratio of Al_2O_3/water was evaluated using three approaches: self-organizing map (SOM), LSSVM-GA, and multilayer perceptron (MLP) ANN [62]. Three suggested inputs were used for the modeling, the same as in the earlier experiments. For the SOM, ANN, and LSSVM-based models, the corresponding correlation coefficients were 0.88125, 0.87575, and 0.89999. On the basis of these values, it can be noted that using LSSVM had better precision compared to the two other approaches, modeled TC of Fe_2O_3/water NF using different intelligent techniques including ANFIS, Conjugate Hybrid Particle Swarm Optimization ANFIS (CHPSO-ANFIS), GA-Radial Basis Function (GA-RBF), and LSSVM [63].

Average Absolute Relative Deviation

Average Absolute Relative Deviation—used in statistics and data analysis to measure the average relative error between predicted and actual values (often seen in ML model evaluation).

$$AARD = \frac{1}{n}\sum_{i=1}^{n}\left|\frac{y_i - \hat{y}_1}{y_1}\right| \times 100$$

The aforementioned models' respective AARDs were 3.6484, 1.0348, 0.4425, and 0.8843. Although it seems that using GA-RBF produces predictions with more accuracy in terms of AARD, LASSVM's performance is noteworthy and dependable. In a research, particle swarm optimization (PSO) was applied using an LSSVM approach. Other intelligent techniques, such as MLP-ANN, RBF-ANN, and ANFIS, were used in addition to PSO-LSSVM to calculate the TC of TiO_2/water. With an R2 of almost 1, the use of PSO-LSSVM yielded the best precision among the methods discussed. The study's LSSVM model's highest absolute relative deviation was less than 0.6%.

Certain TC models of NFs are more thorough than others. In these situations, several more inputs must be taken into account. ANNs are one method that certain scholars utilize to create these kinds of comprehensive models. SVM-based models work similarly to ANNs in this regard. For example, a research predicted the TC of NFs with Fe_2O_3 particles in water-EG base fluids. In addition to temperature and VD, the mass ratio of water/EG was included in their models, PSO-ANN and GA-LSSVM. GA-LSSVM was shown to be more accurate than the PSO-ANN model, with an AARD of 2.3809% as opposed to 10.761% [64, 65].

Dynamic Viscosity Modeling by SVM

Another feature that affects heat transfer and fluid flow is DV, which several researchers have studied and predicted. Like the TC, there are some basic connections for an approximate estimation of DV based mostly on the VF of nanostructures; however, more complicated techniques and additional important elements must be considered for more accuracy. In this context, clever techniques are created to simulate this NF characteristic [66].

Metal oxide NFs have been studied extensively because of their widespread use. DV of Fe_2O_3/water NF was predicted in a research using a number of clever approaches, including ANFIS, CHPSO-ANFIS, GA-RBF, and LSSVM. The DV of the NF was estimated using the following models: 0.9816, 0.9974, 0.9982, 0.9970, and 0.9930. Even though the LSSVM model was less accurate than GA-RBF and CHPSO-ANFIS, it was nonetheless noteworthy with an AARD of 3.2194. In one research, the effectiveness of Coupled Simulated Annealing (CSA)-LSSVM in estimating the DV of TiO2/SAE 50 NF was evaluated and contrasted with two other techniques, namely GA-RBF and Gene Expression Programming (GEP). In terms of the NF's non-Newtonian behavior, shear rate was evaluated in addition to temperature and VF. Although GA-RBF performed better with an R2 of 0.99998, CSA-LSSVM outperformed GEP with an R2 of 0.99777 and demonstrated noteworthy performance with an R2 of 0.99991.

The DV of NFs with hybrid nanostructures may be predicted using SVM methods. The LSSVM approach's viability for predicting shear rate on the DV of MWCNT-MgO/oil was examined in a research. Shear rate, temperature, and VF were all inputs in their model, and a Bayesian technique was used for optimization. The estimated data, as shown, are quite similar to the experimental data, with an R2 of 0.9999, while the value of R2 was 0.9991 when correlation was used for prediction, demonstrating the better performance of the used intelligent technique. In another study, different intelligent methods such as SVM, RBF [66], decision tree (DT), and random forest (RF) were used to estimate the DV of various hybrid NFs, including COOH-functionalized MWCNT/Al_2O_3, COOH-functionalized MWCNT/ZnO, COOH-functionalized MWCNT/MgO, and COOH-functionalized MWCNT/SiO2. VF and temperature served as the models' inputs. Based on predicted values, DT and SVM outperform two other methods in terms of accuracy [67].

The accuracy and results of the intelligent approaches may be impacted by applied functions and procedures. For example, when using MLP-ANN, the kind of training functions may affect accuracy, and when utilizing the ANFIS approach, the membership function used affects how well the predictive models perform. Different optimization techniques may be used to tune the hyperparameters of intelligent approaches. The performance of the suggested models may be impacted by the LSSVM approaches' optimization methodology [68].

In a research on the DV modeling of Al_2O_3/water NFs, multiple optimization techniques such as GA, Imperialist Competitive Algorithm (ICA), PSO, and Hybrid GA PSO (HGAPSO) were utilized in the LSSVM. The model that was created by utilizing GA as the optimization tool had the greatest R2 value; however, HGAPSO/LSSVM had the lowest AARD value among the used techniques, as seen in Fig. 2 [69].

Additional inputs might be added to the models to increase their comprehensiveness. For example, to estimate the DV of NFs with Al_2O_3 particles and different base fluids, apply VF, lower the temperature inverse, the base fluid acentric factor, the particle diameter, and reduce the pressure. Their study used methods such as Cascade Feedforward ANN (CF ANN), MLP-ANN, RBF-ANN, LSSVM, and correlation. Even while MLP-ANN had the lowest AARD of all the approaches used, LSSVM's performance with an AARD of 6.92% was much better than the correlation with an AARD of 14.50% [70].

In another study, the DV of NFs with Fe_2O_3 nanoparticles in water-EG base fluid was calculated using GA-LSSVM and PSO-ANN. Their model's inputs included the mass percent of EG/water, temperature, and VF.

4.2 Optimization of Ionic Liquid-Based Nanocomposites

Multi-component systems known as ionic liquid-based nanocomposites (IL-NCs) blend the functional characteristics of nanomaterials with the diverse physicochemical characteristics of ionic liquids (ILs). From a theoretical perspective, these materials belong to a distinct class of soft matter in which molecular interactions, thermodynamics, and nanostructuring effects regulate their behavior and performance. In this sense, optimization is the methodical adjustment of material composition, microstructure, and interaction parameters to attain certain functional and physicochemical characteristics [70, 71].

Thermodynamic Considerations

The thermodynamic compatibility between the ionic liquid and the nanofiller is the fundamental aspect of IL-NC optimization. The free energy of mixing (ΔGmix) must be negative or low enough to provide interfacial stability and uniform dispersion. An approximation of the Gibbs free energy change is [72]:

$$\Delta G_{mix} = \Delta H_{mix} - T\Delta S_{mix}$$

where ΔH_{mix} is the enthalpic contribution due to interactions between the IL and nanofiller surfaces, ΔS_{mix} is the entropic contribution from the disorder introduced during mixing.

Effective optimization minimizes ΔH_{mix} (via favorable intermolecular forces) and maximizes ΔS_{mix} (by ensuring high dispersibility and low aggregation tendencies).

Molecular Interactions and Interface Engineering

The interaction between IL molecules and nanomaterials is governed by a combination of:

- Electrostatic interactions (ionic pairing and surface charge).
- Van der Waals forces.
- Hydrogen bonding.
- π–π stacking (for carbon-based nanomaterials).

Predicting and optimizing these interactions is done by theoretical models like molecular dynamics (MD) simulations and density functional theory (DFT). Grafting ionic groups or polar moieties onto the surface of nanofillers, for example, improves interfacial adhesion by adding particular IL ion binding sites, which improves miscibility and structural stability [73].

Structural Organization and Phase Behavior

Since ILs are amphiphilic and nanomaterials are anisotropic, IL-NCs often show microphase separation:

- Self-assembly behavior of ILs around nanoparticles.
- Formation of interfacial layers or electric double layers.
- Percolation thresholds, especially in conductive or magnetic nanocomposites [74].

Important characteristics such as ionic conductivity, thermal diffusivity, and mechanical strength are influenced by the degree of organization. Monte Carlo simulations, statistical mechanics, and percolation theory are theoretical methods that aid in the prediction and optimization of these features.

Rheological and Transport Properties

- Viscosity of the IL medium (governed by cation/anion identity).
- Filler aspect ratio and dispersion.
- Interfacial slip conditions and particle–matrix interactions.

The Debye–Hückel theory and the Stokes–Einstein relation are often used to evaluate ionic mobility and diffusivity. Continuum models and multiscale simulations relate molecular interactions to macroscopic transport processes [75].

5 Molecular Interactions and Interface Engineering

The following factors work together to control the interaction between IL molecules and nanomaterials:

- Electrostatic interactions (ionic pairing and surface charge).
- Van der Waals forces.
- Hydrogen bonding.
- π–π stacking (for carbon-based nanomaterials).

Predicting and optimizing these interactions is done by theoretical models like MD simulations and DFT. Surface functionalization of nanofillers (e.g., grafting ionic groups or polar moieties) improves interfacial adhesion by creating specialized binding sites for IL ions, resulting in increased miscibility and structural stability [76].

5.1 ML Models for Stability and Performance Enhancement

In materials science, the use of ML has completely changed how scientists approach the design and optimization of complex systems, especially ionic liquid-based nanocomposites (IL-NCs). Theoretically, ML models are effective instruments that can capture multidimensional interactions and nonlinear correlations between compositional, structural, and environmental factors. These models may greatly help in forecasting the stability and improving the performance of IL-NCs by learning from current data and generalizing to new, undiscovered material combinations [77].

Using molecular descriptors of ILs and nanomaterials, ML techniques may be taught to forecast mechanical, chemical, and thermal robustness. For example, large datasets of IL compositions, nanoparticle surface chemistries, or environmental conditions can be analyzed by models like support vector machines (SVM), choice trees, or random woodlands to predict decomposition temperatures, oxidative degradation behavior, and phase separation tendencies. These prediction skills are based on thermodynamic principles, in which the model learns to estimate enthalpic and entropic contributions to mixing and interaction energies, enabling the selection of compositions that minimize aggregation or chemical breakdown over time [78].

Additionally, ML models play a key role in improving IL-NCs' functional performance. Because performance characteristics including mechanical strength, viscosity, ionic conductivity, dielectric strength, and interfacial charge transfer efficiency are often interdependent, traditional trial-and-error techniques are ineffective. Multiple characteristics may be tuned simultaneously thanks to multi-objective optimization made possible by ML. High-dimensional design spaces may be explored using neural networks, Bayesian optimization, and evolutionary algorithms [79], which provide optimum solutions based on trained models obtained from simulation and observations. These models often include input characteristics like molecular weight, ion-pair polarity, aspect ratio of nanoparticles, or density of surface functional groups, all of which are theoretically connected to performance via well-established physicochemical correlations [80].

Through the discovery of hidden relationships and trends in intricate datasets, ML also makes it possible to theoretically abstract structure-property links. When combined with computational approaches such as density functional theory (DFT) or MD simulations, ML models may be improved with physics-informed characteristics, resulting in more accurate and interpretable predictions. This collaboration helps to build predictive frameworks for rational design, which allow material candidates to be suggested based on performance criteria prior to laboratory synthesis, in addition to speeding up the identification of new IL-NC formulations [81].

ML methods essentially provide a theoretical foundation for negotiating the intricacy of IL-NC systems. These models significantly reduce experimental effort, allow for targeted material creation, and enhance understanding of the fundamental interactions controlling material performance and stability through data-driven learning and accurate forecasting [82].

6 Computational and Experimental Insights

6.1 AI-Assisted Molecular Dynamics and Density Functional Theory (DFT)

Ab initio QM and DFT approaches may also benefit from ML at the Hamiltonian, (Kohn-Sham) molecular orbital, and functional levels. On the DFT side, numerous general-purpose ML-improved DFT functional were presented, including DM2180 and CF22D. Although these methods are undoubtedly slower than SQM techniques enhanced by ML, they may be more resilient due to the less precise physical models at their core. Accurate ab initio techniques, like various forms of linked cluster and configuration interaction, are very slow, therefore, there is a strong motivation to speed them up by, for example, choosing the most crucial configuration and speeding up their integration [83].

A significant divergence from the standard QM techniques is the use of ML to determine the optimal wavefunction for the system. In this sense, impressive findings were obtained by getting more precise values than with a connected cluster, and subsequent investigations have shown that excited-state energies could also be obtained. Despite the high computing cost of these techniques, efforts are being undertaken to develop speedier, more general-purpose solutions. Developments in this subject are vital to keep an eye on [84].

6.2 Data-Driven Experimental Validation of Nanofluid Behavior

Synthesis of Nanoparticles, Characterization, and Nanofluid Preparation

The study's chemical reagents were purchased from Merck and utilized without further purification. Throughout the test, deionized water was utilized regularly. The related study contains information on how the necessary nanoparticles were synthesized. To assess the stability of the nanofluids, a Malvern Instruments Nano-ZS device from the United Kingdom was deployed [85]. X-ray diffraction and field emission scanning electron microscopy were used to examine the nanoparticles' morphological and structural properties. Water-based mono and hybrid nanofluids in a 50:50 ratio were created utilizing a two-step process. Please see our published article for details on how to prepare the nanofluids [86].

Thermal Conductivity and Viscosity

The nanofluids' thermal conductivity was tested using a Decagon Devices Inc. KD2 Pro Thermal Properties Analyzer. This gadget, which has a single KS-1 sensor, uses the transient hot-wire approach. The sensor, which is 60 mm long and 1.3 mm wide, is submerged in the test fluid and serves as both a heat source and a temperature sensor. The transitory hot-wire approach reduces free convection difficulties by creating just a tiny quantity of heat. The instrument may have an error margin of up to $\pm 5\%$. The KD2 Pro was calibrated using a glycerol standard that the manufacturer supplied prior to the studies commencing [87].

To determine the viscosity of the nanofluids, a Brookfield DV-II + Pro viscometer was used, with a maximum inaccuracy of $\pm 2\%$. By measuring the torque applied by a spindle immersed in the fluid, this device determines fluid resistance. A regulated water bath was utilized in conjunction with the viscometer to maintain temperature stability and evaluate the effect of temperature on viscosity in order to guarantee reliable data. Viscosity readings were made in 5-degree increments between 60 °C. To guarantee consistency and reliability, each humidity and temperature measurement was performed five times with a 15-min delay in between [88].

Prognostic Analysis with Machine Learning

Random Forest

An ensemble learning method called RF generates the mean prediction for regression tasks or the mode of categorization by constructing a large number of DTs during training. The model averages the results of individual DTs in the context of regression. Each DT reduces variance and overfitting because it is constructed using a random subset of the data and a random subset of the attributes at each node. During training, the method bootstraps, or selects samples at random with replacement, such that every tree is trained on a different subset. By ensuring that each tree is decorrelated, random feature splitting strengthens the ensemble model as a whole. The final prediction simultaneously minimizes bias and variance, calculated as the average of all tree projections. In datasets where DTs alone may exhibit high variation, this method often works well to prevent overfitting. The training error is often minor because of the average of numerous trees; nonetheless, overfitting on training data subsets may produce a slight rise in the test error, but less noticeably than in single DTs [89].

Gradient Boosting Regression

Gradient boosting (GB) regression develops models one after the other, fixing the mistakes of the models that came before it. For regression issues, it is often mean squared error, a step-by-step additive model that adds new trees to minimize the loss function. After training an initial data-based DT, the process calculates the tree's residuals or mistakes. After that, a new tree is trained to predict these residuals. The projections from this tree are combined with those from previous trees to modify the model's overall prediction [90].

For a certain number of times, this process is repeated, with each subsequent tree correcting the mistakes of the previous one. By minimizing a differentiable loss function, gradient descent aids in model parameter update. Although gradient boosting is known for its excellent predictive power, it may be prone to overfitting, particularly when too many iterations are performed. Because of its iterative nature, which focuses on the most difficult samples at each step, it is often more accurate than random forests and DTs.

Decision Tree

Decision trees (DT) work by recursively dividing data into subsets depending on input feature values, resulting in a decision tree-like model. At each node, the approach selects the feature and threshold that minimizes a specified loss function, such as mean squared error in regression problems. This approach repeatedly creates branches in the tree until the data is adequately divided or a stopping criterion—such as the maximum depth—is satisfied. Without feature scaling or data normalization, DTs may capture complex correlations between attributes. The average of the goal values in each leaf node creates the final prediction. DTs are simple to comprehend because their structure mimics how people make decisions. They are prone to overfitting, too, especially as the trees go deep and start to pick up on noise in the training set. Pruning makes the model simpler by removing branches with little predictive power, which reduces the overfitting problem. Although DTs are strong, their single-tree design makes them more susceptible to significant variation than ensemble techniques like random forests or gradient boosting [91].

SHAP-Based Explainable Machine Learning

SHapley Additive ExPlanations (SHAP) values are a game-theory-based method for evaluating ML models that quantifies each feature's contribution to the model's output. Understanding how variables like temperature, concentration of nanoparticles, and other experimental settings affect the thermal conductivity and viscosity of water-based graphene oxide nanofluids is made easier with the help of SHAP values. SHAP offers insights into the internal dynamics of complex ML models, hence elucidating their output, by calculating the contribution of each feature to each unique prediction. Using the study's top-performing RF [92].

In this effort, ML models will be trained using experimental data to predict TC and viscosity. To see how factors like temperature or nanoparticle concentration affect the expected thermal properties, the researchers would compute SHAP values for each feature after training. Summary plots are useful for visualizing SHAP values; each point in these graphs represents a prediction, and the magnitude of the SHAP values shows the contribution of a feature. The basic physics and data patterns, such as whether and to what extent increasing the quantity of nanoparticles influences heat conductivity or viscosity, may therefore be easily understood. SHAP values' explainability adds a new degree of transparency to ML predictions, allowing domain experts to better comprehend the results [93].

6.3 Case Studies on AI-Driven Nanofluid Applications

AI-Optimized Nanofluids in Solar Thermal Collectors

Thermal efficiency is critical in solar energy systems because it allows for successful energy conversion. Researchers used ANN and GA to enhance the efficiency of a flat-plate solar collector using Al_2O_3 nanoparticles distributed in water.

The research found that the AI model could estimate the collector's thermal efficiency with an average accuracy of 96.8%. Using an optimized nanofluid with a volume concentration of 3% (0.03) under 600 W/m^2 irradiance increased the system's thermal efficiency by 18–22% compared to typical water-based fluids. The AI framework allows real-time adjustments of fluid properties, resulting in improved energy collection [94].

AI-Guided Magnetic Nanofluids for Targeted Drug Delivery

In nanomedicine, targeted drug delivery using magnetic nanofluids is a significant breakthrough. In a biomedical application, reinforcement learning and PSO were utilized to adjust magnetic field settings and guide Fe_3O_4 (magnetite) nanoparticles suspended in a biocompatible fluid.

Through constant manipulation of the external magnetic field, the AI system directed nanoparticles into tumor tissues in a simulated setting. When compared to passive diffusion-based techniques, the method's targeting efficiency increased by 70%. Intelligent nanofluid guiding has also shown the promise of precision medicine by reducing adverse effects by around 45% [95].

AI-Based Contrast Agent Optimization for MRI Imaging

Gadolinium oxide (Gd_2O_3) nanofluids have been studied as MRI contrast agents owing to their improved magnetic characteristics. A research using Support Vector Regression (SVR) to estimate and adjust contrast agent concentration for optimal image clarity.

After training on more than 120 MRI datasets, the AI model was able to predict optimum doses with a coefficient of determination (R2) of 0.95 and a root mean square error (RMSE) of 0.02. The results showed a ~40% enhancement in picture contrast compared to ordinary commercial agents. This method demonstrates how AI may be used to tailor the characteristics of nanofluids for personalized diagnoses [96].

AI-Controlled Nanofluid Cooling in Data Centers

High-performance computing (HPC) and data center cooling systems need effective heat transfer fluids. A deep neural network (DNN) was utilized by a study team to simulate and regulate the thermal behavior of copper oxide (CuO) nanoparticles in ethylene glycol, which is used as a coolant.

The AI model was trained on over 300 experimental test situations and then integrated into a real-time control loop. The results revealed that utilizing the optimized nanofluid lowered CPU running temperatures by up to 15 °C and increased heat dissipation by 25%. The data center's overall energy usage decreased by 11% after a 30-day trial, demonstrating the benefits of AI-optimized cooling systems for both the economy and the environment.

Intelligent Lubricant Design Using Nanofluids in Aerospace Gearboxes

Tribological performance, or wear and friction, is crucial in aeronautical engineering. RF regression was used to build and improve nanolubricants including zirconium oxide (ZrO_2) nanoparticles and ILs.

The AI model predicted optimum formulas that decreased friction by 35% and wear rates by ~40% compared to traditional synthetic lubricants, based on over 2000 testing runs. The dependability of the model was shown by its R2 score of 0.92 [97]. The longevity and functionality of aeronautical components may be improved by using AI-optimized fluids. Other important applications are mentioned in Table 4.

7 Challenges and Future Perspectives

Challenges and future perspectives of applying ML/AI in ILs are discussed in Table 5.

Table 4 Applications in industry and healthcare

Application area	Description	Key benefits	Examples/use cases
AI-driven heat transfer fluids in energy systems	Integrating AI with nanofluids improves heat transfer in power systems by dynamically controlling nanoparticle concentration, flow rate, and type based on operating conditions. This is especially useful in renewable energy systems like solar and geothermal setups, where conditions vary. Neural networks and reinforcement learning algorithms predict optimal thermal performance, enhancing efficiency and reducing energy losses	– Real-time thermal optimization – Increased energy conversion efficiency – Reduced system maintenance and downtime – Lower operational costs	– AI-based thermal management in photovoltaic thermal (PVT) systems – Predictive cooling in nuclear reactors using nanofluids – Dynamic control of thermal fluids in concentrated solar power (CSP) plants [98]
Bionanofluids in drug delivery and medical diagnostics	Bionanofluids contain bio-functionalized nanoparticles in biocompatible fluids, allowing targeted drug delivery and crossing barriers like the blood-brain barrier. They also enhance imaging and sensing in diagnostics due to their superior optical, magnetic, or electrical properties. Common examples are gold nanoparticles, liposomes, and magnetic nanofluids	– Targeted and controlled drug release – Minimization of side effects – Improved diagnostic resolution – Real-time monitoring capabilities	– Nanofluid-based contrast agents in MRI and CT imaging – Targeted chemotherapy using magnetically guided nanofluids – Early detection of cancer biomarkers via biosensor integration [98]

(continued)

Table 4 (continued)

Application area	Description	Key benefits	Examples/use cases
Smart lubricants and advanced cooling technologies	Ionic liquids and nanolubricants are smart fluids that respond to temperature, pressure, or electric fields. They enhance heat dissipation, reduce friction, and adapt to changing conditions. Nanoparticles boost thermal conductivity and load capacity, while ionic liquids offer high thermal and chemical stability, making them crucial in electronics and precision machinery	– Enhanced equipment longevity and decreased mechanical wear – Viscosity and rheological characteristics that can be adjusted; high electrical insulation and thermal conductivity – Formulations that are environmentally friendly	– Smart engine oils in aerospace and automotive industries – High-performance lubricants in robotics and microelectromechanical systems (MEMS) – Nanofluid-based cooling in data centers and high-performance computing (HPC) systems [98]

Table 5 Challenges and future perspectives of applying ML/AI in ionic liquids

No.	Topic	Detailed description	Challenges/limitations	Future perspectives
1.	Lack of interpretability in AI models	Many AI models (e.g., deep neural networks) function as "black boxes," making it difficult to extract underlying scientific mechanisms	Scientific trust is reduced due to a lack of explainability	Development of explainable AI (XAI) and physics-informed models to improve transparency [99]
2.	Insufficient domain-specific data	High-quality, labeled datasets in nanofluidics and nanomaterials are rare due to costly and time-consuming experiments	Data scarcity hinders model generalization and validation	Collaborative data-sharing initiatives, open-access databases, and synthetic data generation using simulations [99]
3.	Computational cost of simulation-based data	Data from first-principles simulations like density functional theory (DFT) or molecular dynamics (MD) is computationally expensive	Resource-intensive simulations limit dataset scale and diversity	Integration of surrogate models or meta-modeling techniques to reduce simulation load
4.	Overfitting in small-scale experimental datasets	AI models may overfit to small datasets common in experimental nanoscience, leading to poor predictive performance	Reduces reliability in real-world applications	Use of regularization techniques, data augmentation, and transfer learning to improve model robustness [99]
5.	Lack of standardization in data collection	Experimental protocols and data formats differ widely across laboratories and studies	Limits reproducibility and model transferability	Adoption of standardized protocols and metadata documentation frameworks for reproducibility [99]
6.	Challenges in multiscale modeling	Nanotechnology involves interactions from atomic to macroscale levels, which are difficult to capture in a single AI model	Loss of critical physical interactions or approximations may lead to inaccuracies	Hybrid multiscale models combining AI with physics-based solvers (e.g., CFD + ML) [99]

(continued)

Table 5 (continued)

No.	Topic	Detailed description	Challenges/limitations	Future perspectives
7.	Difficulty in integrating experimental and simulation data	Combining datasets from different modalities (e.g., experimental and simulated) is complex due to discrepancies in scale, noise, and resolution	Limits model training and testing scope	Use of data fusion techniques, Bayesian approaches, and multi-modal ML architectures [99]
8.	Bias in AI algorithms and dataset representation	Training data may not represent the full design space, leading to biased or skewed predictions	AI recommendations may be misleading or incomplete	Incorporating active learning and uncertainty quantification to prioritize data diversity
9.	Slow adoption in experimental labs	Many experimental nanotech labs lack the computational infrastructure or expertise to apply AI methods	Slows innovation and integration of AI insights into practice	Development of user-friendly AI tools, cloud platforms, and interdisciplinary training programs
10.	Security and ethical concerns in autonomous discovery	AI-designed materials or automated labs could raise safety, ethical, or misuse concerns	Risk of unintended consequences or regulatory gaps	Establishing AI safety guidelines, ethics frameworks, and governance policies
11.	Uncertainty in AI model predictions	Most ML models lack a reliable way to quantify prediction uncertainty, which is critical in material design	Affects decision-making in high-stakes applications	Adoption of Bayesian neural networks, ensemble learning, and probabilistic modeling
12	Future of autonomous material design	AI is expected to guide real-time synthesis, characterization, and optimization of materials via autonomous labs	Requires seamless integration of AI, robotics, and real-time analytics	Rise of self-driving laboratories, digital twins, and closed-loop material discovery systems

8 Conclusion

In this chapter, we have discussed the application of AI/ML in the field of nanotechnology. There are different types of ML algorithms, like supervised and unsupervised ML algorithms, CNN, and RNN, which we have used. There are nanofluid classifications, which use the AI/ML algorithms that we covered in the above chapter. This chapter covers the application of nanofluids and bionanofluids that emerged as critical in energy transfer, biomedical, and electrical cooling also the different Challenges in optimizing nanofluid performance. There are different factors that work together to control the interaction between IL molecules and nanomaterials like electrostatic interactions, Van der Waals forces, hydrogen bonding, and π–π stacking.

Term and Definition

AI (artificial intelligence) Simulation of human intelligence processes by machines, especially computer systems, to perform tasks such as decision-making, speech recognition, and visual perception.

ANN (Artificial Neural Network) Computational models inspired by the human brain's network of neurons, used for pattern recognition and predictive modeling.

Bionanofluids Nanofluids incorporating biological materials or biomolecules for biomedical applications such as targeted drug delivery.

CNN (Convolutional Neural Network) A type of DNN primarily used for image and spatial data processing, consisting of convolutional, pooling, and fully connected layers.

DFT (density functional theory) A quantum mechanical method used to investigate the electronic structure of molecules and condensed matter systems.

DL (deep learning) A class of ML using neural networks with multiple layers (deep architectures) to model complex patterns in data.

GA (genetic algorithm) An optimization technique based on natural selection to find approximate solutions to complex problems.

Ionic Liquids Salts that are liquid below 100 °C, known for their unique properties such as negligible vapor pressure and high thermal stability.

MD (molecular dynamics) Simulation method for studying the physical movements of atoms and molecules over time.

ML (machine learning) Subset of AI that uses algorithms and statistical models to enable computers to learn patterns from data and make predictions or decisions without explicit programming.

Nanofluids Fluids engineered by dispersing nanoparticles (1–100 nm) into base fluids to enhance thermal properties.

PSO (particle swarm optimization) An optimization algorithm inspired by the social behavior of birds or fish to find optimal solutions by moving particles in a search space.

RNN (Recurrent Neural Network) Neural network with loops allowing information to persist, effective for sequential data like time series or language processing.

SHAP (SHapley Additive exPlanations) A game-theoretic approach to explain individual predictions of ML models by computing feature contributions.

SVM (support vector machine) A supervised ML algorithm used for classification or regression tasks by finding the optimal hyperplane that separates data points of different classes.

References

1. I.H. Sarker, Machine learning: Algorithms, real-world applications and research directions. SN Comput. Sci. **2**(3), 160 (2021)
2. M. Soori, B. Arezoo, R. Dastres, Artificial intelligence, machine learning and deep learning in advanced robotics, a review. Cogn. Robot. **3**, 54–70 (2023)
3. I.H. Sarker, AI-based modeling: Techniques, applications and research issues towards automation, intelligent and smart systems. SN Comput. Sci. **3**(2), 158 (2022)
4. S. Malik, K. Muhammad, Y. Waheed, Artificial intelligence and industrial applications-A revolution in modern industries. Ain Shams Eng. J. **15**, 102886 (2024)
5. P. Mikalef, M. Gupta, Artificial intelligence capability: Conceptualization, measurement calibration, and empirical study on its impact on organizational creativity and firm performance. Inf. Manag. **58**(3), 103434 (2021)
6. R.A. Bin, A.K. Kausik, AI revolutionizing industries worldwide: A comprehensive overview of its diverse applications. Hybrid Adv. **7**, 100277 (2024)
7. P.A. Okeleke, D. Ajiga, S.O. Folorunsho, C. Ezeigweneme, Predictive analytics for market trends using AI: A study in consumer behavior. Int. J. Eng. Res. Updat. **7**(1), 36–49 (2024)
8. T. Amr, *Hands-on Machine Learning with Scikit-Learn and Scientific Python Toolkits: A Practical Guide to Implementing Supervised and Unsupervised Machine Learning Algorithms in Python* (Packt Publishing Ltd, Birmingham, 2020)
9. A. Tripathy, A.Y. Patne, S. Mohapatra, S.S. Mohapatra, Convergence of nanotechnology and machine learning: The state of the art, challenges, and perspectives. Int. J. Mol. Sci. **25**(22), 12368 (2024)
10. S.A.D.H. Hassan, M.N.S. Almaliki, Z.A. Hussein, H.M. Albehadili, S. Rabeea Banoon, A. Abboodi, et al., Development of nanotechnology by artificial intelligence: A comprehensive review. J. Nanostruct. **13**(4), 915–932 (2023)
11. L.H. Kumar, S.N. Kazi, H.H. Masjuki, M.N.M. Zubir, A review of recent advances in green nanofluids and their application in thermal systems. Chem. Eng. J. **429**, 132321 (2022)
12. I. Papadimitriou, I. Gialampoukidis, S. Vrochidis, I. Kompatsiaris, AI methods in materials design, discovery and manufacturing: A review. Comput. Mater. Sci. **235**, 112793 (2024)
13. Y. Liu, T. Zhao, W. Ju, S. Shi, Materials discovery and design using machine learning. J. Mater. **3**(3), 159–177 (2017)
14. E. Chávez-Angel, M.B. Eriksen, A. Castro-Alvarez, J.H. Garcia, M. Botifoll, O. Avalos-Ovando, et al., Applied artificial intelligence in materials science and material design. Adv. Intell. Syst. **7**, 2400986 (2025)
15. P.K. Kanti, P. Paramasivam, V.V. Wanatasanappan, S. Dhanasekaran, P. Sharma, Experimental and explainable machine learning approach on thermal conductivity and viscosity of water based graphene oxide based mono and hybrid nanofluids. Sci. Rep. **14**(1), 30967 (2024)

16. S. Badini, S. Regondi, R. Pugliese, Unleashing the power of artificial intelligence in materials design. Materials **16**(17), 5927 (2023)
17. Y. Xu, X. Liu, X. Cao, C. Huang, E. Liu, S. Qian, et al., Artificial intelligence: A powerful paradigm for scientific research. Innovation **2**(4), 100179 (2021)
18. V.J. Reddy, M.F. Ghazali, S. Kumarasamy, Innovations in phase change materials for diverse industrial applications: A comprehensive review. Results Chem. **8**, 101552 (2024)
19. L.J. Habeeb, H.S. Majdi, Nanofluids and computational applications in medicine and biology, in *Applications of Nanobiotechnology*, (IntechOpen, London, 2019)
20. W. Safiei, M.M. Rahman, A.R. Yusoff, M.R. Radin, Preparation, stability and wettability of nanofluid: A review. J. Mech. Eng. Sci. **14**(3), 7244–7257 (2020)
21. H. Babar, M.U. Sajid, H.M. Ali, Viscosity of hybrid nanofluids: A critical review. Therm. Sci. **23**(3), 1713–1754 (2019)
22. A. Asadi, F. Pourfattah, I.M. Szilágyi, M. Afrand, G. Żyła, H.S. Ahn, et al., Effect of sonication characteristics on stability, thermophysical properties, and heat transfer of nanofluids: A comprehensive review. Ultrason. Sonochem. **58**, 104701 (2019)
23. W. Chamsa-Ard, S. Brundavanam, C.C. Fung, D. Fawcett, G. Poinern, Nanofluid types, their synthesis, properties and incorporation in direct solar thermal collectors: A review. Nano **7**(6), 131 (2017)
24. C.L.L. Beicker, M. Amjad, E.P. Bandarra Filho, D. Wen, Experimental study of photothermal conversion using gold/water and MWCNT/water nanofluids. Sol. Energy Mater. Sol. Cells **188**, 51–65 (2018)
25. O.J. Wai, P. Gunnasegaran, H. Hasini, A review on experimental and numerical investigations of jet impingement cooling performance with nanofluids. Micromachines **13**(12), 2059 (2022)
26. H. Guan, Q. Su, R. Wang, L. Huang, C. Shao, Z. Zhu, Why can hybrid nanofluid improve thermal conductivity more? A molecular dynamics simulation. J. Mol. Liq. **372**, 121178 (2023)
27. S. Ebrahimi, Thermal conductivity of water base Ni-np@ MWCNT magnetic nanofluid. Mater. Res. Bull. **150**, 111781 (2022)
28. W. Yu, H. Xie, A review on nanofluids: Preparation, stability mechanisms, and applications. J. Nanomater. **2012**(1), 435873 (2012)
29. Y.Z.N. Htwe, H. Mamat, Progress and prospects on stability and thermal properties of surfactant assisted graphene, carbon nanotubes based nanofluid and their hybrid for next-generation thermal management. J. Mol. Liq. **414**, 126235 (2024)
30. J.D. Holbrey, K.R. Seddon, Ionic liquids. Clean Prod. Process. **1**(4), 223–236 (1999)
31. W. Silva, M. Zanatta, A.S. Ferreira, M.C. Corvo, E.J. Cabrita, Revisiting ionic liquid structure-property relationship: A critical analysis. Int. J. Mol. Sci. **21**(20), 7745 (2020)
32. M.M. Huie, R.A. DiLeo, A.C. Marschilok, K.J. Takeuchi, E.S. Takeuchi, Ionic liquid hybrid electrolytes for lithium-ion batteries: A key role of the separator–electrolyte interface in battery electrochemistry. ACS Appl. Mater. Interfaces **7**(22), 11724–11731 (2015)
33. E.A.C. Panduro, F. Finotti, G. Largiller, K.Y. Lervåg, A review of the use of nanofluids as heat-transfer fluids in parabolic-trough collectors. Appl. Therm. Eng. **211**, 118346 (2022)
34. N. Nikfarjam, M. Ghomi, T. Agarwal, M. Hassanpour, E. Sharifi, D. Khorsandi, et al., Antimicrobial ionic liquid-based materials for biomedical applications. Adv. Funct. Mater. **31**(42), 2104148 (2021)
35. J.A.L. Marques, F.N.B. Gois, J.P. do Vale Madeiro, T. Li, S.J. Fong, Artificial neural network-based approaches for computer-aided disease diagnosis and treatment, in *Cognitive and Soft Computing Techniques for the Analysis of Healthcare Data*, (Elsevier, Amsterdam, 2022), pp. 79–99
36. V.K. Ojha, A. Abraham, V. Snášel, Metaheuristic design of feedforward neural networks: A review of two decades of research. Eng. Appl. Artif. Intell. **60**, 97–116 (2017)
37. M. Mohseni-Dargah, Z. Falahati, B. Dabirmanesh, P. Nasrollahi, K. Khajeh, Machine learning in surface plasmon resonance for environmental monitoring, in *Artificial intelligence and data science in environmental sensing*, (Elsevier, Amsterdam, 2022), pp. 269–298

38. W. Han, C. Zheng, R. Zhang, J. Guo, Q. Yang, J. Shao, Modular neural network via exploring category hierarchy. Inf. Sci. **569**, 496–507 (2021)
39. S. Mohanty, P.K. Patra, S.S. Sahoo, Prediction and application of solar radiation with soft computing over traditional and conventional approach–A comprehensive review. Renew. Sust. Energ. Rev. **56**, 778–796 (2016)
40. S. Das, A. Tariq, T. Santos, S.S. Kantareddy, I. Banerjee, Recurrent neural networks (RNNs): Architectures, training tricks, and introduction to influential research. Mach. Learn. Brain Disord. **2023**, 117–138 (2023)
41. L. Alzubaidi, J. Zhang, A.J. Humaidi, A. Al-Dujaili, Y. Duan, O. Al-Shamma, et al., Review of deep learning: Concepts, CNN architectures, challenges, applications, future directions. J. Big Data **8**, 1–74 (2021)
42. M.R. Keyvanpour, M.B. Shirzad, Machine learning techniques for agricultural image recognition, in *Application of Machine Learning in Agriculture*, (Elsevier, Amsterdam, 2022), pp. 283–305
43. N. Aloysius, M. Geetha, A review on deep convolutional neural networks, in *2017 International conference on communication and signal processing (ICCSP)*, (IEEE, New York, 2017), pp. 588–592
44. Z. Rguibi, A. Hajami, D. Zitouni, A. Elqaraoui, A. Bedraoui, Cxai: Explaining convolutional neural networks for medical imaging diagnostic. Electronics **11**(11), 1775 (2022)
45. N. Kabra, P. Bhattacharya, S. Tanwar, S. Tyagi, MudraChain: Blockchain-based framework for automated cheque clearance in financial institutions. Futur. Gener. Comput. Syst. **102**, 574–587 (2020)
46. M.M. Taye, Theoretical understanding of convolutional neural network: Concepts, architectures, applications, future directions. Computation **11**(3), 52 (2023)
47. K. Ashangani, K.U. Wickramasinghe, D.W.N. De Silva, V.M. Gamwara, A. Nugaliyadde, Y. Mallawarachchi, Semantic video search by automatic video annotation using TensorFlow, in *2016 Manufacturing & Industrial Engineering Symposium (MIES)*, (IEEE, New York, 2016), pp. 1–4
48. P. Purwono, A. Maarif, W. Rahmaniar, H. Fathurrahman, A.Z. Frisky, Q.M. Ul Haq, Understanding of convolutional neural network (cnn): A review. Int. J. Robot Contr. Syst. **2**(4), 739–748 (2022)
49. Y. Wang, Y. Li, Y. Song, X. Rong, The influence of the activation function in a convolution neural network model of facial expression recognition. Appl. Sci. **10**(5), 1897 (2020)
50. N.I. Galanis, P. Vafiadis, K.G. Mirzaev, G.A. Papakostas, Convolutional neural networks: A roundup and benchmark of their pooling layer variants. Algorithms **15**(11), 391 (2022)
51. X. Zhao, L. Wang, Y. Zhang, X. Han, M. Deveci, M. Parmar, A review of convolutional neural networks in computer vision. Artif. Intell. Rev. **57**(4), 99 (2024)
52. M.M. Taye, Understanding of machine learning with deep learning: Architectures, workflow, applications and future directions. Computer **12**(5), 91 (2023)
53. I.D. Mienye, T.G. Swart, G. Obaido, Recurrent neural networks: A comprehensive review of architectures, variants, and applications. Information **15**(9), 517 (2024)
54. F.M. Salem, F.M. Salem, Gated RNN: The gated recurrent unit (GRU) RNN, in *Recurrent neural networks*, (Wiley, Hoboken, 2022), pp. 85–100
55. M.A.A. Elmaleeh. The identification and categorization of anemia through artificial neural networks: A comparative analysis of three models. arXiv Prepr arXiv240404690 (2024)
56. M. Cilimkovic, Neural networks and back propagation algorithm. Inst. Technol. **15**(1), 18 (2015)
57. A. Sherstinsky, Fundamentals of recurrent neural network (RNN) and long short-term memory (LSTM) network. Phys. D Nonlinear Phenom. **404**, 132306 (2020)
58. R. Zhang, S. Qing, X. Zhang, J. Li, Y. Liu, X. Wen, Experimental investigation and machine learning modeling of heat transfer characteristics for water based nanofluids containing magnetic Fe3O4 nanoparticles. Mater Today Commun **36**, 106798 (2023)
59. A. Mirahmad, R.S. Kumar, B.P. Doldán, C.P. Rios, J. Díez-Sierra, Beyond thermal conductivity: A review of Nanofluids for enhanced energy storage and heat transfer. Nano **15**(4), 302 (2025)

60. R.V. Pinto, F.A.S. Fiorelli, Review of the mechanisms responsible for heat transfer enhancement using nanofluids. Appl. Therm. Eng. **108**, 720–739 (2016)
61. M.R. Safaei, A. Hajizadeh, M. Afrand, C. Qi, H. Yarmand, N.W.B.M. Zulkifli, Evaluating the effect of temperature and concentration on the thermal conductivity of ZnO-TiO2/EG hybrid nanofluid using artificial neural network and curve fitting on experimental data. Phys. A Stat. Mech. Appl. **519**, 209–216 (2019)
62. A. Alfaleh, K.N. Ben, S.M. Eldin, M. Alturki, I. Elbadawi, R. Kumar, Predicting thermal conductivity and dynamic viscosity of nanofluid by employment of support vector machines: A review. Energy Rep. **10**, 1259–1267 (2023)
63. M. Kahani, M.H. Ahmadi, A. Tatar, M. Sadeghzadeh, Development of multilayer perceptron artificial neural network (MLP-ANN) and least square support vector machine (LSSVM) models to predict Nusselt number and pressure drop of TiO2/water nanofluid flows through non-straight pathways. Numer. Heat Transf. Part A Appl. **74**(4), 1190–1206 (2018)
64. G.G. Tiruneh, A.R. Fayek, V. Sumati, Neuro-fuzzy systems in construction engineering and management research. Autom. Constr. **119**, 103348 (2020)
65. L. Snoussi, O. Fakhfakh, M.I. Khedher, G.A. Khouqeer, K.V. Sharma, F. Hosni, et al., Comparative analysis of machine learning techniques for estimating dynamic viscosity in various nanofluids for improving the efficiency of thermal and radiative systems. J. Radiat. Res. Appl. Sci. **18**(1), 101205 (2025)
66. T. Zhang, A.M.K. Pasha, S.M. Sajadi, D.J. Jasim, N. Nasajpour-Esfahani, H. Maleki, et al., Optimization of thermophysical properties of nanofluids using a hybrid procedure based on machine learning, multi-objective optimization, and multi-criteria decision-making. Chem. Eng. J. **485**, 150059 (2024)
67. A. Alirezaie, S. Saedodin, M.H. Esfe, S.H. Rostamian, Investigation of rheological behavior of MWCNT (COOH-functionalized)/MgO-engine oil hybrid nanofluids and modelling the results with artificial neural networks. J. Mol. Liq. **241**, 173–181 (2017)
68. S.O. Sada, S.C. Ikpeseni, Evaluation of ANN and ANFIS modeling ability in the prediction of AISI 1050 steel machining performance. Heliyon **7**(2), e06136 (2021)
69. M.A. Ahmadi, M. Ebadi, Evolving smart approach for determination dew point pressure through condensate gas reservoirs. Fuel **117**, 1074–1084 (2014)
70. M.H. Esfe, S.M. Motallebi, D. Toghraie, Investigation the effects of different nanoparticles on density and specific heat: Prediction using MLP artificial neural network and response surface methodology. Colloids Surf. A Physicochem. Eng. Asp. **645**, 128808 (2022)
71. Z. He, P. Alexandridis, Ionic liquid and nanoparticle hybrid systems: Emerging applications. Adv. Colloid Interf. Sci. **244**, 54–70 (2017)
72. F. Li, Z. Zuo, B. Cao, X. Ji, Study on the thermodynamic properties of ionic liquids 1-Hexyl-3-methylimidazolium halide with isopropanol mixtures. J. Chem. Eng. Data **69**(11), 3794–3804 (2024)
73. G. García, M. Atilhan, S. Aparicio, Theoretical study on the solvation of C60 fullerene by ionic liquids. J. Phys. Chem. B **118**(38), 11330–11340 (2014)
74. Y. Liu, Y. Li, J. He, K.J. Duelge, Z. Lu, Z. Nie, Entropy-driven pattern formation of hybrid vesicular assemblies made from molecular and nanoparticle amphiphiles. J. Am. Chem. Soc. **136**(6), 2602–2610 (2014)
75. S. Kondrat, G. Feng, F. Bresme, M. Urbakh, A.A. Kornyshev, Theory and simulations of ionic liquids in nanoconfinement. Chem. Rev. **123**(10), 6668–6715 (2023)
76. H. Tang, S. Zhang, T. Huang, F. Cui, B. Xing, pH-dependent adsorption of aromatic compounds on graphene oxide: An experimental, molecular dynamics simulation and density functional theory investigation. J. Hazard. Mater. **395**, 122680 (2020)
77. G. Huang, Y. Guo, Y. Chen, Z. Nie, Application of machine learning in material synthesis and property prediction. Materials **16**(17), 5977 (2023)
78. K. Low, R. Kobayashi, E.I. Izgorodina, The effect of descriptor choice in machine learning models for ionic liquid melting point prediction. J. Chem. Phys. **153**(10), 104101 (2020)
79. A. Yamada, K. Tatebe, Effect of the interior fill percentage on the deterioration of the mechanical properties of FFF-3D-printed PLA structures. Polymers **17**(6), 828 (2025)

80. O. Oktaviani, Nanoparticles: Properties, applications and toxicities. J. Latih. **1**(2), 11–20 (2021)
81. N.N. Nyangiwe, Applications of density functional theory and machine learning in nanomaterials: A review. Next Mater. **8**, 100683 (2025)
82. S. Morandini, F. Fraboni, M. De Angelis, G. Puzzo, D. Giusino, L. Pietrantoni, The impact of artificial intelligence on workers' skills: Upskilling and reskilling in organisations. Inf. Sci. **26**, 39–68 (2023)
83. R.J. Bartlett, I.V. Schweigert, V.F. Lotrich, Ab initio DFT: Getting the right answer for the right reason. J. Mol. Struct. THEOCHEM **771**(1-3), 1–8 (2006)
84. L. Böselt, M. Thürlemann, S. Riniker, Machine learning in QM/MM molecular dynamics simulations of condensed-phase systems. J. Chem. Theory Comput. **17**(5), 2641–2658 (2021)
85. K.S. Pavithra, M.P. Yashoda, S. Prasannakumar, Synthesis, characterisation and thermal conductivity of CuO-water based nanofluids with different dispersants. Part. Sci. Technol. **38**, 559–567 (2020)
86. R.K. Bumataria, N.K. Chavda, H. Panchal, Current research aspects in mono and hybrid nanofluid based heat pipe technologies. Heliyon **5**(5), e01627 (2019)
87. S. Hasbi, N. Norazman, M.S. Saharudin, Effects of titanium oxide and graphene as nano-fillers on the thermal conductivity of biobased phase change materials as latent thermal heat storage. Mater Today Proc **75**, 181–187 (2023)
88. S.M.S. Murshed, K.C. Leong, C. Yang, Investigations of thermal conductivity and viscosity of nanofluids. Int J Therm Sci **47**(5), 560–568 (2008)
89. E. Scornet, A random forest guided tour. TEST **25**(2), 197–227 (2016)
90. C. Bentéjac, A. Csörgő, G. Martínez-Muñoz, A comparative analysis of gradient boosting algorithms. Artif. Intell. Rev. **54**, 1937–1967 (2021)
91. A.J. Myles, R.N. Feudale, Y. Liu, N.A. Woody, S.D. Brown, An introduction to decision tree modeling. J. Chem. **18**(6), 275–285 (2004)
92. Y. Nohara, K. Matsumoto, H. Soejima, N. Nakashima, in Explanation of machine learning models using improved Shapley additive explanation. Proceedings of the 10th ACM International Conference on Bioinformatics, Computational Biology and Health Informatics (2019), p. 546
93. S. Ganga, Z. Uddin, R. Asthana, H. Hassan, A. Bhardwaj, Modelling of viscosity and thermal conductivity of water-based nanofluids using machine-learning techniques. Int. J. Math. Eng. Manag. Sci. **8**(5), 817–840 (2023)
94. Z. Said, P. Sharma, L.S. Sundar, C. Li, D.C. Tran, N.D.K. Pham, et al., Improving the thermal efficiency of a solar flat plate collector using MWCNT-Fe3O4/water hybrid nanofluids and ensemble machine learning. Case Stud. Therm. Eng. **40**, 102448 (2022)
95. M. Sheikhpour, M. Arabi, A. Kasaeian, A. Rokn Rabei, Z. Taherian, Role of nanofluids in drug delivery and biomedical technology: Methods and applications. Nanotechnol. Sci. Appl. **13**, 47–59 (2020)
96. M.A. Ibrahim, B. Hazhirkarzar, A.B. Dublin, Gadolinium magnetic resonance imaging (2018)
97. Y.J.J. Jason, H.G. How, Y.H. Teoh, H.G. Chuah, A study on the tribological performance of nanolubricants. PRO **8**(11), 1372 (2020)
98. B.H. Ben, N. Ullah, A. Hamid, N.A. Shah, A comprehensive review on nanofluids: Synthesis, cutting-edge applications, and future prospects. Int. J. Thermofluids **22**, 100595 (2024)
99. S. Chakraborty, R. Tomsett, R. Raghavendra, D. Harborne, M. Alzantot, F. Cerutti, et al., Interpretability of deep learning models: A survey of results, in *2017 IEEE smartworld, ubiquitous intelligence & computing, advanced & trusted computed, scalable computing & communications, cloud & big data computing, internet of people and smart city innovation*, (IEEE, New York, 2017), pp. 1–6

Innovations of Machine Learning with AI-Driven Nanotechnology in Ionic Liquids, Nanofluids, and Bionanofluids

A. Ashwini and G. Preemi

1 Introduction to AI and Machine Learning in Nanotechnology

Scientists operate through nanotechnology to transform small-scale atomic and molecular matter resulting in industrial transformations across medicine, electronics, energy, as well as materials science. Scientific progress in this field resulted in the exponential growth of complex information and large data collections [1]. The intricate nature of complex systems makes artificial intelligence (AI) together with machine learning (ML) fundamental tools for outcome management, prediction, and analysis purposes. The combination of AI and ML with nanotechnology lets researchers find new discoveries, make innovative nanomaterials, and enhance operational processes more efficiently than traditional methods provide. The special capabilities of AI and ML allow for extensive dataset processing, pattern detection, and prediction through automated systems that operate without written program codes. Nanotechnology benefits from these technologies to reproduce nanoparticle behavior and generate new material property forecasts as well as improve manufacturing optimization [2]. Neural networks along with decision trees operate predictive models that accept experimental and simulation datasets to determine nanomaterial stability and toxicity and conductive properties. The data-driven process not only reduces affluent experimental period but also grows innovative potential through its ability to discover associations that stand beyond outdated analysis approaches.

A. Ashwini (✉)
Department of Electronics and Communication Engineering, Vel Tech Rangarajan Dr. Sagunthala R&D Institute of Science and Technology, Chennai, Tamil Nadu, India

G. Preemi
Department of Electronics and Communication Engineering, PSN College of Engineering and Technology, Tirunelveli, Tamil Nadu, India

S. Kulkarni, S. Das (eds.), *Ionic Liquids, Nanofluids and Nanotechnology*,
Lecture Notes in Nanoscale Science and Technology 42,
https://doi.org/10.1007/978-3-032-04008-4_23

The combination of AI technology with nanotechnology applications brings substantial advantages to actual product development. Medical applications of artificial intelligence nanotechnology comprise advanced targeted medication schemes, early disease recognition tools, and individualized beneficial methods. The application of machine learning models in energy leads to more efficient production of nanomaterials, which are employed in batteries and solar cells. The ongoing field development shows that AI and ML together with nanotechnology create opportunities to advance science and engineering beyond current limits for various industries [3]. Through AI and ML automation and intelligent insights, nanotechnology will capitalize on leading a new scientific era which produces fast-paced and innovative development.

2 Fundamentals of Ionic Liquids, Nanofluids, and Bionanofluids

Routine salts exist in a liquid form using organic cations with organic or inorganic anions as their foundation. The melting behavior of ionic liquids differs from that of sodium chloride salt, since they have lower melting temperatures that reach far below 100 °C. Their unique properties—such as negligible vapor pressure, high thermal stability, wide electrochemical windows, and excellent solvation abilities—make them highly attractive for a range of applications, including green chemistry, electrochemical devices, and advanced material synthesis [4]. Because they can dissolve a wide variety of compounds and facilitate ion transport, ionic liquids are often used in catalysis, battery technology, and nanomaterial fabrication.

Nanofluids are engineered colloidal suspensions of nanoparticles (typically less than 100 nm in size) in a base fluid like water, ethylene glycol, or oil. The addition of nanoparticles such as metals, metal oxides, or carbon-based materials dramatically enhances the thermal conductivity, viscosity, and heat transfer capabilities of the base fluid. Nanofluids have garnered significant attention in fields like electronics cooling, energy systems, and industrial heat exchangers [5]. Their ability to efficiently manage heat and mass transfer processes makes them crucial in high-performance cooling systems, renewable energy technologies, and advanced manufacturing processes.

Nanofluids that contain either biologically relevant materials or nanoparticles are classified as bionanofluids. The fluids combine dispersed biomolecules made from proteins, DNA, and peptides with nanoparticles or bio-functionalized nanoparticles within a fluid medium. Bionanofluids serve in medical applications, including diagnostics, targeted drug delivery systems, biosensing, and tissue engineering. The combination of biocompatibility and biological compatibility with adjustable physical–chemical properties provides various innovative solutions in biotechnology and medical applications [6]. The combination of nanotechnology abilities with biological scheme elements marks bionanofluids as a progressive platform to yield breakthrough healthcare methods and bio-inspired engineering arrangement plans.

3 Machine Learning Techniques for Predicting Thermophysical Properties

The technological field of machine learning has proven itself capable of precisely estimating properties such as thermophysical behavior in complicated fluid systems, including ionic liquids, nanofluids, and bionanofluids. These fluids demonstrate extreme nonlinearities that result from different interacting elements, including particle size, concentration, temperature, and chemical structure. The detection approaches of simulation and experiments preserve their accuracy but they scrap with cost and time requirements and limitations on application scope. The capability of machine learning to classify sophisticated data designs allows revolutionary improvements in both quick and precise material goods calculations and system optimization, leading to quicker finding of new materials [7]. Multiple machine learning methods function for predicting thermophysical measurements, which include thermal conductivity, viscosity, specific heat, and density assessment.

3.1 *Artificial Neural Networks (ANNs)*

ANNs are commonly used models that efficiently process nonlinear connections that determine input–output dependencies (input: nanoparticle type, volume fraction, temperature; output: thermal conductivity and viscosity). Such models prove effective for extensive datasets since they adapt to discover hidden dependencies without requiring conventional physical models.

3.2 *Support Vector Machines (SVMs)*

SVM techniques with SVR as an excellent choice deliver accurate results for datasets that contain between 50 and 1000 data points. This methodology enables proper relationship detection while simultaneously preventing overfitting issues that present when experimental measurements are scarce for new nanofluids and ionic liquids. The SVM technique has demonstrated its usefulness by accurately predicting thermal conductivity along with viscosity according to research findings [8].

3.3 *Decision Trees and Random Forests (RF)*

Random forests (RF) with their tree-based structure build prediction models that deliver both high stability and understandable output for thermophysical measurements. They can analyze data combinations of diverse types, along with establishing which variables lead to the strongest effects. Random forest algorithms prove exceptional in sensitivity analyses to help researchers identify factors that strongly affect fluid behavior.

3.4 Gaussian Process Regression (GPR)

The accuracy of GPR substantially increases when it combines predictions with uncertainty measurements, thus becoming highly beneficial in critical reliability-driven applications. GPR enables researchers to predict thermophysical properties of ionic liquids and bionanofluids using restricted data while simultaneously producing prediction interval estimates.

3.5 Deep Learning (DL) and Convolutional Neural Networks (CNNs)

Today's deep learning (DL) practice includes the evaluation of CNNs for cases that involve input images including nanoparticle dispersion patterns and molecular structure diagrams. The property calculation correctness of deep learning algorithms rises through self-extracting hierarchical topographies, but this asset is limited by the need for extensive training datasets and operational calculating capabilities [9].

The combination of nanoscience with AI and machine learning algorithms appears as shown in Fig. 1. The study of ionic liquids, nanofluids, and bionanofluids becomes faster with machine learning methods because they now enable design of customized fluids with precise attributes. Relevant examples include how inverse-designed nanofluids with optimal thermal conductivities for cooling applications can be built using models along with how models predict the compatibility of new bionanofluids for medical purposes. The coming fusion of AI-based nanotechnology with trial data, physics-based simulations, and machine learning replicas will fundamentally alter thermophysical property calculation to support quicker innovation and improved design methods.

4 AI-Driven Design and Optimization of Nanofluids

Nanofluid optimization through experimental trials and theoretical modeling serves as the traditional approach in designing suspensions of nanoparticles within base fluids. Traditional approaches face major problems when discovering such a complex design space due to the frequent connections between particle size, shape, material

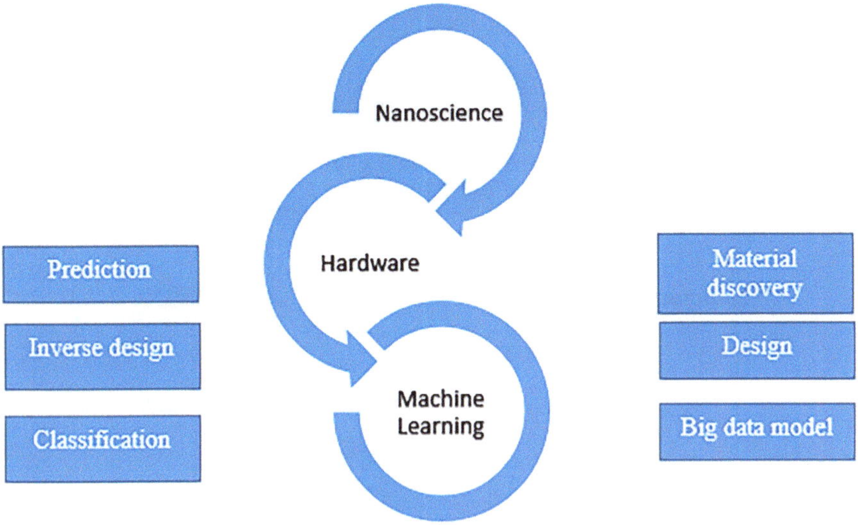

Fig. 1 Nanoscience in AI and machine learning

composition, base fluid properties, and operational circumstances [10]. The claim of artificial intelligence (AI) technologies has altered nanofluid growth through its data-influenced predicting representations and optimization schemes which push the formation of enhanced nanofluids. Accurate forecasting of optimal nanoparticle solutions for heat transfer enhancement becomes possible through training ML models with experimental data [11].

The integration of AI systems for inverse design promises to solve problems by specifying performance parameters first, followed by algorithmic suggestions for matching formulations. Real-time nanofluid preparation procedure monitoring and regulator functions are developed possibly through AI to preserve steady and repeatable fallouts. Recent improvements in machine learning include physical laws during model training so the prediction quality upsurges for data-poor circumstances [12]. AI expansion alongside nanofluid investigation will make purpose-made efficient liquids for progressive technologies to make intelligent engineering solutions that cartel enhanced rapidity and ecological sustainability.

5 Integration of Machine Learning in Bionanofluid Research

Experimental efforts in bionanofluid research show a significant increase in popularity because scientists study and implement biological molecules and biofunctionalized nanoparticles in fluid suspension. The application of bionanofluids presents vast opportunities for medical tests and drugs, as well as diagnostic instruments and tissue repair techniques. The intricate behavior of biological interactions

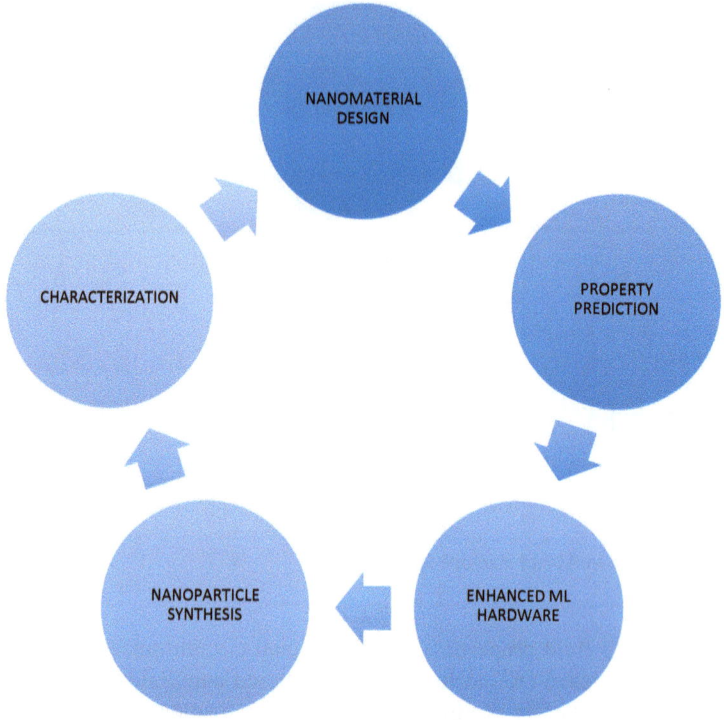

Fig. 2 Collaboration between nanotechnology and ML

together with fluid mechanics and nanoparticle movements creates demanding barriers for bionanofluid system optimization and characterization along with design development [13]. The integration of machine learning (ML) applications in bionanofluid research leads to effective data examination techniques alongside behavioral prediction capabilities and smart fluid development for biomedical needs.

The prediction of bionanofluid behavioral patterns under different physiological and environmental scenarios is now done using machine learning models. Support vector machines (SVM) along with random forests (RF) and neural networks (NNs) generate predictive models to start relationships among nanoparticle size, surface chemistry, concentration, high temperature and subsequent biological consequences with cytotoxicity, cellular acceptance, or drug issue efficacy [14]. Through its application of ML, one can perform fast formulation parameter assessments, which cuts down the expenditures and lengthy biological tests needed to identify prospective candidates, as displayed in Fig. 2.

ML provides design optimization capabilities for bionanofluid systems apart from its predictive functions. Research approaches utilizing Bayesian optimization collected with evolutionary procedures assist in classifying ideal external alterations of nanoparticles along with loading approaches and fluid features which increase treatment results and imaging potentials. Unsupervised learning methods which contain clustering and dimensionality lessening techniques allow scientists to learn concealed patterns within high-dimensional natural datasets, thus creating new scientific findings [15]. When deep learning is combined with imaging analysis, such as microscopy and biosensor outputs, assistants achieve automatic data removal for improved bionanofluid classification in intricate natural systems. The collaborative power of machine learning knowledge and bionanofluid research crops accelerated detection while enabling personal drug development and pioneering bio-nanotechnology advances.

6 Advancements in Ionic Liquid-Based Nanocolloids Through AI

This section clearly explains the AI-driven developments in ionic liquid-based nanocolloids, which are gathered in separate points.

6.1 Predictive Modeling of Thermophysical Properties

The main advancement of AI in ionic liquid-based nanocolloids includes precise predictions of essential thermophysical properties such as thermal conductivity, viscosity, specific heat, and density. Experimental methods taking place in a traditional framework need time along with financial resources to perform measurements that vary conditions. Experimental datasets provide training material for machine learning (ML) models including support vector regression (SVR), random forests (RF), and deep neural networks (DNNs) so they can forecast effects of nanoparticle concentration variations together with particle size and ionic liquid composition changes on these properties [16]. The ability to predict results streamlines formulation optimization and lowers experimental trial efforts.

6.2 Accelerated Material Discovery and Design

AI algorithms have dramatically increased the pace of finding new nanocolloids through inverse design by using ionic liquids. Through optimization approaches involving genetic algorithms (GA) and Bayesian optimization, scientists can deliver target belongings (thermal conductivity with low viscosity profile) to permit AI to style solutions by classifying potent nanoparticle features alongside optimal surface functionalization and ionic liquid coupling systems [17]. The method speeds up the developmental process through which scientists create customized nanocolloids that can be used to produce advanced cooling systems, energy storage technologies, and next-generation lubricants.

6.3 Enhanced Stability Prediction and Optimization

Nanocolloid development faces stability as its main implementation barrier. Nanoparticle agglomeration and sedimentation processes lead to deteriorating long-term performance characteristics of these fluids. The stability prediction of nanocolloids relies on AI models, which evaluate zeta potential measurements together with particle size distribution along with ionic liquid properties. Machine learning outlines deliver optimal approaches for choosing surfactants while enhancing surface change methods of nanoparticles and dispersion methods leading to enhanced chronic colloidal stability.

6.4 Smart Process Control and Quality Monitoring

For the production of ionic liquid-based nanocolloids, it is essential to maintain consistent quality combined with reproducible results. AI-based smart course controls track industrial parameters through real-time sensor variables such as temperature, mixing speed, and particle dispersion grade before adjusting operational situations to steady product quality. These arrangements implement adaptive control and reinforcement learning approaches to acquire dynamic manufacturing skills through real-time data processing which lessens both operator mistakes and giving out material waste. The dispersion of nanoparticles within ionic liquids results in the formation of bio nanofluids that attract research interest for heat transfer applications. Heat transfer performance is enhanced through the combination of ionic liquid unique features including low volatility and high thermal stability with nanoparticles.

6.5 Data-Driven Insights into Structure–Property Relationships

The essential step for making rational designs is to fully understand the complex relationships between structure and properties in ionic liquid-based nanocolloids [18]. Scientific data analysis through these techniques enables researchers to understand the impact of nanoparticle configurations together with the factors involving ionic liquid pairing forces alongside engagement dynamism on macroscopic characteristics such as thermal and electrical conductivity for optimized material designs.

6.6 Personalized and Application-Specific Fluid Formulation

The development of application-specific ionic liquid nanocolloids occurs through AI for particular needs, including high-performance batteries alongside thermal management in electronics and biomedical devices. The application of AI models trained using performance-related datasets can propose optimized formulations, which fulfill specified operating requirements. Research exists on how ML techniques can speed up the development of optimal nanomaterial properties through material optimization processes. Nanotechnology benefits from this method by shortening development timelines while lowering experimental expenses, thus generating greener, more efficient technology practices. AI utilizes its optimization abilities to handle distinct requirements between nanocolloids used in extreme thermal settings and those made for medical implant functions.

7 Machine Learning in Synthesis and Characterization of Nanomaterials

Nanomaterial expansions require synthetic and characterization progressions that build next-generation materials for electronics, medicine, energy, and catalysis subdivisions. Scientific control over nanomaterial goods such as size, shape, surface chemistry, and phase arrangement remains difficult as it requires handling multiple variables instantaneously. The traditional testing technique ends up necessitating long periods of time and using substantial incomes while failing to disclose the complete interdependencies between production variables and material attributes [19]. The field of nanomaterials expansion undergoes a revolutionary transformation through machine learning (ML) since it provides forecasting replicas alongside optimization incomes and automatic analysis purposes, which substantially rise development velocity while enhancing accuracy.

7.1 ML in Nanomaterial Synthesis

The synthesis outcome predictions through machine learning algorithms depend on precursor concentration, temperature, reaction time, pH, and surfactant usage parameters. Random forests, support vector machines (SVM), and neural networks serve as techniques help researchers understand the nonlinear synthesized nanomaterial relationships between conditions and properties. The methodology cuts down both expenses and manufacturing time and enhances the chance of fabricating successful materials.

Predictive Modeling of Synthesis Outcomes

The training phase of support vector machines (SVM), random forests (RF), and artificial neural networks (ANNs) occurs through the analysis of data obtained from past synthesis experiments, which helps the models understand parameter-input-output property relationships, including precursor concentrations, reaction time, temperature, and pH. After training their model, they become competent at forecasting the material outcome from different synthesis protocols while reducing experiments and resource expenditure [20]. Scientists can predict the dimensions of silver nanoparticles by adjusting temperature, reagent volume before executing several experimental tests.

Optimization of Synthesis Parameters

Synthesis conditions are optimized for achieving certain performance levels through the utilization of Bayesian optimization alongside genetic algorithms. Instead of trying all possible combinations (which is impractical), ML models intelligently select the next best experiment to run based on previous results. This closed-loop optimization accelerates the discovery of ideal synthesis recipes. Example: Finding the optimal temperature, time, and surfactant concentration to synthesize uniform gold nanorods with minimal defects.

Inverse Design of Nanomaterials

Traditional synthesis starts with inputs and checks the outputs. Using ML, researchers define the desired material properties first (such as thermal conductivity, bandgap, or magnetic properties). The ML model then predicts the synthesis conditions needed to create a material with those exact properties. Example: Designing a nanoparticle with a target optical absorption wavelength for a biomedical imaging application and then suggesting the synthesis route to achieve it.

Handling and Learning from Sparse or Noisy Data

Experimental results that synthesize materials in real-life environments tend to manifest limited and irregular data patterns caused by fluctuations during actual production. GPR provides advanced ML capability to work effectively with small datasets containing noisy data points. These approaches yield dependable predictions and synthesis directions while working in conditions of incomplete data, thus benefiting research on new nanomaterials with restricted database information. Researchers can predict trusted synthesis results of bionanofluids by analyzing a few experimental results from recently developed fluids.

Real-Time Monitoring and Adaptive Synthesis

The utilization of real-time sensors with ML integration allows synthesis procedures to modify themselves dynamically. Reinforcement learning (RL) uses real-time monitoring data to continuously modify synthesis parameters, which leads to improved experimental results throughout the procedure [21]. The process becomes more reliable through this approach while also enabling instant correction of any synthesis-related errors or deviations. The process regulates flow rate and temperature settings through automatic control systems during nanowire growth with CVD.

7.2 *ML in Nanomaterial Characterization*

The combination of scanning electron microscopy (SEM) with transmission electron microscopy (TEM), X-ray diffraction (XRD), as well as spectroscopy generates huge amounts of difficult data. CNNs process microscopy images by detecting nanoparticle size distributions together with shapes, surface defects, and textural properties at high precision.

Automated Image Analysis (Microscopy Data)

Microscopy techniques like scanning electron microscopy (SEM), transmission electron microscopy (TEM), and atomic force microscopy (AFM) produce thousands of high-resolution images. Deep learning models alongside convolutional neural networks analyze high-resolution images through auto-processing to identify characteristics such as nanoparticle dimension, form, dispersion, cluster formation, and imperfections. Cognitive automation through automated measurements performs much more quickly than manual assessment methods while efficiently minimizing errors made by human operators. A TEM image processing CNN system can quickly determine the diameter of multiple thousands of nanoparticles within a single second while detecting invisible size differences throughout a test sample.

Spectroscopy Data Interpretation

XRD and fourier transform infrared spectroscopy (FTIR), together with Raman spectroscopy, produce spectral signature marks that frequently display overlapping patterns. ML algorithms (e.g., support vector machines, random forests) can be trained to recognize patterns within these spectra to identify materials, analyze phase purity, or detect functional groups. Deep learning models can also denoise raw spectra and extract subtle features invisible to human eyes. Example: Using

ML to classify different types of carbon nanostructures (like graphene vs. carbon nanotubes) based on their Raman spectra.

Property Prediction from Characterization Data

Once characterization data is collected, ML models can predict important properties of the nanomaterials (like electronic conductivity, thermal stability, and optical bandgap) without additional experiments. Models trained on large databases of known materials can infer how minor structural changes affect performance. Example: Predicting the photoluminescence properties of quantum dots based on TEM images and XRD data.

Defect Detection and Quality Assessment

Defects at the atomic or nanoscale often determine the performance of materials but are difficult to identify manually. ML models, especially anomaly detection algorithms, can be trained to recognize "normal" patterns in characterization data and then flag unusual features (defects, cracks, impurities) automatically. Example: Using unsupervised learning to detect grain boundary defects in a nanocrystalline material from electron microscopy data.

Data Fusion for Comprehensive Characterization

Characterization often involves multiple techniques (e.g., combining TEM, XRD, and Raman data). ML enables data fusion, where information from different sources is integrated into a single model for a more comprehensive understanding of the material [22]. Example: Combining Raman spectra and SEM images to predict the mechanical properties of a nanocomposite.

Accelerated High-Throughput Characterization

With ML, high-throughput experiments—generating thousands of samples and measurements—become feasible. ML can quickly screen through massive datasets, prioritize interesting samples, and recommend where to focus more detailed characterization efforts. Example: Automatically analyzing thousands of nanoparticle samples synthesized under different conditions to find those with optimal optical absorption.

7.3 Integration of Synthesis and Characterization through ML

A growing frontier is the integration of ML models that connect synthesis parameters directly with characterization outputs. The entire sequence of models assists users in forecasting changes to synthesis conditions so they can optimize their materials production processes through automated feedback systems. ML functions to find optimal synthesis conditions in addition to performing continuous material evaluation while the model evolves through an iterative process. Rapid material discovery occurs through an automatically improving process that requires minimal human involvement thus being labeled as "autonomous materials research" and "self-driving laboratories."

8 Challenges and Limitations of AI in Nanofluid Research

AI organized with machine learning brings great future prospective to improve nanofluid research. The total realization of these potential benefits depends on solving various restrictive issues and obstacles.

8.1 Data Limitations and Quality Issues

AI systems need large databases containing top-quality information to operate at their best. Studies of nanofluids encounter limited availability of experimental data because it is sometimes inconsistent and without sufficient standardization procedures [23]. The insufficient amount of data affects the development of stable ML models resulting in less accurate predictions and limited applicability. Study data exchange becomes difficult due to the lack of established data format standards throughout research initiatives.

8.2 Complexity of Nanofluid Systems

Nanofluids demonstrate intricate behavior patterns through various connected variables, which include nanoparticle dimensions as well as shapes and concentrations of the base fluid at different temperatures. The modeling of complex associations within ML techniques proves difficult for both advanced algorithms using extensive computational capacity. The nonlinear properties coupled with the dynamic behavior of nanofluids create obstacles during the model convergence process and affect stability.

8.3 Model Interpretability and Transparency

Modern advanced ML models, particularly deep neural networks, operate much like black boxes since they provide minimal insight about the conversion from input to output relationships. Research adoption of AI for nanofluid would likely decrease because of the insufficient clarity regarding decision processes.

8.4 Reproducibility and Validation Challenges

The reproduction of results from AI-powered nanofluid research needs to be established as a fundamental requirement. Research report suggests that performance inconsistencies of models result from changes in experimental practices, measurement approaches, and preprocessing protocols [24]. The enhancement of reproducibility together with cross-study comparisons depends on standardized rules for collecting data and training, and validating models.

8.5 Integration with Experimental Workflows

The integration of AI models into operational experimental processes requires practical installation barriers. The process of developing AI tools that serve nanofluid research requirements needs close partnership work between computational scientists and experimental researchers.

The resolution of these obstacles demands united work toward creating better data quality alongside model interpretation development, together with methodological standardization, along with multidisciplinary group synergy [25]. The combination of AI and ML technologies will greatly improve nanofluid application and comprehension across different technological fields when these challenges are overcome.

9 Case Studies: Successful Applications of AI in Nanotechnology

9.1 AI-Driven Synthesis of Nanoparticles for Drug Delivery

Background

The delivery of drugs to specific locations by nanoscale particles generates extensive use in pharmaceutical applications because of their barrier-crossing ability.

Nanoparticles require precise synthesis conditions for achieving desired dimensions, shape, and surface properties but these requirements make the process very difficult to control.

AI Application

The optimization process of drug delivery AuNPs synthesis utilized machine learning (ML) in a published work [26]. Researchers trained the ML system with experimental data collected from different conditions, including temperature, reaction time, and surfactant concentration together with measured properties of synthesized nanoparticles like size, shape, and surface charge.

Outcome

The AI system effectively identified perfect synthesis parameters needed to create nanoparticles that performed excellently while maintaining a uniform shape and securing both substantial drug retention and stability against degradation. The artificial intelligence model cut down the expenses as well as the experimental times required for nanoparticle synthesis while enabling cost-effective methods for generating nanoparticles that could be used in biomedical applications.

9.2 AI in Nanomaterial Design for Photovoltaics

Background

Quantum dot-based photovoltaics benefit considerably from nanomaterials to improve solar cell effectiveness [27]. Creating new materials that absorb light better and conduct charges effectively presents challenges since researchers must study an extensive array of possible parameters.

AI Application

Scientists at deep learning (DL) developed quantum dot materials evaluation for solar applications. The training procedure involved using a deep neural network (DNN) to assess existing data linking to nanomaterial possessions, which consisted of electron mobility dimensions and light absorption levels composed with bandgap dimensions.

Outcome

Through accurate predictions, AI models enabled scientists to propose materials that computer methods would have discovered much later [28]. Procurement of AI-automated design enabled scientists to develop photovoltaic materials which generated increased solar energy conversions.

9.3 AI-Based Optimization of Nanofluid Thermal Performance

Background

Nanofluids, which are fluids infused with nanoparticles, are known to have enhanced thermal conductivity and heat transfer properties. However, designing a nanofluid with optimal thermal performance requires careful selection of nanoparticle size, material, and concentration.

AI Application

Researchers used Bayesian optimization, a technique that combines prior knowledge with data-driven models, to optimize the thermal conductivity of nanofluids. The AI model considered multiple variables (e.g., particle concentration, size, material, temperature) and predicted the resulting thermal conductivity of various nanofluid formulations.

Outcome

The AI model identified several nanofluid formulations that significantly outperformed traditional fluids in terms of heat transfer efficiency. The technology optimized cooling system efficiency while reducing the development time required to find new efficient industrial nanofluids.

9.4 AI in Nanomedicine: Cancer Diagnosis and Treatment

Background

Nanomedicine employs nanoparticles to bring medical care to cancer patients by transporting diagnostic elements and therapeutic agents to cancer cells. The growth of optimized nanoparticle structures for delivery requires overcoming three fundamental technical hurdles: specific addressing systems, biocompatible assemblies, and measured delivery kinetics.

AI Application

Convolutional neural networks (CNNs) utilize microscopy and imaging data to identify and monitor nanoparticles within biological environments, providing enhanced accuracy in DNA-based analysis and control systems [29]. The CNNs received training that enabled them to identify tumor markers in addition to determining the interaction level between nanoparticles and cancer cells.

Outcome

Controlled enterprises of nanoparticle systems advanced with AI tools led to improved drug-targeting accuracy and delivered more medications successfully while reducing adverse effects. The AI model worked alongside researchers on developing nanoparticles that served as imaging agents for better cancer detection at early stages. The method shortened the development time for improved personalized nanomedicines used for cancer treatment.

9.5 AI in Carbon Nanotube (CNT) Synthesis

Background

The superior combination of mechanical properties, electrical characteristics, and thermal capabilities in carbon nanotubes makes them suitable for various uses within electronics, materials science, and energy storage. Multiple challenges persist during the production of carbon nanotubes (CNTs) because scientists have not mastered their synthetic control for consistent product quality.

AI Application

Researchers deployed reinforcement learning (RL) to improve the production techniques for CNTs. The AI model applied real-time combination data monitoring, including temperatures alongside precursor absorptions for adjusting limits to obtain target CNT belongings such as dimensional and material purity values.

Outcome

The CNT synthesis conditions were adjusted for optimal results, which produced superior quality and more efficient CNTs. The new method minimized CNT production costs and times, which creates more economically feasible opportunities for commercial electronics and energy storage usage.

9.6 AI-Enabled Nanostructure Property Prediction for Batteries

Background

Scientists use nanostructured materials in lithium-ion batteries because they exhibit better storage capabilities and charge/discharge frequency. The behavior of these materials requires an accurate prediction under different conditions for battery performance optimization.

AI Application

The electrochemical routine predictions of battery electrodes with diverse nanostructured materials used random forests and deep learning machine knowledge representations. Past experimental data of silicon and graphene as well as transition metal oxides served as training sets for the developed models.

Outcome

The AI technology made predictions about electrochemical performance characteristics for new nanomaterials, thus identifying materials with extended cycle stability and increased capacity retention. The research resulted in finding innovative materials that delivered superior results for next-generation batteries.

10 Conclusion

Artificial intelligence (AI) incorporated into nanotechnology has established a groundbreaking innovative period by speeding up discoveries and optimizing diverse applications, which span material creation to materials identification. The transformative power of AI in nanotechnology encounters three main obstacles due to sparse datasets, complex physical phenomenon models, and unclear models along with their generalization difficulties. Artificial intelligence continues to revolutionize nanotechnology by delivering tools that help develop innovative materials, which would enable major advancements throughout medicine, energy industry, and environmental sustainability projects. AI integration with nanotechnology brings the approach of developing faster and more efficient cost-effective solutions which will fuel future technological development.

References

1. G.M. Sacha, P. Varona, Artificial intelligence in nanotechnology. Nanotechnology **24**(45), 452002 (2013)
2. W. Liu, Y. Wu, Y. Hong, Z. Zhang, Y. Yue, J. Zhang, Applications of machine learning in computational nanotechnology. Nanotechnology **33**(16), 162501 (2022)
3. D.A. Winkler, Role of artificial intelligence and machine learning in nanosafety. Small **16**(36), 2001883 (2020)
4. D. Li, S.H. Hejazi Dehaghani, A. Karimipour, Developing a novel hybrid nanofluid preparation method using the droplet generation method: Predicting the thermal conductivity, viscosity, and magnetic properties compared to the conventional two-step method. Int. J. Thermophys. **45**(6), 84 (2024)
5. J. Hou, H. Sui, J. Du, L. Sun, Synergistic effect of silica nanofluid and biosurfactant on bitumen recovery from unconventional oil. J. Dispers. Sci. Technol. **43**(5), 704–715 (2022)
6. M.Z. Ghobadlou, N. Asadzadeh, M. Ahmadlouydarab, Impact of cationic surfactant-coated hydrophilic nanoparticles on polymeric solution performance in chemical enhanced oil recovery (CEOR) from a two-dimensional porous medium. Chem. Eng. Res. Des. **204**, 427–437 (2024)
7. F. Jirasek, H. Hasse, Perspective: Machine learning of thermophysical properties. Fluid Phase Equilib. **549**, 113206 (2021)
8. H. Wang, X. Chen, A comprehensive review of predicting the thermophysical properties of nanofluids using machine learning methods. Ind. Eng. Chem. Res. **61**(40), 14711–14730 (2022)
9. A. Maleki, A. Haghighi, I. Mahariq, Machine learning-based approaches for modeling thermophysical properties of hybrid nanofluids: A comprehensive review. J. Mol. Liq. **322**, 114843 (2021)
10. M. Hussain, D. Lin, H. Waqas, Q.M. Al-Mdallal, An artificial intelligence and machine learning-driven CFD simulation for optimizing thermal performance of blood-integrated ternary nano-fluid. Eng. Appl. Comput. Fluid Mech. **19**(1), 2459664 (2025)
11. R. Syah, M. Elveny, M.K. Nasution, V.V. Ponkratov, M.Y. Kuznetsova, A.L. Poltarykhin, M. Babanezhad, Numerical investigation of nanofluid flow using CFD and fuzzy-based particle swarm optimization. Sci. Rep. **11**(1), 20973 (2021)
12. S.A. Abdollahi, A. Basem, A.A. Alizadeh, D.J. Jasim, M. Ahmed, A.J. Sultan, H. Maleki, Combining artificial intelligence and computational fluid dynamics for optimal design of laterally perforated finned heat sinks. Results Eng. **21**, 102002 (2024)
13. P. Chandra, R. Das, B. Manohar, K.M. Reddy, Study of bio-nanofluid flow via vertical surface: A machine learning approach, in *Artificial Intelligence Based Solutions for Industrial Applications*, (CRC Press, Boca Raton, 2024), pp. 267–291
14. Y. Akbar, S. Huang, A. Alshamrani, M.M. Alam, Intelligent computing approach for the bio-convective peristaltic pumping of Powell–Eyring nanofluid: Heat and mass transfer analysis. J. Therm. Anal. Calorim. **149**(15), 8445–8462 (2024)
15. R. Adhikari, S. Das, S. Das, Neural network-based simulation of microbial interactions in tetra-hybrid nanoparticles-enhanced cross blood streaming over stretchy artery stenosis considering magnetic and slip phenomena. Int. J. Model. Simul. (2025). https://doi.org/10.1080/02286203.2025.2478995
16. A.A. Minea, S.M. Sohel Murshed, Ionic liquids-based nanocolloids—A review of progress and prospects in convective heat transfer applications. Nano **11**(4), 1039 (2021)
17. R.A. Khan, H.A. Mohammed, G.M. Sulaiman, A.A. Subaiyel, A. Karuppaiah, H. Rahman, Y.E. Choonara, Molecule (s) of interest: I. Ionic liquids–gateway to newer nanotechnology applications: Advanced nanobiotechnical uses', current status, emerging trends, challenges, and prospects. Int. J. Mol. Sci. **23**(22), 14346 (2022)
18. L.Y. Li, Y.R. Gao, R. Xue, Y. Shu, J.H. Wang, Z.J. Wang, Advances of ionic liquid-based nanohybrids for biomedical applications. J. Mater. Chem. B **11**(28), 6491–6515 (2023)

19. H. Tao, T. Wu, M. Aldeghi, T.C. Wu, A. Aspuru-Guzik, E. Kumacheva, Nanoparticle synthesis assisted by machine learning. Nat. Rev. Mater. **6**(8), 701–716 (2021)
20. A.A. Guda, M.V. Kirichkov, V.V. Shapovalov, A.I. Muravlev, D.M. Pashkov, S.A. Guda, A.V. Soldatov, Machine learning analysis of reaction parameters in UV-mediated synthesis of gold nanoparticles. J. Phys. Chem. C **127**(2), 1097–1108 (2023)
21. J.F. Masson, J.S. Biggins, E. Ringe, Machine learning for nanoplasmonics. Nat. Nanotechnol. **18**(2), 111–123 (2023)
22. P. Sharma, Z. Said, A. Kumar, S. Nizetic, A. Pandey, A.T. Hoang, V.D. Tran, Recent advances in machine learning research for nanofluid-based heat transfer in renewable energy system. Energy Fuel **36**(13), 6626–6658 (2022)
23. I. Gonçalves, R. Souza, G. Coutinho, J. Miranda, A. Moita, J.E. Pereira, R. Lima, Thermal conductivity of nanofluids: A review on prediction models, controversies and challenges. Appl. Sci. **11**(6), 2525 (2021)
24. A. Ullah, M. Kilic, G. Habib, M. Sahin, R.Z. Khalid, K. Sanaullah, Reliable prediction of thermophysical properties of nanofluids for enhanced heat transfer in process industry: A perspective on bridging the gap between experiments, CFD and machine learning. J. Therm. Anal. Calorim. **148**(12), 5859–5881 (2023)
25. S. Talebian, T. Rodrigues, J. Das Neves, B. Sarmento, R. Langer, J. Conde, Facts and figures on materials science and nanotechnology progress and investment. ACS Nano **15**(10), 15940–15952 (2021)
26. O. Adir, M. Poley, G. Chen, S. Froim, N. Krinsky, J. Shklover, A. Schroeder, Integrating artificial intelligence and nanotechnology for precision cancer medicine. Adv. Mater. **32**(13), 1901989 (2020)
27. W. Tawfik, A strategic review of the impact of modern technologies on scientific research: AI, lasers, and nanotechnology. J. Laser Sci. Appl. **1**(2), 38–48 (2024)
28. R. Ravichandran, Nanotechnology applications in food and food processing: Innovative green approaches, opportunities and uncertainties for global market. Int. J. Green Nanotechnol. Phys. Chem. **1**(2), P72–P96 (2010)
29. T. Schnitzler, A. Herrmann, DNA block copolymers: Functional materials for nanoscience and biomedicine. Acc. Chem. Res. **45**(9), 1419–1430 (2012)

Machine Learning and AI in Nanotechnology with Ionic Liquids and Nanofluid and Bionanofluids

M. Manigandan, K. Gayathri, G. Vithiya, N. Vishnuvarthan, and P. Saranraj

1 Introduction

Thanks to AI and machine learning (ML), data gathering has improved, resulting in increased development across all sectors. This is especially proven in the pharma industry, where drug development and discovery processes have improved and enhanced productivity, streamlining clinical trials and minimizing the dependence on human efforts. The interdisciplinary combination of nanotechnology (NT) and artificial intelligence (AI) promises revolutionization in various sectors, yielding significant advantages for the global economy and society. Their extraordinary efforts reveal NT's capability for advanced material manipulation at the atomic level. In nanoscience, the dimensions to adapt or create novel nanostructures are essential. The exceptional electrical properties of nanoparticles offer remarkable adaptability in their electronic, physical, and chemical attributes, which depend on their size and shape. Moreover, nanoparticles' high surface-to-volume ratios make them appealing for varying applications and demanding specific surface chemical or optical interactions. For instance, due to vast surface area and unique exposed atomic structures, diverse nanoparticle types, including metals, alloys, and oxides, have demonstrated remarkable catalytic effects in both thermal and electrochemical conversion processes [1].

M. Manigandan (✉) · K. Gayathri · G. Vithiya · N. Vishnuvarthan · P. Saranraj
PG and Research Department of Microbiology, Sacred Heart College (Autonomous), Tirupattur, Tamil Nadu, India
e-mail: mmanigandan@shctpt.edu

S. Kulkarni, S. Das (eds.), *Ionic Liquids, Nanofluids and Nanotechnology*, Lecture Notes in Nanoscale Science and Technology 42,
https://doi.org/10.1007/978-3-032-04008-4_24

2 Overview of Nanotechnology and Its Applications

Over the last few decades, nanotechnology has rapidly advanced and found in applications across various industries. It plays a crucial role, especially in the pharmaceutical sector, in introducing new medical devices, drug delivery systems, including imaging detectors and diagnostic biosensors. This field discovers how to manipulate the matter at nano scales between 1 and 100 nm, where unique physical properties enable the development of innovative applications. Nanotechnology encompasses the creation of engineered particles within this nanoscale range. One specialization focuses on materials and products that possess nanoscale structures. As a result of advancements in networking, computation, and data processing, nanotechnology has become a vital element in electronic devices [2]. It has improved power densities, resulting in less flammable alternatives and the capability to convert waste heat in nanomachines into usable energy, which enhances the storage capacity utilized by various battery types. Researchers are engineering lightweight piezoelectric nanowires and solar films that can be attached to machinery to harness usable energy from body heat, friction, and wind to power portable electronics [3]. Infrastructure such as roads, bridges, railroads, parking facilities, and floors can harness long-lasting, cost-effective structural solutions made possible by nanoscale and sensor technologies. In the medical field, nanotechnology is under investigation for repairing damage to brain cells, nerves, and spinal cord cells. Despite being regulated by organizations like the food and drug administration (FDA), the field of nanotechnology has been extensively studied for its potential to enhance biotechnology and medical advancements. As the benefits of nanotechnology become apparent in daily life, its multifaceted impact on societies is evident [4]. Moreover, nanotechnology holds significant promise in science and environmental cleanup efforts against continuous challenges such as remaining hazardous waste. In essence, nanotechnology focuses on the study of minuscule particles and aims to design new products at atomic and molecular dimensions [5].

3 Fundamentals of Nanofluids, Ionic Liquids, and Bionanofluids

3.1 Nanofluids

The demand for innovative solutions to enhance heat transmission has escalated, notably with the advent of high-temperature flow systems. For instance, as microprocessors have evolved to become more powerful and compact, the requirements for effective heat flow have intensified, leading to the emergence of new thermal management challenges. Furthermore, there is a growing interest in developing the efficiency of the existing heat transfer technologies. In automotive applications, for example, advancements in heat transfer can yield smaller cooling heat exchangers, subsequently reducing vehicle weight. Additionally, enhancements in heat transfer efficiency can be accomplished by increasing the heat transfer coefficient (h) through the application of more effective heat transfer methods or by optimizing the exchange properties of the exact materials involved. For instance, systems utilizing forced convection of gases demonstrate an increased heat transfer coefficient compared to those relying on free convection. Alternatively, improving the properties of coolants designed for specific heat transfer processes can also enhance the heat transfer coefficient. Typically, liquid coolants are supplemented with additives to boost certain performance characteristics. For instance, glycols are mixed with water to raise their boiling point and lower their freezing point. Adding solid particles, known as nanofluids, to liquid coolants has been shown to greatly enhance the heat transfer coefficient. Nanofluids are utilized across various industries, including energy generation, transportation, electronics like microprocessors and micro electro mechanical systems (MEMS), and biotechnology. Recently, many companies have acknowledged the promise of nanofluid technology and are focusing on specific industrial applications [6, 7].

3.2 Ionic Liquids

Ionic liquids primarily comprise anions and cations that bond via various mechanisms. They exhibit a wide array of physical and chemical properties, including exceptional thermal stability, low vapor pressure, and the ability to function as eco-friendly solvents or catalysts. This adaptability was found to be increased in their use across multiple disciplines such as physics, engineering, and biochemistry. For investigators focusing on ionic liquids (ILs), the environmentally beneficial properties are particularly attractive. However, uncertainties remain about the large-scale manufacturing and application of ILs, encompassing apprehensions like environmental toxicity, potential hazards, cost efficiency, purity, and stability. Additionally, ILs can be measured as green solvents or catalysts by analyzing the life cycle of synthesized products and the production techniques of ILs compared to other liquids [8].

The integration of ionic fluids and machine learning (ML) in nanotechnology characterizes a notable advancement in materials science. Ionic liquids (ILs) are unique solvents created by low vapor pressure and high thermal stability, making them suitable for a wide range of applications, particularly in energy storage and nanomaterial synthesis. Machine learning enhances the design and optimization of these ionic fluids, facilitating rapid identification of their real and unique properties for specific applications [9].

Machine learning (ML) has been found to advanced as an essential tool in designing ionic liquids (ILs), enabling property predictions and significantly reducing the costs and time of experimental testing. For instance, deep learning models have been created to predict IL conductivity, smoothing the development of high-performance

ILs. Additionally, ML techniques are applied rapidly in nanotechnology to improve the synthesis parameters of nanomaterials, improving their thermal and dynamic characteristics. Computer-aided molecular design (CAMD) methods use ML to systematically identify and customize ILs for specific applications [8, 9]. However, considerable challenges are found in validating predictive models' accuracy and ensuring that the synthesized materials are useful for industrial use in all aspects [10].

3.3 Bionanofluids

Bionanofluids were nanomaterials that mix nanoparticles with biomaterials such as enzymes, DNA, antibodies, or cells that live for sophisticated uses. While the nanoparticles improve transport, kinetic performance, and control, these materials provide specific functionality, catalytic activity, and selective recognition. Novel developments in a number of fields, including biosensing, the delivery of drugs, tissue engineering, biocatalysis, and bioimaging, are made possible by these characteristics. High-sensitivity biomarker detection, effective drug and gene delivery to cancerous cells, stable and then reusable nano biocatalyst like green synthesis, 3D culture facilitation for regenerative medicine, and even bioimaging agents for magnetic resonance imaging (MRI) are all made possible by bionanofluids. In their study [11], used homotopy analysis to examine the 3D bioconvective flow of Casson nanofluid in a rotating frame with non-linear heat radiation and slip circumstances.

In fact, it was shown that by increasing the thickness of the boundary layer and decreasing the rate of heat transfer over the surface, slippage effects and thermal radiation would both significantly affect flow patterns and, consequently, heat transfer rates. A study [12] examined the bioconvection of a three-dimensional Casson nanofluid on a heated extendable surface, emphasizing the impacts of heat generation and absorption. The thermal boundary layer and total heat transfer rate, which are essential for both enhancing the efficiency of thermal systems and achieving the ideal temperature distribution, were found to be significantly impacted by both processes. The beginning of bioconvection in a nanofluid that includes contained gyrotactic bacteria was theoretically examined [13].

4 The Emergence of Machine Learning and AI in Nanotechnology

Machine learning evolves with computing power, energy storage, medical technology, and nanotechnology, which consistently improve each other within their individual domains. By using nanotechnology to design specialized chips and systems, we can create a new generation of neural networks that mimic the human brain, thereby improving the effectiveness of machine learning models. The combination of all of these fields has the potential to transform automation across medical, agricultural, and aviation technologies with remarkable strength and efficiency, protecting in a new era of scientific advancements. The integration of artificial intelligence (AI), machine learning, and nanotechnology represents a major discovery in scientific and technical advancements. These two fields complement each other well, as AI's strength lies in analyzing huge quantities of datasets and optimizing complex systems, while nanotechnology offers remarkable precision at the atomic and molecular scales. This interaction creates an exciting possibility in materials science, medicine, energy solutions, and environmental sustainability. However, this convergence also presents challenges. Beyond technical challenges, promoting better collaboration among AI analyzers, nanotechnology experts, and industry stakeholders is essentially connecting these disciplines is crucial for responsible advancement [14].

5 Machine Learning and Artificial Intelligence: An Overview

The advancement of dermatological imaging is significantly shaped by artificial intelligence (AI) and its diverse applications, including image capture, processing, interpretation, reporting, and follow-up planning [15]. Furthermore, benefits such as data mining, storage, and integration are also relevant. The multitude of potential AI applications indicates it will soon become a vital resource for dermatologists. Nevertheless, many dermatologists still possess a limited understanding of AI. As our knowledge of AI expands, two other concepts, machine learning (ML) in particular, are also drawing attention in academic discussions. At present, most medical schools do not integrate AI into their curricula, and the interchangeable use of these terms in academic literature does not aid emerging researchers. To spearhead this transformation, the younger generation must possess a clear understanding of these three concepts and their distinctions. AI, as a subset of computer applications, targeting on creating machines that perform tasks especially require human intelligence. Essentially, it embodies the integration of human intelligence into automation [16].

In the realm of artificial intelligence, machines execute tasks based on predefined rules and algorithms. Any computer program featuring ML, coupled with elements of human intelligence, is classified as artificial intelligence (AI). Machine learning (ML) is a subset of AI that comprises techniques enabling computers to learn from data without explicit programming. The goal of machine learning is to prepare machines ideally from the provided data and methods effectively. In short, ML serves as a pathway to achieve AI [17].

6 Common ML Algorithms Used in Nanotechnology

6.1 Supervised Learning

Supervised machine learning (SML) trains algorithms on labeled datasets to create predictive models that classify future occurrences. One of the most prevalent tasks in intelligent systems is supervised classification. This study quantifies and compares the performance of seven notable classification algorithms—decision table, Naïve Bayes (NB), support vector machine (SVM), random forest (RF), neural networks (Perceptron), decision tree (J48), and JRip, utilizing the Waikato Environment for Knowledge Analysis (WEKA) tool. An analysis was performed by a scientist on a diabetes dataset of patients, with 768 examples with eight independent variables, and a target variable. The results indicated that SVM exhibited other algorithms in terms of accuracy, with Naïve Bayes and random forest following closely behind.

The evaluation helped analyze the key performance metrics like model-building time, the kappa statistic, and mean absolute error (MAE). The algorithms over this provided high accuracy, trade-offs concerning computational efficiency, and error rates. The findings of this study underscore how importance to choose the right machine learning algorithms depending on the particular features of the dataset being analyzed. What really matters in supervised predictive modeling is achieving the highest possible accuracy while keeping errors to an absolute minimum. This research has helped advance our knowledge about fine-tuning machine learning approaches to handle classification problems we encounter in practical, real-life scenarios. At its core, the learning mechanism works by constantly comparing the algorithm's predictions against known correct answers. When discrepancies are found, the system makes careful adjustments to reduce these errors, gradually improving its performance over time. To give a concrete example, we can look at how supervised learning helps predict housing prices. The algorithm studies historical data about home sizes and their corresponding sale prices, then uses these patterns to estimate values for new properties coming onto the market. This practical application demonstrates the method's effectiveness when working with well-structured data [18].

6.2 Unsupervised Learning

In their groundbreaking study, Guyon et al. [19] demonstrated how unsupervised learning enables systems to independently identify meaningful patterns within complex datasets. This approach organizes information into meaningful groups through clustering techniques, much like how our brains naturally categorize similar objects or concepts. The medical field is currently undergoing a remarkable transformation through the combination of AI and nanotech. This powerful combination is paving the way for innovative diagnostic tools and personalized treatment options. By harnessing AI's remarkable data processing capabilities alongside the precision of nanoscale engineering, scientists are developing safer and more effective therapies tailored to individual patients' unique needs. These advancements represent a significant leap forward in precision medicine, offering hope for more targeted and successful treatment outcomes [20].

Traditional drug delivery techniques frequently cause unwanted side effects by affecting healthy cells. AI-guided nanoparticles, on the other hand, are now utilizing this strategy. These tiny carriers, which are equipped with intelligent algorithms, navigate the complex pathways of the body to deliver the medication precisely where it is needed, and their action is like a microscopic GPS for medicine. Scientists enhanced the nanoparticles with the distinctive surface markers that specifically bind to the cancerous cells, ensuring the treatment targets only the affected areas. This targeted approach on cells increases the drug concentrations at the diseased site while reducing side effects. Notably, some AI-driven nanosystems now provide real-time monitoring, enabling the continuous assessment of treatment effectiveness and automatic dosage adjustments, providing doctors unprecedented control over therapy. It is widely recognized that the patients are responding variably to the medications. Here, the combination of AI and nanotechnology raises the development of personalized medicine. By analyzing a patient's genetic information, the AI can predict which treatments will be the most effective, though nanotechnology ensures precise drug delivery [21].

Recent advancements include AI-analyzed genetic profiling that identifies the biomarkers predicting the drug responses for optimal initial treatments. Smart nanocarriers are tiny capsules that release medicine only upon detecting specific signals, such as tumor acidity or certain cancer markers. Cancer-targeting nanoparticles are AI-engineered particles that locate the tumor cells based on genetic signatures, minimizing harm to the healthy cells. As AI and nanotechnology evolve, they promise treatments that are not only more effective but also uniquely suited to each individual's biological characteristics [22].

6.3 Reinforcement Learning

Reinforcement learning is founded on output that guides an agent's behavior in order to optimize long-term rewards. Correct actions yield rewards, while incorrect ones incur penalties. Unlike supervised learning, reinforcement learning lacks explicit corrections for suboptimal actions or direct input/output pairing.

Recommender systems enable users to tailor their online experiences to align with personal preferences. For example, when conducting online searches, users can receive ratings for products or related items due to the established recommender systems. This innovation has transformed how people find products, information, and connections. There are mainly two techniques to help users collect and assess data, generating valuable and unique recommendations: Content-based recommendations and collaborative recommendations. Most e-commerce platforms utilize this approach [19].

Reinforcement learning (RL) is a division of machine learning where an autonomous agent develops optimal decision-making strategies through repeated interactions with a changing environment. The agent tests every action, receives rewards as feedback, and adjusts the policy on it to enhance cumulative rewards over time. Two key strategies for tackling RL challenges include behavior search methods, such as using techniques like genetic algorithms, and statistical methods, which assess state-action utilities via dynamic programming. Reinforcement learning within multi-agent systems employs diverse algorithms to improve decision-making in both competitive and cooperative environments. The Nash learning algorithm modifies the traditional learning for multi-agent scenarios by considering the joint actions and Nash equilibria, enabling them to maintain Q-values for themselves and their peers [23]. Likewise, the Friend or Foe learning (FFQ) algorithm distinguishes the agents as friends or foes, providing better convergence guarantees than Nash learning by leveraging the coordination or adversarial equilibria. Additionally, rQ-learning addresses large search spaces through the use of generalized states (r-states) and the actions (r-actions), although the creation of these can create challenges with the limited prior information [24]. These approaches highlight the complex nature of multi-agent learning, where equilibrium strategies and joint action spaces play a crucial role in achieving successful convergence.

Alternative techniques, the fictitious play algorithm, aimed to address the shortcomings of the Nash equilibrium by simulating opponents' strategies using empirical distributions and adjusting Q-values based on the belief weights. This approach works well with both the stationary and non-stationary opponents, making it suitable for both adversarial and cooperative situations. In contrast to the multi-agent SARSA learning, which is an on-policy method, resulting in optimal Q-values for the current policy, overcoming the off-policy limitations has been seen in Nash Q and Minimax Q-learning. These algorithms highlight the trade-offs between convergence assurances, computational demands, and flexibility within the multi-agent reinforcement learning. While Nash Q and FFQ focus on the equilibrium strategies, the rQ-learning and fictitious play address scalability and opponent modeling. On-policy methods such as the state-action-reward-state-action (SARSA) and policy hill climbing (PHC) prioritize real-time policy optimization, with the WoLF-PHC enhancing resilience via adaptive learning. The choice of algorithmic process depends on the specific context, whether adversarial, cooperative, or a combination of both, necessitating stability in equilibrium versus the need for exploration. Future innovations might integrate these approaches to harmonize the theoretical assurances with practical scalability in intricating the multi-agent scenarios [25].

7 Predicting Thermophysical Properties of Nanofluids Using ML

The current section describes a thorough analysis of the recent literature that uses computational intelligence for modeling the thermophysical properties of nanofluids. This section has been preceded by a thorough analysis of the basic concepts of nanofluid thermophysical properties, taking into account a wide readership with varying levels of scientific ability. The addition of nanoparticles significantly raised in thermal conductivity (TC) of the result the nanofluids [26]. The study demonstrated that the TC of nanofluids increases with nanoparticle concentration, and the silver nanoparticles had a higher TC than copper and titanium oxide nanoparticles. The findings have demonstrated the potential use of nanofluids to improve heat transmission in a variety of settings in electronic cooling systems. ANNs can be used to approximate any function; they must also provide relevant information about the attachments among the inputs and the outputs. The combination of fuzzy logic with ANNs is another common hybrid approach [27]. A discussion of nanofluid prediction has advanced significantly, and as a result [28], three different connectionist approaches were applied and contrasted well. The data set has volumetric concentration ranges from 0.25% to 6%, temperature ranges between 10 °C and 70 ° C, and nanoparticle sizes between 5 and 282 nm. SVM and genetic algorithms (GA) have been employed as innovative methods for forecasting the thermal characteristics of nanofluids. Alade et al. [29] carried out a comprehensive study on applying the SVR to model the TC enhancement of metal and metallic oxide nanofluids [30].

8 AI-Driven Optimization of Nanoparticle Synthesis

To enhance the properties of nanoparticles and investigate the synthesis condition parameters, numerous experimental steps can be automated using robotic synthesis and automated materials characterization. A simple yet effective method for exploring this synthesis space in situ with the robotic system involves performing a grid search that examines every possible combination of experimental conditions [31]. However, this approach can lead to significant wastage, and search efficiency would improve dramatically with an algorithm that intelligently proposes the next experiments based on current data or prior testing. Consequently, in self-driving labs, machine learning-assisted global optimization is crucial for reducing the number of tests, thereby saving time and costs while accelerating the optimization of overall properties. Effectively integrating data from diverse sources in ML-driven optimization tasks is vital for a closed-loop method, particularly with multi-modal data arising from various techniques and experiments, as seen in the current nanoparticle synthesis. To establish a global feature for property optimization, one potential approach is independently constructing encoding schemes for each data type (for instance, using recurrent neural networks for process condition data and convolutional neural networks for image data) and then concatenating these outputs [32]. Through machine learning, nanoparticles can be created with optimal qualities by fulfilling three primary roles.

Regression analysis connects the experimental parameters of nanoparticle production (input) with their resulting characteristics (output), such as size and optical properties, which is a straightforward process [33]. Although various modern neural network architectures and traditional machine learning methods are worth considering, compiling a large dataset of high-quality experimental data is a more vital prerequisite for developing a reliable machine learning model. In 2019, Voznyy et al. [34] observed that, given the extensive history of nanoparticle synthesis, a model for the relationships between conditions and properties in nanoparticle synthesis can be formulated by meticulously curating data and exploring the literature for relevant information and associations [35].

This could be accomplished by using a generative model or conducting a grid search across the experimental parameter space that encompasses most relevant synthesis regions, utilizing a rapid regression model [36] that correlates the properties of synthesized nanoparticles with the distribution of production conditions [37].

9 AI for Toxicity Prediction and Safer Nanomaterials

Assessing nanoparticle safety requires toxicity evaluations. Research using cell cultures provides a practical, scalable, and budget-friendly approach for investigating nanotoxicity in numerous animal and human cell lines. However, these studies do not fully encompass the complex physiology of animal models, which hinders their capacity to accurately predict nanotoxic effects in humans and other species. Although animal models can consider complex physiological circumstances, they frequently fail to accurately forecast toxic responses and adverse conditions in humans. The computational nanotoxicity approaches seek to connect insights from human participants, animal models, and cell cultures, potentially improving nanotoxicity modeling and prediction in the future, provided that the underlying assumptions and models remain sound. Additionally, the computational studies reduce the necessity, costs, and duration of testing for nanotoxicity in animals and cells. However, published research reveals substantial variability in nanoparticle characterization, experimental techniques, dosage metrics, and data completeness, attributed to the absence of standardized testing protocols for nanotoxicity. This variability detracts from the statistical accuracy of computational models were further when predicting nanotoxicity [38, 39].

10 Challenges and Advancements in the AI-Nanotechnology Interface

The integration of AI with human oversight in processing structural and functional images, enhancing human knowledge, and accelerating the intricate procedures involved in cell and tissue cultures could greatly benefit. Given the current rate of AI medical device advancements and the worldwide demand for advanced AI therapies, nanomaterials play a vital role in AI applications, especially in the pharma and healthcare industries. AI holds promise for developing smarter, more effective, and personalized nano therapies for human diseases in the future [40].

The rapidly evolving field of nanotechnology is based on innovative nanomaterials, nanodevices, and nanosystems [41]. Since its conception, nanotechnology has advanced significantly, with important breakthroughs in domains such as electronics, biosensors, field effect transistors, and nanoparticles. When bringing artificial intelligence (AI) to nanosciences, there are a number of significant issues that must be resolved. These include making sure that methods and outcomes are reliable, balancing computational and physical models, making materials flexible, and forecasting the accuracy and uncertainty of generative models.

Novel nanomaterials, nanodevices, and nano systems are the foundation of the quickly expanding field of nanotechnology, with substantial advancements in fields including electronics, field effect transistors, biosensors, nanoparticles, and nanotechnology was inception. When nanosciences and artificial intelligence (AI) are combined, a number of significant issues must be resolved. These materials are adaptable, balancing computational and physical models, anticipating the accuracy and uncertainty of generative models, and making sure that processes are reliable. Researchers and pathologists can solve medical obstacles by fusing AI and nanotechnology [42].

Nanotechnology, an interdisciplinary field, employs various techniques to accurately identify and manipulate matter at the atomic and molecular levels. AI has the potential to elevate nanotechnology by predicting material properties and phenomena, thereby streamlining the necessary experimental efforts [43]. This includes using AI frameworks to explore the morphology of the nanoparticles and the biomolecules' influence, addressing preprocessing challenges encountered in AI for cancer cell segmentation, which demands specific image contrast techniques for optimal analysis. The outcome will be the creation of more targeted and effective image contrast agents. Scanning transmission electron microscopy (STEM) imaging relies heavily on phase contrast detection to produce high-contrast images within theoretical limits. This can be achieved by integrating deep learning methods with conventional physical models in a unified AI simulation framework. The collaboration between AI and electron microscopy can enhance our understanding of nanoparticle and cell interactions. Recent studies utilizing hybrid approaches by Mobarak et al. [44] indicated that AI can predict the interactions between surface, interface, and environment of any nanomaterial, thus managing the complexity of dynamic biological characteristics.

11 Conclusion

Recent advancements in nanotechnology and machine learning hold the promise of radically transforming energy storage, computational power, and medical technologies. By using nanotechnology to create tailored chips and systems, we can produce a new generation of neural networks that mimic the human brain, thereby enhancing the effectiveness of machine learning (ML) models. Conversely, machine learning accelerates nanotechnology's progress by streamlining the discovery and synthesis of new nanomaterials, utilizing previously attained knowledge. The synergy between these two disciplines enables unparalleled strength and efficiency in automating medical, agricultural, and aviation technologies. For instance, in agriculture, ML algorithms paired with nanotechnology-enabled sensors can monitor soil health and boost crop yields. Additionally, optimization techniques can refine the hyperparameters of developed models, facilitating the enhancement of synthetic nanoparticles. To achieve optimization, a loss function must be established to ensure its minimum value aligns with the goal function.

References

1. T.L. Moore, C. Pesce, A. Greco, C. Pisante, P. Decuzzi, A machine learning workflow to enhance microfluidic development of nanomedicines. ChemRxiv (2025). https://doi.org/10.26434/chemrxiv-2025-2lb14
2. J. R. Jeremy, Nanotechnology, 2nd ed. (Springer, Cham, 2016)
3. G.A. Divesh, Literature review of nanotechnology. J. Emerg. Technol. Innov. Res. **6**, 1–9 (2019)
4. S.H. Matthew, *Management of Emerging Public Health Issues and Risks* (Academic Press, New York, 2019)
5. H. Kumar, K. Kuča, S.K. Bhatia, K. Saini, A. Kaushal, R. Verma, D. Kumar, Applications of nanotechnology in sensor-based detection of foodborne pathogens. Sensors **20**, 1–22 (2020)
6. P. Keblinski, S.K. Nayak, P. Zapol, Charge distribution and stability of charged carbon nanotubes. Phys. Rev. Lett. **89**, 255503 (2002). https://doi.org/10.1103/PhysRevLett.89.255503
7. M. Beck, *Thermal Conductivity of Metal Oxide Nanofluids* (Georgia Institute of Technology, Atlanta, 2008). https://doi.org/10.1016/j.physrep.2018.11.004
8. Z. Gadzhimagomedova, Artificial intelligence for nanostructured materials. Nanobiotechnol. Rep. **17**, 1–9 (2022). https://doi.org/10.1134/S2635167622010049
9. R. Gu, Z. Song, Machine learning-aided rational screening of task-specific ionic liquids, in *Ionic Liquids in the Chemical Sciences*, (Wiley, Hoboken, 2024), pp. 79–103. https://doi.org/10.1002/9783527845491.ch3
10. S. Koutsoukos, F. Philippi, F. Malaret, T. Welton, A review on machine learning algorithms for the ionic liquid chemical space. Chem. Sci. **12**, 6820–6843 (2021). https://doi.org/10.1039/D1SC01000J
11. Y. Li, M. Imtiaz, W. Jamshed, S. Rehman, M.R. Eid, N.A.A.M. Nasir, R.W. Ibrahim, Nonlinear thermal radiation and the slip effect on a 3D bioconvection flow of the Casson nanofluid in a rotating frame via a homotopy analysis mechanism. Nanotechnol. Rev. **12**, 20230161 (2023). https://doi.org/10.1016/j.rineng.2024.102652
12. I. Ullah, W.A. Khan, W. Jamshed, A. Abd-Elmonem, N.S.E. Abdalla, R.W. Ibrahim, M.R. Eid, Heat generation (absorption) in 3D bioconvection flow of Casson nanofluid via a convective heated stretchable surface. J. Mol. Liq. **392**, 123503 (2023). https://doi.org/10.1016/j.molliq.2023.123503
13. K. Zaimi, A. Ishak, I. Pop, Stagnation-point flow toward a stretching/shrinking sheet in a nanofluid containing both nanoparticles and gyrotactic microorganisms. J. Heat Transf. **136**, 041101 (2014). https://doi.org/10.1115/1.4026008
14. M. Mousavizadegan, A. Firoozbakhtian, M. Hosseini, H. Ju, Machine learning in analytical chemistry: From synthesis of nanostructures to their applications in luminescence sensing. Trends Anal. Chem. **167**, 117216 (2023). https://doi.org/10.1016/j.trac.2023.117216

15. S. Russell, J. Bohannon, Artificial intelligence. Fears of an AI pioneer. Science **349**, 252 (2015). https://doi.org/10.1126/science.349.6243.252
16. G. Chartrand, P.M. Cheng, E. Vorontsov, et al., Deep learning: A primer for radiologists. Radiographics **37**, 2113–2131 (2017). https://doi.org/10.1148/rg.2017170077
17. P. Lakhani, A.B. Prater, R.K. Hutson, et al., Machine learning in radiology: Applications beyond image interpretation. J. Am. Coll. Radiol. **15**, 350–359 (2018). https://doi.org/10.1016/j.jacr.2017.09.044
18. S. Shen, H. Jiang, T. Zhang, Stock market forecasting using machine learning algorithms. J. Comput. **4**, 1–6 (2012). https://doi.org/10.5120/20182-2402
19. I. Guyon, A. Elisseeff, An introduction to variable and feature selection. J. Mach. Learn. Res. **3**, 1157–1182 (2003)., https://www.jmlr.org/papers/v3/guyon03a.
20. A. Yadav, K. Yadav, R. Ahmad, K.A. Abd-Elsalam, Emerging frontiers in nanotechnology for precision agriculture: Advancements, hurdles, and prospects. Agrochemicals **2**, 220–256 (2023). https://doi.org/10.3390/agrochemicals2020016
21. F.Y. Osisanwo, J.E.T. Akinsola, O. Awodele, J.O. Hinmikaiye, O. Olakanmi, J. Akinjobi, Supervised machine learning algorithms: Classification and comparison. Int. J. Comput. Trends Technol. **48**, 128–138 (2017)., https://www.ijcttjournal.org.
22. O. Adir, M. Poley, G. Chen, S. Froim, N. Krinsky, J. Shklover, J. Shainsky-Roitman, T. Lammers, A. Schroeder, Integrating artificial intelligence and nanotechnology for precision cancer medicine. Adv. Mater. **32**, e1901989 (2020). https://doi.org/10.1002/adma.201901989
23. J. Hu, M.P. Wellman, Nash Q-learning for general-sum stochastic games. J. Mach. Learn. Res. **4**, 1039–1069 (2003)., https://www.jmlr.org/papers/v4/hu03a.html.
24. E.F. Morales, Scaling up reinforcement learning with a relational representation, in Workshop on Adaptability in Multi-Agent Systems (2003)
25. M.H. Bowling, M.M. Veloso, Multiagent learning using a variable learning rate. Artif. Intell. **136**, 215–250 (2002). https://doi.org/10.1016/S0004-3702(02)00121-2
26. J. Esfahani, M.R. Safaei, M. Goharimanesh, L.R. de Oliveira, M. Goodarzi, S. Shamshirband, E.P.B. Filho, Comparison of experimental data, modelling and non-linear regression on transport properties of mineral oil based nanofluids. Powder Technol. **317**, 458–470 (2017). https://doi.org/10.1016/j.powtec.2017.05.003
27. W. Yu, S. Choi, The role of interfacial layers in the enhanced thermal conductivity of nanofluids: A renovated Maxwell model. J. Nanopart. Res. **5**, 167–171 (2003). https://doi.org/10.1023/A:1024438603801
28. M.H. Ahmadi, M.A. Nazari, R. Ghasempour, H. Madah, M.B. Shafii, M.A. Ahmadi, Thermal conductivity ratio prediction of Al_2O_3/water nanofluid by applying connectionist methods. Colloids Surf. A Physicochem. Eng. Asp. **541**, 154–164 (2018). https://doi.org/10.1016/j.colsurfa.2018.01.030
29. I.O. Alade, T.A. Oyehan, I.K. Popoola, S.O. Olatunji, A. Bagudu, Modeling thermal conductivity enhancement of metal and metallic oxide nanofluids using support vector regression. Adv. Powder Technol. **29**, 157–167 (2018). https://doi.org/10.1016/j.apt.2017.10.026
30. S.H. Rostamian, M. Biglari, S. Saedodin, M.H. Esfe, An inspection of thermal conductivity of CuO-SWCNTs hybrid nanofluid versus temperature and concentration using experimental data, ANN modelling and new correlation. J. Mol. Liq. **231**, 364–369 (2017). https://doi.org/10.1016/j.molliq.2017.02.015
31. H. Tao, T. Wu, M. Aldeghi, T.C. Wu, A. Aspuru-Guzik, E. Kumacheva, Nanoparticle synthesis assisted by machine learning. Nat. Rev. Mater. **6**, 701–716 (2021). https://doi.org/10.1038/s41578-021-00337-5
32. M.H.S. Segler, M. Preuss, M.P. Waller, Planning chemical syntheses with deep neural networks and symbolic AI. Nature **555**, 604–610 (2018). https://doi.org/10.1038/nature25978
33. H.A. Nguyen, F.Y. Dou, N. Park, S. Wu, H. Sarsito, B. Diakubama, H. Larson, E. Nishiwaki, M. Homer, M. Cash, B.M. Cossairt, Predicting indium phosphide quantum dot properties from synthetic procedures using machine learning. Chem. Mater. **34**, 6296–6311 (2022). https://doi.org/10.1021/acs.chemmater.2c01099

34. O. Voznyy, L. Levina, J.Z. Fan, M. Askerka, A. Jain, M.-J. Choi, O. Ouellette, P. Todorovic, L.K. Sagar, E.H. Sargent, Machine learning accelerates discovery of optimal colloidal quantum dot synthesis. ACS Nano **13**, 11122–11128 (2019). https://doi.org/10.1021/acsnano.9b03864
35. K. Cruse, A. Trewartha, S. Lee, Z. Wang, H. Huo, T. He, O. Kononova, A. Jain, G. Ceder, Text-mined dataset of gold nanoparticle synthesis procedures, morphologies, and size entities. Sci. Data **9**, 234 (2022). https://doi.org/10.1038/s41597-022-01343-0
36. B. Sanchez-Lengeling, A. Aspuru-Guzik, Inverse molecular design using machine learning: Generative models for matter engineering. Science **361**, 360–365 (2018). https://doi.org/10.1126/science.aat2663
37. J. Noh, J. Kim, H.S. Stein, B. Sanchez-Lengeling, J.M. Gregoire, A. Aspuru-Guzik, Y. Jung, Inverse design of solid-state materials via a continuous representation. Matter **1**, 1370–1384 (2019). https://doi.org/10.1016/j.matt.2019.08.017
38. H.S. Leong, K.S. Butler, C.J. Brinker, M. Azzawi, S. Conlan, et al., On the issue of transparency and reproducibility in nanomedicine. Nat. Nanotechnol. **14**, 629–635 (2019). https://doi.org/10.1038/s41565-019-0496-9
39. A. Afantitis, G. Melagraki, P. Isigonis, A. Tsoumanis, D.D. Varsou, et al., NanoSolveIT project: Driving nanoinformatics research to develop innovative and integrated tools for in silico nanosafety assessment. Comput. Struct. Biotechnol. J. **18**, 583–602 (2020). https://doi.org/10.1016/j.csbj.2020.02.023
40. M.G. Ammendolia, B. De Berardis, Nanoparticle impact on the bacterial adaptation: Focus on nano-titania. Nano **12**, 3616 (2022). https://doi.org/10.3390/nano12203616
41. A. Holzinger, K. Keiblinger, P. Holub, K. Zatloukal, H. Müller, AI for life: Trends in artificial intelligence for biotechnology. New Biotechnol. **74**, 16–24 (2023). https://doi.org/10.1016/j.nbt.2023.02.001
42. L.K. Vora, A.D. Gholap, K. Jetha, R.R.S. Thakur, H.K. Solanki, V.P. Chavda, Artificial intelligence in pharmaceutical technology and drug delivery design. Pharmaceutics **15**, 1916 (2023). https://doi.org/10.3390/pharmaceutics15071916
43. L. Goswami, M.K. Deka, M. Roy, Artificial intelligence in material engineering: A review on applications of AI in material engineering. Adv. Eng. Mater. **25**, 2300113 (2023). https://doi.org/10.1002/adem.202300113
44. M.H. Mobarak, M.A. Mimona, M.A. Islam, N. Hossain, F.T. Zohura, I. Imtiaz, M.I.H. Rimon, Scope of machine learning in materials research—A review. Appl. Surf. Sci. Adv. **18**, 100523 (2023). https://doi.org/10.1016/j.apsadv.2023.100523

Challenges, Opportunities, Emerging Trends, Case Studies, Teaching–Learning, Protocols, Ethical Issues, and Prospects

Anurima Majumdar, Antara Ghosal, Palasri Dhar, and Sunipa Roy

1 Introduction

1.1 Overview of Ionic and Nanofluid Nanotechnologies

Ionic and nanofluids are two innovative branches of nanotechnology that have increasingly come into the forefront on account of their transformative applications in thermal management, energy systems, and biomedical devices, with potential industrial applications. These premium fluids exist to counter a few disadvantages of traditional heat-transfer fluids such as water, ethylene glycol, and oils, thus providing better thermal conductivity and stability and efficiency in different systems.

Nanofluids are dispersion systems containing nanoparticles in conventional base fluids. They have enhanced thermal conductivity and applications concerning alternative heat transfer mediums [1]. Long-term stability and consistent behavior constitute the major problems. Present novel studies focus on hybrid nanofluids and new methods for stabilization against sedimentation and agglomeration. Ionic liquids (ILs) and deep eutectic solvents (DESs) show a bright future owing to their convertible characteristics and environmentally friendly nature [2, 3]. Application of these solvents to nanofluid systems is at the frontiers of green energy, catalysis, and electronics cooling. Research needs to be continued to optimize formulations that offer the best thermal efficiency and stability.

A. Majumdar (✉) · A. Ghosal · P. Dhar · S. Roy
Guru Nanak Institute of Technology, Kolkata, India
e-mail: anurima.majumdar@gnit.ac.in

S. Kulkarni, S. Das (eds.), *Ionic Liquids, Nanofluids and Nanotechnology*,
Lecture Notes in Nanoscale Science and Technology 42,
https://doi.org/10.1007/978-3-032-04008-4_25

1.2 Ionic Fluids

Ionic fluids, particularly the more prominently considered ionic liquids (ILs), are salts in the liquid state composed entirely of ions. These are salts generally melting below 100 °C and, due to their extremely low vapor pressure, high thermal and chemical stability, and great solvating powers, are considered as services for liquid-phase reactions. Their nature is tuned depending on which cations and anions are put together; hence, they are called "designer solvents." They are applied in electrochemical devices, electrolytes to batteries, capturing CO_2, and catalysis [4]. In thermal management, ionic liquids offer better thermal conductivity than organic solvents. Their capability of staying stable through very high temperatures is leveraged in high-performance cooling systems for the electronics, aerospace, and nuclear industries. Another recent trend that researchers are looking into is the dispersion of nanoparticles in ionic liquids for Io nanofluids, which have demonstrated synergistic thermal attributes from the two counterparts.

1.3 Nanofluids

All kinds of nanoparticles can be added to a base fluid to generate a nanofluid essentially. We are talking about common metals like copper or silver, metal oxides (like titanium dioxide or aluminum oxide), exotic carbon materials like graphene or carbon nanotubes, or those crazy magnetic particles people like to play about with. This magic dust helps the fluid move to and from hot or cold surfaces and transfer heat considerably more effectively. It's like really enhancing your dull, old liquid.

Nanofluids were introduced by Choi and Eastman in 1995 and have proved to have the capability to revolutionize heat transfer systems [5]. They bring about more efficient energy heat exchangers, electronic coolers, automotive radiators, and solar thermal collectors. Ingredients are a synergy of thermal conductivity, Brownian motions, particle-fluid interaction, and micro-convection mechanisms.

The major nanofluid issues include nanoparticle agglomeration studies, long-term stability, material wear, and viscosity increase. The researchers address these undesirable concepts through a discussion of surface nanoparticle modification, use of a surfactant, and nanofluid concentration rates.

1.4 Synergistic Potential and Applications

Hybrid nanofluids—that is, Io nanofluids—result from combining the benefits of ionic fluids and nanofluids. These are under investigation for use in advanced heat exchangers, high-capacity batteries, smart cooling systems, and even biomedical uses including drug delivery and hyperthermia treatment; they show even more enhancement in thermal and electrical properties.

All things considered, groundbreaking developments in fluid dynamics and thermal sciences are provided by ionic and nanofluid nanotechnologies. Their application is likely to be a major contributor to next-generation electronic systems, effective industrial processing, and sustainable energy solutions. The future of nanotechnology and material science depends on these fluids since research keeps addressing the constraints related to stability, toxicity, and cost.

2 Constraints in the Field of Ionic and Nanofluids Nanotechnologies

Many researchers in recent years have documented several restrictions and difficulties in this field of study. Focusing on synthesis approaches, thermophysical property enhancement, and stability mechanisms, Rahman et al. [1] offer a thorough review of nanofluids. In useful applications, they stress difficulties, including poor dispersion, sedimentation, and agglomeration of nanoparticles. Singh et al. [2] address the several physical and chemical elements influencing nanofluid stability: zeta potential, pH, and surfactant use among others. The author also lays out rules for enhancing dispersion methods. Long-term nanofluid stability is investigated by Chakraborty and Kumar [3] under consideration of temperature, concentration, and particle material. They stress optimizing the surface modification of nanoparticles for stable formulations. Reviewing nanofluid synthesis methods, Puree et al. [4] pay particular attention to commercialization issues. Especially with regard to long-term storage stability and scalability. They also look at the thermal potential of hybrid nanofluids. Mehta et al. [5] deal with problems related to nanoparticle clustering and optimization of thermophysical properties. To improve stability, they suggest surface coatings and sophisticated dispersion techniques. The impact of fluid type, synthesis techniques, and particle size on nanofluid performance is investigated by Okello et al. [6]. They emphasize that in order to preserve homogeneity and shelf life, consistent protocols are required. In order to comprehend their application in nanofluidic systems, Nesterova et al. [7] examine the behavior of ionic liquids under nanoconfinement, paying particular attention to interfacial instability and degradation pathways in confined systems. A fundamental review of nanofluid stability mechanisms is given by Chakraborty and Panigrahi [8], who emphasize experimental techniques for measuring zeta potential, sedimentation, and rheological behavior. Rahman et al. [9] revisit hybrid nanofluid preparation and emphasize methods to mitigate incompatibility issues between different nanoparticle types and the base fluid.

2.1 *Synthesis and Stability Problems*

The synthesis procedure and the fluids' long-term stability are crucial to the creation and use of ionic and nanofluids in cutting-edge technologies. The development of stable, effective, and scalable formulations still faces a number of obstacles, despite their encouraging thermal and functional characteristics. To increase the effectiveness and dependability of these nanotechnologies in practical applications, it is imperative to comprehend these problems.

2.2 *Synthesis Challenges in Ionic Fluids and Nanofluids and Stability Problems*

The production of ionic fluids and nanofluids presents numerous technical challenges. Purity is crucial. The chemical properties of the ionic fluids can alter and may have an adverse effect on their thermal performance if even a tiny quantity of moisture or residual reactants manages to get into these materials.

Why is scalability such a big issue with ionic fluids then? Because they are made through complex organic synthesis pathways that are difficult to scale up. For nanofluids, high surface energy leads to agglomeration, and that alone makes it virtually impossible to disperse the nanoparticles uniformly throughout the fluid. A common way of making a nanofluid is to use a two-step method. In practice, this stepwise process usually results in sedimentation and a loss of efficiency at some point.

Yet another significant concern is stability. Nanoparticles tend to settle over time, which affects the thermal conductivity and flow behavior of the fluids in which they are dispersed. To increase the dispersion of these nanoparticles, we employ surface functionalization; however, this may change the characteristics of the base fluid, and the functionalization may deteriorate at high temperatures. Consequently, achieving and maintaining long-term chemical compatibility and thermal stability is a real challenge.

Yet another significant concern is stability. Nanoparticles tend to settle over time, which affects the thermal conductivity and flow behavior of the fluids in which they are dispersed. We use surface functionalization to improve the dispersion of these nanoparticles, but doing so may alter the properties of the base fluid, and at high temperatures, the functionalization may degrade. Consequently, achieving and maintaining long-term chemical compatibility and thermal stability is a real challenge.

2.3 *Environmental and Ecotoxicological Implications*

Ionic fluids and nanoparticles/nanofluids, despite imparting attractive thermal and chemical attributes, pose major environmental and ecotoxicological challenges. If leaked into the environment, they may bioaccumulate and cause toxicity to aquatic organisms, disturbing the ecosystems permanently. Ionic fluids might also be persistent and have a low biodegradation tendency, and nanoparticles in Cd nanofluids are able to traverse biological membranes that lead to cellular effects. Stability problems such as agglomeration and sedimentation decrease productivity and hinder the manipulation and disposal thereof. In addition, instability could cause an accidental release in the environment, during storage or operation. Sustainable use of these materials and minimizing threats to the environment need comprehensive risk assessment, better formulation for biodegradability, and long-term ecotoxicity studies.

2.4 Scalability and Commercial Path

There are pros and cons to the scalability and commercial (SC) of ionic fluids and nanofluids. Ionic liquids are appealing materials for energy storage and separation applications because of their designable properties; however, they suffer from expensive synthesis and limited large-scale synthesis methods. Nanofluids from advanced cooling and heat transfer applications are, however, limited by the cost-effective mass production and the higher inter-person variability of the quality. These stability problems—such as nanoparticle agglomeration, sedimentation, and chemical decomposition—are an impediment to commercialization, as they jeopardize the reliability and efficiency of semiconducting devices. To break these barriers, formulation, the development of standardized manufacturing processes, and life cycle assessment are needed to achieve high performance and cost-effectiveness that is environmentally compatible with industrial requirements.

2.5 Technical Barriers for a Range of Applications

Ionic fluids and nanofluids share great potential for applications in energy, electronics, and thermal management, yet there exist many technical obstacles to retard their wide application. For ionic liquids, the high viscosity, corrosivity, and low thermal stability limit their use in high-temperature or high-flow systems. The development of nanofluids is hindered by the problem of obtaining a uniform distribution as well as long-term stabilization, since nanoparticles have a tendency to form aggregates or precipitate, which causes a deterioration in thermal conductivity and the flow properties. Additionally, compatibility with the existing system material and fouling/erosion is also a serious concern. Solving these challenges requires advancements in surface modification, stabilization methods, and application-specific formulations to drive increased reliability and sustained performance under varied operating conditions.

3 Opportunities in the Field of Ionic and Nanofluid Nanotechnologies

3.1 Green Chemistry and Sustainable Technology Novelties

Green chemistry for ionic liquids/nanofluids environmentally benign progress. Ionic liquids that are easily biodegradable and nontoxic have been developed for catalysis, storage energy, and separation. Low-cost, green solutions in the form of deep eutectic solvents are also available. Eco-friendly nanoparticles and water-based dispersions help in minimizing the environmental hazard in nanofluids. New surface treatments improve stability without any nasty additives. These inventions advance sustainable technologies such as solar thermal systems and energy-efficient cooling, consistent with objectives of waste reduction, environmental safety, and long-term environmental compatibility.

3.2 Better Energy Storage and Conversion Efficiency

Ionic fluids and nanofluids present good energy storage and conversion performances because of their exceptional thermal, electrical, and transport properties. Ionic liquids have high ionic conductivity, wide electrochemical windows, and heat stability and are promising materials for future batteries and supercapacitors. Nanofluids, which have a higher thermal conductivity and heat transfer performance, increase coolants effectiveness in energy systems and increase their overall efficiency. Their incorporation in fuel cells, solar collectors, or thermal storage systems enables more efficient, lightweight, and sustainable energy devices for applications in the future.

3.3 Advances in Biomedical and Pharmaceutical Areas

Recent developments in biopharmaceutical sectors indicate that bio-ionic (material) and nanofluids are promising in drug delivery and diagnosis and therapy. Biocompatible ILs for improved solubility, stability, and targeting of drugs. Nanoparticles, when functionalized, can be used as nanofluids for imaging contrast and drug delivery systems. Their controllable features promote minimally invasive therapy and precision medicine. Current research works on contending with the problem between low toxicity and high biocompatibility, which promotes safer and more efficient biomedical applications by developing fluid systems.

3.4 Applications in Industry and Engineering

Ionic fluids (also called electrolytes) and nanofluids (composed of nanoparticles) have gained attention in industrial and engineering processes because of their better thermal, transport, and chemical characteristics. Highly stable and non-volatile, ionic fluids are accessed for lubrication, electroplating, and chemical separations. Nanofluids improve heat transfer in a cooling system, liquid cooling systems used in motor cooling, electronic cooling, and air-conditioning applications. For machining, energy systems, and material processing, they increase efficiency and precision. Current developments continue to address scalability, cost efficiency, and environmental security for wider commercial utilization.

4 New Trends in Ionic and Nanofluid Nanotechnologies

4.1 Smart and Functionalized Ionic Liquids

Rational design of smart and functionalized ILs, through the chemical modification of ILs, is enabling the optimized performance of ILs in a wide range of applied areas. These ILs possess features such as adjustable viscosity, conductivity, and solubility, which render them perfect candidates for applications such as separation processes, catalysis, and energy storage. Functionalized ILs can be responsive to external stimuli (temperature, pH, electric field) allowing for dynamic regulation in drug delivery and sensing applications. These advances improve the flexibility and performance of ionic liquids for editorial in dual-use applications.

4.2 Advanced Nanofluids for Thermal and Energy Systems

With their great increase in thermal conductivity, innovative nanofluids are able to bring forth a revolution in energy and thermal systems performance. The thermal conductivity of these nanofluids is far superior to that of ordinary fluids because they are formed by dispersing metal oxides, carbon nanotubes, or graphene particles in base fluids. In the electronics, power generation, and automotive sectors, this kind of cooling improvement is especially advantageous and can lead to more efficient energy solutions. Furthermore, nanofluid systems are being researched as an environmentally friendly and renewable energy source for solar thermal collectors by scientists.

4.3 Most up-to-Date Nanotechnology in Medicine and Electronics

The latest nanotechnology developments are changing medicine and electronics and coming up with innovations that supercharge precision and efficiency. In the field of medicine, the application of nanoparticles for targeted drug delivery is the focus of attention, since this method not only increases the cure rate but also significantly reduces side effects. The benefit of using nanomaterials for diagnostic tools independently of other things, such as cancer screening, is that these materials in the contrast agents and biosensor areas of the nanoscale can detect the disease in its early stages. In the realm of electronics, nanotechnology is instrumental in making smaller, faster, and more energy-efficient equipment. Flexible electronics, nanotransistors, and quantum dots are the ways of the future in relation to wearable tech, sensors, and energy-saving devices. Healthcare and electronics both continue to benefit from these innovations that are used for solutions promising a major impact on society.

4.4 AI-Machine Learning Integration in Nanoscience

Nanotechnology, the latest breakthroughs have had a drastic impact on the medical and electronics industries and are also driving further exciting developments in precision and efficiency. When it comes to medical science, the use of nanoparticles for targeted drug delivery—transportation of an active substance to a specific part of the body—is and has been the main focus of research because it is a new method that can not only increase the cure rate but also significantly reduce the side effects. One of the advantages of nanomaterials for diagnostic tools and cancer screening as an individual case is the possibility of the same materials of the nanoscale also being in the contrast agents and biosensors areas and that the disease can be detected at the primary stage. Thus, the picture of electronics is that with nanotechnology, it is practically possible to create smaller, faster, and less power-consuming devices. The wearable tech, sensors, and energy-saving devices that the future will see as the most popular ones will be characterized by the presence of flexible electronics, nano transistors, and quantum dots. Both the health and electronic fields are becoming the future of using these exact technologies, which are one of the great promises of societal impact.

5 The Mystery of Case Studies

5.1 Industrial Applications of Ionic Liquids that Have Succeeded

Ionic liquids (ILs) are electrically conducting salts that remain in the liquid state below the temperature of 100 °C. A mixture of various properties sets ILs into motion especially as they demonstrate high thermal and chemical stability, large electrochemical potential, and at the same time are able to dissolve a wide range of substances [10, 11]. For the last 20 years, ILs have ceased to be just a thing of the lab to become widely used in industrial processes that belong to heterogeneous industries like petrochemicals, metallurgy, textiles, environmental technologies, energy, and pharma. In this review article, the author identifies nothing less than the success stories from the real industrial world, through practical uses of ILs that are no longer only found in the realm of ideas.

Petrochemical Refining and Alkylation

The first big and obvious use of ionic liquids in manufacturing was and still is the alkylation of isobutane with olefins, which is mainly done in the petrochemical industry to produce high-octane gasoline. The conventional methods of alkylation with acids such as hydrofluoric and sulfuric ones are harmful, which results in a threat to security and the environment. This prompted Chevron, in a partnership with Queen's University Ionic Liquid Laboratories (QUILL), to develop the ISOALKY™ process that uses chloroaluminate-based ILs [12]. The ISOALKY™ technology became the opportunity for safer, more selective, and more environmentally friendly processes due to the reduced acid consumption and improved product selectivity [12]. The initial large-scale use of the ISOALKY™ process in Chevron's Salt Lake City refinery in 2020 demonstrated mass production using catalysis through ILs and operations that are highly reliable [12].

Metal Recovery and Recycling

The need for recycling technologies with environmentally friendly traits and efficient service has driven the determination toward ILs for metal recovery. ILs have proven to show unexpected potential in extracting real earth elements (REEs), noble metals, and base metals from electronic waste and other materials. We can see this when the researchers at QUILL came up with a method to retrieve neodymium and dysprosium from spent magnets using task-specific ILs with good selectivity and recovery efficiency [13]. This success inspired the establishment of Ionic Technologies, a company that runs a 30-ton demonstration facility and serves the market of REE-containing waste annually in Belfast [13]. In such a context, ILs offer a non-volatile, reusable, and green alternative to traditional leaching agents, thus minimizing secondary waste as well as occupational risks.

Electrochemical Applications

ILs are indispensable in the growth of the electrochemical industry as they have been instrumental in metal electrodeposition and electropolishing. By employing ILs to replace water-based systems, it becomes viable to carry out the electrodeposition of metals of high reactivity or those that withstand great heat and pressure, for example, such metals as aluminum, magnesium, and uranium [14]. Scionix Ltd. developed a low-temperature ionic liquid (a subclass of ILs) from choline chloride and chromium(III) chloride, which was used in the electroplating process of chromium. This process is more environmentally friendly and safer compared to the traditional chromium (VI) method, which is toxic and carcinogenic [14]. Furthermore, these systems offer the benefits of high current efficiency, low energy consumption, and the best quality metal finishing.

Textile and Material Processing

The textile industry has benefited from ILs in the context of dissolving and restructuring natural fibers like cellulose, silk, and wool. Natural Fiber Welding, Inc. developed proprietary IL-based processes for creating high-performance textiles from natural and waste fibers [15]. These processes enable the recycling and reconstitution of fibers without toxic solvents or aggressive chemicals. ILs allow the precise control of fiber morphology and alignment, leading to mechanically superior and more sustainable textiles. The scalability of these methods has led to partnerships with major clothing brands, integrating ILs into global supply chains [15].

Carbon Capture and Environmental Applications

ILs can also be used to great advantage in the field of the environment, namely in the area of carbon capture and storage (CCS). In comparison to other forms of CO_2 solubility, ILs may be formulated in a way that they have high CO_2 solubility without having to care too much about the concentrations and temperatures. Their low vapor pressure and high thermal stability are suitable for their use in a continuous gas absorption process. ILs were synthesized by inserting amine, ether, or carbonate groups to enhance the CO_2 capture capacity [16]. Certainly, some ILs display similar chemisorptive behavior to traditional amines but do not encounter decomposition problems. There are no commercial CCS factories that use ILs, yet pilot projects have been launched to test this approach and to carry out extensive simulations to confirm its industrial flue gas treatment usability [16].

Energy Storage and Conversion

In energy systems, ionic liquids are involved as electrolytes in such functions as batteries, supercapacitors, and hydrogen compressors to provide the respective power. Their stability with respect to temperature and electrochemical energy allows them to be appropriately applied to the requirement indicated, thereby ensuring sustained efficiency and availability of the device. For example, ILs are used as electrolytes for vanadium redox flow batteries (VRFBs), lithium-ion batteries, and sodium-ion batteries [17]. This means one should not go over the points already clearly set forth in the paper. The industry is very likely to benefit from such a kind of application in order to dramatically reduce the flammability and the safety hazards of the system operated at high T or high V. Moreover, the use of materials similar to ILs in hydrogen piston compressors would meet the demands for oil-free compression with minimal leakage. Hence, they are considered promoters for the hydrogen economy, as they guarantee that the storage and conversion of hydrogen will be efficient.

Pharmaceutical and Biomedical Applications

In the pharmaceutical field, ILs are applied as solubilizers, drug carriers, and green reaction media. These can notably increase the solubility and hence bioavailability of poorly soluble drugs. ILs have also been involved in the green extraction of active pharmaceutical ingredients (APIs) from natural sources. A case in point would be the IL-assisted extraction of artemisinin from Artemisia annua, a core antimalarial compound [18]. ILs can also assist in the preparation of pharmaceutical intermediates at lower temperatures and in a more efficient way with good yield and selectivity [10, 18]. Certain ILs display intrinsic antimicrobial action, which can lead to the application of dual-function solvents in drug formulation.

The utilization of ionic liquids in industries ranging from petrochemicals and metallurgy to energy and biomedicine proves their transformative potential. Most of the green technologies have been limited to academic labs; however, ILs have managed to overcome that barrier and have been used in many successful commercial platforms with guaranteed economic and environmental benefits. The research field is still on its way to designing the most effective IL structures for specific functions, and hence, the role of ILs in sustainable industrial processes will most likely be enlarged. Yet, issues such as the cost, toxicity of some ILs, and recycling still pose great challenges and demand continuous attention if we want these to become the sustainable way for the future and be accepted by standards.

5.2 Nanofluids in Heat Transfer Applications in Reality

Nanofluids are synthetic sols of nanoparticles (1–100 nm) in regular base fluids like water, ethylene glycol, or oils and have been shown to be highly effective for heat transfer in real-world applications. The use of such highly conductive materials as Al_2O_3, CuO, TiO_2, or carbon-based nanomaterials has been found to bring significant improvement to the heat transfer characteristics of the base fluids in terms of thermal conductivity and convective heat transfer. This innovation has had the result of them being adapted to various industrial systems starting from electrical and electronic cooling systems and going to renewable energy, automotive, and heating, ventilation and air conditioning (HVAC) systems.

Automotive and Engine Cooling

In automobiles, the proper temperature for operating stably is necessary for productivity, emission reduction, and energy saving. The use of nanofluids as coolants in car radiators or intercoolers has shown an increase in heat exchange efficiency of up to 15–20%. In real vehicle tests, it has been proved that the modified thermal regulation leads to better fuel economy and hence reduced exhaust pollution. To regulate the temperature of the batteries and motors in hybrid and electric cars, vehicle makers have already taken steps to use nanofluid-enhanced coolants.

Renewable Energy Systems

Solar thermal collectors in today's world fully make use of nanofluids due to their better absorption and thermal transfer properties. Apart from that, nanofluids were used in actual installations for the purpose of increasing the absorption and decreasing the heat loss in the case of flat plate and parabolic trough collectors. To illustrate, quite a number of laboratory studies have demonstrated carbon nanotube (CNT) and graphene-based nanofluids as having the best heat transfer and storage properties for concentrating solar power (CSP) plants. In photovoltaic thermal (PV/T) systems nanofluids are the best solution to carry the excessive heat away from the PV panels and at the same time improve both electrical and thermal conversion efficiencies.

Heat Exchangers and Industrial Cooling

Nanofluid has become a very popular kind of fluid preferred for compact heat exchangers in industrial sectors like energy, chemical, and air conditioning. Nanofluids not only have made it possible to reduce the size of the heat exchangers but also have replaced the traditional fluids so that the performance is improved. Companies in the chemical and food industries have used nanofluids in shell-and-tube and plate heat exchangers to make it easier for the heat to go through and decrease fouling.

5.3 Nanotechnology Revolutions in Drug Delivery

Nanotechnology is a breakthrough in the drug delivery field and has provided a way for precise, countable, and efficient treatments, which the traditional pharmaceutical approaches do not have. The use of nanoscale materials allows nanocarriers facilitate the process. At the same time, they keep the drug solubility, stability, and bioavailability and minimize the drug side effects.

Targeted delivery is one of the benefits that is most talked about in nanotechnology in the drug delivery sector. Nanoparticles can be made in such a way to specifically gather into the tissues that are affected, i.e., a tumor, either passively using the electron paramagnetic resonance (EPR) effect or actively by ligands that attract only those receptors on the diseased cells. As a result, drug concentration at the site of action is increased, and the body's toxic reaction is decreased.

Liposomes, polymeric nanoparticles, solid lipid nanoparticles, dendrimers, and nano micelles are nanocarriers that have been used for the treatment of drugs. These carriers are able to carry a variety of drugs such as chemotherapeutics, antibiotics, and anti-inflammatory agents successfully. Doxil, for instance, is a liposomal doxorubicin that comes in clinical application for the treatment of cancer, improving the survival rate of patients with the longer circulation time of the nanoparticles and lower side effects to the heart.

Besides that, nanotechnology is able to cause controlled release of drugs in a particular location through various stimuli-responsiveness situations like pH, temperature, enzymes, or light-induced triggers merely, hence, such methods are great for the release of drugs at the target site.

Besides, the help of nanocarriers to get through biological barriers, including the blood–brain barrier, is also an innovation that opens up interesting ways of treating diseases of the nervous system.

Nanotechnology, in turn, has been changing drug delivery methods gradually, ultimately leading to more individualized, successful, and harmless treatments. Confirmed by the numerous breakthroughs, it can be said that modern medicine will grow and that one of the ways to reach it is via nanotechnology, which holds great promise for assigning a positive outcome to diverse diseases.

6 Teaching Learning Approaches

The teaching of ionic fluids, nanofluids, and nanotechnology at an advanced level commits educators to form new pedagogical approaches, which transfer the foundational knowledge to real-world applications. In modern materials, teaching strategies have shifted toward active learning that uses case studies as problem-based learning as well as introduces the students to interdisciplinary courses. The teachers of today are likely to exploit case studies, research-based projects, and flipped classrooms as well as facilitate students' analytical thinking and real-time problem-solving by asking them, for example, to scrutinize the thermal behavior of nanofluids in energy systems.

It turned out that laboratory and practical applications play a significant role in providing a strong foundation for the theoretical knowledge. The students performed the synthesis of ionic liquids, thermal conductivity tests of nanofluids, and nanoparticle characterization using instruments such as scanning electron microscope (SEM) and UV-Vis spectroscopy. Many institutions have undergone a significant transformation by the addition of real-time lab modules, such as indian institute of technology (IIT) Bombay, where students are able to carry on with the learning process—designing, testing, and optimizing nanofluid properties for heat exchangers—and thus they can simulate exactly the same activities involved in the industry.

E-Learning and virtual simulations have become one solution to the problem. They came in very handy to ensure that access to some kind of knowledge and understanding, particularly during COVID-19, was guaranteed. Online platforms like NanoHub.org present interactive modules and simulation tools that students can use for the purpose of modeling the behavior of nanomaterials. Consequently, without the expensive lab infrastructure, students could deal with nanomaterials quite confidently. The use of virtual reality-based labs is a remarkable invention; in fact, it serves as a support system for gaining an immersive environment where they learn the processes of particle interaction, drug delivery mechanisms, and molecular design of ionic fluids.

Definitely, industry-academia collaboration has a very important impact on skill development. Relationships with industries like Tata Chemicals and badische anilin- und sodafabrik (BASF) are the ones that enable us to adjust the curriculum in accordance with the market. Besides internship programs, research projects and speeches by industrial guests, who are experts in the field, are provided to the students so they can generate ideas for themselves.

They can witness real-life applications like ionic fluids in CO_2 capture or nanofluids in automotive cooling systems.

The above-mentioned teaching methods combine different aspects and give students practical skills for careers in materials science and emerging nanotechnologies.

7 Protocols and Standardization in Ionic Fluids, Nanofluids, and Nanotechnology

The expanding influence of ionic fluids, nanofluids, and nanotechnology in the academic and industrial fields places more and more stress on the need for strong protocols and standardization to maintain safety, consistency, and reproducibility.

For the suitable processing of ionic liquid, purity control, moisture sensitivity management, and waste disposal are the best practices. In most cases, glovebox handling and vacuum drying are inevitable to carry out the syntheses of the ionic liquids without the risk of hydrolysis or contamination. The procedures for pinpointing the task of ionic liquid synthesis, prepared by the QUILL Research Centre and BASF, allow for thermal stability and solvent recovery to be among the most eco-friendly processes. Also, these processes are the most efficient for the purpose of flowing the tasks at hand through the research.

In the electronic/vehicle systems sphere, to ensure that nanofluids are used safely, one must be in possession of clear guidelines on concentration, dispersion techniques, and the compatibility with a given system. The problems of agglomeration and erosion are the focus here. Ultrasonication, surfactant use, and checking if the suspension is stable, in real-time, are the solution. If, for example, Al_2O_3 nanofluids were to be used in automotive radiator applications, the manufacturers would adhere to the ISO/TR 13014:2012 guidelines to gauge the magnitude of exposure and risk. These protocols play a huge role in preventing the sedimentation of nanoparticles, which could, otherwise, block heat exchangers or harm the pump systems.

Regulations in the field of nanotechnology research and manufacturing are crucial; thus, borracelle. They are vital for cross-institutional consistency and safety. The OECD and the ISO as the regulatory bodies have the task of developing the ISO/TS 80004 standard for nanomaterial terminology and the ISO/TR 18401 standard for life cycle risk assessment. For example, multinational technology companies like IBM and Samsung, in their efforts to manufacture nanomaterials that will be used in the production of electronics, draw guidance from these standards.

Adhering to these real-world standards is not only a guarantee of compliance, security, and successful business but is also a way of fastening commercial usage by reducing the risks in both operational and environmental areas of the project, on one side, and of the technology innovation-acceptance in several sectors through trust establishment, on the other.

8 Ethical Issues in Ionic Fluids, Nanofluids, and Nanotechnology

The rapid progress of ionic fluids, nanofluids, and nanotechnology is very useful but also raises some significant ethical issues. One major concern is the possible side effects of nanomaterials on environmental and human health because these particles can enter cells, accumulate in organs, and cause toxicity. For example, the use of silver nanoparticles in consumer products has contaminated the water, bringing danger to the ecosystem.

The ethical dilemmas in the business sector are shown in the early selling of products without proper safety validation; it is also pursued due to the competitive market pressure. If profit becomes the main priority, long-term safety may be overlooked, as happened with the unregulated use of carbon nanotubes in some coatings.

AI integrated into nanotechnology research raises ethical questions, including algorithm bias, data privacy, and the substitution of human roles in research and manufacturing through automation.

Also, public knowledge and the policy around nanotechnologies in society are still minimal. The different regulations among countries, for example, the lack of strictness in the policies of developing nations, set the stage for some nations to exploit these people but also for the implementation of uneven standards of safety. Certainly, informing the stakeholders is the only rational way to avoid any negative consequences of the respective innovation.

9 Future Prospects

The future of ionic fluids, nanofluids, and nanotechnology is moving forward with the help of the most specialized research directed toward smart materials, multifunctional fluids, and AI-based systems. The field of health care is a lot about new nanocarriers that can deliver drugs with the utmost precision and bio-imaging, which can impressively open new frontiers for the greater good. Moreover, nanofluids are making a difference in the thermal systems of solar and battery technologies, and ionic fluids are essential in green electrolytes for the next generation of batteries. Electronics will be faster, smaller, and their energy efficiency will increase thanks to nanoscale components. The new direction of the development priority refers to sustainable development, the one that puts minimal environmental impact and maximal performance at the forefront of the technologies, thus making them universal enablers of cleaner, smarter, and more inclusive global advancement.

References

1. M.A. Rahman et al., Review on nanofluids: Preparation, properties, stability, and thermal performance augmentation in heat transfer applications. ACS Omega **9**(30), 32328–32349 (2024). https://doi.org/10.1021/acsomega.4c03279
2. S.K. Singh, Review on the stability of the nanofluids, in *Pipeline Engineering - Design, Failure, and Management*, ed. by S. Rushd, M.A. Ismail, (IntechOpen, London, 2022). https://doi.org/10.5772/intechopen.107154
3. S. Chakraborty, P. Kumar, Nanofluids long-term stability challenges and guidelines, in *Fundamentals and Transport Properties of Nanofluids*, ed. by S.M.S. Murshed, (Royal Society of Chemistry, London, 2022), pp. 71–146. https://doi.org/10.1039/9781839166457-00071
4. S. Purreea, M. Nadeem, A. Shahzad, H.M. Ali, Synthesis, heat transfer properties and stability of nanofluids for commercialization: A review. Chem. Eng. Commun. **210**(5), 814–836 (2023). https://doi.org/10.1080/00986445.2021.1976163
5. B. Mehta, D. Subhedar, H. Panchal, Z. Said, *Synthesis, Stability, Thermophysical Properties and Heat Transfer Applications of Nanofluid—A Review* (Elsevier, Amsterdam, 2023)
6. J.A. Okello, W.N. Mutuku, A.O. Oyem, A review of nanofluids synthesis, factors influencing their thermophysical properties and applications. J. Eng. Res. Rep. **17**(4), 1–17 (2020). https://doi.org/10.9734/jerr/2020/v17i417192
7. I. Nesterova et al., The role of surface material properties on the behavior of ionic liquids in nanoconfinement: A critical review and perspective. arXiv preprint, arXiv:2410.08721 (2024). https://arxiv.org/abs/2410.08721
8. S. Chakraborty, P.K. Panigrahi, Stability of nanofluid: A review. Appl. Therm. Eng. **174**, 115259 (2020). https://doi.org/10.1016/j.applthermaleng.2020.115259
9. M.A. Rahman et al., An extensive review of preparation, stabilization, and application of single and hybrid nanofluids. J. Therm. Anal. Calorim. **149**, 9523–9557 (2024). https://doi.org/10.1007/s10973-024-13449-1
10. J.P. Hallett, T. Welton, Room-temperature ionic liquids: Solvents for synthesis and catalysis. Chem. Rev. **111**(5), 3508–3576 (2011). https://doi.org/10.1021/cr1003248
11. N.V. Plechkova, K.R. Seddon, Applications of ionic liquids in the chemical industry. Chem. Soc. Rev. **37**(1), 123–150 (2008). https://doi.org/10.1039/B006677J

12. ISOALKY™ technology: Safer alkylation using ionic liquids. Chevron Corporation, https://pmc.ncbi.nlm.nih.gov/articles/PMC7664896/
13. L.T. Stephenson, The challenge of recycling rare earth metals. The Guardian (2025), https://www.theguardian.com/research-to-reality/2025/jan/24/the-challenge-of-recycling-rare-earth-metals-and-the-scientists-making-it-happen
14. A.P. Abbott, G. Capper, D.L. Davies, R.K. Rasheed, V. Tambyrajah, Novel solvent properties of choline chloride/urea mixtures. Chem. Commun. **9**(1), 70–71 (2003). https://doi.org/10.1039/B210714G
15. Natural fiber welding: Sustainable textile solutions. Natural Fiber Welding, Inc., https://www.naturalfiberwelding.com/
16. J.R. Harjani et al., Functionalized ionic liquids for CO^2 capture. J Mater Chem A **9**, 14856–14870 (2021). https://doi.org/10.1039/D1TA03656D
17. M. Armand et al., Ionic-liquid materials for the electrochemical challenges of the future. Nat. Mater. **8**, 621–629 (2009). https://doi.org/10.1038/nmat2448
18. T. Welton, Ionic liquids in pharmaceutical applications. Green Chem. **13**(1), 19–23 (2011). https://doi.org/10.1039/C0GC90049A

Conclusion

As we arrive at the final chapter of *Ionic Liquids, Nanofluids, and Nanotechnology: Innovations, Applications, and Sustainability*, it is clear that we are standing at the threshold of a new era in materials science—one defined by synergy, intelligence, and purpose. The convergence of ionic liquids, nanofluids, and nanotechnology is no longer a speculative venture; it is a living, evolving field that is actively redefining how we approach some of the most pressing challenges of our time. From advancing green chemistry and energy storage to revolutionizing healthcare and smart manufacturing, these materials are enabling transformative solutions across diverse sectors.

Throughout this volume, we have explored how these advanced materials are not only functionally potent on their own, but exponentially more impactful when integrated. Their adaptability, tunability, and compatibility with artificial intelligence (AI) and machine learning (ML) systems have positioned them as foundational elements of a data-driven scientific future. Indeed, AI and ML are now indispensable tools, enhancing our capacity to model, design, and deploy materials with unprecedented speed and precision.

But this journey has also underscored critical responsibilities. As researchers and innovators, we must remain conscious of the ethical, environmental, and societal implications of our work. The potential toxicity of nanoparticles, the ecological footprint of synthesis processes, and the need for inclusive and equitable technology access are not peripheral issues—they are central to the sustainable advancement of the field. True progress lies not just in discovery, but in stewardship.

This book has sought to present more than just a compendium of knowledge. It has aimed to chart a vision—one where innovation is balanced with sustainability, and where scientific inquiry serves both technological progress and human well-being. It is our hope that the insights, case studies, and interdisciplinary perspectives presented in these chapters will not only inform, but inspire a new generation of researchers, educators, and industry leaders to pursue innovation that is both bold and responsible.

S. Kulkarni, S. Das (eds.), *Ionic Liquids, Nanofluids and Nanotechnology*,
Lecture Notes in Nanoscale Science and Technology 42,
https://doi.org/10.1007/978-3-032-04008-4

As we look ahead, the path forward demands collaboration across disciplines, institutions, and borders. The integration of ionic liquids, nanofluids, and nanotechnology with computational intelligence offers vast potential—but realizing it will require shared vision, rigorous standards, and a commitment to ethics and equity.

To all our readers, we extend our deepest gratitude for engaging with this work. May it serve as a launching point for new discoveries, meaningful dialogue, and enduring impact in the dynamic and vital domain of advanced materials science.

Index

S. Kulkarni, S. Das (eds.), *Ionic Liquids, Nanofluids and Nanotechnology*, Lecture Notes in Nanoscale Science and Technology 42,
https://doi.org/10.1007/978-3-032-04008-4